W9-AWU-780

Study Guide

to accompany

PRINCIPLES OF LIFE

Hillis • Sadava • Heller • Price

BETTY MCGUIRE
Cornell University

ED DZIALOWSKI
University of North Texas

DENISE LELLO
Smith College

JACALYN NEWMAN
University of Pittsburgh

Sinauer Associates, Inc.   W. H. Freeman and Company

Cover photograph © Fred Bavendam/Minden Pictures.

Study Guide to accompany ***Principles of Life***

Copyright © 2012 Sinauer Associates, Inc. All rights reserved.
This book may not be reproduced in whole or in part without permission of the publisher.

Address editorial correspondence to:
Sinauer Associates, Inc.
23 Plumtree Road
Sunderland, MA 01375 U.S.A.
Fax: 413-549-1118
Internet: www.sinauer.com; publish@sinauer.com

Address orders to:
MPS/W.H. Freeman & Co. Order Department
16365 James Madison Highway, U.S. Route 15
Gordonsville, VA 22942 U.S.A.
Examination copy information: 1-800-446-8923
Orders: 1-888-330-8477

ISBN 978-1-4292-7930-7
Printed in U.S.A.

4 3 2 1

To the Student

Biology is an incredibly exciting field of study, but in order to appreciate new discoveries and discussions, it is necessary to have a firm grasp of the underlying concepts and ideas. Your textbook is designed to give you a comprehensive overview of important biological phenomena. It will also serve as a resource to you in future studies. Together with your instructor, your textbook will provide you with invaluable information for beginning your study of biology.

This Study Guide is designed to supplement, not replace, your textbook and your instructor. It was written for you, the student, in language that you can understand, but it does emphasize proper usage of biological terminology. Important concepts and ideas have been synthesized into easy-to-read text that provide an overview of the biological concepts discussed in your textbook. Each Study Guide chapter includes three review elements: (1) The Big Picture, an introductory overview of the topics covered in the chapter; (2) Study Strategies, tips for effective ways to study the material and master common problem areas; and (3) Key Concept Review, a detailed outline of each numbered concept in the chapter interspersed with short answer and diagram questions. These will help you preview a chapter before reading it, check your understanding, and review the chapter later.

Each Study Guide chapter also includes a series of questions that have been grouped into two categories: Key Concept Review questions and Test Yourself Questions. Key Concept Review questions are integrated into the summary section, allowing you to test your factual and conceptual knowledge of each chapter while reading the pertinent review text. In some cases, diagrams are provided and you are asked to label them or answer questions based on them, while in other chapters, you are asked to create your own diagrams based on a question or series of questions. Key Concept Review questions ask you to apply the knowledge you have gleaned from a chapter to answer questions that are more open-ended. These questions require you to have assimilated several concepts and to think beyond what you have just read. Your instructor may ask questions similar to these on exams, or may use an entirely different approach to assess your knowledge, but these questions will be a good check of how well you understand the material. Test Yourself questions are multiple choice style questions, designed to determine if you have retained information from a chapter, and if you can put together various concepts in order to answer questions. Answers to both types of questions are provided at the end of each chapter of the Study Guide. These answers are not exhaustive, but are instead are designed to point you to the correct concepts in the textbook. Because of the nature of many of the Key Concept Review questions, your answers should be more expansive than the short explanations given in this Study Guide.

Strategies for Studying Biology

Each individual has his or her own unique study pattern. However, there are some successful study strategies that are universal. We recommend that you first preview a chapter in your textbook. In this initial preview, it is important to note the organization of the chapter and the main points, and to go over the chapter summary at the end. By referring to the Study Guide at this point, you will further understand the organization of the material. If you follow these steps, you might have some specific questions in your mind that you will expect your reading to answer.

Reading a textbook is an active process. We recommend that you always have a pencil and paper available. Jotting notes in margins serves as an excellent mental trigger when it comes time to review. As you read each section of your textbook, see if you can summarize it in your own words. Compare your summaries to those provided at the end of each section and at the end of the chapter. You want to assure yourself that the main points you are noting match those that the author has selected. Refer back to those questions that came up as you previewed the chapter.

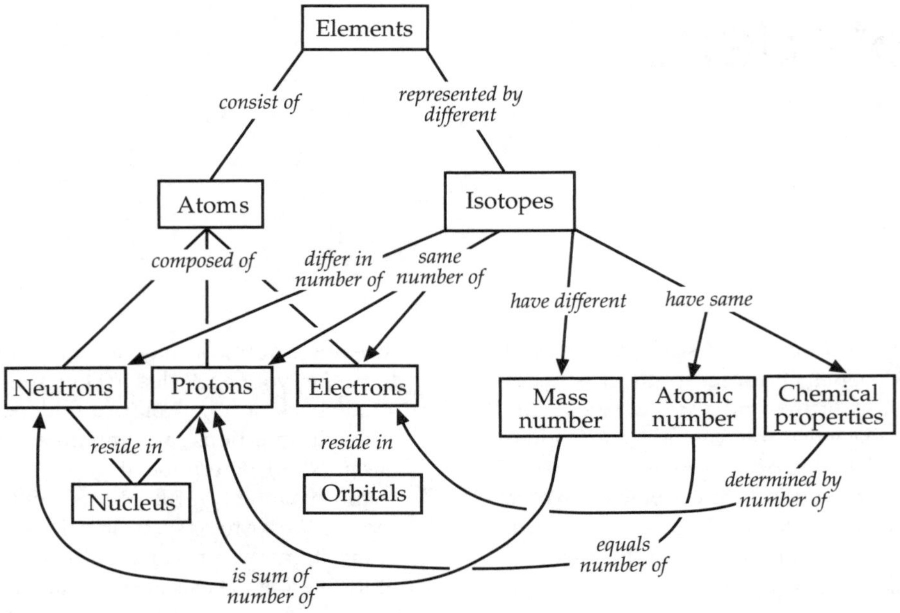

You should be able to answer your own questions by the time you have finished your reading.

As you are reading, take time to review all the figures and tables. Your textbook makes use of figures to illustrate points, describe pathways, and give visual representation to complex topics. Often these figures are more helpful than the paragraphs of written explanation. In places where figures might be beneficial to you but are not provided in the textbook, try to draw them yourself. This is especially important when attempting to understand structures and pathways. You will find that the Study Guide also refers you to specific animated tutorials, Web activities, and working with data exercises, all of which are available within BioPortal, at yourBioPortal.com. Each of these tools will aid in your understanding of the material. In addition, BioPortal includes many other study and review resources, some of which your instructor may assign. (Many of these resources are also available on the textbook's free Companion Website, at: www.whfreeman.com/hillis1e)

Once you have read the material, summarized it in your own words, and reviewed all of the figures and tables, it is time to do another quick review. Scan the summary in your textbook again and read over the Key Concept Review section of this Study Guide. Additionally, the textbook includes many features to help you learn the material as you read and review, including Do You Understand the Concept? questions, Apply the Concept questions, Analyze the Data questions, and Chapter Summaries. (Refer to the inside front cover of the textbook for more about these features.) We recommend further that you construct a "concept map" of the main points of a chapter. This will assure you that you see how concepts are interrelated and that you understand the organization of the material. An example of a concept map is shown above. Main concepts are in boxes, and arrows are used to connect concepts and show relationships.

Finally, when you are comfortable with chapter content, work on the questions section of this Study Guide (both from the Key Concept Review and Test Yourself sections). We recommend working through all of the questions before checking your answers. Any questions that you do not understand or that you answer incorrectly should be flagged as material you need to review. Remember that your goal is to understand the material completely, not just to answer these particular questions.

Experiments have suggested that the average attention span of a reader is approximately 12 minutes. You cannot expect to sit down and master an entire chapter in one sitting. Break up your study into short segments of approximately 30 minutes each. This will give you time to get down to business, maximize your attention, and learn without becoming frustrated or drained. At the end of this time, move to some other activity or area of study. When you return to biology, you will find your mind is ready to absorb more.

This entire process should take place well in advance of your exams. You cannot learn even one chapter adequately the day before an exam. Mastering biology requires daily study and review. If you keep up with learning the concepts as your course proceeds, you will find that reviewing the textbook summaries and the Study Guide, along with the online review tools, is adequate preparation. The key to learning biology well is slow and diligent daily work.

Doing Well on Biology Exams

Your instructors are your best resource for doing well on exams. They are there to assist you in learning the material that is presented in your textbook. They are experts in their fields and understand how the concepts presented are vital to your biology education. Follow their lead in preparing for exams. Exam performance is directly linked to classroom attendance and daily study. Sit front and center in your lecture hall and pay close attention to what your instructor writes and provides in presentations or overhead displays. Your instructor will give you guidance about the most important points to study. You should take careful notes and ask for clarification when necessary.

You should always compare your classroom and lecture notes to your textbook and the Study Guide. Take note of the points your instructor emphasizes. Were there specific questions that the instructor brought up in class? Do these correspond to the questions in the Study Guide? If so, chances are you will see them again on an exam. This practice should be part of your daily study regimen.

Approach the exam itself with confidence. Read the directions carefully and read the questions completely. One of the biggest mistakes students make on exams has to do with not following instructions. We recommend that you read carefully through the entire exam even before you begin to answer the first question. You may find that early questions are answered partially in later ones, and that an overview of the whole exam helps to trigger your memory of the material.

If your exam contains multiple choice questions, treat each of the answer options as if it were a separate true/false question. In other words, mentally fill in the question with each answer and ask yourself if the resulting statement is true or false. Be sure to read each of the options, even if the first one strikes you as the correct one. You may be dealing with a question that has multiple correct answers or asks you to select "all of the above" or "none of the above." If you are unsure about how to answer a question, begin by ruling out answers that you know are incorrect. By eliminating some possibilities, your chance of selecting the right one is greatly improved! If you narrow the choices down to two possible answers and just can't decide between them, go with your gut response. You just may be right. And when you go over your answers, change your response only if you know you answered a question incorrectly. If you aren't sure, it may be best to leave the answer as-is. Your first instincts are often correct.

Essay questions require you not only to think through the concepts, but also to decide how you will present them. Organized, concise essay answers are always preferred to rambling answers with little direction. Take a look at the point value of each question. This is often an indication of how many "points" you need to make in your discussion. A 10-point question rarely can be answered with a single sentence. Look carefully at what your instructor is asking. If you are required to "discuss" a concept or a problem, do not present a list or some scattered phrases. Write in complete sentences and paragraphs unless you are asked specifically to list or itemize the material. Be sure that you address all points in the question. Essay questions frequently have multiple parts to test your understanding of the links between various concepts.

Always be aware of the time you have to complete your exam. Answer the questions you know swiftly, and allow yourself time to go back and really think about those you are struggling with. Be sure to concentrate on the questions with the highest point value. Those are the questions testing the most critical concepts and they are the ones that will have the most influence on your grade.

Once you think you have completed the exam, take a few minutes to go back over it. If there are any questions about which you are still unclear, be sure to ask your instructor for clarification.

Learn from Your Mistakes

After your instructor returns your exam or posts the answers, review the mistakes you have made. Take this opportunity to clarify misunderstandings and re-learn material when necessary. You will find this helpful not only for the final exam, but also as you approach upper-level courses. Material reviewed, corrected, and re-learned is more likely to be retained in the long run.

Get Help before It Is Too Late

Many students come to college without adequate study habits and/or are ill prepared for the rigor of biology at the college level. But this is a problem that can be solved fairly easily. Many colleges and universities have academic centers that assist with locating tutors, teaching study habits, and acquainting you with a variety of other resources. Make good use of these facilities. They are there for you. Also, for many additional study tips and strategies, see the *Survival Skills* document that is available within BioPortal.

Good luck in your study of introductory biology!

Table of Contents

Table of Contents (continued)

1 Principles of Life

The Big Picture

- Biology is the study of living things. Living things are composed of cells, which contain genetic material, require nutrients for energy, maintain a constant internal environment, and interact with one another. All available evidence points to an origin of life some 4 billion years ago. Natural selection and evolution have led to the current host of organisms inhabiting the planet. Life has specific characteristics, and the evolutionary roots of these characteristics can be traced through time.
- Based on molecular evidence, all life on Earth is organized as a tree of life. When scientists study organisms from an evolutionary perspective, they are concerned with evolutionary relatedness and common ancestors.
- Scientific inquiry is pursued according to a specific hypothesis–prediction method, which sets it apart from other methods of inquiry. This method requires a testable and falsifiable hypothesis, control of variables, and repeated testing.

Study Strategies

- The tenets of evolutionary biology may conflict with some of your personal beliefs. However, the textbook is concerned with science and scientific evidence, which ideally are impersonal, objective, and value-neutral.
- The hypothesis–prediction method should not be thought of or memorized merely as a series of steps. Think about how each step in the process follows from the preceding step. Two characteristics of scientific investigation that set it apart from other modes of inquiry are the setting of hypotheses and the continuous nature of the process. With each conclusion leading to a new set of hypotheses, science has the capacity to alter our overall base of knowledge.
- Familiarity with the hypothesis–prediction method will help you understand the evidence behind evolution. It may make sense to turn to Section 1.5 ("Science Is Based on Quantifiable Observations and Experiments") before studying the rest of the chapter.
- Drawing a timeline or examining the calendar model of Figure 1.1 will help you understand the chronology of the major evolutionary events discussed in the textbook.
- Go to yourBioPortal.com to review the following tutorial and activities:

  Animated Tutorial 1.1 Using Scientific Methodology

  Web Activity 1.1 The Major Groups of Organisms

  Web Activity 1.2 The Hierarchy of Life

  Working with Data 1.1 Feminization of Frogs

Key Concept Review

1.1 Living Organisms Share Common Aspects of Structure, Function, and Energy Flow

Life as we know it had a single origin

Life arose from non-life via chemical evolution

Cellular structure evolved in the common ancestor of life

Photosynthesis allowed living organisms to capture energy from the sun

Eukaryotic cells evolved from prokaryotes

Multicellularity allowed specialization of tissues and functions

Biologists can trace the evolutionary tree of life

Discoveries in biology can be generalized

All species on Earth share a common ancestor. The study of the evolution of life on Earth involves both the fossil record and modern molecular methods for comparing genomes. Life arose approximately 4 billion years ago. A random aggregation of complex chemicals led to the existence of the first biological molecules. These initial simple molecules led to molecules that could reproduce themselves and act as templates for larger, more complex molecules. The next step in the origin of life probably involved the enclosure of these molecules within a membrane, a development that created an environment favorable for the necessary biological molecules to interact in more complete ways. Early cells were simple, ocean-living prokaryotes with no internal membrane-enclosed compartments. Metabolism, which is the sum of all chemical reactions that occur within a cell, requires a source of energy. The ability to carry out photosynthesis arose approximately 2.7 billion years ago. Photosynthesis involves converting light energy from the sun into usable chemical energy with oxygen as a by-product. Large numbers of photosynthetic organisms increased atmospheric oxygen levels, allowing for the evolution of aerobic metabolism. Photosynthetic organisms also contributed to the insulating ozone layer that shields Earth from radiation and modifies temperatures. This shield eventually made life on land possible.

Eukaryotic cells have intracellular membrane-enclosed compartments known as organelles that carry out specific cellular functions. These compartments are thought to have arisen when larger prokaryotic cells engulfed and assimilated smaller cells. Multicellularity, which evolved about 1 billion years ago, made it possible for cells to specialize and for organism size to increase. Different species come about when two groups of organisms in one species are isolated from each other so that they can no longer mate and produce viable offspring. As mutations in the separated genomes accumulate, these two groups become new species.

Organisms are referred to by their binomial genus and species names (e.g., humans are *Homo sapiens*). By examining gene sequences, biologists can compile phylogenetic trees that show the evolutionary relationships among species. All life is thereby placed into one of three major domains: the prokaryotic Archaea and Bacteria and the eukaryotic Eukarya. Plantae, Fungi, and Animalia are three major groups of the Eukarya. Because all life is related, scientists work with model systems to help them study general patterns in life.

Question 1. Scientists interested in human biology typically perform experiments with other model systems. Why do scientists work with model systems in this way?
Textbook Reference: *1.1 Living Organisms Share Common Aspects of Structure, Function, and Energy Flow, p. 5*

Question 2. What is the hypothesis for the evolution of eukaryotic cells?
Textbook Reference: *1.1 Living Organisms Share Common Aspects of Structure, Function, and Energy Flow, p. 4*

1.2 Genetic Systems Control the Flow, Exchange, Storage, and Use of Information

Genomes encode the proteins that govern an organism's structure

Genomes provide insights into all aspects of an organism's biology

The genome of a cell contains all the genetic information in the form of deoxyribonucleic acid (DNA). DNA contains the code to make the proteins required for a cell to function. Segments of DNA make up genes, which are used to produce the proteins necessary for life. Proteins regulate the chemical reactions occurring within the cell. All of the cells in an organism contain the same genome, but they express different genes. Mutations occur in the DNA sequence and this can alter protein function; these mutations are the basis for evolution. Sequencing the genome of organisms allows biologists to look for the genetic basis for disease; it also allows them to compare the evolution of genes among species. A relatively new field, bioinformatics, is involved in studying the large amount of data generated from genome sequencing.

Question 3. Your genome has over 3 billion nucleotides. Does each cell use every one of the 3 billion nucleotides in your genome?
Textbook Reference: *1.2 Genetic Systems Control the Flow, Exchange, Storage, and Use of Information, p. 6*

Question 4. Assume you have sequenced the genome of 15 different species of turtle. What can this vast amount of information tell you?
Textbook Reference: *1.2 Genetic Systems Control the Flow, Exchange, Storage, and Use of Information, p. 7*

1.3 Organisms Interact with and Affect Their Environments

Organisms use nutrients to supply energy and to build new structures

Organisms regulate their internal environment

Organisms interact with one another

All cells require nutrients for survival. These nutrients are used to carry out cellular work, such as the work of building the cell, or mechanical work, such as the moving of molecules from one location to another. Metabolism is a measure of all the chemical transformations and work done in a cell or organism. Cells continuously carry out multiple chemical reactions and thus must carefully regulate their internal environment. A group of similar cells that work together is known as a tissue. A number of different tissues can be organized into an organ, and multiple organs work together to form an organ system. Multicellular organisms must also regulate the extracellular fluid that bathes each cell, which requires the specialization of their cells. Organisms use feedback loops to help regulate their physiology and maintain homeostasis.

Organisms interact with other organisms along many levels of organization. Individuals from one species living together make up a population. The many populations of different species in an area make up a community. Communities and their abiotic environment make up ecosystems. Ecology studies how species function within their communities and ecosystems.

Question 5. The study of biology can be organized from the most basic unit, the molecule, up to the biosphere. Create a flow chart describing how each level is connected with the level below it.
Textbook Reference: *1.3 Organisms Interact with and Affect Their Environments, pp. 7–8*

1.4 Evolution Explains Both the Unity and Diversity of Life

Natural selection is an important mechanism of evolution

Evolution is a fact, as well as the basis for broader theory

Charles Darwin proposed that all living things are related to one another and that species have evolved through the process of natural selection. Natural selection is the differential survival and reproduction among individuals in a population. Natural selection can produce adaptations, which are structural, physiological, or behavioral traits that enhance the chance of survival of an organism in a given environment. Scientists can directly measure and observe past and present evolution.

Question 6. Create a scenario for how natural selection might act upon a population of rabbits with fur colors that vary from white to brown.
Textbook Reference: *1.4 Evolution Explains Both the Unity and Diversity of Life, p. 9*

1.5 Science Is Based on Quantifiable Observations and Experiments

Observing and quantifying are important skills

Scientific methods combine observation, experimentation, and logic

Getting from questions to answers

Good experiments have the potential to falsify hypotheses

Statistical methods are essential scientific tools

Not all forms of inquiry into nature are scientific

Biologists study life by means of observation and experimentation. Biologists observe organisms in order to understand their natural history, and they try to quantify these observations. The scientific method, or the hypothesis–prediction method, is the basis of most scientific investigations. This method involves making observations and asking questions that lead to the formation of hypotheses, which are provisional answers to the questions. Predictions are made in regard to the hypotheses by means of deductive logic, and then they are tested.

The predictions derived from a hypothesis can be tested by comparative or controlled experiments. A controlled experiment involves isolating variables of interest (independent variables) while keeping other variables that may influence the outcome as steady as possible. Observations are then made of the dependent variables to test the hypothesis. A comparative experiment involves gathering data from multiple groups or conditions to examine the patterns found in nature. Statistical methods are used to determine if the results of the comparison are significant. Typically, these statistical tests start with a null hypothesis stating that there are no differences.

Science depends on a hypothesis that is testable and that can be rejected by direct observation and experiments. Observations must also be reproducible and quantifiable for a conclusion to be considered scientific.

Question 7. The study of biology is advanced by using the scientific method. Create a flow chart showing the steps involved in the scientific method.
Textbook Reference: *1.5 Science Is Based on Quantifiable Observations and Experiments, p. 11*

Question 8. Look over a newspaper to see how many articles are directly related to biology. Select one article and discuss how the researchers followed or did not follow the hypothesis–prediction method.
Textbook Reference: *1.5 Science Is Based on Quantifiable Observations and Experiments. pp. 10–12*

Test Yourself

1. Life arose on Earth approximately _______ years ago.
   a. 4 billion
   b. 4 million
   c. 4,000
   d. 1.5 billion
   e. 400,000
   Textbook Reference: *1.1 Living Organisms Share Common Aspects of Structure, Function, and Energy Flow, p. 2*

2. Populations of organisms have been able to inhabit a wide variety of environments on Earth because they
   a. have a genome.
   b. contain organelles.
   c. carry out photosynthesis.
   d. adapt through evolution.
   e. are similar to model organisms.
   Textbook Reference: *1.4 Evolution Explains Both the Unity and Diversity of Life, p. 9*
3. Which of the following is *not* a characteristic of most living organisms?
   a. Regulation of internal environment
   b. One or more cells
   c. Ability to produce biological molecules
   d. Ability to extract energy from the environment
   e. All of the above are characteristics of living organisms.
   Textbook Reference: *1.1 Living Organisms Share Common Aspects of Structure, Function, and Energy Flow, pp. 3–4*
4. Photosynthesis was a major evolutionary milestone because
   a. photosynthetic organisms contributed oxygen to the environment, which led to the evolution of aerobic organisms.
   b. photosynthesis led to conditions that allowed life to arise on land.
   c. photosynthesis is the only metabolic process that can convert light energy to chemical energy.
   d. photosynthesis provides food for organisms.
   e. All of the above
   Textbook Reference: *1.1 Living Organisms Share Common Aspects of Structure, Function, and Energy Flow, p. 3*
5. Homeostasis involves the regulation of a
   a. variable internal environment.
   b. constant external environment.
   c. constant internal environment.
   d. constant rate of natural selection.
   e. variable rate of natural selection.
   Textbook Reference: *1.3 Organisms Interact with and Affect Their Environment, p. 8*
6. A group of cells that work together to carry out a similar function is known as a(n)
   a. tissue.
   b. organ system.
   c. unicellular organism.
   d. protein.
   e. gene.
   Textbook Reference: *1.3 Organisms Interact with and Affect Their Environment, p. 8*
7. In order for natural selection to occur,
   a. certain traits must provide greater chances for survival and reproduction than other traits.
   b. random survival of all organisms must occur.
   c. sexual selection must occur.
   d. All of the above
   e. None of the above
   Textbook Reference: *1.4 Evolution Explains Both the Unity and Diversity of Life, p. 9*
8. Which of the following is *not* a domain on the tree of life?
   a. Archaea
   b. Plantae
   c. Eukarya
   d. Bacteria
   e. All of the above are domains.
   Textbook Reference: *1.1 Living Organisms Share Common Aspects of Structure, Function, and Energy Flow, p. 5*
9. The information needed to produce proteins is contained in
   a. nutrients.
   b. tissues.
   c. evolution.
   d. organs.
   e. genes.
   Textbook Reference: *1.2 Genetic Systems Control the Flow, Exchange, Storage, and Use of Information, p. 6*
10. Evolution is
    a. only relevant to the study of biology.
    b. the change in the genetic makeup of a population through time.
    c. the change in protein expression of a population through time.
    d. not influenced by natural selection.
    e. None of the above
    Textbook Reference: *1.4 Evolution Explains Both the Unity and Diversity of Life, p. 9*
11. In a model experiment, researchers subjected frogs to various levels of atrazine while keeping all other variables constant. This is an example of a _______ experiment.
    a. controlled
    b. repeated
    c. laboratory
    d. comparative
    e. None of the above
    Textbook Reference: *1.5 Science Is Based on Quantifiable Observations and Experiments, p. 12*
12. For a hypothesis to be scientifically valid, it must be _______ and it must be possible to _______ it.
    a. testable; prove
    b. testable; reject
    c. controlled; prove
    d. controlled; reject
    e. testable; control
    Textbook Reference: *1.5 Science Is Based on Quantifiable Observations and Experiments, pp. 13–14*

13. Eukaryotic cells differ from prokaryotic cells in that eukaryotic cells have
    a. genes.
    b. proteins.
    c. organelles.
    d. membranes.
    e. All of the above
    Textbook Reference: *1.1 Living Organisms Share Common Aspects of Structure, Function, and Energy Flow, p. 4*
14. In the scientific names of organisms, the _______ is placed first and the _______ is placed second.
    a. species; genus
    b. genus; domain
    c. domain; genus
    d. genus; species
    e. domain; species
    Textbook Reference: *1.1 Living Organisms Share Common Aspects of Structure, Function, and Energy Flow, p. 4*
15. Metabolism refers to
    a. natural selection.
    b. the chemical transformations and work of a cell.
    c. communities.
    d. mutations in DNA.
    e. cellular structure.
    Textbook Reference: *1.3 Organisms Interact with and Affect Their Environment, p. 8*

Answers

Key Concept Review

1. Model systems are useful in the study of biology because all organisms evolved from a common ancestor. Therefore, cellular pathways in a zebrafish are very similar to those found in the human. Model systems are also valuable because in many cases they can be manipulated experimentally.
2. Eukaryotic cells contain a number of membrane-bound organelles. It is hypothesized that organelles such as mitochondria and chloroplasts evolved from engulfed prokaryotic organisms that were not digested, but began a mutual relationship with the new host cell.
3. Genes are made of specific DNA nucleotide sequences in your genome. Your cell uses these genes to code for the proteins necessary for that cell type to function. Not all cell types need all the proteins that can be coded for by your genome, therefore, they do not use all of genome.
4. With the advent of new sequencing techniques, the genomes of different species can be sequenced more easily than in the past. The sequences from the 15 turtles could be used to help determine the evolutionary relationship among these species. They might also allow you to look at specific genes of interest to see how they differ among the species and how this relates to differences in their physiology or morphology.
5. The study of biology involves the smallest molecules to the entire biosphere. See Figure 1.6.
6. Natural selection refers to the differential survival and reproduction among individuals in a population according to their differing traits. In one possible scenario, these rabbits could be living in high mountains that receive a great deal of snow. The white rabbits would be most successful in hiding from predators, and thus the white coloration would increase an animal's likelihood of living and contributing offspring to the next generation.
7.

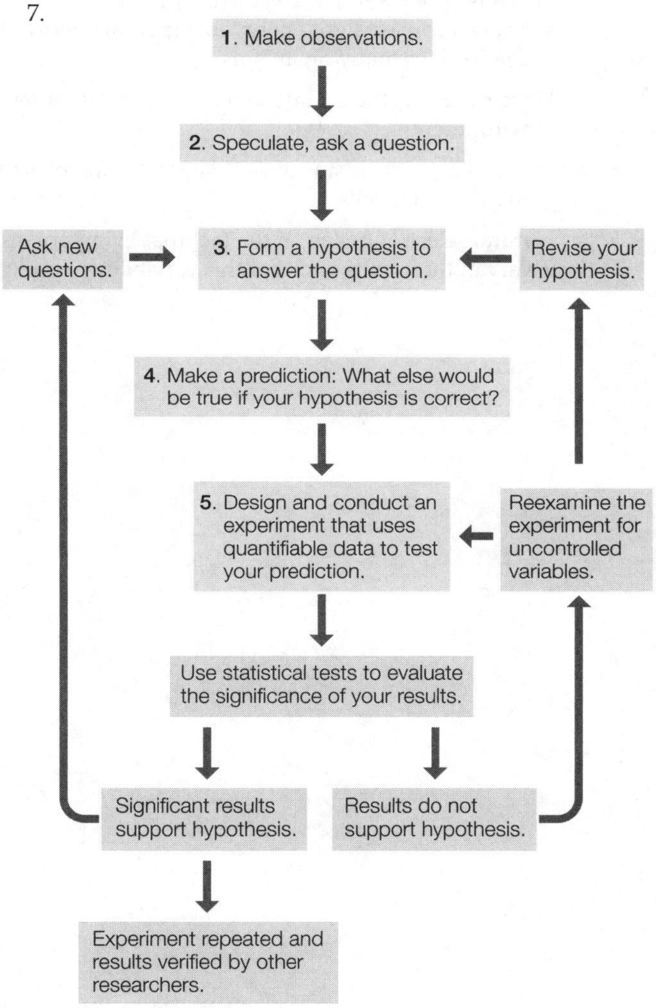

8. Often an article reporting on science news does not provide enough information for the reader to know whether a particular study followed the hypothesis–prediction approach. It pays to be a smart reader and consumer in this respect.

Test Yourself

1. **a.** Available evidence places the beginning of life at about 3.8 billion years ago.

2. **d**. Adaptations are the differences found in organisms that allow them to live in an environment.
3. **e.** Most living organisms are composed of cells, have the ability to make biological molecules, can regulate their internal environment, and get energy from the external environment.
4. **e.** Photosynthesis caused the accumulation of oxygen in the atmosphere and contributed to the formation of an ozone layer. The presence of oxygen made the evolution of aerobic organisms possible, and the ozone layer shielded Earth from harmful radiation. These two phenomena contributed to the evolution of terrestrial life. Because photosynthesis is the only metabolic process that can convert light energy to chemical energy, it provides food for many organisms.
5. **c.** Homeostasis is the maintenance of a constant internal environment.
6. **a.** Multicellular organisms have tissues that are formed from many similar cells.
7. **a.** Natural selection relies on the fact that living organisms vary in their traits and that this variability provides some individuals with an advantage in terms of survival and reproduction.
8. **b.** Plantae is a kingdom found in the domain Eukarya.
9. **e.** Genes are specific sequences of DNA that contain the information used to make proteins.
10. **b.** Evolution is a change in the frequency of genes within a population over time and can occur through natural selection. This is the major unifying principle of biology.
11. **a.** For experiments to be scientifically valid, they must be controlled.
12. **b.** Scientific hypotheses are set apart from mere conjecture by being testable and falsifiable.
13. **c.** Both eukaryotic and prokaryotic cells contain genes, proteins, and some sort of membrane. Only the eukaryotes have organelles.
14. **d.** An example is *Homo sapiens*.
15. **b.** The metabolism, or metabolic rate, of an organism is the sum of all the chemical transformations and work that it performs.

Life Chemistry and Energy 2

The Big Picture

- Understanding the chemical building blocks of all matter is essential to understanding the biology of organisms. Atoms, which contain protons, neutrons, and electrons, form elements or combine to form molecules. Molecules may be held together with covalent or ionic bonds and stabilized in three-dimensional conformations by hydrogen bonds. Hydrogen bonds are important for determining the properties of water such as its high heat capacity and cohesion.
- Two classes of biologically important molecules are carbohydrates and lipids. Carbohydrates are used for energy, storage, structure, and signaling. Polysaccharides are complex sugars made up of multiple simple monosaccharides linked by glycosidic linkages. Lipids are insoluble in water and make up a large component of the cell membrane phospholipid bilayer. Their properties depend on the length and number of double bonds of the hydrocarbon chains.
- Chemical reactions change reactants to products without creating or destroying matter. Metabolism is a series of energy-transferring reactions that fuel the processes of the cell. All reactions either give off energy (exergonic) or require energy to proceed (endergonic). Energy may be stored in chemical bonds (potential energy) and released when the bonds are broken to do work (kinetic energy).

Study Strategies

- Learning quantitative terminology and how solutions are made is often difficult. Completing practice problems is the best way to learn this material.
- It is easy to become overwhelmed by the different chemical bonds. Consult Table 2.1 in the textbook to make sure you understand the components and how they interact. Keep it handy as a reference.
- Use the figures in your book to help you visualize what atoms and molecules look like and how they interact.
- Be careful not to get lost in terminology. Think about what each term means, and focus on understanding the concept while memorizing the term.
- Go to yourBioPortal.com to review the following tutorials and activities:

  Animated Tutorial 2.1 Chemical Bond Formation

  Animated Tutorial 2.2 Macromolecules: Carbohydrates and Lipids

  Animated Tutorial 2.3 Synthesis of Prebiotic Molecules

  Web Activity 2.1 Functional Groups

  Web Activity 2.2 Forms of Glucose

  Working with Data 2.1 Synthesis of Prebiotic Molecules in an Experimental Atmosphere

Key Concept Review

2.1 Atomic Structure Is the Basis for Life's Chemistry

An element consists of only one kind of atom

Electrons determine how an atom will react

Atoms combine to form all matter. The nucleus of an atom contains a defined number of positively charged protons and neutral neutrons. Negatively charged electrons move in their outer shells. When an atom has equal numbers of electrons and protons it is electrically neutral. A dalton is the mass of one proton. An element is made up of only one type of atom. There are 94 natural elements and 24 or more man-made elements. Six elements compose the majority of every living organism: carbon, hydrogen, oxygen, nitrogen, phosphorus, and sulfur. The number of protons in an element determines its type and is known as its atomic number. Neutrons are found in every element except hydrogen. The total number of protons and neutrons make up the mass number of the atom.

The Bohr model shows the interaction between electrons in orbits around the central nucleus of the atom. Interactions between atoms involve the associations of their electrons. Electrons continuously orbit the nucleus of an atom in a defined space. The orbital of an atom is the space in which an electron is found. These patterns of orbitals compose a series of electron shells or energy levels, each with a specific number of electrons. The first shell can contain up to two electrons. The second and subsequent shells can have as many as eight electrons. The number of electrons in the outermost shell determines how the atom interacts with other atoms. Two or more linked atoms compose a molecule, resulting in stabilization of the outer electron shell of the atoms. Many biologically important atoms, such as carbon and nitrogen, are stable when they have eight electrons in the outermost shell. This phenomenon is referred to as the octet rule.

Question 1. Draw the electron shells and place electrons in the appropriate shells based on the atomic numbers of the elements carbon ($_6C$), nitrogen ($_7N$), sodium ($_{11}Na$), chlorine ($_{17}Cl$), argon ($_{18}Ar$), and oxygen ($_8O$).
Textbook Reference: *2.1 Atomic Structure Is the Basis for Life's Chemistry, pp. 17–19*

2.2 Atoms Interact and Form Molecules

Ionic bonds form by electrical attraction

Covalent bonds consist of shared pairs of electrons

Hydrogen bonds may form within or between molecules with polar covalent bonds

Polar and nonpolar substances: Each interacts best with its own kind

Functional groups confer specific properties to biological molecules

Chemical bonds link two atoms together to make a molecule. The bonds include ionic bonds, covalent bonds, and hydrogen bonds. Ions are formed when atoms lose or gain electrons, resulting in a net positive or negative charge. Cations are positively charged and have fewer electrons than protons. Anions are negatively charged and have more electrons than protons. Ionic bonds form between ions of opposite charge and are often called salts. When in solution, these bonds are weak.

Atoms share pairs of electrons to stabilize their outer shells in covalent bonding. This type of bond is very stable and strong and can be broken only with a great deal of energy. All molecules have a three-dimensional shape, and the interactions between a given pair of atoms always have the same length, angle, and direction. A molecule's shape affects how it behaves. Multiple covalent bonds may exist. Although a single covalent bond involves one pair of shared electrons, double bonds involve two pairs of shared electrons. Triple bonds share three pairs, but are rare.

Covalent bonding between atoms of the same element results in equal sharing of electrons. The attractive force that an atom exerts on electrons is known as electronegativity. Two atoms that have the same electronegativity share electrons equally and form nonpolar covalent bonds. Bonds between different elements generally result in polar covalent bonds with unequal sharing of electrons. Unequal sharing of electrons results in partial (δ) charges in molecules because one nucleus is more electronegative than the other, attracting the electrons more strongly. One atom of a molecule may be partially negative, whereas the other is partially positive. This balance of partial charges results in a polar molecule with a δ^- pole and δ^+ pole.

Hydrogen bonds form when the positively charged hydrogen atom on one polar molecule attracts a negatively charged atom on another polar molecule. This occurs between molecules of water where the positive hydrogen is attracted to the negative oxygen. This is a weak bond that is easily broken because no electrons are shared, but this type of bond is important in stabilizing the three-dimensional shape of large molecules such as proteins and DNA.

Hydrogen bonds in water provide for its heat capacity and cohesion. About 3.4 hydrogen bonds are formed by each water molecule. Water has a high heat of vaporization, so lots of heat is required to change water from a liquid to a gaseous state. Much of the energy is used to break hydrogen bonds. This makes evaporating water an effective coolant. The polar nature of water and the formation of hydrogen bonds contribute to the cohesive strength or cohesion of water. Polar molecules are hydrophilic, or "water loving," because of the partial charges and the hydrogen bonds they form. Nonpolar molecules are hydrophobic, or "water hating," and they have hydrophobic interactions. Polar molecules tend to aggregate with other polar molecules and nonpolar molecules aggregate with other nonpolar molecules.

Functional groups are small groups of atoms that provide specific properties to the molecules they are attached to. In chemical structures, the functional group is often shown bonded to an R (for "remainder") to indicate that the functional group can be attached to a wide variety of carbon skeletons. Biological molecules typically contain hydrophobic, polar, and charged functional groups which determine shape and function.

All organisms are composed of four major biological macromolecules: proteins, nucleic acids, carbohydrates, and lipids. In all life forms, the specific types of macromolecules share similar roles that are dependent on the chemical properties of their component monomers.

Polymer macromolecules are made by the covalent bonding of smaller monomers. In condensation, the removal of water links monomers together. (see Figure 2.8A).

Polymers may be broken down into their constituent monomers through hydrolysis (see Figure 2.8B). The size and structure of each polymer is related to its biological function.

Question 2. Calcium has an atomic number of 20. Draw structures for Ca and Ca^{2+}. What is the difference between these structures?
Textbook Reference: *2.2 Atoms Interact and Form Molecules, p. 19*

Question 3. Water is a polar molecule. This property contributes to cohesion and surface tension. Draw six water molecules. In your drawing, indicate how hydrogen bonding between molecules contributes to cohesion and surface

tension. Be sure to include appropriate covalent bonds in each molecule.
Textbook Reference: *2.2 Atoms Interact and Form Molecules, pp. 21–22*

2.3 Carbohydrates Consist of Sugar Molecules

Monosaccharides are simple sugars

Glycosidic linkages bond monosaccharides

Polysaccharides store energy and provide structural materials

Carbohydrates have the general formula $C_n(H_2O)_n$. They are used to store energy, to transport energy, in carbon skeletons, and as signaling molecules in organisms. They can be small monomers or large polymers. Monosaccharides are simple sugar monomers that are used in the synthesis of complex carbohydrates. Simple sugars typically have three to six carbon atoms and may exist as different isomers. Examples include six-carbon hexoses such as glucose, which is used as an energy source, or five-carbon pentoses such as ribose and deoxyribose, which make up the structural backbones of RNA and DNA, respectively.

Disaccharides, oligosaccharides, and polysaccharides are constructed from monosaccharides covalently bonded by condensation reactions that form glycosidic linkages. Disaccharides are formed by two linked monosaccharides and oligosaccharides are formed by several linked monosaccharides. Polysaccharides are very large polymers of monosaccharides that provide energy storage or structural support. The specific structure of polysaccharides contributes to their function. For instance, starch, glycogen, and cellulose are all chains of glucose, but the glycosidic bonds are in different orientations and yield very different chemical properties.

Question 4. Complex carbohydrates should be a mainstay of one's diet. What properties of carbohydrates make them excellent food sources?
Textbook Reference: *2.3 Carbohydrates Consist of Sugar Molecules, p. 24*

2.4 Lipids Are Hydrophobic Molecules

Fats and oils are triglycerides

Phospholipids form biological membranes

Lipids are insoluble in water due to many nonpolar covalent bonds. Lipids aggregate because of their nonpolar nature and are held together by van der Waals interactions. They play a variety of roles in the biology of organisms, including energy storage, structure in cell membranes and on body surfaces, and thermal insulation. Fats and oils, composed of fatty acids and glycerol, are also known as triglycerides or simple lipids, and function primarily in the storage of energy. At room temperature, fats are solid and oils are liquid. Each triglyceride contains a glycerol bound to three fatty acids with carbon atoms that are all single-bonded (saturated fatty acids) or contain double bonds (unsaturated fatty acids). During condensation, three fatty acids are covalently linked to the glycerol by ester linkages. The characteristics of the carbon bonds of fatty acids (i.e., whether they are saturated or unsaturated) influences the shape of the molecule and determines its melting point. The greater the saturation of the fatty acid chains, the higher the melting point.

Phospholipids form cellular membranes. In phospholipids, one of the hydrophobic fatty acids of a typical lipid is replaced with a hydrophilic phosphate group. This allows phospholipids to be amphipathic: they have a hydrophilic "water-loving" head and hydrophobic "water-hating" tails. In an aqueous environment, the hydrophobic tails of the phospholipids tend to aggregate, with the phosphate heads facing out. This bilayer effect allows for the establishment of a hydrophobic inside surrounded by an aqueous environment.

Question 5. Consider the following triglyceride:

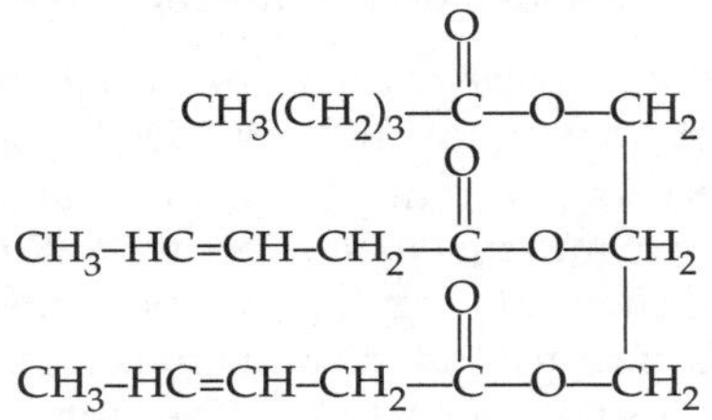

a. Circle the remnant of the glycerol portion of the triglyceride.

b. How many water molecules result from the formation of this triglyceride from glycerol and three fatty acids?
Textbook Reference: *2.4 Lipids Are Hydrophobic Molecules, p. 27*

Question 6. Which triglyceride (*A* or *B*) is probably a solid at room temperature? Explain your answer.
Textbook Reference: *2.4 Lipids Are Hydrophobic Molecules, p. 27*

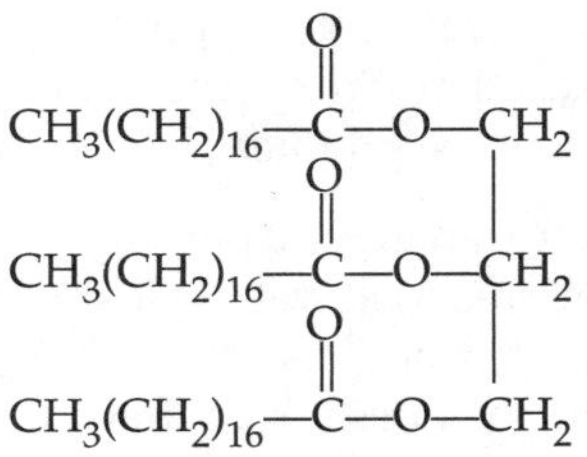

A

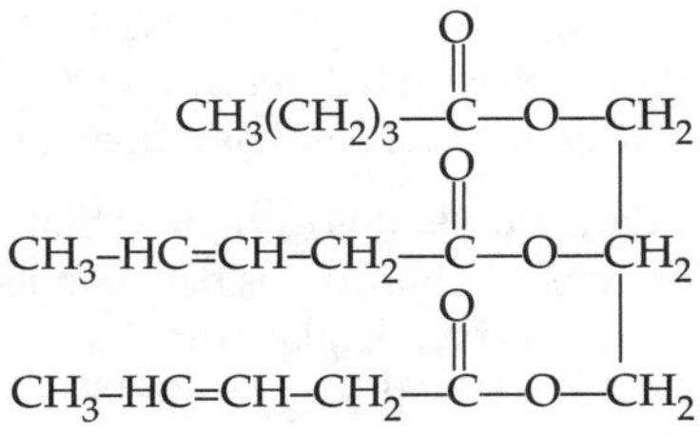

B

Question 7. Draw a phospholipid and a bilayer. What characteristics of phospholipids make them perfectly suited for membranes? What do you think might happen if phospho-

lipids did not form a bilayer? How might they arrange themselves in an aqueous environment?
Textbook Reference: *2.4 Lipids Are Hydrophobic Molecules, pp. 27–28*

Question 8. Dietary guidelines encourage people to stay away from saturated fats. What is meant by the term "saturated fat"? Why is this type of fat of more concern than unsaturated fats in the diet? What is the structural difference between saturated and unsaturated fats?
Textbook Reference: *2.4 Lipids Are Hydrophobic Molecules, p. 27*

2.5 Biochemical Changes Involve Energy

There are two basic types of energy
There are two basic types of metabolism
Biochemical changes obey physical laws

Chemical reactions involve reactants that combine or change bonding partners to produce products. Metabolism is all of the chemical reactions occurring in a living organism. Kinetic energy is energy of movement and does work. Potential energy is stored energy. Energy can be stored biologically in chemical bonds of fatty acids and other molecules. Breaking the bonds converts the energy to kinetic energy. Anabolic reactions or anabolism link together simple molecules to create complex molecules and store energy in the resulting bonds. Catabolic reactions or catabolism break down complex molecules and release stored energy.

The laws of thermodynamics apply to all matter and energy transformations. The first law of thermodynamics states that energy is neither created nor destroyed. Energy can be converted from one form to another. The second law of thermodynamics states that not all energy can be used to do work in a given reaction. When energy is converted from one form to another, some becomes unusable and there is an increase in entropy. The change in available energy that occurs during a reaction is called free energy (G). Reactions are spontaneous when the reactant has a higher G than the products.

Question 9. It is estimated that approximately 90 percent of energy that passes between levels in a food web is "lost" at each level. Explain the first law of thermodynamics and discuss why this apparent loss of energy does not contradict the law.
Textbook Reference: *2.5 Biochemical Changes Involve Energy, pp. 29–30*

Question 10. Explain how free energy, total energy, temperature, and entropy are related.
Textbook Reference: *2.5 Biochemical Changes Involve Energy, pp. 30–31*

Question 11. You decide to purchase a new water heater and start looking at the energy efficiency ratings. You find one unit that is labeled as 100 percent energy efficient, and the salesperson says that the more efficient the appliance is, the more money you will save. However, you don't trust that the store is providing accurate information, and you do not buy the product. Was this decision correct?
Textbook Reference: *2.5 Biochemical Changes Involve Energy, pp. 30–31*

Test Yourself

1. The stability of the three-dimensional shape of many large molecules is dependent on
   a. covalent bonds.
   b. ionic bonds.
   c. hydrogen bonds.
   d. van der Waals attractions.
   e. hydrophobic interactions.
   Textbook Reference: *2.2 Atoms Interact and Form Molecules, p. 21*
2. Which of the following statements about water is true?
   a. Water has a low heat of vaporization.
   b. Water has a high specific heat.
   c. When water freezes, it gains energy from the environment.
   d. All of the above
   e. None of the above
   Textbook Reference: *2.2 Atoms Interact and Form Molecules, p. 22*
3. Which of the following statements about chemical reactions is true?
   a. The bonding partners of atoms remain constant.
   b. All reactions release energy as they proceed.
   c. The bonding partners of atoms change.
   d. All reactions consume energy as they proceed.
   e. None of the above
   Textbook Reference: *2.5 Biochemical Changes Involve Energy, p. 29*
4. Which of the following is *not* a polymer?
   a. A protein
   b. A nucleic acid, such as DNA
   c. A polysaccharide carbohydrate

d. A lipid
e. All of the above are polymers.
Textbook Reference: *2.2 Atoms Interact and Form Molecules, pp. 23–24*

5. The atomic number is determined by the number of _______ in an atom.
a. protons and neutrons
b. electrons
c. neutrons
d. protons
e. protons, neutrons, and electrons
Textbook Reference: *2.1 Atomic Structure Is the Basis for Life's Chemistry, p. 17*

6. Cellulose and starch are composed of the same monomers but have structural and functional differences. Which of the following is the characteristic that accounts for those differences?
a. Different types of glycosidic linkages
b. Different numbers of glucose monomers
c. Different types of bonds holding them together
d. A linear shape in one versus a ring shape in the other
e. None of the above
Textbook Reference: *2.3 Carbohydrates Consist of Sugar Molecules, p. 25*

7. Triglycerides are synthesized from ______ and _______.
a. glycerol; amino acids
b. amino acids; cellulose
c. steroid precursors; starch
d. cholesterol; glycerol
e. fatty acids; glycerol
Textbook Reference: *2.4 Lipids Are Hydrophobic Molecules, p. 27*

8. Which of the following characteristics distinguishes carbohydrates from other macromolecule types?
a. Carbohydrates are constructed of monomers that always have a ring structure.
b. Carbohydrates never contain nitrogen.
c. Carbohydrates consist of a carbon bonded to hydrogen and a hydroxyl group.
d. Carbohydrates contain glycerol.
e. None of the above
Textbook Reference: *2.3 Carbohydrates Consist of Sugar Molecules, p. 24*

9. Which of the following statements about carbohydrates is *false*?
a. Monomers of carbohydrates have six carbon atoms.
b. Monomers of carbohydrates are linked together during dehydration.
c. Carbohydrates are energy-storage molecules.
d. Carbohydrates can be used as carbon skeletons.
e. All of the above are true.
Textbook Reference: *2.3 Carbohydrates Consist of Sugar Molecules, pp. 25–26*

10. One characteristic of phospholipids that allows them to form a bilayer is their
a. hydrophilic fatty acid tail.
b. hydrophobic head.
c. hydrophobic fatty acid tail.
d. hydrophilic glycogen acid tail.
e. All of the above
Textbook Reference: *2.4 Lipids Are Hydrophobic Molecules, p. 28*

11. A five-carbon sugar is known as a
a. glutamine.
b. glucose.
c. hexose.
d. pentose.
e. None of the above
Textbook Reference: *2.3 Carbohydrates Consist of Sugar Molecules, p. 24*

12. Covalent bonds form when
a. atoms of opposite charge are attracted to each other.
b. hydrogen and oxygen interact.
c. hydrophilic molecules bind hydrophobic molecules.
d. electrons of nonpolar substances interact.
e. atoms share electrons.
Textbook Reference*: 2.2 Atoms Interact and Form Molecules, p. 18*

13. ______ energy is the energy of movement.
a. Potential
b. Kinetic
c. Entropic
d. Enthalpic
e. Heat
Textbook Reference: *2.5 Biochemical Changes Involve Energy, p. 29*

14. ______ energy is the energy of state or position.
a. Potential
b. Kinetic
c. Entropic
d. Enthalpic
e. Physical
Textbook Reference: *2.5 Biochemical Changes Involve Energy, p. 29*

15. Which of the following statements concerning energy transformations is true?
a. Increases in entropy reduce usable energy.
b. Energy may be created during transformation.
c. Potential energy increases with each transformation.
d. Increases in temperature decrease total amount of energy available.
e. Decreases in entropy reduce usable energy.
Textbook Reference: *2.5 Biochemical Changes Involve Energy, p. 29*

16. A reaction has a ΔG of –20 kcal/mol. This reaction is
a. endergonic, and equilibrium is far toward completion.
b. exergonic, and equilibrium is far toward completion.

c. endergonic, and the forward reaction occurs at the same rate as the reverse reaction.
d. exergonic, and the forward reaction occurs at the same rate as the reverse reaction.
e. of an indeterminate nature according to the information supplied.

Textbook Reference: *2.5 Biochemical Changes Involve Energy, pp. 29–31*

Answers

Key Concept Review

1.

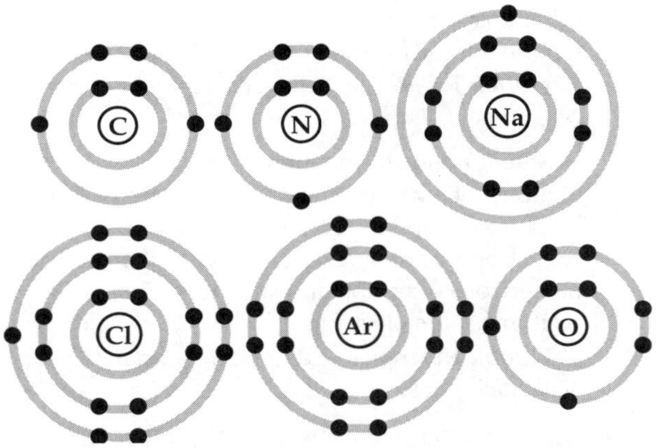

2.

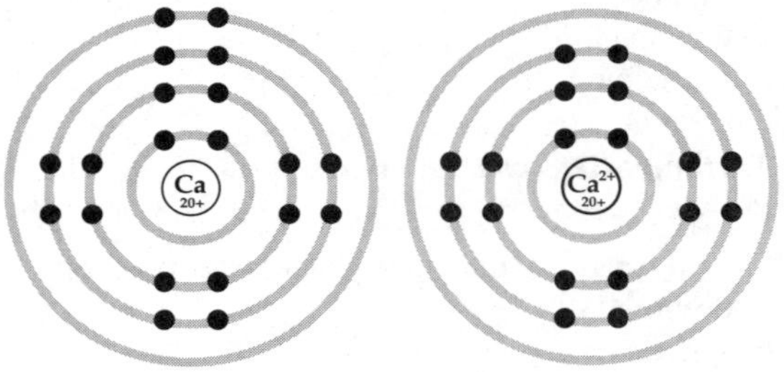

The difference between these structures is that calcium (Ca) has two electrons in its outer shell. Calcium ion (Ca^{2+}) has lost its two outer electrons and therefore has a positive charge of 2 because it has two more protons than electrons.

3. The partially positive hydrogens of one water molecule are attracted to the partially negative oxygens of another molecule of water. This attraction tends to cause water molecules to "stick" together, creating surface tension.

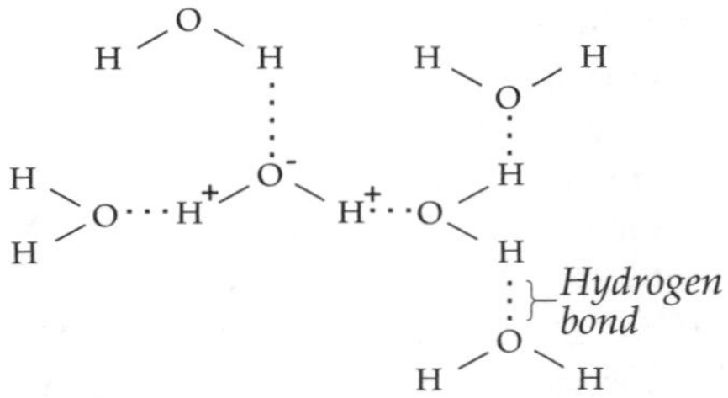

4. Complex carbohydrates are easily broken down into glucose monomers, which provide nearly all cellular energy. By storing glucose monomers in large carbohydrates, the osmotic strain on any given cell is reduced without sacrificing availability of energy.

5. a.

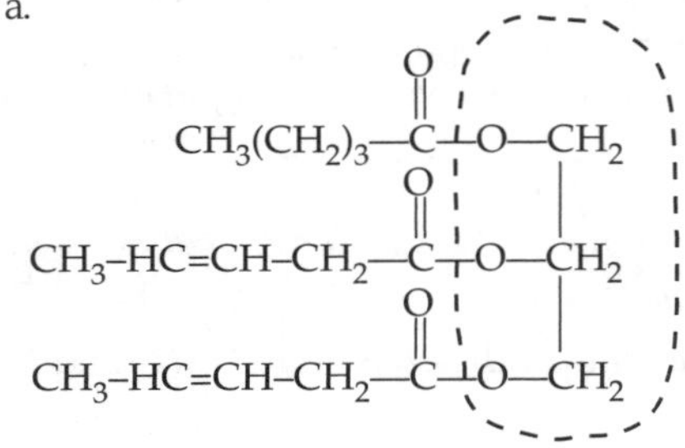

b. Three water molecules will result. One water molecule results for each of the three fatty acids added to glycerol by a condensation reaction.

6. Triglyceride A is probably solid at room temperature. Its fatty acid chains are saturated (no double bonds) and relatively long: both of which are characteristics of solid, animal-derived triglycerides.

7. The hydrophilic head and the hydrophobic tail of phospholipids allow them to have an "inside" that resists an aqueous environment and an "outside" that can reside in such an environment. When they exist as a bilayer, the hydrophobic tails aggregate. If they did not exist in two layers, the tails would still try to aggregate.

   This would result in a spherical aggregation of phospholipids called a micelle in which the tails are arranged toward the center of the sphere, away from the aqueous environment, and the heads are immersed in the aqueous environment.

Phosphatidylcholine

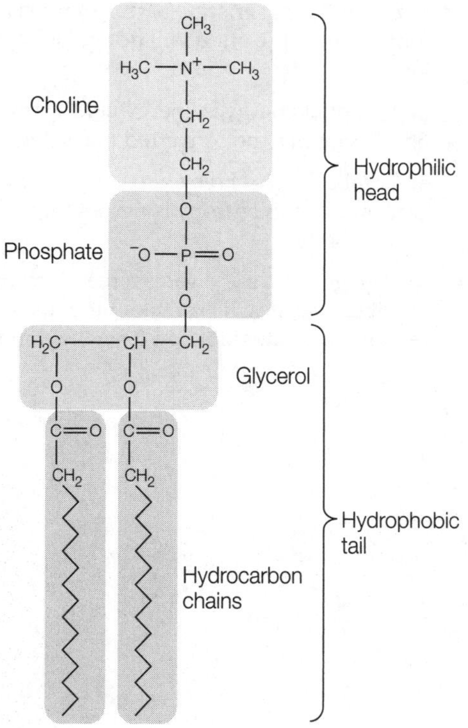

Phospholipid bilayer

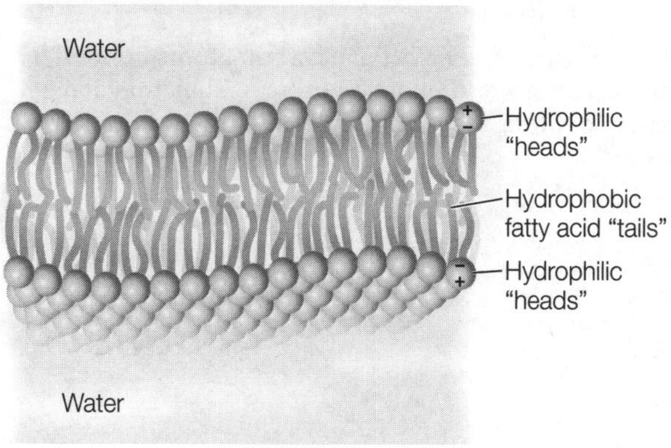

8. Saturated fats contain only single bonds and are "saturated" with hydrogen, which allows fat molecules to pack together densely. This characteristic is easily seen in that most saturated fats are solid at room temperature. Unsaturated fats contain double bonds that affect the shape of the molecule and are not "saturated" with hydrogen. This characteristic keeps them from packing together tightly, and they tend to be liquid at room temperature. Saturated fats are dangerous because of this "packing" ability, which can affect membrane function.

9. The first law of thermodynamics states that energy cannot be created or destroyed, but that it may be converted from one form to another. In the transfer between levels of a food web, approximately 90 percent of the energy is converted to unusable heat energy. There is a net loss of usable energy during each conversion, but the total amount of energy (usable and unusable) remains the same.

10. Total energy = free energy + unusable energy + absolute temperature.

11. The decision was correct. An appliance with 100 percent energy efficiency is not possible. The second law of thermodynamics indicates that every time energy is transformed, some is lost in the form of entropy. An appliance with no energy lost to entropy therefore does not exist.

Test Yourself

1. **c.** Hydrogen bonds, though weak individually, are quite effective in large numbers and are responsible for maintaining the structural integrity of many large molecules such as proteins and DNA.
2. **b.** Water has both a high heat of vaporization and a high specific heat. In addition, when water freezes it gives off a lot of energy to the environment.
3. **c.** Chemical reactions involve the combining or changing of bonding partners of the reactants to produce the products. During this process matter is neither created nor destroyed. These reactions can either release energy or require energy.
4. **d.** Proteins, polysaccharides,and nucleic acids are all polymers made by condensation to link monomers together. Lipids are monomers that do not form polymers when they interact.
5. **d.** The atomic number of an atom is the number of protons in the nuclei.
6. **a.** Starch and cellulose have different types of glycosidic linkages. This difference accounts for structural and functional differences between the two macromolecules.
7. **e.** Triglycerides are formed from one glycerol and three fatty acid molecules.
8. **c.** Carbohydrates always have carbon atoms bonded to hydrogen atoms and hydroxyl groups. They may have a variety of other associated molecules in addition to these.
9. **a.** Carbohydrate monomers may have different numbers of carbon atoms. Five-carbon monomers (pentoses) and six-carbon monomers (hexoses) are both common.
10. **c.** Phospholipids are composed of a hydrophilic head and a hydrophobic tail. When they are placed in water, the hydrophobic tails come together in the interior of the bilayer, surrounded by the hydrophilic heads facing outward.
11. **d.** Five-carbon sugars are known as pentoses and form the backbones for RNA and DNA.
12. **e.** Covalent bonds form when atoms share electron pairs.
13. **b.** The released energy is available to do work; therefore, it is kinetic energy. Some energy is lost in the form of entropy, but it will not be used by the cell to do work.
14. **a.** Potential energy is energy held within chemical bonds that may be converted to working kinetic energy.
15. **a.** Total energy = Free energy + entropy × temperature. Any increase in entropy is necessarily going to reduce free energy.
16. **b.** A negative ΔG indicates a reaction with energy being liberated; it will be a spontaneous reaction, and will tend to go in the direction from reactants to products.

Nucleic Acids, Proteins, and Enzymes

3

The Big Picture

- Sequences of nucleic acids are either single-stranded RNA or double-stranded DNA and contain the genetic information needed to code for the proteins required for life. Because DNA holds essential information, it must be reproduced exactly, so that each new cell or organism receives a complete set of DNA. The information stored in DNA is transcribed into RNA and then translated into specific proteins. Monomer amino acids are joined together to produce proteins. Each amino acid contains an R group that provides it with specific chemical characteristics. The sequence of amino acids in a protein determines its shape and ultimate function within the cell. Many proteins act as enzymes within the cell.
- Enzymes aid biological reactions by lowering the activation energy required to start the reaction. Each enzyme has a specific three-dimensional conformation that interacts specifically with the substrate. Interaction between the enzyme and substrate results in an optimal orientation for a reaction to take place. Enzymes are affected by temperature, pH, reactants, activators, and inhibitors.
- Metabolism is regulated by enzymes. Most metabolic pathways are under allosteric control. Enzyme complexes allow for interactions and regulation of adjacent active sites. Often the final product of a specific pathway regulates the commitment step of the pathway itself by feedback inhibition.

Study Strategies

- Do not get overwhelmed by the different types of macromolecules of biological systems. Remember that each macromolecule is synthesized from smaller components that give each class of molecule its specific function.
- Make a table of macromolecule characteristics. This will help you see the patterns of similarity and the differences. Avoid the temptation to memorize structures. If you understand condensation and hydrolysis and know the basic components of the macromolecules, memorizing the structures of the large macromolecules is unnecessary.
- Allosteric regulation can be confusing because the terms "allosteric regulation" and "allosteric enzyme" are related but slightly different. Remember that an allosteric enzyme is an enzyme with more than one binding site: an active site (for binding and acting on substrates) and one or more allosteric sites. When allosteric regulators are bound to the allosteric site, they control whether or not the enzyme can bind substrate.
- With this chapter, the best strategy is to take "small bites." You may find the concepts to be very unfamiliar; therefore, make sure you understand each section before proceeding. This chapter contains a large amount of terminology. Making a terminology/vocabulary list may be helpful.
- When thinking of enzyme-mediated reactions, be sure to consider that alterations on the left side of the equation lead to changes in the amount of product formed. Remember that enzymes are always conserved across the equation.
- Go to yourBioPortal.com to review the following tutorials and activities:

  Animated Tutorial 3.1 Macromolecules: Nucleic Acids and Proteins

  Animated Tutorial 3.2 Enzyme Catalysis

  Animated Tutorial 3.3 Allosteric Regulation of Enzymes

  Web Activity 3.1 Nucleic Acid Building Blocks

  Web Activity 3.2 DNA Structure

  Web Activity 3.3 Features of Amino Acids

  Working with Data 3.1 Primary Structure Specifies Tertiary Structure

Key Concept Review

3.1 Nucleic Acids Are Informational Macromolecules

Nucleotides are the building blocks of nucleic acids

Base pairing occurs in both DNA and RNA

DNA carries information and is expressed through RNA

The DNA base sequence reveals evolutionary relationships

The primary role of nucleic acids is to store and transmit hereditary information. DNA (deoxyribonucleic acid) and RNA (ribonucleic acid) are the two different types of nucleic acids. Nucleic acids are made up of nucleotide monomers. Each nucleotide monomer consists of a pentose, or five-carbon, sugar (either dexoyribose or ribose), a phosphate group, and a nitrogenous base. The bases of nucleotides contain either a single ring with six members (pyrimidine) or a fused double-ring (purine). The backbone of DNA consists of alternating sugars and phosphates with the bases projecting outward.

In both RNA and DNA, nucleotides are joined by phosphodiester linkages between the nucleotide and the phosphate group, growing from the 5′ to the 3′ direction. Nucleic acids are either short oligonucleotides (such as RNA molecules used for DNA replication or regulation of gene expression) or longer polynucleotides (such as DNA). Four types of nucleotides are found in DNA: two purines (adenine and guanine) and two pyrimidines (thymine and cytosine).

In RNA, thymine is replaced by uracil. In DNA, bases are capable of complementary base pairing, with adenine pairing with thymine and cytosine pairing with guanine. Complementary base pairing is facilitated by the size and shape of the molecules, the geometry of the sugar–phosphate backbone, and hydrogen bonding. DNA is typically found as a double helix with two complementary strands paired and twisted together. The two strands in DNA are stabilized by hydrogen bonds and run in opposite directions. Most RNA is a single strand of nucleotides which in some cases can interact with other RNA molecules through hydrogen bonding to produce different three-dimensional structures.

The sequences of bases in DNA provide a means of storing information in the cell. To ensure the correct use of this information, DNA is replicated precisely and can be transcribed into RNA, which in turn is translated into specific polypeptides. Replication and transcription rely on properties of the five base pairs. The base pairs A-T and G-C in DNA are replaced with their appropriate RNA counterpart (A-U and G-C) during transcription. When a molecule of DNA is replicated, the entire DNA molecule is copied. The genome contains a complete set of an organism's DNA. The specific portions of the DNA that encode for proteins are genes; they are transcribed only when the cell requires the specific proteins.

Similarities in base sequences of DNA can help determine evolutionary relationships among organisms because it is a conserved and regular molecule. Closely related species tend to have greater similarity in base sequences compared to distantly related species. The increased ability of scientists to sequence DNA is leading to insights into the evolutionary relationships of organisms that were not available based on anatomical and behavioral comparisons.

Question 1. If you were to transcribe both strands of the DNA sequence below, what two RNA sequences would result?
Textbook Reference: *3.1 Nucleic Acids Are Informational Macromolecules, p. 36*

A A G C G T C

T T C G C A G

Question 2. In the diagram below, use the base-pairing rules for DNA and RNA to label the strand of RNA (right) that is complementary to the single strand of DNA (left), where C = cytosine, G = guanine, A = adenine, T = thymine, and U = uracil. Then circle and label an example of a nucleotide and a nucleoside, showing the double-stranded DNA and RNA hybrid. Based on the orientation of the sugar molecules, label the four ends of the molecule as 3′ or 5′, showing the double-stranded DNA and RNA hybrid.
Textbook Reference: *3.1 Nucleic Acids Are Informational Macromolecules, pp. 36–38*

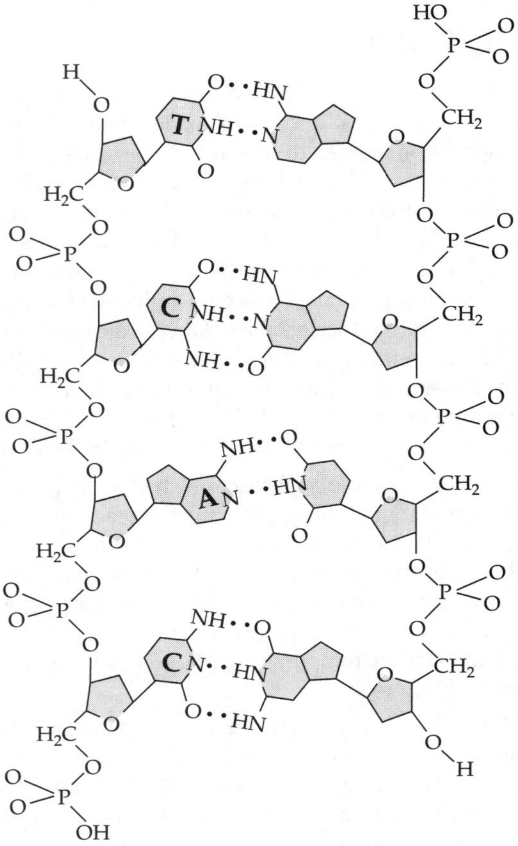

Question 3. What properties of a DNA structure make it well suited for its function as an informational molecule?
Textbook Reference: *3.1 Nucleic Acids Are Informational Macromolecules, p. 37*

3.2 Proteins Are Polymers with Important Structural and Metabolic Roles

Amino acids are the building blocks of proteins

Amino acids are bonded to one another by peptide linkages

Higher-level protein structure is determined by primary structure

Environmental conditions affect protein structure

Proteins have many functions in an organism, including enzymatic activity, defense, hormonal regulation, signaling, storage, structural support, transport, and genetic regulation. Proteins are polymers composed of amino acid monomers, each of which has a central carbon atom with a hydrogen atom, amino group, and carboxyl group attached to it. The shape and structure of proteins are influenced by the properties of the side chains of the constituent amino acids. Proteins contain only 20 amino acids, which are grouped based on their side chains. Five amino acids have charged side chains, which attract water; five have polar, hydrophilic side chains that form hydrogen bonds; seven have hydrophobic, nonpolar side chains; and three have special hydrophobic functions.

The side chain of cysteine, –SH, can form disulfide bridges with other cysteines and influences protein chain folding. Glycine is small and fits into folds of proteins. Proline has limited bonding ability and often contributes to looping in proteins. Proteins range in size from small oligopeptides (also called peptides) to large polypeptides (also called proteins). Amino acids are linked together during condensation. The resulting bonds between amino and carboxyl groups are called peptide bonds or linkages. Peptide chains always begin with the N terminus and end with the C terminus. The subsequent sequence of amino acids is known as the primary structure of the protein.

A protein's primary structure refers to the basic amino acid sequence. The primary structure is held together by covalent bonds, however all higher levels of structure are determined largely by the specific amino acid sequence in the primary structure of the protein. Secondary structure refers to regular, repeated spatial patterns of the amino acid chain stabilized by hydrogen bonds. The α helix consists of a right-handed coiling of a single polypeptide chain. The β pleated sheet is formed from two or more hydrogen-bonded chains. Tertiary structure refers to an amino acid chain that is bent and folded into a more complex pattern. This structure is stabilized by covalent disulfide bridges, hydrophobic interactions, van der Waals forces, ionic interactions, and hydrogen bonds. The sequence of R groups in the primary structure is responsible for this level of folding. Because a protein's three-dimensional structure is stabilized by relatively "weak" bonds, environmental conditions such as temperature, pH (H^+ concentration), and the presence of high concentrations of polar or nonpolar substance can change the shape of the protein into an inactive form. This is called denaturation, and it can be reversible or irreversible depending on the protein. Quaternary structure refers to the three-dimensional interactions of multiple protein subunits. Hydrophobic interactions, hydrogen bonds, and ionic interactions are all involved in maintaining the quaternary structure.

Question 4. Differentiate between primary, secondary, tertiary, and quaternary structures as they relate to proteins. If a protein is immersed in an unfavorable pH solution, which structures are most likely to disassociate first, and why?
Textbook Reference: *3.2 Proteins Are Polymers with Important Structural and Metabolic Roles, pp. 42–44*

Question 5. How do the different chemical properties of amino acid R groups contribute to the final three-dimensional shape of the molecule?
Textbook Reference: *3.2 Proteins Are Polymers with Important Structural and Metabolic Roles, pp. 42–44*

Question 6. Suppose that you have isolated a protein with the following amino acid sequence: RSCFLA. Referring to Table 3.2 of the textbook, draw this protein. In your drawing, label the N terminus and the C terminus and show all peptide linkages. How many water molecules are generated in the synthesis of this protein?
Textbook Reference: *3.2 Proteins Are Polymers with Important Structural and Metabolic Roles, p. 40*

3.3 Some Proteins Act as Enzymes to Speed up Biochemical Reactions

To speed up a reaction, an energy barrier must be overcome

Enzymes bind specific reactants at their active sites

Energy barriers between reactants and products slow down reactions. Activation energy (E_a) must be added to get past this barrier; it may be thought of as a little "shove" to get the reaction going. For reactions to take place, the reactants must be slightly destabilized and turned into a transition state that has higher free energy than that of either the reactants or the products. In biological systems, this is accomplished with the help of a catalytic enzyme. Most biological catalysts are protein enzymes that provide scaffolding upon which reactions necessary for life take place. Enzymes are highly specific biological catalysts that have specific substrate-binding areas on their surfaces called active sites. The enzyme's specificity comes from the shape of its active site. The names for enzymes typically end in the suffix *-ase*. Only specific substrates can fit in and bind with an enzyme's active site. Once bound, the site is referred to as an enzyme–substrate complex (ES). The enzyme alters the conformation of the substrate, lowering the activation energy for the reaction.

The enzyme–substrate complex gives rise to the product and the enzyme remains unchanged. The reaction can be represented as follows:

$$E + S \rightarrow ES \rightarrow E + P$$

Enzymes speed up reactions by orienting substrates for maximum chemical interactions, by inducing strain on the substrate, or by adding chemical groups to substrates. Binding of small molecules to the active sites depends on hydrogen bonds, charge interactions, and hydrophobic interactions. Binding of substrate by an enzyme can change the enzyme's shape to produce an "induced fit." This is made possible by the large size of an enzyme, which helps it position the correct amino acids at the active site and regulate the site's shape. Cofactors, coenzymes, and prosthetic groups all contribute to an enzyme's activity. Cofactors are inorganic ions such as zinc, copper, or iron. Coenzymes are organic molecules that bind in the active site and can be thought of as a co-substrate. Coenzymes are chemically changed during the reaction and regenerated in other pathways. Prosthetic groups are permanently bound to the enzyme. The rate of a reaction increases as the substrate concentration increases, until all enzyme active sites are occupied. At that point, no amount of additional substrate will increase the reaction rate because all of the active sites are saturated and no more enzyme is available for catalysis. Enzyme efficiency is measured in turnover number, or how fast an enzyme can convert substrate to product and free up its active site.

Question 7. Discuss how a protein's three-dimensional structure makes it perfect for acting as a carrier and receptor molecule. In what ways are proteins uniquely suited for this function compared to other macromolecules?
Textbook Reference: *3.3 Some Proteins Act as Enzymes to Speed up Biochemical Reactions, p. 47*

Question 8. Explain how substrate concentration affects the rate of an enzyme-mediated reaction.
Textbook Reference: *3.3 Some Proteins Act as Enzymes to Speed up Biochemical Reactions, p. 49*

Question 9. Label the graph below with the following: Activation energy (E_a) for catalyzed reaction, activation energy for uncatalyzed reaction, ΔG for catalyzed reaction, ΔG for uncatalyzed reaction, free energy of reactants, free energy of products, least stable state on graph, most stable state on graph.
Textbook Reference: *3.3 Some Proteins Act as Enzymes to Speed up Biochemical Reactions, p. 47*

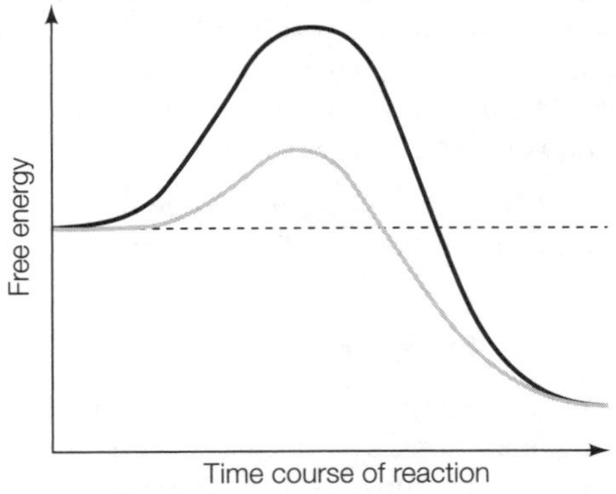

3.4 Regulation of Metabolism Occurs by Regulation of Enzymes

- Enzymes can be regulated by inhibitors
- An allosteric enzyme is regulated via changes in its shape
- Some metabolic pathways are usually controlled by feedback inhibition
- Enzymes are affected by their environments

All life must maintain stable internal conditions, or homeostasis, and this is regulated by enzymes. The chemical reactions occurring in an organism are organized into specific metabolic pathways that are catalyzed by specific enzymes at each step. Enzyme function within a cell is regulated by both the amount of the enzyme and the activity of the enzyme. Inhibitors are substances that bind with enzymes to inhibit or slow their function. Irreversible inhibition occurs when an inhibitor forms a covalent bond with an enzyme and permanently destroys the active site. In reversible inhibition, an inhibitor binds noncovalently to the active site and alters it, but the inhibition is reversible, depending on the concentration of the inhibitor and substrate. Some reversible inhibitors are called competitive inhibitors because they compete with the substrate for the active site. Others are called noncompetitive inhibitors because they bind elsewhere but alter the active site so that the substrate cannot bind, or the rate of binding is reduced.

Allostery is the regulation of an enzyme by a molecule that binds to the enzyme at a site other than the active site, resulting in a change in the enzyme shape. Noncompetitive inhibitors are an example of inhibitors that work allosterically. Most allosteric enzymes exist naturally in an active form and an inactive form, and they can switch back and forth between the two forms. When an allosteric regulator binds to the enzyme, the enzyme becomes locked in either its active form or inactive form. Enzymes' allosteric sites are modified either covalently or noncovalently, causing a change in enzyme shape. Protein kinases are important activation enzymes that regulate other metabolic enzymes.

The first step in an enzyme-mediated pathway is referred to as the commitment step. Once initiated, the pathway is followed to completion. Frequently, the final product allosterically inhibits the enzyme of the commitment step, preventing overproduction of the final product. This process is called end-product inhibition or feedback inhibition.

Enzyme activity is dependent upon the surrounding environment. Changes in pH influence charges of amino and other groups on the enzyme or substrate. This may change the folding of the enzyme or its ability to interact with the substrate and drastically alter its catalytic ability. Enzymes typically have an ideal pH at which they function best. Increases in temperature may aid in reduction of activation energy by adding kinetic energy; however, this may also result in the denaturing of enzymes. Each enzyme has an optimal temperature. Organisms may produce different forms of the same enzyme, called isozymes, which have different optimal temperatures.

Question 10. You discover a fish that possesses two versions of an enzyme. One version functions best at a high temperature, around 37°C, and the other at a lower temperature, around 30°C. Draw a graph showing how temperature affects enzyme activity of these two enzymes. Why would a fish have versions of an enzyme that function optimally at two different temperatures?
Textbook Reference: *3.4 Regulation of Metabolism Occurs by Regulation of Enzymes, p. 52*

Question 11. Amylase is a digestive enzyme that breaks down starch and secreted in the human mouth. Although it functions well in the there, it ceases to function once it contacts the acidic environment of the stomach. Explain why amylase does not function in the stomach.
Textbook Reference: *3.4 Regulation of Metabolism Occurs by Regulation of Enzymes, p. 52*

Question 12. Figure 3.18 shows the behavior of an allosteric enzyme that has binding sites for a positive regulator. Describe the behavior of an enzyme with binding sites for a negative regulator, as opposed to a positive regulator.
Textbook Reference: *3.4 Regulation of Metabolism Occurs by Regulation of Enzymes, p. 51*

Question 13. Enzyme X has an optimal pH of 9.0 and is completely inactive at pH 7.0 and pH 11.0. Draw the predicted enzyme activity and determine if this enzyme prefers an acidic, neutral, or basic environment.
Textbook Reference: *3.4 Regulation of Metabolism Occurs by Regulation of Enzyme, p. 52*

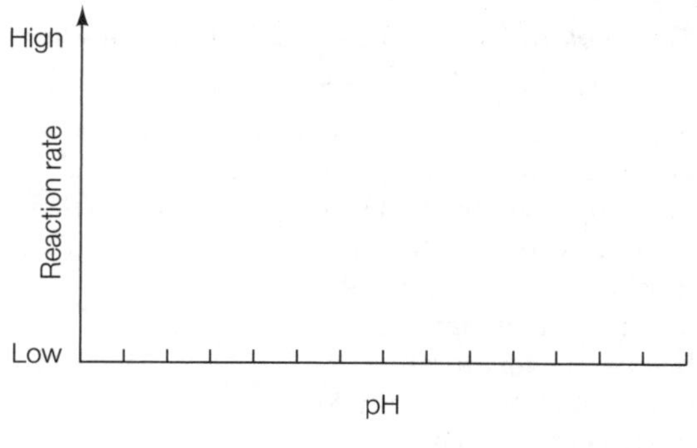

Test Yourself

1. DNA utilizes the bases guanine, cytosine, thymine, and adenine. In RNA, _______ is replaced by _______.
   a. adenine; arginine
   b. thymine; uracil
   c. cytosine; uracil
   d. cytosine; arginine
   e. cytosine; thymine
   Textbook Reference: *3.1 Nucleic Acids Are Informational Macromolecules, p. 35*
2. The pairing of purines with pyrimidines to create a double-stranded DNA molecule is called
   a. complementary base pairing.
   b. phosphodiester linkages.
   c. antiparallel synthesis.
   d. dehydration.
   e. the genome.
   Textbook Reference: *3.1 Nucleic Acids Are Informational Macromolecules, p. 36*
3. The end product of transcription is _______ and the end product of translation is _______.
   a. proteins; DNA
   b. proteins; RNA
   c. RNA; DNA
   d. DNA; RNA
   e. RNA; proteins
   Textbook Reference: *3.1 Nucleic Acids Are Informational Macromolecules, p. 38*
4. The double-helix formation of DNA is caused by
   a. ionic bonds.
   b. covalent bonds.
   c. hydrogen bonds.
   d. hydrophobic side chains.
   e. None of the above
   Textbook Reference: *3.1 Nucleic Acids Are Informational Macromolecules, p. 38*
5. In DNA and RNA, nucleotides are joined by
   a. phosphodiester linkages.
   b. hydrogen bonds.
   c. peptide linkages.
   d. glycosidic linkages.
   e. All of the above
   Textbook Reference: *3.1 Nucleic Acids Are Informational Macromolecules, p. 35*
6. Which of the following statements about proteins is *false*?
   a. Enzymes are proteins.
   b. Proteins are part of the phospholipid bilayer.
   c. Some hormones are proteins.
   d. Proteins are structural components of the cell.
   e. All of the above are true, none is false.
   Textbook Reference: *3.2 Proteins Are Polymers with Important Structural and Metabolic Roles, p. 39*
7. A disulfide bridge is formed by
   a. two cysteine side chains.
   b. two glycerol linkages.
   c. two proline side chains.
   d. condensation.
   e. hydrolysis.

Textbook Reference: *3.2 Proteins Are Polymers with Important Structural and Metabolic Roles, p. 41*

8. Proteins consist of amino acids linked by
a. noncovalent bonds.
b. peptide bonds.
c. phosphodiester bonds.
d. van der Waals forces.
e. Both a and b
Textbook Reference: *3.2 Proteins Are Polymers with Important Structural and Metabolic Roles, p. 41*

9. The R groups of amino acids located on the surface of protein molecules in the interior of biological membranes are
a. hydrophobic.
b. hydrophilic.
c. polar.
d. able to form disulfide bridges.
e. electrically charged.
Textbook Reference: *3.2 Proteins Are Polymers with Important Structural and Metabolic Roles, p. 40*

10. Which of the following represents an enzyme-catalyzed reaction? (E = enzyme, P = product, S = substrate)
a. $E + P \rightarrow E + S$
b. $E + S \rightarrow E + P$
c. $E + S \rightarrow P$
d. $E + S \rightarrow E$
e. $P + S \rightarrow E$
Textbook Reference: *3.2 Proteins Are Polymers with Important Structural and Metabolic Roles, p. 47*

11. Enzymes are biological catalysts that function by
a. increasing free energy in a system.
b. lowering activation energy of a reaction.
c. lowering entropy in a system.
d. increasing temperature near a reaction.
e. altering the equilibrium of the reaction.
Textbook Reference: *3.3 Some Proteins Act as Enzymes to Speed up Biochemical Reactions, p. 46*

12. Which of the following contributes to the specificity of enzymes?
a. Each enzyme is active over a wide range of temperatures and pH conditions.
b. Each enzyme has an active site that interacts with many substrates.
c. Substrates themselves may alter the active site slightly for optimum catalysis.
d. Enzymes are more active at higher temperatures than at lower temperatures.
e. All of the above
Textbook Reference: *3.3 Some Proteins Act as Enzymes to Speed up Biochemical Reactions, p. 47*

13. Coenzymes and cofactors, as well as prosthetic groups, assist enzyme function by
a. stabilizing three-dimensional shape.
b. assisting with the binding of enzyme and substrate.
c. maintaining active sites in an active configuration.
d. reversibly binding to the enzyme to regulate the enzyme's activity.
e. a, b, and c
Textbook Reference: *3.3 Some Proteins Act as Enzymes to Speed up Biochemical Reactions, p. 48*

14. Which of the following statements about enzymes is true?
a. They are consumed by the enzyme-mediated reaction.
b. They are not altered by the enzyme-mediated reaction.
c. They raise activation energy.
d. They can be composed of RNA or proteins.
e. They are rarely regulated.
Textbook Reference: *3.3 Some Proteins Act as Enzymes to Speed up Biochemical Reactions, p. 46*

15. Enzymes alter the _______ of the reaction.
a. ΔG value
b. activation energy
c. equilibrium
d. rate
e. Both a and b
Textbook Reference: *3.4 Regulation of Metabolism Occurs by Regulation of Enzymes, p. 47*

16.–17. Ascorbic acid, which is found in citrus fruits, acts as an inhibitor to catecholase, the enzyme responsible for the browning reaction in fruits such as apples, peaches, and pears.

16. One explanation for the inhibiting function of ascorbic acid could be its similarity, in terms of size and shape, to catechol, the substrate of the browning reaction. If this explanation is correct, then this inhibition is most likely an example of _______ inhibition.
a. competitive
b. indirect
c. noncompetitive
d. allosteric
e. feedback
Textbook Reference: *3.4 Regulation of Metabolism Occurs by Regulation of Enzymes, pp. 50–51*

17. Suppose further studies indicate that ascorbic acid is not similar to catechol in size and shape but that the pH of the ascorbic acid solution alters the protein folding of catecholase. If this is true, then this inhibition is most likely an example of
a. competitive inhibition.
b. enzyme denaturation.
c. noncompetitive inhibition.
d. allosteric regulation.
e. feedback inhibition.
Textbook Reference: *3.4 Regulation of Metabolism Occurs by Regulation of Enzymes, pp. 50–51*

18. Metabolism is organized into pathways that are linked in which of the following ways?

a. All cellular functions feed into a central pathway.
b. All steps in the pathway are catalyzed by the same enzyme.
c. The product of one step in the pathway functions as the substrate in the next step.
d. Products of the pathway accumulate and are secreted from the cell.
e. Different substrates are acted on by the same enzyme.

Textbook Reference: *3.4 Regulation of Metabolism Occurs by Regulation of Enzymes, p. 51*

19. In the pathway A + B → C + D, enzyme X facilitates the reaction. If compound D inhibits enzyme X, you would conclude that
a. enzyme X is an allosteric inhibitor of the reaction.
b. compound D is an allosteric stimulator of the reaction.
c. compound D is a competitive inhibitor of the reaction.
d. enzyme X is subject to feedback stimulation.
e. compound D is a coenzyme in the reaction.

Textbook Reference: *3.4 Regulation of Metabolism Occurs by Regulation of Enzymes, pp. 50–51*

20. You are investingating a newly-discovered species. This organism lives in acidic pools in volcanic craters where temperatures often reach 100°C and normally stay above 90°C. You determine that it has a surface enzyme that catalyzes a reaction leading to its protective coating, and you decide to study this enzyme in the laboratory. At which temperature would you most likely find optimal activity of this enzyme?
a. 0°C
b. 37°C
c. 55°C
d. 95°C
e. 105°C

Textbook Reference: *3.4 Regulation of Metabolism Occurs by Regulation of Enzymes, p. 52*

21. You fill two containers with identical amounts of reactants A and B and enzymes 1–4. In the reactions shown below, if product D inhibits enzyme 2 and product F is an allosteric stimulator of enzyme 1, what will be the final result if you add extra product D to the second container? (Assume that both containers are given enough time for the reactions to go to completion.)

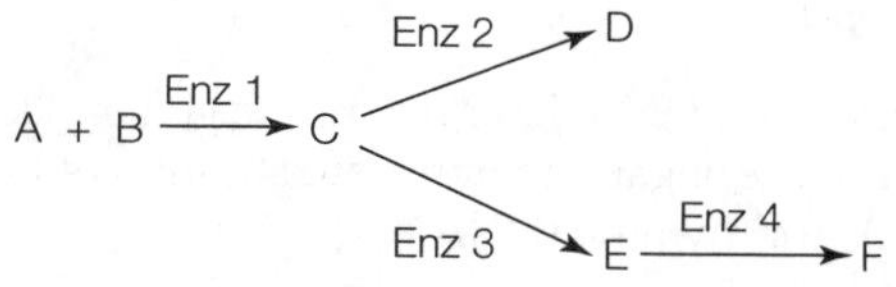

a. The concentration of product C will increase and there will be no change in the concentration of product F compared to that of the first container.
b. The concentration of reactants A and B will increase relative to the first container.
c. The concentration of product F will increase in the second container because more of D is converted back to C.
d. The concentration of products E and F will both increase in the second container, since D inhibits enzyme 2.
e. The concentration of product F will increase relative to the first container, since enzyme 2 will have been inhibited from converting C into D.

Textbook Reference: *3.4 Regulation of Metabolism Occurs by Regulation of Enzymes, pp. 50–51*

22. You are given an unlabeled enzyme in a container and told to add a compound that will bind irreversibly to the enzyme and increase its function. You ask for information about the enzyme, but your instructor simply hands you a list of possible compounds. Based on what you have learned about the enzyme partners below, which one is the best choice?
a. Coenzyme A
b. Zinc (Zn^{2+})
c. Flavin
d. ATP
e. NAD

Textbook Reference: *3.4 Regulation of Metabolism Occurs by Regulation of Enzymes, p. 48*

Answers

Key Concept Review

1. RNA sequence for top: UUCGCAG; RNA sequence for bottom: AAGCGUC.

2.

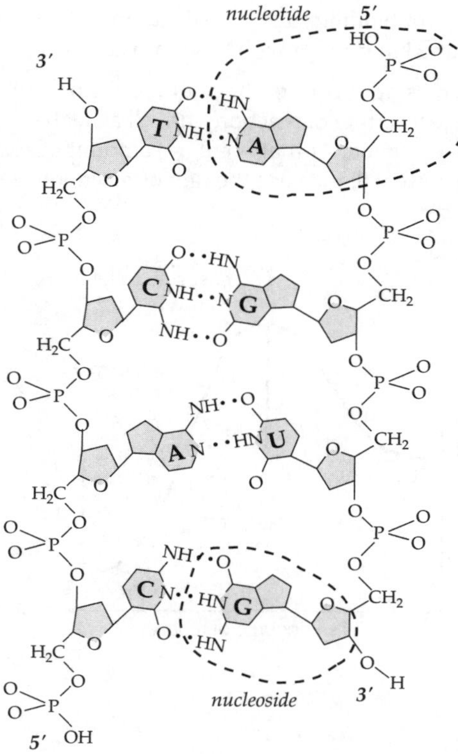

3. DNA structure is highly regular and conserved, which makes it usable by all cells, while the different bases allow it to encode specific genetic information.
4. See Figure 3.7 in the textbook for diagrams of primary, secondary, tertiary, and quaternary structures. The quaternary structure is the least stable and breaks down first in unfavorable conditions. Protein structures continue to denature by unfolding the tertiary structures, then the secondary structures. Primary structure is maintained by covalent bonds and is the last to break down. Disruptions in tertiary and quaternary structures are often reversible. Disruptions in primary or secondary structures are generally irreversible.
5. The size of the R group, the charge of the R group, and any special binding properties all contribute to the final orientation of a protein molecule.
6. Five water molecules are generated.

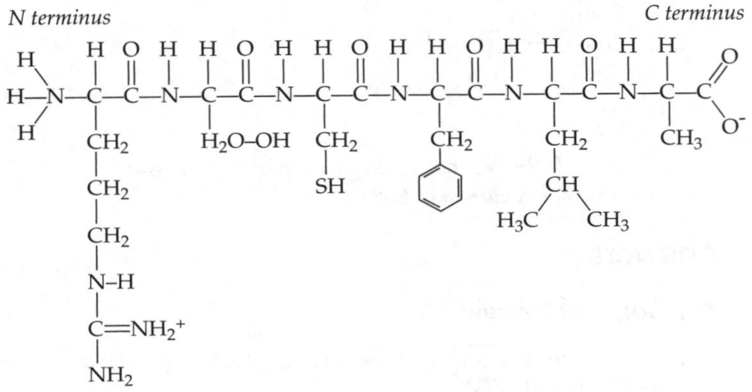

7. The three-dimensional nature of proteins allows them to form binding sites. These binding sites are uniquely shaped to interact with other molecules.
8. Increasing substrate concentration will result in an increased rate of reaction until all available active sites are occupied. At that point, no amount of substrate increase will increase the rate of reaction (see Figure 3.15).
9.

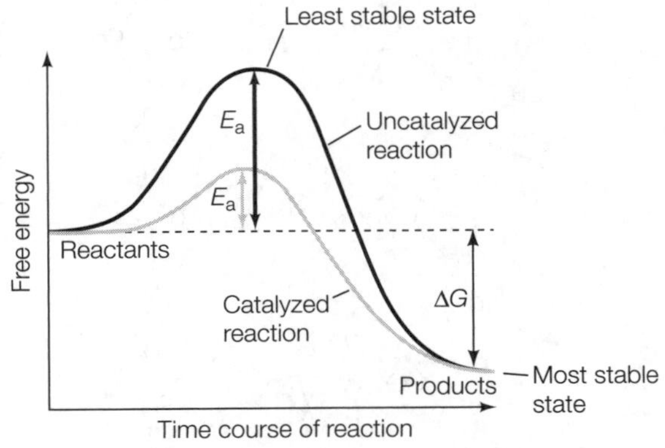

10. It is possible that this fish lives in an environment that changes temperature over some time period, (e.g., daily or seasonally). The two different versions of the enzyme allow it to function well over a range of body temperatures.

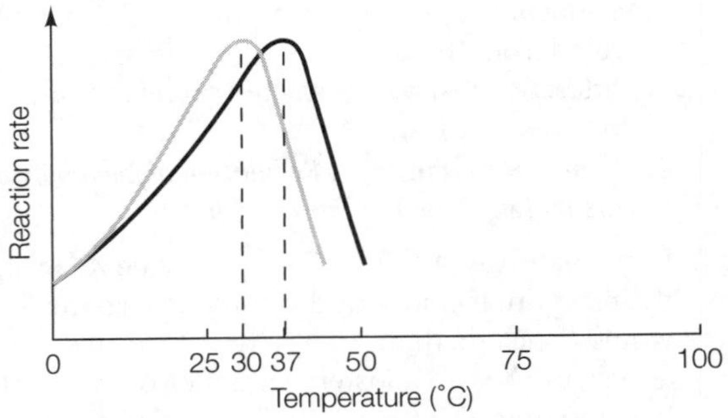

11. The pH optimum of amylase is approximately 7. At this pH, the protein has a three-dimensional shape that allows starch to bind to its active site and catalyze its hydrolysis. When it is at the stomach pH (approximately 2), the protein is denatured, and its three-dimensional shape and active site are lost; therefore, it can no longer catalyze the reaction.
12. A negative regulator would stabilize the enzyme in its inactive form. In the absence of the regulator, the enzyme would alternate between its inactive and active forms. When it encounters substrate, it is able to bind only if it happens to be in its active form. When bound to the negative regulator, the enzyme would be fixed in its inactive form, and it would not be able to the bind substrate.
13.

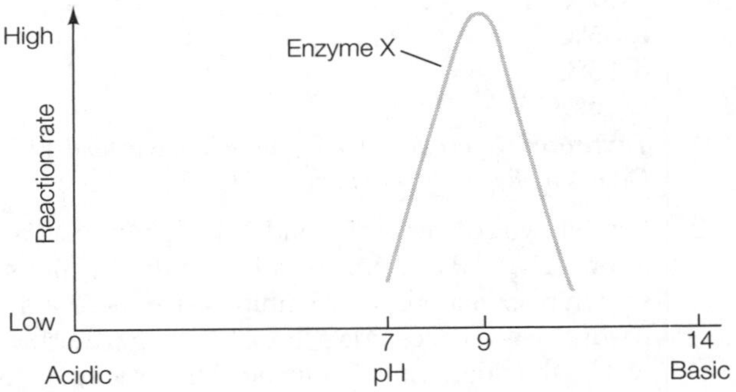

The enzyme prefers a basic environment. Your graph should be symmetrical around a peak at pH 9.0.

Test Yourself

1. **b.** RNA and DNA differ by one oxygen molecule in their ribose sugar and in the substitution of uracil (RNA) for thymine (DNA).
2. **a.** Complementary base pairing results from the attraction of charges and the ability of purines to pair with pyrimidines through hydrogen bond formation.
3. **e.** Transcription involves making RNA from DNA, and translation involves making proteins from RNA.

4. **c.** The double-helix structure of DNA is due to the hydrogen bonding.

5. **a.** The sugar of one nucleotide and the phosphate of the next are joined by phosphodiester linkages in both DNA and RNA.

6. **b.** Proteins do not make up the phospholipid bilayer, which is composed of phospholipids. Proteins do perform all of the other functions.

7. **a.** Disulfide bridges are formed when the –SH groups of two cysteines interact.

8. **b.** The peptide bond that links amino acids to form proteins is a type of covalent bond.

9. **a.** The interior of the membrane is hydrophobic; therefore, an embedded protein would have hydrophobic residues.

10. **b.** Substrate is converted to product and the enzyme is unchanged.

11. **b.** Enzymes reduce activation energy and speed up reactions.

12. **d.** Enzymes are specific to particular substrates that may actually "adjust" the fit of the active site. They also function in specific narrow optimum ranges of pH and temperature.

13. **e.** Cofactors, coenzymes, and prosthetic groups assist with the maintenance of an enzyme's three-dimensional shape, and the conformation of the active prosthetic groups are permanently bound to the enzyme.

14. **b.** Enzymes are not consumed or altered in any way during an enzyme-mediated reaction, and they function to lower the activation energy of a reaction. Ribozymes (composed of RNA) are catalysts, but they are not true enzymes.

15. **b.** The free energy levels of the reactants and products are not changed by enzymes. Only the activation energy is altered. See Figure 3.12.

16. **a.** Competitive inhibitors compete for the active site with the substrate.

17. **b.** Destroying the three-dimensional structure of an enzyme, or denaturing it, is usually irreversible.

18. **c.** Within a given pathway, the products of the preceding step act as substrates for subsequent steps.

19. **c.** Products of a reaction that inhibit the enzymes via feedback inhibition are allosteric inhibitors.

20. **d.** Enzymes typically work at a maximal rate at a particular temperature or within a range of temperatures. This optimal temperature tends to be correlated with the body temperature of the organism.

21. **e.** The addition of more D to the second container will reduce the activity of enzyme 2. The pathway from A + B going all the way to F will be the predominant reaction that takes place, leading to a greater final concentration of F in the second container. The first container will convert some of C into D, and thus the level of intermediate E and final product F will be lower than in the second container. (Hint: Draw a diagram of the experiment.)

22. **c.** Of the five compounds listed, only one is a prosthetic group, which is irreversibly bound to its target enzyme.

Cells: The Working Units of Life

4

The Big Picture

- Cell theory states that all living things are composed of cells and all cells come from preexisting cells. Living things employ one of two basic cell forms; the small prokaryotic cell has no organelles, and the eukaryotic cell has many membrane-enclosed organelles.
- The organelles of eukaryotic cells have a variety of functions, including information storage and transmission, modification and packaging of cellular products, and energy transformation. See Figure 4.7 to understand the organization, structure, and function of each component.
- The cytoskeleton of the cell provides support, structure, and protection. Components of the cytoskeleton also aid in movement of the entire cell or cell components. Cilia and flagella provide cell mobility.

Study Strategies

- Students often are overwhelmed by the number of organelles and their diverse functions in a eukaryotic cell. Remember that each organelle has a distinct location within the cell and that an organelle's structure is related to its function.
- Terminology is the most difficult part of this material. The terminology is easier to learn if you understand the concepts before you memorize the terms.
- Refer to Figure 4.7 while you are studying the organelles and their components in order to visualize how their functions relate to their structure.
- Go to yourBioPortal.com to review the following tutorials and activities:

  Animated Tutorial 4.1 Eukaryotic Cell Tour

  Animated Tutorial 4.2 The Golgi Apparatus

  Web Activity 4.1 The Scale of Life

  Web Activity 4.2 Lysosomal Digestion

  Web Activity 4.3 Animal Cell Junctions

Key Concept Review

4.1 Cells Provide Compartments for Biochemical Reactions

Cell size is limited by the surface area-to-volume ratio

Cells can be studied structurally and chemically

The plasma membrane forms the outer surface of every cell

Cells are classified as either prokaryotic or eukaryotic

The cell theory states that the cell is the basic unit of life, that all organisms are composed of cells, and that all cells come from preexisting cells. This means that when we study cell biology we are studying life and that life is continuous. Cell size is limited by surface area-to-volume ratio. Very large cells are not feasible because as volume increases, surface area increases at a slower rate. A cell's capacity for chemical activity is related to the cell volume. However, cells with large volumes are unable to take up enough material across their surface to support the metabolic activity of the volume. Because cells are small, either light microscopes or electron microscopes must be used to visualize them. Light microscopes, which use lenses and light, can resolve objects as small as 0.2 μm. Electron microscopes, which use electron beams, can resolve objects as small as 0.2 nm. Staining techniques assist in visualizing cellular components. Cells can also be studied by chemical analysis that involves breaking open the cells and examining the cell-free extract (for, e.g., activity of enzymes).

All cells are surrounded by a plasma membrane composed of a phospholipid bilayer with many embedded and protruding proteins. The plasma membrane and the associated proteins are selectively permeable, permitting some small molecules to pass through but preventing others from doing so. The membrane allows for maintenance of the internal cellular environment and is responsible for communication and interactions with other cells. Prokaryotes, which include the Archaea and Bacteria, lack a nucleus and other membrane-bound internal compartments. Eukaryotes, which are composed of the domain Eukarya, have organelles and a nucleus containing the cell's DNA. The organelles compartmentalize the different functions occurring in the cell.

Question 1. What is the primary function of a cell membrane? What characteristics of membranes allow them to contribute to metabolic activity?
Textbook Reference: *4.1 Cells Provide Compartments for Biochemical Reactions, p. 58*

4.2 Prokaryotic Cells Do Not Have a Nucleus

Prokaryotic cells share certain features

Specialized features are found in some prokaryotes

In terms of numbers and diversity, prokaryotes are the most successful cell type on Earth. Prokaryotic cells do not have internal membrane compartments and are generally smaller than eukaryotic cells. Prokaryotes are one-celled organisms found only in domains Archaea and Bacteria. These single cells have a plasma membrane and a nucleoid containing the genetic material, which is not membrane-enclosed. The cytoplasm of the cell consists of the liquid cytosol and insoluble particles that act as subcellular machinery (i.e., ribosomes for protein synthesis). The cytoplasm is in constant motion to ensure that reactants come together to bring about biochemical reactions.

Most prokaryotes have a rigid cell wall located outside the plasma membrane that supports the cell and determines its structure. The cell walls of bacteria contain peptidoglycan, which is relatively permeable; cell wall type is frequently used as a characteristic to identify bacteria. Some prokaryotes produce an outer slimy layer, or capsule, that is rich in polysaccharides and located outside the cell wall. Though it provides protection, it is not necessary for the cell's survival. Some prokaryotes have an internal membrane that provides some compartmentalization. Photosynthetic prokaryotes have stacks of folded membranes on which photosynthesis is carried out. Other prokaryotes have membrane folds that function in energy reactions and cell division. Some prokaryotes have structures to help with movement or support. Flagella are tiny protein machines containing the protein flagellin that facilitate motion of the cell. Actin-like proteins may help the cell maintain its shape by providing a cytoskeleton.

Question 2. In your biology course, your professor gives you a sample of cells and tells you that they can only be prokaryotic cells. What characteristics identify them definitively as prokaryotic cells?
Textbook Reference: *4.2. Prokaryotic Cells Do Not Have a Nucleus, pp. 59–61*

4.3 Eukaryotic Cells Have a Nucleus and Other Membrane-Bound Compartments

Compartmentalization is the key to eukaryotic cell function

Ribosomes are factories for protein synthesis

The nucleus contains most of the DNA

The endomembrane system is a group of interrelated organelles

Some organelles transform energy

Several other membrane-enclosed organelles perform specialized functions

Eukaryotic cells are larger than prokaryotic cells and are distinguished by their membrane-enclosed internal organelles. Each organelle has its own internal environment uniquely suited to its function. Animal and plant cells share common organelles, while other organelles are found only in plant cells. Ribosomes (an RNA molecule) found in the cytoplasm of both eukaryotic and prokaryotic cells act as information transcription centers and guide the synthesis of proteins from the messenger RNA nucleic acid blueprints. The nucleus stores DNA and is also the site of DNA duplication and regulation of DNA transcription into RNA. The nucleolus region of the nucleus is involved in ribosome assembly and RNA synthesis. The nucleus is bounded by a double lipid bilayer called the nuclear envelope. Small openings in the nuclear envelope, called nuclear pores, allow passage of RNA and ribosomes to the cell cytoplasm. Inside the nucleus, DNA and proteins combine to make chromatin. At cell division, chromatin condenses to form chromosomes. The outer membrane of the nuclear envelope is continuous with the endoplasmic reticulum (ER).

The endomembrane system consists of the nuclear envelope, the ER, the Golgi apparatus, and lysosomes. Membrane-bound vesicles move various substances within the endomembrane system. The endoplasmic reticulum is a complex of membrane sacs throughout the cell and is continuous with the nuclear membrane. The endoplasmic reticulum is classified into two types based on the presence or absence of attached ribosomes. Rough endoplasmic reticulum (RER) is studded with active ribosomes and is involved in the synthesis, storage, transport, and modification of new proteins. Many of the cell's membrane-bound proteins are produced in the RER. Carbohydrate groups are added to proteins to make glycoproteins in the RER. These carbohydrate groups help, for example, in indentifying proteins and ensuring that they reach the correct destinations within the cell. Smooth endoplasmic reticulum (SER) has a role in protein modification and transport, but its more important functions are in the modification of chemicals taken in by the cell, such as drugs and pesticides, and as the site of hydrolysis of glycogen and synthesis of steroids and lipids. The Golgi apparatus is an organelle composed of flattened membrane stacks called cisternae and membrane-enclosed vesicles. It contributes to further modification, packaging, and concentration of proteins. In plants, it is the site of polysaccharide synthesis. The three different regions of the Golgi have different enzymes and functions. The *cis* region is closest to the nucleus or RER, the *trans* region is closest to the cell surface, and the medial region lies in between. Proteins are released from the ER in a vesicle and are transported to the *cis* region, where the vesicle membrane fuses with the Golgi membrane and the contents are released into the Golgi. Vesicles containing proteins pinch off of the *trans* Golgi for transport to the plasma membrane or lysosomes. Lysosomes are "digestion centers" within a cell that break down proteins, polysaccharides, nucleic acids, and lipids into their monomer components. Primary lysosomes released from the Golgi contain a host of powerful enzymes in a slightly acidic environment; these break down engulfed molecules and cellular waste. In the process of phagocytosis, the cell takes up material in a phagosome, which fuses with a primary lysosome to make a secondary lysosome. Through the process of autophagy, lysosomes digest organelles such as mitochondria, breaking them down to monomers for reuse in new organelles. Lysosomal storage diseases occur when lysosomes fail to digest internal components. Plant cells do not contain lysosomes.

Two organelles, mitochondria and chloroplasts, are involved in harvesting energy. During cellular respiration, mitochondria convert potential chemical energy stored in glucose into adenosine triphosphate (ATP), a form of energy readily usable by the cell. Cells can contain anywhere from one to a few hundred thousand mitochondria. Mitochondria have two membranes—an outer smooth membrane and a highly folded internal membrane. The folds are called cristae, and the remaining internal space is called the matrix. The matrix holds ribosomes and DNA. Protein complexes used during cellular respiration are embedded in the cristae. Plastids are found in plants and protists, but not in animal cells. Plant cells have several different types of plastids, each with unique functions.

Chloroplasts containing the photosynthetic pigment chlorophyll are the site of photosynthesis, where light energy is converted to chemical energy. Chloroplasts have two membranes. The innermost membrane contains circular compartments, or thylakoids, which are folded into stacks called grana and hold chlorophyll and enzymes for photosynthesis. The liquid content of the chloroplast is called the stroma and contains ribosomes and DNA. Other plastid types include chromoplasts, which contain pigments involved in flower color, and starch-storing leucoplasts. Other specialized organelles are found in some, but not all, cells. Peroxisomes function to break down harmful peroxide by-products. Glyoxysomes, which are found in young plants, convert lipids to carbohydrates. Vacuoles, found in plants and protists, have various functions depending on the organism, including storage of waste products, structural support, reproduction, food storage, and water regulation.

Question 3. Eukaryotic cells possess organelles, where specific metabolic functions occur. What are the benefits of compartmentalization to a cell?
Textbook Reference: *4.3 Eukaryotic Cells Have a Nucleus and Other Membrane-Bound Compartments, p. 64*

Question 4. If we can assume that form follows function, what would be the explanation for the structural similarities between mitochondria and chloroplasts?
Textbook Reference: *4.3 Eukaryotic Cells Have a Nucleus and Other Membrane-Bound Compartments, p. 68*

Question 5. The role of a certain cell in an organism is to secrete a protein. Create a flow chart in which you trace the production of that protein from the nucleus through all necessary organelles to the point of release from the cell.
Textbook Reference: *4.3 Eukaryotic Cells Have a Nucleus and Other Membrane-Bound Compartments, pp. 64–66*

4.4 The Cytoskeleton Provides Strength and Movement

Microfilaments are made of actin
Intermediate filaments are diverse and stable
Microtubules are the thickest elements of the cytoskeleton
Cilia and flagella provide mobility
Biologists manipulate living systems to establish cause and effect

The cytoskeleton is involved in cell support, position and movement of organelles, cytoplasmic streaming, and anchoring of the cell. Microfilaments, composed of the protein actin, assist with contraction of the cell. These filaments are involved in muscle cell contraction, cell division, cytoplasmic streaming, and stabilizing cell shape. The cytoskeleton can lengthen or shorten due to the dynamic instability of actin. In muscle cells, the protein myosin interacts with actin to produce muscle contraction. Intermediate filaments are found only in multicellular organisms and function in stabilizing structure and resisting tension. They also help to anchor the nucleus and maintain rigidity with desmosomes. Hair and fingernails are composed of the intermediate filament keratin.

Microtubules are long hollow tubes of the dimer protein tubulin that contribute to the rigidity of the cell and act as a framework for the movement of motor proteins. They have a very specific structure that can be quickly added to or reduced, showing dynamic instability. Kinesin (a motor protein) delivers vesicles or organelles to various locations in the cell by moving along microtubule "tracks."

Cilia and flagella of eukaryotes are powered by microtubules. Though cilia and flagella differ in size and function, they have the same basic structure: a "9 + 2" arrangement of microtubules. Movement of cilia and flagella occurs when the microtubules slide past one another. The sliding is caused by an ATP-driven shape change in molecules of dynein bound to the microtubules. Dynein binds two microtubule doublets, and nexin cross-links the doublets, allowing the cillium to bend. Motor proteins called kinesin are involved in the transfer of vesicles along microtubules within the cell.

Question 6. Explain how microtubules and dynein function to make cilia and flagella move.
Textbook Reference: *4.4 The Cytoskeleton Provides Strength and Movement, pp. 70–71*

Question 7. As you examine a cell under the microscope, you notice what appear to be small, membrane-bound units moving along a path within the cell. Based on your knowledge of filaments in a cell, describe what you are observing.
Textbook Reference: *4.4 The Cytoskeleton Provides Strength and Movement, pp. 71–72*

4.5 Extracellular Structures Allow Cells to Communicate with the External Environment

The plant cell wall is an extracellular structure

The extracellular matrix supports tissue functions in animals

Cell junctions connect adjacent cells

Plants have semirigid cell walls composed of fibrous cellulose and a gel-like polysaccharide and protein matrix. The cell wall functions to support and protect the cell. Plasmodesmata are small holes in the cell walls that allow connections between plant cells. The extracellular matrix of some animal cells, which is composed of collagen and proteoglycans, functions to hold cells together, filter materials, orient cell movement, and assist with chemical signaling. Some cells, such as bone cells, secrete an elaborate and rigid matrix. The extracellular matrix and plasma membrane are connected by proteins such as integrin.

Cell junctions help to hold the cells of multicellular animals together. The three types of cell junctions are tight junctions, desmosomes, and gap junctions. Tight junctions, which are found in the epithelium of the bladder, do not allow for movement of substances and prevent urine from leaking out into the body. Desmosomes are involved in joining cells strongly while allowing for some movement around the extracellular matrix. Gap junctions connect cells and act as channels between the interiors of the adjoining cells.

Question 8. You are looking at a group of cells from an animal and notice that the cell junctions look like channels. What type of junctions are these, what do they do, and what type of tissue are you most likely examining?
Textbook Reference: *4.5 Extracellular Structures Allow Cells to Communicate with the External Environment, p. 74*

Test Yourself

1. A mass of cells is found in the sediment surrounding a thermal vent in the ocean floor. The salinity in the area is quite high. Microscopic examination of one of the cells reveals no evidence of membrane-enclosed organelles. This cell would be classified as a
   a. eukaryotic cell.
   b. prokaryotic cell.
   c. member of domain Archaea or Bacteria.
   d. Both a and c
   e. Both b and c
   Textbook Reference: *4.1 Cells Provide Compartments for Biochemical Reactions, p. 59*
2. Centrifugation of a cell results in the rupture of the cell membrane and the compacting of the contents into a pellet in the bottom of the centrifuge tube. Bathing this pellet with a glucose solution yields metabolic activity, including the production of ATP. One of the contents of this pellet is most likely which of the following?
   a. Cytosol
   b. Mitochondria
   c. Lysosomes
   d. Golgi bodies
   e. Thylakoids
   Textbook Reference: *4.3 Eukaryotic Cells Have a Nucleus and Other Membrane-Bound Compartments, p. 68*
3. The cell theory states all of the following *except*
   a. all living things are composed of cells.
   b. cells are the fundamental units of life.
   c. all cells come from preexisting cells.
   d. all cells contain mitochondria.
   e. None of the above is an element of the cell theory.
   Textbook Reference: *4.1 Cells Provide Compartments for Biochemical Reactions, p. 57*

4. Though science fiction has produced stories like "The Blob," we do not see very many large single-celled organisms. Which of the following tends to limit cell size?
   a. The difficulty in maintaining a continuous large membrane
   b. The difficulty of reproduction in a large cell
   c. Surface area-to-volume ratios
   d. All of the above
   e. None of the above

   Textbook Reference: *4.1 Cells Provide Compartments for Biochemical Reactions, p. 57*

5. Microscopes are used to resolve images that cannot be seen with the unaided eye. Electron microscopes use _______ to resolve images, whereas light microscopes use _______ to resolve images.
   a. light and lenses; diffraction of electron beams
   b. diffraction of electron beams; light and lenses
   c. lasers; light and lenses
   d. light and lenses; lasers
   e. None of the above

   Textbook Reference: *4.1 Cells Provide Compartments for Biochemical Reactions, p. 58*

6. The cellular function of the RER is
   a. DNA synthesis.
   b. photosynthesis.
   c. cellular respiration.
   d. protein synthesis.
   e. mRNA degradation.

   Textbook Reference: *4.3 Eukaryotic Cells Have a Nucleus and Other Membrane-Bound Compartments, p. 65*

7. Photosynthesis occurs in the
   a. chloroplast.
   b. mitochondria.
   c. Golgi apparatus.
   d. nucleus.
   e. RER.

   Textbook Reference: *4.3 Eukaryotic Cells Have a Nucleus and Other Membrane-Bound Compartments, p. 68*

8. Lysosomes are involved in
   a. DNA synthesis.
   b. the breakdown of phagocytized material.
   c. protein folding.
   d. pigment production.
   e. cell membrane production.

   Textbook Reference: *4.3 Eukaryotic Cells Have a Nucleus and Other Membrane-Bound Compartments, pp. 66–67*

9. The packaging of proteins to be used outside the cell occurs in the
   a. nucleus.
   b. SER.
   c. Golgi apparatus.
   d. chromoplast.
   e. nuclear pore.

   Textbook Reference: *4.3 Eukaryotic Cells Have a Nucleus and Other Membrane-Bound Compartments, p. 66*

10. Which of the following organelles is *not* enclosed in two membranes?
    a. Nucleus
    b. Chloroplast
    c. Mitochondrion
    d. RER
    e. All of the above have two membranes.

    Textbook Reference: *4.3 Eukaryotic Cells Have a Nucleus and Other Membrane-Bound Compartments, pp. 64, 68*

11. Movement of cells in both prokaryotes and eukaryotes is accomplished by which of the following structures?
    a. Cilia
    b. Pili
    c. Dynein
    d. Cell membranes
    e. Flagella

    Textbook Reference: *4.4 The Cytoskeleton Provides Strength and Movement, pp. 70–71*

12. Which of the following statements about mitochondria and chloroplasts is true?
    a. Animal cells produce chloroplasts.
    b. Both mitochondria and chloroplasts may be found in the same cell.
    c. Mitochondria and chloroplasts are not found in the same cell.
    d. In certain conditions, chloroplasts can revert to mitochondria.
    e. None of the above

    Textbook Reference: *4.3 Eukaryotic Cells Have a Nucleus and Other Membrane-Bound Compartments, p. 68*

13. Which of the following statements about ribosomes is true?
    a. Ribosomes guide protein synthesis.
    b. Ribosomes are found only in the nucleus or on the RER.
    c. There are no ribosomes in the mitochondria.
    d. Ribosomes are the site of photosynthesis.
    e. All of the above

    Textbook Reference: *4.3 Eukaryotic Cells Have a Nucleus and Other Membrane-Bound Compartments, p. 64*

14. Nuclear DNA exists as a complex of proteins called _______ that condenses into _______ during cellular division.
    a. chromosomes; chromatin
    b. chromatids; chromosomes
    c. chromophors; chromatin
    d. chromatin; chromosomes
    e. None of the above

    Textbook Reference: *4.3 Eukaryotic Cells Have a Nucleus and Other Membrane-Bound Compartments, p. 64*

15. Rough endoplasmic reticulum and smooth endoplasmic reticulum differ
    a. only in the presence (RER) or absence (SER) of ribosomes.
    b. in their function, and also in terms of the presence (RER) or absence (SER) of ribosomes.

c. only in microscopic appearance.
d. only in their function.
e. None of the above

Textbook Reference: *4.3 Eukaryotic Cells Have a Nucleus and Other Membrane-Bound Compartments, pp. 65–66*

Answers

Key Concept Review

1. The cell membrane allows the enclosure of biochemical functions within a membrane and acts as a selectively permeable barrier. It also allows a cell to maintain homeostasis and is important in communication with adjacent cells and in receiving signals from the environment. In addition, it plays an important structural role and plays a role in cell shape.
2. Since the cell is a prokaryote, it will have a number of features that are not shared with eukaryotic cells, such as a plasma membrane and a nucleoid. It is likely that it will have a cell wall. Unlike eukaryotes, it will not contain any organelles and will be small.
3. Organelles allow different metabolic environments to exist in the same cell. This partitioning of jobs allows for greater specialization.
4. Both mitochondria and chloroplasts are involved in energy-transformation activities that require many enzymes. The stacking or folding of membranes provides enzymatic activity centers for these reactions.
5.

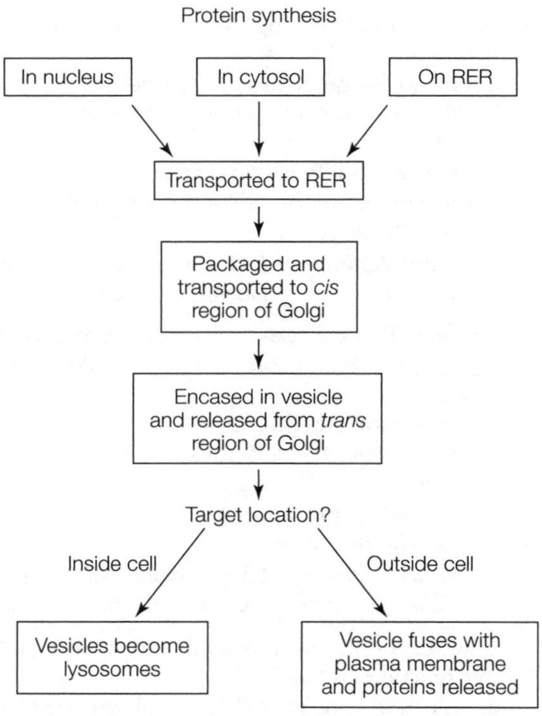

6. Dynein molecules bind to pairs of microtubules in the flagella or cilia. With the addition of cellular energy, the dynein molecules undergo a conformational change that results in the microtubules sliding past one another, resulting in a whiplike motion of the flagella.
7. The moving units are vesicles that are travelling along microtubules within the cell. The vesicles are attached to the microtubules by kinesin. The kinesin "walks" along the microtubule by detaching and reattaching to the microtubule.
8. The junctions between these cells are gap junctions. Gap junctions function as channels to allow two adjoining cells to communicate with each other. Heart muscle is a classic example of a cell type that relies on gap junctions to pass along information.

Test Yourself

1. **e.** Several characteristics suggest that this is a prokaryote. It survives in high salinity and high heat, although the sure indication is that it contains no membrane-enclosed organelles. Prokaryotes are in the domain Archaea and Bacteria.
2. **b.** The pellet is undergoing cellular respiration, a function that occurs in the mitochondria. You can also assume that if the single membrane of the cell itself is ruptured, other organelles enclosed in single membranes would be ruptured as well.
3. **d.** The cell theory states that cells are the fundamental units of life, all living things are composed of cells, and that all cells come from preexisting cells.
4. **c.** As volume increases, the surface area available for exchange does not increase proportionally. Eventually the surface is not large enough for maintenance of the metabolic activity of the cell.
5. **b.** In electron microscopy, a concentrated beam of electrons is focused on an object, allowing resolution of structures as small as 2 nm. Light microscopy, using lights and lenses, can only resolve objects to approximately 0.2 μm.
6. **d.** The RER is the site of protein synthesis.
7. **a.** The chloroplasts are the organelles involved in photosynthesis.
8. **b.** Lysosomes are organelles that contain digestive enzymes used to break down macromolecules taken in by phagocytosis.
9. **c.** The Golgi apparatus packages proteins for both internal and external use.
10. **d.** The nucleus, the mitochondria, and the chloroplasts are the only organelles enclosed in two membranes.
11. **e.** Though the flagella have different structures, they serve the same role in prokaryotes and eukaryotes.

12. **b.** Mitochondria and chloroplasts may be found in the same cell. Almost all eukaryotic cells contain mitochondria.
13. **a.** Ribosomes, found in the nucleus, the cytosol, and in organelles such as the mitochondria, RER, and chloroplasts, complex with RNA to guide protein production.
14. **d.** The complex of proteins and DNA is called chromatin. Chromatin takes the form of chromosomes only during cell division.
15. **b.** Both the structure and the function of RER and SER differ.

Cell Membranes and Signaling

5

The Big Picture

- Cellular membranes are a dynamic composition of a phospholipid bilayer, integral and peripheral proteins, and carbohydrates. The nature of the constituent phospholipids allows for the formation of a barrier that is semipermeable. Small hydrophilic molecules can traverse the membrane via simple diffusion, water can cross the membrane through aquaporins by osmosis, small charged ions can pass through protein channels, and some other molecules can be shepherded through by carrier proteins. Transport of molecules against their concentration gradient is active and requires energy input, either directly from ATP or as coupled to ATP-driven transport. Larger substances depend on endocytosis and exocytosis for transport into and out of the cell.
- Signal transduction is the means by which cells receive information from the environment or other cells and react to those signals. Transduction is a highly regulated series of events that depends on the binding of a signal ligand to a receptor protein. The signal binding must cause a change in the shape of the receptor protein, which causes a responder protein to initiate events in the cell that change its function. The effects of signals are often mediated by secondary messengers.

Study Strategies

- In the study of membranes, the processes of diffusion and osmosis are the most difficult to understand. It is very easy to get the terminology confused, especially the terms "hypertonic" and "hypotonic." A consideration of the Latin roots of the words is helpful. "Hyper-" generally means excess, and "hypo-" generally means "less than." "Tonic" refers to solute concentration. Therefore, "hypertonic" means excess solutes, and "hypotonic" means fewer solutes.
- Secondary active transport also tends to be confusing. Remember that secondary active transport does not use ATP directly, but is tightly coupled to ion transport that does require ATP.
- When studying the fluid mosaic model, think about the properties of the constituent molecules. This will make understanding the membrane's structure much easier. Draw a cartoon of the plasma membrane and all the potential components.
- For the study of diffusion and osmosis, draw diagrams of the movement of water and solutes. Diagrams will help you visualize what is happening across a membrane.
- Create a flow chart or a diagram of the signal transduction pathways.
- Make a list of the different secondary signals and provide an example of how each signal is activated and what it affects.
- Review the figures of the chapter to clarify the different examples of signal transduction.
- It is easy to become overwhelmed by the different examples of signal transduction. Try to focus on particular details of the systems. For instance, distinguish between direct and indirect signal transduction. Compare plasma membrane receptors with cytoplasmic receptors. List the three kinds of secondary messengers for signal transduction. Then, pick one of the signal transduction examples and create a table that includes the signal, the receptor, the transduction (responders and amplification), and the effect in the cell. Expand your table to include other examples of signal transduction.
- G protein action is a difficult concept. Remember that the G protein interacts with the receptor, binds GTP, and then interacts with the effector protein.
- Go to yourBioPortal.com to review the following tutorials and activities:

  Animated Tutorial 5.1 Passive Transport

  Animated Tutorial 5.2 Active Transport

  Animated Tutorial 5.3 Endocytosis and Exocytosis

  Animated Tutorial 5.4 G Protein–Linked Signal Transduction and Cancer

  Animated Tutorial 5.5 Signal Transduction Pathway

  Interactive Tutorial 5.1 Lipid Bilayer: Temperature Effects on Composition

  Web Activity 5.1 Membrane Molecular Structure

  Web Activity 5.2 Concept Matching

Working with Data 5.1 Aquaporins Increase Membrane Permeability to Water

Working with Data 5.2 The Discovery of a Second Messenger

Key Concept Review

5.1 Biological Membranes Have a Common Structure and Are Fluid

Lipids form the hydrophobic core of the membrane

Membrane proteins are asymmetrically distributed

Plasma membrane carbohydrates are recognition sites

Membranes are constantly changing

Biological membranes evolved to provide a barrier for cellular life. A biological membrane is mostly composed of lipids, with phospholipids as the most abundant component. Phospholipids have both hydrophilic ("water-loving") phosphorus heads and hydrophobic ("water-hating") nonpolar fatty tails. They arrange themselves into a bilayer with the hydrophobic tails touching and the heads extending into the aqueous environment inside and outside the cell. This arrangement allows for the fluid movement of the two layers on top of each other and the sealing of any disruptions in the membrane. This membrane is about 8 nm thick. Though the basic structure of a bilayer is always the same, the inner and outer halves differ in lipid composition and thus have slightly different properties. Phospholipids differ in their length, degree of unsaturation, and degree of polarity. Phospholipids that are saturated can be packed together more closely in the membrane, while unsaturated phospholipids have kinks in their fatty acids that make them less dense. Cholesterol can constitute up to 25 percent of the lipid content in the membrane of animals. The amount of cholesterol present and the degree of fatty acid saturation (membrane kinks) influences the fluidity of the membrane. Increases in cholesterol and fatty acid saturation make the membrane less fluid. Decreases in temperature also make the membrane less fluid. Organisms have the ability to change the lipid composition of their membranes in response to a change in temperature.

The typical ratio of proteins to phospholipid molecules in a plasma membrane is 1 to 25. This ratio can vary: in mitochondria the ratio is 1 protein to 5 lipids, while in neurons it is 1 protein to 70 lipids. Proteins may be embedded in or extend across membranes. Regions or domains of a membrane protein with hydrophobic amino acid side chains tend to be found in the hydrophobic environment of the membrane. The regions of a membrane protein containing hydrophilic amino acid side chains extend out from the membrane. The phospholipids and proteins are independent of each other, allowing for movement throughout the membrane. This happens for the most part because the constituents interact only noncovalently. Anchored membrane proteins have lipid groups that anchor them within the bilayer. Proteins that penetrate the phospholipid bilayer are called integral proteins. Special types of integral proteins called transmembrane proteins span the entire bilayer. Those not embedded are referred to as peripheral proteins. Proteins are distributed in membranes asymmetrically "as needed," and the numbers of proteins and their placement vary greatly based on cell type. The "inside" and "outside" of a membrane often have different properties due to the different characteristics of the transmembrane proteins on the two sides. Some membrane proteins may be anchored to cytoskeletal components. Such proteins have very specific functions.

Membrane carbohydrates serve as recognition sites for other cells and molecules. Carbohydrates are frequently bound to lipids or proteins, forming glycolipids or glycoproteins, respectively. Because of the nature of carbohydrates, they contribute to cell signaling. The binding of oligosaccharides on membrane glycoproteins is known as cell–cell adhesion. Membranes are in a constant state of flux. Phospholipids are produced on the surface of the smooth endoplasmic reticulum and distributed to the Golgi as vesicles. From the Golgi they move to the plasma membrane. The membranes of the different organelles differ.

Question 1. Diagram a cell membrane and label the phospholipid bilayer, integral proteins, peripheral membrane proteins, and carbohydrates. Describe the fluid mosaic model with reference to your diagram.
Textbook Reference: *5.1 Biological Membranes Have a Common Structure and Are Fluid, p. 79*

Question 2. You are examining the membrane of a cell and notice many carbohydrates in the plasma membrane. What role do these carbohydrates play in the cell?
Textbook Reference: *5.1 Biological Membranes Have a Common Structure and Are Fluid, p. 81*

5.2 Some Substances Can Cross the Membrane by Diffusion

Diffusion is the process of random movement toward a state of equilibrium

Simple diffusion takes place through the phospholipid layer

Osmosis is the diffusion of water across membranes

Diffusion may be aided by channel proteins

Carrier proteins aid diffusion by binding substances

Passive transport is the movement of a substance across a lipid bilayer by simple diffusion or facilitated diffusion via channel proteins or by carrier molecules. Passive transport does not require an input of energy, but relies on diffusion. Diffusion is the net movement of a substance from an area of greater concentration to an area of lesser concentration.

It is a random process that moves toward equilibrium. A membrane is said to be permeable to those substances that can pass through it and impermeable to those that cannot. If a substance can pass through a membrane, it will diffuse until concentrations on either side of the membrane are equal. At equilibrium, the molecules of the substance continue to move across the membrane, but the net movement of molecules in both directions is equal. The rate of diffusion depends on the size of the diffusing substance, the temperature of the solution, and the concentration gradient. Diffusion within small areas such as single cells may occur rapidly, but diffusion occurs more slowly with increasing distance. In simple diffusion, small molecules pass through a membrane. The more lipid-soluble the molecules are, the faster they diffuse across the membrane. In contrast, charged and polar molecules do not readily pass through a membrane due to the formation of many hydrogen bonds with water and the hydrophobic nature of the internal layer of the membrane.

Osmosis is the diffusion of water across a membrane, typically through channels in the membrane. Water will move across a membrane from areas of low solute concentration to areas of high solute concentration in order to equalize solute concentrations on either side of the membrane. Solute concentrations separated by a membrane are classified as isotonic, hypertonic, or hypotonic. An isotonic solution has the same solute concentrations on both sides of a membrane. A hypertonic solution has a solute concentration that is higher than the concentration on the other side of the membrane, and a hypotonic solution has a concentration that is lower. Environmental and cellular solute concentrations dictate the direction of osmosis in living cells. The pressure within cells, called turgor pressure, changes according to the amount of water taken up by osmosis. Turgor pressure can build up only if there is a cell wall to limit cell expansion.

Facilitated diffusion is the passive movement of a substance across a membrane with the help of membrane-bound proteins that act as channels or carriers. A substance can cross the membrane by facilitated diffusion through protein channels running through the plasma membrane. The pores of these proteins have polar amino acids that allow polar molecules and ions to cross the membrane. The best-studied channels are ion channels. Most ion channels are either ligand-gated or voltage-gated, allowing the passage of ions to be controlled depending on the cellular environment. Specific channels called aquaporins allow water to cross the membrane by osmosis. Water can also diffuse through ion channels. Facilitated diffusion may be aided by carrier proteins that bind a substance and transport it across the membrane. Diffusion in this case is not only limited by the concentration gradient, but also by the number of available membrane carrier proteins. When all of the carrier proteins are bound to the substance, they are saturated, limiting the rate of diffusion.

Question 3. A marathon runner has just arrived in the emergency room with severe dehydration, and the physician must decide which type of solution to pump into his veins: pure water, 0.9 percent saline, or 1.5 percent saline. In order to be certain, blood samples are treated with each solution and observed under a microscope. Describe what is likely to happen to the blood cells when they are exposed to each solution. (Hint: Blood cells are approximately 0.9 percent saline.) Which solution should the physician choose for rehydrating the runner?
Textbook Reference: *5.2 Some Substances Can Cross the Membrane by Diffusion, p. 84*

Question 4. The plasma membrane is a good barrier to the movement of water. Given this, how does water typically cross the plasma membrane into a cell?
Textbook Reference: *5.2 Some Substances Can Cross the Membrane by Diffusion, p. 85*

5.3 Some Substances Require Energy to Cross the Membrane

Active transport is directional

Different energy sources distinguish different active transport systems

Primary active transport requires the energy stored in ATP to move ions against their concentration gradient. This is a directional process in which a molecule is moved either into or out of a cell against its concentration gradient. Primary active transport makes direct use of energy from the hydrolysis of ATP to drive the transporter. The sodium–potassium pump is an integral membrane glycoprotein found in animal cells that uses ATP to transport two K^+ ions into the cell and three Na^+ ions out of the cell. Secondary active transport uses ATP indirectly by coupling solute transport with the ion concentration gradient established by primary transport. It typically moves amino acids and sugars across the membrane with the help of an ion like Na^+.

Question 5. You are examining a membrane that contains a number of proteins that appear to be involved in transport of an ion across that membrane. Design an experiment that will allow you to determine if these transporters are passive transporters or active transporters.
Textbook Reference: *5.3 Some Substance Require Energy to Cross the Membrane, p. 87*

Question 6. The sodium–potassium pump is important for maintaining gradient of Na^+ and K^+ between the inside and outside of the cell. Describe how these gradients function in secondary active transport to move another molecule such as glucose.

Textbook Reference: *5.3. Some Substances Require Energy to Cross the Membrane, pp. 87–88*

5.4 Large Molecules Cross the Membrane via Vesicles

Macromolecules and particles enter the cell by endocytosis

Receptor-mediated endocytosis is specific

Exocytosis moves materials out of the cell

Some macromolecules are unable to cross the plasma membrane because of their size or charge, or because they are polar. Endocytosis is the process by which a cell brings these large substances into the cell. This is accomplished by the cell membrane, which folds around the substance to form an endocytotic vesicle. Large substances and even entire cells are engulfed in the process of phagocytosis. Once inside the cell, the vesicle fuses with a lysosome for digestion. The cell takes up liquids in small vesicles from the outside in the process of pinocytosis. Animal cells use receptor-mediated endocytosis to capture specific macromolecules such as cholesterol from the environment. Receptors for specific macromolecules cluster together on the cell surface in coated pits containing the protein clathrin. Upon binding of the specific molecule to the receptors, the coated pit invaginates to form a vesicle. The resulting vesicle becomes clathrin-coated until it is well inside the cell, where it loses its coat and fuses with a lysosome. Animal cells take up cholesterol by the process of receptor-mediated endocytosis.

Exocytosis moves materials in vesicles out of the cell. Binding proteins found on the surface of the cell-produced vesicles bind with receptor proteins of the cytoplasmic side of the cell membrane. The vesicle membrane fuses with the cell membrane, and the contents of the vesicle are released outside the cell membrane.

Question 7. Cells have the ability to take in large molecules by endocytosis and secrete them to the environment by exocytosis. Describe each process and explain why both are important for the cell.
Textbook Reference: *5.4 Large Molecules Cross the Membrane via Vesicles, pp. 89–90*

5.5 The Membrane Plays a Key Role in a Cell's Response to Environmental Signals

Cells are exposed to many signals and may have different responses

Membrane proteins act as receptors

Receptors can be classified by location and function

Cells receive signals from their environment (chemicals, light, temperature, touch, or sound) and from other signals (usually chemicals). This process of receiving a signal and communicating it to the cell is called signal transduction. Multicellular organisms receive signals from their environment, from other cells, or from extracellular fluid. Autocrine signals are local signals that affect the cells that make them. Paracrine signals are local signals that affect nearby cells. Hormones are circulatory signals that travel through the circulatory system and affect distant cells. Only those cells that have the correct receptor will respond to the chemical signal. The response may occur through a short-acting enzyme or longer-term gene expression.

A cell's response to a signal is dependent upon the presence of membrane proteins that function as receptors. Binding to the receptor requires a fit of the ligand to the receptor binding site. Once the ligand binds with the receptor, the receptor must undergo an allosteric change to have an effect. The ligand is not changed in the binding process, and binding of the signal to the receptor is reversible. Binding sites may be inhibited by competing chemicals. There are two classes of receptors: plasma membrane receptors, which bind large and/or polar ligands that cannot cross the plasma membrane, and cytoplasmic receptors, which bind small nonpolar ligands that diffuse across the plasma membrane.

There are three main types of plasma membrane receptors: ion channel receptors, protein kinase receptors, and G protein–linked receptors. In the case of ion channel receptors ligand binding causes a conformational change to open "gates" that allow ions (Na^+, K^+, Ca^+, or Cl^-) to pass through. An example is the acetylcholine receptor found on muscle cells. When two molecules of acetylcholine bind to the receptor, the channel in the receptor opens and sodium flows through. In the case of protein kinase receptors, ligand binding stimulates the transfer of a phosphate group from ATP to a target protein. Insulin works through a protein kinase receptor. The term "G protein–linked receptor" applies to a group of receptors for which ligand binding changes its shape so that a G protein on the cytoplasmic side can bind. When the G protein is activated (by binding to the cytoplasmic side of the G-linked receptor), it binds GTP and activates an effector protein, changing cellular function. After binding the effector protein, GTP is hydrolyzed to GDP, causing inactivation of the G protein until the receptor binds its ligand again.

Question 8. Acetylcholine is polar ligand that initiates a series of events. Based on the properties of this ligand, what types of receptors would you predict that it could interact with? Why?
Textbook Reference: *5.5 The Membrane Plays a Key Role in a Cell's Response to Environmental Signals, p. 93*

Question 9. Describe the steps involved in a signaling cascade acting through a G protein–linked receptor.
Textbook Reference: *5.5 The Membrane Plays a Key Role in a Cell's Response to Environmental Signals, p. 94*

5.6 Signal Transduction Allows the Cell to Respond to Its Environment

Second messengers can stimulate signal transduction

A signaling cascade involves enzyme regulation and signal amplification

Signal transduction is highly regulated

Cell functions change in response to environmental signals

Second messengers, such as cyclic AMP, help to transduce the message from an activated receptor to the associated events in the cell. The advantage of second messengers is that they distribute and amplify the initial signal. Signaling cascades allow for amplification of the message at each step. Signal transduction pathways are highly regulated to ensure their proper function. Function in a pathway is dependent on the synthesis and breakdown of the enzymes involved and the activation or inhibition of these enzymes.

These signaling transduction pathways have a number of ways in which they can bring about the response in the cell. Some signal transduction pathways open ion channels. Others alter enzyme function, either through inhibition or activation (as with epinephrine stimulation). Signal transduction pathways also influence the expression of genes by regulating transcription.

Question 10. Discuss the role of secondary messengers in a signaling pathway. How are they different from ligands and receptors? What roles do secondary messengers and signals have in common?
Textbook Reference: *5.6 Signal Transduction Allows the Cell to Respond to Its Environment, pp. 94–96*

Question 11. Describe how G protein–linked receptors and protein kinases interact in a signal transduction cascade.
Textbook Reference: *5.6 Signal Transduction Allows the Cell to Respond to Its Environment, p. 96*

Question 12. Why are signal transduction pathways highly regulated? What would happen if they were not?
Textbook Reference: *5.6 Signal Transduction Allows the Cell to Respond to Its Environment, pp. 96–97*

Test Yourself

1. Which of the following statements regarding cellular membranes is *false*?
   a. The hydrophobic nature of the phospholipid tails limits the migration of polar molecules across the membrane.
   b. Integral proteins and phospholipids move fluidly throughout the membrane.
   c. Membrane phospholipids flip back and forth from one side of the bilayer to the other.
   d. Glycolipids and glycoproteins serve as recognition sites on the cell membrane.
   e. All of the above are true; none is false.
   Textbook Reference: *5.1 Biological Membranes Have a Common Structure and Are Fluid, pp. 79–82*
2. Which of the following contributes to differences in the two sides of the cell membrane?
   a. Differences in peripheral proteins
   b. Different domains expressed on the ends of integral proteins
   c. Differences in phospholipid types
   d. Differences in the carbohydrates attached to membrane proteins
   e. All of the above
   Textbook Reference: *5.1 Biological Membranes Have a Common Structure and Are Fluid, pp. 81–82*
3. Which of the following cell membrane components serve as recognition signals for interactions between cells?
   a. Cholesterol
   b. Glycolipids or glycoproteins
   c. Phospholipids
   d. Carrier proteins
   e. All of the above
   Textbook Reference: *5.1 Biological Membranes Have a Common Structure and Are Fluid, p. 82*
4. You are monitoring the diffusion of a molecule across a membrane. An internal concentration of _______ and an external concentration of _______ will result in the fastest rate of diffusion.
   a. 5; 60
   b. 60; 5
   c. 35; 40
   d. 50; 50
   e. Either a or b
   Textbook Reference: *5.2 Some Substances Can Cross the Membrane by Diffusion, p. 83*

5. If a red blood cell with an internal salt concentration of about 0.85 percent is placed in a saline solution that is 4 percent, the
   a. cell will lose water and shrivel.
   b. cell will gain water and burst.
   c. turgor pressure in the cell will increase greatly.
   d. turgor pressure in the cell will decrease greatly.
   e. cell will remain unchanged.
   Textbook Reference: *5.2 Some Substances Can Cross the Membrane by Diffusion, p. 83*
6. Solution X is hypotonic relative to solution Y if solution X has a solute concentration that is _______ solution Y.
   a. greater than that of
   b. lower than that of
   c. the same as that of
   d. All of the above
   e. None of the above
   Textbook Reference: *5.2 Some Substances Can Cross the Membrane by Diffusion, p. 84*
7. Which of the following statements about osmosis is *false*?
   a. Osmosis refers to the movement of water along a concentration gradient.
   b. In osmosis, water moves to equalize solute concentrations on either side of the membrane.
   c. The movement of water across a membrane can affect the turgor pressure of some cells.
   d. If osmosis occurs across a membrane, then diffusion is not occurring.
   e. During osmosis, water is moving through membrane channels.
   Textbook Reference: *5.2 Some Substances Can Cross the Membrane by Diffusion, p. 84*
8. Channel proteins allow ions that would not normally pass through the cell membrane to pass through via the channel. What property of the channel proteins makes this possible?
   a. A pore of polar amino acid groups
   b. A pore of hydrophobic amino acid groups
   c. A pore of Ca^{2+}
   d. All of the above
   e. None of the above
   Textbook Reference: *5.2 Some Substances Can Cross the Membrane by Diffusion, p. 85*
9. Which of the following limits the movement of molecules by means of carrier-mediated facilitated diffusion?
   a. The concentration gradient
   b. The availability of carrier molecules
   c. Temperature
   d. Both a and b
   e. All of the above
   Textbook Reference: *5.2 Some Substances Can Cross the Membrane by Diffusion, pp. 85–86*
10. Active transport differs from passive transport in that active transport
   a. requires energy.
   b. never requires direct input of ATP.
   c. moves molecules with a concentration gradient.
   d. Both b and c
   e. Both a and c
   Textbook Reference: *5.3 Some Substances Require Energy to Cross the Membrane, p. 86*
11. Single-celled animals, such as amoebas, engulf entire cells for food. This manner of "eating" is called
   a. exocytosis.
   b. endocytosis.
   c. facilitative transport.
   d. active transport.
   e. osmosis.
   Textbook Reference: *5.4 Large Molecules Cross the Membrane via Vesicles, pp. 88–89*
12. Many cells have a sodium–potassium pump. In order to function, sodium–potassium pumps require
   a. ATP.
   b. a channel protein.
   c. the absence of a concentration gradient.
   d. ADP.
   e. All of the above
   Textbook Reference: *5.3 Some Substances Require Energy to Cross the Membrane, p. 87*
13. Bacterial cells are often found in very hypotonic environments. Which of the following characteristics keeps them from taking in too much water from their environment?
   a. The presence of a cell wall, which allows for a build-up of turgor pressure, preventing additional water from entering the cell
   b. The presence of a cell wall, which allows for a build-up of tonic pressure, preventing additional water from entering the cell
   c. The capacity of the cell to expel water as quickly as it takes it up
   d. The presence of an active water pump
   e. None of the above
   Textbook Reference: *5.2 Some Substances Can Cross the Membrane by Diffusion, p. 84*
14. The rate of diffusion can be affected by
   a. temperature.
   b. molecule size.
   c. the concentration gradient.
   d. the electrical charge.
   e All of the above
   Textbook Reference: *5.2 Some Substances Can Cross the Membrane by Diffusion, p. 84*
15. Which of the following does *not* occur during signal transduction?
   a. Binding of ligand to receptor
   b. Conformational change to the receptor protein
   c. Conformational change of the signal
   d. Alteration of cellular activity
   e. Allosteric regulation

Textbook Reference: *5.5 The Membrane Plays a Key Role in a Cell's Response to Environmental Signals, p. 92*

16. Which of the following statements about secondary messengers is true?
    a. They amplify the signal.
    b. They bind to the active site of the receptor.
    c. They result in multiple effects from a single signal.
    d. Both a and c
    e. All of the above

    Textbook Reference: *5.6 Signal Transduction Allows the Cell to Respond to Its Environment, pp. 94–97*

17. Caffeine is a stimulant that works because it acts as a(n) _______ to the adenosine receptors in a person's brain, and stimulates _______ in that person's heart and liver that increases blood flow and blood glucose.
    a. effector; a pathway
    b. inhibitor; a cascade pathway
    c. inhibitor; a ligand
    d. signal; inhibitors
    e. pathway; a ligand

    Textbook Reference: *5.5 The Membrane Plays a Key Role in a Cell's Response to Environmental Signals, p. 92*

18. Cytoplasmic receptors bind
    a. small signals that can diffuse through the plasma membrane.
    b. secondary messengers, such as cAMP.
    c. hydrophilic molecules.
    d. ligands.
    e. All of the above

    Textbook Reference: *5.5 The Membrane Plays a Key Role in a Cell's Response to Environmental Signals, p. 93*

19. Which of the following statements about protein kinase cascades is true?
    a. Amplification can occur at each step in the path.
    b. Information at the plasma membrane is communicated to the nucleus.
    c. The multiple steps allow for the specificity of the process.
    d. Different targets can produce variation in the cellular response.
    e. All of the above

    Textbook Reference: *5.5 The Membrane Plays a Key Role in a Cell's Response to Environmental Signals, p. 93*

20. Which of the following statements about receptors is true?
    a. Receptors are found only on the surface of cells.
    b. Receptors are specific to the signal ligand.
    c. Most receptors can bind with many signal ligands.
    d. All receptors can act as ion channels.
    e. All of the above

    Textbook Reference: *5.5 The Membrane Plays a Key Role in a Cell's Response to Environmental Signals, pp. 92–93*

21. Paracrine signals
    a. act on the cells that made them.
    b. move through the blood and act on cells far from their source.
    c. act on cells that are near to those that secrete them.
    d. do not act through receptors.
    e. require large concentrations of the signaling molecule to function.

    Textbook Reference: *5.5 The Membrane Plays a Key Role in a Cell's Response to Environmental Signals, p. 91*

22. Which of the following represents the correct order of signal transduction?
    a. Binding of signal, release of secondary messenger, alteration of receptor conformation, alteration of cellular function
    b. Binding of signal, release of secondary messenger, alteration of receptor conformation, transcription of gene
    c. Binding of signal, activation of target protein by responder, alteration of receptor conformation, release of secondary messenger; transcription of gene
    d. Binding of signal, alteration of receptor conformation, alteration of cellular function, release of second messenger
    e. Binding of signal, alteration of receptor conformation, activation of target protein by responder, alteration of cellular function

    Textbook Reference: *5.6 Signal Transduction Allows the Cell to Respond to Its Environment, pp. 94–97*

Answers

Key Concept Review

1.

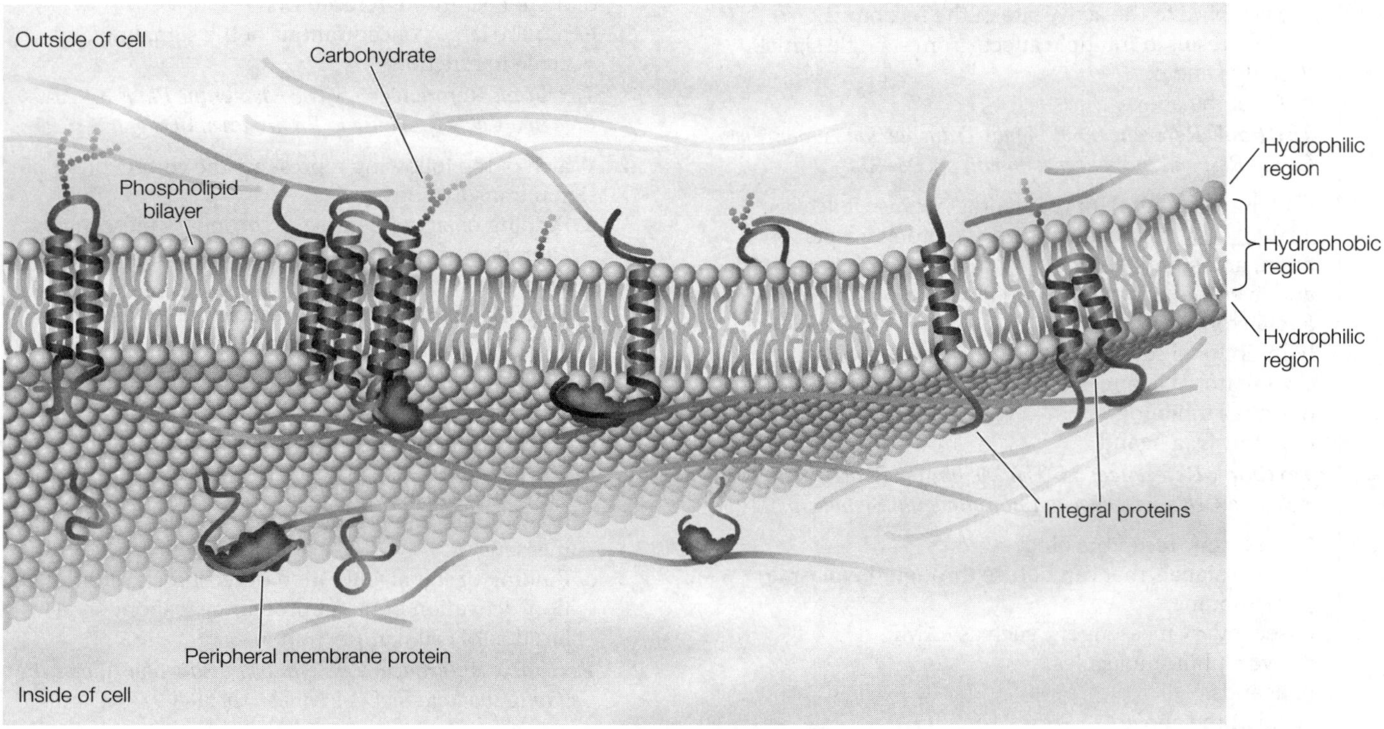

The fluid mosaic model states that the phospholipid bilayer allows the embedded proteins to float freely through the bilayer.

2. Carbohydrates located on the surface of the plasma membrane are typically involved in recognition. This can be recognition between cells or as signaling sites.
3. In pure water, the blood cells will take on water through osmosis, swell, and eventually rupture. In 0.9 percent saline, the cells should neither gain nor lose a significant amount of water. In a 1.5 percent saline solution, the cells should lose water and shrivel. In order to rehydrate the runner, a solution isotonic to the patient's blood cells, 0.9 percent, should be infused into his bloodstream. A hypotonic solution would end up rupturing the patient's cells.
4. Some plasma membranes contain special protein channels known as aquaporins, which allow for the movement of water by osmosis into and out of the cell.
5. The main difference between active and passive transport is that active transport goes against a concentration gradient and requires energy, whereas passive transport diffuses passively and does not require energy. To test this, you would want to measure the movement of the ions across a membrane. One side of the membrane should have a high concentration of the ion and the other a low concentration. If ions move from the area of high concentration to low concentration, then the transporter is a passive transporter. If instead, the addition of ATP is needed for movement to occur, it is an active transporter.
6. The sodium–potassium pump sets up a gradient of Na^+ and K^+, with more Na^+ outside the cell and more K^+ inside the cell. A secondary active transport mechanism has a transport protein for the specific molecule, which in this case is glucose. This transporter passively transports Na^+ into the cell by means of the Na^+ gradient and brings along the glucose.
7. Cells take up large particles, foreign cells, and food sources by endocytosis, in which the plasma membrane of the cell surrounds the particle to form an endocytotic vesicle. Substances such as undigested material, digestive enzymes, neurotransmitters, and material for plant wall construction are secreted by the cell by means of exocytosis. During exocytosis, the membrane of a secretory vesicle fuses with the plasma membrane and the contents are released to the outside of the cell.
8. Polar molecules such as acetylcholine are unable to cross the plasma membrane and therefore cannot enter the cell freely. Therefore, they require a receptor that is

located on the outside of the plasma membrane. This receptor must span the membrane and initiate a signaling cascade upon the binding of acetylcholine.

9. A G protein–linked receptor is stimulated by hormone binding to its external binding site. This in turn activates the G protein, which activates an effector protein. The effector protein converts reactants into signaling products, amplifying the signal from the single hormone.
10. Secondary messengers function to amplify and spread signals. They do not bind to receptors and do not act like signals in the cascade. Signals result in a change in cell function; secondary messengers are part of the pathway that results in the change in cell function.
11. G protein–linked receptors frequently expose the protein kinase activities of effector molecules.
12. Signal transduction pathways are highly regulated because they are temporary events within a cell that help to regulate the function of the cell. If they were not highly regulated, then the response of the cell would lag behind the changes in the environment. This would potentially lead to cell death.

Test Yourself

1. **c.** Because of the hydrophobic tails and hydrophilic heads of the phospholipids, it is impossible for them to flip back and forth from one side of the membrane bilayer to the other.
2. **e.** The cell membrane is asymmetric and has different properties and functions on the cytoplasmic side versus the extracellular side. These properties arise from differences in the constituents of the membrane.
3. **b.** Both glycolipids and glycoproteins serve as recognition signals.
4. **e.** Diffusion may take place in either direction across a membrane and always follows a concentration gradient. The larger the gradient, the faster the diffusion will occur.
5. **a.** The cell will lose water as solute concentrations on both sides of the membrane equalize.
6. **b.** A solution that has a lower solute concentration than another is hypotonic in comparison to the other one.
7. **d.** Diffusion and osmosis are not mutually exclusive and may take place at the same time.
8. **a.** The charged or polar lining of the channel proteins allows passage of polar and charged molecules.
9. **e.** Anything that affects the rate of diffusion will affect carrier-mediated facilitated diffusion. Carrier-mediated facilitated diffusion also relies on the availability of carrier molecules. Temperature can also be a limiting factor.
10. **a.** Active transport works against a concentration gradient and requires energy to do so. That energy does not always have to be directly supplied in the form of ATP.
11. **b.** Cells carry out cellular eating by phagocytosis, which is a type of endocytosis.
12. **a.** Sodium–potassium pumps are forms of primary active transport and require energy in the form of ATP.
13. **a.** Turgor pressure limits osmosis, and once a cell is turgid, no more water may be taken on.
14. **e.** Temperature, molecule size, molecule charge, and concentration gradients all affect the rate at which diffusion takes place.
15. **c.** In order for signal transduction to take place, the ligand must bind to the receptor, the receptor must undergo a conformational change, and the activity of the cell must be altered.
16. **d.** Secondary messengers both amplify the signal and have multiple effects within a cell.
17. **b.** Caffeine and adenosine bind to the same receptor protein. The binding of caffeine prevents adenosine from binding in the brain. In other organs, a series of cascades begin in response to caffeine binding.
18. **a.** To bind to a cytoplasmic receptor, the ligand must be able to pass through the plasma membrane.
19. **e.** Amplification, regulation, specificity, and variation are all roles of protein kinase cascades.
20. **b.** Receptors interact only with specific ligands to bring about a response in the cell.
21. **c.** Autocrine signals affect the cells that make them, while paracrine cells affect nearby cells. A receptor in the plasma membrane binds hydrophilic signals.
22. **e.** Signals must bind their receptors, the conformation of the receptor is altered, a responder activates a target protein, and cell function changes.

Pathways that Harvest and Store Chemical Energy

6

The Big Picture

- Cells require energy to carry out their functions. Metabolism is a series of energy-transferring reactions that fuel the processes of the cell. All reactions either give off energy (exergonic) or require energy (endergonic) to proceed. Energy may be stored in chemical bonds (potential energy) and released when the bonds are broken to do work (kinetic energy). ATP serves as the energy shuttle for many metabolic processes.
- Metabolism is regulated by enzymes. Most metabolic pathways are under allosteric control. Enzyme complexes allow for interactions and regulation of adjacent active sites. Often the final product of a specific pathway regulates the commitment step of the pathway itself.
- Glucose provides cellular energy. It may be metabolized in the absence of oxygen through glycolysis and fermentation. In the presence of oxygen, it is metabolized through glycolysis and cellular respiration in the form of the citric acid cycle, electron transport, and oxidative phosphorylation. Compared to aerobic respiration, anaerobic respiration produces significantly less cellular energy (in the form of ATP) from the same amount of glucose.
- Photosynthesis allows organisms with the appropriate pigments and metabolic processes to convert light energy from the sun to chemical energy that can be used in the cell or stored. The process of photosynthesis occurs in two steps. The first step utilizes light energy to produce ATP and NADPH with the help of the electron transport system. The second step uses the products of the light reactions to fix CO_2 in the Calvin cycle. Photosynthesis forms the basis of all food webs and is vital to life on Earth.

Study Strategies

- The chemistry presented in this chapter is often difficult, and you may question why you are studying it in a biology course. Remember that biological processes are based on physical and chemical properties. To understand the function of organisms, it is necessary to understand the basics of energy transfer. If you do not have a firm understanding of topics covered in previous chapters—such as potential energy, entropy, and enzyme activity—it would be helpful for you to review these before learning more about them in the context of cellular metabolism.
- The biggest mistake would be to memorize the pathways without understanding them. When you begin your study, do not focus on the pathways themselves; instead, focus on the beginning and end products and why each pathway does what it does. Once you understand conceptually why the pathway is present, move on to studying the steps in the pathway itself. You may find the concepts to be very unfamiliar; therefore, make sure you understand each section before proceeding. Spend a significant amount of time on the diagrams so you can visualize the pathways' purposes and interactions.
- This chapter contains a great deal of new terminology. Making a vocabulary list may be helpful. You might also benefit from writing the various reactants, products, key intermediates, and electron carriers on index cards. You can then use the cards to arrange the relevant molecules into the proper sequence for the pathway you are studying (i.e., glycolysis, cellular respiration, photosynthesis). Pay attention to multiple pathways that use the same molecules but in different ways, and also note which molecules are limited to a single pathway.
- Remember that ATP and electron carriers such as NAD^+ are energy currencies; they move energy from pathway to pathway. In order to understand the pathways completely, always think about the roles of ATP and electron carriers like NAD^+ in redox reactions. Other electron carriers mentioned in this chapter are $NADP^+$ and FAD. It may help to mentally link the P in $NADP^+$ with photosynthesis so you do not confuse it with the electron carriers used in glycolysis (NAD^+) and cellular respiration (NAD^+ and FAD).
- The properties of chlorophyll and light itself greatly affect how well a plant carries out photosynthesis. Take some time to understand the properties of light and how pigments interact with light.

- A common mistake is to think that the light reactions occur in the light and the Calvin cycle occurs in the dark. This is not the case! The light reactions must occur simultaneously with the Calvin cycle to supply the needed energy in the form of ATP and NADPH + H^+.
- When you are studying energy pathways in the cell, it is easy to forget that all living plant cells use cellular respiration. Cellular respiration takes the energy stored in photosynthesis and makes it available to drive other cellular processes. Photosynthesis is limited to photosynthetic cells in the plant; cellular respiration is more widespread in the organism.
- Go to yourBioPortal.com to review the following tutorials and activities:

  Animated Tutorial 6.1 Two Experiments Demonstrate the Chemiosmotic Mechanism

  Animated Tutorial 6.2 Electron Transport and ATP Synthesis

  Animated Tutorial 6.3 Photophosphorylation

  Animated Tutorial 6.4 The Source of the Oxygen Produced by Photosynthesis

  Animated Tutorial 6.5 Tracing the Pathway of CO_2

  Web Activity 6.1 ATP and Coupled Reactions

  Web Activity 6.2 The Citric Acid Cycle

  Web Activity 6.3 Respiratory Chain

  Web Activity 6.4 Glycolysis and Fermentation

  Web Activity 6.5 Energy Levels

  Web Activity 6.6 Regulation of Energy Pathways

  Web Activity 6.7 The Calvin Cycle

  Working with Data 6.1 An Experiment Demonstrates the Chemiosmotic Mechanism

Key Concept Review

6.1 ATP, Reduced Coenzymes, and Chemiosmosis Play Important Roles in Biological Energy Metabolism

ATP hydrolysis releases energy

Redox reactions transfer electrons and energy

Oxidative phosphorylation couples the oxidation of NADH to the production of ATP

ATP and reduced coenzymes link catabolism and anabolism: An overview

This chapter builds on your understanding of energy, enzymes, and metabolism to explain how they all occur in living cells. The specificity of enzymes and their regulation allows a cell to tightly control its metabolic activity. It is important to keep in mind that a cell cannot create energy; it can only transform it from one form to another. Photosynthesis transforms light energy into chemical energy and stores it in covalent bonds in carbohydrates. In contrast, glucose catabolism transforms the potential energy in the carbohydrates into potential energy in ATP.

ATP is the "energy currency" of the cell. When the bond between the second and third phosphate groups is hydrolyzed, energy is released. This linkage of ATP hydrolysis with an endergonic reaction is called reaction coupling. Redox reactions refer to the coupled reduction and oxidation reactions, and ATP hydrolysis is just one example of this process. It is important to remember that when the electrons are transferred, protons are also transferred; for this reason, some prefer to think of redox reactions as transferring hydrogen atoms.

A common electron carrier in redox reactions is nicotinamide adenine dinucleotide (NAD^+/NADH). While NAD^+ is a common electron carrier in cells, there are other molecules with the same role. Flavin adenine dinucleotide (FAD) plays a role in the citric acid cycle and nicotinamide adenine dinucleotide phosphate ($NADP^+$) has a function in photosynthesis.

The concepts of redox reactions, electron carriers, and ATP synthesis all come together when the oxidation of NADH is used to produce ATP. NADH is oxidized and ATP is synthesized. This process is called oxidative phosphorylation.

An understanding of redox reactions, electron carriers, and ATP synthesis is needed for an understanding of the process of oxidative phosphorylation, by which NADH is oxidized and ATP is synthesized.

The mechanism for this energy transfer is dependent on a proton gradient that drives the synthesis of ATP via the enzyme ATP synthase embedded in the membrane; this process is called chemiosmosis. Thus, chemiosmosis makes oxidative phosphorylation possible. While this chapter will illustrate chemiosmosis in the inner mitochondrial membrane and the thylakoid membrane of the chloroplast, chemiosmosis can also occur across a plasma membrane of a prokaryotic cell.

Question 1. What is the result of the hydrolysis of ATP? Is there an alternative pathway that results in the release of more energy? Draw the products for both reactions using the reactants structures given in the diagram below as a guide.
Textbook Reference: *6.1 ATP, Reduced Coenzymes, and Chemiosmosis Play Important Roles in Biological Energy Metabolism, pp. 101–102*

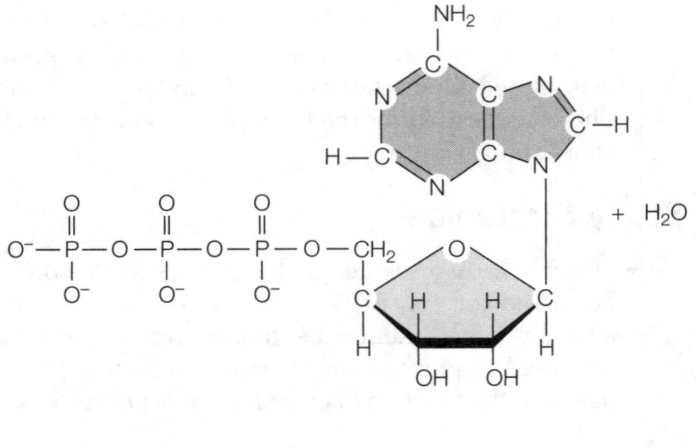

Question 2. It is estimated that approximately 90 percent of the energy that passes between levels in a food web is "lost" at each level. Explain this in the context of ATP usage in the cell.
Textbook Reference: *6.1 ATP, Reduced Coenzymes, and Chemiosmosis Play Important Roles in Biological Energy Metabolism pp. 102–103; see also 2.5 Biochemical Changes Involve Energy pp. 29–30*

Question 3. In the diagram below, label each compound as either oxidized or reduced. Circle the two lowest energy compounds. Which is the oxidizing agent and which is the reducing agent?
Textbook Reference: *6.1 ATP, Reduced Coenzymes, and Chemiosmosis Play Important Roles in Biological Energy Metabolism, p. 103*

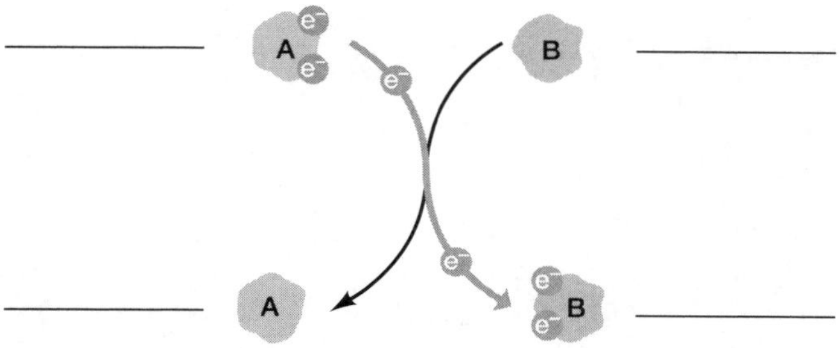

Question 4. Explain how the proton-motive force drives chemiosmosis.
Textbook Reference: *6.1 ATP, Reduced Coenzymes, and Chemiosmosis Play Important Roles in Biological Energy Metabolism, pp. 103–104*

6.2 Carbohydrate Catabolism in the Presence of Oxygen Releases a Large Amount of Energy

- In glycolysis, glucose is partially oxidized and some energy is released
- Pyruvate oxidation links glycolysis and the citric acid cycle
- The citric acid cycle completes the oxidation of glucose to CO_2
- NADH is oxidized by the respiratory chain, and ATP is formed by chemiosmosis

Glucose metabolism is one source of cellular energy. Glucose may be metabolized in the absence of oxygen through glycolysis and fermentation. In the presence of oxygen, it is metabolized through glycolysis and cellular respiration in the form of the citric acid cycle, electron transport, and oxidative phosphorylation. Compared to aerobic respiration, anaerobic respiration produces significantly less cellular energy (in the form of ATP) from the same amount of glucose.

Both metabolic pathways using glucose (glycolysis followed by fermentation and glycolysis followed by cellular respiration) are regulated by enzymes. These pathways may be further controlled by compartmentalization into organelles in eukaryotic cells. Because metabolic processes are energy-transferring reactions, ATP and NAD^+ are necessary for shuttling energy and electrons between steps of the pathways. Redox reactions are the basis of metabolism. As one compound is oxidized, another is reduced.

Question 5. In the diagram below, name the processes that are bracketed, enter the products in the blanks indicated, and indicate how many ATP molecules are generated.
Textbook Reference: *6.2 Carbohydrate Catabolism in the Presence of Oxygen Releases a Large Amount of Energy, pp. 106–110*

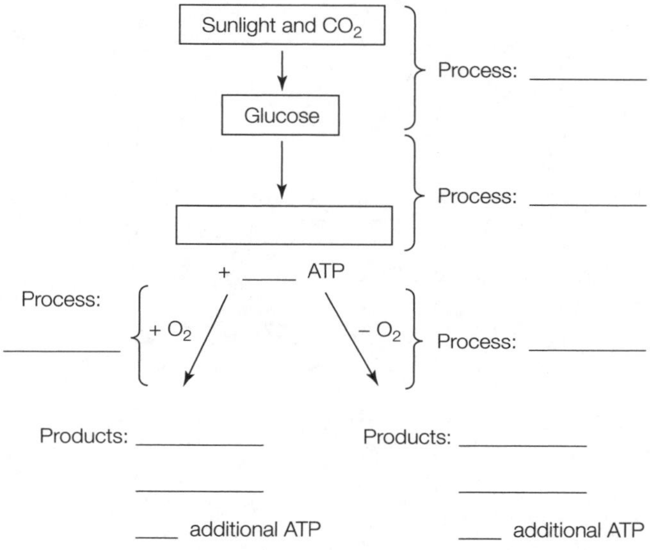

Question 6. In the diagram below, fill in the numbers of reactants and products on each side of the equation.
Textbook Reference: *6.2 Carbohydrate Catabolism in the Presence of Oxygen Releases a Large Amount of Energy, pp. 106–110*

1 Glucose { + ____ ADP, + ____ NAD^+ + H^+, + ____ P_i, + ____ ATP } → ____ Pyruvate { + ____ ATP, + ____ NADH, + ____ H_2O, + ____ ADP, + ____ P_i }

Cross out anything appearing on both sides of the equation to get the net reaction shown below.

1 Glucose { + ____ ADP, + ____ NAD^+ + H^+, + ____ P_i } → ____ Pyruvate { + ____ ATP, + ____ NADH, + ____ H_2O }

Question 7. In the diagram below, label all of the structures and indicate the necessary reactants and products for the different steps of the pathway.

Textbook Reference: *6.2 Carbohydrate Catabolism in the Presence of Oxygen Releases a Large Amount of Energy, p. 108*

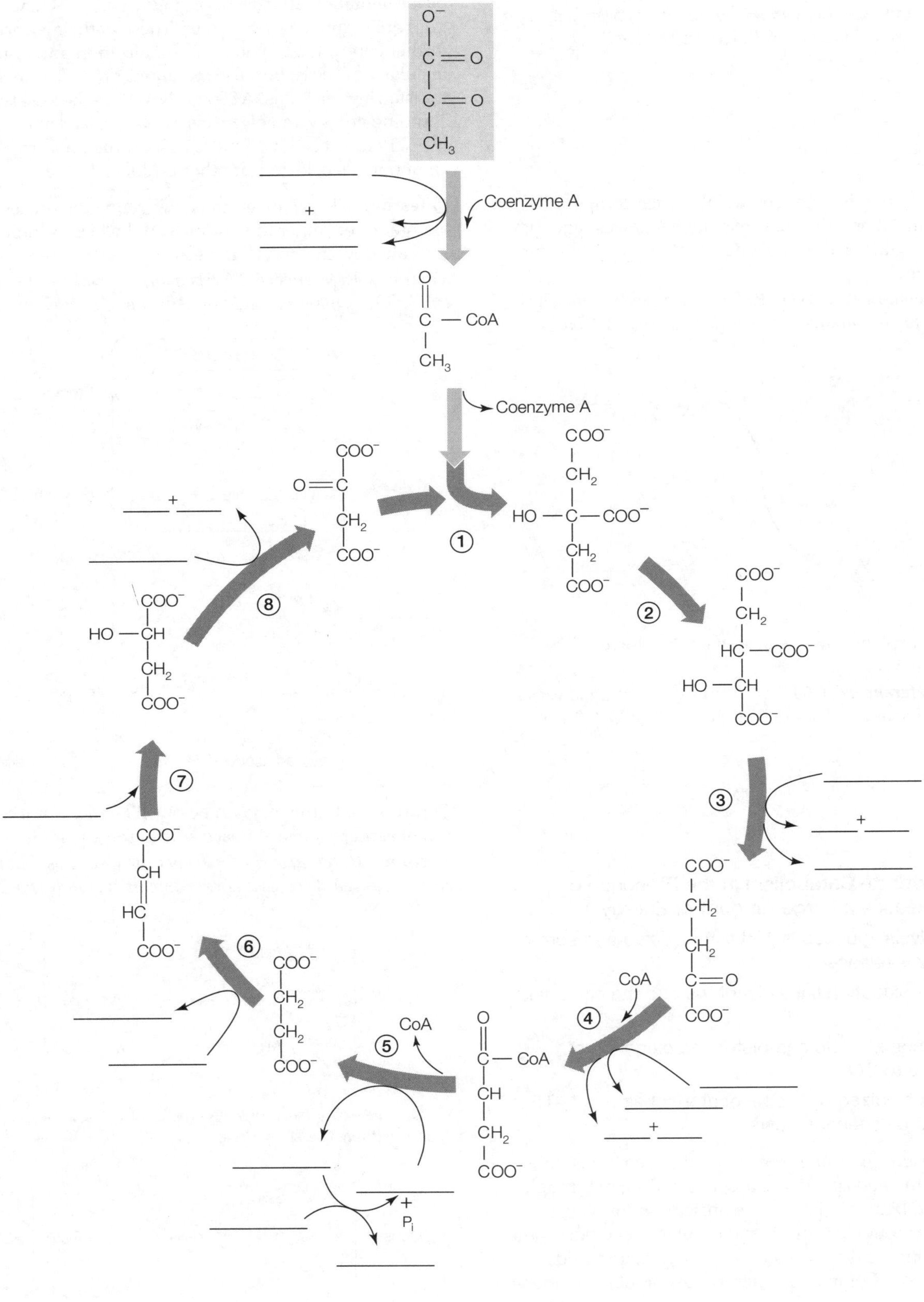

Question 8. Label the diagram below of the respiratory chain and chemiosmosis in the mitochondria. Be sure to label the mitochondrial parts as well.
Textbook Reference: *6.2 Carbohydrate Catabolism in the Presence of Oxygen Releases a Large Amount of Energy, p. 109*

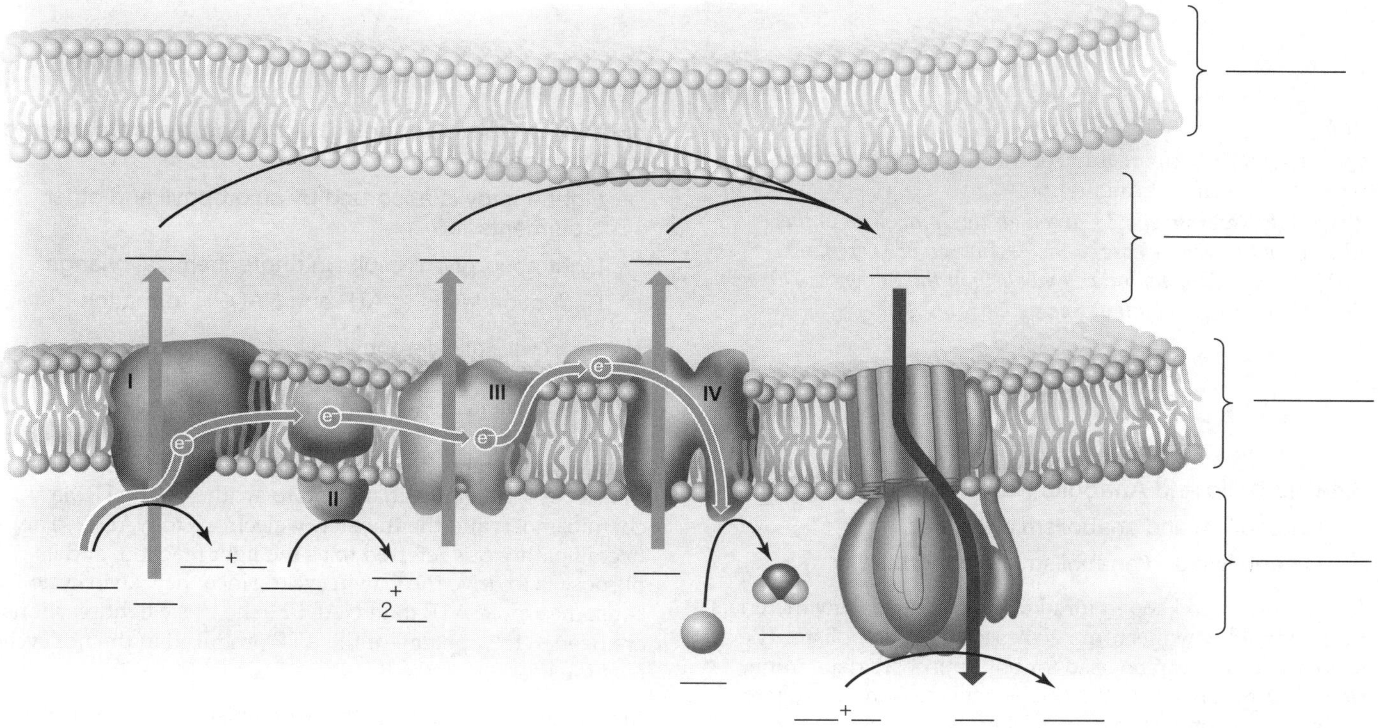

Question 9. Why is oxygen necessary for aerobic respiration?
Textbook Reference: *6.2 Carbohydrate Catabolism in the Presence of Oxygen Releases a Large Amount of Energy, pp. 109–110*

Question 10. One of the by-products of aerobic cellular respiration is carbon dioxide. Assume that you are working with labeled glucose, and trace the fate of that molecule until carbon dioxide is released.
Textbook Reference: *6.2 Carbohydrate Catabolism in the Presence of Oxygen Releases a Large Amount of Energy, pp. 107–108*

Question 11. Cyanide kills by inhibiting cytochrome oxidase in the mitochondria so that oxygen can no longer be utilized and the electron transport chain is halted. However, many cells in the human body are capable of lactic acid fermentation. Since cyanide does not inhibit glycolysis and fermentation, what could explain cyanide's lethal affect on humans?
Textbook Reference: *6.2 Carbohydrate Catabolism in the Presence of Oxygen Releases a Large Amount of Energy, pp. 109–110*

Question 12. Where do the bubbles in beer come from?
Textbook Reference: *6.2 Carbohydrate Catabolism in the Presence of Oxygen Releases a Large Amount of Energy, p. 110*

6.3 Carbohydrate Catabolism in the Absence of Oxygen Releases a Small Amount of Energy

There are two fermentation pathways—lactic acid formation and alcoholic fermentation—which regenerate NAD^+ so that glycolysis can continue. There is no additional ATP obtained from the fermentation reactions, but the overall glycolysis-to-fermentation pathway does yield 2 ATPs per glucose molecule.

Question 13. Glycolysis yields two molecules of pyruvate, two ATP, and two NADH + H^+, regardless of whether oxygen is present or not. What are the fates of these molecules in the absence of oxygen? What would happen if NADH + H^+ were not recycled?
Textbook Reference: *6.3 Carbohydrate Catabolism in the Absence of Oxygen Releases a Small Amount of Energy, p. 111*

Question 14. Compare and contrast energy yields from aerobic respiration and fermentation.
Textbook Reference: *6.3 Carbohydrate Catabolism in the Absence of Oxygen Releases a Small Amount of Energy, pp. 110–111; 6.2 Carbohydrate Catabolism in the Presence of Oxygen Releases a Large Amount of Energy, pp. 106–110*

6.4 Catabolic and Anabolic Pathways Are Integrated

Catabolism and anabolism are linked

Catabolism and anabolism are integrated

It is important to keep in mind that a cell needs raw materials both to build new macromolecules and for catabolism. The same molecules can be used for both purposes depending on the current needs of the cell. For this reason, the different pathways all connect to one another. As Figure 6.14 shows, all four groups of macromolecules (lipids, carbohydrates, nucleic acids, and proteins) can feed into the glycolysis and cellular respiration pathway. Likewise, intermediates of glycolysis and cellular respiration can be diverted from energy production and used to build new macromolecules. When these intermediates are used to form glucose, the process is called gluconeogenesis.

Figure 6.7 (in Concept 6.2) shows how photosynthesis ties in to the pathways discussed so far in this chapter. A single photosynthetic plant cell can be actively converting light energy into carbohydrates while also catabolizing some of those carbohydrates for energy to power other cellular activities. Metabolic enzymes are regulated to ensure that the appropriate pathways are active for the cell's needs at all times.

Question 15. The fate of acetyl CoA differs according to how much ATP is present in the cell. Explain what happens to acetyl CoA when ATP is limited, and compare that to what happens when acetyl CoA is abundant. How do these processes help regulate metabolism?
Textbook Reference: *6.4 Catabolic and Anabolic Pathways Are Integrated, pp. 111–112*

Question 16. When a person consumes a packet of pure sugar and burns it for energy, where does the carbon in the sugar ultimately go? Is the same true of the carbons in fat molecules when a person loses weight?
Textbook Reference: *6.4 Catabolic and Anabolic Pathways Are Integrated, pp. 111–112*

6.5 During Photosynthesis, Light Energy Is Converted to Chemical Energy

Light energy is absorbed by chlorophyll and other pigments

Light absorption results in photochemical change

Reduction leads to ATP and NADPH formation

The properties of chlorophyll, which are organized into two photosystems, allow photons of light to excite electrons. This process transfers the light energy to the molecule, and this energy, in the form of energized electrons, is then transferred to antenna systems in a redox reaction. Redox reactions in the electron transport chain lead to synthesis of ATP via chemiosmosis and the transfer of electrons to $NADP^+$. The overall pathway is referred to as the light reaction, and its purpose is to drive the Calvin cycle. Since the Calvin cycle consumes more ATP than NADPH, the cyclic light reactions are needed to supplement the ATP produced in the noncyclic reactions.

In an ecosystem, photosynthetic organisms provide energy in a usable form to heterotrophic organisms. This is why photosynthetic organisms are referred to as the foundation of most food webs.

Question 17. Label the photosystems, pathways, reactants, and products in the diagram below. Include both electron pathways in your answer.
Textbook Reference: *6.5 During Photosynthesis, Light Energy Is Converted to Chemical Energy, p. 116*

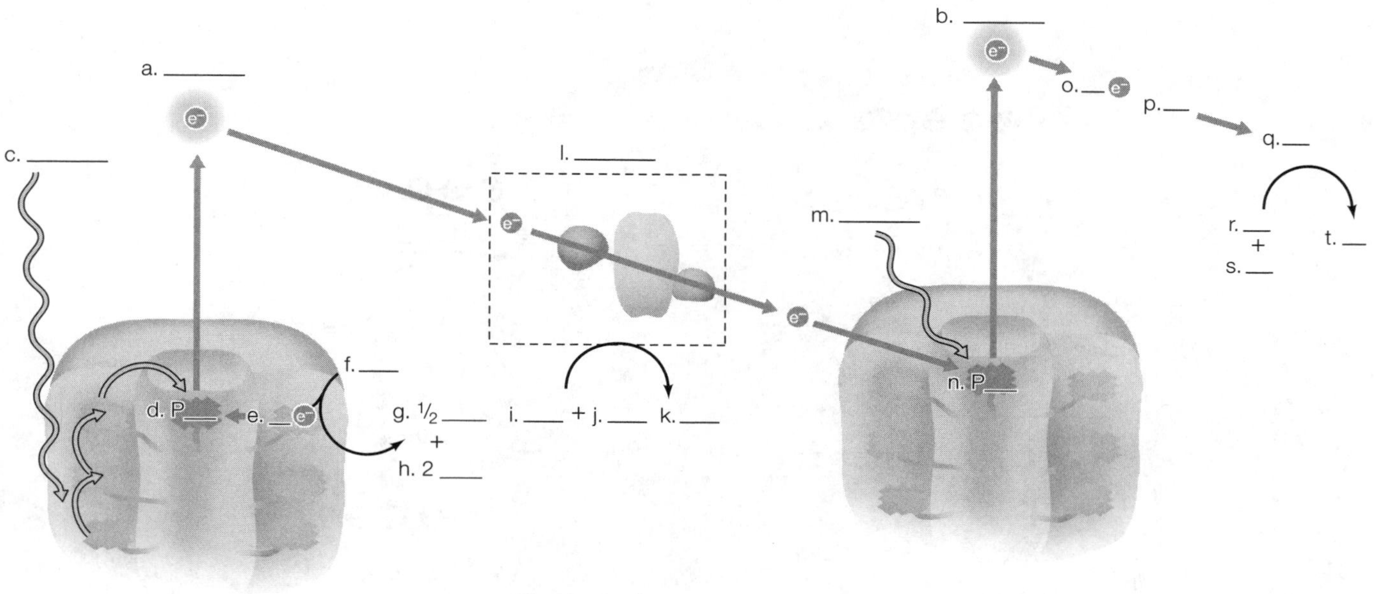

Question 18. Plants consume CO_2 and give off O_2. How is this possible if plants must also undergo cellular respiration?
Textbook Reference: *6.5 During Photosynthesis, Light Energy Is Converted to Chemical Energy p. 113*

Question 19. Explain the differences between cyclic and noncyclic electron flow. Why are both processes necessary?
Textbook Reference: *6.5 During Photosynthesis, Light Energy Is Converted to Chemical Energy, pp. 116–117*

Question 20. How do accessory pigments enhance photosynthetic activity in plants?
Textbook Reference: *6.5 During Photosynthesis, Light Energy Is Converted to Chemical Energy, p. 115*

Question 21. Why are plants green?
Textbook Reference: *6.5 During Photosynthesis, Light Energy Is Converted to Chemical Energy, p. 114*

6.6 Photosynthetic Organisms Use Chemical Energy to Convert CO_2 to Carbohydrates

The Calvin cycle takes place in the stroma and has three stages: carbon fixation, reduction, and RuBP regeneration. The first stage uses energy in the 5-carbon RuBP molecule to add a molecule of CO_2. This reaction yields two 3-carbon molecules. The second stage reduces the 3-carbon molecules, yielding a sugar precursor called G3P. Approximately one-sixth of the G3P product is used by the cell either to synthesize other molecules or catabolized as an energy source. The remaining G3P is recycled into RuBP. Although the process by which G3P is synthesized and recycled into RuBP looks inefficient, the energy driving the reactions is essentially "free" from the plant's perspective. It originally enters the pathway in the form of light photons, so as long as there is sunlight, there is energy available to drive the light reactions and ultimately the Calvin cycle.

Question 22. Label the stages, reactants, and products in the diagram below. Which enzyme is most important in this system?
Textbook Reference: *6.6 Photosynthetic Organisms Use Chemical Energy to Convert* CO_2 *to Carbohydrates, pp. 118–119*

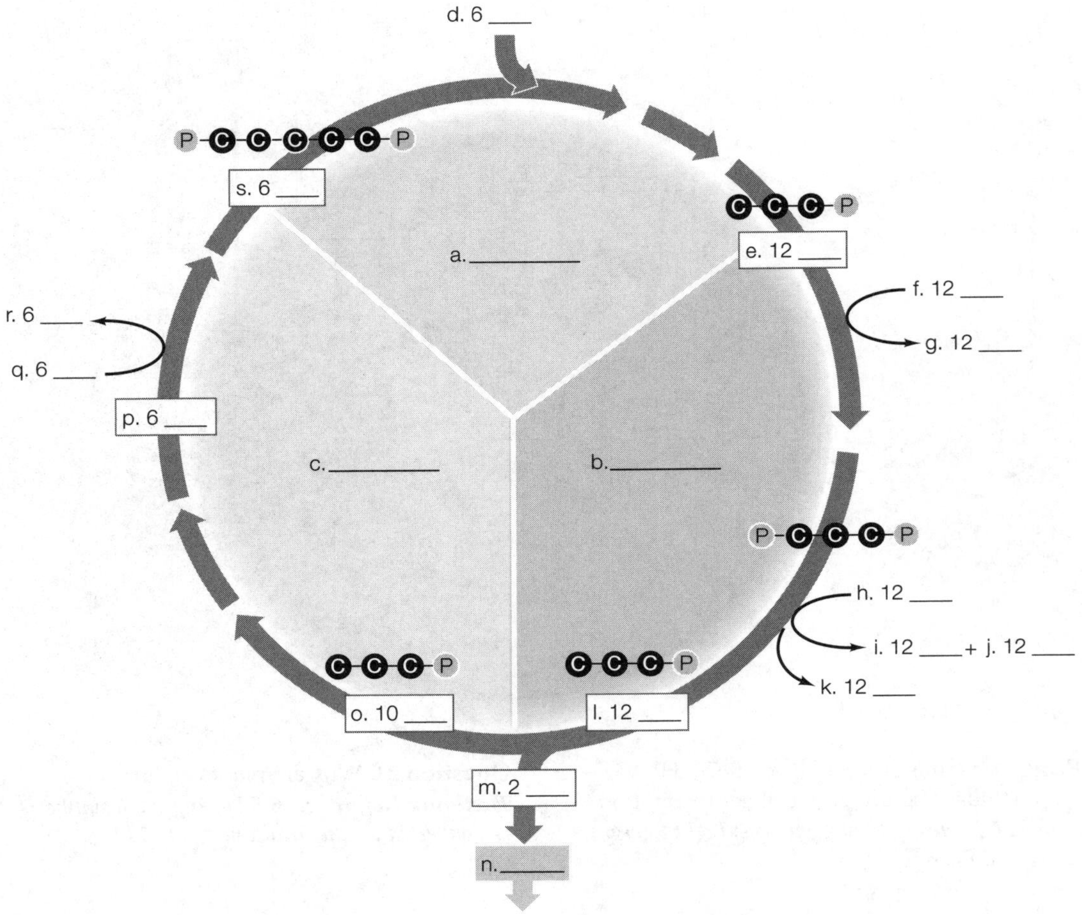

Question 23. Label the cycles, pathways, reactants, and products in the diagram below. As you work, focus on the connections between the two systems. Note how the products of one system provide the raw materials for the other system.
Textbook Reference: *6.6 Photosynthetic Organisms Use Chemical Energy to Convert CO_2 to Carbohydrates, pp. 118–119, Figure 6.7; 6.5 During Photosynthesis, Light Energy Is Converted to Chemical Energy, pp. 113–118*

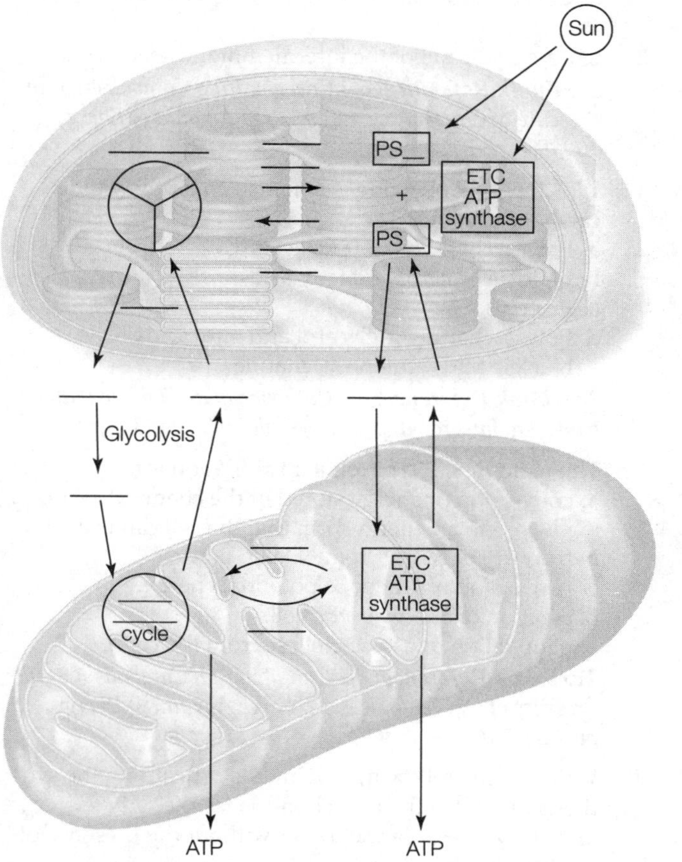

Question 24. Cellular respiration occurs simultaneously with many other cellular processes. Describe, in general, how cellular respiration interacts with other cellular metabolic events in plant and animal cells.
Textbook Reference: *6.6 Photosynthetic Organisms Use Chemical Energy to Convert CO_2 to Carbohydrates, p. 118*

Question 25. Why do plants undergo both photosynthesis and cellular respiration, even in the daytime? Why don't they simply use the ATP produced in the light reactions of photosynthesis to drive cellular processes?
Textbook Reference: *6.6 Photosynthetic Organisms Use Chemical Energy to Convert CO_2 to Carbohydrates, p. 118*

Question 26. The Calvin cycle was once referred to as the "dark" reactions of photosynthesis. Why is this a misnomer?
Textbook Reference: *6.6 Photosynthetic Organisms Use Chemical Energy to Convert CO_2 to Carbohydrates, p. 118*

Test Yourself

1. Before ATP is split into ADP and P_i, it holds what type of energy?
   a. Potential
   b. Kinetic
   c. Entropic
   d. Enthalpic
   e. Physical
   Textbook Reference: *6.1 ATP, Reduced Coenzymes, and Chemiosmosis Play Important Roles in Biological Energy Metabolism, p. 101*
2. ATP hydrolysis is
   a. endergonic.
   b. exergonic.
   c. chemoautotrophic.
   d. anabolic.
   e. None of the above
   Textbook Reference: *6.1 ATP, Reduced Coenzymes, and Chemiosmosis Play Important Roles in Biological Energy Metabolism, pp. 101–102*
3. Which of the following cellular metabolic processes is active in *all* cells, regardless of the presence or the absence of oxygen?
   a. The citric acid cycle
   b. Electron transport
   c. Glycolysis
   d. Fermentation
   e. Pyruvate oxidation
   Textbook Reference: *6.2 Carbohydrate Catabolism in the Presence of Oxygen Releases a Large Amount of Energy, pp. 106–107, Figure 6.10*
4. Which of the following statements regarding glycolysis is *false*?
   a. A 6-C sugar is broken down to two 3-C molecules.
   b. Two ATP molecules are consumed.
   c. Glycolysis requires oxygen.
   d. A net sum of two ATP molecules is generated.
   e. Glycolysis occurs in the cytosol.
   Textbook Reference: *6.2 Carbohydrate Catabolism in the Presence of Oxygen Releases a Large Amount of Energy, pp. 106–107, Figure 6.10*
5. During which process is most ATP generated in the cell?
   a. Glycolysis
   b. The citric acid cycle
   c. Electron transport coupled with chemiosmosis
   d. Fermentation
   e. Pyruvate oxidation

Textbook Reference: *6.2 Carbohydrate Catabolism in the Presence of Oxygen Releases a Large Amount of Energy, pp. 109–110*

6. Which of the following is a function of the electron transport chain?
   a. Cycling NADH + H^+ back to NAD^+
   b. Using the intermediates from the citric acid cycle
   c. Breaking down pyruvate
   d. Increasing the number of protons in the mitochondrial matrix
   e. Consuming excess ATP

   Textbook Reference: *6.2 Carbohydrate Catabolism in the Presence of Oxygen Releases a Large Amount of Energy, pp. 108–109*

7. Which of the following describes the role of the inner mitochondrial membrane?
   a. It acts as an anchor for the membrane-associated enzymes of cellular respiration.
   b. It allows for the establishment of a proton gradient.
   c. It separates the mitochondria's environment from that of the cytosol.
   d. It anchors enzymes and allows for the establishment of the proton gradient, but it is not involved in separating the contents of the mitochondria from the cytosol.
   e. It anchors enzymes, allows for the establishment of the proton gradient, and is involved in separating the contents of the mitochondria from the cytosol.

   Textbook Reference: *6.2 Carbohydrate Catabolism in the Presence of Oxygen Releases a Large Amount of Energy, pp. 108–109, Figure 6.12*

8. In the following redox reaction, _______ is oxidized and _______ is reduced.

   Glyceraldehyde 3-phosphate (G3P) + NAD^+ + H^+ + P_i → 1,3-Bisphosphoglycerate (BPG) + NADH
   a. G3P; NAD^+
   b. BPG; NADH + H^+
   c. G3P; NADH + H^+
   d. NAD^+; NADH + H^+
   e. None of the above; the equation does not show a redox reaction.

   Textbook Reference: *6.1 ATP, Reduced Coenzymes, and Chemiosmosis Play Important Roles in Biological Energy Metabolism, p. 102*

9. Which of the following statements about redox reactions is true?
   a. Oxidizing agents accept electrons.
   b. Oxidizing agents donate electrons.
   c. A molecule that accepts electrons is said to be oxidized.
   d. A molecule that donates electrons is said to be reduced.
   e. Oxidizing agents accept electrons and are reduced in the process.

   Textbook Reference: *6.1 ATP, Reduced Coenzymes, and Chemiosmosis Play Important Roles in Biological Energy Metabolism, p. 102*

10. Cyanide poisoning inhibits aerobic respiration at cytochrome *c* oxidase. Which of the following is *not* a result of cyanide poisoning at the cellular level?
    a. Reduction of oxygen to water
    b. Cessation of ATP synthesis in the mitochondria, because electron transport is never completed
    c. Switching of cells to anaerobic respiration and fermentation if possible
    d. Continuation of glycolysis as long as NAD^+ is available
    e. Less acidic pH of the intermembrane space

    Textbook Reference: *6.2 Carbohydrate Catabolism in the Presence of Oxygen Releases a Large Amount of Energy, p. 109, Figure 6.12*

11. Which of the following is matched correctly with its catabolic product?
    a. Polysaccharides – amino acids
    b. Lipids – glycerol and fatty acids
    c. Proteins – glucose
    d. Polysaccharides – glycerol and fatty acids
    e. Nucleic acids – monosaccharides

    Textbook Reference: *6.4 Catabolic and Anabolic Pathways Are Integrated, p. 112, see also Chapter 3*

12. The main function of cellular respiration is the
    a. conversion of energy stored in the chemical bonds of glucose to an energy form that the cell can use.
    b. recovery of NAD^+ from NADPH.
    c. conversion of kinetic to potential energy.
    d. creation of energy in the cell.
    e. elimination of excess glucose from the cell.

    Textbook Reference: *6.2 Carbohydrate Catabolism in the Presence of Oxygen Releases a Large Amount of Energy, pp. 109–110*

13. Which of the following statements concerning the synthesis of ATP in the mitochondria is *false*?
    a. ATP synthesis cannot occur without the presence of ATP synthase.
    b. The proton-motive force is the establishment of a charge and concentration gradient across the mitochondrial membrane.
    c. The proton-motive force drives protons back across the membrane through channels established by the ATP synthase channel protein.
    d. The ATP synthase protein is composed of two units.
    e. The intermembrane space is more basic than the mitochondrial matrix.

    Textbook Reference: *6.2 Carbohydrate Catabolism in the Presence of Oxygen Releases a Large Amount of Energy, pp. 109–110*

14. Which of the following does *not* occur in the mitochondria of eukaryotic cells?
    a. Fermentation
    b. Oxidative phosphorylation
    c. Citric acid cycle
    d. Electron transport chain
    e. Creation of a proton gradient

Textbook Reference: *6.2 Carbohydrate Catabolism in the Presence of Oxygen Releases a Large Amount of Energy, p. 108*

15. Which of the following is recycled and reused in cellular metabolism?
    a. ADP
    b. NAD^+
    c. FAD
    d. P_i
    e. All of the above
    Textbook Reference: *6.1 ATP, Reduced Coenzymes, and Chemiosmosis Play Important Roles in Biological Energy Metabolism, p. 103; 6.2 Carbohydrate Catabolism in the Presence of Oxygen Releases a Large Amount of Energy, pp. 106–107*

16. For each molecule of glucose, how many ATPs are synthesized in fermentation?
    a. 0
    b. 1
    c. 2
    d. 3
    e. 4
    Textbook Reference: *6.2 Carbohydrate Catabolism in the Presence of Oxygen Releases a Large Amount of Energy, p. 110*

17. The main function of photosynthesis is the
    a. consumption of CO_2.
    b. production of ATP.
    c. conversion of light energy to chemical energy.
    d. production of starch.
    e. production of O_2.
    Textbook Reference: *6.5 During Photosynthesis, Light Energy Is Converted to Chemical Energy, p. 114*

18. Which of the following best represents the components that are necessary for photosynthesis to take place?
    a. Mitochondria, accessory pigments, visible light, water, and CO_2
    b. Chloroplasts, accessory pigments, visible light, water, and CO_2
    c. Mitochondria, chlorophyll, visible light, water, and O_2
    d. Chloroplasts, chlorophyll, visible light, water, and CO_2
    e. Chlorophyll, accessory pigments, visible light, water, and O_2
    Textbook Reference: *6.5 During Photosynthesis, Light Energy Is Converted to Chemical Energy, p. 113, Figure 6.15*

19. Chlorophyll is suited for the capture of light energy because
    a. certain wavelengths of light raise it to an excited state.
    b. in its excited state it gives off electrons.
    c. its structure allows it to attach to thylakoid membranes.
    d. it can transfer absorbed energy to another molecule.
    e. All of the above
    Textbook Reference: *6.5 During Photosynthesis, Light Energy Is Converted to Chemical Energy, p. 115*

20. Plants give off O_2 because
    a. O_2 results from the incorporation of CO_2 into sugars.
    b. they do not respire; they photosynthesize.
    c. water is the initial electron donor, leaving O_2 as a photosynthetic by-product.
    d. electrons moving down the electron chain bind to water, releasing O_2.
    e. O_2 is synthesized in the Calvin cycle.
    Textbook Reference: *6.5 During Photosynthesis, Light Energy Is Converted to Chemical Energy, p. 116*

21. In plants, cyclic and noncyclic electron flow
    a. meet the ATP demands of the Calvin cycle.
    b. produce excess NADPH + H^+.
    c. synthesize proportional amounts of ATP and NADPH + H^+ in the chloroplast.
    d. consume the products of the Calvin cycle.
    e. produce O_2 for the atmosphere.
    Textbook Reference: *6.5 During Photosynthesis, Light Energy Is Converted to Chemical Energy, p. 117*

22. Which of the following statements about the light reactions of photosynthesis is true?
    a. Photosystem I cannot operate independently of photosystem II.
    b. Photosystems I and II are activated by different wavelengths of light.
    c. Photosystems I and II transfer electrons and create proton equilibrium across the thylakoid membrane.
    d. Photosystem I is more significant than photosystem II.
    e. Oxygen gas is a product of photosystem I.
    Textbook Reference: *6.5 During Photosynthesis, Light Energy Is Converted to Chemical Energy, pp. 116–117*

23. ATP is produced during the light reactions via
    a. CO_2 fixation.
    b. chemiosmosis.
    c. reduction of water.
    d. glycolysis.
    e. noncyclic electron flow from photosystem I.
    Textbook Reference: *6.5 During Photosynthesis, Light Energy Is Converted to Chemical Energy, p. 116*

24. Because of the properties of chlorophyll, plants need adequate _______ light to grow properly.
    a. green
    b. blue and red
    c. infrared
    d. ultraviolet
    e. blue and blue-green
    Textbook Reference: *6.5 During Photosynthesis, Light Energy Is Converted to Chemical Energy, p. 114, Figure 6.17*

25. Which of the following statements concerning the Calvin cycle is *false*?
   a. Light energy is not required for the cycle to proceed.
   b. CO_2 is assimilated into sugars.
   c. RuBP is regenerated.
   d. It makes use of energy stored in ATP and NADPH + H^+.
   e. All of the above are false.
   Textbook Reference: *6.6 Photosynthetic Organisms Use Chemical Energy to Convert CO_2 to Carbohydrates, p. 118*

26. Which of the following initiates the Calvin cycle and results in the entire pathway being carried out under environmental conditions?
   a. 3PG is reduced to G3P using ATP and NADPH + H^+.
   b. RuBP is regenerated.
   c. CO_2 and RuBP join, forming 3PG.
   d. G3P is converted into glucose and fructose.
   e. Any of the above; since it is a cycle, it can start at any point.
   Textbook Reference: *6.6 Photosynthetic Organisms Use Chemical Energy to Convert CO_2 to Carbohydrates, p. 118*

27. The Calvin cycle results in the production of
   a. glucose.
   b. starch.
   c. rubisco.
   d. G3P.
   e. ATP.
   Textbook Reference: *6.6 Photosynthetic Organisms Use Chemical Energy to Convert CO_2 to Carbohydrates, p. 118*

28. Which of the following statements regarding the relationship between photosynthesis and cellular respiration in plants is true?
   a. Photosynthesis occurs in specialized photosynthetic cells.
   b. Cellular respiration occurs in specialized respiratory cells.
   c. Cellular respiration and photosynthesis can occur in the same cell.
   d. Photosynthesis is limited to specialized plant cells and cellular respiration does not occur in plant cells.
   e. Both a and c
   Textbook Reference: *6.1 ATP, Reduced Coenzymes, and Chemiosmosis Play Important Roles in Biological Energy Metabolism, p. 106*

29. Photosynthesis occurs
   a. in all plant cells.
   b. only in photosynthetic plant cells.
   c. only in plant cells lacking mitochondria.
   d. only in the stroma.
   e. only in the thylakoid membrane.
   Textbook Reference: *6.5 During Photosynthesis, Light Energy Is Converted to Chemical Energy, p. 113*

30. Activities such as amino acid synthesis and active transport in plant cells are powered by
   a. the light-dependent and light-independent reactions of photosynthesis.
   b. ATP from the light reactions of photosynthesis.
   c. ATP from fermentation.
   d. ATP from glycolysis and cellular respiration.
   e. All of the above
   Textbook Reference: *6.4 Catabolic and Anabolic Pathways Are Integrated, p. 112*

Answers

Key Concept Review

1. Products of reaction 1:

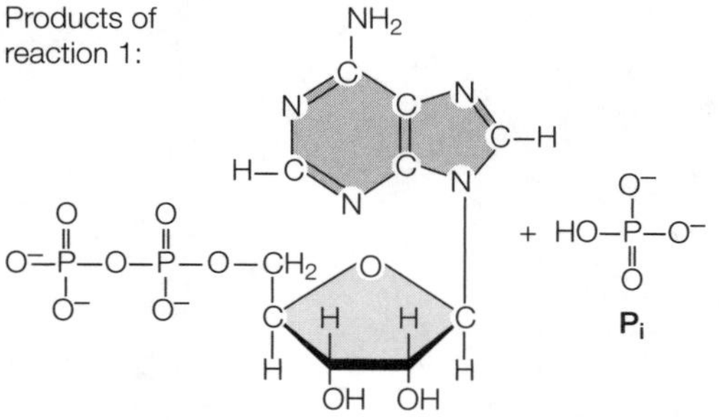

Products of reaction 2:

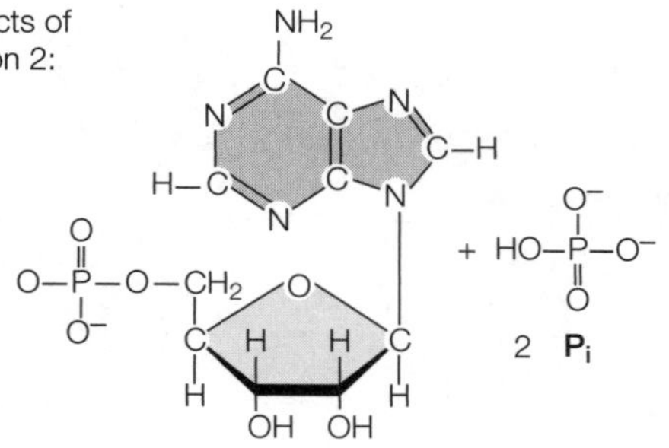

2. Recall that energy cannot be created or destroyed, but that it may be converted from one form to another. Many of these energy transformations use ATP as the energy carrier. As each molecule of ATP is hydrolyzed, some energy is converted to unusable heat energy. This phenomenon is not limited to ATP hydrolysis, but applies to every energy conversion reaction in the cell. Thus, there is a net loss of usable energy among different members of the food web.

3.

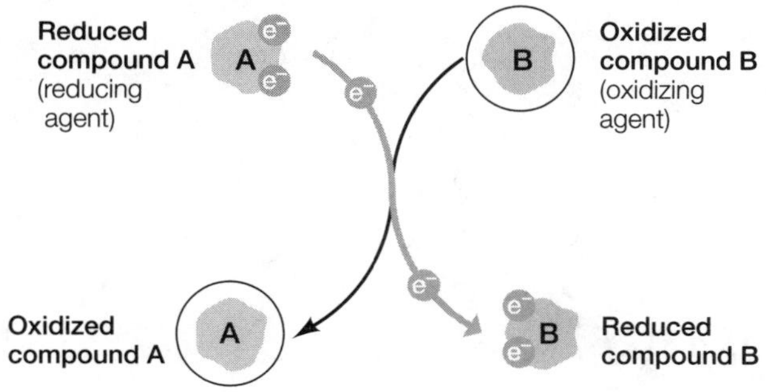

4. The proton-motive force results in a concentration and charge gradient across the mitochondrial membrane. For that gradient to equalize, the protons must flow through a channel protein. If this channel protein has an associated ATP synthase, ATP is generated as protons flow through.

5.

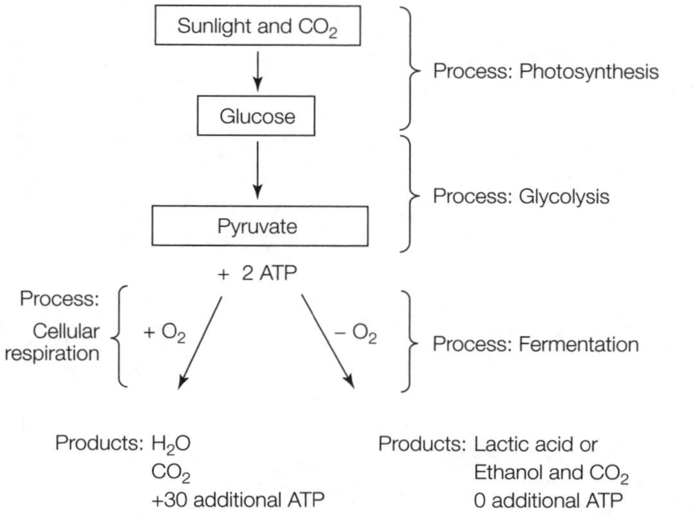

6.

1 Glucose + ~~4~~ ADP (becomes 2 ADP) + 2 NAD^+ + H^+ + 4 P_i (becomes 2 P_i) + ~~2 ATP~~ → 2 Pyruvate + ~~4~~ ATP (becomes 2 ATP) + 2 NADH + 2 H_2O + ~~2 ADP~~ + ~~2 P_i~~

Net Reaction:

1 Glucose + 2 ADP + 2 NAD^+ + H^+ + 2 P_i → 2 Pyruvate + 2 ATP + 2 NADH + 2 H_2O

7.

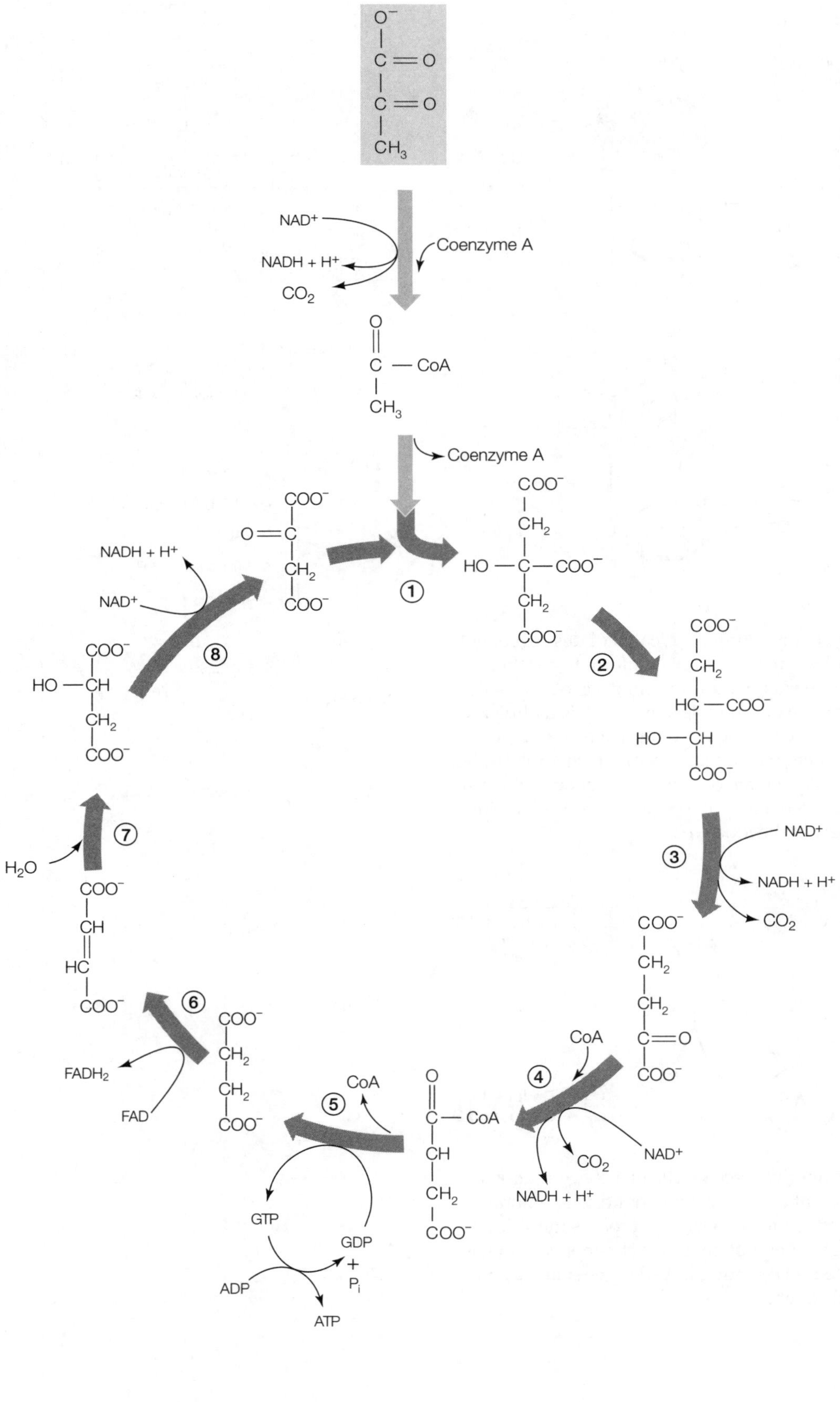
NAD+
NADH + H+
CO2
Coenzyme A
C — CoA
CH3
Coenzyme A
NADH + H+
NAD+
1
2
3
4
5
6
7
8
NAD+
NADH + H+
CO2
CoA
CoA
NAD+
CO2
NADH + H+
GTP
GDP
+
Pi
ADP
ATP
FADH2
FAD
H2O

8.

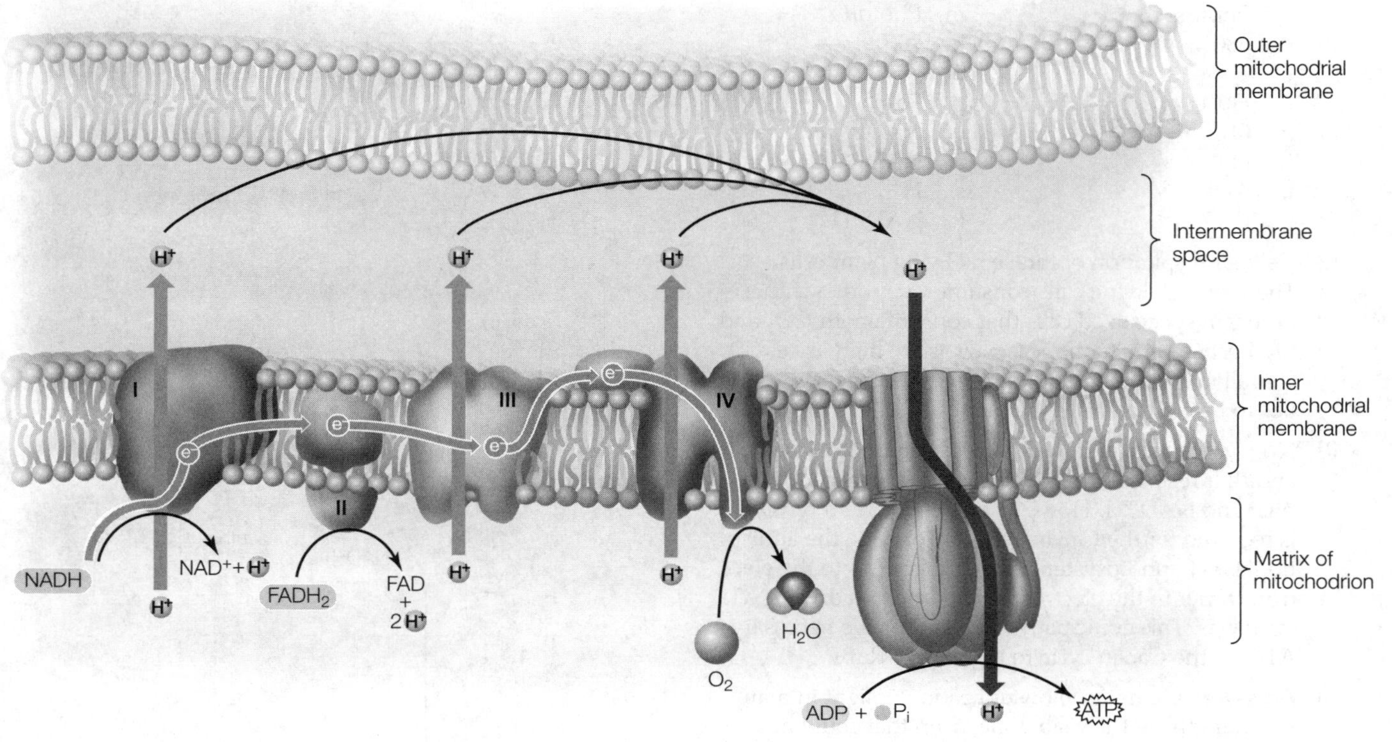

9. Oxygen acts as the terminal electron acceptor in the electron transport pathway. Without it, NADH + H^+ cannot be cycled back to NAD^+. The accumulated NADH + H^+ acts as an inhibitor to the citric acid cycle and effectively shuts it down. Therefore, in the absence of oxygen, a cell can undergo glycolysis only.

10. Carbon dioxide is liberated when pyruvate is oxidized to acetate (In-Text Art, p. 108) and in the citric acid cycle (Figure 6.11).

11. The reduced efficiency of glycolysis and fermentation for ATP synthesis is one reason why oxygen deprivation is so deadly to humans. (And, as later chapters will show, the resulting lactic acid buildup also causes problems.) In addition, not all cells are capable of carrying out the reactions of fermentation. Brain cells, for example, will simply die in the absence of oxygen.

12. The bubbles in beer are bubbles of CO_2 released during the fermentation of pyruvate into ethyl alcohol (see Figure 6.13.)

13. In the absence of oxygen, pyruvate is either reduced to lactate (in lactic acid fermentation) or it is metabolized and its metabolites are reduced to ethyl alcohol (in alcoholic fermentation). In either case, NADH + H^+ is the reducing agent, and it is oxidized back to NAD^+ in the process. The two molecules of ATP are used to power cellular activities. If NADH + H^+ were not oxidized to NAD^+, there would eventually be no NAD^+ available for glycolysis.

14. Fermentation yields only 2 ATP. Cellular respiration yields 32 ATP.

15. If ATP is limited, acetyl CoA enters the citric acid cycle and cellular respiration utilizes it to produce ATP. If ATP is abundant, acetyl CoA is shuttled to fatty acid synthesis, thus storing the energy in chemical bonds. This enables the cell to store energy for times when energy sources are abundant and consume the stored energy when external sources are limited.

16. The carbon in the sugar is exhaled in the form of CO_2. When someone loses weight, the carbon from the fat molecules is also exhaled in the form of CO_2.

17.

a. Photosystem II	k. ATP
b. Photosystem I	l. Electron transport
c. Photon	m. Photon
d. 680	n. 700
e. 2	o. 2
f. H_2O	p. Fd
g. O_2	q. $NADP^+$ reductase
h. H^+	r. $NADP^+$
i. ADP	s. H^+
j. P_i	t. NADPH

18. Cellular respiration occurs in all living plant cells. Therefore, all living cells consume O_2. Photosynthesis occurs in specialized cells that consume both CO_2 and O_2. Because atmospheric O_2 levels are high, excess O_2 is available for the plant to utilize; therefore, O_2 continues to be emitted from the plant.

19. Noncyclic electron flow involves both photosystems I and II. It results in the synthesis of equal amounts of ATP and NADPH. However more ATP than NADPH is required for the Calvin cycle. To provide the additional ATP, photosystem I sends electrons to the electron carrier in the electron transport chain driving ATP synthesis. This cyclic pathway provides the necessary ATP for the Calvin cycle to regenerate RuBP.

20. Accessory pigments allow utilization of light in many wavelengths of the visible spectrum that could not be used by chlorophyll alone. The energy absorbed is channeled to the reaction centers of the photosystems.

21. The primary pigments in plants are chlorophylls. Chlorophylls absorb blue and orange-red wavelengths of light and reflect green light, thus making plants appear green. See the absorption spectra and action spectra of chlorophyll in Figure 6.17.

22. The most important enzyme in this system is rubisco, which is active in the first stage.

a. Carbon fixation	j. H^+
b. Reduction and sugar production	k. P_i
	l. G3P
c. Regeneration of RuBP	m. G3P
d. CO_2	n. Sugars
e. 3PG	o. G3P
f. ATP	p. RuMP
g. ADP	q. ATP
h. NADPH	r. ADP
i. $NADP^+$	s. RuBP

23.

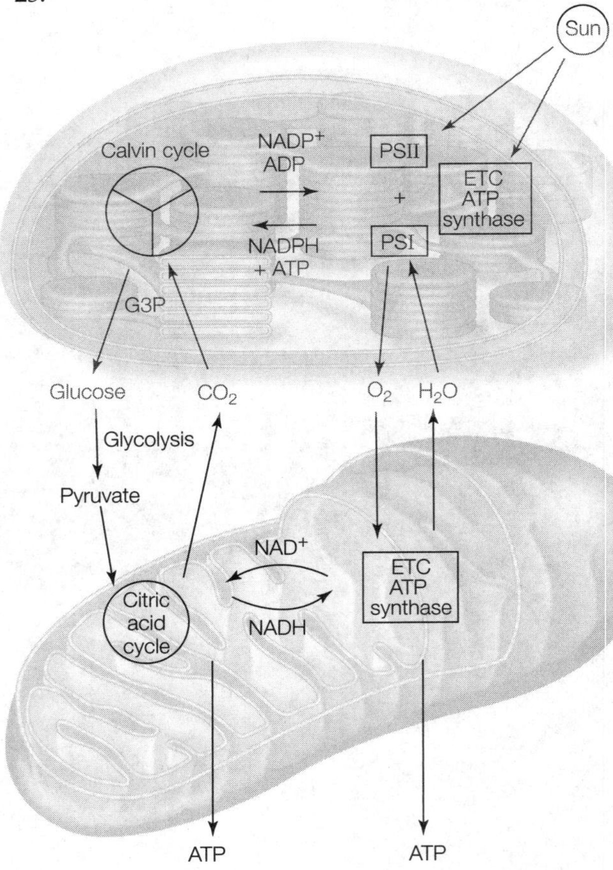

24. Consult Figures 6.7 and 6.14 to see where different metabolic pathways in the cell interact. It is crucial to remember that photosynthetic plant cells contain both chloroplasts and mitochondria. Living nonphotosynthetic plant cells, have only mitochondria and lack chloroplasts. Animals, in contrast, have only mitochondria. The pathways seen in Figure 6.14 are common to organisms in both kingdoms.

25. The light reactions of photosynthesis produce ATP. ATP cannot be stored for use later (such as when light is not available) or transported to nonphotosynthetic cells. Therefore, there has to be a mechanism for that energy to be stored in a more stable form. The Calvin cycle stores the energy in the chemical bonds of G3P, which can be incorporated into carbohydrates for longer-term storage. When the cell needs energy for cellular processes, the carbohydrate is processed through the catabolic pathways (glycolysis, cellular respiration) already discussed.

26. Light is required for both the light reactions of photosynthesis and the Calvin cycle. The Calvin cycle depends on the ATP generated during the light-dependent reactions.

Test yourself

1. **a.** Potential energy is energy held within chemical bonds that may be converted to working kinetic energy.
2. **b.** ATP hydrolysis is exergonic.
3. **c.** Glycolysis proceeds during both fermentation and cellular respiration. Only in cellular respiration is oxygen needed as the terminal electron acceptor of the pathway.
4. **c.** During glycolysis, 6-C glucose is broken down into two 3-C pyruvate molecules. In the process, four total ATP are produced, but two are consumed, leaving a net production of two ATP molecules. No oxygen is required in glycolysis.
5. **c.** Most of the ATP produced during cellular respiration is produced by electron transport and chemiosmosis coupled in oxidative phosphorylation.
6. **a.** The electron transport chain is responsible for oxidizing NADH + H^+ back to NAD^+.
7. **c.** The mitochondrial membrane is necessary for the anchoring of proteins as well as the establishment of a barrier across which a gradient can be established.
8. **a.** A molecule is oxidized when it loses electrons or protons and is reduced when it gains electrons or protons. In this reaction, G3P donates electrons and therefore is oxidized, while NAD^+ accepts them and thus is reduced.
9. **e.** Oxidizing agents accept electrons and cause oxidation of another molecule. Reducing agents donate electrons and cause the reduction of another molecule.
10. **a.** Cyanide stops aerobic cellular respiration because cytochrome *c* oxidase loses the ability to reduce oxygen to water. Those cells that can switch to anaerobic respiration (fermentation) do so and use their remaining glucose more quickly. Cells that cannot make that switch die.
11. **b.** Lipids are broken down into glycerol and fatty acids; polysaccharides are broken down into glucose; proteins are broken down into amino acids. Nucleic acids are converted into some amino acids and fed into the citric acid cycle.
12. **a.** Cellular respiration is the cell's way of converting potential energy in the chemical bonds of glucose to potential energy that the cell ultimately can use.
13. **a.** Substrate level phosphorylation occurs in step 5 of the citric acid cycle in the mitochondria. The intermembrane space is more acidic than the matrix.
14. **a.** Fermentation occurs in the cytosol, whereas all the other processes occur in the mitochondria of eukaryotic cells.
15. **e.** ADP, NAD^+, FAD, and P_i are all recycled and reused in the process of cellular respiration.
16. **a.** Fermentation regenerates NAD^+ so that glycolysis can continue.
17. **c.** Photosynthetic organisms, including but not limited to plants, are the only life forms capable of trapping light energy and converting it to chemical energy. Because of this they form the basis of many of Earth's food webs.
18. **d.** Chloroplasts are the site of the photosynthetic reactions; chlorophyll is excited by photons of light and serves as reaction centers for the photosystems; visible light is necessary to excite chlorophyll and accessory pigments; water is the initial electron donor for the pathway; and CO_2 is necessary to make precursor molecules for energy storage.
19. **e.** The "tails" of chlorophyll molecules are associated with the thylakoid membranes of the chloroplasts. This close membrane association assists with establishing the proton-motive force that will drive ATP synthesis. When excited by light, the chlorophyll moves into an excited state and passes electrons to acceptor molecules. This begins to set up the proton gradient across the membrane that will drive ATP synthesis.
20. **c.** Water is split at photosystem II to donate electrons to the reaction center. The resulting protons are moved across the membrane to establish the proton-motive force, and O_2 is given off as a by-product.
21. **a.** ATP is required in greater amounts in the Calvin cycle than NADPH + H^+ is; therefore, there must be a mechanism for producing additional ATP. Cyclic electron flow provides that mechanism. If noncyclic electron flow were to be sped up to meet ATP needs, an excess of NADPH + H^+ would result. Shifting between cyclic and noncyclic flow balances ATP/NADPH + H^+ ratios. Oxygen gas is a by-product of the light reactions, but its production is not the purpose of the reactions.
22. **b.** Photosystems I and II operate depending on whether electron flow is cyclic or noncyclic. The ATP levels in the chloroplast control activity. Compared to photosystem I, photosystem II is activated by light of a higher energy level. Both photosystems transfer electrons and create proton gradients across the thylakoid membranes; photosystem I does this via the cyclic pathway. Water is split by a structure embedded in the photosystem II complex.
23. **b.** In the light reactions, ATP synthesis occurs when protons flow through an ATP synthase channel protein in the thylakoid membrane. This is a chemiosmotically driven process. Photosystem II is always involved, while photosystem I operates via cyclic electron transport only.
24. **b.** Chlorophyll and accessory pigments absorb light in the blue and red wavelengths of visible light. Green light is reflected; therefore, plants appear green. (Accessory pigments allow energy from additional wavelengths to be absorbed as well.)
25. **a.** Light energy is required for the Calvin cycle to proceed. ATP synthesis is dependent on light energy, and the Calvin cycle is dependent on ATP.

26. **c.** The first step of the Calvin cycle is the fixation of CO_2 into 3PG. This is the regulatory step, and it requires ATP and NADPH + H^+. While it is true that the Calvin cycle is a cycle, there is a net consumption of CO_2 for the purpose of building carbohydrates.

27. **d.** The Calvin cycle produces only G3P, which can then be metabolized into storage products, such as sugars and starch.

28. **e.** Photosynthesis occurs only in plant cells that have the necessary structures, but cellular respiration occurs in every living plant cell that has mitochondria and O_2.

29. **b.** Photosynthesis is limited to photosynthetic plant cells. There are many plant cells that are not exposed to light or that lack chloroplasts; these cells rely on cellular respiration.

30. **d.** Plant cells have mitochondria and rely on the processes of glycolysis and cellular respiration to provide ATP for cellular activities. Photosynthesis converts light energy into potential energy stored in chemical form, but then the cells must make the energy usable. Plant cells release this stored energy via glycolysis and cellular respiration.

The Cell Cycle and Cell Division

7

The Big Picture

- During the life cycle of a multicellular organism, cells reproduce in a controlled process for both sexual and asexual reproduction.
- Prokaryotic cells reproduce by cell division, specifically binary fission.
- In the presence of an internal or external environmental signal, cells undergo a specific set of steps to replicate their DNA, segregate the newly replicated chromosomes, and divide their cytoplasm into two daughter cells. These steps ensure that the daughter cells receive a complete set of genetic instructions.
- Diploid organisms produce haploid gametes by meiosis, which fuse during fertilization to form new diploid organisms.
- During the generation of these gametes, crossing over between homologous chromatids and random segregation of those chromatids create genetically diverse gametes, leading to increased species diversity and survival.
- The cell cycle is a tightly regulated process involving many regulatory molecules and checkpoints.
- Many cells undergo programmed cell death (apoptosis) when they are no longer needed or are likely to have accumulated so much damage they might lead to the development of cancer.

Study Strategies

- Be sure you understand the differences between mitosis and meiosis. After every DNA replication, there is at least one cell division. In mitosis, the replicated chromosomes are segregated to the two daughter cells, resulting in daughter cells that each have two sets of chromosomes (diploid), like the parent cell. In meiosis, the replicated chromosomes undergo two cell divisions, resulting in four cells that each have only one set of chromosomes (haploid). Furthermore, the homologous chromosomes pair in prophase of meiosis I to allow crossing over to occur.
- It is helpful to work through the processes of mitosis and meiosis with various props. For instance, yarn of different colors, different flavors of licorice, or Play-Doh "snakes" can represent cells with at least two different chromosome pairs. Build a parent cell and then use additional objects to represent cells at each stage of the cell division process. The hands-on work will allow you to see more clearly what is taking place in the drawings in the book. After you understand the overall process you can record it in sequential diagrams. Alternatively, you can use a digital camera to photograph your models in different stages, print them out, and then shuffle them. Practice putting the images into the proper order and explain the purpose of each phase of mitosis and meiosis.

  Once you understand how cell division in mitosis and meiosis works, explore the processes of crossing over and nondisjunction in order to see how these influence genetic composition of daughter cells.
- It also might be helpful to work with a study partner. For instance, you can diagram the starting cell, some of the intermediate steps, and the resulting cells and then challenge your partner to fill in the missing steps. You and your study partner can also take turns introducing different errors in cell division and challenging each other to explain how these errors will influence the genetic composition of the daughter cells.
- Draw a picture of the cell cycle and indicate the points at which the cyclin–CDK complexes act during that cycle. Use paper cups to represent CDK. Replicate the synthesis of cyclin by adding candies to the cups. Draw a large circle and label the stages of the cell cycle. Then place the cyclin–CDK complexes in their proper places on the circle. What would happen if you were unable to degrade cyclin (cup stays full) or unable to synthesize cyclin (cup stays empty)? Repeat these activities with a study partner so you can practice breaking the cell cycle in different ways while you challenge each other to explain the outcomes.
- Always keep in mind that the "stages" of the cell cycle, like the stages of mitosis and meiosis, are artificial labels used to help us understand the overall process.

The cells do not lurch from stage to stage in incremental jumps! But thinking in terms of phases of the cycle allows us to identify key features of the overall process as it flows from beginning to end.

- Remember that programmed cell death is beneficial. The sacrifice of individual cells aids in the development and/or survival of the multicellular organism.
- Go to yourBioPortal.com to review the following tutorials and activities:

  Animated Tutorial 7.1 Mitosis

  Animated Tutorial 7.2 Meiosis

  Web Activity 7.1 Sexual Life Cycle

  Web Activity 7.2 The Mitotic Spindle

  Web Activity 7.3 Images of Mitosis

  Web Activity 7.4 Images of Meiosis

Key Concept Review

7.1 Different Life Cycles Use Different Modes of Cell Reproduction

Asexual reproduction by binary fission or mitosis results in genetic constancy

Sexual reproduction by meiosis results in genetic diversity

Organisms can reproduce themselves by means of asexual reproduction and sexual reproduction. Asexual reproduction of cells allows for far greater genomic stability than sexual reproduction does, although evolution occurs in asexually reproducing organisms, too; DNA replication is not always 100 percent accurate and the molecule is somewhat prone to breakage, although the errors occur only rarely. Sexual reproduction guarantees that each generation has a genome that is different from that of the parent generation.

Sexual reproduction by means of the diplontic life cycle is most familiar to students since that is the system of human reproduction. Some organisms are diploid only in their zygote stage and are haploid in the mature form; we call this a haplontic life cycle. The third system is called alternation of generations, in which both the haploid and dipoid generations have multicellular stages.

Question 1. In the diagrams below, label the sexual life cycle strategy that is represented (diplontic, haplontic, or alternation of generations). For each, provide one example of an organism that demonstrates that life cycle strategy.
Textbook Reference: *7.1 Different Life Cycles Use Different Modes of Cell Reproduction, p. 126*

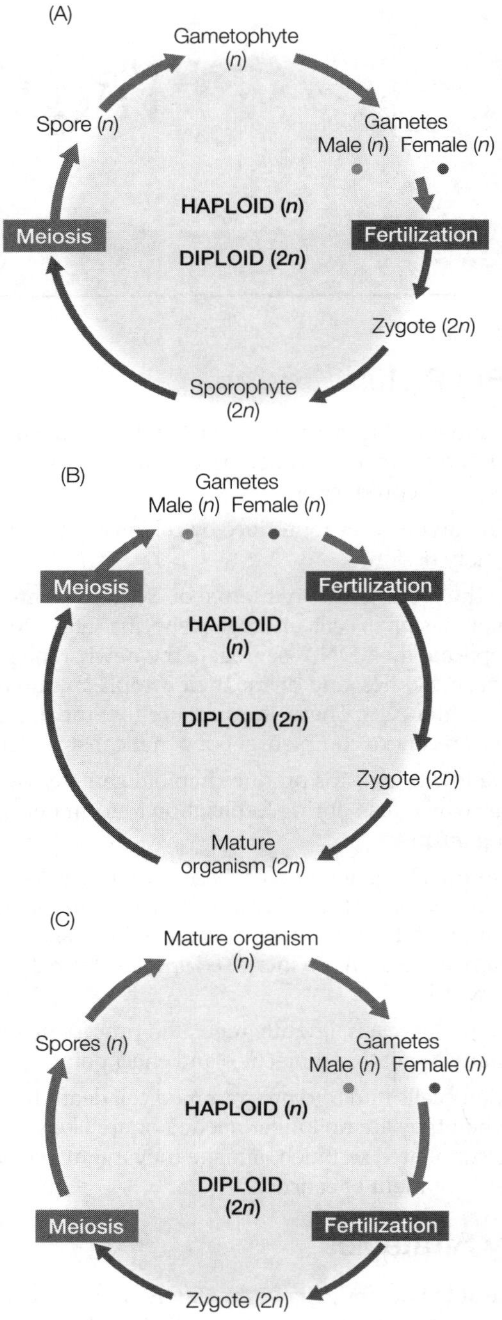

(A) Life cycle ________

Sample organism ________

(B) Life cycle ________

Sample organism ________

(C) Life cycle ________

Sample organism ________

7.2 Both Binary Fission and Mitosis Produce Genetically Identical Cells

Prokaryotes divide by binary fission

Eukaryotic cells divide by mitosis followed by cytokinesis

Prophase sets the stage for DNA segregation

Chromosome separation and movement are highly organized

Cytokinesis is the division of the cytoplasm

Cell division in both prokaryotic and eukaryotic cells consists of three stages: DNA replication, DNA segregation, and cytokinesis. Prokaryotic cells divide by means of binary fission. Since the cell has only a single, circular chromosome, there is no need to have a system to organize multiple chromosomes prior to division of the cell. Eukaryotic cells, by contrast, have multiple chromosomes, each of which needs to be replicated and distributed to the daughter cells. This is why you see chromosomes aligning at the equatorial plate in mitosis (and meiosis). Mitosis is a more complex process than binary fission. As such it is helpful to think of it in terms of stages, with each stage flowing directly into the next. Interphase is the stage of the cell cycle when most metabolic functions operate, and DNA is replicated in the S phase of interphase.

By the time the cell enters mitosis, the DNA is condensed, and replication would no longer even be possible. In mitosis, the division of the nucleus takes place in a number of stages: prophase, prometaphase, metaphase, anaphase, and telophase. This is followed by cytokinesis, which is the division of the cytoplasm and the organelles it contains. In prophase, the DNA is packaged into a very condensed state. The kinetochores assemble on the centromeres and the spindle fibers form. In prometaphase, the nuclear envelope breaks down and the chromosomes attach to the spindles extending to each pole. In metaphase, the chromosomes line up at the midline in preparation for moving away from each other in anaphase. In telophase, the nuclear envelope re-forms around the nucleus and the process of nuclear division is terminated.

Question 2. How does the process of cell division differ in prokaryotic cells and eukaryotic cells? Which elements are the same?
Textbook Reference: *7.2 Both Binary Fission and Mitosis Produce Genetically Identical Cells, pp. 127–128*

Question 3. How does binary fission permit the evolution of new organisms when the benefit of genetic diversity from sexual reproduction is not possible?
Textbook Reference: *7.2 Both Binary Fission and Mitosis Produce Genetically Identical Cells, pp. 127–128*

Question 4. Diagram normal mitosis in a diploid cell ($n = 2$). Use either colors or sizes to differentiate between homologous and nonhomologous chromosomes.
Textbook Reference: *7.2 Both Binary Fission and Mitosis Produce Genetically Identical Cells, pp. 128–130*

Question 5. Describe how two meters of DNA in a typical human cell can fit into the nucleus, which is 5 μm in diameter.
Textbook Reference: *7.2 Both Binary Fission and Mitosis Produce Genetically Identical Cells, p. 129*

Question 6. How does cytokinesis differ in animal and plant cells?
Textbook Reference: *7.2 Both Binary Fission and Mitosis Produce Genetically Identical Cells, p. 131*

Question 7. Suppose that by using a chemical that inhibits cytokinesis, you have created peaches that are tetraploid. How many sets of chromosomes do these peaches have? What will be the ploidy of the gametes?
Textbook Reference: *7.2 Both Binary Fission and Mitosis Produce Genetically Identical Cells, p. 131*

7.3 Cell Reproduction Is Under Precise Control

The eukaryotic cell division cycle is regulated internally

The cell cycle is controlled by cyclin-dependent kinases

Cell reproduction is a process that is carefully controlled and includes multiple regulatory molecules and checkpoints. Cyclin-depending kinases (CDKs) interact with the protein cyclin, and via allosteric regulation cyclin is converted to the active form. The individual checkpoints each have their own specific CDK molecule. For example, one type of CDK moves a cell from G2 phase into S phase by inactivating the protein RB, which blocks such movement.

Question 8. What would happen to a cell if it was unable to synthesize cyclin (as described in the textbook)?
Textbook Reference: *7.3 Cell Reproduction Is Under Precise Control, pp. 133–134*

Question 9. In the cell cycle shown in the diagram below, where is the checkpoint that determines whether the cell will proceed from G1 to S? Label G1, S, and G2 phases. Then add CDK and cyclin to the diagram in the appropriate places and indicate if they are present in an intact or degraded form.
Textbook Reference: *7.3 Cell Reproduction Is Under Precise Control, pp. 133–135*

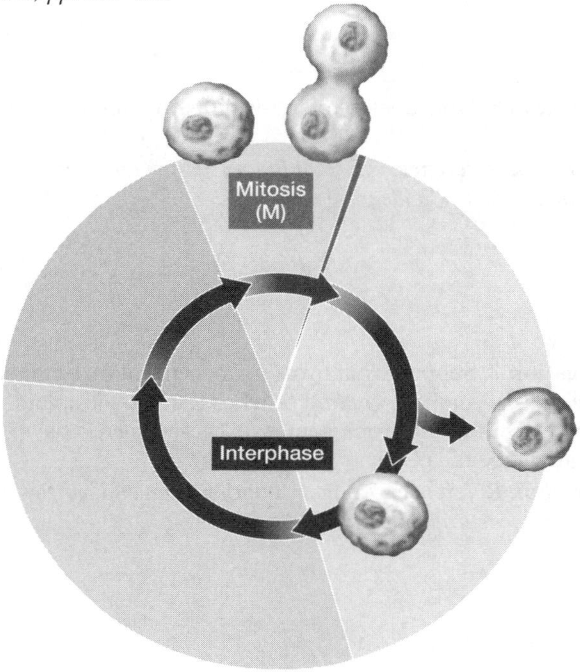

7.4 Meiosis Halves the Nuclear Chromosome Content and Generates Diversity

Meiotic division reduces the chromosome number

Crossing over and independent assortment generate diversity

Meiotic errors lead to abnormal chromosome structures and numbers

Meiosis requires two sequential nuclear divisions (referred to as meiosis I and II) in order to generate haploid gametes. The first round of nuclear divisions separates homologous chromosomes, while the second round separates sister chromatids. As in mitosis, DNA replication occurs in the S phase of interphase and the DNA is replicated a single time.

In meiosis I, the replicated homologous chromosomes pair up at the equator instead of lining up randomly as occurs in mitosis. The pairing of homologous chromosomes results in a tetrad. The chromosomes in the tetrads are aligned along their lengths in a process referred to as synapsis. The chromosomes in the tetrad exchange material, resulting in recombinant chromatids. This is one source of genetic diversity in the next generation.

A second source of genetic diversity during meiosis is the phenomenon known as independent assortment. In essence, each tetrad and its combination of parental chromosomes separates independently of other tetrads. Neither chromosome is a match to the parent because of crossing over, and no gamete has a chromosome assortment identical to a parent because of independent assortment. These are key sources of genetic diversity in sexually reproducing organisms.

Meiosis, like mitosis, is tightly regulated, but errors are possible. One type of error is a result of chromosome breakage and incorrect repair, resulting in the movement of a genetic sequence to a different location on a chromosome or even to a different chromosome altogether. This is called translocation. A second error can occur when homologous chromosomes fail to separate (nondisjunction) in meiosis I or II; the gametes will either have two copies or zero copies of that chromosome. Many of these irregularities are fatal, but there are examples in which the zygote survives with the extra or missing chromosome (aneuploidy). Trisomy 21 in humans is not lethal like many other aneuploidy cases, but it does result in phenotypic changes compared to diploid 21 individuals; trisomy 21 is referred to as Down syndrome.

Polyploid organisms may have three, four, or even more sets of homologous chromosomes. As long as the ploidy number is even, meiosis can continue normally and the organism is fertile. Nuclei with odd ploidy cannot divide their chromosomes evenly and the organism is sterile. In plants, evenly polyploid crops are more robust and may produce larger fruits or flowers, while oddly polyploid crops are seedless.

Question 10. Starting with a diploid cell ($n = 2$), diagram normal meiosis. How is this process different from mitosis?
Textbook Reference: *7.4 Meiosis Halves the Nuclear Chromosome Content and Generates Diversity, pp. 134–137*

Question 11. A cell at the start of meiosis has two $2n$ chromosomes, representing the replicated DNA of a diploid cell. If the cell has 1 microgram of DNA in G1, it would have 2 micrograms of DNA at the start of meiosis. At the end of meiosis I, there is 1 microgram of DNA in a cell, but the cell is not diploid. Why not?
Textbook Reference: *7.4 Meiosis Halves the Nuclear Chromosome Content and Generates Diversity, pp. 134–137*

Question 12. Starting with a diploid cell ($n = 2$), diagram meiosis with a nondisjunction in meiosis I, and then meiosis with a nondisjunction in meiosis II. Label the gametes that would result from each meiotic event. Describe the types of zygotes that would be produced from these gametes.
Textbook Reference: *7.4 Meiosis Halves the Nuclear Chromosome Content and Generates Diversity, pp. 134–139*

Question 13. Describe two ways in which the genetic diversity of organisms is increased during meiosis.
Textbook Reference: *7.4 Meiosis Halves the Nuclear Chromosome Content and Generates Diversity, p. 138*

Question 14. Are the tetraploid peaches described in Question 7 fertile? Would triploid peaches be fertile? Explain.
Textbook Reference: *7.4 Meiosis Halves the Nuclear Chromosome Content and Generates Diversity, pp. 138–139*

7.5 Programmed Cell Death Is a Necessary Process in Living Organisms

Cell death can occur in cells in two ways: by necrosis (cells are damaged by toxins or starved of oxygen or essential nutrients) or by apoptosis (programmed cell death). Apoptosis is a normal process during development that eliminates unneeded tissue. Apoptosis also occurs in older cells in which genetic damage may be present.

Signals for cell death include a lack of mitotic signals, recognition of DNA damage, or changes in a receptor protein that activate signal transduction. This in turn leads to the activation of caspases, which are proteases that hydrolyze target molecules leading to cell death.

Question 15. Describe how a cell exposed to radiation for a prolonged period of time might trigger and proceed with apoptosis.
Textbook Reference: *7.5 Programmed Cell Death Is a Necessary Process in Living Organisms, p. 141*

Test Yourself

1. Which of the following statements about mitosis is true?
   a. Cytokinesis follows mitosis.
   b. DNA replication is completed prior to the beginning of this phase.
   c. The chromosome number of the resulting cells is the same as that of the parent cell.
   d. The daughter cells are usually genetically identical to the parental cell.
   e. All of the above
   Textbook Reference: *7.2 Both Binary Fission and Mitosis Produce Genetically Identical Cells, p. 128*

2. Which of the following statements about meiosis is true?
   a. The chromosome number in the resulting cells is halved.
   b. DNA replication occurs before meiosis I and again before meiosis II.
   c. The homologs pair during prophase II.
   d. The daughter cells are genetically identical to the parental cell.
   e. The chromosome number of the resulting cells is the same as that of the parent cell.
   Textbook Reference: *7.4 Meiosis Halves the Nuclear Chromosome Content and Generates Diversity, pp. 134–137*

3. Which of the following statements about kinetochores on mitotic chromosomes is true?
   a. They are located at the centromere of each chromosome.
   b. They are the sites where microtubules attach to separate the chromosomes.
   c. They are organized so that there is one per sister chromatid.
   d. Kinetochore microtubules from opposite poles attach to each sister chromatid.
   e. All of the above
   Textbook Reference: *7.2 Both Binary Fission and Mitosis Produce Genetically Identical Cells, p. 129*

4. Which of the following statements about the mitotic spindle is true?
   a. It is composed of polar and kinetochore microtubules, both of which attach to chromosomes.
   b. It is composed of actin and myosin microfilaments.
   c. It is composed of kinetochores at the metaphase plate.
   d. It is composed of microtubules, which help separate the chromosomes to opposite poles of the cell.
   e. It originates only at the centrioles in the centrosomes.
   Textbook Reference: *7.2 Both Binary Fission and Mitosis Produce Genetically Identical Cells, p. 130*

5. Imagine that there is a mutation in the cyclin gene such that its gene product is nonfunctional. What kind

of effect would this mutation have on a skin cell in the area of a cut?
a. Cdk would not be synthesized.
b. There would be no effect, because skin cells do not replicate.
c. The cell would be stuck in S phase and unable to replicate.
d. The cell would not be able to enter G1.
e. The cell would be unable to reproduce itself.

Textbook Reference: *7.3 Cell Reproduction Is Under Precise Control, p. 134*

6. Imagine that there is a mutation in a Cdk gene such that its gene product is present but nonfunctional. What kind of effect would this mutation have on a mammalian white blood cell based on what you know about Cdk and protein RB interactions?
a. The cell would be unable to replicate its DNA.
b. The cell would be unable to enter mitosis.
c. The cell would be unable to reproduce itself.
d. The cell would not be able to phosphorylate its associated cyclin.
e. All of the above

Textbook Reference: *7.3 Cell Reproduction Is Under Precise Control, pp. 132–134*

7. Which of the following statements about DNA replication and cytokinesis in *Escherichia coli* is true?
a. DNA replication occurs in the nucleus.
b. Cytokinesis is facilitated by microfilaments of actin and myosin.
c. DNA replication occurs during the S phase of the cell cycle.
d. Cell reproduction is initiated by reproductive signals, which lead to DNA replication, DNA segregation, and cytokinesis.
e. The *E. coli* chromosome is linear.

Textbook Reference: *7.2 Both Binary Fission and Mitosis Produce Genetically Identical Cells, pp. 127–128*

8. Which of the following statements about chromatids is true?
a. They are replicated chromosomes still joined together at the centromere.
b. They are identical in mitotic chromosomes.
c. They undergo recombination in mitosis.
d They are identical in meiotic chromosomes.
e. Both a and b

Textbook Reference: *7.2 Both Binary Fission and Mitosis Produce Genetically Identical Cells, p. 129*

9. Chromosome movement during anaphase is the result of
a. the hydrolysis of ATP by dynein.
b. molecular motors at the kinetochores that move the chromosomes toward the poles.
c. molecular motors at the centrosome that pull the microtubules toward the poles.
d. shortening of the microtubules at the centrosome that pull the chromosomes toward the poles.
e. a, b, and d

Textbook Reference: *7.2 Both Binary Fission and Mitosis Produce Genetically Identical Cells, p. 130*

10. Programmed cell death (apoptosis)
a. occurs in cells that have been deprived of essential nutrients.
b. occurs only in cells that have damaged DNA.
c. is a natural process during development.
d. is signaled by the initiation of mitosis.
e. is well controlled in cancer cells.

Textbook Reference: *7.5 Programmed Cell Death Is a Necessary Process in Living Organisms, p. 141*

11. If the *ori* site on the *E. coli* chromosome is deleted,
a. nothing will happen.
b. replication will start but not be able to continue.
c. replication will not start.
d. replication will be initiated at another *ori* site on the chromosome.
e. the chromosome will be replicated but the cell will not be able to divide.

Textbook Reference: *7.2 Both Binary Fission and Mitosis Produce Genetically Identical Cells, pp. 127–128*

12. Which of the following statements about chiasmata is true?
a. They are sites where nonsister chromatids can exchange genetic material during meiosis and increase genetic variation in gametes.
b. They are sites where sister chromatids can exchange genetic material during meiosis and increase genetic variation in gametes.
c. They increase genetic variation among the gametes due to exchange between all members of the tetrad during meiosis.
d. They increase genetic variation among the gametes due to exchange between all members of the tetrad during mitosis.
e. All of the above

Textbook Reference: *7.4 Meiosis Halves the Nuclear Chromosome Content and Generates Diversity, p. 138*

13. The difference between asexual and sexual reproduction is that
a. asexual reproduction occurs only in bacteria, whereas sexual reproduction occurs in plants and animals.
b. asexual reproduction results from meiosis, whereas sexual reproduction results from mitosis.
c. asexual reproduction results in an organism that is identical to the parent, whereas sexual reproduction results in an organism that is not identical to either parent.
d. asexual reproduction results from the fusion of two gametes, whereas sexual reproduction produces clones of the parent organism.
e. asexual reproduction occurs only in haplontic organisms, whereas sexual reproduction occurs only in diplontic organisms.

Textbook Reference: *7.1 Different Life Cycles Use Different Modes of Cell Reproduction, pp. 125–126*

14. A chromatid is
    a. a chromosome before it has undergone DNA replication.
    b. one of the pairs of homologous chromosomes.
    c. a homologous chromosome.
    d. a newly replicated bacterial chromosome.
    e. one-half of a newly replicated eukaryotic chromosome.

    Textbook Reference: *7.2 Both Binary Fission and Mitosis Produce Genetically Identical Cells, p. 129*

15. A diploid cell in G1 has _______ micrograms of DNA as one of its daughter cells at the end of meiosis I and _______ number of chromosomes as one of its daughter cells at the end of meiosis II.
    a. twice the number of; the same
    b. one-half the number of; twice the
    c. the same number of; twice the
    d. one-half the number of; the same
    e. the same number of; one-half the

    Textbook Reference: *7.4 Meiosis Halves the Nuclear Chromosome Content and Generates Diversity, pp. 134–137*

Answers

Key Concept Review

1.

   (A) Life cycle: Alternation of generations; Sample organism: Most plants, some fungi

   (B) Life cycle: Diplontic; Sample organism: Animals, brown algae, some fungi

   (C) Life cycle: Haplontic; Sample organism: Most protists, fungi, some green algae

2. All cells divide by replicating their DNA, segregating the DNA, and then splitting the cytoplasm by cytokinesis. In most prokaryotic cells there is only one circular chromosome. As the cell enlarges to prepare for division, the newly replicated daughter chromosomes are separated at opposite sides of the cell. During fission, the cell membrane pinches in, and cell wall components are synthesized between the daughter cells. In eukaryotic cells, there are more chromosomes, and they are linear. The cell undergoes a sequential set of steps called the cell cycle, in which the chromosomes are replicated and then separated to opposite poles of the cell. The chromosomes are segregated equally into the daughter cells by means of microtubules, and actin filaments and myosin cause the cell membrane to form a contractile ring (in animals) and separate to form two daughter cells.

3. Binary fission typically results in two daughter cells that are genetically identical to the parent cell. Errors in DNA replication, while rare, can introduce changes to the genetic code. These changes are then passed on in subsequent rounds of cell division.

4.

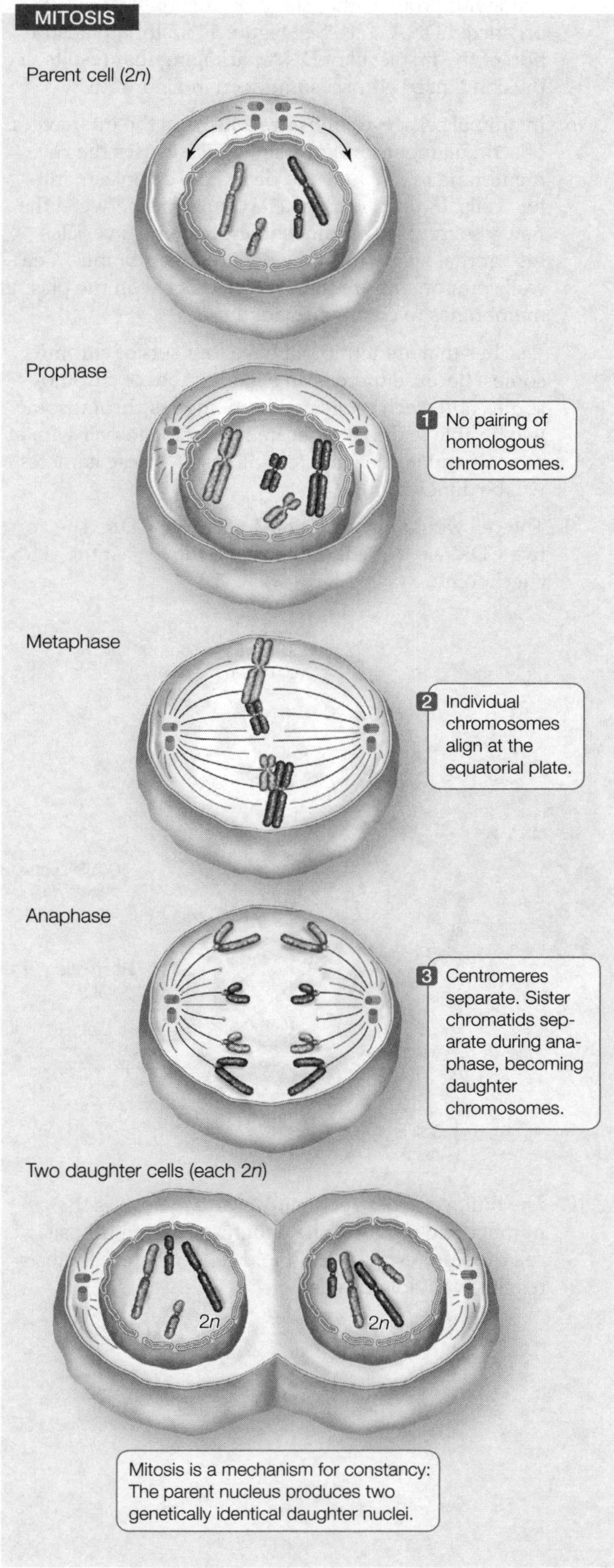

MITOSIS
Parent cell (2n)
Prophase
1 No pairing of homologous chromosomes.
Metaphase
2 Individual chromosomes align at the equatorial plate.
Anaphase
3 Centromeres separate. Sister chromatids separate during anaphase, becoming daughter chromosomes.
Two daughter cells (each 2n)
2n
2n
Mitosis is a mechanism for constancy: The parent nucleus produces two genetically identical daughter nuclei.

5. The DNA is very tightly condensed because it is packaged with proteins that stabilize the double helix in organized DNA coils. See Figure 11.15 for an illustration of the first level of DNA packaging that results in the condensed chromosome seen in Figure 7.5.
6. In animal cells, cytokinesis results from the interaction of actin filaments and myosin, which causes the cell membrane to pinch in and divide the cytoplasm into two cells. In plant cells, a cell plate forms between the newly segregated chromosomes, and Golgi vesicles fuse at that site to form the new cell membranes. Cell wall components are then secreted between the plasma membranes to complete cytokinesis.
7. Peaches that are tetraploid have four sets of chromosomes. Because there is an even number of chromosomes (4*n*), each replicated homologous chromosome will be able to find a replicated homolog to pair with at meiosis and fertile gametes will result. These gametes will be diploid (2*n*).
8. The cell would never be able to activate CDK. The active CDK is required for the cell to move past the G1/S checkpoint.
9.

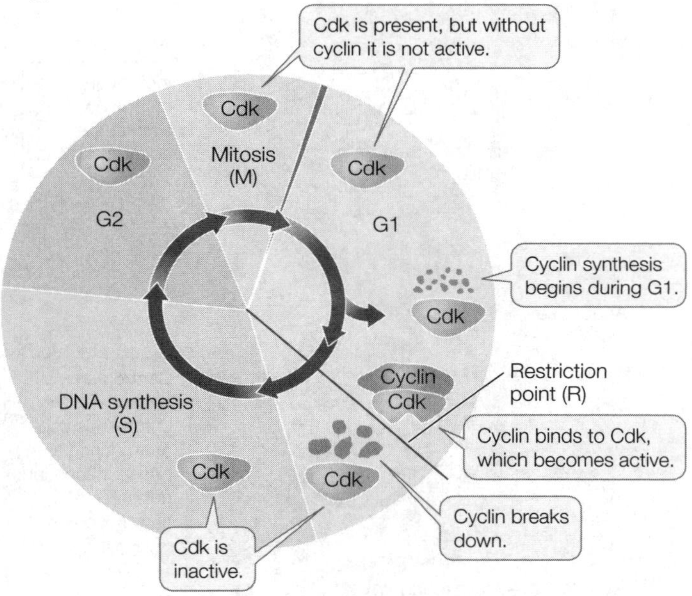

10. The difference between mitosis and meiosis is the number and ploidy of the daughter cells. In mitosis, cell division results in 2 diploid cells, whereas meiosis results in 4 haploid cells.

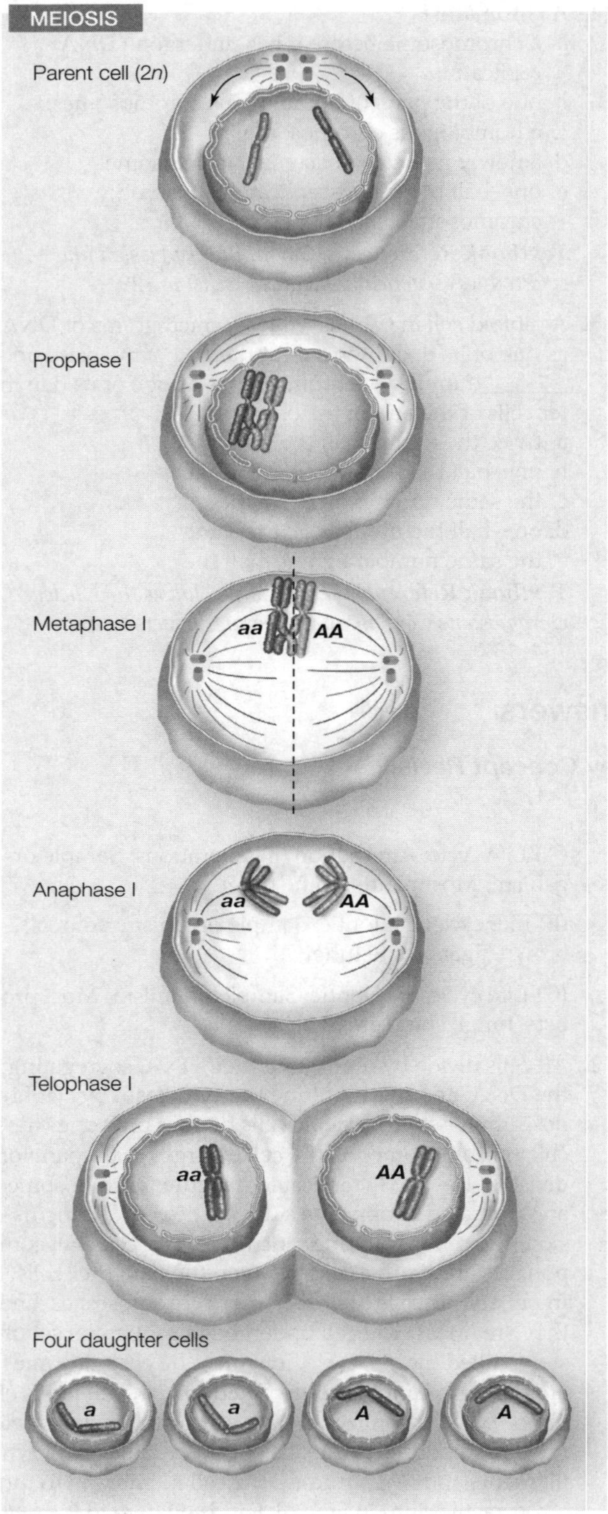

11. At the start of meiosis, the 2 micrograms of DNA are approximately divided among a total of $2n = 2$ (4 total, or 0.5 micrograms each) chromosomes. At the end of meio-is I, each of the cells has $n = 2$ chromosomes (2 total) because the homologous pairs have separated but not the sister chromatids, so each cell has one-half the amount of DNA, 1 micrograms, but is haploid, not diploid.

12.

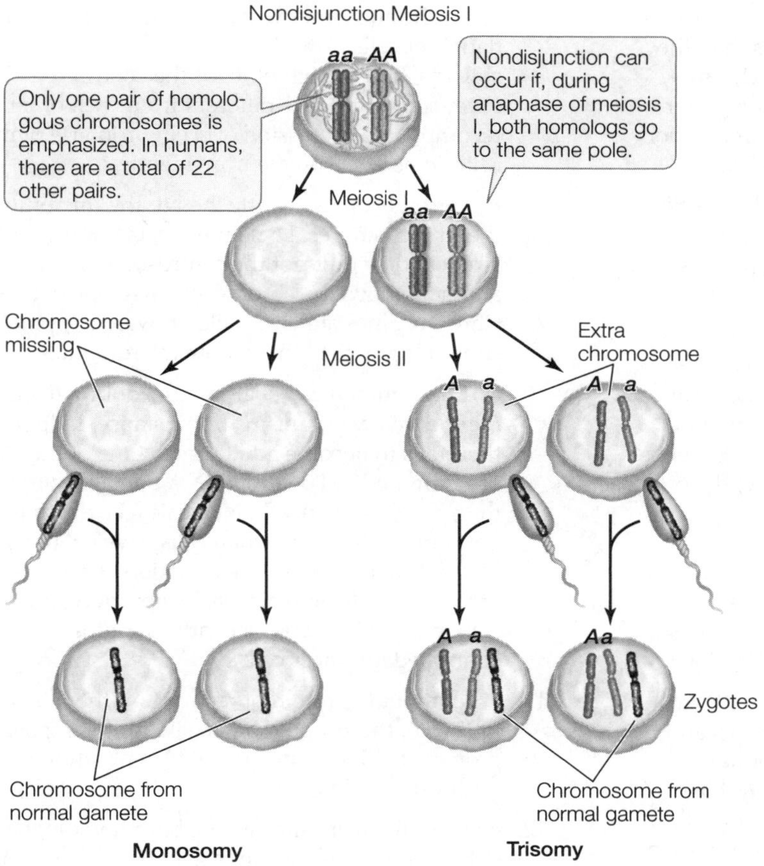
Nondisjunction Meiosis I
aa AA
Only one pair of homologous chromosomes is emphasized. In humans, there are a total of 22 other pairs.
Nondisjunction can occur if, during anaphase of meiosis I, both homologs go to the same pole.
Meiosis I
aa AA
Chromosome missing
Meiosis II
Extra chromosome
A a
A a
A a
Aa
Zygotes
Chromosome from normal gamete
Chromosome from normal gamete
Monosomy
Trisomy

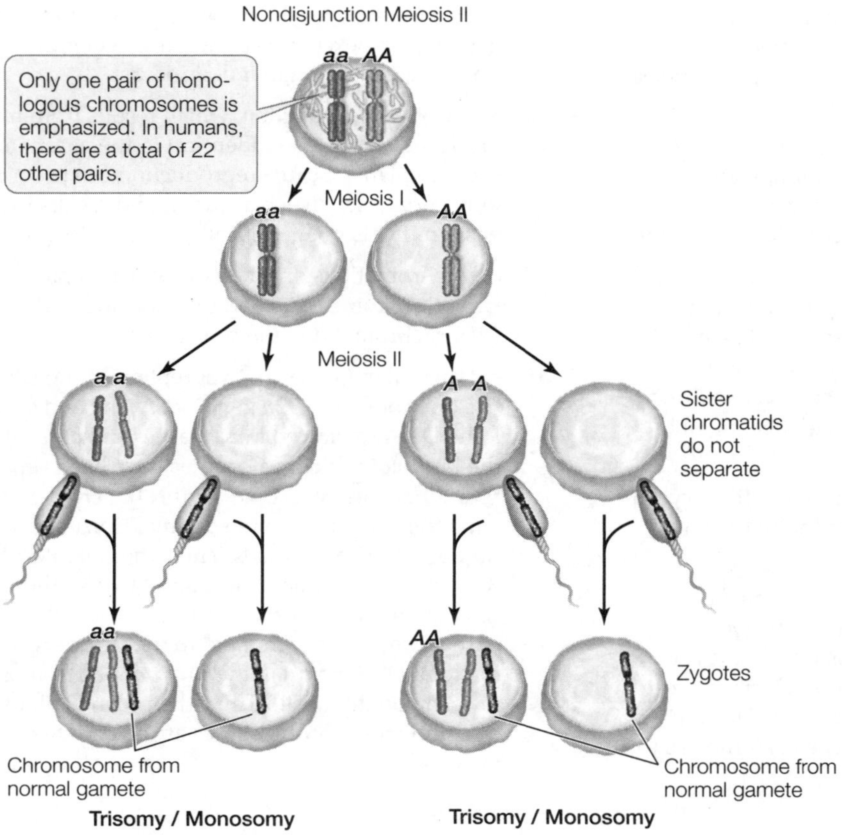
Nondisjunction Meiosis II
aa AA
Only one pair of homologous chromosomes is emphasized. In humans, there are a total of 22 other pairs.
Meiosis I
aa
AA
Meiosis II
a a
A A
Sister chromatids do not separate
aa
AA
Zygotes
Chromosome from normal gamete
Chromosome from normal gamete
Trisomy / Monosomy
Trisomy / Monosomy

13. Genetic diversity is increased during crossing over of prophase I of meiosis so that each gamete has chromosomes with different combinations of alleles than the parents. During meiosis, each homologous chromosome is randomly sorted to one of the four daughter cells. This results in 223 different possible combinations of homologous chromosomes per human gamete!
14. Because these gametes are diploid, they will be fertile. Triploid peaches will not be fertile because one of the three homologs will not find its pair during prophase of meiosis I, and the single homologs will be sorted randomly into the daughter cells.
15. A cell that has been damaged by radiation likely has DNA damage. This is the internal signal for a signal transduction pathway that leads to the activation of caspases. When the caspases degrade the nuclear envelope, nucleosomes, and plasma membrane, the cell dies.

Test Yourself

1. **e.** Mitosis occurs after DNA replication and results in cells with the same number of genetically identical chromosomes as the parent cell. Cytokinesis follows mitosis.
2. **a.** Meiosis occurs after one round of DNA replication. Homologous chromosomes pair during prophase I of meiosis, and after meiosis II the resulting cells have half the number of chromosomes as the parent cell. Those chromosomes are not genetically identical to the parent cells.
3. **e.** Kinetochores, one per sister chromatid, are assembled at the centromere of each chromosome and are the sites in which microtubules from opposite poles attach to segregate the chromosomes.
4. **d.** The mitotic spindle is composed of microtubules, not actin and myosin filaments. The spindle originates from the centrosome, which may or may not have centrioles, and only the kinetochore microtubules attach to the chromosomes.
5. **e.** While many cells lose the ability to divide when they mature, you know from experience that skin wounds heal. Without the presence of cyclin, however, there will be no signal to the cell to replicate, and that particular skin cell will not participate in healing of the wound.
6. **e.** Functional Cdk is required to complex with its cyclin and inactivate protein RB, permitting the cell to enter the cell cycle.
7. **d.** *Escherichia coli* is a prokaryote. It lacks a nucleus and does not undergo the cell cycle seen in eukaryotes. It has a circular chromosome, and does not synthesize actin or myosin proteins. Cytokinesis in *E. coli* is a result of a reproductive signal that causes the DNA to be replicated and segregated, and finally causes the cell to divide.
8. **e.** Chromatids are highly condensed, newly replicated chromosomes which will be segregated to the daughter cells. After DNA replication, chromatids are still attached to each other at the centromere. Meiotic sister chromatids are different from each other due to recombination (crossing over) in prophase of meiosis I. Mitotic sister chromatids are identical.
9. **e.** Chromosomes are attached to the microtubules at their kinetochores. Dynein molecular motors at the kinetochores (but not the centrosome) hydrolyze ATP and help move the chromosomes to opposite poles. Chromosomes are also pulled toward the poles by the shortening of the kinetochore microtubules.
10. **c.** Programmed cell death occurs during the development of many organisms (for instance, tadpoles lose their tails to become adult frogs). One of the stimuli for programmed cell death is DNA damage, but it is not the only cause of death. Necrosis (cell death that is not programmed) occurs when cells have been deprived of essential nutrients. The initiation of mitosis is part of the cell cycle, in which cells reproduce, and is not a step in programmed cell death. Apoptosis is not well controlled in cancer cells.
11. **c.** Without the origin of replication (there is only one in *E. coli*), there would be no site for the replication proteins to bind to initiate DNA replication, so DNA synthesis would not start.
12. **c.** Chiasmata are sites where both homologous and sister chromatids exchange genetic material during meiosis (not mitosis). The exchange between homologous chromosomes increases genetic variation in the gametes (the exchange between sister chromatids has no effect, even though it does occur).
13. **c.** Asexual reproduction, which results from mitosis, produces cells that are identical to the parent and can occur in plants. Sexual reproduction can occur in haplontic organisms (such as fungi) and results in an organism that is not genetically identical to either parent.
14. **e.** A chromatid is one-half of a newly replicated eukaryotic chromosome and is connected to the other (sister) chromatid at the centromere.
15. **c.** The cell in G1 has not yet replicated the DNA, so it has one copy of the $2n$ genome. At the end of meiosis I, the DNA replicated in S phase is still present, but the homologous chromosomes have been separated. Thus, there are two copies of the $1n$ genome in each daughter cell, or twice the original DNA but half the number of chromosomes. Thus, the cell in G1 had the same number of microgams and DNA as the daughter cell after meiosis I. (In meiosis II, the result is one copy of the $1n$ genome compared to the original cell, or half the total DNA and half the number of chromosomes in each daughter cell). Thus, the cell in G1 had twice the number of chromosomes as the daughter cells after meiosis II.

Inheritance, Genes, and Chromosomes

8

The Big Picture

- Genes are units of inheritance that are passed down to progeny by the fusion of gametes to form a new organism.
- Phenotypes are observable traits and are the result of gene expression. An individual may have different alleles for a trait, but only a single phenotype results.
- Meiosis generates gametes that contain only one allele for each gene, and those alleles are segregated independently (unless they are linked).
- By studying the expression of particular phenotypes in successive generations, one can begin to determine, according to Mendel's laws, which alleles gave rise to those phenotypes (i.e., rare recessive, rare dominant, sex-linked).
- Exceptions to Mendel's laws also help explain the interactions of gene products and the interaction of alleles that lead to the observed phenotype.
- Bacteria can transfer genes to other recipient cells during conjugation.

Study Strategies

- It is easy to be overwhelmed by all the exceptions to Mendel's laws. Initially, make sure you understand the laws of segregation and independent assortment. The law of independent assortment focuses on different genes, while the law of segregation applies to different alleles of the same gene. Examine these laws by thinking about the molecular role of these alleles. For example, what is the explanation for a codominant phenotype at the molecular level, such as the AB blood type in humans? Both gene products are synthesized by the cell and modify the cell surface glycoproteins, producing both A and B antigens. What is the molecular explanation for epistasis? The enzyme required to produce an initial substrate necessary for pigmentation is missing, even though the other gene products (farther down the biochemical pathway) for producing a particular color are present in the cell.
- Review the figures of meiosis in Chapter 7 to see how Mendel's laws function in meiosis. Draw the chromosomes with different alleles (*S* on one chromosome, *s* on the other) and follow how those alleles are segregated to the gametes during meiosis.
- Use a Punnett square to predict the outcome of different monohybrid and dihybrid crosses. Use the probability rules to predict the genotypes and phenotypes of these same crosses and compare the two methods.
- After reviewing Mendel's laws of independent assortment and allele segregation, make a list of the exceptions to those laws with specific examples for each exception.
- Go to yourBioPortal.com to review the following tutorials and activities:

  Animated Tutorial 8.1 Independent Assortment of Alleles

  Animated Tutorial 8.2 Alleles That Do Not Assort Independently

  Interactive Tutorial 8.1 Pedigree Analysis

  Web Activity 8.1 Homozygous or Heterozygous?

  Web Activity 8.2 Concept Matching I

  Web Activity 8.3 Concept Matching II

Key Concept Review

8.1 Genes Are Particulate and Are Inherited According to Mendel's Laws

Mendel used the scientific method to test his hypotheses

Mendel's first experiments involved monohybrid crosses

Mendel's first law states that the two copies of a gene segregate

Mendel verified his hypotheses by performing test crosses

Mendel's second law states that copies of different genes assort independently

Probability is used to predict inheritance

Mendel's laws can be observed in human pedigrees

The scientific method was used by Mendel and by those who came after him, who analyzed his results in the context of new evidence. Mendel's work was conducted long before the structure of DNA was known or the process of meiosis understood. As we have gained an understanding of cell division, genes, and chromosomes, we have been able to further verify Mendel's original hypotheses and refine his original ideas.

Mendel's experiments with peas revealed the way multiple alleles for a trait are expressed in an individual and passed on to the next generation, often in a new combination and resulting in a phenotype that is different from that of the parent. Mendel demonstrated that different alleles of the same gene are sorted into different gametes in meiosis (the law of segregation) and that different genes for different traits move to gametes independently of one another (the law of independent assortment). We know that the latter law is affected by the physical placement of genes on a single chromosome, and is referred to as linkage.

These two laws allow us to predict the probability of a certain progeny's having a particular phenotype and genotype when the parents' genotypes are known. This ability to make predictions can be particularly useful when potential parents are counseled about the probability that a certain trait will be passed on to their children.

Question 1. Diagram two separate pairs of chromosomes, each bearing a different allele; for instance, *Ss* and *Yy*, as seen in the dihybrid cross with seed shape (*Ss*) and seed color (*Yy*). Show how these alleles assort independently during meiosis to produce haploid gametes (see Figure 8.3).
Textbook Reference: *8.1 Genes Are Particulate and Are Inherited According to Mendel's Laws, pp. 147–148*

Question 2. Draw a diagram starting with the same parent (*SsYy*), but assume that *S* and *Y* are linked and are 20 map units apart. For this parent, the *S* and *Y* alleles are on one chromosome and the *s* and *y* alleles are on the other chromosome. Draw the gametes you would expect if there is a crossover between *S* and *Y*.
Textbook Reference: *8.1 Genes Are Particulate and Are Inherited According to Mendel's Laws, pp. 148–149*

Question 3. Suppose you are a genetics counselor and you are working with a 21-year-old pregnant woman who has just discovered that her father has Huntington's chorea, a rare dominant autosomal trait. This disease usually develops in middle age, so without genetic testing people carrying this trait do not know about this until midlife. What are the chances that the child she is carrying will develop the disease? (Assume that her husband's family has no history of the disease.) What is the chance that she has Huntington's chorea?
Textbook Reference: *8.1 Genes Are Particulate and Are Inherited According to Mendel's Laws, p. 151*

8.2 Alleles and Genes Interact to Produce Phenotypes

New alleles arise by mutation

Dominance is not always complete

Genes interact when they are expressed

The environment affects gene action

A genetic code is much like a blueprint; it provides directions but must be translated into a product in order to be useful. As alleles for traits are expressed, they may interact with each other or mask the expression of one of the traits. The environment may also influence the expression of the genes. The final outcome is the phenotype, and this is what we are able to observe and measure, much like a finished building.

In some cases, the expression of one allele masks the other. This is often true if one allele does not code for a functional product and the other does. In other cases, both alleles code for a product and neither product can mask the presence of the other, leading to a blending of traits we call codominance. Alternately, the products of both alleles may be visible as separate traits, as seen in human blood types.

Not all traits are the result of a single gene product. In many cases, there are multiple genes that influence the final phenotype. Examples of this include epistasis, in which one set of genes may affect the expression of other genes (as in Labrador retriever coat color), and quantitative traits, in which multiple genes all contribute to a measurable outcome such as height or grain yield.

In all cases, is it important to remember that genes are not expressed in a vaccuum. The environment plays a very important role in determining the final phenotype of an individual. In order to meet one's full genetic potential, the environment must have sufficient resources. A plant without water will not produce a large crop any more than a child with insufficient food will reach his or her maximum height. We now know that phenotype is not determined by either nature or nurture alone, but by how they work together, and result in a particular phenotype.

Question 4. A woman who has O type blood has a son with A type blood and a daughter with O type blood. What is the genotype of each parent? If both of the children have A type blood, would that change your answer, and if so, how?
Textbook Reference: *8.2 Alleles and Genes Interact to Produce Phenotypes, pp. 153–154*

Question 5. In mice, agouti (fur with dark and light coloration) and color are both dominant traits, while black and albino are recessive traits. We know that coat color is an epistatic trait. If you cross a population of true-breeding albino mice with a population of true-breeding agouti mice, you get progeny that are all agouti. If you then cross members of the agouti population with each other, you get the following results: 9/16 agouti, 4 albino, 3 black. How do you explain this? Diagram the parent and agouti crosses.
Textbook Reference: *8.2 Alleles and Genes Interact to Produce Phenotypes, pp. 154–155*

8.3 Genes Are Carried on Chromosomes

Genes on the same chromosome are linked, but can be separated by crossing over in meiosis

Linkage is also revealed by studies of the X and Y chromosomes

Some genes are carried on chromosomes in organelles

The DNA sequences coding for particular products are located on chromosomes in both prokaryotes and eukaryotes. Since chromosomes are much longer than individual genes, some traits are linked and tend to travel together to gametes in meiosis. The farther apart the genes are on the chromosome, the more likely it is that crossing over in meiosis I will separate them to different gametes.

The sex chromosomes also carry linked genes, and due to the different sizes of the X and Y chromosomes, an allele is sometimes found on one sex chromosome and not on the other. Since a single allele on the X chromosome has no homologue on the Y chromosome, males are said to be hemizygous for these traits. Sex-linked traits are inherited according to Mendel's laws, but the phenotypes are more likely to be observed in males.

Chromosomes in the organelles are not subject to the recombination seen in meiosis and are passed virtually unchanged from one parent to the offspring. In humans, cytoplasmic chromosomes are inherited from the mother.

Question 6. Draw a pedigree for three generations in which the father has red–green color blindness, his daughter is a carrier, and this daughter has four sons. Predict how many of these grandsons of the man with red–green color blindness will be color blind themselves.
Textbook Reference: *8.3 Genes Are Carried on Chromosomes, pp. 155–159; 8.1 Genes Are Particulate and Are Inherited According to Mendel's Laws, p. 150*

Question 7. Draw a sample pedigree with three generations in which the paternal grandfather has a rare dominant autosomal trait. What is the probability that one of his children will have the disease? What is the probability that one of his grandchildren will have the disease?
Textbook Reference: *8.3 Genes Are Carried on Chromosomes, pp. 155–159; 8.1 Genes Are Particulate and Are Inherited According to Mendel's Laws, p. 150*

Question 8. Draw a sample pedigree with three generations in which the maternal grandmother and paternal grandfather are carriers of a rare recessive autosomal trait. What is the probability that one of their children will be carriers of this trait? What is the probability that a grandchild will have the disease?
Textbook Reference: *8.3 Genes Are Carried on Chromosomes, pp. 155–159; 8.1 Genes Are Particulate and Are Inherited According to Mendel's Laws, p. 150*

Question 9. Cytoplasmic traits in certain species of trees are passed from the male plant to all of its progeny. Compare this phenomenon to that of cytoplasmic inheritance in humans.
Textbook Reference: *8.3 Genes Are Carried on Chromosomes, pp. 159–160*

8.4 Prokaryotes Can Exchange Genetic Material

Bacteria exchange genes by conjugation

Plasmids transfer genes between bacteria

Prokaryotic organisms have single, haploid chromosomes but divide by means of binary fission. This deprives prokaryotes of the benefits of sexual reproduction, namely the guarantee of genetic diversity from one generation to the next. While mutations do occur, their occurrence is relatively rare.

The solution to this problem is that many prokaryotes have developed ways of sharing genetic information. The first is though conjugation, by which a sequence of chromosomal DNA is transferred to a cell, recombines with the host chromosome, and results in a chromosome with a genotype different from that of either parent cell. The second is through the transfer of plasmid DNA. Plasmids do not incorporate into the bacterial chromosome, but they are replicated during cell division and transmitted to the next generation. Plasmids also move through the conjugation tube and transfer new genes to the recipient cell.

Question 10. Two strains of *E. coli*, each having alleles *Abcde* or *abcDEF* are grown together in the laboratory for several weeks. Most of the resulting cells are genetically identical to the parent strains, but some have the genotypes *ABCDef*, *ABCDEf*, *abcdEF*, and *abcdeF*. Explain these results.
Textbook Reference: *8.4 Prokaryotes Can Exchange Genetic Material, p. 161*

Question 11. Many genes for antibiotic resistance are located on plasmids, and many antibiotics are given to healthy livestock in order to prevent illness and increase productivity. Explain why the use of antibiotics in livestock has raised concern about use of antibiotics in humans.
Textbook Reference: *8.4 Prokaryotes Can Exchange Genetic Material, p. 162*

Test Yourself

1. Hemophilia is a trait carried by the mother and passed to her sons. The allele for hemophilia, therefore,
   a. is carried on one of the mother's autosomal chromosomes.
   b. is carried on the Y chromosome.
   c. can be carried on the X or Y chromosome.
   d. is on the X chromosome and can be inherited by the son only if the mother is a carrier (heterozygous).
   e. is carried in the mitochondrial genome because a son inherits this allele from his mother.
   Textbook Reference: *8.3 Genes Are Carried on Chromosomes, pp. 157–159*
2. Before Mendel, genetic inheritance was thought to be a function of the blending of traits from the two parents. Which exception to Mendel's laws is in fact an example of blending?
   a. X linkage
   b. Polygenic inheritance
   c. Incomplete dominance
   d. Codominance
   e. Pleiotropism
   Textbook Reference: *8.2 Alleles and Genes Interact to Produce Phenotypes, p. 153*
3. Which of the following statements about true-breeding plants is true?
   a. When a true-breeding plant with a particular trait is crossed with another plant of the same variety, all of their offspring produce plants with that same trait.
   b. When a true-breeding plant with a particular trait is crossed with another plant of the same variety, all of their offspring will be sterile.
   c. They result from a monohybrid cross.
   d. They result from a dihybrid cross.
   e. They result from crossing over during prophase I of meiosis.
   Textbook Reference: *8.1 Genes Are Particulate and Are Inherited According to Mendel's Laws, p. 145*
4. What is the probability that a cross between a true-breeding pea plant with spherical seeds and a true-breeding pea plant with wrinkled seeds will produce F_1 progeny with spherical seeds?
   a. ½
   b. ¼
   c. 0
   d. ⅛
   e. 1
   Textbook Reference: *8.1 Genes Are Particulate and Are Inherited According to Mendel's Laws, p. 150*
5. Which of the following statements about the pattern of inheritance for a rare recessive allele is true?
   a. Every affected person has an affected parent.
   b. Unaffected parents can produce children who are affected.
   c. Affected parents do not produce affected children.
   d. Unaffected mothers have affected sons and daughters who are carriers.
   e. None of the above
   Textbook Reference: *8.1 Genes Are Particulate and Are Inherited According to Mendel's Laws, p. 148*
6. Which of the following statements about the pattern of inheritance for a rare dominant allele is true?
   a. Every affected person has an affected parent.
   b. Unaffected parents can produce children who are affected.
   c. Affected parents do not produce affected children.
   d. Unaffected mothers have sons who are affected and daughters who are carriers.
   e. None of the above
   Textbook Reference: *8.1 Genes Are Particulate and Are Inherited According to Mendel's Laws, p. 152*
7. Which of the following statements about the pattern of inheritance for a sex-linked allele is true?
   a. Every affected person has an affected parent.
   b. Unaffected parents can produce children who are affected.
   c. Affected parents do not produce affected children.
   d. Unaffected mothers have sons who are affected and daughters who are carriers.
   e. None of the above
   Textbook Reference: *8.3 Genes Are Carried on Chromosomes, pp. 157–158*
8. The terms "penetrance" and "expressivity" refer to
   a. the increased expression of a particular trait when a hybrid species is formed.
   b. quantitative traits that diminish or intensify a particular phenotype.

c. the influence of environment on the expression of a particular genotype.
d. the expression of one gene masking the effects of another gene.
e. the expression of a dominant phenotype in a heterozygote.

Textbook Reference: *8.2 Alleles and Genes Interact to Produce Phenotypes, p. 155*

9. Sex determination is similar in humans and *Drosophila* because in both species
a. females are hemizygous.
b. males have one X chromosome and females have two X chromosomes.
c. all males have one Y chromosome.
d. secondary sex characteristics are determined by genes on the X chromosome.
e. the ratio of X chromosomes to sets of autosomes determines maleness or femaleness.

Textbook Reference: *8.3 Genes Are Carried on Chromosomes, pp. 158–159*

10. Linked genes are genes that
a. assort independently.
b. segregate equally in the gametes during meiosis.
c. always contribute the same trait to the zygote.
d. are found on the same chromosome.
e. recombine during mitosis.

Textbook Reference: *8.3 Genes Are Carried on Chromosomes, pp. 156–157*

11. Cytoplasmic inheritance
a. results from polygenic nuclear traits.
b. is determined by nuclear genes.
c. is the result of the gametes' contributions of equal amounts of cytoplasm to the zygote.
d. is determined by genes on DNA molecules in mitochondria and chloroplasts.
e. follows Mendel's law of segregation.

Textbook Reference: *8.3 Genes Are Carried on Chromosomes, pp. 159–160*

12. Epistasis is
a. the degree to which a particular genotype is expressed in an individual.
b. the proportion of individuals within a group that have a particular genotype and show the expected phenotype.
c. a situation in which a heterozygotic individual expresses phenotypic traits that are intermediate between those of the parents.
d. a situation in which one gene masks the expression of another gene.
e. a situation in which both alleles are expressed equally.

Textbook Reference: *8.2 Alleles and Genes Interact to Produce Phenotypes, p. 154*

13. Quantitative traits are traits
a. that are affected by the environment.
b. that affect the same physical characteristic.
c. in which each allele intensifies or diminishes the phenotype.
d. All of the above
e. None of the above

Textbook Reference: *8.2 Alleles and Genes Interact to Produce Phenotypes, p. 154*

14. A test cross
a. is used to determine if an organism that is displaying a dominant trait is heterozygous or homozygous for that trait.
b. is used to determines if an organism that is displaying a recessive trait is heterozygous or homozygous for that trait.
c. causes the loss of hybrid vigor.
d. results in an F_2 generation with a phenotypic ratio of ¾ dominant to ¼ recessive.
e. results in the transfer of the same alleles from generation to generation.

Textbook Reference: *8.1 Genes Are Particulate and Are Inherited According to Mendel's Laws, p. 148*

15. A bacterial cell's genotype can be altered by
a. the introduction of a plasmid that carries some of the bacteria's genes.
b. mating with a bacterial cell with the same genotype.
c. homologous recombination with human DNA.
d. the forming of a conjugation tube with another bacterial cell.
e. the transferring of genetic material from a different strain of bacteria.

Textbook Reference: *8.4 Prokaryotes Can Exchange Genetic Material, p. 161*

Answers

Key Concept Review

1.

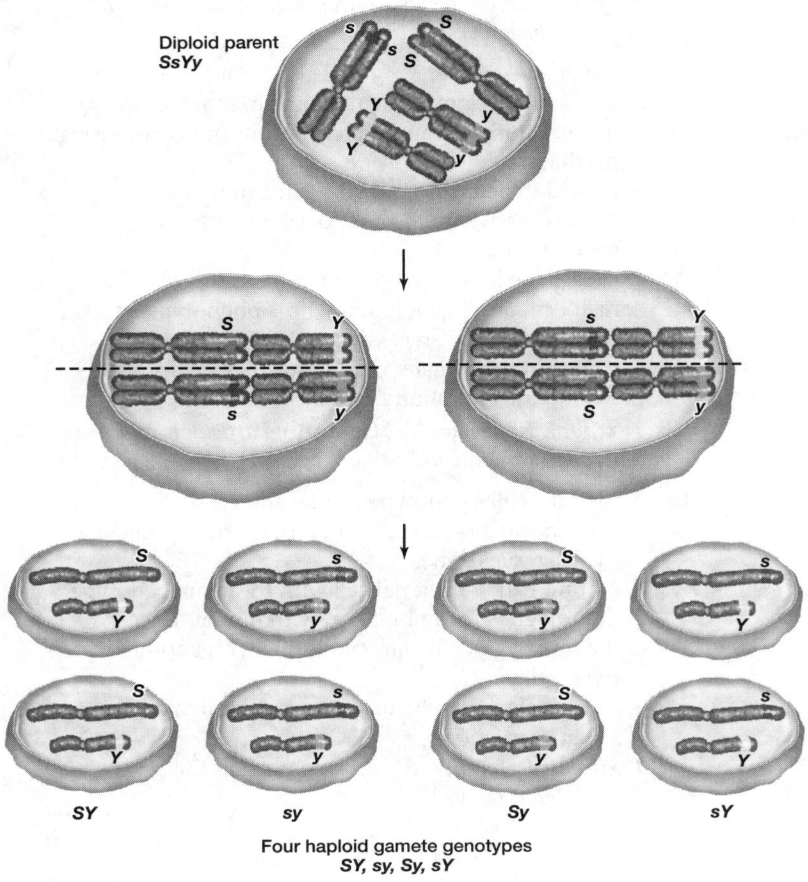

2.

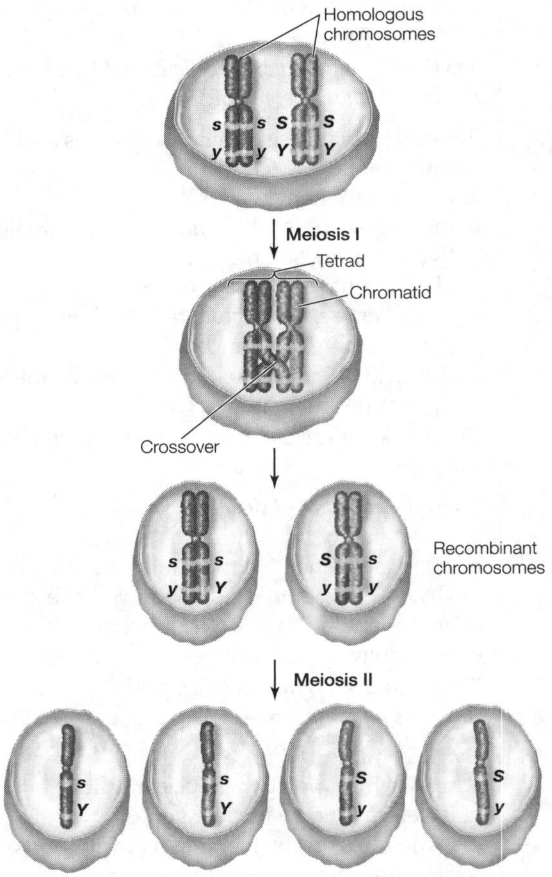

3. There is a 50 percent chance that she will develop Huntington's chorea. Because the trait is an autosomal dominant allele, half of her father's gametes will contain the homologous chromosome carrying that allele and half of his gametes will contain the homologous chromosome that carries the wild-type allele. If she received the Huntington's allele, her child has a 50 percent chance of receiving this allele from her. The product rule is used to predict the probability that her child will inherit the Huntington's allele: ½ (the probability that she has the Huntington's allele) × ½ (the probability that her child will inherit this allele from her) = ¼ (the probability that her child has the allele). Her child has a 25 percent chance of carrying the Huntington's chorea allele and thus of developing the disease.

4. In the first case, in which some children have O type blood and others have A type blood, the father has a genotype of I^AI^O while the mother has a genotype of I^OI^O. Because I^O is recessive to I^A, the children could have O type blood only if they inherited the I^O allele from each parent. If all of the children have A type blood, then the father likely has an I^AI^A genotype. However, we do not know how many children these parents have. Since the sample size is small, it is possible that the children all inherited the I^A allele from an I^AI^O parent.

5. Agouti = *A*, Black = *a*, Color = *G*, Albino = *g*. The true-breeding agouti parent must be homozygous for agouti (*AA*) and for color expression (*GG*). The true-breeding albino parent must be homozygous for albinism: *gg*. Because all of the progeny are agouti, you don't know from that generation what the albino parents' color alleles are. The albino parents might have alleles that are *aa*, *Aa*, or *AA*. Write it as *ggXX* for now, using *X*'s as placeholders for the unknown alleles. So *ggXX* = true = breeding albino parent, *GGAA* = true-breeding agouti parent. All F_1 progeny are *GgAX*. Continue to use *X* as a placeholder for the F_1 cross:

	GA	**gA**	**GX**	**gX**
GA	GGAA	GgAA	GGA	GgAX
gA	GgAA	ggAA	GgA	ggAX
GX	GGAX	GgAX	GGXX	GgXX
gX	GgAX	ggAX	GgXX	ggXX

Resulting phenotypes: albino *gg* (light gray), agouti *AA* and *Aa* (dark gray), three are unknown (white).

Now go back to the original problem and see what else you can determine once you have the Punnett square filled out. You are told that the progeny ratio is 9 agouti, 4 albino, 3 black. The black progeny must be the unknown phenotypes above and have genotype *aa*. Fill in those three squares. This information can then be used to figure out the genotypes of the parents. Below is the Punnett square with the alleles all filled in.

	GA	**gA**	**Ga**	**ga**
GA	GGAA	GgAA	GGA	GgAa
gA	GgAA	ggAA	GgA	ggAa
Ga	GGAa	GgAa	GGaa	Ggaa
ga	GgAa	ggAa	Ggaa	ggaa

Resulting phenotypes: albino *gg* (light gray), agouti *AA* and *Aa* (dark gray), black *Ggaa* (lightest gray).

Now you know that the F_1 parents were actually all *GgAa* because it is the only way to get three black mice.

Going back to the original albino parents, the F_1 mice had to get the *A* allele from the true-breeding agouti parent (*GGAA*), so the *a* allele must have come from the albino parent (*ggaa*).

6. One-half of her sons could be color-blind.

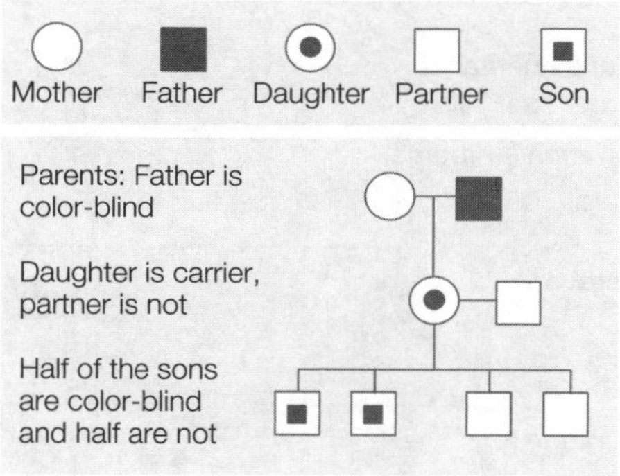
Mother
Father
Daughter
Partner
Son
Parents: Father is color-blind
Daughter is carrier, partner is not
Half of the sons are color-blind and half are not

7. One-half of his children could get the disease; one-quarter of the grandchildren could get the disease.

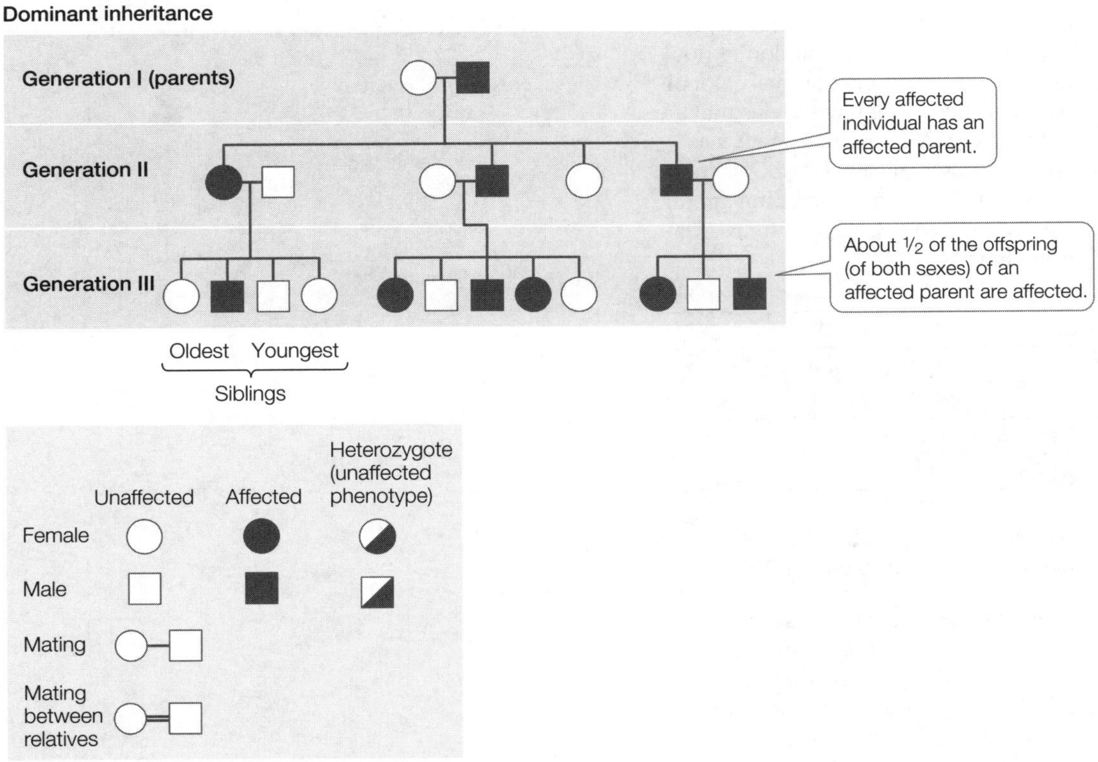

8. One-half of the children of these grandparents could be carriers. One-sixteenth of the grandchildren could have the disease.

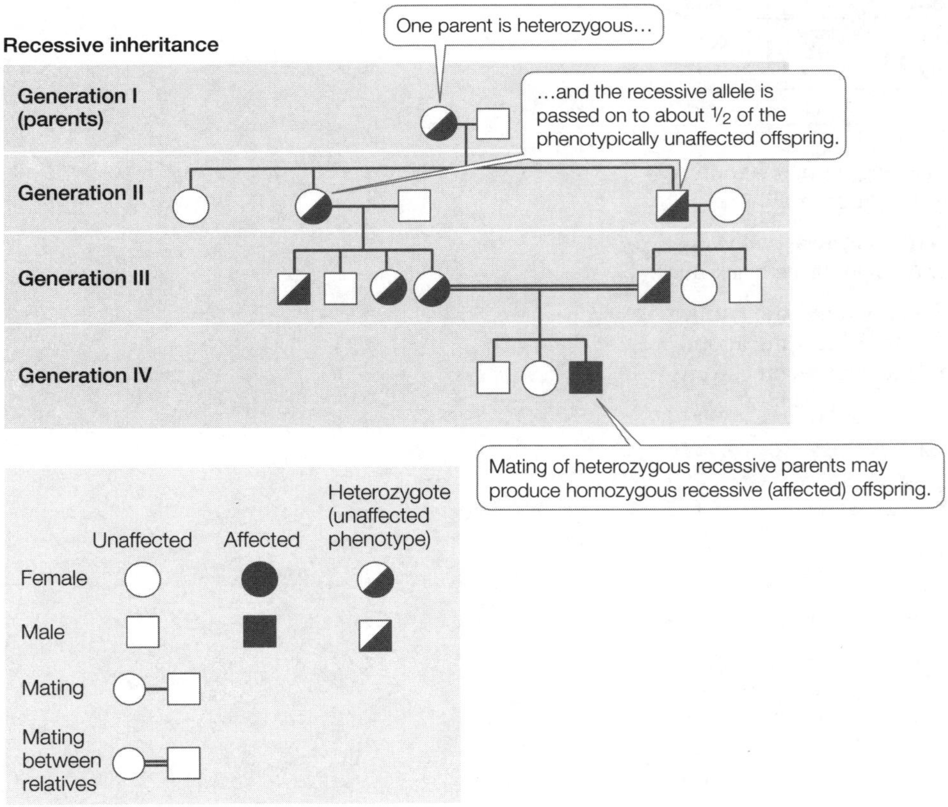

9. In humans, the gamete with the largest cytoplasmic contribution is the egg, so cytoplasmic inheritance is passed from the female parent to all her children. In certain tree species, the male gamete contributes most of the cytoplasm to the zygote, so all the mitochondria and chloroplasts in the zygote are inherited from the male parent.
10. Conjugation allows for recombination between the DNA transferred though the pilus and the recipient chromosome. The process of conjugation can be disrupted if the cells are pulled apart. In addition, because crossing over is random, there is no guarantee that all of the transferred DNA will be exchanged into the new chromosome. Either or both of these events could result in the genotypes seen in the results.
11. We know that plasmids are transferred between bacteria, including bacteria of different strains. Antibiotics in the livestock feed select for those bacteria containing antibiotic resistance genes. This makes it more likely that they will be transferred to strains of bacteria that cause disease in humans and be resistant to the antibiotics normally used to treat the infection. (Alternatively, the bacteria themselves might cause diseases in humans that are untreatable with the antibiotics normally used.)

Test Yourself

1. **d.** Hemophilia is an X-linked trait and can only be inherited by the son from his mother's X chromosome (and not her mitochondrial chromosome). The father contributes the Y chromosome to his son (not his X chromosome) and thus cannot pass any of his X-linked alleles to his son.
2. **c.** Incomplete dominance results in the progeny's expressing an intermediate form of the two parental alleles. (In a cross between red-flowered plants and white-flowered plants, the expression of pink-flowered plants would be a "blend" of the parental traits.) Codominance is not an example of blending because both alleles are fully expressed in the individual.
3. **a.** Monohybrid and dihybrid crosses produce heterozygous individuals; true-breeding individuals are always homozygous.
4. **e.** This is an example of a monohybrid cross. All of the F_1 progeny would have spherical seeds. (The F_1 generation would all have the genotype *Ss*, producing the phenotype of spherical seeds because the spherical allele, *S*, is dominant to the wrinkled allele, *s*.)
5. **b.** Rare recessive alleles can be carried by both parents but not expressed in those parents. If the parents are heterozygous for this allele (*Aa*), their children will have a ¼ probability of expressing that recessive allele (*aa*). If both parents are affected (*aa*), their children will also be affected (*aa*).
6. **a.** If an allele is dominant, every affected individual has at least one dominant allele. An affected individual must have received that allele from one of his or her parents. Because the allele is dominant, that parent must also be affected.
7. **d.** The most common sex-linked alleles are X-linked and are passed from a mother to her son (because the mother always donates one of her X chromosomes to her son, and the father always donates the Y chromosome to his son). Daughters can also receive the X-linked allele from their mothers, but the father donates the other X chromosome, so daughters can be carriers.
8. **c.** "Penetrance" and "expressivity" refer to the effects of the environment on a particular phenotype. Answer a. refers to hybrid vigor, answer b. refers to quantitative traits, answer d. refers to epistasis, and answer e. refers to expression of a dominant allele.
9. **b.** In these species, females have two X chromosomes and males have one X chromosome. The alleles for secondary sex characteristics are found on the X chromosome and the autosomes.
10. **d.** Linked genes, by definition, are on the same chromo some and thus do not assort independently, do not contribute the same trait to the zygote, and do not recombine during mitosis or segregate equally to the gametes during meiosis.
11. **d.** The genes on the mitochondria and chloroplast chromosomes, which are cytoplasmically inherited (unlike nuclear genes), are passed on to all of the progeny from the gamete that contributes most of the cytoplasm.
12. **d.** Answer a. refers to expressivity, answer b. refers to penetrance, answer c. refers to incomplete dominance, and answer e. refers to codominance.
13. **d.** Quantitative traits are traits that are affected by the environment and can either diminish or intensify one phenotype.
14. **a.** A test cross is used to determine if an organism that is expressing a dominant trait is homozygous or heterozygous for that trait. True-breeding individuals continue to express the same alleles generation after generation.
15. **e.** A bacterial cell's genotype changes when new genetic information is introduced either on a plasmid or by conjugation (which requires a conjugation tube and the transfer of DNA). If conjugation occurs, those genes need to recombine into the recipient cell's chromosome in order to be maintained. Transferring the same genetic information on a plasmid or mating with a bacterial cell with the same genotype will not change the genotype of the recipient bacteria. Human DNA cannot recombine with a bacterial chromosome unless there are identical (homologous) DNA sequences in both DNA molecules.

DNA and Its Role in Heredity 9

The Big Picture

- The identification of DNA as the genetic material has allowed scientists to understand how hereditary information is passed on at the molecular level and how mutations can alter that hereditary information.
- Advances in our knowledge of DNA replication have led to new technologies for understanding genes, their function, their expression, and genetic relatedness in different organisms.

Study Strategies

- Be careful to keep track of the DNA's orientation. The 5′ and 3′ ends are often confused when parental (template) strands and daughter (newly synthesized) strands are compared. Remember that on both strands, the 3′ end corresponds to the site (or former site) of a hydroxyl group (—OH), and the 5′ end corresponds to the site (or former site) of a phosphate tail.
- Study the figures in the textbook to visualize what is occurring during DNA replication. Make your own diagrams of these processes or arrange simple manipulatives. For instance, to visualize the differences between leading and lagging strands, lay out colored candies on a strip of paper. Separate the strips, and "synthesize" a new strand of DNA using the complementary colors of candy (set up a pairing scheme of your choosing). Make sure you work from a 5′ to 3′ direction.
- Take advantage of laboratory activities that simulate or allow you to experience sequencing or PCR. Nearly all molecular research labs utilize one or both of these techniques.
- Draw a picture of the replication fork. Position the primers on the fork, and then draw in the leading and lagging strands. Label all of the 3′ and 5′ ends. Draw a box to indicate where ligase will seal up the ends.
- Many different mutations can occur in a gene that will alter the function of the gene product. Some of those mutations may have more deleterious effects than others. Make a table listing the different types of mutations, the scale of the changes, and where appropriate, a simple illustration of the change compared to a template DNA sequence you design or to a template chromosome.
- Go to yourBioPortal.com to review the following tutorials and activity:

  Animated Tutorial 9.1 The Hershey–Chase Experiment

  Animated Tutorial 9.2 Experimental Evidence for Semiconservative DNA Replication

  Animated Tutorial 9.3 DNA Replication Part 1: Replication of a Chromosome and DNA Polymerization

  Animated Tutorial 9.4 DNA Replication Part 2: Coordination of Leading and Lagging Strand Synthesis

  Interactive Tutorial 9.1 Polymerase Chain Reaction (PCR): Identifying Pathogens

  Web Activity 9.1 DNA Polymerase

Key Concept Review

9.1 DNA Structure Reflects Its Role as the Genetic Material

Circumstantial evidence suggested that DNA was the genetic material

Experimental evidence confirmed that DNA is the genetic material

The discovery of the three-dimensional structure of DNA was a milestone in biology

The nucleotide composition of DNA was known

Watson and Crick described the double helix

Four key features define DNA structure

The double-helical structure of DNA is essential to its function

Today we take for granted our ability, given the investment of some time and money, to sequence the genome of a given species. But the role of DNA in a cell was not always understood and the structure of the DNA molecule was also a mystery. Before the mid-twentieth century the technology available made it difficult for scientists to directly verify that DNA was the genetic material. The issue was finally resolved when bacteria, but not proteins, were observed to acquire viral DNA when infected by bacteriophage T2. Once DNA was established as the genetic material, it was important to identify its structure. It had already been observed that the ratios of adenine to thymine and cytosine to guanine are always fixed in the DNA polymer. Watson and Crick made use of Rosalind Franklin's work in X-ray crystallography as well as models to determine that the DNA molecule is a double helix with antiparallel strands. This structure proved to be an elegant solution to question of how DNA stores genetic information in a form that can be efficiently copied and "read" by the cell machinery.

Question 1. Diagram the double helix. Be sure to label those properties that make it most suited as the genetic material. Think about DNA replication and encoding of information.
Textbook Reference: *9.1 DNA Structure Reflects Its Role as the Genetic Material, p. 170*

Question 2. Given the following DNA sequence, what would the sequence of the complementary strand look like?

5′-GCTAACTGTGATCGTATAAGCTGA-3′
Textbook Reference: *9.1 DNA Structure Reflects Its Role as the Genetic Material, p. 172*

9.2 DNA Replicates Semiconservatively

- DNA polymerases add nucleotides to the growing chain
- The two DNA strands grow differently at the replication fork
- Telomeres are not fully replicated in most eukaryotic cells
- Errors in DNA replication can be repaired
- The basic mechanisms of DNA replication can be used to amplify DNA in a test tube

A key feature of DNA is that it replicates semiconservatively. This means that each "half" of the double helix serves as the template for a new, complementary DNA strand. The process of replication requires several proteins as well as an abundant supply of energy and nucleotides. Starting at the origin of replication, the hydrogen bonds between the complementary bases are broken and the strands are pulled apart. Replication begins with an RNA primer, and DNA polymerase synthesizes the new strand of DNA by adding nucleotides to the 3′ end of the primer. One of the strands (the leading strand) is in the proper orientation to allow for continuous synthesis. However, the other strand (the lagging strand) is antiparallel but still must be replicated by means of nucleotides added to the 3′ end of the new strand. This necessitates a discontinuous form of DNA synthesis in which short stretches of DNA (Okazaki fragments) are synthesized from a series RNA primers as single-stranded template becomes available. Later, the cell removes the RNA primers, fills in the resulting gaps in the sequence, and ligates the backbone so the lagging strand is complete.

In linear chromosomes, the directional nature of DNA results in a continued shortening of the lagging strand with each round of DNA replication. To ensure that this shortening does not affect coding regions, and to mark the ends of a chromosome as "proper" ends and not the result of breakage, telomeres are located at the ends of chromosomes. These stretches of repeating sequences can be lengthened with the action of the enzyme telomerase, which is not expressed in most somatic cells. The absence of telomerase acts as a "countdown clock" to tell a cell when to stop dividing. Cells that should continue to divide contain active telomerase that keeps the telomeres from being lost over time.

DNA replication needs to be accurate. To help ensure the accuracy, the cell has several strategies to monitor the process. DNA polymerase itself has a proofreading capability and will remove an incorrectly selected nucleotide if it makes a mistake. There are other proteins that monitor the DNA strand for mismatched base pairs and will remove the incorrect nucleotides. (DNA polymerase will then replace them.)

DNA replication is possible in a test tube. By selecting primers for a DNA region of interest, scientists can amplify a target sequence by means of PCR (the polymerase chain reaction). This is a powerful tool that has revolutionized molecular biology research.

Question 3. Diagram the replication complex and label the following: DNA polymerase, RNA primer, Okazaki fragment, the leading and lagging strands, and the leading and lagging template strands. Be sure to label the 5′ and 3′ ends of the parental and newly replicated DNA.
Textbook Reference: *9.2 DNA Replicates Semiconservatively, pp. 172–176, Figures 9.11–9.12*

Question 4. Explain the role of Okazaki fragments in the synthesis of the lagging strand.
Textbook Reference: *9.2 DNA Replicates Semiconservatively, pp. 172–176*

Question 5. Differentiate between proofreading and mismatch repair.
Textbook Reference: *9.2 DNA Replicates Semiconservatively, p. 177*

Question 6. Explain how PCR amplifies a particular sequence of DNA. Is the replication process linear or exponential? Explain your answer.

Textbook Reference: *9.2 DNA Replicates Semiconservatively, pp. 178–179*

Question 7. Refer to the following DNA sequence. Note that only the "top" strand is provided here.

TATCGTCAGA TTTCAATCTA TTGGCGTTGT TAAAAAACTA

TGGTTATACT AACGGCAAAA ACGCTCTGAA ACTAGATCCT

AATGAAGTCT TCAACGTGAC TTTTGACCGT TCAATGTTCA

The nucleotides can be referred to by number (i.e., the first one is number 1 and the last is 120), design PCR primers that would amplify nucleotides 31–80. Your primers should be beyond the region you are amplifying and should be 15 nucleotides long. What do you need in the test tube to make the reaction work?

Textbook Reference: *9.2 DNA Replicates Semiconservatively, pp. 178–79*

9.3 Mutations Are Heritable Changes in DNA

Mutations can have various phenotypic effects

Point mutations change single nucleotides

Chromosomal mutations are extensive changes in the genetic material

Mutations can be spontaneous or induced

Some base pairs are more vulnerable than others to mutation

Mutagens can be natural or artificial

Mutations have both benefits and costs

Mutations are alterations to the genetic code. Mutations in somatic cells are not confined to the individual whereas mutations in germline cells are passed to the next generation. Regardless of the cell type, mutations may be limited to a single nucleotide change or they may be chromosomal mutations in which entire chromosome fragments are moved, lost, duplicated, or flipped. Whether or not the mutation is beneficial, harmful, or neutral to the organism is dependent on how the mutation affects the phenotype.

The causes of mutations are numerous and may be natural or artificial. DNA replication itself can introduce changes to the genetic code (spontaneous mutation), and this happens naturally in a cell. Chromosomes may break or fail to separate, or they can be repaired incorrectly. Exposure to radiation and mutagenic chemicals can lead to induced mutations. While an individual cannot prevent spontaneous mutations, limiting one's exposure to mutagens can reduce the incidence of induced mutations.

Question 8. Two individuals have a mutation in gene *X* but at different sites. The mutation affects the first individual adversely, and the second individual experiences no effect. Explain this observation.

Textbook Reference: *9.3 Mutations Are Heritable Changes in DNA, pp. 179–180*

Question 9. Are all mutations inherently bad? What about those in somatic cells?

Textbook Reference: *9.3 Mutations Are Heritable Changes in DNA, pp. 179–184*

Question 10. Starting with chromosome ABCDEFG and the nonhomologous chromosome LMNOPQR, illustrate the following:

a. Deletion of segment C
b. A duplication of segment C
c. A duplication and a deletion of a segments D and E, respectively
d. An inversion of a segment, going from A to C
e. A reciprocal translocation between CDEFG and LMNO

Textbook Reference: *9.3 Mutations Are Heritable Changes in DNA, pp. 181–182, Figure 9.17*

Test Yourself

1. Chargaff observed that the amount of _______ was roughly equal to the amount of _______ in all tested organisms.
   a. purines; pyrimidines
   b. A; T
   c. A + T; G + C
   d. A + G; T + C
   e. a, b, and d

   Textbook Reference: *9.1 DNA Structure Reflects Its Role as the Genetic Material, p. 169*

2. Watson and Crick's model allowed them to visualize
   a. the molecular bonds of DNA.
   b. the sugar and phosphate component of the DNA molecule's surface.
   c. how the purines and pyrimidines fit together in a double helix.
   d. the antiparallel design of two strands of the DNA double helix.
   e. All of the above

   Textbook Reference: *9.1 DNA Structure Reflects Its Role as the Genetic Material, p. 169*

3. A fundamental requirement for the functioning of genetic material is that it must be
   a. conserved among all organisms with very little variation.
   b. passed intact from one species to another.

c. replicated accurately and passed from a parent to its offspring or from a cell to its daughter cells.
d. found outside the nucleus.
e. replicated accurately over many millions of years of evolution.

Textbook Reference: *9.1 DNA Structure Reflects Its Role as the Genetic Material, pp. 170–171*

4. The primary function of DNA polymerase is to
a. add nucleotides to the growing daughter strand.
b. seal nicks along the sugar–phosphate backbone of the daughter strand.
c. unwind the parent DNA double helix.
d. generate primers to initiate DNA synthesis.
e. prevent reassociation of the denatured parental DNA strands.

Textbook Reference: *9.2 DNA Replicates Semiconservatively, pp. 174–175*

5. When the lagging daughter strand of DNA is synthesized, unreplicated gaps are formed on the parental DNA. Lagging strand synthesis fills these gaps by
a. synthesizing short Okazaki fragments in a 5′-to-3′ direction.
b. synthesizing multiple short RNA primers to initiate DNA replication.
c. using DNA polymerase I to remove RNA primers from Okazaki fragments.
d. filling in those gaps with new strands of complementary DNA as the replication fork proceeds.
e. All of the above

Textbook Reference: *9.2 DNA Replicates Semiconservatively, p. 176*

6. RNA primers are necessary in DNA synthesis because
a. DNA polymerase is unable to initiate replication without an origin.
b. the DNA polymerase enzyme can catalyze the addition of deoxyribonucleotides only onto the 3′ (—OH) end of an existing strand.
c. RNA primase is the first enzyme in the replication complex.
d. primers mark the sites where helicase has to unwind the DNA.
e. All of the above

Textbook Reference: *9.2 DNA Replicates Semiconservatively, p. 174*

7. Proofreading and repair occur
a. at any time during or after synthesis of DNA.
b. only before mitosis.
c. only in the presence of DNA polymerase.
e. only during replication.

Textbook Reference: *9.2 DNA Replicates Semiconservatively, pp. 177–178*

8. If 30 percent of the bases in a sample of DNA extracted from eukaryotic cells are adenine, what percentage of the bases in this DNA are cytosine?
a. 10 percent
b. 20 percent
c. 30 percent
d. 40 percent
e. 50 percent

Textbook Reference: *9.1 DNA Structure Reflects Its Role as the Genetic Material, p. 169*

9. Which of the following statements about DNA replication is *false*?
a. Okazaki fragments are synthesized as part of the leading strand.
b. Replication forks represent areas of active DNA synthesis on the chromosomes.
c. Error rates for DNA replication are reduced by proofreading of the DNA polymerase.
d. Ligases and polymerases function in the vicinity of replication forks.
e. The sliding clamp protein increases the rate of DNA synthesis.

Textbook Reference: *9.2 DNA Replicates Semiconservatively, pp. 175–176*

10. The PCR technique
a can amplify only very small samples of DNA.
b. amplifies several random DNA sequences within a genome.
c. requires synthetic primers to flank the regions of interest.
d. is accomplished in three sequential steps: binding, denaturation, and replication.
e. generates DNA molecules that all have variable sequences.

Textbook Reference: *9.2 DNA Replicates Semiconservatively, p. 178*

11. Which of the following represents a bond between a purine and a pyrimidine, respectively?
a. C–T
b. G–A
c. G–C
d. T–A
e. A–G

Textbook Reference: *9.1 DNA Structure Reflects Its Role as the Genetic Material, p. 169*

12. Which of the following would *not* be found in a DNA molecule?
a. Purines
b. Ribose sugars
c. Phosphates
d. Sulfur
e. Nitrogenous bases

Textbook Reference: *9.1 DNA Structure Reflects Its Role as the Genetic Material, pp. 169–170, Figure 9.5*

13. If a high concentration of a particular nucleotide lacking a hydroxyl group at the 3′ end is added to a PCR reaction,
a. no additional nucleotides will be added to a growing strand containing that nucleotide.
b. strand elongation will proceed as normal.
c. nucleotides will be added only at the 5′ end.

d. *T. aquaticus* DNA polymerase will become nonfunctional.
e. the primer will be unable to bind with the DNA template.

Textbook Reference: *9.2 DNA Replicates Semiconservatively, p. 176*

14. The telomeres at the ends of linear chromosomes allow
a. the 5′ ends of the chromosomes to undergo recombination.
b. the gaps left by primer removal of lagging strands to be repaired by telomerase.
c. DNA repair enzymes to recognize those ends and remove them.
d. normal cells to divide continuously.
e. DNA breaks to be examined at cell division checkpoints.

Textbook Reference: *9.2 DNA Replicates Semiconservatively, pp. 176–177*

15. Which of the following chromosomal mutations would still allow protein X to be functional?
a. Deletion of the last 100 codons of gene *X*
b. A duplication of gene *X*
c. An inversion of the last 100 codons of gene *X*
d. A translocation of the last 100 codons of gene *X* to another chromosome
e. None of the above

Textbook Reference: *9.3 Mutations Are Heritable Changes in DNA, pp. 181–182*

16. Gain-of-function mutations
a. are dominant mutations that are expressed in wild-type cells.
b. are dominant mutations that are expressed in mutant cells.
c. are the cause of continuous division in cancer cells.
d. can be analyzed only under restrictive conditions.
e. are expressed only in response to the appropriate environmental signals.

Textbook Reference: *9.3 Mutations Are Heritable Changes in DNA, p. 180*

Answers

Key Concept Review

1. See Figure 9.5 in the textbook. Did you indicate that nucleotides can only hydrogen bond with their complement (C and G, A and T)? Did you note the antiparallel structure of double stranded DNA and indicate that new nucleotides are only added to the 3′ end? Did you note that the nucleotides can appear in any order on a stand of DNA, which allows the DNA sequence to be unique when the nucleotides are "read" by cellular machinery?

2. 3′-CGATTGACACTAGCATATTCGACT-5′

3.

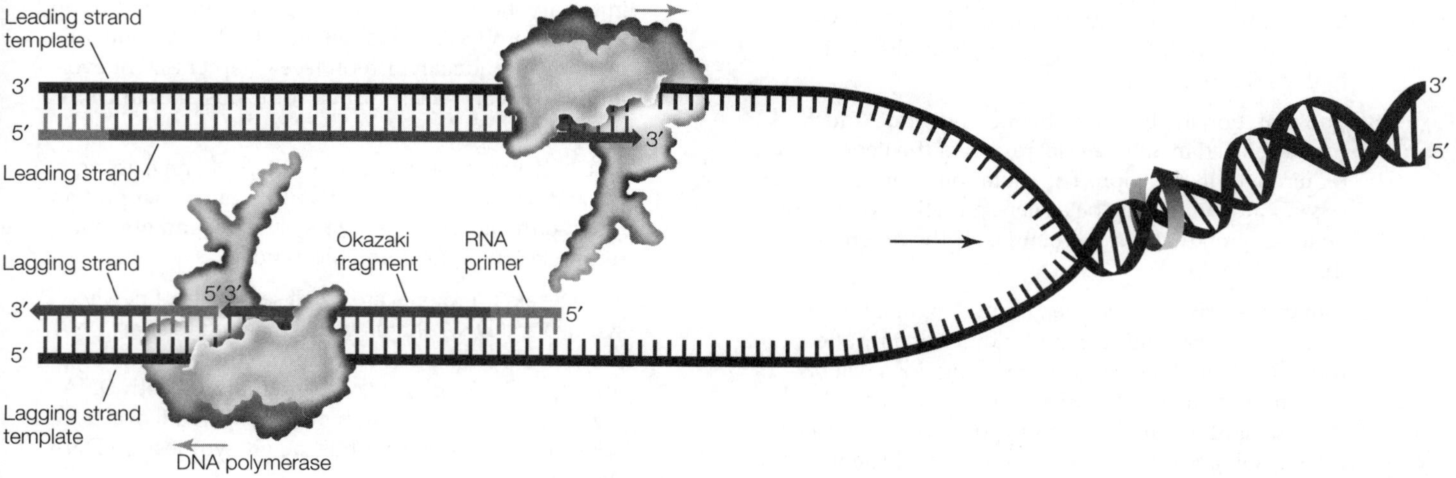

4. Okazaki fragments are short, discontinuous stretches of sequence that are formed when the lagging strand is replicated. Because only a portion of the DNA is opened up at the replication fork at any given time, and because DNA polymerase adds nucleotides to the 3′ end of the previous nucleotide, continuous growth is not possible on the lagging strand. Therefore, replication of the lagging strand produces short, discontinuous stretches of DNA that are synthesized in the 5′-to-3′ direction—the opposite direction from the movement of the replication fork.

5. Proofreading occurs during synthesis of the DNA molecule, whereas mismatch repair occurs after synthesis.
6. Refer to Figure 9.15 in the textbook. PCR amplifies a sequence of DNA by means of a laboratory technique that uses a sample of double-stranded DNA, two artificially synthesized primers that are complementary to the ends of the sequence to be amplified, the four dNTPs, and a DNA polymerase. During each cycle of a PCR reaction, the double-stranded DNA is separated and replicated, producing two double-stranded copies. These copies are themselves copied in the next cycle of the PCR reaction. Since every cycle of the reaction results in a doubling of the DNA, the amount of DNA increases exponentially.
7. Any 15 nucleotide sequence identical to the sequence shown between 1 and 30 would work for one of the primers. For example,

   5′-ATCTATTGGCGTTGT-3′ goes from 16 to 30. It will bind to the complementary sequence and allow DNA to be synthesized from position 31 to 120. To amplify the double-stranded DNA sequence, a primer is also needed to bind to the other strand of DNA and synthesize a strand complementary to the one shown. For example, the sequence between 81 and 95 is

   5′-AATGAAGTCTTCAAC-3′

   3′-TTACTTCAGAAGTTG-5′

   with the bottom strand showing the primer sequence. Convention has us write single-stranded DNA from 5′ to 3′, so the two primers will be

   5′-AATGAAGTCTTCAAC-3′ (top), and

   5′-GTTGAAGACTTCATT-3′ (bottom).

   The test tube must also contain the template DNA, both primers, DNA polymerase that is heat stable, dATP, dTTP, dCTP, dGTP nucleotides, salts, and a buffer solution.
8. The mutation in gene *X* in the first individual must have occurred in an essential region of the gene that is required for its function. The mutation in gene *X* in the second individual may be a silent mutation, or it may be in a region that is nonessential for the function of that protein.
9. Mutations may be beneficial, neutral, or deleterious to an organism. Somatic cell mutations will only affect the individual, whereas germ cell mutations will be passed to the next generation. If a mutation results in the loss of function, the phenotype may be altered if the homologous gene's expression is insufficient. In the case of a germ cell loss-of-function mutation, the mutation may be deleterious in the next generation if two nonfunctional genes are inherited. Some mutations do not alter the function of genes and are called silent. Still other mutations may result in gene products with altered function. Depending on the environment, this altered function may provide some benefit for survival. It is also possible that this altered function is deleterious. On a species level, mutations in germ cells are one source of genetic diversity. In organisms that replicate asexually, all cell mutations are a potential source of genetic diversity. Genetic diversity helps ensure that at least some organisms will have a chance of surviving in spite of changing environmental conditions.
10.
    a. ABDEFG
    b. ABCCDEFG
    c. ABCDDFG
    d. CBADEFG
    e. ABLMNO and CDEFGPQR

Test Yourself

1. **e.** Chargaff found that in most DNA sampled, the amount of A equaled the amount of T and the amount of G equaled the amount of C. It follows that the amount of purines (A + G) equals the amount of pyrimidines (T + C). The same does not hold for the amount of A + T versus the amount of G + C.
2. **e.** Model-building by Watson and Crick created a three-dimensional visualization of the size, bond angles, base pairings, and overall structure of the DNA molecule. The information that contributed to their model came from a variety of experiments performed by other scientists.
3. **c.** Replication of DNA is a fundamental requirement for its function as the genetic material. DNA must be correctly replicated in each cell of an organism and passed from parent to offspring or from cell to daughter cell. DNA is not passed from one species to another. There are variations in DNA sequences from different organisms. DNA is found outside the nucleus in eukaryotic cells (specifically in the mitochondria, and in the chloroplasts of plants), but this DNA is inherited in a non-Mendelian fashion (see Chapter 8). For evolution to occur, there must be some mistakes in DNA replication over time.
4. **a.** DNA polymerase adds nucleotides to an existing nucleotide strand, ligase seals nicks, and other proteins unwind the DNA, separate the strands, and prevent reassociation of the parental strands.
5. **e.** Okazaki fragments are small segments of newly synthesized DNA that have been added to short RNA primers. The RNA is removed from these small fragments by DNA polymerase I and replaced with DNA. The remaining DNA fragments are ligated (covalently bound) to form a continuous, newly synthesized DNA strand.
6. **b.** DNA polymerase cannot initiate synthesis of a nucleotide strand; it can only add on to an existing strand of RNA primer or DNA. Primers provide DNA polymerase with the 3′ (—OH) end required for the addition of deoxyribonucleotides. The origin is also required for DNA synthesis, but it is the site where the

replication complex is initially assembled. Primase is not the first enzyme in the replication complex.

7. **a.** The mismatch repair mechanism operates before the newly synthesized DNA strand is methylated, but for the integrity of DNA to be maintained, repair mechanisms must be active during synthesis, modification, and utilization of DNA.
8. **b.** If 30 percent of DNA is adenine, then according to Chargaff's rule, 30 percent must be thymine. The remaining 40 percent of the DNA is cytosine and guanine. Because the ratio of cytosine to guanine must be equal, the percentage of cytosine in this DNA must be 20 percent.
9. **a.** Okazaki fragments are involved in synthesis of the lagging strand.
10. **c.** PCR amplifies specific DNA sequences from small or large samples of DNA and requires three sequential steps: denaturation, binding, and replication. The amplified DNA molecules do not have variable DNA sequences.
11. **c.** Guanine is a purine, and its paired pyrimidine is cytosine.
12. **d.** Sulfur is a constituent of many protein molecules, but it is not found in DNA.
13. **a.** A hydroxyl group at the 3′ position of a nucleotide is necessary for the binding of any additional nucleotides. If this hydroxyl group is absent, no other nucleotides can be added to a growing strand.
14. **b.** Telomerase binds the telomeres at the ends of the chromosomes, protects them from being degraded or recombined with other chromosomes, and ensures that these ends will be replicated after primer removal. Telomeres are not associated with cell division checkpoints or continuous cell division.
15. **c.** A duplication of gene *X* might allow the protein to be functional. The other chromosomal mutations would remove large regions of the coding sequence or would position part of that sequence in the opposite direction (inversion), which would produce a nonfunctional protein X.
16. **b.** Gain-of-function mutations are dominant and are expressed in mutant cells. They do not respond to the appropriate environmental signals and do not have a restrictive condition under which they are expressed. Some gain-of-function mutations are seen in cancer cells, but mutations in tumor supressors are also responsible for unregulated cell division in cancer cells.

From DNA to Protein: Gene Expression

10

The Big Picture

- Every protein and RNA in the cell has a DNA blueprint (the gene sequence) that specifies the amino acid or nucleotide sequence of that gene product. Those sequences determine how that gene product will fold up three-dimensionally and ultimately function in the cell.
- Alterations in those gene products, which are caused by mutations in the genes, help us understand how those proteins and RNAs function in the cell.
- Some gene products are RNA (e.g., rRNA and tRNA) while others are proteins. Many proteins are modified after translation or combined with other peptides to generate a complete, functional product.

Study Strategies

- Don't try to understand every detail of gene expression without first comprehending the larger picture: the way genetic information is accessed in the DNA and expressed in the cell so that the cell can function. Familiarize yourself with the details of the central dogma, but then take a step back and understand that all of these details describe the steps required to synthesize gene products.
- It is easy to confuse transcription and translation. Draw diagrams of the processes of transcription and translation including initiation, elongation, and termination. Carefully review the template, the product, and the sites of initiation and termination. Label the components that play a role in each process. Be sure to orient the 5′ and the 3′ ends of the nucleic acid, and the N and C terminus of the protein.
- Choose a hypothetical gene that encodes a peptide of four amino acids and draw a flowchart that shows how that gene in the chromosome is made into protein. Next, follow the details of gene expression: the synthesis of the RNA from the DNA, and the synthesis of the protein from the mRNA. Be sure to include the start and stop signals in each of these processes. Then alter one of those nucleotides in the DNA sequence, and transcribe and translate the gene. Is the protein sequence changed? Repeat this process using a different mutation in the DNA.
- Review the experiments that revealed the presence of introns in the primary RNA transcript and describe how those introns are removed during splicing. Develop analogies to explain the editing process to a non-biology student.
- Introns and exons are easily confused. It may help to remember that Introns Interrupt and Exons are Expressed.
- When taking notes in class and from the book, you will find you need to write the words transcription and translation again and again. Many scientists use "txn" as shorthand for transcription and "tln" as shorthand for translation. The abbreviations for ribosomal RNA (rRNA), transfer RNA (tRNA), and messenger RNA (mRNA) can also be added to your repertoire.
- Go to yourBioPortal.com to review the following tutorials and activities:

  Animated Tutorial 10.1 Transcription

  Animated Tutorial 10.2 RNA Splicing

  Animated Tutorial 10.3 Deciphering the Genetic Code

  Animated Tutorial 10.4 Protein Synthesis

  Interactive Tutorial 10.1 Genetic Mutations

  Web Activity 10.1 Eukaryotic Gene Expression

  Web Activity 10.2 The Genetic Code

  Working with Data 10.1 Deciphering the Genetic Code

Key Concept Review

10.1 Genetics Shows That Genes Code for Proteins

Observations in humans led to the proposal that genes determine enzymes

The concept of the gene has changed over time

Genes are expressed via transcription and translation

In metabolic pathways, the product of one reaction becomes the substrate for the next reaction. If we understand how the larger pathways work, we can figure out which step failed in a metabolic disorder such as alkaptonuria or phenylketonuria. In both of these disorders, the mutation in genes coding for enzymes leads to the buildup of an undesirable intermediate. Study of this type of disorder led to the idea of one gene correlating to one enzyme.

Further work revealed that some proteins comprise multiple peptide subunits. This led to the refinement of the idea that a single gene coded for a single protein and the relationship was modified to one gene–one polypeptide. While still useful for many gene products, this relationship oversimplifies how genes work. For example, RNA is a gene product that may or may not be translated into a protein. Even so, the transcription of the genome into an RNA sequence followed by translation into a polypeptide chain is true in many cases in a cell and in all instances of protein synthesis.

Question 1. Suppose that two different mutant strains of a bacterium are unable to grow on a minimal medium without the addition of the amino acid lysine, even though the "normal" strain does not need a lysine supplement. Explain how different mutations in each strain, either on the same gene or in two different genes, might have resulted in this lysine-requiring phenotype.
Textbook Reference: *10.1 Genetics Shows That Genes Code for Proteins, pp. 188–189*

Question 2. Imagine that you are trying to explain the process of translation and transcription to a non-biologist, and you decide to use the analogy of a builder's construction of a new house. How would you create this analogy? What are the biological equivalents of the architect's blueprints, the construction site, the builder's photocopies of the blueprints, the actual building materials, the Bobcat loader, and the finished building? Be sure to detail the roles of DNA, mRNA, rRNA, tRNA, amino acids, and protein.
Textbook Reference: *10.1 Genetics Shows That Genes Code for Proteins, pp. 188–190*

10.2 DNA Expression Begins with Its Transcription to RNA

- RNA polymerases share common features
- Transcription occurs in three steps
- Eukaryotic coding regions are often interrupted by introns
- Eukaryotic gene transcripts are processed before translation

In both eukaryotes and prokaryotes, transcription occurs in three steps: initiation, elongation, and termination. The process of initiation begins with a promoter sequence, which orients RNA polymerase and indicates which strand of DNA is to be transcribed. RNA polymerase does not bind all by itself; rather, it is one of many proteins that are recruited to the promoter region in order to initiate transcription. In the second stage, elongation, RNA polymerase unwinds the DNA double helix and synthesizes a single strand of mRNA in a 5′-to-3′ direction while reading the DNA template in the 3′-to-5′ direction. The complementary triphosphate nucleotides are added using energy contained in the bonds between the three phosphates. When RNA polymerase encounters the termination sequence in the DNA, either the RNA polymerase falls off the DNA template or additional proteins remove it from the DNA template, and transcription ends.

In eukaryotes, the coding sequences (exons) often have noncoding "interruptions" called introns. The product of transcription includes all of these sequences, and before translation begins the cell works to remove the introns by a process called RNA splicing. Splicing is completed by an RNA protein complex called a spliceosome, which reads consensus sequences at the boundaries of introns and exons in order to splice the mRNA properly. Once splicing is complete, a GTP cap is added to the 5′ end of the pre-mRNA and a poly A tail is added to the 3′ end. These modifications help mRNA bind to the ribosome, protect the mRNA from premature degradation, and assist in the transport of mRNA from the nucleus to the cytosol.

Question 3. Suppose that you place a double-stranded DNA sequence in a test tube, along with the following components: RNA polymerase, dATP, dCTP, dGTP, dUTP, and the required transcription factors. A coworker uses the same solution with a different DNA template sequence and finds that mRNA is the result. Your procedure, however, does not result in mRNA, even though you are sure that the reagents are sound. What went wrong?
Textbook Reference: *10.2 DNA Expression Begins with Its Transcription to RNA, p. 191*

Question 4. On the DNA and mRNA fragments below, label the following: 5′ GPT cap, 3′ poly A tail, introns, exons, promoter, terminator, and splice sites.
Textbook Reference: *10.2 DNA Expression Begins with Its Transcription to RNA, p. 192, Figure 10.6*

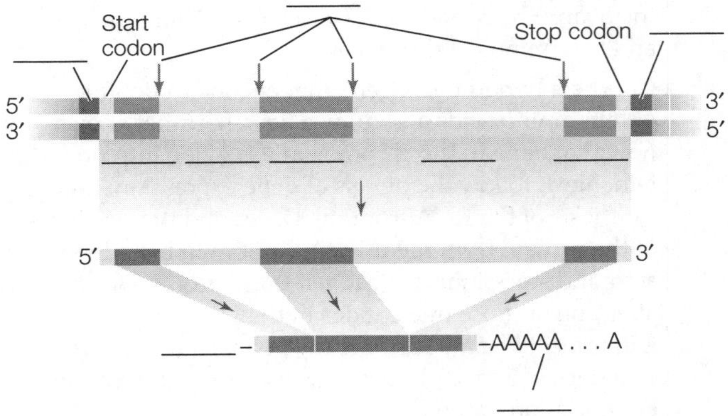

10.3 The Genetic Code in RNA Is Translated into the Amino Acid Sequences of Proteins

The information for protein synthesis lies in the genetic code

Point mutations confirm the genetic code

Trios of nucleotides (called a codon) code for individual amino acids. There is more than one codon for many amino acids, as well as a start codon and three stop codons. The genetic code is very highly conserved in living things, indicating that it arose in a common ancestor early in the evolution of living things. Alterations to the genetic code may not alter the amino acid, especially if the change is in the third nucleotide of the codon (a silent mutation). In other cases, a single nucleotide change may change the codon to that of a different amino acid (a missense mutation) or a premature stop codon (a nonsense mutation). If an extra nucleotide is added to a DNA sequence, or one is lost, the resulting shift in reading frame will result in the production of a different polypeptide (a frame-shift mutation).

Question 5. Starting with the following mature mRNA sequence, identify the start codon, stop codon, and amino acid sequence.

10 20 30 40 50 60

5′–UUAGCTAUCC UAAAGUAUGC GUCAUUCUCA AAUCGUUUGG GGUUGUUAAU GUAAACGUCA–3′

Textbook Reference: *10.3 The Genetic Code in RNA Is Translated into the Amino Acid Sequences of Proteins, pp. 196–198*

Question 6. Using the mature mRNA sequence shown in Question 5 as a starting point, determine the amino acid sequences that would result from the following changes and classify them as silent, missense, nonsense, or frameshift mutations:

a. Nucleotide 29 is replaced with guanine
b. Nucleotide 29 is replaced with uracil
c. Nucleotide 37 is replaced with adenine
d. One uracil is inserted between nucleotides 10 and 11.
e. The nucleotide at position 20 is deleted
f. One guanine is inserted between nucleotides 17 and 18.

Textbook Reference: *10.3 The Genetic Code in RNA is Translated into the Amino Acid Sequences of Proteins, pp. 196–198, Figure 10.11*

10.4 Translation of the Genetic Code Is Mediated by tRNA and Ribosomes

Transfer RNAs carry specific amino acids and bind to specific codons

Each tRNA is specifically attached to an amino acid

Translation occurs at the ribosome

Translation takes place in three steps

Polysome formation increases the rate of protein synthesis

Translation is the process of reading the codons in the mRNA to properly link amino acids in a peptide chain. The job of tRNA is to transport specific amino acids to the ribosome. By reading the codon, the tRNA is able to bind to the mRNA at the appropriate time, allowing the attached amino acid to be linked to the peptide chain. The other main structure involved is rRNA, which is composed of two subunits. These subunits position the tRNAs along the mRNA sequence and catalyze peptide bond formation.

Like transcription, translation has three steps: initiation, elongation, and termination. The initiation process involves binding of the small ribosomal subunit and a charged tRNA to the start codon on the mRNA. The large subunit of the ribosome then joins the complex. Elongation breaks the bond between the tRNA and the amino acids while forming peptide bonds between amino acids already present at the P site and those delivered to the A site. The ribosome and mRNA shift relative to each other by one codon and the process is repeated until a termination codon is reached. At this point, a protein release factor frees the peptide chain and translation is complete.

Question 7. Draw a diagram of a eukaryotic cell and show where in the cell the gene is transcribed and translated. In your diagram, indicate how this particular gene product is targeted to a compartment in the cell such as the endoplasmic reticulum. Create a stepwise list of all the important proteins and enzymes required for this process.

Textbook Reference: *10.4 Translation of the Genetic Code Is Mediated by tRNA and Ribosomes, pp. 199–203*

Question 8. What would happen if the tRNA synthase for tryptophan added a phenylalanine to the tryptophan tRNAs instead of tryptophan?
Textbook Reference: *10.4 Translation of the Genetic Code Is Mediated by tRNA and Ribosomes, pp. 199–203*

10.5 Proteins Are Modified after Translation

Signal sequences in proteins direct them to their cellular destinations

Many proteins are modified after translation

The polypeptide released from a ribosome is often not a complete, functional protein. Proteins are sent to different organelles and they may be modified after translation is complete.

Proteins are sent to the proper destination by use of signal sequences at the N terminus of the peptide chain. They may be modified with the addition of phosphates or sugars. They may be synthesized in an inactive form and become active only after being cleaved by proteases. The combination of signal sequences and posttranslational modifications may confer additional functions to the proteins.

Question 9. Suppose that a protein is supposed to go to the endoplasmic reticulum and the DNA encoding the signal sequence for that gene product is deleted. What would be the result?
Textbook Reference: *10.5 Proteins Are Modified after Translation, pp. 204–205*

Test Yourself

1. Transcription in prokaryotic cells
   a. occurs in the nucleus, whereas translation occurs in the cytoplasm.
   b. is initiated at a start codon with the help of initiation factors and the small subunit of the ribosome.
   c. is initiated at a promoter and uses only one strand of DNA (the template strand) to synthesize a complementary RNA strand.
   d. is terminated at a stop codon.
   e. is initiated at an *ori* site on the chromosome.
   Textbook Reference: *10.2 DNA Expression Begins with Its Transcription to RNA, p. 190*

2. Which of the following statements about RNA polymerase is *false*?
   a. It synthesizes mRNA in a 5′-to-3′ direction, reading the DNA strand 3′ to 5′.
   b. It synthesizes mRNA in a 3′-to-5′ direction, reading the DNA strand 5′ to 3′.
   c. It binds at the promoter and unwinds the DNA.
   d. It does not require a primer to initiate transcription.
   e. It uses only one strand of DNA as a template for synthesizing RNA.
   Textbook Reference: *10.2 DNA Expression Begins with Its Transcription to RNA, p. 192*

3. Translation of messenger RNA into protein occurs in a _______ direction, and from the _______ terminus to the _______ terminus.
   a. 3′-to-5′; N; C
   b. 5′-to-3′; N; C
   c. 3′-to-5′; C; N
   d. 5′-to-3′; C; N
   e. 3′-to-5′; C; C
   Textbook Reference: *10.3 The Genetic Code in RNA Is Translated into the Amino Acid Sequences of Proteins, pp. 197, Figure 10.11*

4. If codons were read two bases at a time instead of three bases at a time, how many different possible amino acids could be specified?
   a. 16
   b. 64
   c. 8
   d. 32
   e. 128
   Textbook Reference: *10.3 The Genetic Code in RNA Is Translated into the Amino Acid Sequences of Proteins, p. 197*

5. Translate the following mRNA:
   3′-GAUGGUUUUAAAGUA-5′
   a. NH_2 met—lys—phe—leu—stop COOH
   b. NH_2 met—lys—phe—trp—stop COOH
   c. NH_2 asp—gly—phe—lys—val COOH
   d. NH_2 met—gly—phe—lys—val COOH
   e. NH_2 asp—gly—phe—lys—stop COOH
   Textbook Reference: *10.3 The Genetic Code in RNA Is Translated into the Amino Acid Sequences of Proteins, pp. 196–198*

6. What would happen if a mutation occurred in DNA such that the second codon of the resulting mRNA was changed from UGG to UAG?
   a. Translation would continue and the second amino acid would be the same.
   b. Nothing. The ribosome would skip that codon and translation would continue.
   c. Translation would continue, but the reading frame of the ribosome would be shifted.
   d. Translation would stop at the second codon, and no functional protein would be made.
   e. Translation would continue, but the second amino acid in the protein would be different.
   Textbook Reference: *10.3 The Genetic Code in RNA Is Translated into the Amino Acid Sequences of Proteins, pp. 197–198*

7. If the following synthetic RNA were added to a test tube containing all the components necessary for protein translation to occur, what would the amino acid sequence be?
5′-AUAUAUAUAUAU-3′
a. Polyphenylalanine
b. Isoleucine–tyrosine–isoleucine–tyrosine
c. Isoleucine–isoleucine–isoleucine–isoleucine
d. Tyrosine–tyrosine–tyrosine–tyrosine
e. Aspargine–aspargine–aspargine–aspargine
Textbook Reference: *10.3 The Genetic Code in RNA Is Translated into the Amino Acid Sequences of Proteins, pp. 197–198*

8. Which part of the tRNA base-pairs with the codon in the mRNA?
a. The 3′ end, where the amino acid is covalently attached
b. The 5′ end
c. The anticodon
d. The start codon
e. The promoter
Textbook Reference: *10.3 The Genetic Code in RNA Is Translated into the Amino Acid Sequences of Protein, pp. 199–200, Figure 10.13*

9. Peptidyl transferase is an
a. enzyme found in the nucleus of the cell that assists in the transfer of mRNA to the cytoplasm.
b. enzyme that adds the amino acid to the 3′ end of the tRNA.
c. enzyme found in the large subunit of the ribosome that catalyzes the formation of the peptide bond in the growing polypeptide.
d. RNA molecule that is catalytic.
e. Both c and d
Textbook Reference: *10.3 The Genetic Code in RNA Is Translated into the Amino Acid Sequences of Proteins, p. 202*

10. Termination of translation requires
a. a termination signal, RNA polymerase, and a release factor.
b. a release factor, initiator tRNA, and ribosomes.
c. initiation factors, the small subunit of the ribosome, and mRNA.
d. elongation factors and charged tRNAs.
e. a stop codon positioned at the A site of the ribosome, peptidyl transferase, and a release factor.
Textbook Reference: *10.3 The Genetic Code in RNA Is Translated into the Amino Acid Sequences of Proteins, pp. 202–203*

11. If the DNA encoding a nuclear signal sequence were placed in the gene for a cytoplasmic protein, the protein would
a. be modified in the Golgi.
b. be directed to the lysosomes.
c. be directed to the nucleus.
d. be directed to the cytoplasm.
e. stay in the endoplasmic reticulum.
Textbook Reference: *10.5 Proteins Are Modified after Translation, pp. 204–205*

12. The central dogma of molecular biology states that _______ is transcribed into _______, which is (are) translated into _______.
a. a gene; polypeptides; a gene product
b. protein; DNA; RNA
c. DNA; mRNA; tRNA
d. DNA; RNA; protein
e. RNA; DNA; protein
Textbook Reference: *10.1 Genetics Shows That Genes Code for Proteins, p. 190*

13. A gene product can be
a. an enzyme.
b. a polypeptide.
c. RNA.
d. microRNA.
e. All of the above
Textbook Reference: *10.1 Genetics Shows That Genes Code for Proteins, p. 189*

14. The enzyme that catalyzes the synthesis of RNA is
a. peptidyl transferase.
b. DNA polymerase.
c. tRNA synthase.
d. ribosomal RNA.
e. RNA polymerase.
Textbook Reference: *10.2 DNA Expression Begins with Its Transcription to RNA, p. 191*

15. A mutation occurs such that a spliceosome cannot remove one of the introns in a gene. What effect will this have on that gene?
a. It will have no effect; the gene will be transcribed and translated into protein.
b. Transcription will terminate early and the protein will not be made.
c. Transcription will proceed, but translation will stop at the site where the intron remains.
d. Translation will continue, but a nonfunctional protein will be made.
e. Translation will continue and will skip the intron sequence.
Textbook Reference: *10.3 The Genetic Code in RNA Is Translated into the Amino Acid Sequences of Proteins, pp. 192–193*

Answers

Key Concept Review

1. The mutations in these two strains of bacteria apparently interfere with lysine synthesis. Both mutations might be in the same gene coding for an enzyme necessary for lysine synthesis, but one could be a nonsense mutation in the fifth codon, for example, and the other could be a frame-shift mutation in the twenty-third codon (the number of mutations that can disable a gene is enormous). If lysine synthesis in this bacterium requires more than one enzyme (as is likely), the two mutations could be in different genes coding for different enzymes. In this case, the phenotypes would not be strictly identical; it would be possible to distinguish between the two by trying to grow them on minimal media to which different intermediates in the synthesis of lysine were added
2. The master blueprint is DNA, which is too precious to be taken to the actual worksite and possibly damaged. Instead, the DNA is photocopied into mRNA. Multiple copies of the same blueprint (mRNA) may be generated to allow the same blueprint to be read multiple times in multiple manufacturing sites in the cell. Ribosomal RNA serves as the main workbench where all of the raw materials are assembled. It brings together the mRNA template along with the amino acids in order to build a protein. The amino acids can be thought of as the actual building materials. They are brought to the workbench in the appropriate order by the tRNA, which serves as a kind of Bobcat loader, fetching the appropriate materials and delivering them at the proper time and in the proper amount. The mRNA blueprint is carefully read as each building block is put into place on the ribosome. Finally, the building (the protein) is constructed according to the plans (mRNA) and the construction crew (RNAs, energy, amino acids) can leave the scene and go to a new site.
3. Your DNA sequence does not have a promoter. The process of transcription depends on the promoter sequence to recruit RNA polymerase to the DNA, orient the RNA polymerase, and indicate which DNA strand is the template strand. Your coworker's DNA template included a promoter sequence.

4.

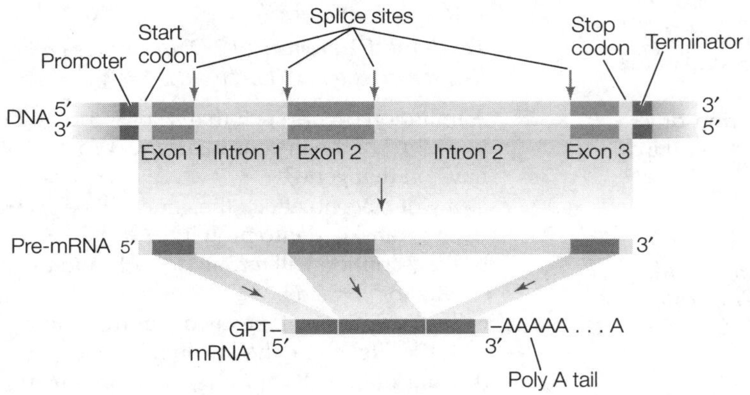

5. The amino acid sequence is coded from nucleotides 17–46. The stop codon is from nucleotides 47–49. The amino acid sequence is Met-Arg-His-Ala-Gln-Ile-Val-Trp-Gly-Cys.

10 20 30 40 50 60
5′–UUAGCTAUCC UAAAGU**AUG**C GUCAUGCUCA AAUCGUUUGG GGUUGU**UAA**U GUAAACGUCA–3′
Met Arg His Ala Gln Ile Val Trp Gly Cys **Stop**

6.

a. Met Arg His Ala Glu Ile Val Trp Gly Cys (Missense, amino acid 5)
b. Met Arg His Ala (Nonsense; premature stop after serine)
c. Met Arg His Ala Gln Ile Val Trp Gly Cys (Silent)
d. Met Arg His Ala Gln Ile Val Trp Gly Cys (Silent; the insertion was before the start codon, so the reading frame was not changed.)
e. Met Val Met Leu Lys Ser Phe Gly Val Val Asn Val Asn Val (Frame-shift; there is no stop codon in this sequence in frame with the new reading frame.)
f. Met Leu Lys Ser Phe Gly Val Val Asn Val Asn Val. (Frame-shift; the first start codon was disrupted, so the second Met in answer e. is the new start codon here. There is no stop codon in this sequence in frame with the new reading frame.)

7.

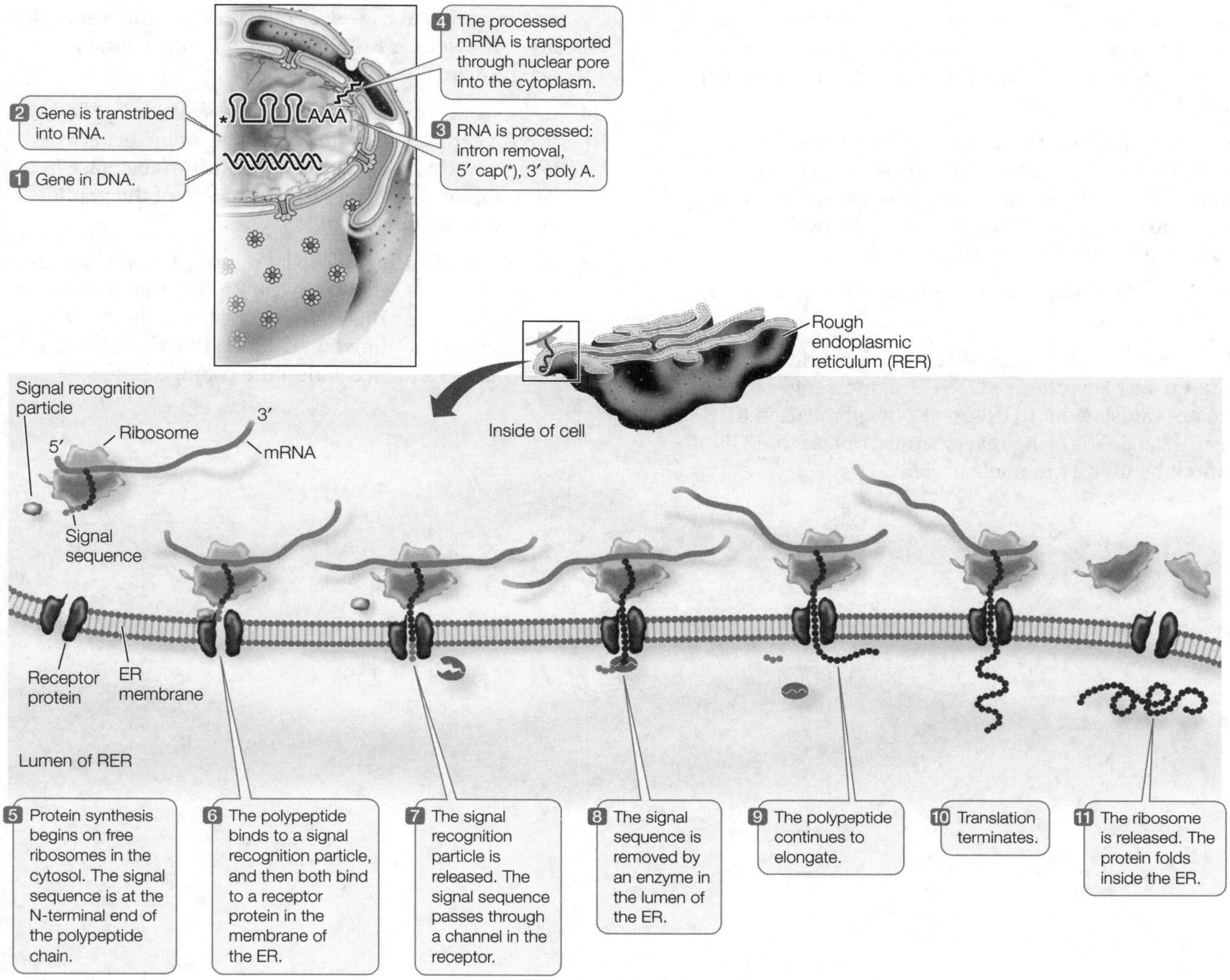

4 The processed mRNA is transported through nuclear pore into the cytoplasm.
2 Gene is transtribed into RNA.
3 RNA is processed: intron removal, 5′ cap(*), 3′ poly A.
1 Gene in DNA.
AAA
Rough endoplasmic reticulum (RER)
Signal recognition particle
3′
Ribosome
5′
mRNA
Inside of cell
Signal sequence
Receptor protein
ER membrane
Lumen of RER
5 Protein synthesis begins on free ribosomes in the cytosol. The signal sequence is at the N-terminal end of the polypeptide chain.
6 The polypeptide binds to a signal recognition particle, and then both bind to a receptor protein in the membrane of the ER.
7 The signal recognition particle is released. The signal sequence passes through a channel in the receptor.
8 The signal sequence is removed by an enzyme in the lumen of the ER.
9 The polypeptide continues to elongate.
10 Translation terminates.
11 The ribosome is released. The protein folds inside the ER.

8. If the tRNA synthetase for tryptophan added phenylalanine to the tryptophan tRNAs, whenever a tryptophan codon was read by these tryptophan tRNAs, phenylalanine would be added to the polypeptide. This would create proteins that were nonfunctional, and the cell would die.

9. Deleting the signal sequence would not affect the transcription or translation of this gene, but it would affect the targeting of the gene product. During translation of the mRNA, no signal peptide would be made and the signal recognition particle would be unable to bind. The mRNA would continue to be translated in the cytoplasm and remain there.

Test Yourself

1. **c.** Transcription occurs in the nucleus and translation occurs in the cytoplasm of eukaryotic cells, not of prokaryotic cells. Translation, rather than transcription, is initiated at start codons and terminates at a stop codon. Replication of the circular chromosome starts the *ori* site.

2. **b.** RNA polymerase binds at a promoter, unwinds the DNA, synthesizes mRNA in a 5′-to-3′ (not 3′-to-5′) direction, and does not require a primer to synthesize the RNA.

3. **b.** Translation of messenger RNA occurs 5′ to 3′, and the polypeptide is synthesized from the N terminus to the C terminus.

4. **a.** Four possible bases read two at a time would yield 4^2, or 16, different codons.

5. **b.** See the codon table (Figure 10.11). Recall that translation occurs in the 5′-to-3′ direction.

6. **d.** UAG is a stop codon, so translation would terminate at that site.

7. **b.** See the codon table (Figure 10.11).

8. **c.** Neither the 3′ end nor the 5′ end of the tRNA is part of the anticodon. The promoter is a DNA sequence, to which RNA polymerase binds to initiate transcription. The start codon is found in the mRNA.

9. **e.** Peptidyl transferase is the enzyme that catalyzes the formation of the peptide bond, and it is located in the large subunit of the ribosome. Its catalytic activity is due to ribosomal RNA found in the large subunit of the ribosome.
10. **e.** Termination of translation requires a stop codon positioned at the A site of the ribosome, peptidyl transferase, and a release factor. Peptidyl transferase hydrolyzes the last amino acid attached to the tRNA in the P site, creating the C terminus.
11. **c.** The nuclear sequence would direct this protein to the nucleus.
12. **d.** Genes are not transcribed into polypeptides, protein is not used to synthesize DNA, and messenger RNAs are not translated into tRNAs. RNA can be used to synthesize DNA using reverse transcriptase, but DNA cannot be utilized to make protein.
13. **e.** Gene products can be RNAs (such as rRNA, tRNA, and microRNA) as well as enzymes and other polypeptides. Messenger RNA is translated into a gene product, protein.
14. **e.** DNA polymerase catalyzes the synthesis of DNA, tRNA synthase covalently attaches amino acids to tRNAs, and ribosomal RNA (peptidyl transferase in the large subunit) catalyzes the formation of the peptide bond during translation.
15. **d.** When an intron fails to be removed, that noncoding sequence is retained in the RNA within the coding sequence. When this RNA is translated, the protein will likely be nonfunctional due to the insertion of a noncoding sequence within the coding sequence.

Regulation of Gene Expression 11

The Big Picture

- Gene expression is highly regulated, allowing an organism to respond to external and internal cues. Genes can be expressed in response to specific environmental or developmental conditions (inducible) or all the time (constitutive). Regulation of genes can potentially occur at many points: before transcription, during transcription, after transcription but before translation, at translation, and after translation. Gene expression is controlled at transcription by activators (proteins that stimulate transcription) or repressors (proteins that prevent transcription).
- Viruses and bacteriophage are acellular and depend on host cell metabolism to carry out their gene expression and reproduction. Viruses can have a DNA or RNA genome, and inject their genome into the host cell when they infect it. They may enter a lytic or lysogenic phase once in the cell. HIV is an example of a retrovirus.
- Prokaryotes transcriptionally regulate gene expression at promoters and posttranscriptionally by mRNA degradation, translational regulation, and protein degradation. Prokaryotic cells use operons to coordinately regulate the expression of several genes.
- Eukaryotes regulate gene expression at the transcriptional level with general transcription factors as well as specific repressors and activators, and they coordinate that regulation by placing regulatory sequences in front of different genes. Epigenetic modifications also affect gene expression. Posttranscriptional regulation includes alternate splicing, modification of the 5′ cap, control of translational initiation, and protein degradation.

Study Strategies

- It is easy to confuse prokaryotic inducible operons (the *lac* operon) with repressible operons (the *trp* operon) because they both use repressors to regulate gene expression. Review the environmental conditions that must exist for each repressor to bind its operator site and what causes each repressor to release the operator site.
- It may seem puzzling at first that some viruses insert their chromosome into the host chromosome, because the goal of viral infection seems to be rapid multiplication and infection of adjacent cells. However, lysogeny allows the viral genome to persist while local environmental resources are abundant. When local resources are depleted or cell damage occurs, the virus can enter the lytic cycle to escape and infect other cells.
- Gene expression in eukaryotes can be regulated epigenetically by DNA methylation and chromatin remodeling. RNA polymerase and transcription factors require access to the DNA sequences before transcription can begin. Diagram the process of transcriptional activation for a gene that can be modified by methylation, and for a gene that undergoes histone modification. When is the gene active? When is it silent?
- Gene expression often occurs in response to environmental signals. Make a list of the environmental signals that a prokaryote receives and then outline the steps initiated by the cell in response to those signals. Include the following signals: lactose in the cell, glucose and lactose in the cell, high levels of glucose in the cell, high or low levels of tryptophan in the cell, bacteriophage infection under poor or rich environmental conditions.
- Outline the different viral life cycles, from the start of infection through multiplication of virus to viral release (see Figure 11.3). Compare the lytic and lysogenic life cycles of bacteriophage.
- Gene regulation can occur at many steps in a eukaryotic cell. Starting with chromatin remodeling, list the steps that must occur for a gene to be made into functional protein in the cytoplasm of the cell.
- Go to yourBioPortal.com to review the following tutorials and activities:

  Animated Tutorial 11.1 The *lac* Operon

  Animated Tutorial 11.2 The *trp* Operon

  Animated Tutorial 11.3 Initiation of Transcription

  Web Activity 11.1 Eukaryotic Gene Expression Control Points

  Web Activity 11.2 Concept Matching

Key Concept Review

11.1 Several Strategies Are Used to Regulate Gene Expression

Genes are subject to positive and negative regulation

Viruses use gene regulation strategies to subvert host cells

Genes can be negatively or positively regulated at many points: before transcription, during transcription, after transcription but before translation, at translation, and after translation. At transcription, gene expression can be negatively regulated by a repressor that inhibits transcription or positively regulated by an activator protein that stimulates transcription.

Viruses are acellular and perform no metabolic functions. They develop and reproduce only inside the cells of specific hosts. Outside the host cell, viruses exist as virions, which consist of nucleic acid and proteins. By means of gene regulation, they manipulate host cell to replicate the virus particles (virions). Some viruses are lytic (they immediately activate the host cell to replicate the virus and lyse, releasing the virions) and some alternate between lytic and lysogenic periods. The latter is a dormant period during which the virus genome is replicated with the cell genome until induced to begin the lytic stage.

Many bacteriophages are lytic. Lytic viral cycles have two stages, early and late. During the early phase, the phage injects its nucleic acid into the host cytoplasm, and viral genes are transcribed and translated. These early gene products shut down host transcription, degrade host DNA, and stimulate viral genome replication and viral gene transcription. Late gene products include viral capsid proteins and proteins that lyse the cell at the end of the lytic cycle. The lytic cycle takes about 30 minutes and produces hundreds of bacteriophage per cell (see Figure 11.3).

Human immunodeficiency virus (HIV) is a retrovirus (RNA genome) that infects only immune system cells and alternates between lytic and lysogenic stages. HIV is an enveloped virus, enclosed in a plasma membrane derived from the previous host cell. HIV copies its RNA genome into the host DNA using reverse transcriptase. The DNA copy of the viral genome is then integrated into a host cell chromosome where it becomes a provirus (see Figure 11.4). The provirus may remain dormant in the host virus for years.

When activated, the provirus is transcribed as mRNA and translated into protein by the host cell's translation machinery. Usually, the cell has regulatory mechanisms to terminate expression of invader virus genes. The HIV viral protein Tat (*t*rans*a*ctivation of *t*ranscription) prevents termination, allowing viral gene transcription by the host RNA polymerase (see Figure 11.5).

Question 1. Which type of gene regulation do viral genomes employ: positive, negative, or a combination of both?
Textbook Reference: *11.1 Several Strategies Are Used to Regulate Gene Expression, pp. 210–212*

Question 2. Some people who are exposed to the HIV virus do not become infected. Describe two possible ways their cells might resist the HIV invasion.
Textbook Reference: *11.1 Several Strategies Are Used to Regulate Gene Expression, pp. 211–212*

11.2 Many Prokaryotic Genes Are Regulated in Operons

Regulating gene transcription conserves energy

Operons are units of transcriptional regulation in prokaryotes

Operator–repressor interactions regulate transcription in the *lac* and *trp* operons

RNA polymerase can be directed to a class of promoters

Prokaryotes conserve energy by making proteins only when they need them. The most efficient means of regulating gene expression is at the level of transcription, because protein synthesis is energetically expensive. Another method of gene regulation (allosteric regulation) allows fine tuning of metabolism.

An example of transcription-level gene regulation is the response to food availability of the gut bacteria, *E. coli*. Lactose uptake by *E. coli* involves three proteins: β-galactoside permease, which transports lactose into the cell; β-galactosidase, which breaks down lactose into glucose and lactose; and β-galactoside transacetylase, which transfers acetyl groups to certain β-galactosides.

The genes that encode these three enzymes are called structural genes, and they reside close to each other in the *E. coli* genome. A cluster of genes regulated by a single promoter is called an operon. *E. coli* responds to changes between glucose and lactose availability using the *lac* operon. The *lac* operon includes the sequences for the promoter, the operator, and the structural genes for the enzymes involved in lactose metabolism (see Figure 11.7). The operator controls transcription of the structural genes.

The *lac* operon is an inducible operon and is transcribed only in the presence of the inducer, β-galactoside (e.g., lactose; see Figure 11.8). Other operons (e.g., *trp* operon) are repressible operons that are turned off only in response to a repressor.

In the inducible *lac* operon, a repressor protein is normally bound to the operator, preventing transcription. When the inducer is abundant, it binds to the repressor, causing it to change shape so it can no longer bind to the DNA and block transcription.

In summary, inducible operons have the following features: In the absence of the inducer, the operator is bound by a regulatory protein (the repressor) to prevent transcription. In the presence of the inducer, the repressor binds to the inducer so that it changes shape and no longer binds with the DNA, so transcription of the operon occurs.

The *trp* operon is a repressible operon. Like an inducible operon, the repressible operon is switched off when the repressor is bound to the operator. However, in the case of the repressible operon, the repressor is *not* bound to the operator unless another molecule, the co-repressor (tryptophan, in this case), first binds to the repressor. When the co-repressor binds to the repressor, the repressor becomes active and binds to the operator, preventing transcription.

In repressible systems (like that of the *trp* operon), the product of the metabolic pathway (the co-repressor) interacts with the regulatory protein, enabling it to bind the promoter and block transcription. The presence of the product thus turns *off* the transcriptional machinery necessary to produce more of that product.

In general, inducible systems like the *lac* operon regulate catabolic pathways that turn on only when the substrate is available, while repressible systems like the *trp* operon regulate anabolic pathways that are always turned on unless the concentration of the product becomes excessive.

Other proteins in prokaryotes called sigma factors regulate transcription by binding to RNA polymerase and guiding it to specific promoters.

Question 3. Suppose that a cell has a mutation that deletes the gene encoding the repressor for a certain operon, and a plasmid is introduced into the host cell that carries a wild-type copy of the gene for the repressor. Is normal regulation of this operon restored in the presence of this plasmid?
Textbook Reference: *11.2 Many Prokaryotic Genes Are Regulated in Operons, pp. 213–214*

Question 4. Suppose that a cell has a mutation that deletes the gene encoding the operator for a certain operon, and a plasmid is introduced into the host cell that carries a wild-type copy of the operator. Is normal regulation of this operon restored in the presence of this plasmid?
Textbook Reference: *11.2 Many Prokaryotic Genes Are Regulated in Operons, pp. 214–215*

11.3 Eukaryotic Genes Are Regulated by Transcription Factors and DNA Changes

- Transcription factors act at eukaryotic promoters
- The expression of sets of genes can be coordinately regulated by transcription factors
- Epigenetic changes to DNA and chromatin can regulate transcription
- Epigenetic changes can be induced by the environment

Eukaryotic promoters have three important sequences, a regulatory binding site, a transcription binding site that often has a TATA box, and an RNA polymerase binding site. Initiation of transcription by RNA polymerase II in eukaryotic cells requires regulatory proteins called general transcription factors (TFIID). These are proteins that bind to the promoter and form a transcription complex to which RNA polymerase II can bind in order to initiate transcription. General transcription factors are different from transcription factors that are specific to promoters or classes of promoters.

TFIID binds the TATA box, changing the shape of the DNA and the transcription factor itself. Other transcription factors then bind the complex on the promoter, and RNA polymerase II binds to initiate transcription (see Figure 11.10).

Other short DNA sequences bind regulatory proteins that affect transcription positively (enhancers, which bind activator proteins) and negatively (silencers, which bind repressors). Regulatory DNA sequences can be located close to or far away from the gene they affect. Often, many binding factors are involved and transcription of any eukaryotic gene is determined by the combination of transcription factors, repressors, and activators.

Recognition of specific nucleotide sequences by transcription factors involves available sites for hydrogen bonding, hydrophobic interactions, and induced fit.

Coordinated gene regulation is achieved when genes share regulatory sequences that bind the same transcription factors. An example of coordinated gene regulation is seen in plant drought response genes. Regulatory elements known as dehydration response elements (DREs) are found near the promoters of genes that respond to drought (see Figure 11.12).

The term "epigenetics" refers to reversible changes in the DNA structure of a gene or genes that occurs without any change in the gene sequence (no mutation). These are heritable. Epigenetic processes include DNA methylation and chromatin remodeling.

In methylation, cytosine residues are methlyated by DNA methyl transferase and the methlyated state can be inherited. In mammals this usually happens in regions rich in CG-adjacent pairs called CpG islands. Maintenance methylases catalyze the formation of 5-methylcytosine on the replicated DNA strand. Repressor proteins bind methylated regions on DNA, so methylated genes are usually silenced. Reversal of methylation is accomplished when demethylases remove the methyl group from the cytosine.

DNA methylation occurs in male and female genomes upon fertilization to silence duplicate genes on the extra X chromosome, and in tumor suppressor genes of cancer cells. When a long stretch or entire choromosome is methylated it appears as heterochromatin (darkly staining). The X chromosome in female animals is an example of heterochromatin. The genotype for the sex chromosomes in mammals is XY in males and XX in females. However, the expression of X-linked genes (the gene dosage) is the same in both sexes, due to X-inactivation in females. Early in embryonic development, one of the two X chromosomes is randomly inactivated, producing a highly condensed heterochromatic chromosome called a Barr body (see Figure 11.14).

Another mechanism for gene regulation is chromatin remodeling. The DNA in eukaryotic chromosomes is attracted to positively charged histone proteins of the nucleosomes that can inhibit the initiation and elongation steps of transcription. Chromatin remodeling must occur to make the DNA accessible to transcription complexes (see Figure 11.15).

Epigenetic changes that occur in germline cells can be inherited, but later modified by environmental factors including stress experienced by offspring.

Question 5. Suppose you are engineering gene *Y*, such that when it is inserted into a eukaryotic chromosome it will be expressed continuously. Which specific sequences must be part of this gene so that it will be expressed?
Textbook Reference: *11.3 Eukaryotic Genes Are Regulated by Transcription Factors and DNA Changes, p. 217, Figure 11.10*

Question 6. Suppose that you are engineering a new plant in which gene *Y* will be activated under drought conditions. What kinds of DNA sequences have to be present to ensure activation of the gene under these conditions?
Textbook Reference: *11.3 Eukaryotic Genes Are Regulated by Transcription Factors and DNA Changes, p. 218, Figure 11.12*

Question 7. Diagram a gene in both a prokaryotic cell and a eukaryotic cell. Include the types of sequences that are important for transcriptional regulation, the location of those sequences, and the proteins that bind those regions.
Textbook Reference: *11.3 Eukaryotic Genes Are Regulated by Transcription Factors and DNA Changes, pp. 216–217*

11.4 Eukaryotic Gene Expression Can Be Regulated after Transcription

- Different mRNAs can be made from the same gene by alternative splicing
- MicroRNAs are important regulators of gene expression
- Translation of mRNA can be regulated
- Protein stability can be regulated

Alternate splicing of mRNA is a mechanism for generating different proteins from the same DNA (see Figure 11.16). There are only about 24,000 protein coding genes in the human genome, but there are many more human mRNAs than human genes and about 80 percent human genes are alternatively spliced.

Some "non-coding" regions of DNA code microRNAs (miRNAs about 22 bp long) that bind specific mRNAs and block their translation. Each miRNA has dozens of target mRNAs. MiRNAs are transcribed as longer precursors that can fold into double-stranded hairpin structures. The folded miRNAs are then cleaved by dicer protein to produce short double-stranded miRNAs (see Figure 11.17). Those miRNAs are converted to single-stranded RNA by a protein complex, guided by proteins to a target where they bind to and inhibit translation of mRNAs and target them for degradation.

The amount of protein in a cell does not always correlate with the amount of mRNA, so translation of mRNAs in the cytoplasm must sometimes be regulated, or some other process must regulate how long the proteins persist in the cell.

Two other ways that translation of mRNA can be regulated is by modification of the 5′ cap and binding of translational repressor proteins. If the mRNA 5′ cap is not modified, the mRNA will not be translated, and these unprocessed mRNAs can be stored until needed and their 5′ cap will then be modified by GTP. Translational repressor proteins block translation by binding to mRNAs and preventing them from binding to ribosomes (see Figure 11.18).

Protein content of cell is a balance of synthesis and degradation. Proteins are targeted for degradation when an enzyme attaches ubiquitin to a lysine residue of the protein to be destroyed. Subsequently, more ubiquitin chains attach, forming a polyubiquitin chain. This polyubiquitin complex binds to a proteasome complex.

When a protein–ubiquitin complex enters the proteasome, ubiquitin is removed using ATP, the protein is unfolded, and three proteases digest the protein (see Figure 11.19).

Question 8. Although miRNA appear to be evolutionarily ancient and biologically important, what is a drawback to this form of gene regulation? What might favor its persistence?
Textbook Reference: *11.4 Eukaryotic Gene Expression Can Be Regulated after Transcription, pp. 221–222*

Question 9. Describe how a translational repressor functions.
Textbook Reference: *11.4 Eukaryotic Gene Expression Can Be Regulated after Transcription, pp. 222–223, Figure 11.18*

Test Yourself

1. Viruses consist of
   a. a protein core and a nucleic acid capsid.
   b. a cell wall surrounding nucleic acid.
   c. RNA and DNA enclosed in a membrane.
   d. a nucleic acid core surrounded by a protein capsid, and in some cases, a membrane.
   e. a nucleic acid core surrounded by a cell membrane.

   Textbook Reference: *11.1 Several Strategies Are Used to Regulate Gene Expression, pp. 210–211*

2. Lytic bacterial viruses
   a. infect the cell, replicate their genomes, and lyse the cell.
   b. infect the cell, replicate their genomes, transcribe and translate their genes, and lyse the cell.
   c. infect the cell, replicate their genomes, transcribe and translate their genes, package those genomes into viral capsids, and lyse the cell.
   d. infect the cell, transcribe and translate their RNA, replicate their genomes, package those genomes into viral capsids, and lyse the cell.
   e. insert their chromosome into the host chromosome.

   Textbook Reference: *11.1 Several Strategies Are Used to Regulate Gene Expression, pp. 210–211*

3. Retroviruses such as HIV
   a. have DNA as their genome.
   b. are prophages.
   c. copy their RNA genome into DNA using reverse transcriptase.
   d. replicate their genome using RNA polymerase.
   e. can only undergo a lytic infection cycle.

   Textbook Reference: *11.1 Several Strategies Are Used to Regulate Gene Expression, pp. 210–211*

4. An operon
   a. is regulated by a repressor binding at the promoter.
   b. has structural genes that are all transcribed from same promoter.
   c. has several promoters, but all of the structural genes are related biochemically.
   d. is a set of structural genes that are all under the same translational regulation.
   e. is transcribed when RNA polymerase binds the operator.

   Textbook Reference: *11.2 Many Prokaryotic Genes Are Regulated in Operons, p. 213*

5. If the gene encoding the *lac* repressor is mutated so that the repressor can no longer bind the operator, will transcription of that operon occur?
   a. Yes, because the repressor transcriptionally activates the *lac* genes.
   b. Yes, but only when lactose is present.
   c. No, because RNA polymerase is needed to transcribe the genes.
   d. Yes, because RNA polymerase will be able to bind the promoter and transcribe the operon.
   e. No, because cAMP levels are low when the repressor is nonfunctional.

   Textbook Reference: *11.2 Many Prokaryotic Genes Are Regulated in Operons, pp. 213–214, Figure 11.8*

6. If the gene encoding the *trp* repressor is mutated such that it can no longer bind tryptophan but can still bind the operon, will transcription of the *trp* operon occur?
   a. Yes, because the *trp* repressor can bind the *trp* operon and block transcription only when it is bound to tryptophan.
   b. No, because this mutation does not affect the part of the repressor that can bind the operator.
   c. No, because the *trp* operon is repressed only when tryptophan levels are high.
   d. Yes, because the *trp* operon can allosterically regulate the enzymes needed to synthesize the amino acid tryptophan.
   e. No, because the repressor will be continuously bound to the operator.

   Textbook Reference: *11.2 Many Prokaryotic Genes Are Regulated in Operons, p. 214, Figure 11.9*

7. Transcriptional regulation in prokaryotes can occur by
   a. a repressor binding an operator and preventing transcription.
   b. proteins that direct the RNA polymerase to specific promoters.
   c. activator proteins that bind to DNA elements near the promoter and promote transcription.
   d. the control of promoter efficiency.
   e. All of the above

   Textbook Reference: *11.2 Many Prokaryotic Genes Are Regulated in Operons, pp. 214–215*

8. How does the Tat protein in HIV disable the host cell negative regulatory system?
   a. It binds to the RNA polymerase
   b. It binds to the host mRNA to block the host terminator genes.
   c. It binds to the host terminator protein complex.
   d. It binds to the viral mRNA to block the terminator protein
   e. It binds to the viral mRNA and to proteins associated with the RNA polymerase.

   Textbook Reference: *11.1 Several Strategies Are Used to Regulate Gene Expression, pp. 211–212*

9. Imagine that the TATA box for gene *X* becomes highly methylated. How will this affect the expression of gene *X*?
   a. There will be no effect.
   b. Gene *X* will be transcribed but not translated.
   c. Gene *X* will be transcribed if the transcription factors receive the appropriate environmental signal.
   d. Gene *X* will not be transcribed or translated.
   e. Gene *X* will be transcribed if the histones become acetylated.

   Textbook Reference: *11.3 Eukaryotic Genes Are Regulated by Transcription Factors and DNA Changes, p. 219*

10. Imagine that gene *X* is moved to a part of the chromosome where the histones are highly acetylated. How will this affect its expression?
  a. There will be no effect.
  b. It will be transcribed but not translated.
  c. It will be transcribed if the transcription factors receive the appropriate environmental signal.
  d. It will not be transcribed or translated.
  e. It will be transcribed if the histones become deacetylated.
  Textbook Reference: *11.3 Eukaryotic Genes Are Regulated by Transcription Factors and DNA Changes, p. 220*
11. Which of the following is an example of regulation of eukaryotic transcription?
  a. Iron binding the repressor protein for the ferritin mRNA and increasing ferritin expression
  b. Proteostome breakdown of protein–ubiquitin complexes
  c. MicroRNAs binding their target mRNA and causing their degradation
  d. Alternate splicing of an mRNA transcript
  e. Activator proteins binding an enhancer
  Textbook Reference: *11.3 Eukaryotic Genes Are Regulated by Transcription Factors and DNA Changes, pp. 216–221*
12. Which of the following statements about histone modifications is *false*?
  a. They cause some genes to be transcriptionally activated.
  b. They can result in the repression of gene transcription.
  c. They are reversible.
  d. They cause Barr bodies to form.
  e. All of the above are false, none is true.
  Textbook Reference: *11.3 Eukaryotic Genes Are Regulated by Transcription Factors and DNA Changes, p. 219*
13. What would happen initially to cells that lack a functional ubiquitin?
  a. Translation of proteins would be more efficient.
  b. Transcriptional initiation would increase.
  c. Protein degradation would decrease.
  d. Histone modifications would increase.
  e. Nothing would happen.
  Textbook Reference: *11.4 Eukaryotic Gene Expression Can Be Regulated after Transcription, p. 223*
14. Which of the following is *not* a means by which translation can be regulated in a eukaryotic cell?
  a. Inhibition of translation with miRNAs
  b. Modification of the 5′ cap.
  c. Translational repressor proteins
  d. Ferritin mRNA
  e. Alternative splicing
  Textbook Reference: *11.4 Eukaryotic Gene Expression Can Be Regulated after Transcription, pp. 221–223*
15. Which of the following statements about miRNA is *false*?
  a. They are usually about 22 nucleotides long.
  b. They are double stranded RNA's.
  c. They are complementary to their target mRNA.
  d. They are translation inhibitors
  e. They each have dozens of mRNA targets.
  Textbook Reference: *11.4 Eukaryotic Gene Expression Can Be Regulated after Transcription, p. 222*

Answers

Key Concept Review

1. Viral genomes use positive regulation of transcription including a promoter that binds host RNA polymerase. They also act at the posttranscription level by using virally encoded enzymes to break down host mRNA before it can be translated.
2. The surface receptor on their immune system cells may possess a variant of the CD4 receptor that the virus does not recognize. The terminator protein of the host human may not be recognizable to the viral Tat and thus may be able to terminate the transcription of viral DNA.
3. Yes. The repressor gene can be transcribed and translated from the plasmid DNA, and normal regulation will be restored.
4. No. The operator site on the plasmid cannot restore regulation unless it recombines with the host operator site in such a way that it replaces the mutant operator on the host chromosome. The DNA site on the plasmid would bind repressor, but because that site is not adjacent to the promoter or the structural genes on the chromosome, normal regulation of those genes cannot occur.
5. You will need a promoter that binds transcription factors (such as a TATA box), an RNA polymerase, and regulatory binding sites that bind activator proteins. It will also have to be put it into a chromosomal region that has not been silenced by condensed nucleosomes.
6. The plant will require stress response elements (SREs) in front of the promoter for gene *Y*. SREs are bound by transcription factors that are sensitive to drought, and genes with SRE sequences in front of their promoters can be coordinately regulated.

7.

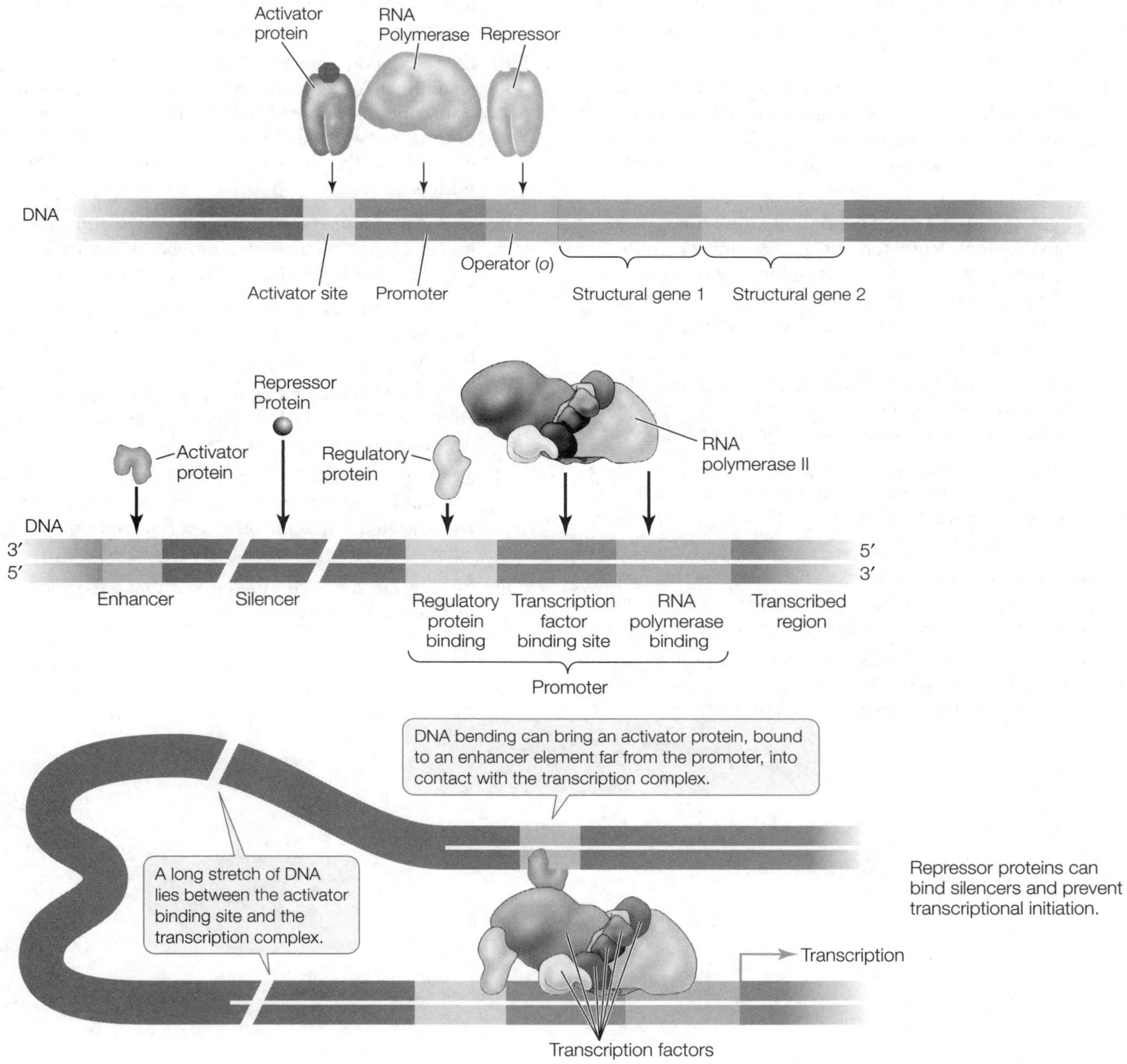

8. miRNA degrades mRNA which means that transcription is already complete, so the cell has wasted energy. However, because of its small size and the potential for many combinations of miRNA and protein complex guides, it provides a system for fine-tuning regulation at the posttranscription level.

9. A translational repressor binds to the mRNA and prevents it from attaching to the ribosome. The repressor can be removed via allosteric regulation. For example, free iron ions that are necessary in low concentrations but toxic in high concentrations, will bind to the repressor when in high concentrations, thereby causing it to release the mRNA and permitting translation of ferritin. Ferritin binds and sequesters excess iron ions, and once they are in low concentrations, there will be few to bind the repressor and translation will cease.

Test Yourself

1. **d.** Nucleic acids do not form capsids; cell walls are found in bacterial and plant cells, not viruses. Viruses are organized so that the nucleic acid is surrounded by protein (not membranes), and a membrane may surround the protein capsid.

2. **d.** This sequence includes the most complete details of the viral life cycle. Viral transcription and translation have to occur first so that viral gene products needed for viral replication will be synthesized.

3. **c.** Answer b. describes a provirus, which is a bacterial virus that has inserted its genome into a host chromosome. Animal viruses replicate their RNA genomes using reverse transcriptase and do not lyse the cells they infect.

4. **b.** An operon is a set of genes that are all transcribed from the same promoter, which is the site where RNA polymerase binds. The repressor binds at the operator site, which overlaps the promoter. Answer d is not correct because the operon is regulated transcriptionally, not translationally.

5. **d.** If the *lac* repressor is nonfunctional, it cannot bind the operator site, and transcription of the *lac* operon will occur at all times, whether or not lactose is present.

6. **a.** If the repressor can no longer bind tryptophan, then it cannot bind the operator, and transcription of the *trp* operon will always be on, whether tryptophan levels in the cell are high or low.

7. **e.** Answer a. refers to the *lac* and *trp* repressors, answer b. to the sigma factors, and answer c. to promoters proteins. Answer d. refers to the *lac* operon.

8. **c.** The Tat complex binds the terminator proteins so they cannot interfere with the RNA polymerase and it can complete transcription of the HIV DNA.

9. **d.** Gene *X* will not be transcribed or translated, since methylation sites on DNA are transcriptionally inactive.

10. **c.** Acetylation leads to loosening of the nucleosomes, resulting in more DNA sequence being accessible for transcription. If transcription factors receive the appropriate environmental signal, transcription will occur. Deacetylation leads to tighter packing of the nucleosomes.

11. **e.** Answer a. refers to translational regulation; answer b. refers to the regulation of protein longevity; answers c. and d. refer to posttranscriptional regulation.

12. **d.** The Barr body is heterochromatin that is heavily methylated DNA. All of the other statements about histone modifications are true.

13. **c.** In the absence of ubiquitin, protein degradation would decrease, since ubiquitin targets proteins for degradation in the proteasome.

14. **e.** Alternative splicing occurs after primary transcription but before mature messenger RNA is ready to be translated. All of the other processes are means by which translation is regulated.

15. **b.** They are single stranded RNAs that are complementary to their target RNAs.

Genomes 12

The Big Picture

- The science of genomics is the study and comparison of genomes from different organisms and is used to identify organisms, genes, potential genes, and their regulatory sequences. Information from these studies has been used to design new medicines to combat pathogens and to understand the minimal requirements for a cell to sustain life. Knowing the DNA sequence of an organism and the genes it contains permits comparisons of normal and mutant genes and the study of gene product variants (usually proteins).
- The science of proteomics focuses on the regulation of protein production and the consequences of changes to the proteins expressed by different genes and alternative splicing.
- Metabolomics is the study of the active metabolites in the cell (primary and secondary metabolites) that control cell functions, and how metabolite pools change under different environmental and developmental conditions.
- There are many motivations for these sub-disciplines including better understanding of diseases and disabilities as well as improvement of agriculture, pharmacology, and environmental reclamation. Some diseases in humans can be directly related to a change in the genome that causes the production of an abnormal gene product. Other molecular diseases are affected by the individual's environment. In agriculture, understanding the production of disease resistance compounds or how to enhance expression of inserted genes are areas of research. Understanding the molecular alterations in a particular gene product and their effects can lead to therapeutic approaches, new diagnostic tools, and preventative measures in health care, agriculture and many other areas.

Study Strategies

- It may be difficult to understand why so little of the eukaryotic genome codes for proteins. However, noncoding regions (transposons, introns, and repetitive regions of the chromosome) may have played a role in creating new functional genes over the course of evolution.
- There are many methods for analyzing genomes and proteomes, including those of pharmagenomics and metagenomics. Review what kinds of sequences are analyzed by each of these methods and give examples of how those analyses have furthered our understanding of the functions of genes, proteins, and cells in different organisms.
- Describe the steps that must be followed in order to generate a genome sequence from a particular organism.
- Comparing genomes between different organisms has allowed us to determine which genes are essential for all cells and which genes are designated for specific adaptive functions. Compare prokaryotic and eukaryotic genomes and list classes of genes that are essential for both types of organisms and classes of genes that are specialized for each type of organism.
- Compare the information and applications of genome and proteome data.
- List the agricultural, medical, and environmental benefits of genome sequencing.
- Go to your BioPortal.com to review the following tutorial and activity:

  Animated Tutorial 12.1 Next-Generation Sequencing

  Web Activity 12.1 Concept Matching

Key Concept Review

12.1 There Are Powerful Methods for Sequencing Genomes and Analyzing Gene Products

New methods have been developed to rapidly sequence DNA

Genome sequences yield several kinds of information

Phenotypes can be analyzed using proteomics and metabolomics

The Human Genome Project, completed in 2003, sequenced all 3.3 billion base pairs in haploid cells. At the same time, genomes from model organisms across all domains of life have been sequenced allowing general comparisons.

To sequence genomic DNA using current methods, it must first be cut into smaller (100-bp) fragments. DNA fragments are denatured using heat, creating single-strand DNA templates. These fragments are attached to short adapter sequences that are mounted on a solid surface. Each fragment is then amplified by PCR. Synthesis of DNA complementary to each tethered, amplified fragment is carried out one nucleotide at a time using universal primers, DNA polymerase and the four nucleotides. Each nucleotide is labeled with a different color fluorescent tag. Unused nucleotides are removed after each nucleotide addition. A camera records the color of the fluorescent tag identifying the nucleotide after each addition, building a color-coded sequence of the DNA fragment. This powerful method is fully automated, can process millions of fragments simultaneously (is massively parallel), and is fast and cheap. Determining the entire genome sequence from these fragments is possible because the fragments are overlapping.

Bioinformatics was developed to reconstruct whole genome sequences from DNA fragment sequences and to analyze sequence information using complex mathematics and computer programs.

Two fields of research utilize genome sequence data. Functional genomics involves identification of functional areas of the genome. In comparative genomics the genomes of different organisms are compared to understand evolutionary relationships among the organisms as well as to discover their similar areas of function.

Proteomics is the study of the variation and interactions of proteins translated in an organism. The total amount of proteins produced by an organism is called its proteome. Two common methods used together to study the proteome are separation through gel-electrophoresis and mass spectrometry, which determines mass and structure of proteins. Both methods are used to isolate, analyze and characterize expressed proteins.

Metabolomics is the study of the metabolites in a cell or organism, focusing on their function in variable cellular environments. Primary metabolites are involved in normal cell processes like glycolysis and hormone signaling, and secondary metabolites are involved in special responses to the environment and can be unique to the organism. These include antibiotics made by bacteria and disease resistance and other defense compounds made by plants.

Together, genomics, proteomics and metabolomics increase our understanding of the genotype reaction to the environment that produces the phenotype.

Question 1. Diagram the steps required to sequence a DNA molecule using the next generation sequencing method described in the textbook. In your diagram be sure to list all of the necessary components and technologies that are utilized for DNA sequencing.
Textbook Reference: *12.1 There Are Powerful Methods for Sequencing Genomes and Analyzing Gene Products, pp. 227–228*

Question 2. Suppose you possess the sequenced genome for a prokaryote and you are looking for the gene(s) that code for a given protein. How would you begin your search?
Textbook Reference: *12.1 There Are Powerful Methods for Sequencing Genomes and Analyzing Gene Products, pp. 229–230*

12.2 Prokaryotic Genomes Are Relatively Small and Compact

Prokaryotic genomes are compact

Metagenomics allows us to describe new organisms and ecosystems

Will defining the genes required for cellular life lead to artificial life?

The genomes of prokaryotes and archaea are generally small (160,000 to 2 million bp). They are organized in a single circular chromosome and are compact (85 percent of the DNA codes proteins or RNA). Smaller circular DNA molecules, called plasmids, are often present in addition to the main chromosome. Most prokaryote genes do not possess introns, but archaeal rRNA and tRNA genes often have introns. Other than these general similarities, these groups possess varied genomes, reflecting the broad range of environments they inhabit.

Functional genomic analysis has been used to assign functions to gene products. Genes can be identified that are involved in the prokaryote's metabolism, transport, and its infectious properties. Prokaryotes with smaller genomes tend to have fewer types of proteins dedicated to a given function (see Table 12.1, p. 232).

Comparative genomics is being used to examine the differences among the genomes of prokaryotes and correlate that information with functional differences between the species. This provides insight about requirements for gene regulation and function, and ultimately how these organisms adapt to different environments.

Genome sequencing has enabled scientists to understand more about transposons (transposable elements), which are small (1,000–2,000 bp) mobile genetic elements that can move from one site in the genome to another. They are often called "jumping genes." Transposons can affect the phenotype of the organism when they insert into a gene's coding region, which alters the coding sequence. Some transposons carry the genes for antibiotic resistance.

Metagenomics is used to analyze gene sequences from complex samples without isolating individual organisms. The PCR is used to amplify specific sequences from an environmental sample. Using this method, thousands of new viruses and bacteria have been found in seawater, marine sediment, and mine water runoff (see Figure 12.7). It is estimated that 90 percent of the microbial world was invisible to microbiologists prior to metagenomic sequence analysis methods.

Some genes or gene segments appear universal, or nearly so, among all genomes compared to date. An example is the sequence that codes for ATP binding sites on proteins. Comparative genomics can be used to determine the minimal number of genes needed for life. The *Mycoplasma genitalium* genome has been mutated to determine the smallest number of genes needed for survival, which is 382 genes, and experiments are now under way to make synthetic genomes based on that of *M. discoides.* This new knowledge has the potential to help us create new microbes to degrade oil spills, reduce tooth decay, or convert cellulose to ethanol for use as fuel.

Question 3. Genomics has revealed that the bacterium that causes tuberculosis has more than 250 genes that metabolize lipids. What does this finding suggest about the bacterium and about medical approaches to fight this disease?
Textbook Reference: *12.2 Prokaryotic Genomes Are Relatively Small and Compact, pp. 231–232*

Question 4. Some transposons carry the genes for antibiotic resistance when they move. By what mechanism could this happen?
Textbook Reference: *12.2 Prokaryotic Genomes Are Relatively Small and Compact, p. 232*

Question 5. You have isolated a new strain of bacteria from a contaminated piece of Swiss cheese. Diagram how you would use transposon mutagenesis to find out the minimal number of genes in the bacterium's genome that are necessary for its survival on Swiss cheese. Include the results you would expect for an essential gene and for a nonessential gene.
Textbook Reference: *12.2 Prokaryotic Genomes Are Relatively Small and Compact, p. 234*

12.3 Eukaryotic Genomes Are Large and Complex

Model organisms reveal many characteristics of eukaryotic genomes

Gene families exist within individual eukaryotic organisms

Eukaryotic genomes contain many repetitive sequences

Eukaryotic genomes are larger, and have more protein-coding genes and regulatory sequences, than prokaryotic genomes. This reflects the greater cellular and organismal complexity of eukaryotes compared to prokaryotes. Eukaryotes have multiple chromosomes. Much of eukaryotic DNA is noncoding, which may be introns, regulatory sequences, or repetitive sequences.

Model eukaryotic organisms include *Saccharomyces cerevisiae, Caenorhabditis elegans, Drosophila melanogaster, Arabidopsis thaliana*, and *Oryza sativa* (see genome comparisons in Table 12.2).

Saccharomyces cerevisiae (budding yeast) is a single-celled eukaryote that has 16 linear chromosomes. The genomes of *S. cerevisiae* and *E. coli* have the same number of genes for the basic functions of cell survival, but *S. cerevisiae* has 5,770 total genes while *E. coli* has 4,377. Many of the additional genes possessed by *S. cerevisiae* code for products involved in protein targeting and organelle function required for eukaryotic compartmentalization.

The soil dwelling nematode *Caenorhabditis elegans* is a simple multicellular organism used to study development. The worm's body is transparent, and its growth from a fertilized egg to a differentiated adult with 959 cells takes just three days. The genome of *C. elegans* is eight times larger than that of yeast and has 3.5 times the number of protein-coding genes (19,427 proteins). Gene inactivation studies have revealed that the worm can survive in laboratory culture with only 10 percent of those genes.

The fruit fly *Drosophila melanogaster* has ten times more cells than *C. elegans*. Its genome has more DNA but it contains fewer genes than *C. elegans*. *D. melanogaster* has many genes that code for transcription factors needed to regulate its more complex development, which includes egg, larva, pupa, and adult stages (see Figure 12.9 for distribution of gene functions in *D. melanogaster*).

The genome of the *Arabidopsis thaliana* plant has 28,000 protein-coding genes. Many of these genes are duplicates of each other. When the duplicates are removed, only 15,000 unique genes remain and many of these are very similar to genes found in nematodes and fruit flies. *Arabidopsis* contains genes unique to plants, including genes involved in photosynthesis, water transport into the root, cell-wall synthesis, uptake and metabolism of inorganic substances, and defense against herbivores. Many of the genes in *Arabidopsis* can be found in rice, *Oryza sativa,* and poplar, *Populus trichocarpa*. (see Figure 12.10 for overlap among plant genomes.)

Eukaryotes have gene families—sets of duplicate or closely related genes (such as the globin genes). Gene families provide the organism with a functional gene while allowing mutations in other members of the gene family. Some of these mutations may create new gene variants that are advantageous to the organism while others may create pseudogenes. Pseudogenes are nonfunctional because they lack promoters and/or recognition sites for intron removal.

In humans, the globin gene family consists of three α-globin genes and five β-globin genes (see Figure 12.11). Different globins are expressed at different times in development. λ-globin, which is expressed in the fetus, binds oxygen more tightly to ensure oxygen transfer from the mother across the placenta to the fetus.

Highly repetitive sequences are short (less than 100 bp) and are repeated thousands of times in tandem in the genome. They can be densely packed in heterochromatin regions or scattered around the chromosome (as seen for short tandem repeats, STRs). The number of repeats at a given location in the genome varies among individual organisms and is heritable, which provides unique molecular markers that can be used to identify individuals.

Moderately repetitive DNA sequences include sequences that are repeated 10–1,000 times and include tRNA and rRNA genes. Multiple copies of the tRNA and rRNA genes are needed to provide the cell with high concentrations of components needed for protein translation.

Most moderately repetitive sequences are not stably integrated into the DNA but can move from place to place in the genome via transposons. Transposons make up about 40 percent of the human genome. There are two main types of transposons in eukaryotes: retrotransposons and DNA transposons (see Table 12.2).

Retrotransposons make an RNA copy of themselves which is then re-copied into DNA before it is inserted into a different site in the genome. There are two kinds of retrotransposons: LTR and non-LTR. LTR retrotransposons have long terminal repeats (LTR) of DNA at each end. Non-LTR retrotransposons lack these repeats.

Non-LTR retrotransposons are further subdivided into SINEs and LINEs. SINEs are *s*hort *in*terspersed *e*lements of up to 500 bp; they are transcribed but not translated. They include the 300-bp *Alu I* element, of which there are a million copies that accounts for 11 percent of the human genome. LINEs are *l*ong *in*terspersed *e*lements of up to 7,000 bp and some *are* transcribed and translated. LINEs make up about 17 percent of the human genome.

DNA transposons do not replicate when they move to a new site on the chromosome, nor do they use RNA as an intermediate. Like some prokaryote transposons, they may be excised and moved to a new location without being replicated.

Question 6. Compare and contrast gene families with repetitive sequences in eukaryote genomes.
Textbook Reference: *12.3 Eukaryotic Genomes Are Large and Complex, pp. 236–238*

Question 7. Which model organism would you use to study the functional genomics of neuron signaling and why?
Textbook Reference: *12.3 Eukaryotic Genomes Are Large and Complex, pp. 235–236*

12.4 The Human Genome Sequence Has Many Applications

The human genome sequence held some surprises

Human genomics has potential benefits in medicine

DNA fingerprinting uses short tandem repeats

The human genome contains about 24,000 protein-coding genes. This is a small number compared with the number of proteins in humans, so posttranscriptional mechanisms such as alternative splicing must provide the additional proteins.

The average gene has 27,000 base pairs although the size ranges from 1,000 to 2.4 million bp. Virtually all human genes have numerous introns and at least 3.5 percent of the genome is functional but noncoding, having roles in gene regulation. Over 50 percent of the genome contains highly repetitive sequences.

Even though almost all genes (97 percent) are the same in all people, scientists have mapped over 7 million single nucleotide polymorphisms (SNPs) in humans. Single nucleotide polymorphisms are when two or more sequences differ by one nucleotide; for example, if one section of a non-coding DNA strand is ATTGCGC in one person and ATAGCGC in another person, that locus is polymorphic in the population.

Comparisons of genes from different organisms have revealed evolutionary relationships (see Figure 12.12). 95 percent of the human genome is shared with the chimpanzee.

Because complex phenotypes such as susceptibility to disease are determined by multiple genes interacting with the environment, human genomics has potential benefits in medicine. One approach taken by medical genetics researchers is haplotype mapping. This method uses SNPs that are linked (inherited together because they are located close together on individual chromosomes). The piece of chromosome with the SNPs is called a haplotype.

Over 500,000 SNPs from different individuals can be placed on a chip and used to analyze disease states. Statistical measures of association of SNP data can be used to determine increased risk for particular diseases (see Figure 12.13). SNP testing will eventually be replaced with DNA sequencing.

Pharmagenomics is the study of how an individual's genome can affect his or her response to drugs or other outside agents (see Figure 12.14). The enzyme variants produced by different individuals may impact the activity of drugs that are processed by the enzyme, so doctors can use this information to tailor drug dose to the individual.

The proteome is more complex than the genome. Because of alternative splicing and posttranslational modifications, the sum total of proteins produced (the proteome) is more complex than the genome. Comparative proteome analysis has revealed a common set of proteins that provide the basic metabolic functions of a eukaryotic cell in humans, worms, flies, and yeast. The unique proteins in each organism may result from a reshuffling of the same domains that exist in proteins from other organisms.

Metabolomics is being used to learn about how the levels of different metabolites influence human physiology, and may help in development of new tools for disease diagnosis and management.

Question 8. Analysis of the human genome has revealed many genes that cause disease when mutated. Now that we know the genes that are involved, why aren't these diseases eliminated?
Textbook Reference: *12.4 The Human Genome Sequence Has Many Applications, pp. 239–240*

Question 9. Why do different organisms have more similarities in their proteomes than in their genomes?
Textbook Reference: *12.4 The Human Genome Sequence Has Many Applications, pp. 240–241*

Test Yourself

1. Functional genomics
   a. assigns functions to the products of genes.
   b. assigns functions to regulatory sequences.
   c. compares genes in different organisms to see how those organisms are related physiologically.
   d. Both a and c
   e. All of the above
   Textbook Reference: *12.2 Prokaryotic Genomes Are Relatively Small and Compact, pp. 231–232*
2. Comparative genomics
   a. assigns functions to the products of genes.
   b. assigns functions to regulatory sequences.
   c. compares genes in different organisms to see how those organisms are related physiologically.
   d. Both a and c
   e. All of the above
   Textbook Reference: *12.2 Prokaryotic Genomes Are Relatively Small and Compact, pp. 231–232*
3. Which of the following is *not* a way that proteomic studies identify proteins that are coded by specific genes?
   a. Reading the genetic code that specifies the amino acid sequence
   b. Using PCR to amplify the protein
   c. Separation of the expressed protein by gel electrophoresis and then analyzing and sequencing the protein
   d. Knock out studies that prevent gene expression
   e. Mass spectrometry characterization of the expressed protein and comparisons with known proteins
   Textbook Reference: *12.2 Prokaryotic Genomes Are Relatively Small and Compact, p. 230*
4. Genes that cause cancer when mutated
   a. can be analyzed by means of linked markers such as SNPs.
   b. can be analyzed to determine which cancer treatment will work the best for an individual.
   c. do not have any homologs in other organisms.
   d. can help us predict which individuals are more likely to develop cancer.
   e. a, b, and d
   Textbook Reference: *12.4 The Human Genome Sequence Has Many Applications, pp. 240–241*
5. Sequencing of the human genome has allowed scientists to
   a. understand regulatory sequences that are important for gene expression.
   b. locate genes that cause disease.
   c. understand evolutionary relationships by comparing human genes to genes in other organisms.
   d. investigate gene families and their origins.
   e. All of the above
   Textbook Reference: *12.4 The Human Genome Sequence Has Many Applications, pp. 240–241*
6. Proteomics has been used to compare
   a. DNA sequences between closely related species.
   b. gene expression during embryonic development.
   c. protein sequences between closely related species.
   d. shotgun cloned sequences.
   e. prokaryotic genomes.
   Textbook Reference: *12.4 The Human Genome Sequence Has Many Applications, p. 240*
7. Which of the following is *not* a component of next generation sequencing?
   a. Cutting DNA into 100-bp fragments
   b. cDNA cloning
   c. Computer alignment of overlapping pieces of chromosomes
   d. Sequencing of DNA with dideoxy nucleotides
   e. Bioinformatics
   Textbook Reference: *12.1 There Are Powerful Methods for Sequencing Genomes and Analyzing Gene Products, p. 227*
8. Which of the following statements about transposable elements is *false*?
   a. They can inactivate genes into which they are inserted.
   b. They can contain gene sequences.
   c. They are mobile genetic elements that move from RNA molecule to RNA molecule.
   d. They may be spliced out of one region of the genome and inserted into another.
   e. They replicate themselves before moving to another site on the genome.
   Textbook Reference: *12.2 Prokaryotic Genomes Are Relatively Small and Compact, pp. 231–232*

9. Gene inactivation studies have allowed us to
   a. determine the minimal number of genes humans need to survive.
   b. demonstrate that *C. elegans* needs most of its genes.
   c. investigate the minimal number of genes needed to sustain life.
   d. create artificial life in a test tube.
   e. All of the above

   Textbook Reference: *12.2 Prokaryotic Genomes Are Relatively Small and Compact, p. 234*

10. Comparisons of yeast and bacterial cell genomes have revealed that
    a. yeast cells have more genes devoted to the basic functions of survival than bacteria do.
    b. eukaryotic cells are structurally similar to bacterial cells in terms of complexity.
    c. there are more genes for targeting proteins to organelles in yeast than in bacteria.
    d. the histones of bacteria are very similar to those of yeast.
    e. bacteria and yeast are both haploid.

    Textbook Reference: *12.3 Eukaryotic Genomes Are Large and Complex, pp. 235–236*

11. Which one of the following does *not* represent information that we have gained from genome sequencing?
    a. Mycobacteria have many genes devoted to metabolizing fats.
    b. There is extensive genetic exchange between different kinds of bacteria.
    c. Genomes can be sequenced even when organisms cannot be cultured.
    d. Much of the eukaryotic genome contains coding sequences.
    e. The *Drosophila* genome contains many genes encoding transcription factors.

    Textbook Reference: *12.3 Eukaryotic Genomes Are Large and Complex, pp. 235–236*

12. Which of the following was *not* one of the discoveries that resulted from the genome sequencing of plants?
    a. There are more protein-coding genes in animals than plants.
    b. Many *Arabidopsis* genes are duplicated due to chromosomal rearrangements.
    c. There are more genes in plants that are similar to each other than there are genes that are unique.
    d. Plants have many genes whose products are used for defense against microbes and herbivores.
    e. All of the above were discoveries.

    Textbook Reference: *12.3 Eukaryotic Genomes Are Large and Complex, pp. 235–236*

13. Which of the following statements about eukaryote retrotransposons is *false*?
    a. They are translated before they are transcribed.
    b. They use an RNA intermediate to move from one region of the genome to another.
    c. They are highly repetitive sequences found throughout the genome.
    d. They can encode gene products that are required for their own transposition.
    e. All of the above are true.

    Textbook Reference: *12.3 Eukaryotic Genomes Are Large and Complex, p. 230*

14. Which of the following statements about the human genome is *false*?
    a. Over 50 percent of the genome contains transposons.
    b. Almost every gene has introns.
    c. About 97 percent of the genome is shared among human individuals while about 90 percent is shared with chimpanzees.
    d. About 2 percent of the genome codes for genes.
    e. Humans have about the same number of genes as fruit flies have.

    Textbook Reference: *12.4 The Human Genome Sequence Has Many Applications, p. 239*

15. Single nucleotide polymorphisms (SNPs)
    a. can be used to map unlinked genes in order to follow the inheritance of disease traits.
    b. can be used to predict if a patient is at risk for a particular disease.
    c. can be linked in order to generate gene sequences.
    d. have limited sequence variations.
    e. All of the above

    Textbook Reference: *12.4 The Human Genome Sequence Has Many Applications, pp. 239–240*

16. The study of proteomes allows scientists to compare
    a. the proteome with the genome to see if the gene sequences are correct.
    b. proteome sequences between species to see if similar proteins are expressed in all species.
    c. transcriptional patterns in different organisms.
    d. highly repetitive DNA sequences in different organisms.
    e. how noncoding regions of the genome differ in different organisms.

    Textbook Reference: *12.1 There Are Powerful Methods for Sequencing Genomes and Analyzing Gene Products, p. 230*

17. Which of the following is *not* a characteristic you would look for when choosing the next "model" plant genome to study agriculturally important traits?
    a. A small genome
    b. One with little redundancy
    c. Ease of laboratory culture
    d. Rapid life cycle
    e. Large numbers of transposons

    Textbook Reference: *12.3 Eukaryotic Genomes Are Large and Complex, pp. 235–237*

18. Which of the following is *not* something you would expect about susceptibility to a disease revealed by haplotype mapping?
    a. The SNP patterns would have a good chance of being altered by transposable elements.

b. They might involve interactions with environmental cues.
c. The susceptibility would be shared by genetically related individuals.
d. Some of the SNP patterns would impact post transcription events.
e. The susceptibility might be greater if the individual was homozygous for the haplotype.

Textbook Reference: *12.4 The Human Genome Sequence Has Many Applications, pp. 239–240*

Answers

Key Concept Review

1. Your list should include the following steps:
   1. Cut the DNA into 100-bp fragments physically or using enzymes to hydrolyze phosphodiester bonds.
   2. Heat the DNA to break the hydrogen bonds holding the two strands together.
   3. Attach each end to a short adapter sequence that is anchored to a bead or flat surface.
   4. Amplify using PCR.
   5. Heat the fragments to denature them.
   6. Add a universal primer complementary to one of the adaptor sequences, DNA polymerase, and the four nucleotides each with an identifying fluorescent dye.
   7. DNA synthesis stops when a nucleotide is incorporated into the replicating strand.
   8. All unincorporated nucleotides are removed.
   9. A camera photographs the color of the added nucleotide.
   10. The dye is removed from the added nucleotide and the process is repeated from step 5.
   11. The sequence of nucleotides is recorded by the color of the nucleotides in the series of photographs taken.
   12. Now use a computer program to analyze the photographs to determine how the sequences of the fragments overlap and to reconstruct the genome sequence.
2. Examine the genome sequence for start and stop codons for translation, which indicate the location of genes. Then check each gene sequence for correct codons to produce the protein's sequence of amino acids.
3. The tuberculosis bacterium must use lipids as a source of energy-rich compounds; a drug that inhibits lipid uptake or metabolism in this bacterium may inhibit its growth.
4. Some transposons become duplicated, with two copies flanking one or more genes. The duplicated transposons together with the genes form a single transposable element. If the flanked genes contain antibiotic resistance factors, they could be transported along with the new large transposable element to different parts of the genome or onto plasmids.
5.

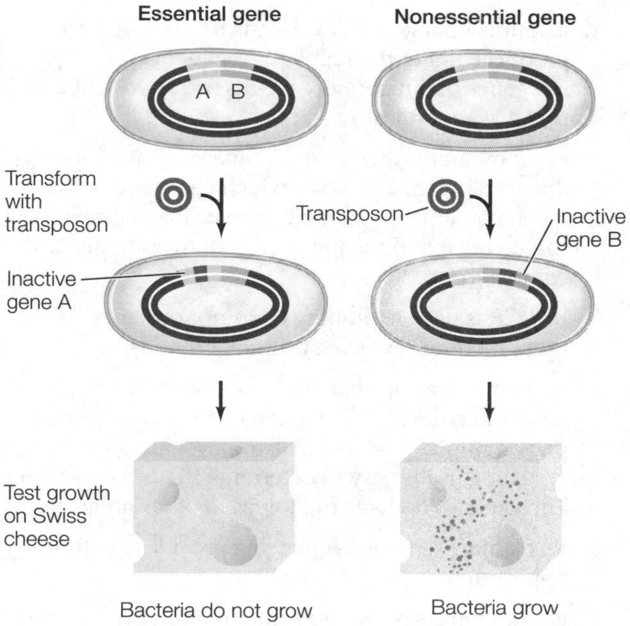

   1. Transform the bacteria with a transposon that can insert into random sites in the genome.
   2. Test subsamples of each transposon-mutated bacteria culture for growth and survival on Swiss cheese.
   3. Isolate DNA from bacteria cultures that show defective growth and survival to determine which gene was inactivated by the transposon. Such genes are likely to be important for growth and survival on Swiss cheese.
6. Gene families are duplicated genes that code for proteins and can include a few to hundreds of copies of the genes in a single genome. Usually different copies contain slight mutations that code for a functional variant of the gene product. Pseudogenes are nonfunctional duplicate genes. Repetitive sequences may be short non-transcribed sequences that exist in thousands of tandem copies or they may be moderately repetitive (10–1,000 repeats) that code for tRNAs and rRNAs.
7. The simplest model organism that includes genes for a nervous system and intercellular communication is the nematode *Caenorhabditis elegans* so this would be a good choice.
8. We have not been able to eliminate genetic diseases because, except in a few rare cases, we are not able to replace the mutated genes in the diseased individual.
9. More sequence similarities are revealed when proteomes are compared than when genomes are compared because the genetic code is redundant. There can be more than one codon for a particular amino acid, so the DNA sequences can vary and still generate the same amino acid sequences.

Test Yourself

1. **a.** Functional genomics assigns functions to gene products, not to the regulatory sequences. Comparing the genes of different organisms is the work of comparative genomics.
2. **c.** Comparative genomics compares genes between different organisms to see which genes are possessed by one organism but not another. These comparisons can be related to the physiology of the organisms being compared.
3. **b.** PCR is used to amplify DNA not proteins. All of the other methods are used to study proteins.
4. **e.** Mutated genes that are linked to cancer can be linked to particular SNPs, can be used to predict if a person might be susceptible to developing cancer, and are being used to try to determine the best treatment. Those genes do have homologs in other organisms.
5. **e.** Human genome sequencing provides all these opportunities.
6. **c.** Proteomics is used to compare protein sequences between different organisms. Answers a, b, and d all refer to techniques used in genomics.
7. **b.** cDNA cloning is used to determine what mRNAs are being expressed in cells. All of the other techniques were utilized to sequence the human genome and order those sequences.
8. **c.** Some transposons do contain gene sequences. They can be spliced out of one region and inserted into another and also replicate themselves and then move to another site. They do not move from one RNA molecule to the next.
9. **c.** Gene inactivation studies have been used in some organisms (not humans) to determine minimal number of genes needed by cells in order to survive. *C. elegans* needs only about 10 percent of its genes to survive. Artificial life has not yet been created in a test tube, but scientists are actively pursuing this possibility.
10. **c.** A comparison of yeast and bacterial cell genomes revealed that the number of genes needed for survival in these two organisms was roughly the same but that yeast cells had many genes for targeting proteins to organelles. Yeast cells are structurally more complex than bacterial cells. Bacterial cells do not have histones. Yeast cells can be haploid or diploid; bacteria are haploid (which had been known long before comparative genomics was developed).
11. **d.** Only 2 percent of the human genome codes for protein-coding genes.
12. **a.** The reverse is true: There are more protein-coding genes in plant cells than in animal cells.
13. **a.** Retrotransposons can move from one region of the genome to another using an RNA intermediate to make a DNA copy of themselves. They are highly repetitive sequences found throughout the genome. Some transposons encode gene products needed for their transposition. Retrotransposons are not translated before they are transcribed.
14. **c.** Humans and chimpanzees share 95 percent of their genomes while individuals within humans share 97 percent.
15. **b.** SNPs that are linked to a disease gene can be used to analyze patients' DNA samples to determine if they are at risk for that disease. SNPs are quite variable. They are small sequences and are not used to generate gene sequences.
16. **b.** Proteomics studies can be used to compare proteins in different organisms. The gene sequence will determine the protein sequence. Proteomic studies cannot be used to study transcriptional patterns, which are cellular processes involving RNA. Non-coding regions of the genome do not specify proteins and highly repetitive sequences do not code for protein.
17. **e.** A plant with large numbers of transposons will be subject to gene interruptions and movement which could complicate studies. A small genome is more manageable and means less redundancy and non-coding DNA and this simplifies search for genes of interest. The ability to be cultivate many generations of a research organism easily minimizes time and effort needed to observe expressed phenotypes.
18. **a.** Because the haplotype is a set of closely linked SNPs on a chromosome, they would not be likely to be interrupted by a transposable element.

Biotechnology 13

The Big Picture

- The ability to isolate DNA from any organism, ligate it to vector DNA, introduce that DNA into host cells, and propagate those cells has had an enormous impact on our understanding of genetics, molecular biology, and cell function and development. These techniques are being used to elucidate evolutionary relationships among different organisms and to understand gene regulation and function in greater depth. These techniques are being used to develop new medicines and diagnostics, agricultural products, and powerful forensic tools.

Study Strategies

- There are a variety of ways to clone DNA fragments, and trying to remember all the different cloning methods can be challenging. Consider the following as you review the different cloning procedures: the size of the cloned DNA, the host cell in which the DNA can be cloned, and the expression of the cloned DNA in the host cell. Different vectors can be used in different host cells to address each of these considerations.
- Many different experimental questions can be answered using cloning. Ask yourself what cloning strategies could be used to answer the following questions, and review the textbook for answers: What is the sequence of a gene? What sequences are important for regulation of that gene? What sequences are important for targeting that gene to a particular site in a eukaryotic cell? What is the difference in function between a mutant gene product and a wild-type gene? What kinds of genes are expressed during the development of an organism? How can cloned genes be expressed in the seeds of plants or in the milk of mammals?
- Outline the specific steps needed to clone a gene in a bacterial cell. Include how the gene is initially isolated, what kind of vectors can be used, how to introduce the recombinant DNA into the cell, and how to confirm that the recombinant molecule is in the host cell. Then outline the steps needed to clone a gene in a eukaryotic cell.
- Complementary base pairing is important for many aspects of biotechnology. Describe each technique in gene cloning that uses complementary base pairing, detailing specifically how base pairing is involved.
- Go to yourBioPortal.com to review the following tutorials and activites:

  Animated Tutorial 13.1 Separating Fragments of DNA by Gel Electrophoresis

  Animated Tutorial 13.2 DNA Chip Technology

  Web Activity 13.1 Expression Vectors

  Working with Data 13.1 Recombinant DNA

Key Concept Review

13.1 Recombinant DNA Can Be Made in the Laboratory

Restriction enzymes cleave DNA at specific sequences

Gel electrophoresis separates DNA fragments

Recombinant DNA can be made from DNA fragments

Recombinant DNA, single molecule DNA sequences made from at least two different organisms, is used for genetic modification of organisms. There are three widely used tools used in construction of recombinant DNA: restriction enzymes for cutting DNA into fragments; gel electrophoresis for the analysis and purification of fragments; and DNA ligase for joining DNA fragments to create novel combinations.

Restriction endonucleases are enzymes that function as defense molecules of bacteria. They target invader bacteriophage double-stranded DNA and inactivate it by cutting it at particular recognition sequences or restriction sites. Each restriction endonuclease recognizes a specific palindromic sequence in DNA and cuts at or near those sites. Some restriction endonucleases cut between the same bases on both DNA strands, producing blunt ends. When restriction endonucleases cut DNA, they often leave ends that have 5′ or 3′ overhangs of single-stranded DNA. These ends are called "sticky ends," and they can form complementary base pairs with other DNA molecules that have the same sticky ends (see Figure 13.3).

The bacterial cell protects itself from these enzymes by methylating their own restriction sites.

DNA fragments cut with the same restriction enzyme can be joined together by ligase, even if they are from different species. DNA ligase (the enzyme that covalently joins the Okazaki fragments during DNA replication and mends broken DNA; see Chapter 9) is used to form a covalent bond on each DNA strand of the recombinant molecule (see Figure 13.3).

The recombinant DNA molecule can be incorporated, using these tools, into the DNA of a vector (such as a plasmid used for bacterial gene insertions). The genes inserted by the vector into a host genome may be expressed by the host organism (see Figure 13.4).

Question 1. Sometimes laboratory technicians may try several restriction enzymes before they find a satisfactory one for the genome of interest. What are two reasons that a specific restriction enzyme might not cleave a given genome?
Textbook Reference: *13.1 Recombinant DNA Can Be Made in the Laboratory, pp. 245–246*

Question 2. Describe how you would create a recombinant plasmid DNA, that includes the gene for production of an enzyme used in drug therapy for insertion into a bacterial host cell.
Textbook Reference: *13.1 Recombinant DNA Can Be Made in the Laboratory, pp. 245–248*

13.2 DNA Can Genetically Transform Cells and Organisms

- Genes can be inserted into prokaryotic or eukaryotic cells
- Recombinant DNA enters host cells in a variety of ways
- Reporter genes are used to identify host cells containing recombinant DNA

Cloned genes are useful for sequence analysis, to produce quantities of their protein products for medicinal or agricultural use, or to create new transgenic organisms.

Recombinant DNA is inserted into a host cell by transformation (or transfection, if the host cell is an animal cell), creating a transgenic cell or organism. Because only a few of the cells exposed to the recombinant DNA are transformed, selectable markers such as genes that confer resistance to antibiotics are often included on the recombinant DNA molecule.

Theoretically, genes can be cloned into any cell or organism. Most research has centered on model organisms. Prokaryotes, especially *E. coli*, have been used to clone many genes and have plasmids that can easily insert genes into the host organism genome. Because of differences in gene expression between prokaryotes and eukaryotes, they are not a suitable research organism for eukaryote genes. Eukaryote gene expression studies are often carried out in yeast cells such as *Saccharomyces* that have a rapid cell division cycle (2–8 hours), are easy to grow, have a small genome size (12 million base pairs), and have been used to clone many eukaryotic genes.

Many plant cells are totipotent, so unspecialized stem cells can be cultured from mature cells. They can be grown in culture, transformed with recombinant DNA, and manipulated to form an entire new transgenic plant containing the recombinant DNA molecule. Recombinant DNA that is incorporated into germ cells will be passsed on to the next generation in the seeds. Recombinant DNA can also be incorporated into cultured animal cells and whole transgenic animals.

Methods for inserting recombinant DNA into host cells vary and can be challenging. The DNA must not only be incorporated into the cell, but also into a replicon, a replication unit containing an origin of replication. This can be accomplished either by inserting the recombinant DNA into the host chromosome, where it is replicated when the chromosome is replicated, or recombinant DNA molecules can enter the host cell as part of a vector that already has an origin of replication. Vectors can be plasmids or viruses. Plasmids are useful because they can replicate independently, have one or more restriction sites where DNA can be cut to insert genes, often contain selectable markers such as bacterial resistance, and are small in size.

Plasmids are capable of making many copies of plasmid DNA per cell. Scientists have produced desirable combinations of unique restriction enzyme sites, origins of replication for specific host organisms, and a variety of reporter and selectable marker genes on strains used in the laboratory.

Agrobacterium tumefaciens, a bacterium that causes crown gall disease in plants, harbors a Ti (tumor inducing) plasmid. This plasmid contains a segment, T DNA, which inserts itself into the host plant cell's DNA when the bacteria infect the plant. This plasmid has been modified to be non-pathogenic and is widely used to insert desirable genes into plant genomes.

Plasmid replication limits size of DNA insertions to about 10,000 base pairs. Most eukaryotic genes are larger than this. For larger DNA sequences (up to 20,000 bp of inserted DNA), virus vectors are used. Viruses infect cells naturally, allowing easy entry of cloned sequences into the cytoplasm of the cell.

Reporter genes identify host cells that contain recombinant DNA. These are important because not all vector copies contain the recombinant DNA and only a small proportion of potential host cells actually takes up the vector. Selectable markers such as antibiotic resistance can be used to determine if the host cell contains the recombinant DNA molecule (see Figure 13.5). Other types of reporter genes include those for β-galactosidase and green fluorescent protein (GFP), which can be used as visual markers to detect uptake of recombinant DNA molecules in host cells. Reporter genes can also be attached to promoters of gene coding regions to visualize transcription or protein localization in cells.

Question 3. What are the critical elements needed for a useful plasmid vector for recombinant DNA?
Textbook Reference: *13.2 DNA Can Genetically Transform Cells and Organisms, p. 249*

Question 4. What are the constraints on the use of plasmids as vectors for inserting recombinant DNA into cells and organisms?
Textbook Reference: *13.2 DNA Can Genetically Transform Cells and Organisms, p. 250*

13.3 Genes and Gene Expression Can Be Manipulated

- DNA fragments for cloning can come from several sources
- DNA mutations can be made in the laboratory
- Genes can be inactivated by homologous recombination
- Complementary RNA can prevent the expression of specific genes
- DNA microarrays reveal RNA expression patterns

One of the reasons for cloning DNA is to study its function, including the proteins it codes and their regulatory sequences. Sources of DNA for cloning include genomic libraries, cDNA libraries, and synthetic data.

Genomic libraries are a collection of DNA fragments from the entire genome of an organism. After the DNA is cut or broken into fragments, each fragment is inserted into a vector, which is used to insert the gene into bacteria. DNA hybridization with colonies of transformed bacteria can identify fragments that contain particular coding sequences.

cDNA libraries consist of all the genes transcribed in a particular tissue. To make cDNA (complementary DNA), the messenger RNA must first be isolated from the cell. A DNA molecule complementary to that RNA is synthesized using reverse transcriptase, and the new molecule can then be cloned (see Figure 13.7). cDNA clones are used to compare gene expression in different tissues at different stages of development.

Synthetic DNA can be made using PCR to amplify a specific sequence. Artificial genes can be made if the amino acid sequence of the gene is known. Sequences for transcriptional and translational initiation and termination can be added to the gene sequence.

Recombinant technology permits generation of mutant genes in the lab. Mutants can be synthesized and compared to wild-type genes to analyze gene function.

Another method of studying gene function is to eliminate gene expression by creating knockout genes. This can be done in a variety of ways, one of which is called homologous recombination. In this case (used in the mouse model) a normal gene is replaced by an inactivated form of the gene.

In constructing transgenic mice to test the function of a gene, the gene of interest is disrupted by the insertion of a reporter or marker gene into the middle of the native gene. A plasmid containing the inactivated gene with the marker is transfected into a mouse stem cell (see Figure 13.8). If recombination occurs, the marker gene will be expressed. The transfected stem cell is then transplanted into an early mouse embryo, and the resulting phenotype is analyzed.

Antisense messenger RNA and RNAi can prevent the translation of specific genes. Antisense RNA will base pair with mRNA in the cytoplasm and form a double-stranded RNA molecule, which cannot be translated and will be degraded by the cell. Interference RNAs, known as small interfering RNAs (siRNAs), are short (about 20 nucleotides) double-stranded RNA molecules that bind specific mRNAs and target them for degradation. Because siRNAs are more stable than antisense RNA, they are the preferred method in medicine and in the laboratory for inhibiting translation.

DNA microarrays reveal RNA expression patterns using hybridization in a large number of sequences simultaneously. It is possible to examine patterns of expression in different tissues, under different conditions, and in individuals with known mutations. Microarrays are small glass chips containing thousands of copies of each DNA sequence per chip. These sequences (greater than 20 bp) are attached to the chip in spots in a precise order. Each spot on the chip contains a unique sequence. Chip technology has been used to look at gene expression from different breast cancers to predict the prognosis for patients and determine treatments (see Figure 13.10).

Question 5. What techniques can be used to study gene expression during development?
Textbook Reference: *13.3 Genes and Gene Expression Can Be Manipulated, pp. 252–255*

Question 6. Hybridization is a useful technique in biotechnology. Describe three ways in which hybridization is used experimentally in DNA recombinant technology.
Textbook Reference: *13.3 Genes and Gene Expression Can Be Manipulated, pp. 252–256*

Question 7. Suppose that your lab assistant has cloned gene *X* into yeast and confirmed that the recombinant DNA molecule is present in the yeast cells. However, the yeast cell is unable to synthesize protein X. Suggest why this part of the experiment is not working and what modifications are needed in the cloning procedure.
Textbook Reference: *13.3 Genes and Gene Expression Can Be Manipulated, p. 256*

13.4 Biotechnology Has Wide Applications

Expression vectors can turn cells into protein factories

Medically useful proteins can be made by biotechnology

DNA manipulation is changing agriculture

There is public concern about biotechnology

Biotechnology is the use of living cells to produce or modify useful materials for people, including food, medicines, and chemicals. This includes turning organisms into factories by manipulating them to express genes at high levels.

For cells to produce a cloned gene product, the vector must have appropriate DNA sequences that allow the cloned gene to be expressed in the host organism. Prokaryotic expression vectors require a promoter, a termination site for transcription, and a ribosome-binding site. Eukaryotic expression vectors require a poly A–addition site, transcription factor binding sites, and enhancers. Modifications of expression vectors include the addition of inducible promoters (which respond to a specific signal), tissue-specific promoters to localize the expression to a specific tissue or time in development, and signal sequences that direct the product to the appropriate destination.

Medically useful products that have been cloned in expression vectors include tissue plasminogen activator, human insulin, and vaccine proteins (see Table 13.1). Bacteria or other organisms can be transformed using recombinant DNA to produce the desired substance or its subunits that can then be extracted for human use.

Recombinant DNA offers breeders the opportunity to choose specific genes that will be incorporated into an organism, to introduce any gene into a plant or animal species, and to generate new organisms quickly.

Transgenic plants have been created that express toxins for insect larva (using the gene for the toxin naturally produced by the bacteria *Bacillus thuringiensis)*, that produce extra nutrients (adding genes for β-carotene to rice to reduce vitamin A deficiencies in human populations), or that can tolerate drought or high-salt conditions (adding salt-tolerance genes to tomatoes).

The creation of transgenic plants has raised some concerns that these crops could be unsafe for human consumption, that it is unnatural to interfere with nature, and that transgenes could escape into other noxious plants. Most scientists agree that transgenic plants should be extensively field tested and that the technology should proceed cautiously.

Recombinant bacteria have been used to clean up the environment in composting programs, wastewater treatments, oil spills, and other such efforts.

Question 8. Describe three useful products that have been produced using biotechnology. Outline two specific dangers that could result from producing organisms that contain foreign genes.
Textbook Reference: *13.4 Biotechnology Has Wide Applications, pp. 259–261*

Question 9. You are working at a biotech company and your project is to clone a eukaryotic gene (*X*) so that you can isolate large amounts of protein X. First, diagram the steps you would take to successfully complete this project. Next, assume that you have your clone but you realize that protein X, which is being made from this clone, is nonfunctional. What do you have to change in your procedure to clone gene *X* so that the expressed protein will be functional?
Textbook Reference: *13.4 Biotechnology Has Wide Applications, pp. 255–257*

Test Yourself

1. Cloning a gene may involve
   a. restriction endonucleases and ligase.
   b. plasmids and bacteriophage λ.
   c. transformation or transfection.
   d. selectable markers and/or reporter genes.
   e. All of the above
   Textbook Reference: *13.1 Recombinant DNA Can Be Made in the Laboratory, pp. 245–248*

2. Complementary base pairing is important for
   a. ligation reactions with blunt-end DNA molecules.
   b. hybridization between DNA and transcription factors.
   c. restriction endonucleases for cutting cell walls.
   d. synthesizing cDNA molecules from mRNA templates.
   e. the transcriptional activation of expression vectors.
   Textbook Reference: *13.3 Genes and Gene Expression Can Be Manipulated, p. 252*

3. For a prokaryotic vector to be propagated in a host bacterial cell, the vector needs
   a. an origin of replication.

b. telomeres.
c. centromeres.
d. drug-resistance genes.
e. reporter genes.
Textbook Reference: *13.2 DNA Can Genetically Transform Cells and Organisms, p. 249*

4. For a Ti plasmid to be propagated in a host plant cell, the vector needs
a. telomeres.
b. centromeres.
c. an origin of replication.
d. a reporter gene.
e. a, b, and c
Textbook Reference: *13.2 DNA Can Genetically Transform Cells and Organisms, p. 250*

5. Reporter genes include genes for
a. drug resistance.
b. bioluminescence.
c. DNA origins.
d. restriction endonucleases.
e. Both a and b
Textbook Reference: *13.2 DNA Can Genetically Transform Cells and Organisms, p. 251*

6. Vectors include all of the following *except*
a. bacterial plasmids.
b. viruses.
c. plant plasmids.
d. bacteriophage λ.
e. All of the above are vectors.
Textbook Reference: *13.2 DNA Can Genetically Transform Cells and Organisms, pp. 249–250*

7. A cDNA clone is
a. mostly cytosine.
b. a copy of the DNA identical to the nuclear gene.
c. a copy of noncoding DNA.
d. a DNA molecule complementary to an mRNA molecule.
e. a fragment of DNA inserted into the host chromosome.
Textbook Reference: *13.3 Genes and Gene Expression Can Be Manipulated, p. 252*

8. Gene expression can be inhibited by
a. antisense RNA.
b. knockout genes.
c. DNA microarrays.
d. microRNA.
e. a, b, and d
Textbook Reference: *13.4 Biotechnology Has Wide Applications, pp. 253–254*

9. Expression vectors are different from other vectors because they contain
a. drug-resistance markers.
b. telomeres.
c. regulatory regions that permit the cloned DNA to produce a gene product.
d. DNA origins.
e. reporter genes that are expressed in the host.
Textbook Reference: *13.4 Biotechnology Has Wide Applications, p. 256*

10. RNAi
a. is more effective than antisense RNA in inhibiting translation.
b. inhibits transcription in eukaryotes.
c. is produced only by viruses.
d. requires other proteins to modify the RNAi.
e. Both a and d
Textbook Reference: *13.4 Biotechnology Has Wide Applications, p. 254*

11. DNA chip technologies can be used to
a. predict an individual's likelihood of getting cancer.
b. show transcriptional patterns in an organism during different times of development.
c. clone DNA.
d. make transgenic plants.
e. inhibit transcription of disease genes.
Textbook Reference: *13.4 Biotechnology Has Wide Applications, pp. 254–255*

12. Which of the following is *not* an application of recombinant DNA technology?
a. Generating large amounts of tissue plasmonigen factor to help dissolve blood clots
b. Making plants more resistant to insect larva
c. Creating vaccines for pathogens
d. Reducing the salt in environmental soil so that plants can grow
e. Creating bacteria that can accelerate the breakdown of wood chips and paper
Textbook Reference: *13.4 Biotechnology Has Wide Applications, p. 255*

13. A potential disadvantage of recombinant DNA technology is that it
a. has the capacity to spread transgenes from crops to other species.
b. could make weeds herbicide-resistant.
c. could create transgenic plants that kill beneficial insects.
d. creates transgenic plants that could be harmful for human consumption.
e. All of the above
Textbook Reference: *13.4 Biotechnology Has Wide Applications, pp. 259–261*

14. Expression vectors
a. are useful for analyzing RNA transcription patterns.
b. are useful for isolating large amounts of DNA.
c. can be used to make large amounts of protein in *E. coli* cells.
d. are useful only in prokaryotic cells and cannot be expressed in eukaryotic cells.
e. are used to create SNPs for DNA microarrays.
Textbook Reference: *13.4 Biotechnology Has Wide Applications, p. 256*

15. Which of the following is required for a transgene to be expressed in a eukaryotic host?
   a. A transcriptional termination site and a ribosome binding site
   b. A DNA origin and an antibiotic resistance marker
   c. Transcription factor binding sites, enhancers, and a poly A–recognition sequence
   d. An inducible promoter and repressor proteins
   e. a, c, and d

   Textbook Reference: *13.4 Biotechnology Has Wide Applications, p. 256*

16. Gel electrophoresis
   a. causes DNA to be pulled through the gel toward the negative end of the field.
   b. causes larger DNA fragments to move more quickly through the gel than smaller DNA fragments.
   c. is required for PCR reactions.
   d. is used to identify and isolate DNA fragments.
   e. is used in allele-specific oligonucleotide hybridization.

   Textbook Reference: *13.1 Recombinant DNA Can Be Made in the Laboratory, p. 246*

17. Which of the following is *not* a step in the process of creating a transgenic mouse with a nonfunctional gene for a protein of interest?
   a. Incorporation of an inactivated copy of the gene into a vector
   b. Addition of an NLS to the gene sequence
   c. Homologous recombination
   d. Inactivation of the gene by insertion of a reporter gene
   e. Insertion of the vector into a mouse stem cell

   Textbook Reference: *13.3 Genes and Gene Expression Can Be Manipulated, p. 253*

18. Which of the following statements about typical useful recombinant gene vectors is *false*?
   a. They are about 60,000 bp long
   b. They have several restriction sites.
   c. They can reproduce independently.
   d. Some potentially harmful genes must be removed before it can be useful.
   e. They contain reporter genes.

   Textbook Reference: *13.2 DNA Can Genetically Transform Cells and Organisms, pp. 249–250*

Answers

Key Concept Review

1. A given restriction endonuclease may not be able to cleave the genome if the target sequences are methylated. The recognition sequence may not appear at all in a very small genome.
2. First, the gene for production of the enzyme must be identified. A first step is by creating a restriction enzyme digest to create a DNA map. The resulting fragments are then analyzed using gel electrophoresis to separate fragments by size and abundance. Next, genes in each fragment can be determined by purifying and analyzing its sequence. Once the appropriate sequence is identified, the fragment must be inserted into a vector organism. This is accomplished by first cleaving the vector genome using an appropriate endonuclease and then using DNA ligase to join the fragment containing the gene coding the enzyme with the vector genome. Now the vector is ready to insert the enzyme gene into a host bacterial genome.
3. A successful plasmid vector must include 20 or more unique restriction sites, origins of replication (*ori*) sites, and reporter genes and selectable marker genes.
4. Constraints on plasmid replication limit the size of the new DNA that can be inserted to 10,000 bp, while most eukaryotic genes together with their introns and flanking sequences are bigger. In addition, plasmids do not naturally infect cells, and sometimes it is difficult to get them inside the desired host cells.
5. cDNA libraries can be made from different developing tissues in order to see what genes are being expressed. DNA microarrays can also be used to analyze transcriptional patterns during development.
6. Hybridization of complementary base pairs allows ligase to seal DNA fragments with complementary sticky ends to create a recombinant molecule. Hybridization of newly made DNA to mRNA allows reverse transcriptase to generate cDNAs. Hybridization is also used in DNA microarray technology.
7. The *X* gene can be cloned into a yeast cell, but unless the vector has the appropriate regulatory signals (promoters, poly A–addition sites, translational initiation, and termination signals), no expression of gene *X* will occur. Recloning gene *X* in a specialized yeast expression vector will result in the expression of the *X* gene.
8. Useful products include rice grains that produce β-carotene, plants that are resistant to herbicides and insect larvae, the production of human growth hormone in cow's milk, and others (see Tables 13.1 and 13.2). Dangers include the creation of genetically engineered foods that could adversely affect human nutrition, the transfer of herbicide- and insect-resistant genes from crop plants to noxious weeds, and the introduction into the wild of new organisms that might have unforeseen ecological consequences.

9. You would use an expression vector that has a promoter, a transcriptional termination site, and a ribosome binding site that will be recognized by the host cell (*E. coli*), so that the protein will be expressed in that host cell.

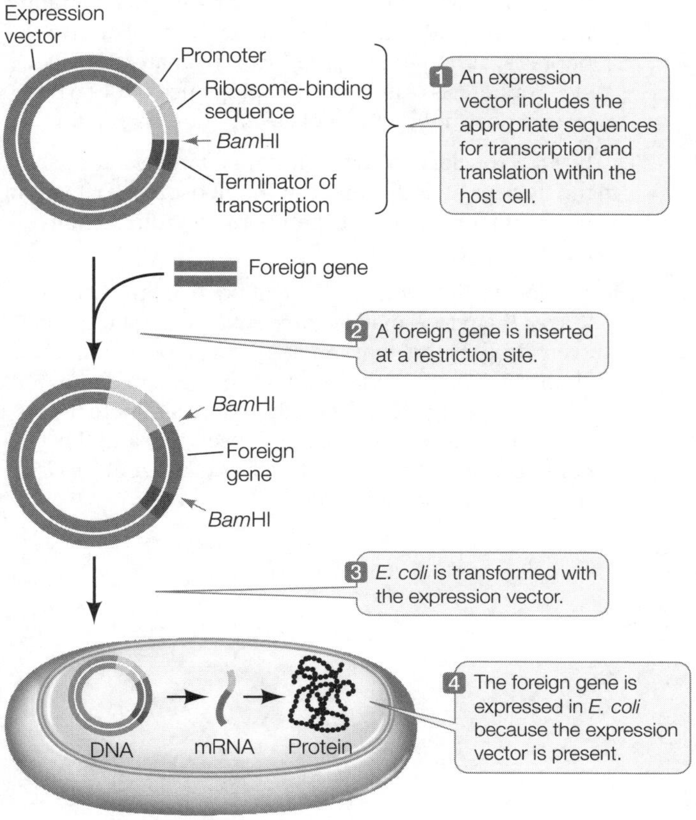
Expression vector
Promoter
Ribosome-binding sequence
BamHI
Terminator of transcription
1 An expression vector includes the appropriate sequences for transcription and translation within the host cell.
Foreign gene
2 A foreign gene is inserted at a restriction site.
BamHI
Foreign gene
BamHI
3 E. coli is transformed with the expression vector.
DNA
mRNA
Protein
4 The foreign gene is expressed in E. coli because the expression vector is present.

There could be two possible procedures. The steps in one procedure would be the following: (1) isolate DNA containing gene *X* from your cell sample; (2) create a genomic library with DNA; (3) using a probe for gene *X*, select a colony containing gene *X* from the genomic library; and (4) isolate DNA from the selected clone and cut with restriction endonuclease. Another procedure would have the following steps: (1) isolate DNA containing gene *X* from your cell sample; (2) utilizing the PCR technique and primers flanking gene *X*, amplify gene *X*; (3) purify the PCR fragment and cut with restriction endonuclease; (5) cut expression vector with same restriction endonuclease as gene *X*; (6) ligate gene *X* into expression vector; (7) transform *E. coli* with expression vector; and (8) transcriptionally activate gene *X* and isolate protein X from *E. coli* cells.

The reason that protein X is nonfunctional is that gene *X* was cloned from genomic DNA, which still had introns in the gene sequence. A cDNA clone (which is made from the mRNA from gene *X* and thus lacks introns) should have been used as a source for gene *X*.

Test Yourself

1. **e.** Cloning a gene requires restriction endonucleases to cut the gene of interest, vectors (including plasmids and bacteriophage λ), and ligase to covalently join the DNA to the vector. Transformation or transfection is needed to introduce the vector into the host cells and reporter genes and selectable markers can be used to detect the presence of the gene in the host cells.

2. **d.** No complementary base pairing can occur between blunt-end cut DNA molecules. Transcription factors are proteins that bind DNA through interactions with their side chains (which are amino acids) and the nucleotides of the DNA. Restriction endonucleases cut double-stranded DNA, not cell walls. The activation of vectors does not require complementary base pairing

3. **a.** A prokaryotic vector needs an origin of replication to be propagated in a prokaryotic cell.

4. **c.** The Ti plasmid requires an origin of replication but not telomeres or centromeres to propagate itself inside the plant cell.

5. **e.** Reporter genes include genes for drug resistance and bioluminescence.

6. **c.** There are no plant plasmids; the vector for carrying DNA into the plants the the bacterium *Agrobacterium tumefaciens*.

7. **d.** cDNA clones are not clones that contain mostly cytosine, nor are they copies of noncoding genes. A cDNA clone is generated by making a DNA copy of a particular messenger RNA using reverse transcriptase. The cDNA clone is not identical to the nuclear gene because in the messenger RNA (which served as a template for the cDNA) the introns have been removed, leaving coding sequence and 5′ and 3′ flanking sequences.

8. **e.** Gene expression can be inhibited by antisense RNA, which complementarily base pairs with the target messenger RNA, making it inaccessible to the translation machinery in the cell. MicroRNAs bind mRNAs and target them for degradation. Knockout genes are genes that have been inactivated by the insertion of DNA into their coding sequences. DNA microarrays are used in hybridization experiments and do not inhibit gene expression.

9. **c.** Expression vectors may contain drug-resistance markers and reporter genes, and they must contain DNA origins, but none of these features distinguishes them from other vectors. Expression vectors are unique because they contain regulatory sequences that allow the cloned gene to be expressed in the host cell.

10. **e.** RNAi is more effective than antisense RNA at inhibiting translation (not transcription). It is produced by viruses and by eukaryotic cells in small amounts and requires other modifying proteins.

11. **b.** DNA chips can be used to analyze gene expression at different times in development and to predict if an individual is at risk for developing cancer. Other factors (e.g., environmental) also determine if a person will get cancer. They are not used to make transgenic plants, to clone DNA, or to inhibit transcription of disease genes.
12. **d.** Soils have not been made less salty using biotechnology; plants have been made more salt-tolerant by means of recombinant DNA techniques.
13. **e.** All of the problems listed are disadvantages of recombinant DNA technology.
14. **c.** Expression vectors have the appropriate control regions that allow them to be expressed (make protein) in both prokaryotic and eukaryotic host cells. They are not used to create SNPs, nor are they used to isolate DNA or study transcription.
15. **c.** Transcription factor binding sites, enhancers, and a poly A–sequence are required for an expression vector to produce its cloned gene product in a host eukaryotic cell. Inducible promoters and repressors will regulate gene expression but are not required for expression. A tissue-specific promoter will cause the gene to be expressed in particular cells (and would be required for the expression vector to work in a particular tissue), and signal sequences will target the protein to particular compartments in the cell, but they are not part of the minimal requirements for expression.
16. **d.** DNA fragments migrate toward the positive end of the electric field, with the smallest fragments migrating the fastest. PCR reactions do not require gel electrophoresis, although their products are analyzed by gel electrophoresis. Allele-specific oligonucleotide hybridization does not require gel electrophoresis.
17. **b.** NLS (nuclear localization signal) targets a protein to the nucleus after it is made at the ribosome, so it would not be important in an experiment regarding a nonfunctional gene.
18. **a.** Most useful vectors are small (<6,000 bp). This makes them too small to carry eukaryotic genes, but even virus vectors that can carry eukaryotic genes are about 45,000 bp. Some vectors like Ti plasmid and viruses have their genomes altered to eliminate the genes that cause tumors (in the case of Ti) and the genes that cause host cells to die and lyse, but this is not typical of a useful vector.

Genes, Development, and Evolution

14

The Big Picture

- The development of a mature organism from a fertilized egg involves the patterned activation of genes in response to environmental signals. In many organisms, positional determinants are present in various regions of the egg due to maternal factors. The asymmetric distribution of these factors will activate genes differentially, and as cell division proceeds, that asymmetric gene expression continues. As a result, different cells experience different environmental signals based on their position in the developing embryo, and they respond by activating different genes. This sort of gene activation determines the fate of the cell.
- The zygote and early embryonic cells are totipotent; they can develop into any structure in the adult organism. As development proceeds, the developmental potential of embryonic cells narrows; for example, cells of the inner cell mass of the blastocyst are pluripotent. Adult stem cells are multipotent. In the adult organism, differentiated cells express particular genes that give those tissues a particular structure and function, even though all the genes are still present in the nucleus of those cells.
- Positional information, often coming from inducers called morphogens, leads to changes in the expression of key developmental genes, which in turn control morphogenesis, the creation of body form.
- Evolutionary developmental biology is the study of how organisms evolved by examining the differences in developmental gene expression and regulation between species. Animals have highly conserved genes, such as homeobox genes, that control development.
- Genetic switches determine where and when genes will be expressed. Changes in these switches can result in the evolution of species differences.
- Heterochrony is the phenomenon whereby differences in organisms have arisen because of differences in the timing of developmental processes. The timing of gene expression can differ between organisms because of the modular makeup of organisms. Species differences can also result from changes in the spatial expression pattern of developmental genes.

Study Strategies

- Sometimes it is difficult to visualize how a single cell can divide and grow to produce a mature functional organism with highly differentiated tissues. At some very early point in development (either before fertilization or in one of the first set of divisions), different genes begin to be expressed in different cells. This gene expression can be in response to cytoplasmic factors or signals from other cells. Early gene expression sets up positional determinants that activate another wave of genes that further divides regions of the embryo into different developmental areas. As cell division proceeds in the embryo, these developmental genes continue to be differentially expressed, resulting in differentiation and morphogenesis in the organism. Review the four key processes of development: (1) determination; (2) differentiation; (3) morphogenesis; and (4) growth.
- Make sure that you understand the different potentials of totipotent, pluripotent, and multipotent stem cells. Review these in the context of human development, focusing on when and where each type of stem cell occurs.
- Review some of the important genes whose products direct development in the model organisms discussed in the chapter. Identify orthologous genes from different organisms.
- The concept of heterochrony can be difficult to grasp. Try to remember that when we discuss heterochrony, we are talking about differences among species, not within species.
- Go to yourBioPortal.com to review the following tutorials and activities:

  Animated Tutorial 14.1 Embryonic Stem Cells

  Animated Tutorial 14.2 Early Asymmetry in the Embryo

  Animated Tutorial 14.3 Pattern Formation in the *Drosophila* Embryo

  Animated Tutorial 14.4 Modularity

  Interactive Tutorial 14.1: Cell Fates: Genetic vs. Environmental Influences

Web Activity 14.1 Stages of Development

Web Activity 14.2 Plant and Animal Development

Working with data 14.1 Cloning a Mammal

Key Concept Review

14.1 Development Involves Distinct but Overlapping Processes

Four key processes underlie development

Cell fates become progressively more restricted during development

Cell differentiation is not irreversible

Stem cells differentiate in response to environmental signals

During development, an organism progresses through successive forms as it moves through its life cycle (see Figure 14.1). The fertilized egg is called a zygote, and from the zygote an embryo develops. Development includes four processes: determination, differentiation, morphogenesis, and growth. Development ends with death.

The developmental fate of a cell is set during determination and is influenced by internal and external conditions. More specifically, differential gene expression and the extracellular environment set the fate of a cell. During differentiation, cells become specialized to contain specific structures and to perform particular functions. Morphogenesis is the creation of form as seen in body shape and organs. It can occur by cell division (an increase in the number of cells), cell expansion (an increase in the size of existing cells), cell movements, and programmed cell death (apoptosis). Growth is an increase in size due to cell division or cell expansion. Cell division is important in the development of plants and animals; cell expansion is particularly important in the development of plants.

Transplantation experiments have shown that the environment in which early embryonic cells exist can redirect them along different developmental paths (see Figure 14.2). Nevertheless, embryonic cells eventually become committed to a particular developmental fate, even though they are not yet differentiated. This is because their fate has been determined. When these cells from older embryos are transplanted, they continue to develop into the original differentiated tissue, regardless of their environment. Cell fate becomes apparent as cells differentiate.

Some cells, such as a zygote, are totipotent, meaning they can develop into any of the different kinds of cells in the mature organism. Developmental possibilities narrow as cell determination and differentiation occur.

In plants, differentiation is reversible in some cells (see Figure 14.3). Under certain conditions, plant cells in culture can give rise to a new, genetically identical plant (a clone). Plant cells first dedifferentiate, and then produce a callus (mass of cells), which develops into a plant embryo when placed in the appropriate medium. Thus, the original differentiated cells contained all the genetic information needed to express genes in the correct sequence to generate a whole plant; this is evidence of genomic equivalence.

Mammals can be cloned by fusing somatic cells with enucleated eggs and implanting the resulting embryos in surrogate mothers (see Figure 14.4). Several mammals, including sheep, mice, and cattle, have been cloned by the technique of nuclear transfer. Practical uses of cloning include increasing the number of valuable animals (for example, genetically engineered animals) and preservation of endangered species and pets.

Development occurs in adults as well as in embryos. In adult plants, the areas of undifferentiated cells in the growing tips of roots and stems are known as meristems. These cells can become any cell type found in the plant.

Undifferentiated dividing cells in mammals are known as stem cells. Stem cells in adult mammals are specific for the kinds of tissue they replace, typically skin, the lining of the intestines, and blood cells. These stem cells are multipotent, meaning that they can differentiate into a limited number of cell types. Signals from adjacent cells or from circulation influence the differentiation of stem cells.

Bone marrow contains two types of multipotent stem cells: hematopoietic and mesenchymal. Hematopoietic stem cells produce red and white blood cells, whereas mesenchymal stem cells produce bone and muscle cells. In hematopoietic stem cell transplantation, stem cells from a patient about to undergo high doses of cancer treatment are removed from the blood, stored, and encouraged to increase in number, and then they are returned to the depleted bone marrow once the treatment has been completed.

In embryonic mammals, cells from a part of the blastocyst (the inner cell mass) are pluripotent, which means they are capable of forming nearly every type of cell. These pluripotent embryonic stem cells can be taken directly from human embryos made available through in vitro fertilization or by making induced pluripotent stem cells from skin cells. The latter technique does not destroy human embryos or provoke an immune response in recipients.

Question 1. Create a flow chart detailing the three different types of stem cells present in a developing human. Name and define each type of stem cell, and include the developmental stage (and location, when possible) at which each type of stem cell is present.
Textbook Reference: *14.1 Development Involves Distinct but Overlapping Processes, pp. 265–269*

Question 2. Describe one application for human health of multipotent stem cells and one application for human health of cloned mammals.
Textbook Reference: *14.1 Development Involves Distinct but Overlapping Processes, pp. 267–268*

Question 3. When mammals are cloned by the fusing of whole cells with enucleated eggs, why is the original nucleus removed from the egg?
Textbook Reference: *14.1 Development Involves Distinct but Overlapping Processes, pp. 265–267*

14.2 Changes in Gene Expression Underlie Cell Differentiation in Development

- Differential gene transcription is a hallmark of cell differentiation
- Gene expression can be regulated by cytoplasmic polarity
- Inducers passing from one cell to another can determine cells fates

Differentiated cells retain all their original genetic content, even though they express a very small subset of their genes. Mechanisms that control gene expression leading to cell differentiation typically work at the level of transcription.

Transcription factors are proteins that bind DNA and regulate the expression of genes. In response to the transcription factor MyoD (myoblast-determining gene), undifferentiated muscle precursor cells stop dividing; this step is necessary for the eventual differentiation of these cells into mature muscle cells (see Figure 14.7). Sometimes a single transcription factor causes a cell to differentiate. In other cases, several different transcription factors promote differentiation.

Cells can be made to transcribe different sets of genes in two ways: (1) by asymmetrical distribution of cytoplasmic factors; and (2) by differential exposure to an external inducer. In the first way, cytoplasmic segregation of factors in the egg results in an unequal distribution of maternal elements in each of the cells of the embryo (see Figure 14.8). Each cell's fate and pattern of gene expression is determined by the amount of cytoplasmic determinants received. Unequal distribution of cytoplasmic determinants directs embryonic development and controls the polarity of the organism. In the second way, cells can induce other cells to differentiate by secreting chemical signals called inducers.

The nematode *Caenorhabditis elegans* is a model organism used in developmental studies, including those that demonstrate induction. The egg of *C. elegans* develops into a larva in 8 hours and an adult in 3.5 days. The animal has a transparent body, making it easy to follow the developmental fate of its cells. The adult is hermaphroditic, having female and male reproductive organs. Eggs are laid through a pore called the vulva, which is induced to form from a single cell called an anchor cell. If this cell is destroyed, no vulva develops. The anchor cell determines the fates of six cells on the ventral surface of *C. elegans* by producing a primary inducer (LIN-3 protein) that diffuses toward adjacent cells, establishing a concentration gradient. The closest cell is exposed to the highest concentration of LIN-3 and becomes the primary precursor cell. It produces a secondary inducer that causes the next closest cells to become secondary precursors. Descendants of primary and secondary precursor cells form the vulva. The final three cells (which are farthest from the anchor cell) become epidermal cells (see Figure 14.9). When LIN-3 binds to a receptor on the surface of the closest cell, it causes a signal transduction cascade, the end result of which is the differentiation of vulval cells in the nematode.

Question 4. Activated MyoD has been found in muscle stem cells of adult vertebrates. What might its role be in this situation?
Textbook Reference: *14.2 Changes in Gene Expression Underlie Cell Differentiation in Development, pp. 269–270*

14.3 Spatial Differences in Gene Expression Lead to Morphogenesis

- Multiple genes interact to determine developmental programmed cell death
- Expression of transcription factor genes determines organ placement in plants
- Morphogen gradients provide positional information during development
- A cascade of transcription factors establishes body segmentation in the fruit fly

Pattern formation results in the spatial organization of a tissue or organism, and morphogenesis is the creation of body form. These two closely linked processes arise from spatial differences in gene expression. For spatial differences in gene expression to occur, cells must know their relative locations in the body and must activate the pattern of gene expression appropriate for their location.

During morphogenesis, some cells are programmed to die through apoptosis. In human embryos, apoptosis occurs in the webs of skin that initially form between fingers and toes. Apoptosis also occurs in *C. elegans*. As the nematode develops from a fertilized egg into an adult, 1,090 cells are produced. However, 131 of these cells are programmed to die due to the sequential expression of *ced-4* and *ced-3* genes. The genes controlling apoptosis are crucial for proper development in many organisms. These genes have been conserved in organisms separated by 600 million years of evolution (nematodes and humans).

During plant development, organs such as leaves, roots, and flowers are produced. Flowers have four types of organs: sepals, petals, stamens, and carpels. Stamens are male reproductive organs and carpels are female reproductive organs. Flower organs occur in whorls organized around a central axis and are derived from meristematic tissue on the plant (see Figure 14.11).

Plant geneticists have studied the development of flower organs in *Arabidopsis*. There are four whorls of organs in *Arabidopsis.* Three genes expressed in the whorls act as organ identity genes to guide differentiation in each whorl. Gene A is expressed in whorls 1 and 2, which form the sepals and petals, respectively; gene B is expressed in whorls 2 and 3, which form petals and stamens, respectively; and gene C is expressed in whorls 3 and 4, which form stamens and carpels, respectively. The three genes—A, B, and C—all encode transcription factors that are active as dimers (proteins with two polypeptide subunits). Gene regulation is combinatorial, and the combination of different dimers determines which genes are activated. A dimer of the transcription factor encoded by gene A will activate genes that make sepals. A dimer of transcription factor A with transcription factor B will result in petals. LEAFY is a transcription factor that regulates the transcription of the genes A, B, and C.

Positional information allows cells to determine where they are in the developing organism. Morphogens are signals that establish positional information. They act directly on the target cell, and different concentrations within the embryo cause different effects. Concentration gradients of the morphogen Sonic hedgehog, secreted by the zone of polarizing activity in the limb bud, determine the anterior–posterior axis of the developing vertebrate limb. According to the "French flag" model, cells in the limb bud form different digits depending on the concentration of the morphogen.

A cascade of transcription factors controls development in the fruit fly (*Drosophila melanogaster*). The body of *Drosophila* is segmented, consisting of a head (which is formed from fused segments), three thoracic segments, and eight abdominal segments. Three classes of genes, expressed in sequence, define the body segments: (1) maternal effect genes; (2) segmentation genes; and (3) Hox genes.

The first step in body segment definition in *Drosophila* is the establishment of anterior–posterior and dorsal–ventral polarity. Polarity is based on the cytoplasmic distribution of mRNA and proteins produced by maternal effect genes. Mutations in maternal effect genes (*bicoid* and *nanos*) create larvae lacking anterior structures *(bicoid*) or abdominal segments (*nanos*). *Bicoid* encodes a transcription factor that positively affects some genes (*hunchback*) and negatively affects others. Nanos protein inhibits the translation of *hunchback.*

Segmentation genes are expressed when the embryo is at the 6,000-nuclei stage and determine the number, boundaries, and polarity of the segments. Three types of segmentation genes regulate segmental development: gap genes, pair rule genes, and segment polarity genes. Gap genes organize large areas along the anterior–posterior axis of the developing embryo. Gap mutants produce larvae that are missing consecutive larval segments. Pair rule genes divide the larva into units of two segments each. Mutations in the pair rule genes produce larvae that are missing every other segment. Segment polarity genes determine the boundary and the anterior–posterior organization of each segment. Mutations in segment polarity genes produce larvae in which the posterior structures in the segments have been replaced by reversed anterior structures.

Differences between segments are encoded by Hox genes, which give each segment an identity. For instance, in response to Hox gene expression, cells in a thoracic segment produce legs and cells in the head segment produce eyes. Homeotic mutants include antennapedia (producing legs in place of antennae; see Figure 14.14) and bithorax (producing a fly with an extra set of wings). A DNA sequence, the homeobox, encodes a 60-amino acid sequence, the homeodomain. The homeodomain includes a DNA-binding motif and acts as a transcription factor.

Question 5. The textbook describes development in several model organisms, including thale cress (*Arabidopsis thaliana*), fruit fly (*Drosophila melanogaster*), and nematode (*Caenorhabditis elegans*). How would you define "model organism"? What traits do you think characterize model organisms?
Textbook Reference: *14.3 Spatial Differences in Gene Expression Lead to Morphogenesis, pp. 273, 275 (see also p. 272)*

Question 6. Develop a flow chart showing the gene cascade that controls body segmentation in the fruit fly. For each class of genes, include a brief description of general function.
Textbook Reference: *14.3 Spatial Differences in Gene Expression Lead to Morphogenesis, pp. 276–277*

14.4 Gene Expression Pathways Underlie the Evolution of Development

Developmental genes in distantly related organisms are similar

Genetic switches govern how the genetic toolkit is used

Modularity allows for differences in the pattern of gene expression among organisms

Evolutionary developmental biology (often shortened to evo-devo) is the contemporary study of evolution and development. The major ideas of this field include: (1) the molecular mechanisms for morphogenesis and pattern formation are shared across diverse organisms; (2) the molecular pathways for different developmental processes operate independently from one another in modules; (3) new structures can evolve through changes in the spatial location and timing of expression of certain genes; and (4) structures change over time, largely as the result of modifications to existing developmental genes and pathways.

Many of the genes controlling development are conserved across different organisms. Thus, these genes can be very similar in organisms that have very different appearances. As an example, the development of an anterior–posterior axis in insects and mammals is due to the same types of homeobox-containing genes that are expressed at either the anterior or posterior end of the developing embryo (see Figure 14.15). The major differences in body form arise because genes are turned on and off at different times and locations during development. The organism's genes and transcription factors, along with the way they interact with extracellular signals, can be thought of as a genetic toolkit used to assemble the organism.

Genetic switches, which consist of promoters and the transcription factors that bind them, control the way in which a gene is used during development. Each gene is controlled by multiple switches that influence where and when it is expressed. For example, genetic switches control spatial development of the embryo. The pattern of development depends on the Hox genes that are expressed within the module. This can be seen in the development of wings in *Drosophila*, in which Hox genes are differentially expressed in the three thoracic segments of the embryo (see Figure 14.16).

The many parts of the developing embryo can change independently from one another because the organism consists of developmental modules. A module, such as the genes and signaling pathways that lead to development of the heart, can change independently of the other modules because some developmental genes affect only one module. On the time scale of evolution, modularity means that changes in the timing or spatial position of a particular developmental process can occur without disrupting the entire organism.

Heterochrony occurs when the relative timing of developmental processes independently shifts in different species. Like almost all mammals, the giraffe has seven cervical vertebrae in its neck. Thus, its long neck cannot be explained by the presence of additional vertebrae. Instead, its long neck results from a delay in the signaling process that stops bone growth, and this delay allows the vertebrae of the neck to grow longer (see Figure 14.17). In other words, the giraffe's long neck results from changes in the timing of expression of genes that control bone formation.

Evolutionary change can also occur as a result of changes in the spatial expression of developmental genes. Bird embryos have webbed feet. The webbing is retained in adult ducks but not in chickens. The signaling protein bone morphogenetic protein 4 (BMP4) causes the cells of the webbing to undergo apoptosis. Although BMP4 is present in the webbing of embryonic ducks and chickens, these species differ in whether the gene *Gremlin* is expressed. *Gremlin* encodes a protein that inhibits BMP4. In ducks, *Gremlin* is expressed, encoding a protein that inhibits BMP4, thereby preventing apoptosis and producing webbed feet.

Question 7. There are morphological differences between adult chickens and ducks that relate to the duck's aquatic lifestyle. Why were these adaptations able to occur without influencing the development of the rest of the duck's morphology?
Textbook Reference: *14.4 Gene Expression Pathways Underlie the Evolution of Development, p. 280*

Question 8. Evaluate the following statement: "Most changes in morphology over evolutionary time occur as a result of the introduction of radically new developmental mechanisms."
Textbook Reference: *14.4 Gene Expression Pathways Underlie the Evolution of Development, p. 278*

14.5 Developmental Genes Contribute to Species Evolution but Also Pose Constraints

- Mutations in developmental genes can cause major evolutionary changes
- Evolution proceeds by changing what's already there
- Conserved developmental genes can lead to parallel evolution

Major evolutionary changes can occur as a result of mutations in developmental genes. The insect *Ultrabithorax* (*Ubx*) homeotic gene has a mutation that results in the repression of the gene involved in limb formation, the *Distal-less* gene. The *Ubx* gene is expressed in the abdomen during development, leading to the absence of limbs on the abdomen in insects. Other arthropods, such as centipedes, do not have this mutation and have abdominal legs. Thus, a mutation in a Hox gene changed the number of legs in insects relative to other arthropods (see Figure 14.19).

Evolution works on existing genes and their expression and can occur through changes in Hox gene expression in different modules. Thus, most evolutionary innovations are modifications of previously existing structures. For example, the wings of birds and bats are not new structures; rather, they are modified limbs (see Figure 14.20). Evolutionary losses are also controlled by Hox genes, such as the loss of forelimbs in the ancestors of present-day snakes.

Similar traits evolve repeatedly because of the highly conserved nature of the genetic code. This can lead to parallel phenotypic evolution of a trait, such as the evolutionary loss of body armor in numerous populations of freshwater stickleback fish. Freshwater populations of sticklebacks are descended from marine populations. In marine sticklebacks, the gene *Pitx1* codes for a transcription factor that stimulates the production of protective plates and spines. The *Pitx1* gene has changed in various freshwater populations of sticklebacks, leading to the loss of body armor in these populations.

Question 9. Why do insects have legs only on thoracic segments, whereas centipedes have legs on thoracic and abdominal segments?
Textbook Reference: *14.5 Developmental Genes Contribute to Species Evolution but Also Pose Constraints, p. 282*

Question 10. Amphipods are small aquatic crustaceans. Researchers have discovered the independent reduction of eyes in different populations of a species of cave-dwelling amphipod. What phenomenon does this represent?
Textbook Reference: *14.5 Developmental Genes Contribute to Species Evolution but Also Pose Constraints, pp. 282–283*

Test Yourself

1. Cell differentiation
   a. results from the loss of particular genes from the nucleus of the differentiated cell.
   b. results from the differential expression of genes that are responsive to environmental signals.
   c. involves the persisting totipotency of early embryonic cells in the mature organism.
   d. results from mutations in genes that control the synthesis of DNA.
   e. precedes cell determination.
   Textbook Reference: *14.1 Development Involves Distinct but Overlapping Processes, pp. 264–265*

2. A totipotent cell is a cell
   a. whose developmental fate has been decided.
   b. that has differentiated into a specialized tissue.
   c. that is fated to form a particular structure.
   d. whose developmental potential is extremely broad.
   e. whose developmental potential is narrower than that of a pluripotent cell.
   Textbook Reference: *14.1 Development Involves Distinct but Overlapping Processes, p. 265*

3. Heterochrony
   a. can explain the long neck of the giraffe.
   b. involves shifts in the relative timing of developmental processes.
   c. can lead to major morphological changes.
   d. does not explain the reduction in body armor of freshwater sticklebacks.
   e. All of the above
   Textbook Reference: *14.4 Gene Expression Pathways Underlie the Evolution of Development, pp. 280–281*

4. The fate of a cell
   a. refers to the cell's original type.
   b. refers to its genetic makeup.
   c. refers to the type of cell into which it will differentiate.
   d. describes its death.
   e. cannot by modified by its cytoplasmic environment.
   Textbook Reference: *14.1 Development Involves Distinct but Overlapping Processes, p. 264*

5. Genes that regulate development are highly conserved. This means that
   a. large differences have evolved among multicellular organisms.
   b. they have changed very little over the course of evolution.
   c. they are always turned on.
   d. they have undergone mutation.
   e. they have been lost in some lineages.
   Textbook Reference: *14.4 Gene Expression Pathways Underlie the Evolution of Development, p. 278*

6. The protein MyoD
   a. is a transcription factor that controls the expression of genes involved in the differentiation of muscle cells.
   b. controls segment identity in mice.
   c. is a transcription factor that activates a gene causing the cell cycle to start in muscle precursor cells.
   d. causes the induction of vulval cells in nematodes.
   e. is the only transcription factor involved in differentiation of mesoderm cells into mature muscle cells.
   Textbook Reference: *14.2 Changes in Gene Expression Underlie Cell Differentiation in Development, pp. 269–270*

7. Induction occurs when
   a. nuclear genes are lost in some tissues.
   b. one cell contacts another and fails to alter its developmental fate.
   c. a cell or tissue sends a chemical signal to another, causing differentiation of that cell or tissue.
   d. two tissues come into contact, causing the release of transcription factors.
   e. factors are unequally distributed in the cytoplasm of an egg.
   Textbook Reference: *14.2 Changes in Gene Expression Underlie Cell Differentiation in Development, pp. 271–272*

8. Which of the following statements about plant development is *false*?
   a. Both cell division and cell expansion contribute to growth in early plant embryos.
   b. Differentiation in plant cells is irreversible.
   c. Meristems at the tips of the roots and stems consist of undifferentiated cells.

d. Flower development in a plant involves organ identity genes.
e. In plants, the fertilized egg is called a zygote.
Textbook Reference: *14.1 Development Involves Distinct but Overlapping Processes, pp. 264–266; 14.3 Spatial Differences in Gene Expression Lead to Morphogenesis, pp. 273–274*

9. It may be possible to genetically engineer plants to produce more seeds and fruits by
a. providing more fertilizer in the spring.
b. manipulating their gap genes.
c. altering the homeotic genes that control carpel development.
d. inducing a mutation that eliminates LEAFY gene function.
e. altering the homeotic genes that control stamen development.
Textbook Reference: *14.3 Spatial Differences in Gene Expression Lead to Morphogenesis, pp. 273–274*

10. Morphogens
a. diffuse within the embryo to set up a concentration gradient.
b. are expressed as positional signals in developing embryos.
c. direct differentiated cells to form organs.
d. must specifically affect target cells.
e. All of the above
Textbook Reference: *14.3 Spatial Differences in Gene Expression Lead to Morphogenesis, pp. 274–275*

11. Maternal effect genes
a. begin setting up positional axes in the egg prior to fertilization.
b. are the same as segmentation genes.
c. are masked by paternal genes.
d. operate late in the gene cascade regulating development of *Drosophila* embryos.
e. determine cell fate within each segment.
Textbook Reference: *14.3 Spatial Differences in Gene Expression Lead to Morphogenesis, p. 276*

12. Hox genes
a. encode protein domains that are important in development and have been highly conserved over evolutionary time.
b. are found in diverse organisms.
c. can produce the wrong structure in the wrong place when mutated.
d. help determine cell fate within each segment of a developing *Drosophila* embryo.
e. All of the above
Textbook Reference: *14.3 Spatial Differences in Gene Expression Lead to Morphogenesis, pp. 276–277*

13. What principle of development states that the nuclei of cells do not lose any genetic information during the stages of development?
a. Genomic equivalence
b. Apoptosis
c. Totipotency
d. Transcription
e. Fate mapping
Textbook Reference: *14.1 Development Involves Distinct but Overlapping Processes, p. 265*

14. What is the role of cytoplasmic segregation in determining the fate of a cell?
a. It keeps cells totipotent.
b. It determines polarity within the organism.
c. It ensures that gradients do not develop in the developing organism.
d. It stops cell division.
e. It has no role in determining cell fate.
Textbook Reference: *14.2 Changes in Gene Expression Underlie Cell Differentiation in Development, pp. 270–271*

15. Module development has allowed for evolutionary changes to occur while still resulting in a viable organism because
a. all modules must change together.
b. modules can change independently of one another.
c. modules are unimportant for development.
d. modules prevent heterochrony.
e. developmental genes always exert their effects on more than one module.
Textbook Reference: *14.4 Gene Expression Pathways Underlie the Evolution of Development, p. 280*

16. Which of the following statements is *false*?
a. Body armor is absent from sticklebacks living in freshwater environments due to a genetic mutation.
b. The *Pitx1* gene promotes the production of protective plates and spines in marine populations of sticklebacks.
c. Similar traits are likely to evolve repeatedly because of the highly conserved nature of developmental genes.
d. The *Pitx1* gene is inactive in freshwater stickleback populations.
e. Environmental conditions induce the loss of body armor in freshwater sticklebacks.
Textbook Reference: *14.5 Developmental Genes Contribute to Species Evolution but Also Pose Constraints, pp. 282–283*

17. Which of the following statements is true?
a. Wings have evolved independently four times in vertebrates.
b. Wings arose as a result of new "wing genes."
c. Wings arose as modifications to already existing structures.
d. Wings have different skeletal components in birds and bats.
e. Like birds, pterosaurs have a flight surface made of feathers.
Textbook Reference: *14.5 Developmental Genes Contribute to Species Evolution but Also Pose Constraints, pp. 282–283*

Answers

Key Concept Review

1.

Name: Totipotent stem cell
Definition: Can specialize to be any type of cell
Developmental stage: Zygote

↓

Name: Pluripotent stem cell
Definition: Can specialize to be nearly every type of cell (cannot form cells of the placenta)
Location and developmental stage: Inner cell mass of the blastocyst

↓

Name: Multipotent stem cell
Definition: Can specialize to be many types of cells
Location and developmental stage: Tissues that need frequent cell replacement such as bone marrow, brain, and skin of an adult

2. Hematopoietic stem cells are multipotent stem cells that produce red and white blood cells. If left in the body, these dividing cells are killed during cancer treatment along with cancer cells. In hematopoietic stem cell transplantation, these stem cells are removed prior to cancer treatment, stored in a laboratory where they are provided with growth factors, and then replaced in greater numbers once treatment has been completed. Through cloning we can increase the number of particularly valuable animals, such as transgenic cows that produce human growth hormone, and thereby increase the supply of growth hormone needed for treating hormone deficiencies.

3. The nucleus of the egg contains a haploid set of chromosomes. Fusing a diploid somatic cell with an egg cell that still contained a nucleus would result in a cell with three copies of each chromosome and little chance of developing normally.

4. The discovery of activated MyoD in muscle stem cells of adult vertebrates suggests that this transcription factor might also be involved in muscle repair. In other words, its role is not restricted to stimulating muscle cell differentiation from mesoderm in vertebrate embryos.

5. Model organisms are research organisms that are expected to yield insights into how other organisms work. Model organisms generally share genetic, cellular, and/or molecular characteristics with these other organisms. They also have short generation times, are easy to obtain and maintain, are amenable to experimental manipulation, and are not of concern from a conservation standpoint.

6.

Maternal effect genes
Establish the major axes of the egg

↓

Gap genes (segmentation genes)
Organize broad areas along the anterior–posterior axis

↓

Pair rule genes (segmentation genes)
Divide embryo into two-segment units

↓

Segment polarity genes (segmentation genes)
Determine the boundaries and anterior–posterior organization of individual segments

↓

Hox genes
Determine what organs will develop in particular segments

7. Organisms are made up of modules. Change can occur in a module independent of other modules. The webbing found in duck feet is due to expression of the *Gremlin* gene in the foot module. Expression of *Gremlin* in the webbing cells of the feet results in a protein that inhibits the BMP4 protein responsible for prompting apoptosis.

8. The statement is incorrect. In fact, most evolutionary changes in morphology occur by modification of existing development genes and developmental pathways.

9. Leg development in arthropods involves the homeotic gene *Ultrabithorax* (*Ubx*) and the gene *Distal-less* (*Dll*). Expression of *Dll* in a segment results in the development of a pair of legs in that segment. In insects, the *Ubx* gene has a mutation that suppresses the expression of *Dll* in the abdomen. As a result, no legs will develop in the abdomen. Centipedes do not have this mutation. Therefore, legs develop in thoracic and abdominal segments.

10. The independent reduction of eyes in several populations of a cave-dwelling amphipod is an example of parallel phenotypic evolution, the phenomenon whereby similar traits evolve repeatedly across multiple populations of a species.

Test Yourself

1. **b.** Genes are not lost from differentiated cells. There are no cells left from the embryo in a mature organism. Mutations in cells that affect DNA replication would result in cells that were unable to replicate their DNA. Cell differentiation occurs after cell determination.

2. **d.** A totipotent cell is a cell that can develop in any number of ways. Its fate has not been determined, it

has not differentiated, and it can develop into more cell types than can a pluripotent cell.

3. **e.** Heterochrony is the process by which the timing of gene expression differs between two or more species. Heterochrony can explain the long neck of the giraffe (delays occur in the signaling process that stops bone growth, so the neck vertebrae grow longer) but not the reduction in body armor in freshwater populations of sticklebacks.

4. **c.** The fate of a cell is the type of cell it will eventually become once it has differentiated.

5. **b.** Conserved genes are genes that are found in many organisms and have undergone very little change.

6. **a.** MyoD is a transcription factor that controls the expression of genes involved in the differentiation of mesoderm cells into mature muscle cells. It activates a gene that causes the cell cycle to stop, allowing differentiation can begin. The transcription factor MyoD is one of several involved in the differentiation of muscle cells.

7. **c.** Induction occurs when one cell or tissue secretes factors that influence the developmental fate of other cells or tissues. Nuclear genes are not lost in developing cells (red blood cells are an exception). Transcription factors are found in the nucleus and are not released to adjacent tissues.

8. **b.** In general, it is easier to reverse differentiation in plant cells than it is in animal cells, as demonstrated by the cloning of a carrot from a differentiated storage cell in its root.

9. **c.** Fertilization is not genetic engineering. Gap genes are found in *Drosophila,* not plants. The LEAFY protein transcriptionally activates the homeotic genes that control organ development (and subsequent seed and fruit formation). In its absence, no fruit or seeds would develop. Altering the homeotic genes that control carpel (not stamen) development could lead to more fruit and seeds.

10. **e.** Morphogens diffuse in the embryo to form concentration gradients that act as positional signals to direct differentiated tissues to form organs. They must specifically affect target cells.

11. **a.** Beginning before fertilization, maternal effect genes determine the anterior–posterior axis. These genes, which operate early in the gene cascade, also induce segmentation genes. The paternal genotype does not affect the expression of maternal gene products in the egg. Hox genes, and not maternal effect genes, determine cell fate within each segment.

12. **e.** Hox genes encode proteins that are highly conserved and important for directing development in many different organisms. Mutations in these genes can produce an abnormal developmental event in the organism. In *Drosophila* embryos, Hox genes determine segment identity.

13. **a.** Genomic equivalence is a fundamental principle of developmental biology. It states that none of the genetic information contained in the original zygote is lost during subsequent cell division and growth.

14. **b.** Cytoplasmic segregation refers to the unequal distribution of factors within the developing embryo. This asymmetry helps set up, for example, the polarity (distinct top and bottom) of the developing organism.

15. **b.** Embryos are composed of self-contained units called modules that can change independently of one another. This allows for one module to evolve with no disruption of the other modules of the animal.

16. **e.** Absence of body armor in freshwater sticklebacks results from a mutation in the *Pitx1* gene; the absence of armor is not induced by environmental conditions.

17. **c.** Wings arose as modifications of preexisting structures. In vertebrates, wings are modified forelimbs.

Mechanisms of Evolution 15

The Big Picture

- Most biologists rank Darwin's *The Origin of Species* as the most significant book ever written about their subject. Its importance is twofold. First, it presents a vast amount of evidence for the fact of evolution. Second, it describes a mechanism—natural selection—to explain evolutionary change. Evolution through natural selection remains the most important unifying concept in biology.
- Mutations in the genome produce the genetic variation upon which natural selection acts. From generation to generation, there is an increase in the frequency of beneficial mutations within a population if natural selection and evolution are occurring. The allele frequency of a population can be altered by gene flow, genetic drift, the founder effect, and sexual selection. The changes in allele frequency can be measured to provide an assessment of the evolution occurring in a population. The Hardy–Weinberg equilibrium principle describes a population in which allele frequencies are not changing across generations.
- Selection can be stabilizing, directional, or disruptive. These types of selection alter the distribution of phenotypes in a population. Selection can act only on missense substitutions and not on synonymous substitutions or silent substitutions. The neutral theory states that the majority of variants in a population are selectively neutral. Heterozygotes can be at a selective advantage over homozygotes.
- New features in a population can arise from a number of other means. Sexual recombination increases the number of possible genotypes in a population. New functions can be obtained by lateral gene transfer from other organisms. In addition, gene duplication can result in genes with new functions.
- Multiple applications exist for evolutionary theory. Protein function can be studied with the knowledge of gene evolution. New molecules are being produced by in vitro evolution. Finally, medical sciences are benefiting from our knowledge of molecular evolution.

Study Strategies

- A common mistake when attempting to solve Hardy–Weinberg problems is the use of the wrong equation. For example, if you are given phenotypic frequencies for a trait that is at genetic equilibrium and shows dominance, remember that the frequency of the recessive phenotype (and genotype) is equal to q^2, not q. By taking the square root of q^2, you can obtain the frequency of the recessive allele and then determine the frequency of the dominant allele by subtraction ($p = 1 - q$).
- It is important to bear in mind that for any gene locus with two alleles, the allele frequencies must total 1 (i.e., it is always the case that $p + q = 1$). The equation that specifies genotype frequencies ($p^2 + 2pq + q^2 = 1$) is valid only for a population at genetic equilibrium.
- The different means by which evolution can occur can become overwhelming. To help organize them, construct a table listing the different means by which a population can evolve.
- Go to yourBioPortal.com to review the following tutorials and activities:

  Animated Tutorial 15.1 Natural Selection

  Animated Tutorial 15.2 Hardy–Weinberg Equilibrium

  Interactive Tutorial 15.1 Genetic Drift

  Web Activity 15.1 Gene Tree Construction

  Working with Data 15.1 Testing for Significant Differences

  Working with Data 15.2 Determining the Paternity of Butterfly Larvae

Key Concept Review

15.1 Evolution Is Both Factual and the Basis of Broader Theory

Darwin and Wallace introduced the idea of evolution by natural selection

Evolutionary theory has continued to develop over the past century

Evolutionary theory encompasses our understanding of the mechanisms that lead to biological changes in populations over time. It provides biologists with a means to understand how life diversified. Geological, morphological, and molecular data provide strong support for evolution.

Charles Darwin studied natural history during his five-year voyage on the HMS *Beagle*. Darwin's main contribution to biology was the theory of evolution by natural selection. This is based on the evidence that species change over time, that divergent species share a common ancestor, and that differential survival and reproduction in the population are based on variation of traits. In short, natural selection is the differential contribution of offspring to the next generation by various genetic types belonging to the same population. Russel Wallace independently came up with the concept of natural selection at roughly the same time as Darwin. The work of Gregor Mendel, Watson and Crick, and E. O. Wilson has led to continued development of evolutionary theory.

Question 1. We call the mechanisms of evolution the "theory" of evolution. Strictly speaking, however, the term "theory" is incorrect. Why?
Textbook Reference: *15.1 Evolution Is Both Factual and the Basis of Broader Theory, p. 289*

15.2 Mutation, Selection, Gene Flow, Genetic Drift, and Nonrandom Mating Result in Evolution

- Mutation generates genetic variation
- Selection on genetic variation leads to new phenotypes
- Natural selection increases the frequency of beneficial mutations in populations
- Gene flow may change allele frequencies
- Genetic drift may cause large changes in small populations
- Nonrandom mating can change genotype or allele frequencies

Biologically, evolution results in a change in the genetics of a population over time. A population is a group of individuals of a single species living and interbreeding in a particular area at the same time. Populations evolve, whereas individuals do not.

The ultimate source of genetic variation in a population is mutation. Because mutations are random changes in genetic material, most are harmful or neutral, but the environment determines whether a particular mutation is disadvantageous or adaptive. Rates of mutation can be high, as in a virus, or they can be low. However, even low rates of mutation can lead to genetic variation in a population. Mutations lead to the production of different forms a gene or alleles. The allele frequency is the proportion of each allele in a population, while the genotype frequency is the proportion of each genotype.

New phenotypes can arise as a result of selection on genetic variation. When selection of individuals with desirable traits is carried out by humans, such as plant and animal breeders, it is referred to as artificial selection. Slight differences in characteristics among individuals may favor survival and reproduction over others. An adaptation is a characteristic that helps its bearer survive and reproduce; the term also refers to the evolutionary process that produces such characteristics.

A number of other mechanisms can also result in evolutionary change within populations. Gene flow occurs when individuals migrate from one population to another and breed in their new location. Gene flow can add new alleles to a population's gene pool, change the frequencies of alleles already present, or both. Genetic drift is caused by chance events that alter allele frequencies in a population. It has its greatest impact on small populations, in which it may even cause harmful alleles to increase in frequency. During a population bottleneck, when a large population is severely reduced in size, allele frequencies may shift drastically, and genetic variation may be reduced as a result of genetic drift. The change in genetic variation that occurs when a few individuals originate a new population is called the founder effect. As in the case of a population bottleneck, some alleles found in the source population will be missing from the founding population, and others will occur with altered frequencies. The preferential mating of individuals with others, either of the same genotype or of a different genotype, is called nonrandom mating. The effect of self-fertilization, another form of nonrandom mating, is a reduction in the frequency of heterozygotes. Sexual selection favors traits that benefit their bearers (generally males) in the competition for access to members of the other sex, or make their bearers more attractive to members of the other sex. Sexual selection often results in sexually dimorphic species, in which males and females differ in appearance.

Question 2. Genetic drift, population bottlenecks, and the founder effect are all means by which allele frequencies can change. Are these processes random or nonrandom? Why?
Textbook Reference: *15.2 Mutation, Selection, Gene Flow, Genetic Drift, and Nonrandom Mating Result in Evolution, pp. 295–296*

Question 3. Describe what is meant by sexual selection and discuss what behavioral ecologists discovered about sexual selection of tail length from studies of male widowbirds.
Textbook Reference: *15.2 Mutation, Selection, Gene Flow, Genetic Drift, and Nonrandom Mating Result in Evolution, pp. 295–296*

15.3 Evolution Can Be Measured by Changes in Allele Frequencies

Evolution will occur unless certain restrictive conditions exist

Deviations from Hardy–Weinberg equilibrium show that evolution is occurring

Evolution can be measured as the changes in allele frequencies in a population. The sum of all allele frequencies at a locus is equal to 1, as is the sum of all genotype frequencies. For a locus with two alleles, the frequencies of the dominant and recessive alleles typically are represented by p and q, respectively; thus $p + q = 1$.

Biologists estimate allele frequencies by measuring numbers of alleles in a sample of individuals from the population. A given allele frequency is then calculated by taking the number of copies of the allele in the population and dividing it by the total number of copies of that allele in the population. If there is only one allele at a locus, the population is *monomorphic* and the allele is *fixed*. If two or more alleles exist, the population is *polymorphic* at that locus. The genetic structure of a population is described by the frequencies of different alleles at each locus and the frequencies of different genotypes.

A population at Hardy–Weinberg equilibrium is not changing genetically and hence is not evolving. To be at Hardy–Weinberg equilibrium, a population must meet five conditions: (1) no mutation; (2) no differential selection among genotypes; (3) no gene flow; (4) the population must be infinite; and (5) random mating. If these conditions are met, allele frequencies at a locus remain the same from generation to generation. Moreover, after one generation of random mating, the genotype frequencies will remain in the proportions $p^2 + 2pq + q^2 = 1$, where p^2, $2pq$, and q^2 represent the frequencies of the homozygous dominant, heterozygous, and homozygous recessive genotypes, respectively. Though populations in nature can never fully meet the conditions of the Hardy–Weinberg equilibrium, this equation is often useful for predicting the approximate genotype frequencies in a population. It is also important because deviations from it may show that evolution is occurring. Moreover, the pattern of deviations is useful in identifying the agents of evolutionary change operating on the population.

Question 4. Discuss the main application of the Hardy–Weinberg rule in evolutionary biology.
Textbook Reference: *15.3 Evolution Can Be Measured by Changes in Allele Frequencies, pp. 298–299*

Question 5. In a population with 600 members, the numbers of individuals of three different genotypes are *AA* = 350, *Aa* = 100, *aa* = 150. Answer the following questions about this population:

(1) What are the genotype frequencies of *AA*, *Aa*, and *aa*?

(2) What are the frequencies of the *A* and *a* alleles?

(3) What would be the expected genotype frequencies of *AA*, *Aa*, and *aa* if this population were in genetic equilibrium?

(4) Is this population in genetic equilibrium? Explain.
Textbook Reference: *15.3 Evolution Can Be Measured by Changes in Allele Frequencies, pp. 298–299*

15.4 Selection Can Be Stabilizing, Directional, or Disruptive

Stabilizing selection reduces variation in populations

Directional selection favors one extreme

Disruptive selection favors extremes over the mean

Natural selection, unlike other agents of evolution, adapts organisms to their environment. In cases in which the distribution of a phenotype approximates a bell-shaped curve because it is controlled by many gene loci, selection can produce any one of three results.

Stabilizing selection reduces variation in the population by favoring average individuals. This can result in the selection against deleterious mutations resulting in purifying selection.

Directional selection changes the mean value for a character by favoring individuals that vary in one direction. This can result in positive selection of a genetic variant at a single locus. Many generations of directional selection will result in an evolutionary trend.

Disruptive selection, by favoring both extremes, leads to a population with two peaks in the distribution of the character. This bimodal distribution will be maintained within the population by disruptive selection.

Question 6. Construct a concept map with the theme of "evolutionary agents." Include in your map the following terms: evolutionary agents, allele frequencies, directional, disruptive, founder effects, gene flow, genetic drift, genetic structure of a population, genotype frequencies, mutation, natural selection, nonrandom mating, phenotypic variation, population bottlenecks, and stabilizing. Connect these terms with verbs or short phrases to indicate the relationships among them.
Textbook Reference: *15.4 Selection Can Be Stabilizing, Directional, or Disruptive, pp. 300–302; 15.2. Mutation, Selection, Gene Flow, Genetic Drift, and Nonrandom Mating Result in Evolution, pp. 292–297*

15.5 Genomes Reveal Both Neutral and Selective Processes of Evolution

Much of molecular evolution is neutral

Positive and purifying selection can be detected in the genome

Heterozygote advantage maintains polymorphic loci

Genome size and organization also evolve

Molecular evolution occurs by diverse mechanisms. A point mutation of a single nucleotide is known as a nucleotide substitution. A synonymous (silent) mutation replaces a nucleotide base in a codon but does not change the amino acid specified by the codon. Because synonymous mutations do not affect the functioning of a protein, they are unlikely to be affected by natural selection. A nonsynonymous mutation changes the amino acid specified by the codon. Though such mutations are likely to be harmful, they are sometimes selectively neutral, or nearly so, and are occasionally advantageous.

Within functional genes, nucleotide substitution rates are highest at nucleotide positions that do not change the amino acid being expressed. The rate of substitution is higher in pseudogenes—duplicate copies of genes that are never expressed—than in functional genes. The neutral theory of molecular evolution postulates that most evolutionary change in macromolecules, as well as much of the genetic variation within species, is the result of random genetic drift, rather than natural selection. The rate of fixation of neutral mutations is theoretically constant and equal to the mutation rate. The rate of evolution of genes or proteins can thus be used as a molecular clock of evolution.

According to the neutral theory of molecular evolution, it is possible to distinguish among evolutionary processes by comparing the rates of synonymous and nonsynonymous substitutions in a protein-coding gene. If an amino acid substitution is neutral in its effect on fitness, then the rates of synonymous and nonsynonymous substitutions in the corresponding DNA sequences are expected to be very similar. If an amino acid position is under strong selection for change, then the rate of nonsynonymous substitutions in the corresponding DNA is expected to exceed the rate of synonymous substitutions. If an amino acid position is under purifying selection, then the rate of synonymous substitutions in the corresponding DNA is expected to be much higher than the rate of nonsynonymous substitutions.

The enzyme lysozyme, while serving in almost all animals as an important first line of defense against invading bacteria, also has evolved to take on an essential role in the digestive process of several groups of foregut fermenters. By comparing the lysozyme-coding sequences in foregut fermenters with several of their non-fermenting relatives, molecular evolutionists have discovered that neutral evolution, purifying selection, and selection for change have all occurred as lysozyme evolved to take on its new function. The independent evolution in several groups of foregut fermenters of a type of lysozyme adapted to its new environment and function shows that convergent evolution occurs at the molecular level.

In some cases, heterozygous individuals have an advantage over homozygous individuals. Heterozygous individuals may be polymorphic for one or more genes. This polymorphism will allow the heterozygous individual to produce two different forms of the same protein, thereby allowing them to cope with a wider range of environmental conditions.

Genome size and organization also evolve. The size of the coding portion of the genome is larger in more complex organisms. Thus, eukaryotes have many times more genes than prokaryotes, and multicellular eukaryotes with tissue organization have more genes than single-celled eukaryotes. Most of the variation in genome size of various organisms is due not to differences in the number of functional genes, but in the amount of noncoding DNA. Although much of the noncoding DNA appears to be nonfunctional, it may alter the expression of surrounding genes. Important categories of noncoding DNA include pseudogenes and parasitic transposable elements.

Question 7. Explain why the majority of mutations are selectively neutral.

Textbook Reference: *15.5 Genomes Reveal Both Neutral and Selective Processes of Evolution, p. 303*

15.6 Recombination, Lateral Gene Transfer, and Gene Duplication Can Result in New Features

Sexual recombination amplifies the number of possible genotypes

Lateral gene transfer can result in the gain of new functions

Many new functions arise following gene duplication

Sexual recombination generates variety in the genotype combinations in a population. Sexual reproduction has a number of disadvantages over asexual reproduction. Adaptive combinations of genes can be broken up, the rate of gene flow to the next generation is slower, and reproduction rate is lower. However, sexual reproduction has a number of benefits as well. It may have evolved to facilitate DNA repair, promote the elimination of deleterious mutations, or to increase the variety of genetic combinations. Sexual reproduction generates new combinations of alleles upon which natural selection can act.

Lateral gene transfer occurs when a species picks up fragments of foreign DNA directly from the environment, or via a virus, or through hybridization with another species. This process increases the genetic variability of the species and thus provides additional raw material on which natural selection can act. It can also result in the spread of genetic functions between distantly related species.

Gene duplication can produce new functions. When a gene is duplicated, four evolutionary outcomes are possible: (1) both copies can retain the gene's original function; (2) both copies can retain the ability to produce the original gene product, but the expression of the genes may diverge in different tissues or at different times of development; (3) one copy can become a functionless pseudogene; or (4) one copy can retain its original function, while the second mutates so extensively that it can perform a different function. When an entire genome is duplicated (as in polyploid organisms), there are major opportunities for new gene functions to evolve.

Genome duplication events that occurred in the ancestor of the jawed vertebrates have permitted many individual vertebrate genes to become highly tissue-specific in their expression. Successive rounds of gene duplication and mutation can result in a gene family—a group of homologous genes with related functions. The globin gene family shows that gene diversification can produce molecules with different functions (hemoglobin and myoglobin) as well as functionless pseudogenes.

Question 8. Describe the advantages and disadvantages of sexual and asexual reproduction.
Textbook Reference: *15.6 Recombination, Lateral Gene Transfer, and Gene Duplication Can Result in New Features, pp. 308–309*

15.7 Evolutionary Theory Has Practical Applications

Knowledge of gene evolution is used to study protein function

In vitro evolution produces new molecules

Evolutionary theory provides multiple benefits to agriculture

Knowledge of molecular evolution is used to combat diseases

The principles of molecular evolution help us understand function and diversification of function in many proteins. For example, detection of strong selection for change in a nucleotide sequence can help us identify molecular changes that have resulted in functional changes. Molecular evolutionary principles underlie the field of in vitro evolution, in which new molecules are produced in the laboratory to perform particular desired functions. The basis of in vitro evolution is the creation of random molecular variation followed by selection by the experimenter. Agriculture has used evolutionary theory to combat pesticide resistance. Biomedical scientists are using principles of molecular evolution to identify and combat human diseases.

Question 9. Compare in vitro evolution with molecular evolution in organisms.
Textbook Reference: *15.7 Evolutionary Theory Has Practical Applications, p. 311*

Question 10. How is the study of molecular evolution important in efforts to combat HIV and other viral pathogens that have recently emerged?
Textbook Reference: *15.7 Evolutionary Theory Has Practical Applications, p. 312*

Test Yourself

1. Evolution occurs at the level of
   a. the individual genotype.
   b. the individual phenotype.
   c. environmentally based phenotypic variation.
   d. the population.
   e. the species.
   Textbook Reference: *15.2. Mutation, Selection, Gene Flow, Genetic Drift, and Nonrandom Mating Result in Evolution, p. 292*

2. Natural selection acts on
   a. the gene pool of the species.
   b. the genotype.
   c. the phenotype.
   d. multiple gene inheritance systems.
   e. the environment.
   Textbook Reference: *15.2. Mutation, Selection, Gene Flow, Genetic Drift, and Nonrandom Mating Result in Evolution, p. 293*

3. In comparing several populations of the same species, the population with the greatest genetic variation will have the
   a. greatest number of genes.
   b. greatest number of alleles per gene.
   c. greatest number of population members.
   d. largest gene pool.
   e. None of the above
   Textbook Reference: *15.2. Mutation, Selection, Gene Flow, Genetic Drift, and Nonrandom Mating Result in Evolution, p. 293*

4. The ability to taste the chemical PTC (phenylthiocarbamide) is determined in humans by a dominant allele *T*, with tasters having the genotypes *Tt* or *TT* and nontasters having *tt*. If 36 percent of the members of a population cannot taste PTC, then according to the Hardy–Weinberg rule, the frequency of the *T* allele should be
   a. 0.36.
   b. 0.4.
   c. 0.6.
   d. 0.64.
   e. 0.8.
   Textbook Reference: *15.3 Evolution Can Be Measured by Changes in Allele Frequencies, p. 298*

5. A gene in humans has two alleles, *M* and *N*, that code for different surface proteins on red blood cells. If you know that the frequency of allele *M* is 0.2, according to the Hardy–Weinberg rule, the frequency of the genotype *MN* in the population should be
   a. 0.16.
   b. 0.2.
   c. 0.32.
   d. 0.64.
   e. 0.8.
   Textbook Reference: *15.3 Evolution Can Be Measured by Changes in Allele Frequencies, p. 298*

6. Random genetic drift would probably have its greatest effect on a
   a. small, isolated population.
   b. large population in which mating is nonrandom.
   c. large population in which mating is random.
   d. large population with regular immigration from a neighboring population.
   e. large population with a high mutation rate.
   Textbook Reference: *15.2 Mutation, Selection, Gene Flow, Genetic Drift, and Nonrandom Mating Result in Evolution, pp. 294–295*

7. Allele frequencies for a gene locus are *least* likely to be significantly changed by
   a. mutation.
   b. the founder effect.
   c. self-fertilization.
   d. gene flow.
   e. natural selection.
   Textbook Reference: *15.6 Recombination, Lateral Gene Transfer, and Gene Duplication Can Result in New Features, p. 308*

8. Which of the following evolutionary agents would produce nonrandom changes in the genetic structure of a population?
   a. Self-fertilization
   b. Population bottlenecks
   c. Mutation
   d. Natural selection
   e. Both a and d
   Textbook Reference: *15.2 Mutation, Selection, Gene Flow, Genetic Drift, and Nonrandom Mating Result in Evolution, pp. 293–295*

9. Suppose that a particular species of flowering plant lives only one year and can produce red, white, or pink blossoms, depending on its genotype. Biologists studying a population of this species count 300 red-flowering, 500 white-flowering, and 800 pink-flowering plants. When a census of the population is taken the following year, 600 red-flowering, 900 white-flowering, and 1,000 pink-flowering plants are observed. Which color has the highest fitness?
   a. Red
   b. White
   c. Pink
   d. All colors are equally fit.
   e. The answer cannot be determined from this information.
   Textbook Reference: *15.5 Genomes Reveal Both Neutral and Selective Processes of Evolution, p. 304*

10. The graph below shows the range of variation among population members for a trait determined by multiple genes.

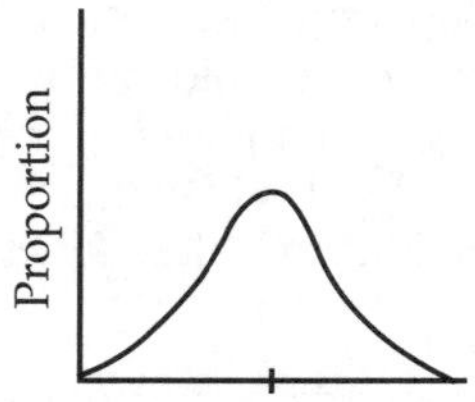

If this population is subject to stabilizing selection for several generations, which distribution(s) would most likely result?

a.
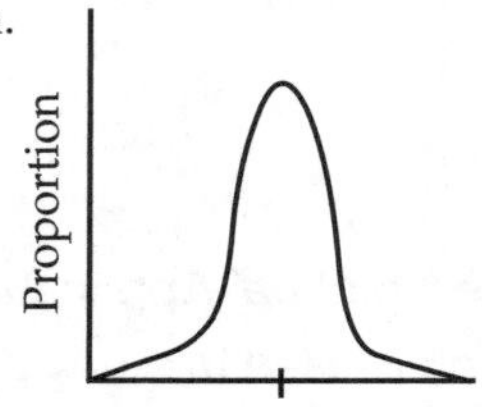

b.
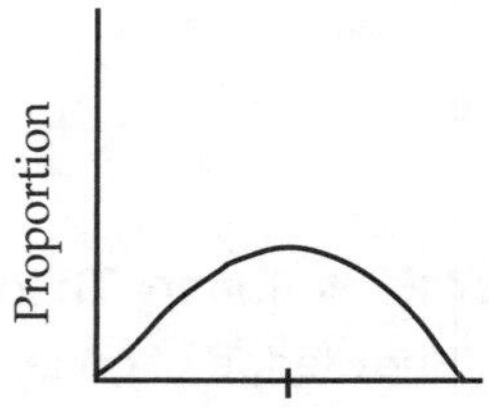

c.
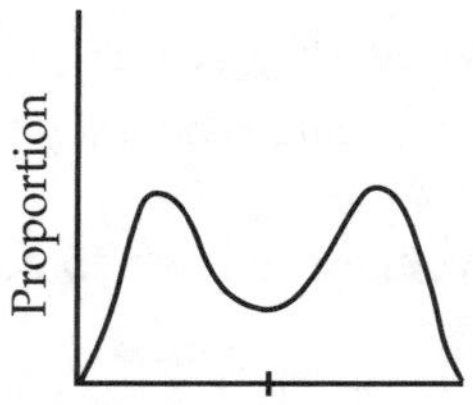

d.
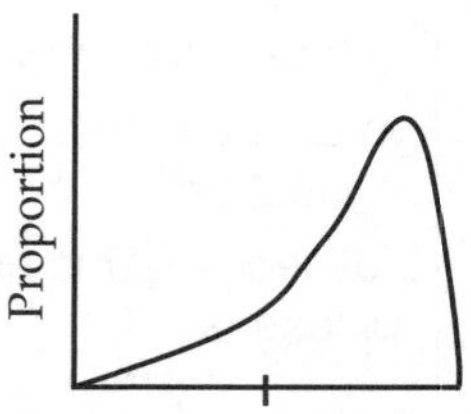

   e. Both a and b
   Textbook Reference: *15.4 Selection Can Be Stabilizing, Directional, or Disruptive, p. 300*

11. In areas of Africa in which malaria is prevalent, many human populations exist in which the allele that produces sickle-cell disease and the allele for normal red blood cells occur at constant frequencies, despite the fact that sickle-cell disease frequently causes death at an early age. This phenomenon is an example of
   a. the founder effect.
   b. a stable polymorphism.
   c. mutation.
   d. nonrandom mating.
   e. Both b and c
   Textbook Reference: *15.3 Evolution Can Be Measured by Changes in Allele Frequencies, p. 297*

12. Which of the following is *not* a disadvantage of sexual reproduction?
   a. When it involves separate genders, it reduces the overall reproductive rate.
   b. It breaks up adaptive combinations of genes.
   c. It reduces the rate at which females pass genes on to their offspring.
   d. It increases the difficulty of eliminating harmful mutations from the population.
   e. All of the above are disadvantages of sexual reproduction.

Textbook Reference: *15.6 Recombination, Lateral Gene Transfer, and Gene Duplication Can Result in New Features, p. 308*

13. In a eukaryote, one would expect to find the lowest rate of nonsynonymous nucleotide substitutions in an
    a. intron of a protein-coding gene.
    b. exon of a protein-coding gene.
    c. intron of a pseudogene.
    d. exon of a pseudogene.
    e. intron or exon of a protein-coding gene.
    Textbook Reference: *15.5 Genomes Reveal Both Neutral and Selective Processes of Evolution, pp. 302–303*

14. Which of the following statements about mutations is *false*?
    a. A silent mutation results in no change in the amino acid sequence of a protein.
    b. According to the neutral theory of molecular evolution, most substitution mutations are selectively neutral and accumulate through genetic drift.
    c. A base substitution mutation in the third codon position is more likely to be neutral than a substitution at the first or second codon position.
    d. Nonsynonymous mutations are virtually always deleterious to the organism.
    e. The rate of fixation of neutral mutations is equal to the mutation rate and is independent of population size.
    Textbook Reference: *15.5 Genomes Reveal Both Neutral and Selective Processes of Evolution, pp. 302–303*

15. Which of the following statements about the enzyme lysozyme is true?
    a. A small group of closely related mammals has evolved a special form of lysozyme that functions in digestion.
    b. The lysozymes found in the foregut fermenters resulted from convergent evolution.
    c. Lysozyme could not have evolved a secondary function if it had been an enzyme with a vital primary function.
    d. A higher mutation rate in the foregut fermenters allows their lysozymes to evolve rapidly.
    e. Lysozyme first evolved as a defense against bacteria in the common ancestor of mammals.
    Textbook Reference: *15.5 Genomes Reveal Both Neutral and Selective Processes of Evolution, p. 304*

16. Which of the following ranks the organisms correctly in terms of the expected total amount of coding DNA in their genomes (from least coding DNA to most coding DNA)?
    a. Bacterium, single-celled eukaryote, *Drosophila*, bird
    b. Bacterium, *Drosophila*, bird, single-celled eukaryote
    c. Single-celled eukaryote, bacterium, *Drosophila*, bird
    d. *Drosophila*, single-celled eukaryote, bird, bacterium
    e. *Drosophila*, bacterium, single-celled eukaryote, bird
    Textbook Reference: *15.5 Genomes Reveal Both Neutral and Selective Processes of Evolution, pp. 306–307*

17. Which of the following would be the *least* likely result of gene duplication?
    a. Production of less of the gene product than was produced before duplication.
    b. Expression of the two copies of the gene at different stages of the organism's development
    c. One copy retaining its original function and the other copy acquiring a different function
    d. One copy of the gene remaining functional and the other copy evolving into a functionless pseudogene
    e. All of the above are about equally likely results.
    Textbook Reference: *15.6 Recombination, Lateral Gene Transfer, and Gene Duplication Can Result in New Features, p. 309*

18. In vitro evolution
    a. can produce both nucleic acid and protein molecules unknown in living organisms.
    b. requires many rounds of production of variant molecules and the selection of those that look promising in terms of the desired properties.
    c. often involves techniques and molecules employed in recombinant DNA technology, such as PCR and cDNA.
    d. Both b and c
    e. All of the above
    Textbook Reference: *15.7 Evolutionary Theory Has Practical Applications, p. 311*

Answers

Key Concept Review

1. A theory typically refers to an untested hypothesis in science. The concept of evolutionary theory is not limited to a single hypothesis, but refers to many different concepts. Additionally, there are vast amounts of data that support evolutionary theory.

2. These processes are random processes that typically occur in small populations. Genetic drift refers to random changes in allele frequencies. A population bottleneck occurs after an environmental event in which a large percentage of a population has died. In this case, the survival of an individual is due solely to chance, as are the resulting allele frequencies among these survivors. In the founder effect, the small number of individuals that found a population is unlikely to have the same allele frequency as the source population. As with the other two mechanisms, the alleles present in the founding population is going to be random.

3. Sexual selection is the spread of a trait that improves the reproductive success of an individual. The trait may improve the ability of its bearer to compete with other members of its sex for access to mates, or it may make its bearer more attractive to members of the opposite sex. By artificially lengthening or shortening the tails of male widowbirds, behavioral ecologists were able to show that increased tail length in males made them more attractive to females but did not give the long-

tailed males an advantage in their interactions with other males.

4. No population meets the Hardy–Weinberg conditions for genetic equilibrium. However, by determining how a population deviates from the expectations of Hardy–Weinberg, evolutionary biologists can identify which evolutionary agents are affecting the population.

5.

(1) The genotype frequencies in this population are the following:

$AA = 350/600 = 0.58$

$Aa = 100/600 = 0.17$

$aa = 150/600 = 0.25$

(2) The allele frequencies in this population are the following:

$A = (700 + 100)/1200 = 0.67$

$a = (300 + 100)/1200 = 0.33$

(3) The expected genotype frequencies of this population in genetic equilibrium would be the following:

$AA = p^2 = (0.67)^2 = 0.45$

$Aa = 2pq = 2 \times 0.67 \times 0.33 = 0.44$

$aa = q^2 = (0.33)^2 = 0.11$

(4) The population is not in genetic equilibrium. Using the observed allele frequencies for p and q, the observed genotypic frequencies differ from those predicted by the Hardy–Weinberg rule.

6.

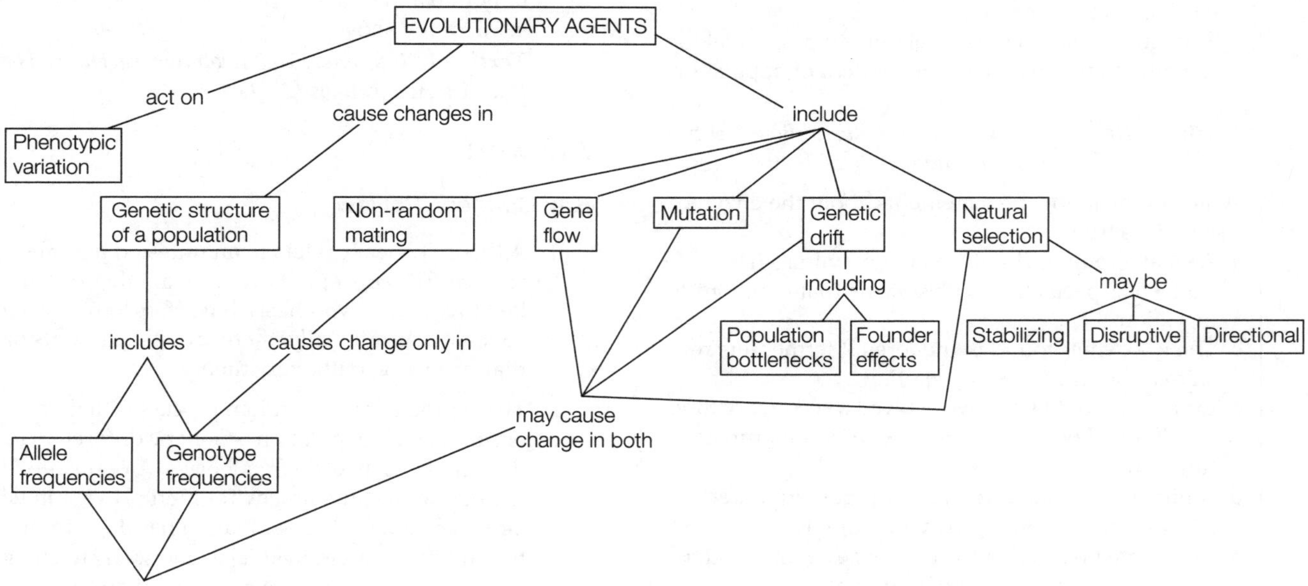

7. When a mutation occurs it can be either synonymous (silent), with no change in the encoded amino acid, or nonsynonymous, with a change in the encoded amino acid. One type of nonsynonymous substitution results in a protein with an altered or different function. Some nonsynonymous substitutions of amino acids are neutral because the resulting protein function has not been altered. When a synonymous substitution or neutral nonsynonymous substitution occurs, there is neither an advantage nor disadvantage to the change. Therefore, there is no selection against these mutations and they tend to accumulate in the population. In the case of deleterious mutations, they are rapidly selected against and removed from the population.

8. In asexual reproduction, an organism's offspring is genetically identical to the parent. One advantage is the rapid rate of reproduction possible in asexual species. The major disadvantage is that deleterious mutations cannot be easily eliminated from the population. Sexual reproduction has the disadvantage of being slower and potentially breaking up advantageous combinations of genes. It has the advantage of creating greater genetic variation and a means to eliminate deleterious alleles from the population.

9. Variation and selection are involved in both in vitro evolution and in molecular evolution in organisms. With in vitro evolution, a huge number of variant molecules are created in the laboratory, and the human experimenters select those that show any sign of having the desired property. Many repetitions of these two steps—production of variants and selection—eventually produce the targeted molecule. In molecular evolution in organisms, naturally occurring mutations in previously existing DNA sequences are the ultimate source of variation, and the environment determines through natural selection which mutations are the favorable variants. However, according to the neutral theory of molecular evolution, much evolutionary change in DNA (and in the encoded proteins) is not adaptive but rather the result of genetic drift.
10. The principles of molecular evolution are critical for understanding the origin and evolutionary development of such pathogenic viruses as hantaviruses, the SARS virus, and HIV. By studying the evolutionary changes that occur in such viruses, medical researchers gain valuable insights into such questions as how some viruses are able to switch from an animal to a human host and how vaccines can remain effective as viruses evolve. In the future, as more extensive genomic databases and evolutionary trees for viruses are developed, it will be possible to identify and treat a much wider array of human diseases.

Test Yourself

1. **d.** Evolution is defined as changes in the genetic structure of a population over time, so evolution occurs at the level of the population.
2. **c.** Natural selection acts on phenotypes, not genotypes. For example, a harmful recessive allele is "invisible" to natural selection when it occurs in a heterozygote, where its harmful effect is masked by the dominant allele.
3. **b.** Genetic variation is related to the number of different alleles per gene. Answer a is incorrect because all members of a species have the same number of genes. Population size per se has little to do with genetic variation, so choices c and d are also incorrect.
4. **b.** The frequency of the *tt* genotype is $q^2 = 0.36$, so the frequency of the *t* allele is q (0.6, the square root of 0.36). If there are only two alleles for this trait, then $T + t = 1$. The frequency of the *T* allele is therefore $1 - t = 1 - 0.6 = 0.4$.
5. **e.** Because $p = 0.2$, $q = 1 - 0.2 = 0.8$, the frequency of the *MN* genotype is $2pq$, or $2 \times 0.2 \times 0.8 = 0.32$.
6. **a.** Genetic drift is most significant in small populations.
7. **c.** Unlike the founder effect, mutation, gene flow, and natural selection, all of which may change allele frequencies in a population, self-fertilization (like other types of nonrandom mating, with the exception of sexual selection) only causes a deviation from the frequency of heterozygotes predicted by Hardy–Weinberg equilibrium.
8. **e.** Self-fertilization (a) leads to increased numbers of homozygous individuals, and natural selection (d) is also nonrandom.
9. **a.** Fitness measures the relative contribution of a genotype or phenotype to subsequent generations. The red-flowering plants, which doubled in number, had the greatest percentage increase of any of the plants and thus had the highest fitness.
10. **a.** Stabilizing selection results when individuals that are intermediate in phenotype make a larger contribution to future generations than individuals of a more extreme phenotype. This leads to reduced variation for the trait and causes the curve to be higher and narrower. Curve *b* shows greater variation, curve *c* would result from disruptive selection, and curve *d* would result from directional selection.
11. **b.** Polymorphism in a population is the existence of two or more alleles at a particular gene locus. If phenotypes are stable through time, then the underlying alleles will also be constant. In this instance, the polymorphism is stable because malaria is a significant cause of mortality in some parts of Africa, and heterozygotes have greater resistance to this disease than individuals with a "normal" phenotype. Thus the superior fitness of the heterozygotes maintains both alleles in the population.
12. **d.** Sexual recombination produces some individuals in a population who are less fit than others because they carry a greater than average number of deleterious mutations. Because these individuals are selected against, sexual reproduction is able to reduce the number of deleterious mutations in the population over time.
13. **b.** Eukaryotic genes usually contain both protein-coding regions (exons) and noncoding regions (introns). A substantial proportion of nonsynonymous substitutions occurring in an exon of a gene would most likely be deleterious and therefore be eliminated by selection. Because introns and pseudogenes are not expressed, nonsynonymous substitutions occurring in them cannot be selected against.
14. **d.** A nonsynonymous substitution mutation can be selectively neutral if, as sometimes happens, it results in an amino acid change that has no significant effect on the shape (and hence the functional properties) of the protein.
15. **b.** Because the animals in which a similar lysozyme has evolved do not share a recent common ancestor, the mechanism involved is convergent evolution. All the other statements are false.
16. **a.** There is a rough relationship between the amount of coding DNA and organismal complexity.

17. **a.** If there are two copies of the gene, it is more likely to produce more of its product than less of it.

18. **e.** In vitro evolution can create novel molecules not known to occur in living organisms. The process starts with a very large pool of variant molecules (nucleic acid or protein) and involves selection on the part of the experimenters of those molecules that show some promise in terms of their eventual possession of the desired properties. Many rounds of production of variant molecules and selection are typically needed to produce the final product.

Reconstructing and Using Phylogenies

16

The Big Picture

- Several factors have combined in recent years to bring new excitement to the field of systematics. First was the development of a new, rigorous approach known as cladistics. More recently, new methods of sequencing DNA and RNA, coupled with enormous increases in computer power, have enabled biologists to apply sophisticated mathematical techniques, such as maximum likelihood analyses, to phylogenetic studies. As a result, phylogenetics is now contributing to a wide array of biological studies, including subjects of medical importance such as the origins and types of human immunodeficiency virus (HIV) and other infectious organisms.
- The parsimony principle, which holds that one should prefer the simplest hypothesis capable of explaining the known facts, is fundamental not only to the reconstruction of phylogenies but also to every other field of scientific research.
- The concept of evolution has revolutionized taxonomy. Before Darwin's time, Linnaeus and others had developed "natural" systems of classification based primarily on similarity of morphology. Modern biologists view similarities of organisms (other than homoplasies) as the result of descent from a common ancestor and expect a system of classification to reflect evolutionary relationships.
- Knowing that organisms are evolutionarily related enables biologists to make predictions about their characteristics. This knowledge can provide important hints in the search for organisms with valuable properties, such as the ability to produce medically useful drugs.

Study Strategies

- The concept of homology can be confusing. Figure 16.2 provides a helpful image of the difference between homologous and homoplastic traits.
- Distinguishing monophyletic, paraphyletic, and polyphyletic groups is difficult for many students. Bear in mind that a monophyletic group is analogous to a branch (or twig) of a tree: a single "cut" can remove it from a phylogenetic tree. This analogy should help you work out which kinds of "cuts" would result in paraphyletic and polyphyletic groups.
- Here is a mnemonic for the hierarchy of taxa (kingdom, phylum, class, order, family, genus, species) in the Linnaean system of classification: "Kindly Professors Cannot Often Fail Good Students." Note that the plural of genus is genera; the words *general* and *generic* come from the same root; this can help you remember that the genus is the more general taxon in the genus/species binomial. *Species* and *specific* also come from the same root, and the species is, of course, the most specific taxon.
- Go to yourBioPortal.com to review the following tutorials and activities:

  Animated Tutorial 16.1 Using Phylogenetic Analysis to Reconstruct Evolutionary History

  Interactive Tutorial 16.1 Phylogeny and Molecular Evolution

  Web Activity 16.1 Constructing a Phylogenetic Tree

  Web Activity 16.2 Types of Taxa

  Working with Data 16.1 Constructing a Phylogenetic Tree

Key Concept Review

16.1 All of Life Is Connected through Its Evolutionary History

Phylogenetic trees are the basis of comparative biology

Derived traits provide evidence of evolutionary relationships

A phylogeny is a description of the evolutionary history of relationships among organisms or their genes. Phylogenetic trees display the order in which lineages are hypothesized to have split. Each split (or node) in a phylogenetic tree represents a point at which lineages diverged in the past. The common ancestor of all the organisms in the tree forms the root of the tree. The timing of separations between lineages of organisms is shown by the positions of nodes on a time or divergence axis. A taxon is any named group of species.

A taxon consisting of all the evolutionary descendants of a common ancestor is called a clade. Just as species that are each other's closest relatives are called sister species, so clades that are each other's closest relatives are called sister clades. Systematics is the study and the classification of the diversity of life.

All of life is connected through its evolutionary history, known as the tree of life. The evolutionary relationships among species, as shown by the tree of life, form the basis for biological classification. Knowledge about these relationships is important when one makes comparisons of species, populations, or genes. Homologous traits are keys to reconstructing phylogenetic trees because they are features that are shared by members of a lineage because of their descent from a common ancestral trait. A trait that differs from its ancestral form is called a derived trait. Conversely, a trait that was present in the ancestor of a group is known as an ancestral trait for that group. Synapomorphies are derived traits that are shared among a group of organisms and are viewed as evidence of the common ancestry of the group. Homoplasies (homoplastic traits) create confusion in reconstructing the evolutionary history of a lineage because they are features that are similar for some reason other than descent from a common ancestral trait. Two processes generate homoplasies: convergent evolution and evolutionary reversals. In convergent evolution, features that evolved independently become superficially similar. In an evolutionary reversal, a character reverts from a derived state to an ancestral one.

Question 1. All but one of the trees shown in the diagram below portray the same phylogenetic relationships among taxa A, B, C, D, E, F, and G. Which one depicts a different phylogeny?
Textbook Reference: *16.1 All of Life Is Connected through Its Evolutionary History, pp. 316–317*

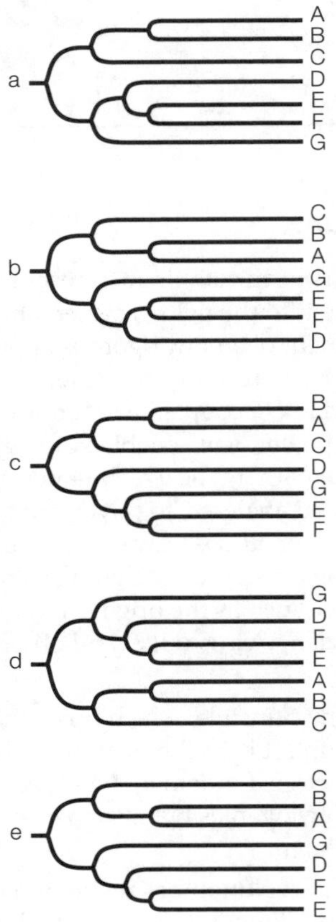

Question 2. In the phylogeny shown below, which group of taxa would *not* constitute a clade?
Textbook Reference: *16.1 All of Life Is Connected through Its Evolutionary History, pp. 316–317*

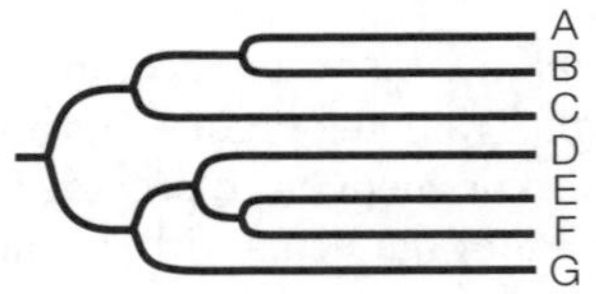

a. A, B, C, D, E, F, and G and their common ancestor
b. E, F, and G and their common ancestor
c. A, B, and C and their common ancestor
d. C, D, E, and F
e. Both b and d

Question 3. You are constructing a phylogenetic tree for several animals, including two flying species. You determine that the presence of wings in the two flying species can be explained by convergent evolution. How can this trait help you in creating the phylogenetic tree for the species.
Textbook Reference: *16.1 All of Life Is Connected through Its Evolutionary History, p. 318*

Question 4. Discuss the implications of the following statement for the field of systematics: "DNA is the genetic material for all prokaryotes and eukaryotes."
Textbook Reference: *16.1 All of Life Is Connected through Its Evolutionary History, p. 318*

16.2 Phylogeny Can Be Reconstructed from Traits of Organisms

- Parsimony provides the simplest explanation for phylogenetic data
- Phylogenies are reconstructed from many sources of data
- Mathematical models expand the power of phylogenetic reconstruction
- The accuracy of phylogenetic methods can be tested

Reconstructing accurate phylogenies involves several steps. The first step is the choice of the ingroup and the appropriate outgroup. The ingroup is an assemblage of organisms whose phylogeny is to be determined. The outgroup can be any species or group of species outside the group of interest. One method of distinguishing ancestral and derived traits is to compare the ingroup to an outgroup. Ancestral traits should be present in both, whereas derived traits should occur only in the ingroup. The root of the tree is determined by the relationship of the ingroup to the outgroup.

In reconstructing phylogenies, systematists are guided by the parsimony principle, which states that the preferred explanation of the observed data is the simplest explanation. In practice, this means that the best phylogenetic reconstruction is the one that minimizes the number of evolutionary changes that need to be assumed over all characters in all groups in the tree. In other words, the best hypothesis is the one that requires the fewest homoplasies.

Phylogenies are constructed from many sources of data. Morphology—the presence, size, shape, and other attributes of body parts—is an important source of traits for phylogenetic analysis. The morphology of fossils is particularly useful in helping to distinguish ancestral and derived traits. The fossil record also reveals when lineages diverged. In groups with few living representatives, information on extinct species may be critical to an understanding of the large divergences among the surviving species.

Similarities in developmental pattern also may reveal evolutionary relationships, since structures in early developmental stages may reveal relationships that are not evident in adults. Behavior, if genetically determined, is another useful source of information.

The complete genome of an organism contains an enormous set of traits (the individual nucleotide bases of DNA) that can be used to analyze phylogenies. Like morphological characters, molecules are heritable characteristics that may diverge among lineages. Both nuclear and organelle DNA sequences are used in phylogenetic studies, as are sequences in gene products (such as the amino acid sequences of proteins). Because the chloroplast genome has changed slowly over evolutionary time, it is often used for the study of relatively ancient phylogenetic relationships among plants. Animal mitochondrial DNA has changed more rapidly, making it useful for studies of evolutionary relationships among closely related animal species.

Biologists have conducted experiments both in living organisms and with computer simulations that have demonstrated the effectiveness and accuracy of phylogenetic methods. The maximum likelihood method uses computer analysis for the reconstruction of phylogenies. A likelihood score of a tree is based on the probability that the observed data evolved on the specified tree, given an explicit mathematical model of evolution for the characters. An advantage of maximum likelihood analyses is that they incorporate more information about evolutionary change than parsimony methods do.

Question 5. The phylogenetic tree below shows the evolutionary relationships of five species (A–E) relative to five traits (1–5). Based on this tree, fill in the table below it, using 1 to indicate the presence of a derived trait and 0 to indicate the presence of an ancestral trait.
Textbook Reference: *16.2 Phylogeny Can Be Reconstructed from Traits of Organisms, p. 319*

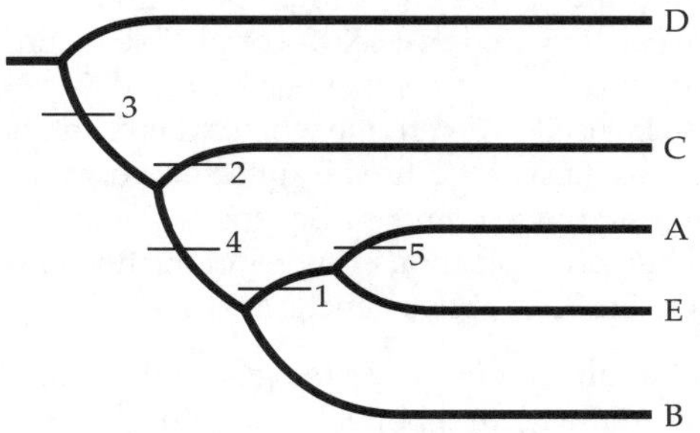

SPECIES	TRAIT 1	2	3	4	5
A					
B					
C					
D					
E					

Question 6. The following table shows the ancestral and derived traits of five species (A–E). Based on the table, and following conventions presented in the textbook, construct a phylogenetic tree that represents the evolutionary relationships of this group. In this table, the ancestral state of each trait is indicated by 0 and the derived state is indicated by 1.
Textbook Reference: *16.2 Phylogeny Can Be Reconstructed from Traits of Organisms, p. 319*

	TRAIT				
SPECIES	1	2	3	4	5
A	1	1	1	0	0
B	0	0	0	0	0
C	1	1	0	1	0
D	0	1	0	0	0
E	1	1	0	1	1

Question 7. Discuss the application of the parsimony principle in the construction of phylogenetic trees.
Textbook Reference: *16.2 Phylogeny Can Be Reconstructed from Traits of Organisms, p. 320*

16.3 Phylogeny Makes Biology Comparative and Predictive

Reconstructing the past is important for understanding many biological processes

Phylogenies allow us to understand the evolution of complex traits

Ancestral states can be reconstructed

Molecular clocks help date evolutionary events

Phylogenetic trees provide information that helps biologists answer many kinds of questions. Phylogenetic trees are used to determine how many times a particular trait has evolved within a lineage. Using the phylogeny of angiosperm fertilization, scientists have determined that self-incompatible mating is the ancestral state and that self-compatibility evolved three times. They can also be used to determine when, where, and how zoonotic diseases (diseases caused by infectious organisms that have been transferred to humans from another animal host) first entered human populations. Studies of the transmission of HIV have benefited from phylogenetic analysis.

Phylogenetic analysis is used by biologists to help them make comparisons among genes, populations, and species, and they also provide an understanding of how complex traits evolve. Phylogenetic methods are used not only to discover the evolutionary relationships among lineages of living organisms, but also to reconstruct morphological, behavioral, and molecular characteristics of ancestral species. In a molecular clock analysis, biologists assume that particular DNA sequences evolve at a reasonably constant rate and can beused as a metric to gauge the time of divergence for a particular split in a phylogeny. Molecular clocks must be calibrated with independent data such as the fossil record, known times of divergence, or biogeographic dates.

Question 8. How does knowledge the phylogeny of a group of organisms one is studying contribute to one's biological understanding of the organisms?
Textbook Reference: *16.3 Phylogeny Makes Biology Comparative and Predictive, pp. 324–327*

16.4 Phylogeny Is the Basis of Biological Classification

Evolutionary history is the basis for modern biological classification

Several codes of biological nomenclature govern the use of scientific names

Biological classification systems are designed to express evolutionary relationships among organisms. The system of binomial nomenclature developed by Linnaeus in 1758 assigns two names to each species, one identifying the species itself and the other the genus to which it belongs. The generic name (always capitalized) is followed by the species name (lower case), and both are italicized. In the Linnaean system of classification, species are grouped into higher-order taxa. The hierarchy of taxa, ranked from most to least inclusive, is kingdom, phylum, class, order, family, genus, species. Thus, a genus includes one or more species, a family includes one or more genera, and so forth.

Biologists today recognize the tree of life as the basis for classification and often name clades without placing them into any Linnaean rank. Taxa in biological classifications are expected to be monophyletic. A monophyletic taxonomic group (a clade) contains an ancestor, all descendants of that ancestor, and no other organisms. A polyphyletic group does not include its common ancestor, whereas a paraphyletic group includes some, but not all, descendants of a particular ancestor.

Polyphyletic and paraphyletic groups are inappropriate as taxonomic units. Several sets of rules govern the use of scientific names, with the goal of providing unique and universal names for biological taxa. One such rule is that if a species is named more than once, the valid name is the first name proposed. In the past, different sets of taxonomic rules were developed by zoologists, botanists, and microbiologists, resulting in many duplicated names. Today, taxonomists are developing rules to ensure that every taxon has a unique name.

Question 9. Why are polyphyletic and paraphyletic groups inappropriate as taxonomic units?
Textbook Reference: *16.4 Phylogeny Is the Basis of Biological Classification, p. 329*

Test Yourself

1. A group that consists of all the evolutionary descendants of a common ancestor is called a(n)
   a. grade.
   b. taxon.
   c. homology.
   d. ingroup.
   e. clade.

   Textbook Reference: *16.1 All of Life Is Connected through Its Evolutionary History, p. 317*

2. A synapomorphy is
   a. the product of convergent evolution.
   b. the result of an evolutionary reversal.
   c. a shared derived characteristic.
   d. a trait that was present in the ancestor of a group.
   e. a phylogenetic tree.

   Textbook Reference: *16.1 All of Life Is Connected through Its Evolutionary History, p. 318*

3. Members of genus *X*, a hypothetical taxon of invertebrates, have antennae with a variable number of segments. Species A and B have 10 segments; species C and D have 9 segments; and species E has 8 segments. In all other genera in this family (including genus *Y*), all species have antennae with 10 segments. Which of the following character states is a synapomorphy that would be useful for determining evolutionary relationships within genus *X*?
   a. 10-segment antennae in species A and B
   b. 10-segment antennae in genus *Y* and in two species of genus *X*
   c. Antennae with fewer than 10 segments in species C, D, and E
   d. 8-segment antennae in species E
   e. Both c and d

   Textbook Reference: *16.1 All of Life Is Connected through Its Evolutionary History, p. 318*

4. Which of the following would *not* be expected to result in homoplasy?
   a. Convergent evolution
   b. The independent evolution of similar structures in different lineages
   c. Selection for traits that perform similar functions
   d. The inheritance of ancestral traits
   e. An evolutionary reversal

   Textbook Reference: *16.1 All of Life Is Connected through Its Evolutionary History, p. 318*

5. A derived trait is one that
   a. differs from its ancestral form.
   b. is homologous with another trait found in a related species.
   c. is the product of an evolutionary reversal.
   d. has the same function, but not the same evolutionary origin, as a trait found in another species.
   e. is found only in members of the outgroup.

   Textbook Reference: *16.1 All of Life Is Connected through Its Evolutionary History, p. 318*

6. Which of the following statements about reconstructing phylogenies is *false*?
   a. Traits found in the outgroup as well as in the ingroup are likely to be ancestral traits.
   b. Shared traits are generally assumed to be homoplastic until they can be proven to be homologous.
   c. Phylogenetic trees do not always provide an explicit time scale by which to date the splits between lineages.
   d. In a phylogenetic tree, branches can be rotated around any node without changing the meaning of the tree.
   e. A particular trait may be either ancestral or derived depending on the point of reference of the phylogeny.

   Textbook Reference: *16.2 Phylogeny Can Be Reconstructed from Traits of Organisms, pp. 318–320*

7. Which of the following is the most significant limitation of fossils as a source of information about evolutionary history?
   a. It is sometimes impossible to determine when a fossil organism lived.
   b. The fossil record for many groups is fragmentary or even nonexistent.
   c. Most fossils contain no nucleic acids or proteins and therefore are worthless for studies of molecular evolution.
   d. It is impossible to determine if morphologically similar fossils belong to the same species, because one cannot know if the fossil species interbred.
   e. Most fossils provide no information about the morphology of soft anatomical structures or about external characteristics such as color.

   Textbook Reference: *16.2 Phylogeny Can Be Reconstructed from Traits of Organisms, p. 321*

8. Which of the following sources of molecular data would be most helpful for a study of the evolutionary relationships of closely related animal species?
   a. Chloroplast DNA
   b. Mitochondrial DNA
   c. The amino acid sequences of a protein found in all animals, such as cytochrome *c*
   d. Ribosomal RNA sequences
   e. Both b and d

   Textbook Reference: *16.2 Phylogeny Can Be Reconstructed from Traits of Organisms, p. 321*

9. Which of the following statements about the role of molecular clocks in phylogenetic analyses is true?
   a. A given gene usually evolves at the same rate in two different species regardless of differences in generation time of the species.
   b. Because changes in DNA sequences occur very slowly, molecular clocks can be used only to date evolutionary divergences that occurred millions of years ago.

c. Molecular clocks must be calibrated with independent data, such as the fossil record.
d. Even in a group of closely related species, different genes have been found to evolve at different rates.
e. Both c and d
Textbook Reference: *16.3 Phylogeny Makes Biology Comparative and Predictive, p. 327*

10. Which of the following statements describes a purpose for which biologists use phylogenetic trees?
a. Phylogenetic trees are helpful in determining when and where an infectious organism once found in other animals first entered human populations.
b. Phylogenetic trees are useful for determining how many times a particular trait may have evolved independently within a lineage.
c. Phylogenetic trees can be used to reconstruct ancestral traits.
d. Phylogenetic trees can be used in conjunction with molecular clocks to estimate the timing of evolutionary events.
e. All of the above
Textbook Reference: *16.3 Phylogeny Makes Biology Comparative and Predictive, pp. 324–327*

11. The *most* important attribute of a biological classification scheme is that it
a. avoids the ambiguity created by the use of common names.
b. reflects the evolutionary relationships among organisms.
c. helps us organize our knowledge about organisms and their traits.
d. improves our ability to make predictions about the morphology and behavior of organisms.
e. groups together organisms with similar traits.
Textbook Reference: *16.4 Phylogeny Is the Basis of Biological Classification, p. 329*

12. Suppose you are writing a scientific paper about a unicellular green alga called *Chlamydomonas reinhardtii.* What is the proper way to refer to this species after the full binomial has been used once?
a. *Chlamydomonas reinhardtii*
b. *Chlamydomonas* spp.
c. *Chlamydomonas* sp.
d. *C. reinhardtii*
e. *Chlamydomonas r.*
Textbook Reference: *16.4 Phylogeny Is the Basis of Biological Classification, p. 328*

13. The organisms that make up a class are _______ diverse and _______ numerous than those in a family within that class. The organisms that make up a phylum all diverged from a common ancestor _______ recently than did the organisms in an order within that phylum.
a. more; more; less
b. more; more; more
c. more; less; less
d. less; less; more
e. less; less; less
Textbook Reference: *16.4 Phylogeny Is the Basis of Biological Classification, p. 328*

14. Which of the following lists of taxonomic categories represents the correct ordering, from most inclusive to least inclusive?
a. Phylum, order, family, genus
b. Class, phylum, order, species
c. Order, class, family, genus
d. Family, order, class, kingdom
e. Kingdom, class, species, genus
Textbook Reference: *16.4 Phylogeny Is the Basis of Biological Classification, p. 328*

15. The ratites are a group of flightless birds comprising the ostrich, emu, cassowaries, rheas, and kiwis. All share certain morphological similarities (such as a breastbone without a keel) not found in other birds, but they live on different continents. In the past, some ornithologists regarded their similarities as homoplasies, but they are now thought to be synapomorphies. Based on this information, you would conclude that the ratites were once regarded as a _______ group but are now believed to be _______.
a. polyphyletic; paraphyletic
b. paraphyletic; monophyletic
c. polyphyletic; monophyletic
d. monophyletic; polyphyletic
e. monophyletic; paraphyletic
Textbook Reference: *16.4 Phylogeny Is the Basis of Biological Classification, p. 329*

Answers

Key Concept Review

1. **c.** Recall that branches of a phylogenetic tree can be rotated around any node without changing the meaning of the tree. Tree "c" is different from all the others because it shows G (rather than D) as the sister taxon to taxa E and F.

2. **e.** E, F, and G and their common ancestor constitute a paraphyletic group, to be a clade, D would have to be included in the group. C, D, E, and F represent a polyphyletic group because the common ancestor of these taxa is not included in the group. (If the common ancestor of these four taxa were included, they would make up a paraphyletic group.)

3. When two animals have traits that can be explained by convergent evolution, these traits cannot be used to construct a phylogenetic tree. This is because they have evolved independently of each other and were not shared by a common ancestor of the two species.

4. Some of the implications of this statement are that DNA evolved as the genetic material before eukaryotes had diverged from prokaryotes, that DNA is an ancestral and general homologous trait, and that all surviving eukaryotes have DNA as their genetic material.

5.

	TRAIT				
SPECIES	1	2	3	4	5
A	1	0	1	1	1
B	0	0	1	1	0
C	0	1	1	0	0
D	0	0	0	0	0
E	1	0	1	1	0

6.

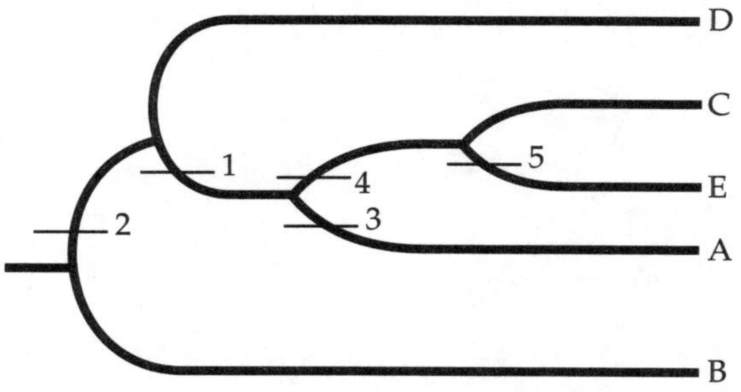

7. In the construction of a phylogenetic tree, the initial assumption is that derived traits appear only once and never disappear. Given a set of traits for a group of species, these restrictions sometimes must be relaxed to produce a phylogenetic tree for the group. Parsimony involves arranging the species so that the number of required reversals and multiple origins is minimized. Generally, the simplest explanation is most likely to be the most accurate.

8. Knowledge of the phylogeny of a group of organisms can provide information about the origin and evolution of different traits. In medicine, the history of infectious pathogens can be determined using a phylogeny. Phylogenies can be used to determine ancestral traits and the evolution of complex adaptations.

9. Monophyletic groups are taxa that contain all the ancestors and all the descendants of that ancestor. It consists of all of the related organisms on a tree. A polyphyletic group does not contain the common ancestor, and a paraphyletic group does not include all of the descendants of a common ancestor. The problems with both of these groups is that they provide an incomplete picture of the phylogenetic tree.

Test Yourself

1. **e.** A clade can be thought of as a complete branch on the tree of life. It includes the ancestor of a group, all the ancestor's descendants, and no other organisms.

2. **c.** Synapomorphies are traits that are not found in the ancestor of a group (hence they are derived) and that are found in more than one member of a group (hence they are shared).

3. **c.** Because antennae with 10 segments are found in all genera in this family except for genus X, the trait of 10-segment antennae is best regarded as ancestral; hence, having fewer than 10 segments in the antennae is a derived trait. The presence of 8-segment antennae in species E is not a synapomorphy because it is a trait found only in that species.

4. **d.** Homoplasy is the appearance of similar structures in different lineages that were not present in the common ancestor.

5. **a.** Derived traits are those that have undergone a change during evolution from the ancestral (original) character state.

6. **b.** Most shared traits, especially in species with a recent common ancestor, are likely to be homologous, not homoplastic. Therefore, the assumption that traits are homologous until proven homoplastic is more consistent with the parsimony principle than the reverse assumption is.

7. **b.** The incompleteness of the fossil record is by far the greatest limitation of its usefulness in determining phylogenies.

8. **b.** Mitochondrial DNA in animals changes rapidly over evolutionary time and hence would be most useful in determining evolutionary relationships among species that have diverged from one another only recently.

9. **e.** It is true that molecular clocks must be calibrated with independent data, and it is true that different genes and other DNA sequences evolve at different rates. But it is not true that the rate of gene evolution is independent of generation time or several other biological factors, or that molecular clocks cannot be used to date comparatively recent events.

10. **e.** The textbook describes specific examples of all four of the purposes listed as answers.

11. **b.** Although all of the statements listed are important attributes of biological classification schemes, the most important attribute is that biological classification reflects evolutionary relationships.

12. **d.** After a scientific name is referenced once in a text, the genus typically is abbreviated but the species name is given in full.

13. **a.** Organisms in a higher taxon are *less* similar, have diverged from a common ancestor *less* recently, and include *more* species than organisms in a lower included taxon.

14. **a.** The complete hierarchy of taxonomic categories, from most to least inclusive, is: kingdom, phylum, class, order, family, genus, species.

15. **c.** If the shared characteristics of the ratites are homoplasies (meaning that they evolved independently by convergent evolution), then the group did not have a single common ancestor and is polyphyletic. If (as is now accepted) the shared traits are synapomorphies shared among all ratites and not found in other birds, then the group is monophyletic.

Speciation 17

The Big Picture

- Speciation is the process that has produced the millions of life forms—each adapted to a particular environment and way of life—that constitute life on Earth. A number of concepts have been proposed to define a species, including the morphological species concept, the biological species concept, and the lineage species concept. Speciation occurs when there is reproductive isolation of one population into new lineages that cannot interbreed successfully.
- Allopatric speciation involves separation of populations due to geographical barriers and isolation of populations. Sympatric speciation occurs without any physical isolation and can involve polyploidy. When divergent populations come into contact, reproductive isolation is reinforced by prezygotic and postzygotic isolating mechanisms.

Study Strategies

- You may find have difficulty understanding why hybrids between species that have different numbers of chromosomes are inevitably sterile unless the hybrid is an allopolyploid. Recall that in meiosis I, homologous chromosomes undergo synapsis. This process cannot occur properly if the haploid sets of chromosomes inherited from the parents contain different numbers of chromosomes, because it is then impossible for every chromosome to have a homolog. As a consequence, meiosis does not proceed normally and few if any normal gametes will be produced. Because allopolyploids possess four sets of chromosomes (two from each parent), their chromosomes can synapse normally and they can produce viable gametes.
- The heart of this chapter is the discussions of allopatric (geographic) speciation and reproductive isolating mechanisms, so be sure to focus on them.
- Go to yourBioPortal.com to review the following tutorials and activities:

  Animated Tutorial 17.1 Founder Events and Allopatric Speciation

  Animated Tutorial 17.2 Speciation Mechanisms

  Interactive Tutorial 17.1 Speciation: Trends

  Web Activity 17.1 Concept Matching

  Working with Data 17.1 Flower Color and Reproductive Isolation

Key Concept Review

17.1 Species Are Reproductively Isolated Lineages on the Tree of Life

We can recognize many species by their appearance

Reproductive isolation is key

The lineage approach takes a long-term view

The different species concepts are not mutually exclusive

Speciation is the process by which one species splits into two species. Determining whether two populations constitute different species may be difficult because speciation is frequently a gradual process. The morphological concept of species used by early biologists grouped organisms into species on the basis of their appearance. This concept has limitations, because in some instances not all members of a species look alike and in other instances two or more cryptic species are morphologically indistinguishable but do not interbreed. The biological species concept defines a species as a group of actually or potentially interbreeding natural populations that are reproductively isolated from other such groups. It emphasizes the significance of reproductive isolation in keeping sexual lineages separated from one another. This definition cannot be applied to organisms that reproduce asexually, and it is limited to a single point in evolutionary time. The lineage species concept regards species as the smallest branches on the tree of life. The lineage splitting may be sudden or gradual, but in either case the lineages are thereafter independent of each other, allowing biologists to consider species over evolutionary time. These concepts of species are not mutually exclusive.

Question 1. There are a number of competing concepts that describe a species. What are the limitations of each concept? ***Textbook Reference:*** *17.1 Species Are Reproductively Isolated Lineages on the Tree of Life, p. 333*

17.2 Speciation Is a Natural Consequence of Population Subdivision

Incompatibilities between genes can produce reproductive isolation

Reproductive isolation develops with increasing genetic divergence

Evolutionary change can occur without speciation. A single lineage may change through time without diverging into two species. Speciation requires that the gene pool of the original species divide into two isolated gene pools. According to the Dobzhansky–Muller model, after speciation the isolated populations will accumulate allelic differences at gene loci (or chromosomal differences) that eventually will make it impossible for members of the two populations to successfully interbreed if they come together again. The rate of reproductive isolation can be gradual, over millions of years, or as short as a few generations.

Question 2. Describe how centric fusion can play a role in speciation.
Textbook Reference: *17.2 Speciation Is a Natural Consequence of Population Subdivision, p. 336*

17.3 Speciation May Occur through Geographic Isolation or in Sympatry

Physical barriers give rise to allopatric speciation

Sympatric speciation occurs without physical barriers

In allopatric speciation, the population is initially divided by a geographic barrier. Evidence suggests that allopatric speciation is the most common mechanism of speciation in most groups of organisms. The barrier can result from a geological or climatic change, and it results in two isolated populations that often are large and genetically similar. These populations diverged not only because of genetic drift, but also because the environments in which they live are, or became, different. Alternatively, separation may occur when some members of a population cross a barrier and found a new, isolated population. The speciation of the 14 species of Darwin's finches in the Galápagos are an example of speciation through geographic isolation.

Sympatric speciation occurs without geographic subdivision of the gene pool of the original species. Disruptive selection, in which different genotypes have high fitness with one of two different food resources, may be a widespread mechanism of sympatric speciation among insects. Sympatric speciation by polyploidy, the production of duplicate sets of chromosomes within an individual, is common in plants. Polyploidy produces new species because the polyploid organisms cannot interbreed with members of the parent species. Polyploid species that have a single ancestor are called autopolyploids, whereas those that have resulted from the hybridization of two species are referred to as allopolyploids. New species may arise by polyploidy much more easily among plants than among animals because plants of many species can reproduce by self-fertilization.

Question 3. Discuss the conditions on the Galápagos Islands that led to the evolution of the birds known as Darwin's finches.
Textbook Reference: *17.3 Speciation May Occur through Geographic Isolation or in Sympatry, pp. 338–339*

Question 4. In autopolyploidy, a new species of plant can arise by the doubling of chromosome numbers in a single individual of one species (provided that the individual is capable of self-fertilization). Why is it virtually impossible for such a tetraploid plant to interbreed successfully with diploid individuals of the "same" species?
Textbook Reference: *17.3 Speciation May Occur through Geographic Isolation or in Sympatry, p. 340*

17.4 Reproductive Isolation Is Reinforced When Diverging Species Come into Contact

Prezygotic isolating mechanisms prevent hybridization between species

Postzygotic isolating mechanisms result in selection against hybridization

Hybrid zones may form if reproductive isolation is incomplete

If two populations reestablish contact before reproductive isolation is complete, several results are possible. If hybrid offspring are not at a selective disadvantage, they may spread through both populations with the result that the gene pools of the populations combine. Thus, no new species would result from the period of isolation. If hybrid offspring are less successful, reinforcement may strengthen prezygotic reproductive barriers.

Prezygotic isolating mechanisms prevent members of different species from mating. Differences in reproductive organs may prevent interbreeding (mechanical isolation). Species may not be able to interbreed because they are fertile at different times (temporal isolation). The two species may not recognize or respond to each other's mating behaviors, or in flowering plants, the behavioral preferences of the pollinating animals may prevent interbreeding (behavioral isolation). Species may simply mate in different areas or different parts of a habitat (habitat isolation). The sperm and egg may be chemically incompatible (gametic isolation). Postzygotic barriers can prevent effective gene flow between species, even if mating occurs. Hybrid zygotes may not mature normally (low hybrid zygote viability). Hybrids may survive less well than either parent species (low hybrid adult viability). Hybrids may be infertile (hybrid infertility).

The evolution of more effective prezygotic reproductive barriers is known as reinforcement. It may occur if the hybrid offspring of two species survive poorly. If hybrid offspring are at a disadvantage but reinforcement fails to occur, a stable, narrow hybrid zone may form.

Question 5. Suppose that members of two populations are separated by a geographic barrier and begin to diverge genetically. Many generations later, when the barrier is removed, the two populations can interbreed, but the hybrid offspring do not survive and reproduce well. Explain how natural selection might lead to the evolution of more effective prezygotic barriers in these species.
Textbook Reference: *17.4 Reproductive Isolation Is Reinforced When Diverging Species Come into Contact, pp. 341–342*

Question 6. The yellow-rumped warbler was formerly split into two species (Myrtle warbler and Audubon's warbler), but in 1973 they were reclassified as a single species. Myrtle warblers and Audubon's warblers have largely allopatric ranges but hybridize where they are sympatric in the Canadian Rockies. They are similar in appearance but are readily distinguished by experienced birders. What further data about these two forms should ornithologists collect and analyze in order to decide whether they should continue to be classified as a single species?
Textbook Reference: *17.4 Reproductive Isolation Is Reinforced When Diverging Species Come into Contact, pp. 341–345*

Question 7. Construct a concept map whose theme is "species." Include in your map the following terms: species, allopatric, allopolyploidy, autopolyploidy, biological, concepts, founder events, independent evolution, interruption of gene flow, lineage, morphological, physical similarity, postzygotic barriers, prezygotic barriers, reproductive isolation, speciation, and sympatric. Connect these terms by verbs or short phrases to indicate the relationships among them.
Textbook Reference: *17.1 Species Are Reproductively Isolated Lineages on the Tree of Life, pp. 333–334; 17.2 Speciation Is a Natural Consequence of Population Subdivision, pp. 335–336; 17.3 Speciation May Occur through Geographic Isolation or in Sympatry, pp. 337–340; 17.4 Reproductive Isolation Is Reinforced When Diverging Species Come into Contact, pp. 340–345*

Test Yourself

1. It is difficult to apply the biological species concept to groups of organisms that
   a. are asexual.
   b. produce hybrids only in captivity.
   c. show little morphological diversity.
   d. exist only in the fossil record.
   e. Both a and d
   Textbook Reference: *17.1 Species Are Reproductively Isolated Lineages on the Tree of Life, pp. 333–334*
2. Which of the following statements about allopatric speciation is *false*?
   a. It can sometimes involve small populations.
   b. It occurs only in species that are widely distributed.
   c. It always involves a physical barrier that interrupts gene flow.
   d. It sometimes can involve chance events.
   e. It is the dominant mode of speciation in most groups of organisms.
   Textbook Reference: *17.3 Speciation May Occur through Geographic Isolation or in Sympatry, pp. 337–338*
3. A long, narrow hybrid zone exists in Europe between the ranges of the fire-bellied toad and the yellow-bellied toad. The persistence of this zone can be attributed to which of the following factors?
   a. Reinforcement strengthens the prezygotic barriers between the two species.
   b. Hybrid offspring have the same fitness as nonhybrid offspring.
   c. Both species travel long distances over the course of their lives.
   d. Individuals from outside the hybrid zone regularly move into the hybrid zone.
   e. None of the above
   Textbook Reference: *17.4 Reproductive Isolation Is Reinforced When Diverging Species Come into Contact, pp. 344–345*
4. Which type of speciation is most common among flowering plants?
   a. Geographic
   b. Sympatric
   c. Allopatric
   d. Disruptive
   e. None of the above
   Textbook Reference: *17.3 Speciation May Occur through Geographic Isolation or in Sympatry, p. 340*
5. Which of the following would *not* be considered an example of a prezygotic reproductive isolating mechanism?
   a. One bird species forages in the tops of trees for flying insects, whereas another forages on the ground for worms and grubs.
   b. The males of one species of moth cannot detect and respond to the sex attractant chemicals produced by the females of another species.

c. Sperm of one species of sea urchin are unable to penetrate the egg plasma membrane of another species.
d. Mosquitoes of one species are active in foraging and searching for mates at dusk, whereas those of another species are active at dawn.
e. Flowers of one orchid species mimic female bees of species A, whereas flowers of another orchid species mimic female bees of species B.

Textbook Reference: *17.4 Reproductive Isolation Is Reinforced When Diverging Species Come into Contact, pp. 341–342*

6. The species definition that includes similarity of appearance among individual members is the basis for
   a. the lineage species concept.
   b. reproductive isolation.
   c. the morphological species concept.
   d. centric fusion.
   e. the biological species concept.

   Textbook Reference: *17.1 Species Are Reproductively Isolated Lineages on the Tree of Life, p. 333*

7. The Dobzhansky–Muller model describes how
   a. two independent lineages come together to form an ancestral population.
   b. an ancestral population separates to become two independent lineages.
   c. geographical barriers lead to speciation.
   d. species should be defined.
   e. allopatric speciation occurs.

   Textbook Reference: *17.2 Speciation Is a Natural Consequence of Population Subdivision, p. 335*

8. More than 800 species of *Drosophila* occur in the Hawaiian Islands, representing 30 to 40 percent of all the species in this genus. The occurrence of so many *Drosophila* species in this island chain is
   a. the result of many founder events followed by genetic divergence.
   b. an example of an evolutionary radiation.
   c. largely the result of sympatric speciation.
   d. evidence that the genus *Drosophila* first evolved in the Hawaiian Islands.
   e. Both a and b

   Textbook *Reference: 17.3 Speciation May Occur through Geographic Isolation or in Sympatry, pp. 337–338*

9. Which of the following statements about speciation is *false*?
   a. A small founding population can be involved in speciation.
   b. Speciation always involves interruption of gene flow between different groups of organisms.
   c. The rate of speciation can vary for different groups of organisms.
   d. Speciation always requires many generations.
   e. Speciation may occur because certain genotypes within a population prefer distinct microhabitats where mating takes place.

   Textbook Reference: *17.3 Speciation May Occur through Geographic Isolation or in Sympatry, p. 340*

10. Which of the following observations would constitute conclusive evidence that two overlapping populations that have been geographically separated have *not* diverged into distinct species?
    a. Matings between members of the two populations produce viable hybrids.
    b. A hybrid zone exists where their ranges overlap.
    c. Interbreeding is common between members of the two populations.
    d. All of the above
    e. None of the above

    Textbook Reference: *17.4 Reproductive Isolation Is Reinforced When Diverging Species Come into Contact, p. 344*

11. In allopatric speciation, which process is likely to be *least* important?
    a. A founder event
    b. Allopolyploidy
    c. Behavioral isolation
    d. Genetic drift
    e. Both b and d

    Textbook Reference: *17.3 Speciation May Occur through Geographic Isolation or in Sympatry, p. 337*

12. A field contains two related species of flowering plants. Species A has a diploid chromosome number of 16, and species B has a diploid number of 18. If a third species arises as a result of hybridization between A and B, how many chromosomes will it have?
    a. 17
    b. 32
    c. 34
    d. 36
    e. 68

    Textbook Reference: *17.3 Speciation May Occur through Geographic Isolation or in Sympatry, p. 340*

13. Speciation by polyploidy occurs far more often in plants than in animals because
    a. plants are more likely to be capable of self-fertilization than animals are.
    b. plant cells can tolerate extra sets of chromosomes, whereas animal cells cannot.
    c. plants as a rule have higher reproductive rates than animals do.
    d. many plants are specialized with respect to their pollinating agent.
    e. All of the above

    Textbook Reference: *17.3 Speciation May Occur through Geographic Isolation or in Sympatry, p. 340*

14. Two species of narrowmouth frogs in the United States have mating calls that differ more in their region of sympatry than in those parts of their ranges that do not overlap. If this difference in their vocalizations has the function of preventing hybridization between the two species, it is an example of

a. a hybrid zone.
b. reinforcement.
c. sympatric speciation.
d. a postzygotic reproductive barrier.
e. allopatric speciation.
Textbook Reference: *17.4 Reproductive Isolation Is Reinforced When Diverging Species Come into Contact, p. 340*

15. Centric fusion of acrocentric (one-armed) chromosomes results in
a. mechanical isolation.
b. behavioral isolation.
c. temporal isolation.
d. reproductive isolation.
e. allopatric speciation.
Textbook Reference: *17.2 Speciation Is a Natural Consequence of Population Subdivision, p. 336*

Answers

Key Concept Review

1. Limitations of the morphological species concept include the fact that the different sexes of the same species may not look alike or two species may look very much alike but are unable to interbreed. The limitation of the biological species concept is that it does not apply to asexual species. The limit of the lineage species concept is limited because of lack of knowledge as to when a lineage actually split or when whether the incipient species will continue to diverge or merge at some point in the future.

2. Centric fusion is the fusion of two acrocentric (one-armed) chromosomes to form a metacentric chromosome. If centric fusion becomes fixed at one chromosome in one population of organisms but at a different chromosome in another population, individuals from the two populations would be unable to produce viable offspring. This is because the offspring will not be able to produce normal games in meiosis.

3. The relatively great distance between the Galápagos Islands and the South American mainland and between each of the islands in the archipelago ensured that once immigrants had arrived on an island, they would be genetically isolated for a substantial period of time. Also, because the islands differ greatly in climate and vegetation, the resident birds were subject to different selection pressures. This, in combination with reduced gene flow between the islands, led to a rapid evolutionary radiation of finches.

4. Recall from Chapter 7 that pairs of homologous chromosomes synapse during prophase and metaphase of the first division of meiosis. The homologs then separate, so that each cell resulting from meiosis I is haploid, as are the products of meiosis II. In tetraploids as in diploids, meiosis is normal because pairing of homologs can occur. Any offspring of a cross between tetraploid and diploid individuals will be triploid, however, and therefore sterile because correct synapsis of homologs cannot occur. Because the tetraploid product of autopolyploidy is reproductively isolated from its diploid relatives, it is a new species.

5. Recall that natural selection tends to remove traits that reduce survival or reproductive success from a population. Individuals that interbreed between populations will have lower fitness (they will contribute fewer offspring to future generations) than those who breed within their own population. If the tendency to avoid interbreeding is heritable (and not just the result of chance), the frequency of alleles that prevent interbreeding will increase in each population. How might such a trait be heritable? Any of the prezygotic barriers to interbreeding might be heritable traits. For example, if the species in question are frogs, and if their mating calls started to diverge while the populations were separated, the following traits might be heritable: a tendency to make a call that is more distinct from that of the other population, or the ability to distinguish between the existing calls of the two populations (coupled with a preference for the call of one's own population). As alleles for these traits increased in frequency, they would contribute to the behavioral isolation of the two populations and perhaps eventually to more complete speciation.

6. During allopatric speciation, divergence of two populations often occurs gradually. In such cases it is inevitable that intermediate stages of speciation occur, and it may be a matter of opinion whether two forms have diverged sufficiently to be considered separate species. With respect to these two warbler populations, biologists would seek answers to these questions: Are hybrid offspring as fit as those resulting from mating of individuals of the same population? Is there evidence that the zone of hybridization is expanding, indicating that the gene pools of the populations are combining? Is there evidence of reinforcement of prezygotic barriers to interbreeding (e.g., a greater difference in the songs of the two forms in the area of hybridization than in allopatric parts of their ranges)? Lesser fitness of hybrid offspring, a stable, narrow zone of hybridization, and evidence of reinforcement would favor the conclusion that the populations are best regarded as separate species.

7.

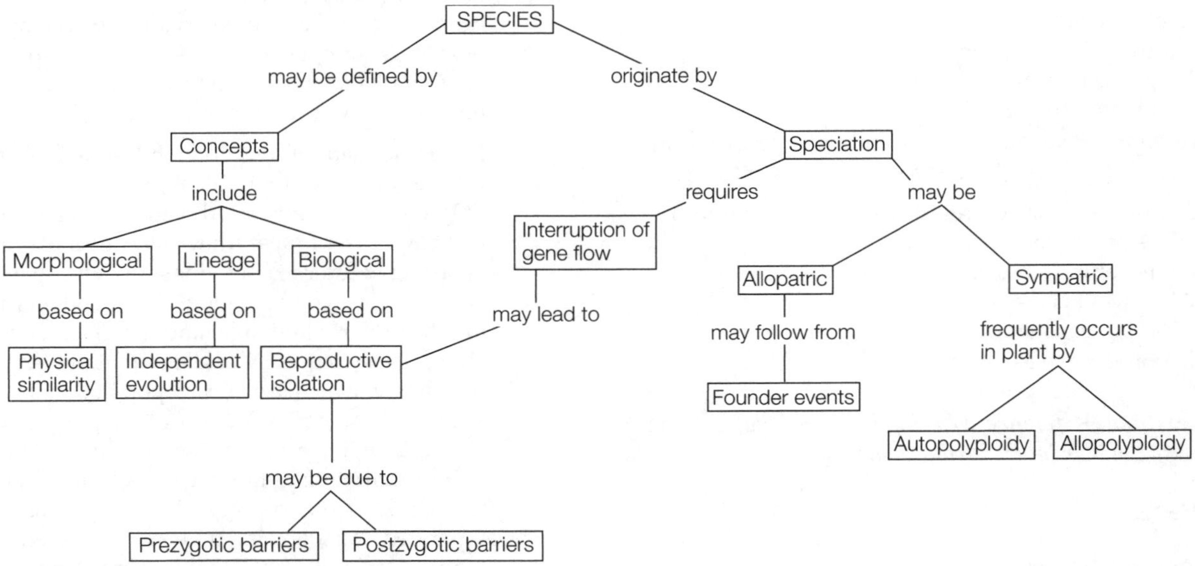

Test Yourself

1. **e.** The key criterion of a biological species is that its members are reproductively isolated from other such groups. This criterion is impossible to evaluate in asexual and fossil species.
2. **b.** A wide distribution is not a requisite for allopatric speciation.
3. **d.** These toads are an example of related species in which reinforcement does not strengthen prezygotic barriers, even though hybrid offspring are only half as fit as nonhybrid offspring. The reason for this is that toads from outside the hybrid zone (not subject to the selective pressure against hybridizing) regularly move into the hybrid zone and mate with members of the other species. The hybrid zone remains narrow because toads do not travel long distances; natural selection removes hybrids from the population before they can disperse very far.
4. **b.** Sympatric speciation is most common among flowering plants. It has been estimated that about 70 percent of all flowering plant species are polyploid.
5. **a.** Provided that the two species are active in the same locality at the same time, a difference in the habitat in which they forage would not in itself be a barrier to interbreeding (though seeking mates in different habitats might well be a barrier to interbreeding). All the other choices describe reproductive barriers that would act prior to fertilization.
6. **c.** The morphological species concept assumes that individuals that look alike can be considered as belonging to the same species.
7. **b.** The Dobzhansky–Muller model describes how changes in different alleles can produce two independent lineages from a single ancestral population.
8. **e.** Island groups are frequently sites of evolutionary radiations through repeated allopatric speciation events that are initiated by individuals (or groups) dispersing from one island to another. The numerous species of *Drosophila* in the Hawaiian Islands are believed to have originated in this way.
9. **d.** New species formed by polyploidy can arise in only two generations.
10. **e.** Interbreeding, production of viable hybrids, and establishment of a hybrid zone do not necessarily mean that speciation is incomplete. If, however, the hybrids were successful, fertile, and bred freely with members of both original populations, their gene pools would merge, and you would conclude that speciation had not taken place.
11. **b.** Genetic drift, founder events, and behavioral isolation may all play a role in allopatric speciation. Allopolyploidy, however, can occur only as the result of hybridization between individuals of different species; hence the two parent species cannot be allopatric.
12. **c.** If haploid gametes of species A and B joined, the result would be a zygote with 17 chromosomes. A mature plant with 17 chromosomes would be sterile, because chromosomes would be unable to pair properly during prophase and metaphase of meiosis I. A fertile allopolyploid would therefore have to have 34 chromosomes so that each chromosome would have a homolog with which to pair. Autopolyploids of species A and B would have 32 and 36 chromosomes, respectively.
13. **a.** If a polyploid plant or animal is capable of self-fertilization, then a new species can arise from a single individual. The ability to self-fertilize is far more common among plants than animals.

14. **b.** Reinforcement is defined as the evolutionary strengthening of prezygotic barriers to interbreeding within the zone of sympatry of two closely related species.
15. **d.** When two chromosomes fuse in one lineage of a population, this lineage is unable to successfully produce viable offspring with the other lineage.

The History of Life on Earth 18

The Big Picture

- To understand evolution, we must have an understanding of the age of rocks and fossils. Radiometric dating provides a means of dating rocks. This provides scientists with the ability to construct a geological time scale and determine the age of fossils. Prior to the development of this technique, no absolute dates could be assigned accurately to any of the important geological and evolutionary events described in this chapter.
- Throughout Earth's history, its physical environment has changed, resulting in a number of mass extinctions. The major land masses have moved as a result of continental drift. The climate has alternated between hot and cold conditions. A number of large volcanic eruptions and extraterrestrial events have influenced the survival of organisms on Earth. Initially, there was no oxygen on Earth, but with the evolution of photosynthetic bacteria the oxygen levels rose, allowing for the evolution of multicellular organisms.
- Life first appeared on Earth about 3.8 bya. In the Precambrian, life was found only in the oceans. The Cambrian explosion resulted in rapid diversification of life, and it continued to diversify from the forms that evolved in the Cambrian. This was punctuated with multiple mass extinctions in the Ordovician, Devonian, and Permian. A single supercontinent, Pangea, was formed during the Permian period and broke apart into two large continents during the Triassic and Jurassic periods. This was also the time of the dinosaur. Modern organisms evolved during the Cenozoic era.

Study Strategies

- The concept of the half-life of radioactive isotopes may seem confusing. Bear in mind that if half of a radioisotope decays in a given time period (its half-life), then at the end of the first half-life only one-half of the original quantity of the isotope is still present, one-half of which will decay in the second half-life. Thus after two half-lives, one-quarter of the original quantity of radioisotope is still present. The same line of reasoning applies to all further half-lives. Study Figure 18.1, which explains this point.
- This chapter contains many names of geological eras and periods and many dates. Instructors vary with respect to the importance they place on memorization of this information. In general, it is best to focus on broad evolutionary patterns and trends. Most instructors will also place more importance on the phylogenetic sequences (e.g., amphibians gave rise to amniotes, which gave rise to reptiles and to mammals) than on the dates at which these events occurred.
- Making a table, chart, or timeline to summarize and organize the events that occurred in each geological era and period is an excellent study strategy.
- Go to yourBioPortal.com to review the following tutorial and activity:

  Animated Tutorial 18.1 Evolution of the Continents

  Web Activity 18.1 Concept Matching

Key Concept Review

18.1 Events in Earth's History Can Be Dated

Radioisotopes provide a way to date rocks

Radiometric dating methods have been expanded and refined

Scientists have used several methods to construct a geological time scale

Evolutionary changes may occur over both short and long time frames. Short-term evolutionary changes occur rapidly enough to be studied directly. Long-term evolutionary changes involve the appearance of new species and evolutionary lineages. The fossil record provides evidence of such changes. Several types of evidence have been used to estimate the age of rocks, of fossils, and of Earth itself. Sedimentary rocks (formed by the accumulation of sediments at the bottom of bodies of water) are deposited in strata (layers). The oldest layers are found at the bottom, and successively higher strata are progressively younger. Fossils, which are the preserved remains of ancient organisms, are useful for establishing the relative ages of the sedimentary rocks in which they occur. The fossil record shows that an organism of any specific type can predictably be found in rocks of a particular age. Living organisms resemble recent fossils more closely than they resemble fossils from more ancient periods.

The regular pattern of decay of radioactive isotopes provides a means of estimating the absolute ages of fossils and rocks. The half-life of a radioisotope is the time period during which half of the remaining radioactive material decays to become a different, stable isotope. The ratio of radioactive carbon-14 to its stable isotope, carbon-12, is used to date fossils less than 60,000 years old. The decay of potassium-40 to argon-40 is widely used to date ancient evolutionary events. Paleomagnetic dating is based on the fact that the age of sedimentary and igneous rocks can be determined because they preserve a record of Earth's magnetic field at the time they were formed.

Earth's geological history is divided into eras, which are subdivided into periods. The boundaries between these divisions are marked by changes in the types of fossils found in successive layers of sedimentary rock. The first forms of life evolved early in the Precambrian era, which lasted for more than three billion years and was marked by enormous physical changes on Earth.

Question 1. For each geological period, make a chart summarizing its major geological and evolutionary events.
Textbook Reference: *18.1 Events in Earth's History Can Be Dated, p. 348*

Question 2. ^{14}C decays to ^{14}N with a half-life of about 5,700 years. Suppose you find a fossil in which the amount of ^{14}C is only $1/16$ of what one would find in a living organism with the same carbon content. Approximately how old is the fossil?
Textbook Reference: *18.1 Events in Earth's History Can Be Dated, p. 349*

Question 3. In 2006, paleontologists announced that fossils of *Tiktaalik roseae*, an important link between fish and tetrapods, had been discovered in freshwater sediments in the Canadian Arctic that were 375 million years old. How were geologists able to determine the age of these sediments?
Textbook Reference: *18.1 Events in Earth's History Can Be Dated, p. 349*

18.2 Changes in Earth's Physical Environment Have Affected the Evolution of Life

The continents have not always been where they are today

Earth's climate has shifted between hot and cold conditions

Volcanoes have occasionally changed the history of life

Extraterrestrial events have triggered changes on Earth

Oxygen concentrations in Earth's atmosphere have changed over time

The theory of plate tectonics proposes that Earth's crust consists of solid lithospheric plates floating on a fluid layer of magma. Convection currents in the magma result from heat emanating from Earth's core. The currents cause continental drift, which is the gradual shift in the position of the plates and the continents they contain. The drifting of continents has had profound effects on climate, sea level, oceanic circulation, and the distributions of organisms.

Changes in the physical parameters on Earth have resulted in numerous mass extinctions. The climate of Earth has alternated between hot/humid and cold/dry conditions. Though most major climatic changes have been gradual, some have occurred within periods of 5,000 to 10,000 years or less. The rapid climate change occurring today is thought to be caused by a buildup of atmospheric CO_2, primarily from the burning of fossil fuels. The climatic shifts caused by massive volcanic eruptions associated with continental drift are implicated in several mass extinctions. The late Permian mass extinction was in part caused by volcanism triggered by the collision of continents that formed the supercontinent Pangaea. Several mass extinctions have probably been caused by collisions of Earth with meteorites or comets, such as the event 65 mya that is thought to have resulted in the extinction of dinosaurs.

No free oxygen was present in the atmosphere until certain bacteria evolved the ability to use water as a source of hydrogen ions for photosynthesis. The gradual increase in oxygen concentration in the atmosphere resulted in the dominance of organisms using aerobic metabolism and in the evolution of larger eukaryotic cells and of multicellular organisms. The exceptionally high oxygen concentrations that occurred during the Carboniferous and Permian periods are associated with the evolution of giant amphibians and flying insects.

Question 4. Scientists believe that if there are no controls on the emission of CO_2 from the burning of fossil fuels, the concentration of this gas could double by the end of the current century, leading to a significant rise in the average temperature of Earth. What would be some of the likely evolutionary effects of this climatic change?
Textbook Reference: *18.2 Changes in Earth's Physical Environment Have Affected the Evolution of Life, p. 352*

Question 5. In what way was the evolution of eukaryotic cells linked to the increase in the oxygen concentration in the atmosphere that occurred during the Precambrian?
Textbook Reference: *18.2 Changes in Earth's Physical Environment Have Affected the Evolution of Life, pp. 353–354*

18.3 Major Events in the Evolution of Life Can Be Read in the Fossil Record

- Several processes contribute to the paucity of fossils
- Precambrian life was small and aquatic
- Life expanded rapidly during the Cambrian period
- Many groups of organisms that arose during the Cambrian later diversified
- Geographic differentiation increased during the Mesozoic era
- Modern biotas evolved during the Cenozoic era
- The tree of life is used to reconstruct evolutionary events

All of the organisms found in a particular place or time constitute its biota. The plant component of a biota is its flora, and the animal component is its fauna. The 300,000 known fossil species represent only a tiny fraction of the species that have ever lived. Some groups, such as hard-shelled marine animals, are much better represented in the fossil record than others. Because an oxygen-rich environment favors rapid decomposition, organisms that become fossils are likely either to have lived in a poorly oxygenated environment or to have been transported to such a site soon after death.

The earliest life on Earth appeared about 3.8 bya, but the fossil record of organisms that lived in the Precambrian is fragmentary. The first organisms were unicellular prokaryotes. For most of the Precambrian, which lasted for over 3 billion years, life consisted of microscopic prokaryotes. The first unicellular eukaryotes evolved about 1.5 billion years ago. The best Precambrian fossil deposits, dating from about 600 mya, contain diverse soft-bodied invertebrates, some of which may represent lineages with no living descendents.

During the Cambrian period, the oxygen concentration in the atmosphere approached its current level, and several large continents formed. The rapid increase in the diversity of multicellular life forms that occurred at this time is known as the "Cambrian explosion."

An evolutionary radiation occurs when there is a rapid diversification of organisms. The Ordovician period was marked by a proliferation of marine filter feeders living on the sea floor. At the end of this period, massive glaciers formed over the southern continents, the sea level and ocean temperatures dropped, and the majority of animal species became extinct. The Silurian period witnessed the evolution of swimming marine animals and the diversification of jawless fish. It also marked the appearance of the first terrestrial arthropods and vascular plants. During the Devonian period, all major groups of fishes evolved. On land, the first insects and amphibians evolved, and forests of club mosses, horsetails, and tree ferns appeared. A mass extinction of approximately three-quarters of all marine species occurred at the end of this period, possibly caused by the collision of two large meteorites with Earth.

In the Carboniferous period, swamp forests consisting largely of giant tree ferns and horsetails became widespread. The fossilized remains of these plants formed coal. The first winged insects evolved during this period, while amphibians became better adapted to life on land and gave rise to the lineage leading to the amniotes (whose eggs can be laid in dry places). The Permian period was marked by the formation of a single supercontinent called Pangaea. As the climate cooled drastically, amniotes split into two lineages, the reptiles and one leading to the mammals. The occurrence of the most extensive of all mass extinctions brought the Permian period to a close.

During the Mesozoic era (251–65 mya), the continents that formed Pangaea slowly separated to form Laurasia and Gondwana, and distinct assemblages of plants and animals evolved on each continent. During the Triassic period, seed ferns and conifers were the dominant forms of terrestrial vegetation, and reptiles were the dominant vertebrates. A mass extinction occurred at the end of the Triassic. The Jurassic period was marked by the complete separation of Pangaea to form Laurasia in the north and Gondwana in the south. It also witnessed the radiation of the dinosaurs, the evolution of flying reptiles, and the first appearance of mammals. The earliest fossils of flowering plants date from late in this period. In the sea, ray-finned fishes also began a great radiation. By the early Cretaceous period, Laurasia and Gondwana had begun to break apart into the present-day continents. The flowering plants began the radiation that led to their current dominance. The dinosaurs continued as the dominant land vertebrates, despite the presence of many groups of mammals. At the end of this period, the collision of a large meteorite with Earth caused the extinction of the dinosaurs and of many other animal and plant lineages.

Modern groups of plants and animals evolved during the Cenozoic era (65 mya–present). During the Tertiary period, the continents drifted toward their present positions. As the climate became cooler and drier, extensive grasslands appeared. Many groups of land vertebrates—especially frogs, snakes, lizards, birds, and mammals—radiated extensively. The current geological period, the Quaternary, is divided into the Pleistocene and Holocene (Recent) epochs. Modern humans evolved during the Pleistocene, which was a time of severe climatic fluctuations, including four major periods of extensive glaciation. As humans spread geographically, many species of large birds and mammals became extinct.

Phylogenetic trees help reconstruct the timing of evolutionary events and clarify relationships among modern species.

Question 6. Below is a chronologically scrambled list of important events in the history of life on Earth. Draw four lines on a sheet of paper to represent the time lines of events occurring in the Precambrian, the Paleozoic era, the Mesozoic era, and the Cenozoic era. Place the listed events on the correct line and in proper sequence.
Textbook Reference: *18.3 Major Events in the Evolution of Life Can Be Read in the Fossil Record, pp. 357–363*

Conifers become dominant
Cambrian "explosion"
Evolution of *Homo*
Fifth mass extinction
First eukaryotes
First flowering plant fossils
First forests ("fern" forests), first jawed fish
First fossils of multicellular animals
First mammals, dinosaurs diversify
First mass extinction
First photosynthetic eukaryotes
First vascular plants and terrestrial arthropods
Fourth mass extinction
Grasslands spread
Origin of amniotes
Origin of life
Origin of photosynthesis
Second mass extinction
Third mass extinction
Rapid radiation of mammals

Test Yourself

1. The half-life of an isotope is the
   a. time it takes a fixed fraction of isotope material to change from one form to another.
   b. time interval during which the isotope is useful for dating rocks.
   c. ratio of one isotope species to another in a sample of organic matter.
   d. Both a and b
   e. None of the above
   Textbook Reference: *18.1 Events in Earth's History Can Be Dated, p. 349*
2. Mountain ranges are fundamentally the result of
   a. plates in Earth's crust that move against one another on top of a fluid layer of molten rock.
   b. climate changes and the movement of glacial ice sheets.
   c. leftover debris from ancient collisions with an asteroid or meteor.
   d. the breakup of Laurasia and Gondwana.
   e. Both a and b
   Textbook Reference: *18.2 Changes in Earth's Physical Environment Have Affected the Evolution of Life, p. 351*
3. Which of the following environments would likely be most conducive to the occurrence of fossilization?
   a. The surf zone along a sandy beach
   b. A shallow, cool swamp with good deposition rates of mud sediments
   c. The bottom of a hot, dry cave with no running water
   d. A fast-running mountain stream
   e. It is not possible to decide based on the information provided.
   Textbook Reference: *18.3 Major Events in the Evolution of Life Can Be Read in the Fossil Record, p. 356*
4. Despite being incomplete as a whole, the fossil record is rather detailed for
   a. soft-bodied insects.
   b. cnidarians and sponges.
   c. most terrestrial animals.
   d. hard-shelled mollusks.
   e. Both c and d
   Textbook Reference: *18.3 Major Events in the Evolution of Life Can Be Read in the Fossil Record, pp. 356–357*
5. During the _______ geological period, Pangaea became fully divided and distinctive assemblages of plants and animals begin to arise on different continents.
   a. Cambrian
   b. Tertiary
   c. Devonian
   d. Permian
   e. Jurassic
   Textbook Reference: *18.3 Major Events in the Evolution of Life Can Be Read in the Fossil Record, p. 362*
6. Which of the following statements about the Ordovician period is *false*?
   a. Marine filter feeders flourished.
   b. The number of classes and orders increased.
   c. Modern mammals appeared.
   d. Many groups became extinct at the end of the period.
   e. The continents were located primarily in the Southern Hemisphere.
   Textbook Reference: *18.3 Major Events in the Evolution of Life Can Be Read in the Fossil Record, p. 358*
7. Most of human evolution has occurred during the
   a. Paleozoic era.
   b. Devonian period.
   c. Quaternary period.
   d. Carboniferous period.
   e. Cretaceous period.
   Textbook Reference: *18.3 Major Events in the Evolution of Life Can Be Read in the Fossil Record, p. 363*
8. For terrestrial animals and plants, the most recent mass extinction event that occurred prior to the evolution of humans took place approximately _______ mya.
   a. 10
   b. 65
   c. 200
   d. 250
   e. 400
   Textbook Reference: *18.3 Major Events in the Evolution of Life Can Be Read in the Fossil Record, p. 362*

9. One of the main factors that distinguishes the Cambrian explosion from all others is that
   a. evolutionarily, it was the most recent explosion.
   b. many new major groups of animals appeared at this time.
   c. it was the time when the dinosaurs became extinct.
   d. it saw a dramatic drop in species diversity, especially among marine organisms.
   e. it was a time of massive volcanic eruptions.

   Textbook Reference: *18.3 Major Events in the Evolution of Life Can Be Read in the Fossil Record, p. 358*

10. During which of the following geological times did the most new kinds of body plans appear?
    a. Carboniferous
    b. Triassic
    c. Jurassic
    d. Devonian
    e. Cambrian

    Textbook Reference: *18.3 Major Events in the Evolution of Life Can Be Read in the Fossil Record, p. 358*

11. The sudden disappearance of the dinosaurs some 65 mya may have been the result of
    a. Earth's collision with a large meteorite.
    b. slow climate changes due to planetary cooling.
    c. competition from better adapted organisms.
    d. the rise of birds and mammals.
    e. the formation of Pangaea.

    Textbook Reference: *18.3 Major Events in the Evolution of Life Can Be Read in the Fossil Record, p. 362*

12. Fossil insects many times larger than any insects alive today have been dated to the Carboniferous and Permian periods. These fossils are evidence that during these periods
    a. the climate was much warmer than it is today.
    b. there were fewer predators on insects than there are today.
    c. the carbon dioxide concentration of the atmosphere was lower than it is today.
    d. insects had fewer competitors for food than modern insects.
    e. the oxygen concentration of the atmosphere was significantly higher than it is today.

    Textbook Reference: *18.2 Changes in Earth's Physical Environment Have Affected the Evolution of Life, pp. 354–355*

13. Which of the following statements about patterns or processes in the evolution of life is *false*?
    a. ^{14}C can be used to date the age of dinosaur bones.
    b. The supercontinent Pangaea formed during the Permian period.
    c. Mass extinctions of marine organisms have coincided with periods of low sea levels.
    d. The ends of five geological periods have been marked by mass extinctions.
    e. All of the above are true.

    Textbook Reference: *18.1 Events in Earth's History Can Be Dated, p. 349*

14. Which of the following paired organisms were *not* present on Earth in their living forms at the same time?
    a. Tree ferns and ray-finned fish
    b. Amphibians and birds
    c. Gymnosperms and insects
    d. Humans and dinosaurs
    e. Jawless fish and vascular plants

    Textbook Reference: *18.3 Major Events in the Evolution of Life Can Be Read in the Fossil Record, pp. 362–363*

15. The most severe mass extinction event, linked to the formation of Pangaea and massive volcanic eruptions, occurred at the end of the _______ period.
    a. Ordovician
    b. Devonian
    c. Permian
    d. Triassic
    e. Cretaceous

    Textbook Reference: *18.3 Major Events in the Evolution of Life Can Be Read in the Fossil Record, p. 362*

Answers

Key Concept Review

1. See Table 18.1 in the textbook.
2. If only $\frac{1}{16}$ of the ^{14}C remains, then four ^{14}C half-lives have passed since the fossil was formed ($\frac{1}{2} \times \frac{1}{2} \times \frac{1}{2} \times \frac{1}{2} = \frac{1}{16}$). Because the half-life of ^{14}C is roughly 5,700 years, the fossil must be about 22,800 years old ($4 \times 5{,}700 = 22{,}800$).
3. Recall that radioisotope dating is used to determine the absolute age of rocks. Igneous rocks, formed from volcanic ash or lava flows, are required for this purpose. Such rocks may have been found near the discovery site in Canada, or they may have been found at one or more other sites that have sediments of the same relative age. The relative age of the sediments can be determined by the types of fossils found in them.
4. The fossil record shows that major environmental changes occurring over a short time interval sometimes lead to large-scale rapid extinctions that appear "instantaneous" in the fossil record. The possible effects of global warming on biodiversity are discussed in more detail in Chapter 46.
5. Small prokaryotic cells can obtain enough oxygen by diffusion even when oxygen concentrations are very low. Since eukaryotic cells are larger, they have a lower surface area-to-volume ratio and hence need a higher concentration of oxygen in their environment for diffusion to meet their requirement for this gas.

6.

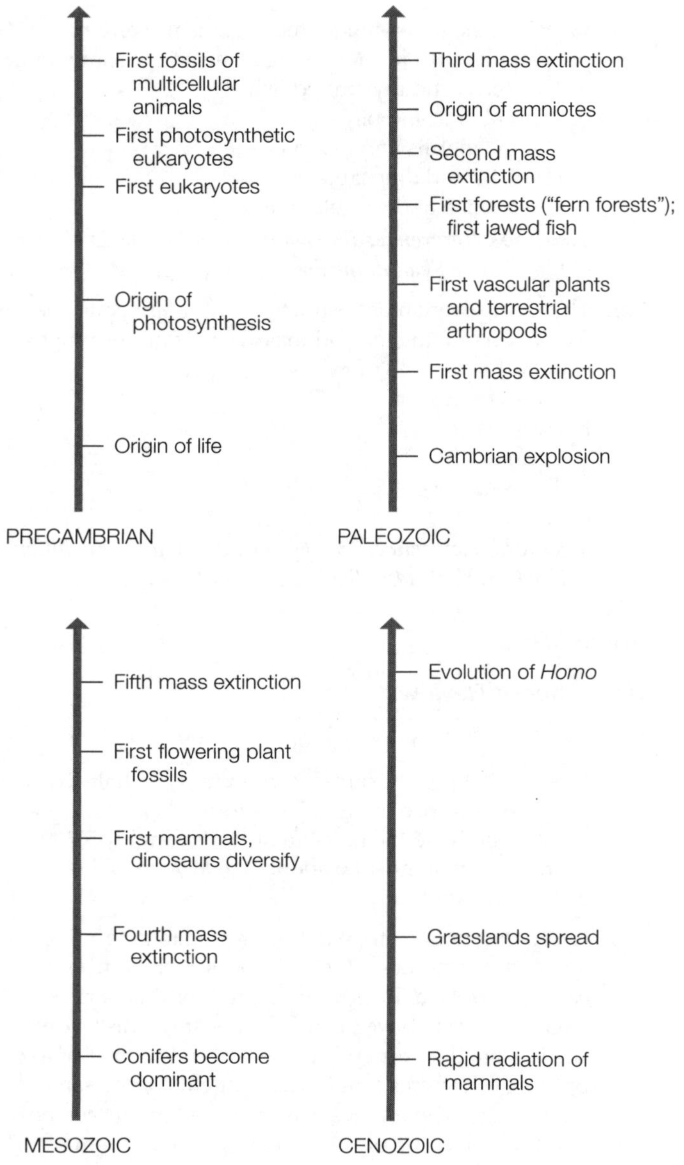

Also see Figures 18.10 and 18.12 in the textbook.

Test Yourself

1. **a.** Radioactive decay is measured as the time it takes for one-half the amount of a substance to spontaneously convert into another substance.
2. **a.** As the crustal plates are pushed together, one may move underneath the other, pushing up mountain ranges.
3. **b.** Fossilization occurs best in areas of low oxygen concentration and rapid sedimentation, where scavengers cannot destroy the body.
4. **d.** Hard-shelled animals such as mollusks are good candidates for fossilization because their shells can withstand decay long enough for them to be buried. Many also tend to live in quiet, shallow waters.
5. **e.** The division of Pangaea began during the Triassic period, but Laurasia and Gondwana did not become completely separated until the Jurassic.
6. **c.** The modern mammals did not appear until the Tertiary, more than 350 million years after the Ordovician.
7. **c.** The Pleistocene epoch, which is part of the Quaternary, was the time of most hominid evolution.
8. **b.** The mass extinction at the end of the Cretaceous, 65 mya, was the most recent mass extinction affecting terrestrial life before the evolution of humans.
9. **b.** The Cambrian explosion produced many novel groups of animals characterized by distinctive body plans. Later explosions caused an increase in diversity only within already existing major lineages.
10. **e.** The Cambrian explosion produced many novel groups of animals characterized by distinctive body plans.
11. **a.** Paleontologists believe that a large meteorite's collision with Earth so altered conditions that the dinosaurs rapidly succumbed during what we know as the great Cretaceous mass extinction.
12. **e.** There is experimental evidence that insects can grow larger when raised in hyperbaric conditions (see Figure 18.8). The current level of atmospheric O_2 appears to limit the evolution in body size of flying insects.
13. **a.** Carbon dating is generally reliable only for fossils less than 60,000 years old.
14. **d.** Dinosaurs disappeared at the end of the Cretaceous, over 60 million years before humans evolved.
15. **c.** Though mass extinctions occurred at the end of all the periods listed, the highest percentage of species became extinct during the Permian period.

Bacteria, Archaea, and Viruses

19

The Big Picture

- All organisms fall within one of three domains: the Archaea, the Bacteria, and the Eukarya. This chapter focuses on the two prokaryotic domains (Archaea and Bacteria) and also on the nonliving viruses. Archaea and Bacteria retain many features of the earliest forms of life on Earth. They have circular chromosomes and cell walls made of peptidoglycans or proteins, and they lack membrane-enclosed organelles. Their structures are simple, and they exist as single cells. Some prokaryotes are stationary, while others move by means of flagella. Their metabolic processes are highly varied, as are their modes for procuring energy. Many prokaryotic organisms make valuable contributions to the environment through nitrogen fixation, photosynthesis, and other metabolic processes. Most are free-living, but some live in mutualistic relationships and others are pathogenic to other organisms.
- The evolutionary relationships among prokaryotes are still highly disputed and best understood at the genetic level. The most ubiquitous extant prokaryotes are the bacteria. They are grouped according to physical and biochemical characteristics, but these groupings do not necessarily reflect their evolutionary relationships. The archaea are more closely related to the eukarya, and exist in some of the harshest known environments.
- While there is no single definition of "life" that all scientists agree upon, many do not consider viruses living organisms, even though they arose from cells and require a host cell to carry out their basic functions. Viral genomes are varied in form and structure. They can have positive- or negative-stranded RNA, single-stranded or double-stranded DNA, and they can be linear or circular. Currently, the genome structure is the basic classification system used, since the evolutionary history of viruses is unknown and unlikely to be determined.

Study Strategies

- Because this chapter covers a large number of diverse microscopic organisms included in two of the three domains of life, the details and exceptions can be overwhelming. Many of the names of these groups are long, complex, and unfamiliar. Because many of the distinguishing features of the different groups of bacteria and archaea are biochemical, you may need to review information from Part I of the textbook. Use flash cards to practice identifying the groups by their defining traits.
- It is tempting to focus on pathogenic organisms, but remember that only a very small percentage of prokaryotes are pathogenic. Pay attention to the ecological services performed by members of different groups as well as the vast array of different conditions in which they thrive. Again, use flash cards to sort different organisms into groups according to where they live or what they do.
- Go to yourBioPortal.com to review the following tutorial and activity:

  Animated Tutorial 19.1 The Evolution of the Three Domains

  Web Activity 19.1 Gram Stain and Bacteria

Key Concept Review

19.1 Life Consists of Three Domains That Share a Common Ancestor

The two prokaryotic domains differ in significant ways

The small size of prokaryotes has hindered our study of their evolutionary relationships.

The nucleotide sequences of prokaryotes reveal their evolutionary relationships

Lateral gene transfer can lead to discordant gene trees

The great majority of prokaryote species have never been studied

Domains Archaea and Bacteria are both prokaryotic and replicate by binary fission rather than mitosis. Both groups of organisms have single chromosomes that are often circular. Additional DNA plasmids may be present in some. Because they lack membrane-bound compartments, they rely on infoldings of the plasma membrane to carry out many metabolic processes

These similarities lead many students to treat Archaea and Bacteria as a single group, but there are some significant differences between them. The cell wall in bacteria is made of peptidoglycan, which is not found in archaea. In addition, the cellular machinery in archaea is more like that of eukaryotes than of bacteria.

The classification of prokaryotes was originally based on shape, color, motility, nutritional requirements, and the amount of peptidoglycan found in the bacterial cell wall. Modern techniques allow for the study of genomes, and many archaea have been identified only though analysis of DNA samples collected from the environment. Analysis of prokaryotic genomes is aiding in our understanding of evolutionary relationships, and rRNA sequences have been particularly useful in phylogenetic studies. We now know that some genes have been transferred between species via lateral gene transfer and that most genomes have a stable core of genes that lets us trace the phylogeny despite this transfer.

Question 1. Runoff into a stream from a wastewater treatment plant has been killing the fish in the stream. The organism responsible has been isolated, and consulting scientists are examining it under a microscope. How can they tell if the organism is a prokaryote or a eukaryote?
Textbook Reference: *19.1 Life Consists of Three Domains That Share a Common Ancestor, pp. 368–369*

Question 2. Describe how Gram staining depends on bacterial cell wall structure. Would Gram staining work for archaea?
Textbook Reference: *19.1 Life Consists of Three Domains That Share a Common Ancestor, pp. 368–369*

Question 3. Label the three basic shapes of bacteria shown in the micrograph below.
Textbook Reference: *19.1 Life Consists of Three Domains That Share a Common Ancestor, p. 369, Figure 19.3*

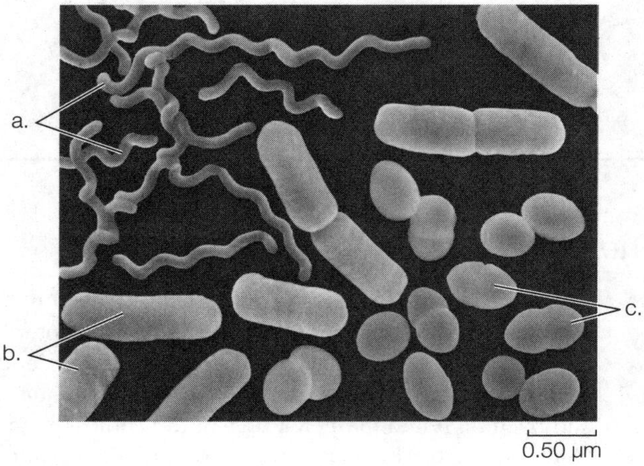

Question 4. In the diagrams below, label the bacteria as Gram-positive or Gram-negative and identify the structures in each.
Textbook Reference: *19.1 Life Consists of Three Domains That Share a Common Ancestor, p. 369, Figure 19.2*

This is a Gram-_______ bacteria.

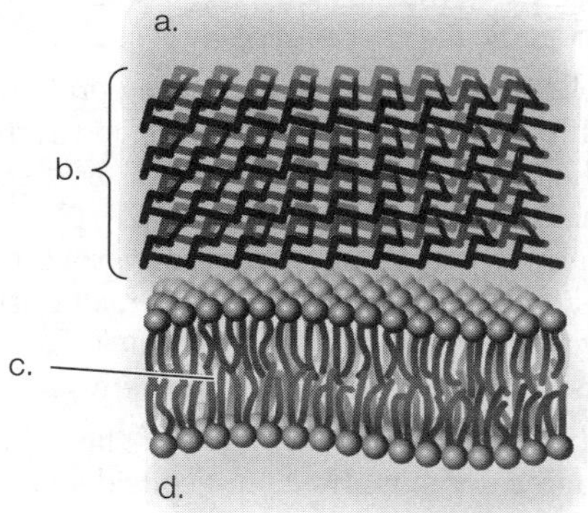

This is a Gram-_______ bacteria.

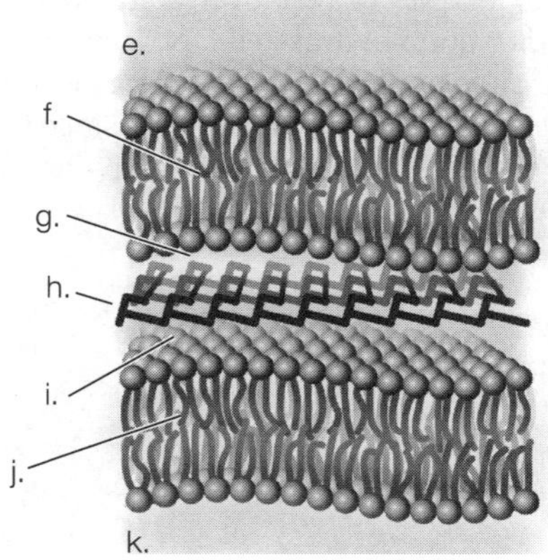

19.2 Prokaryote Diversity Reflects the Ancient Origins of Life

- The low-GC Gram-positives include the smallest cellular organisms
- Some high-GC Gram-positives are valuable sources of antibiotics
- Hyperthermophilic bacteria live at very high temperatures
- Hadobacteria live in extreme environments
- Cyanobacteria were the first photosynthesizers
- Spirochetes move by means of axial filaments
- Chlamydias are extremely small parasites
- The proteobacteria are a large and diverse group
- Gene sequencing enabled biologists to differentiate the domain Archaea
- Most crenarchaeotes live in hot or acidic places
- Euryarchaeotes are found in surprising places
- Korarchaeotes and nanoarchaeotes are less well known

This section focuses on domain Archaea and the eight groups in domain Bacteria. The bacteria groups are fairly well characterized and classified based on nucleotide sequence data; far less is known about archaea.

The eight groups of bacteria are: (1) low-GC Gram-positives, (2) high-GC Gram-positives, (3) hyperthermophilic bacteria, (4) hadobacteria, (5) cyanobacteria, (6) spirochetes, (7) chlamydias, and (8) proteobacteria. Two groups of archaea that have been studied most are Crenarchaeota and Euryarchaeota.

Low-GC Gram-positives include members that can form endospores, mycoplasmas that lack cell walls, and *Staphylococcus*, which is a common cause of skin infections in humans. High-GC Gram-positives are a source of antibiotics and also include the human pathogen *Mycobacterium tuberculosis.* Hyperthermophilic bacteria, hadobacteria, and some archaea thrive in extreme environments where the temperatures, radiation levels, or toxin levels are fatal to other organisms. Cyanobacteria are believed to be the earliest photosynthetic organisms and have the ability to form multicellular colonies. Spirochetes are characterized by axial filaments that enable a corkscrew-like movement. Some species are the cause of syphilis and Lyme disease. Chlamydias are extremely small obligate parasites known to cause some human STDs. Proteobacteria include the nitrogen-fixing *Rhizobium, Escherichia coli,* and many species that infect plants.

Archaea phylogeny is still being developed, but two large groups in the current classification scheme are Crenarchaeota and Euryarchaeota. Crenarchaeotes are generally thermophilic and/or acidophilic, while the euryarchaeotes are methanogens, halophiles, and other varieties of extremophile. Aside from a lack of peptidoglycan, the archaea are characterized by their distinct membrane lipids.

Question 5. Complete the table below.

Domain	Group Name	Key Features	Example Genera or Associated Diseases
	Chlamydias		
	Crenarchaeota		
	Cyanobacteria		
	Euyarchaeota		
	High-GC Gram-positives		
	Korarchaeota		
	Low-GC Gram-positives		
	Nanoarchaeota		
	Proteobacteria		
	Spirochetes		

Textbook Reference: *19.2 Prokaryote Diversity Reflects the Ancient Origins of Life, pp. 371–378*

Question 6. You discover a new prokaryotic organism. What sorts of traits will you look for in order to determine which bacteria or archaea group this organism belongs to? (Assume you cannot sequence the genome but must use other criteria for your initial classification.)
Textbook Reference: *19.2 Prokaryote Diversity Reflects the Ancient Origins of Life, pp. 371–378*

19.3 Ecological Communities Depend on Prokaryotes

Many prokaryotes form complex communities

Prokaryotes have amazingly diverse metabolic pathways

Prokaryotes play important roles in element cycling

Prokaryotes live on and in other organisms

A small minority of bacteria are pathogens

Prokaryotes as a group perform many different roles in the larger ecosystem. Some are able to form biofilms, aggregations of cells that bind to a solid surface and secrete a sticky matrix that protects them from antibiotics and other agents that might harm them. Biofilms are the subject of much current research since many biofilm-forming bacteria are human pathogens or cause corrosion industrial plants.

The long evolutionary history of bacteria and archaea has led to the wide diversity of their metabolic "lifestyles"—their use or nonuse of oxygen, their energy sources, their sources of carbon atoms, and the materials they release as waste products.

Some prokaryotes serve vital roles in the cycling of elements in the biosphere. A limited number of prokaryotes are pathogens and thus are important in terms of human health, agriculture, and animal husbandry. The protein products of prokaryote metabolism released by replicating cells, called exotoxins, are often fatal to the host. Endotoxins, which are components of lipopolysaccharides that are released when cells grow or lyse, are less dangerous to humans. The work of Robert Koch was key in developing a method of determining whether or not a prokaryotic organism is responsible for a particular disease. Even in modern times, when it is well known that microbes can cause disease, Koch's postulates help researchers determine if a specific organism is the cause of a particular disease. For example, Koch's postulates were used to demonstrate that *Helicobacter pylori* causes stomach ulcers.

Question 7. Differentiate between the following terms: obligate anaerobe versus obligate aerobe; photoautotroph versus photoheterotroph; chemolithotroph versus chemoheterotroph.
Textbook Reference: *19.3 Ecological Communities Depend on Prokaryotes, pp. 378–380*

Question 8. A friend with young children tells you that she intends to protect their health by eradicating all bacteria from her home and their immediate environment. What do you tell her?
Textbook Reference: *19.3 Ecological Communities Depend on Prokaryotes, pp. 378–382*

Question 9. Biologists from the Centers for Disease Control in Atlanta have been called to a remote area of Uganda to study a mysterious disease that is causing respiratory ailments in a small village. They isolate a bacterium from several patients that seems to be the likely cause of the illness. How can they determine if this is indeed the pathogen?
Textbook Reference: *19.3 Ecological Communities Depend on Prokaryotes, pp. 381–382*

Question 10. A U.S. Department of Agriculture field representative counsels an inexperienced farmer to plant alfalfa in fields with soils that have low nitrogen levels because alfalfa roots are hosts for nitrogen-fixing bacteria. How will the alfalfa and its associated bacteria help "fertilize" this farmer's soil?
Textbook Reference: *19.3 Ecological Communities Depend on Prokaryotes, pp. 381–382*

Question 11. A patient with a bacterial infection has caused a great deal of concern to his physician because the bacteria responsible for the infection produce an exotoxin. Why are exotoxins often more dangerous than endotoxins?
Textbook Reference: *19.3 Ecological Communities Depend on Prokaryotes, pp. 381–382*

19.4 Viruses Have Evolved Many Times

Many RNA viruses probably represent escaped genomic components

Some DNA viruses may have evolved from reduced cellular organisms

Viruses lack a cellular structure and are not considered living organisms by many scientists. (Note that they are not even named according the binomial nomenclature system used for the three domains of life.) Even so, they have some traits of living things, including a genetic code and the ability to reproduce. Viruses are classified on the basis of their genomes, since their evolutionary history is complex and viruses most likely arose multiple times from multiple branches of the tree of life. RNA viruses can be positive stranded, negatively stranded, retroviruses, or double stranded. DNA viruses are double stranded.

Question 12. Why would a physician refuse to prescribe antibiotics for a cold?
Textbook Reference: *19.4 Viruses Have Evolved Many Times, pp. 134–137; 19.1 Life Consists of Three Domains That Share a Common Ancestor, p. 368*

Test Yourself

1. Which of the following characteristics is unique to prokaryotes?
   a. Lack of membrane-enclosed organelles
   b Presence of cell walls
   c. Presence of plasma membranes
   d. Presence of a cytoskeleton
   e. Absence of ribosomes
   Textbook Reference: *19.1 Life Consists of Three Domains That Share a Common Ancestor, p. 370*

2. Rod-shaped bacteria are referred to as
   a. bacilli.
   b. cocci.
   c. spiral.
   d. helici.
   e. roddi.
   Textbook Reference: *19.1 Life Consists of Three Domains That Share a Common Ancestor, p. 369, Figure 19.3*

3. Which of the following statements concerning prokaryotes is *true*?
   a. Because prokaryotes do not contain organelles, they cannot photosynthesize or carry out cellular respiration.
   b. Prokaryotes have no chromosomes and therefore lack DNA.
   c. Prokaryote flagella are similar in structure to eukaryote flagella.
   d. Prokaryotes undergo mitosis.
   e. None of the above
   Textbook Reference: *19.1 Life Consists of Three Domains That Share a Common Ancestor, p. 367*

4. Gram-negative bacteria stain pink because
   a. they have specialized lipids in their cell walls.
   b. their peptidoglycan layer is thin.
   c. their peptidoglycan layer is thick.
   d. they are receptive to antibiotics.
   e. their cell walls are composed largely of proteins.
   Textbook Reference: *19.1 Life Consists of Three Domains That Share a Common Ancestor, p. 369*

5. Archaea are more closely related to _______ than to any other monophyletic group.
   a. bacteria
   b. eukaryotes
   c. bacteria and eukaryotes
   d. fungi
   e. protists
   Textbook Reference: *19.1 Life Consists of Three Domains That Share a Common Ancestor, p. 367, Figure 19.1*

6. The dense films laid down by many prokaryotes are
   a. endotoxins.
   b. denitrifiers.
   c. biofilms.
   d. pathogens.
   e. endospores.
   Textbook Reference: *19.2 Prokaryote Diversity Reflects the Ancient Origins of Life, p. 372*

7. Which of the following groups of bacteria has the highest proportion of pathogens?
   a. Proteobacteria
   b. Cyanobacteria
   c. Chlamydias
   d. High-GC Gram-positives
   e. Low-GC Gram-positives
   Textbook Reference: *19.2 Prokaryote Diversity Reflects the Ancient Origins of Life, p. 375*

8. The mitochondria of eukaryotes were derived by endosymbiosis from
   a. proteobacteria.
   b. chemoheterotrophs.
   c. eukaryotes.
   d. archaea.
   e. viruses.
   Textbook Reference: *19.2 Prokaryote Diversity Reflects the Ancient Origins of Life, p. 375*

9. Autoclaves, which sterilize medical and laboratory equipment by means of pressurized heat, must pass a "spore test" in many states to demonstrate that they work correctly. The spore-producing bacteria used for this test are most likely taken from which of the following groups?
   a. Low-GC Gram-positive bacteria
   b. Proteobacteria
   c. Cyanobacteria
   d. Chlamydias
   e. High-GC Gram-positive bacteria
   Textbook Reference: *19.2 Prokaryote Diversity Reflects the Ancient Origins of Life, p. 372*

10. Archaea that live in extremely salty conditions are referred to as
    a. thermophiles.
    b. halophiles.
    c. salinophiles.
    d. brinophiles.
    e. sodiophiles.
    Textbook Reference: *19.2 Prokaryote Diversity Reflects the Ancient Origins of Life, p. 373*

11. Methane gas contributes to the greenhouse effect that is raising atmospheric temperatures. A large portion of all methane emission is from grazing cattle because cows harbor methane-producing archaea from the _______ group.
    a. Crenarchaeota
    b. Euryarchaeota
    c. Anarchaeota
    d. Proteobacteria
    e. Nanoarchaeota
    Textbook Reference: *19.2 Prokaryote Diversity Reflects the Ancient Origins of Life, p. 377*

12. One of the major diagnostic characteristics that distinguishes archaea from bacteria is the absence of
    a. peptidoglycan cell walls in archaea.
    b. peptidoglycan cell walls in bacteria.
    c. ribosomes in archaea.

d. chemoautotrophy in bacteria.
e. None of the above
Textbook Reference: *19.2 Prokaryote Diversity Reflects the Ancient Origins of Life, pp. 375–376*

13. A bacterium that requires a carbon source other than carbon dioxide, yet can convert light energy to chemical energy, is called a
a. photoautotroph.
b. photoheterotroph.
c. chemoautotroph.
d. chemoheterotroph.
e. chemolithotroph.
Textbook Reference: *19.3 Ecological Communities Depend on Prokaryotes, pp. 379–380, Table 19.2*

14. A bacterium that *cannot* live in the presence of oxygen is called a(n)
a. obligate aerobe.
b. facultative aerobe.
c. obligate anaerobe.
d. facultative anaerobe.
e. aerotolerant anaerobe.
Textbook Reference: *19.3 Ecological Communities Depend on Prokaryotes, p. 380*

15. Which of the following virus types inserts a double-stranded DNA copy of its genome into the host cell's genome?
a. Double-stranded RNA viruses
b. Double-stranded DNA viruses
c. Retroviruses
d. Negative-sense single-stranded RNA viruses
e. Positive-sense single-stranded RNA viruses
Textbook Reference: *19.4 Viruses Have Evolved Many Times, p. 385*

16. Which of the following is *not* one of Koch's postulates?
a. The introduction of the disease-causing microorganism to a new, healthy host causes the same disease that existed in the original host.
b The disease-causing microorganism is always found in the person with the disease.
c. After the disease-causing microorganism has been introduced into a healthy host who then gets the disease, the same microorganism can be isolated from this patient.
d. The disease-causing microorganism is susceptible to antibiotic treatment.
e. The disease-causing microorganism can be isolated from an infected individual and grown in culture.
Textbook Reference: *19.3 Ecological Communities Depend on Prokaryotes, pp. 380–382*

Answers

Key Concept Review

1. The most immediately apparent difference between a prokaryote and a eukaryote is the much smaller size of the prokaryote and the absence of membrane-enclosed organelles in prokaryotes. With proper staining techniques, the nuclei of any eukaryote would likely be visible with the light microscope.

2. Gram staining results depend on the arrangement and amounts of peptidoglycans in the bacterial cell wall. This technique does not work with archaea because they do not have peptidoglycan in their cell walls.

3.
a. Helices
b. Bacilli
c. Cocci

4. positive; negative
a. Outside of cell
b. Cell wall (peptidoglycan)
c. Plasma membrane
d. Inside of cell
e. Outside of cell
f. Outer membrane of cell envelope
g. Periplasmic space
h. Peptidoglycan layer
i. Periplasmic space
j. Plasma membrane
k. Inside of cell

5.

Domain	Group Name	Key Features	Example Genera or Associated Diseases
Bacteria	Chlamydias	Gram-negative cocci. Very small, obligate parasites. Life cycle involves two forms of cells called elementary bodies and reticulate bodies.	Cause human eye disease, the STD chlamydia, and some types of pneumonia
Archaea	Crenarchaeota	Most live in hot and/or acidic environments such as sulfur hot springs. Internal cellular environment is close to pH 7.	*Sulfolobus*
Bacteria	Cyanobacteria	Photoautotrophs. Use chlorophyll *a* for photosynthesis. May form colonies with differentiated cells (see Figure 26.13). Some also fix nitrogen. Thought to be the source of eukaryotic chloroplasts.	*Anabaena* (see Figure 19.9A)
Archaea	Euyarchaeota	Many are methanogens living in the guts of ruminant herbivores or termites and cockroaches. Others are extreme halophiles. *Thermoplasma* lacks a cell wall and lives in coal deposits.	*Methanopyrus thermoplasma*
Bacteria	High-GC Gram-positives	Gram-positive, GC-rich genomes, may form spores at tips of filaments. Source of most of our antibiotics.	*Mycobacterium tuberculosis,* the cause of tuberculosis and *Streptomyces,* the source of streptomycin.
Archaea	Korarchaeota	Poorly characterized, only DNA has been isolated directly from hot springs.	None available
Bacteria	Low-GC Gram positives	Gram-positive, although some are Gram negative and lack a cell wall. Genome is not GC-rich. May produce endospores capable of surviving extreme conditions. Mycoplasmas are the smallest cellular organisms known.	*Bacillus anthracis:* causes anthrax, *Clostridium:* a cause of food poisoning, *Staphylococcus aureus*: can colonize healthy individuals or cause infection.
Archaea	Nanoarchaeota	Poorly characterized. Lives attached to cells of *Ignicoccus,* a crenarchaeote. Discovered at a deep-sea thermal vent near Iceland's coast.	*Nanoarchaeum equitans*
Bacteria	Proteobacteria	Largest group of identified species. Very diverse group classified into five groups based on their metabolic pathways.	*Rhizobium:* important for nitrogen fixation, *Yersinia pestis:* cause of bubonic plague, *Vibrio cholerae:* cause of cholera, *Salmonella typhinurium*: cause of gastrointestinal disease, and *Escherichia coli:* well characterized.
Bacteria	Spirochetes	Gram-negative, helical structure (See Figure 19.10) hemoheterotrophic. Move via axial filaments.	Some species cause syphilis and Lyme disease.

6. A Gram stain would allow you to determine if the prokaryote is a member of archaea (which will not stain at all) or bacteria (Gram-positive or Gram-negative). You could also use information about the environment where the new organism was found. Some bacteria and archaea are known to live in extreme environments, while other bacteria can form endospores, fix nitrogen, or have unique structures like axial filaments. Using the environmental clues, information from Gram staining, and the shape and motility of the cells, you would probably be able to predict which group this new organism belongs to.

7. Obligate anaerobes die in the presence of oxygen gas, whereas obligate aerobes die without oxygen gas for cellular respiration. Photoautotrophs convert light energy to chemical energy using carbon dioxide as their carbon source; photoheterotrophs rely on an outside carbon source other than carbon dioxide but still can convert light energy to chemical energy. Chemolithotrophs can produce what they need from inorganic molecules, whereas chemoheterotrophs depend on nutrients from external sources.

8. This effort to eradicate all bacteria will not be successful, since humans are hosts to innumerable bacteria both on and in our bodies. Furthermore, most bacteria are not harmful. In fact, many are beneficial. For example, bacteria are essential in nutrient cycling in the ecosystem. In some cases, the attempt to eradicate bacteria has created dangers of its own. The abuse and overuse of antibiotics, for instance, creates selective pressures that lead to an increase in antibiotic-resistant pathogenic bacteria, rendering previously effective treatments useless. Your friend would be better-served by declaring a truce with the bacteria and recognizing them as part of the world we inhabit.
9. Koch's postulates stipulate that in order for an organism to be identified as a disease-causing agent: (1) it must always be found in individuals with the disease; (2) it must be taken from the host and grown in pure culture; (3) a sample of the culture must produce disease if injected into a healthy individual; and (4) the newly infected individual must yield a new pure culture of the same organism. That said, it is unethical to knowingly infect a person with a pathogen just for the purposes of identification! Therefore, a diagnosis can be made based on the first three postulates but not on the fourth one.
10. Nitrogen-fixing bacteria can convert atmospheric nitrogen into ammonia. Other bacteria can then convert the ammonia into nitrates that plants can use. (This topic is covered in greater detail in Chapter 25.)
11. Exotoxins are produced continuously by the infecting bacteria, whereas endotoxins are produced only at cell death. Only a limited number of endotoxin molecules are released, but exotoxins can be released until the host dies.
12. Antibiotics such as penicillin and ampicillin interfere with the synthesis of the peptidoglycan-based cell walls of bacteria. Viruses do not have a cellular structure, and they use the host cell's machinery to replicate and spread. Since the cold virus is replicating in a person's cells, an antibiotic would have no effect on a cold and would indeed kill many of the "good" bacteria living in the body.

Test Yourself

1. **a.** Prokaryotes are differentiated from eukaryotes by their lack of membrane-enclosed organelles. They also have a single circular piece of genomic DNA and lack a cytoskeleton.
2. **a.** A rod-shaped bacterium is known as a bacillus (plural, bacilli). Cocci are roughly spherical, and the helices are corkscrew shaped.
3. **e.** Bacteria do carry out both photosynthesis and multiple forms of cellular respiration. All life forms have DNA. The flagella of bacteria consist of a single protein called flagellin rather than multiple proteins as found in eukaryotes. Bacteria replicate by binary fission.
4. **b.** Gram-negative bacteria have thin peptidoglycan layers that do not retain crystal violet stain.
5. **b.** The Archaea and the Eukarya share a more recent common ancestor with each other than either does with the Bacteria. One piece of evidence for this is that a signature sequence from rRNA has been found in all archaea and eukaryotes tested so far, but in none of the bacteria.
6. **c.** Biofilms are gel-like polysaccharide matrices that are laid down by prokaryotes and trap other bacteria.
7. **c.** Chlamydias are all parasitic and cause several human diseases.
8. **a.** Endosymbiosis of proteobacteria gave rise to the mitochondria of eukaryotes.
9. **a.** Among the groups listed, only low-GC Gram-positive bacteria, high-GC Gram-positive bacteria, and cyanobacteria produce spores. Low-GC Gram-positive spores can withstand extreme environmental condition, so they would be useful for testing an autoclave's ability to sterilize equipment.
10. **b.** The term "halophile" means "salt loving."
11. **b.** Methanogenic bacteria that reside in the guts of cows belong to the Euryarchaeota.
12. **a.** Archaea lack peptidoglycans. Instead they have a unique lipid in their cell walls.
13. **b.** Photoheterotrophs require a carbon source other than carbon dioxide, yet are able to harvest light energy.
14. **c.** Oxygen gas is toxic to obligate anaerobes.
15. **c.** Retroviruses, such as HIV, insert a form of their genome into the host's DNA. This provirus is replicated every time the cell divides by mitosis.
16. **d.** Koch's postulates were developed in the 1880s, a time when even the role of microorganisms in disease was not understood. Antibiotic therapies to target these microorganisms came later.

The Origin and Diversification of Eukaryotes

20

The Big Picture

- The many lineages of protists, as well as all fungi, animals, and plants, are eukaryotic. Eukaryotic cells are characterized by membrane-enclosed organelles, the presence of a cytoskeleton, and a nuclear envelope. The evolution of the eukaryotic cell, which acquired features of both archaea and bacteria, was a major evolutionary milestone that occurred in the Precambrian. The eukaryotic cell allows for compartmentalization of processes and increased adaptability. Though the evolutionary lineage is still debated, it is thought that eukaryotes evolved via multiple symbiotic events.
- Protists as a group are not monophyletic. The most recent common ancestor of protists gave rise to the plants, animals, and fungi, groups that are not included in the protists. Protists are diverse, ranging from microscopic single-celled organisms to complex multicellular organisms. Several subgroups of protists are covered in this chapter, some of which are monophyletic and some that are not. The five major eukaryotic clades designated protists are: alveolates, excavates, stramenophiles, rhizaria, and amoebozoans.
- Certain body plans and modes of nutritional acquisition are repeated throughout the microbial eukaryote groups and are good examples of form following function. Protists as a group play a vast role in the environment and medicine, are economically valuable, and include some of the most challenging disease organisms.

Study Strategies

- It is easy to become overwhelmed with organism names and characteristics. Focus on the trends outlined in the chapter and use lab opportunities to understand the organism groupings. For a start, see the summaries of the five major protist groups at the end of the chapter.
- This chapter contains a great deal of information. It is best not to try to learn it all at once, but to break it up into smaller pieces and study it during several sessions.
- To learn the organism groups, create a chart with key characteristics. This will help you compare and contrast the groups.
- On one side of a 4 × 6 index card, write down a particular clade or subgroup; on the other side, note the defining characteristics. Do the same with sample organisms, with the genus specified on one side of the card and the clade and subgroup on the other side. Then shuffle the deck and organize them into different categories. For example, you can sort all of the subgroups into their proper clades. You can match sample organisms with their subgroups or clades. Or, you can sort the groups according to the presence or absence of morphological characteristics, life cycle strategies, or other criteria of your choosing.
- Go to yourBioPortal.com to review the following tutorials and activity:

  Animated Tutorial 20.1 Family Tree of Chloroplasts

  Animated Tutorial 20.2 Digestive Vacuoles

  Animated Tutorial 20.3 Life Cycle of the Malarial Parasite

  Web Activity 20.1 Anatomy of a *Paramecium*

Key Concept Review

20.1 Eukaryotes Acquired Features from Both Archaea and Bacteria

The modern eukaryotic cell arose in several steps

Chloroplasts have been transferred among eukaryotes several times

Protists are a paraphyletic group of very diverse organisms. The term "protist" lacks any real taxonomic meaning. It is really shorthand for all eukaryotes that are neither land plants, fungi, nor animals. You may also see the term "microbial eukaryotes" used in place of protists. Because of the "catch-all" nature of this grouping, it is extremely diverse, containing many different organisms that are not necessarily closely related to one another. Most are unicellular, but many are colonial and some are multicellular. Most eukaryotes can be classified in one of eight major clades, and five of these clades are collectively referred to as "protists." The five protist clades are: alveolates, excavates, stramenopiles, rhizaria, and amoebozoans.

The origin of the eukaryotic cell is still being debated, but evidence suggests that eukaryotes are monophyletic. The evolution of the eukaryotic cell was revolutionary and occurred at a time of great environmental change. Important events include: (1) origin of a flexible cell surface; (2) origin of a cytoskeleton; (3) origin of a nuclear envelope; (4) appearance of digestive vesicles; and (5) endosymbioses that led to the development of organelles.

The first step in the path toward a eukaryotic cell may have involved loss of the rigid cell wall, which allowed cells to become larger and which also allowed infoldings of the cell surfaces to increase the surface area-to-volume ratio. This also permitted vesicles to form from bits of infolded membrane that pinch off. Such infolding may have been the origin of the nuclear envelope.

New structures then evolved; ribosomes became associated with internal membranes to form the endoplasmic reticulum, a cytoskeleton of actin and microtubules formed, and digestive vesicles developed. The cytoskeleton opened up modes of locomotion by actin- and myosin-based flagella. How these structures formed is currently a matter of conjecture. It is thought that these early motile cells may have been phagocytes that engulfed prokaryotic cells. Subsequent endosymbiotic events may have led to the evolution of mitochondria and chloroplasts (see Figure 20.1).

An understanding of the evolution of eukaryotes is complicated by lateral gene transfer. The prokaryotic genes in many eukaryotes are so numerous that lateral transfer cannot explain them all, nor can endosymbiosis. Additional hypotheses are being tested.

Some chloroplasts are enclosed in double membranes, and others are enclosed in triple membranes. This phenomenon can be explained by endosymbiosis. All chloroplasts can be traced to the engulfment of an ancestral cyanobacterium with a single membrane. Its engulfment in a vesicle of the phagocyte's membrane resulted in a double membrane. This initial event is called primary endosymbiosis. Primary endosymbiosis gave rise to the green and red algae. Photosynthetic euglenoids, with their triple membranes, arose from secondary endosymbiosis (see Figure 20.2). In this case, a chlorophyte was ingested with its housed chloroplast, resulting in three membranes. Ultimately, all of the constituents of the chlorophyte except the chloroplast were lost. Tertiary endosymbiosis occurred when a dinoflagellate lost its chloroplast and took up a protist that had aquired its chloroplast through secondary endosymbiosis.

Question 1. Explain one line of thought regarding the origin of the eukaryotic cell. Why are scientists unsure of the evolutionary relationships among the protists?
Textbook Reference: *20.1 Eukaryotes Acquired Features from Both Archaea and Bacteria, p. 389*

Question 2. Describe the origin of double membrane-enclosed and triple membrane-enclosed chloroplasts.
Textbook Reference: *20.1 Eukaryotes Acquired Features from Both Archaea and Bacteria, p. 391*

20.2 Major Lineages of Eukaryotes Diversified in the Precambrian

- Alveolates have sacs under their plasma membranes
- Excavates began to diversify about 1.5 billion years ago
- Stramenopiles typically have two unequal flagella, one with hairs
- Rhizaria typically have long, thin pseudopods
- Amoebozoans use lobe-shaped pseudopods for locomotion

Radiation of eukaryotes began in the Precambrian. The five major groups of protistan eukaryotes include the alveolates, excavates, stramenopiles, rhizaria, and amoebozoans. Protists are very diverse, even within the major groups. Some protists are autotrophic, some are heterotrophic, and some have the ability to switch between the two modes. Many protists are involved in symbioses, from mutualism to parasitism. Some protists are motile and move via pseudopodia, flagella, or cilia. Cilia and eukaryotic flagella are identical in cross section; they differ only in length. The amoeboid body plan (which includes pseudopods) is highly effective in nutrient-rich environments. Some organisms have an amoeboid body plan at different stages of their life cycle.

Many protists are microscopic, though some algae form aquatic forests of very large, multicellular organisms. Single-celled protists are able to increase their cell size with the presence of vesicles, which effectively increase surface area. Vesicles assist with water regulation (contractile vacuoles) and nutrient procurement (food vacuoles). Many protists have cell walls, or "shells," to protect their cell membranes. The chemical composition of these structures varies among organisms.

Many protists contain endosymbionts. Some harbor other protists that photosynthesize or carry out other metabolic processes. Most protists undergo both sexual and asexual reproduction, but some lack the ability to reproduce sexually. Sexual and asexual reproductive practices are as diverse as the protists themselves.

The synapomorphy that defines the Alveolata is the possession of cavities called alveoli just below their cell surfaces. All alveolates are unicellular.

The majority of dinoflagellates are mostly marine organisms with two flagella. They are golden brown in color due to photosynthetic and accessory pigments and are important primary producers in marine environments. Many live in symbiosis with other organisms. They are common endosymbionts in corals, for example. Dinoflagellates such as *Pfiesteria piscicida* and *Gonyaulax* are responsible for "red tides" in warm marine waters and can damage fish. Toxins produced by *Gonyaulax* species can accumulate in shellfish and be fatal to people eating the shellfish. Dinoflagellates have a distinct appearance due to the special arrangement of their two flagella: one in an equatorial groove around the cell and another that passes through a longitudinal groove before extending beyond the cell into the surroundings.

Apicomplexans are parasitic protists. Their name derives from the cluster of organelles at the apex of their cells, which assists with invasion of their host organism. The life cycles of these parasites are complex and often involve infection of multiple hosts to complete the life cycle. Apicomplexans in the genus *Plasmodium* cause malaria, a disease that claims more than a million lives annually.

Ciliates move via cilia and are distinguished by their possession of two types of nuclei: macronuclei (usually one per cell) and micronuclei (from one to several per cell; see Figure 20.18). Most ciliates are heterotrophic. *Paramecium* is a common and well-studied ciliate genus. Paramecia move by means of cilia, their membranes are protected by a pellicle (a structure composed of an outer membrane and an inner layer of closely packed membrane-enclosed sacs). They also protect themselves with trichocysts present in the pellicle, which act as sharp darts (see Figure 20.6). Paramecia reproduce asexually by binary fission. Genetic recombination is accomplished through conjugation, which involves the exchange of equal amounts of genetic material (see Figure 20.18). The resulting recombined paramecia then go through binary fission.

The excavates include diverse groups that split from one another soon after the eukaryotes originated. Several of these groups lack mitochondria, although there is evidence that they once possessed them. The continuing existence of excavates demonstrates that eukaryotic life is possible without mitochondria. Five major subgroups are aggregated into three main groupings: the diplomonads and parabasalids; the heteroloboseans; and the euglenoids and kinetoplastids.

Diplomonads and parabasalids lack mitochondria. *Giardia lamblia* is a well-known diplomonad that causes the human intestinal disorder giardiasis. *Trichomonas vaginalis* is a parabasalid responsible for a sexually transmitted disease in humans. Both of these organisms have multiple flagella and use an undulating membrane in locomotion.

Heteroloboseans have a life cycle that alternates between an amoeboid stage and a flagellated stage. One species of *Naegleria* can cause a fatal disease of the nervous system in humans. The amoeba enters the body through the nose and moves through the olfactory nerve to the brain where it destroys tissues.

Euglenids and Kinetoplastids together constitute a clade of unicellular excavates with flagella. They reproduce asexually through binary fission. Their mitochondria contain disc-shaped cristae and their flagellae contain a unique crystalline rod (see Figure 20.8). Euglenids are flexible in nutritional requirements and may switch between autotrophism and heterotrophism. Euglenid chloroplasts have triple membranes. The longer of the two flagella possessed by Euglenids may be used for propulsion and as an anchor. Kinetoplastids are parasitic and are characterized by their single large mitochondrion. The mitochondrion is unique in having a kinetoplast housing multiple circular DNA molecules and associated proteins. Many tropical human diseases are caused by kinetoplastids, including African sleeping sickness (see Table 20.1). One of the ways these organisms evade control efforts is by frequently altering their cell surface recognition proteins.

Stramenopiles are the diatoms, brown algae, oomycetes, and a few smaller groups, and are characterized by their rows of tubular hairs on the longer of their two flagella. Those that are not flagellated have lost their flagella over the course of evolution.

Diatoms are yellowish brown and store carbohydrates and oils as photosynthetic products. They are most noted for their silica-containing cell walls, which have two halves that fit together like a petri plate (see Figure 20.9) and have been used as filters, insulation, and insecticides. All diatoms are symmetrical and unicellular. They reproduce both asexually and sexually, but since each asexual reproduction event results in size reduction, occasional sexual reproduction is required. Diatoms are major photosynthetic producers.

Brown algae are multicellular and are the largest of the protists. Brown algae are almost exclusively marine. They produce branched filaments (Figure 20.10A) or leaflike growths (Figure 20.10B). The giant kelp that grow in large oceanic "forests" are brown algae. Brown algae are brown because of chlorophylls *a* and *c* and the carotenoid fucoxanthin. They have specialized regions called holdfasts, which anchor them to a substrate. Brown algae are commercially important for the presence of alginic acid, which is used as a binder in many food and cosmetic products.

Oomycetes are a nonphotosynthetic group of stramenopiles. These are commonly known as water molds and downy mildews, but their resemblance to fungi is only superficial. For example, unlike the cell walls of fungi, which contain chitin, cell walls of oomycetes typically contain cellulose. Oomycetes are coenocytic, and have multiple nuclei per cell. Many are saprobic and feed on dead material. Others are infectious to plants.

Rhizaria comprise three related groups of unicellular aquatic eukaryotes, including the cercozoans, foraminiferans, and radiolarians. Cercozoans are very diverse, with either amoeboid or flagellated forms and occur in aquatic or soil habitats. Limestone deposits are often the result of discarded foraminiferan shells made of calcium carbonate. Some foraminiferans live as plankton, while others dwell on the sea floor and have been found at the deepest parts of the oceans. Their pseudopods are used to trap food. Radiolarians have stiff, microtubule-reinforced pseudopods that help the cells float in their marine environments and provide additional surface area. Their glassy endoskeletons sometimes have elaborate geometric designs and are as varied as snowflakes. They include some of the largest unicellular eukaryotes, up to several millimeters across.

Amoebozoans include the loboseans, the plasmodial slime molds, and the cellular slime molds. Their relationships to other groups of eukaryotes are unclear.

Loboseans live as independent single cells. They feed by phagocytosis. A few produce casings by gluing sand grains together (see Figure 20.15). Slime molds are motile, feed by endocytosis, form spores on erect fruiting bodies, and undergo drastic changes during their life cycle. Plasmodial slime molds form multinucleate masses (see Figure 20.16A). During the vegetative stage, an acellular slime mold is a wall-less mass of cytoplasm with multiple diploid nuclei. It oozes over a network of strands called plasmodium. This phenomenon is called cytoplasmic streaming and is used to move and engulf food. If exposed to adverse environmental conditions, the slime mold will form a dormant sclerotium from which it can turn back into a plasmodium when environmental conditions again become favorable. It may also transform into a fruiting structure called a sporangiophore (see Figure 20.16B). Cellular slime molds are made of large numbers of cells called myxamoebas with single haploid nuclei. They reproduce by mitosis and fission, and this life cycle can continue as long as conditions are favorable. Once conditions become unfavorable, myxamoebas secrete cAMP, aggregate, and form a motile slug that eventually produces fruiting bodies. Spores from the fruiting bodies germinate to form more myxamoebas. Sexual reproduction occurs with the fusion of two myxamoebas, which then undergo meiosis to release haploid myxamoebas (see Figure 20.17).

Question 3. What would be the criteria for placing a newly discovered organism within the protists?
Textbook Reference: *20.2 Major Lineages of Eukaryotes Diversified in the Precambrian, p. 392*

Question 4. Many protists are motile. Describe three types of mobility and the function that each type enhances (e.g., food acquisition, mate acquisition, etc.).
Textbook Reference: *20.2 Major Lineages of Eukaryotes Diversified in the Precambrian, pp. 392–401*

Question 5. Many excavates lack mitochondria yet are very successful. How might you explain this for at least some excavates?
Textbook Reference: *20.2 Major Lineages of Eukaryotes Diversified in the Precambrian, p. 395*

20.3 Protists Reproduce Sexually and Asexually

Some protists have reproduction without sex and sex without reproduction

Some protist life cycles feature alternation of generations

Most protists reproduce both sexually and asexually, although sexual reproduction has not been confirmed for some protists. Modes of asexual reproduction in protists include binary fission, multiple fission, budding, and sporulation. All of these asexual modes of reproduction lead to lines of genetically identical individuals (except for rare mutants that may arise). Sexual reproduction in protists is variable: gametes may be the only haploid cell in the life cycle; the zygote may be the only diploid cell in the life cycle; and multicellular haploid and/or diploid generations may occur between sexual reproduction events.

Conjugation is an elaborate and unusual way of exchanging and rearranging genetic material (see Figure 20.18). Like many other multicellular protists, some fungi, and all land plants, brown algae go through alternation of generations, in which a diploid sporophyte gives rise to haploid spores through meiosis to form the haploid gametophyte, which in turn produces haploid gametes. Heteromorphic alternation of generations occurs when the two generations differ morphologically, and isomorphic alternation of generations occurs when the two generations are morphologically similar. The gamete-producing generation is haploid and produces gametes by mitosis. The spore-producing generation is diploid and its specialized cells (called sporocytes) each produce four haploid spores by dividing meiotically.

Question 6. Explain how the process of sexual reproduction in *Paramecium,* as diagrammed in the figure below, results in genetic change. At which stages are there opportunities for genetic rearrangement? Are the resulting daughter cells genetically identical? Why or why not?
Textbook Reference: *20.3 Protists Reproduce Sexually and Asexually, pp. 401–402*

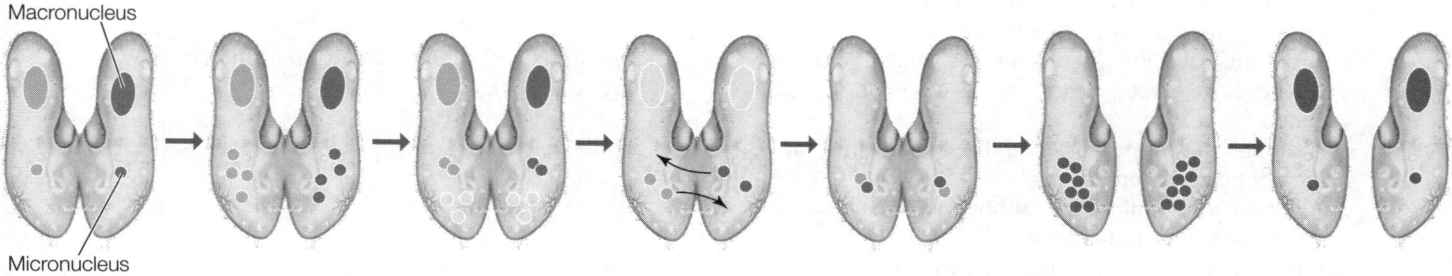

Question 7. Differentiate between asexual and sexual reproduction. Describe one mode of microbial eukaryotic sexual reproduction.
Textbook Reference: *20.3 Protists Reproduce Sexually and Asexually, pp. 401–402*

Question 8. Explain alternation of generations. Why are the protists the first group in which this process could be observed?
Textbook Reference: *20.3 Protists Reproduce Sexually and Asexually, p. 402*

20.4 Protists Are Critical Components of Many Ecosystems

Phytoplankton are primary producers

Some microbial eukaryotes are deadly

Some microbial eukaryotes are endosymbionts

We rely on the remains of ancient marine protists

Diatoms produce about one-fifth of all fixed carbon on Earth, and other protist members of the phytoplankton are important as well. Many important plant and animal pathogens are also microbial eukaryotes. These include *Plasmodium,* the protist that causes malaria, and dinoflagellates that produce neurotoxins that can affect fish and humans (see Table 20.1).

Protists are also important symbionts with many organisms. Many are endosymbionts with other protists and animals, including corals.

Ancient deposits of protists provide oil, natural gas, diatomaceous earth, and limestone. They also provide important clues to Earth's evolutionary history and past climate.

Question 9. Many protists are significant human pathogens. Describe the pathogenic protist that causes malaria, describe its life cycle, and identify the microbial eukaryote group to which it belongs.
Textbook Reference: *20.4 Protists Are Critical Components of Many Ecosystems, p. 403*

Question 10. Apicomplexans, the group containing *Plasmodium,* the disease organism responsible for malaria, contain a much reduced, nonphotosynthetic chloroplast. The chapter notes that researchers are targeting this organelle in efforts to develop anti-malarial drugs. How would you explain the presence of chloroplasts in *Plasmodium*?
Textbook Reference: *20.4 Protists Are Critical Components of Many Ecosystems, p. 403; 20.2 Major Lineages of Eukaryotes Diversified in the Precambrian, pp. 392–393*

Test Yourself

1. Which of the following modes of reproduction can be found in at least some protists?
   a. Binary fission
   b. Sexual reproduction
   c. Spore formation
   d. Multiple fission
   e. All of the above
   Textbook Reference: *20.3 Protists Reproduce Sexually and Asexually, p. 401*
2. During the evolution of eukaryotes from prokaryotes, which of the following did *not* occur?
   a. Infolding of the flexible cell membrane
   b. Loss of the cell wall

c. A switch from aerobic to anaerobic metabolism
d. Endosymbiosis of once free-living prokaryotes
e. Development of a cytoskeleton

Textbook Reference: *20.1 Eukaryotes Acquired Features from Both Archaea and Bacteria, pp. 389–390*

3. Which of the following statements about protists is *false*?
a. Apicomplexans are the only microbial eukaryote group without parasitic representatives.
b. Foraminiferans and radiolarians are shelled protists.
c. Ciliates have great control over the direction of their beating cilia.
d. Although they appear structurally simple, amoebas are not primitive organisms.
e. All diatoms are unicellular, although a few individual species associate in filaments.

Textbook Reference: *20.2 Major Lineages of Eukaryotes Diversified in the Precambrian, pp. 392–393*

4. Ciliates, as represented by *Paramecium*, have defensive organelles in their pellicles called
a. trichonympha.
b. tridents.
c. trichomes.
d. trichocysts.
e. trochlea.

Textbook Reference: *20.2 Major Lineages of Eukaryotes Diversified in the Precambrian, p. 394, Figure 20.6*

5. A major difference between the vegetative states of cellular and plasmodial slime molds is that plasmodial slime molds _______ whereas cellular slime molds _______.
a. have haploid nuclei; have diploid nuclei
b. produce fruiting bodies; do not produce fruiting bodies
c. undergo aggregation under adverse conditions; do not undergo aggregation under adverse conditions
d. exist as a coenocytic mass; exist as individual myxamoebas
e. grow almost indefinitely in favorable conditions; must aggregate into myxamoebas every seven generations

Textbook Reference: *20.2 Major Lineages of Eukaryotes Diversified in the Precambrian, pp. 399–401*

6. Why is sexual reproduction in diatoms required periodically?
a. This is the only way they can evade parasitic excavates.
b. Half of the asexually produced offspring sink to the sea floor to become diatomaceous earth.
c. During asexual reproduction, the cells decrease in size each generation.
d. This allows them to move, since only the male gametes have flagella.
e. Diatoms lose their symmetry after a certain number of generations of asexual reproduction.

Textbook Reference: *20.2 Major Lineages of Eukaryotes Diversified in the Precambrian, pp. 396–397*

7. Red tides often cause massive fish kills and human illness in those eating shellfish. Which group of protists is responsible for red tides?
a. Parabasalids
b. Red algae
c. Euglenozoans
d. Dinoflagellates
e. Radiolarians

Textbook Reference: *20.0 Introduction, p. 388*

8. Holdfasts and alginic acid are characteristic of which group of protists?
a. Parabasalids
b. Red algae
c. Brown algae
d. Stramenopiles
e. Unikonts

Textbook Reference: *20.2 Major Lineages of Eukaryotes Diversified in the Precambrian, p. 397, Figure 20.10*

9. Three kinetoplastid trypanosomes cause diseases that result in more than 150,000 deaths annually. Which of the following characteristics of these excavates has made it difficult to kill them and therefore to eradicate these diseases?
a. Their lack of mitochondria
b. Their ability to frequently change cell surface recognition molecules
c. Their asexual reproduction by binary fission
d. Their nuclear genes, which replace the function of mitochondria
e. Their "guide proteins," which edit mRNA in the mitochondria

Textbook Reference: *20.2 Major Lineages of Eukaryotes Diversified in the Precambrian, pp. 395–396*

10. Which of the following statements regarding conjugation in *Paramecium* is *false*?
a. It results in genetic recombination.
b. It results in clones.
c. It results in offspring.
d. It is a sexual process.
e. It results in the production of no new cells.

Textbook Reference: *20.3 Protists Reproduce Sexually and Asexually, pp. 401–402, Figure 20.18*

11. Which of the following statements regarding protists is true?
a. They are always parasitic.
b. They are all single-celled.
c. They are all heterotrophic.
d. They are always photosynthetic.
e. They are always aquatic for at least at some part of their life cycle.

Textbook Reference: *20.2 Major Lineages of Eukaryotes Diversified in the Precambrian, pp. 392–401*

12. Which of the following statements about the evolution of eukaryotes is true?
a. The alveolates are more closely related to the stramenopiles than to any other group.

b. The most recent common ancestor of the parabasilids and euglenoids also gave rise to the animals.
c. The opisthokonts are polyphyletic.
d. Euglenids and Cercozoans are subgroups of the clade Rhizaria.
e. None of the above is true, all are false.

Textbook Reference: *20.1 Eukaryotes Acquired Features from Both Archaea and Bacteria, p. 393, Figure 20.3*

13. Which of the following statements about oomycetes is *false?*
    a. The oomycetes are more distantly related to fungi than are humans.
    b. Their cell walls are typically made of chitin.
    c. They are absorptive heterotrophs.
    d. Their cell walls are typically made of cellulose.
    e. Some oomycetes are plant parasites.

    Textbook Reference: *20.2 Major Lineages of Eukaryotes Diversified in the Precambrian, pp. 397–398*

14. Which of the following statements about protists is *false?*
    a. They are photosynthetic and are one of the foundations of the marine food web.
    b. They are synthesizers of materials that form many sandy beaches
    c. They neutralize the sulfuric acid in deep sea vents.
    d. They are used as nutritional supplements for humans.
    e. They aid in the formation of what we call crude oil.

    Textbook Reference: *20.4 Protists Are Critical Components of Many Ecosystems, pp. 402–405*

15. Which of the following statements about the protist cytoskeleton is *false?*
    a. It allows for the formation of pseudopods.
    b. It is the main structural component of cilia.
    c. It is the primary component of the outer shell of diatoms.
    d. It is the main structural component of flagella.
    e. It helps in removing excess water via the contractile vacuoles.

    Textbook Reference: *20.1 Eukaryotes Acquired Features from Both Archaea and Bacteria, pp. 390–391*

16. Which of the following statements about coral bleaching is *false?*
    a. It results from the loss of a dinoflagellate.
    b. It results from the loss of an endosymbiont.
    c. It results from food depletion.
    d. It results from environmental conditions such as rising water temperatures or increased water turbidity.
    e. It results from a parasitic invasion.

    Textbook Reference: *20.4 Protists Are Critical Components of Many Ecosystems, p. 404, Figure 20.19*

17. Slime molds were once classified with the fungi. Which of the following characteristics favor their placement among the Amoebozoans?
    a. Cytoplasmic streaming, endocytosis, and alternation of generations
    b. Cytoplasmic streaming, phagocytosis, and pseudopods
    c. Cell walls, sporangia, and pseudopods
    d. Creeping locomotion and engulfing of food particles
    e. Production of thick-walled spore-bearing structures

    Textbook Reference: *20.2 Major Lineages of Eukaryotes Diversified in the Precambrian, pp. 397–398*

Answers

Key Concept Review

1. One hypothesis is that Eukarya split from Archaea and after the split, endosymbioses with bacterial lineages resulted in mitochondria and chloroplasts. Another hypothesis proposes that eukaryotes resulted from the fusion of lineages from Archaea and Bacteria. Regardless of these early relationships, we can infer that the evolution of eukaryotic cells resulted from a number of events including: the origin of a flexible cell surface; the origin of a cytoskeleton; the origin of a nuclear envelope enclosing a chromosomal genome; the development of digestive vacuoles; and the endosymbiotic development of organelles, notably mitochondria and chloroplasts. The evolutionary relationships are difficult to understand due to limited fossilization, lateral gene transfers, and the contributions of symbiotic prokaryotes.

2. Double-membrane-enclosed chloroplasts most likely arose from the endosymbiosis of a cyanobacteria in a process called primary endosymbiosis; triple-membrane-enclosed chloroplasts most likely arose from the endosymbiosis of a chlorophyte or a rhodophyte

3. The organism must be a eukaryote and not fit the criteria for land plants, animals, or fungi.

4. Protists move by flagella, cilia, undulating membranes, pseudopodia, and cytoplasmic streaming.

   Flagella: Many protists possess one to many flagella in at least at one stage of the life cycle. Flagella propel them through the water to locate food and mates or escape enemies. In Euglena, flagella may also serve as an anchor to hold the organism in place.

   Cilia: Cilia, for example, provide *Paramecium* with a form of locomotion that is generally more precise than locomotion by flagella or pseudopods. A *Paramecium* can coordinate the beating of its cilia to propel itself either forward or backward in a spiraling manner. This movement can assist *Paramecium* to align for conjugation.

   Undulating membranes: In addition to flagella and a cytoskeleton, the parabasalids have undulating membranes that also contribute to the cell's locomotion.

Pseudopodia: In foraminiferans, long, threadlike, branched pseudopods extend through microscopic apertures in the shell and interconnect to create a sticky net that the they use to obtain food. The pseudopods also can provide locomotion. Loboseans use their pseudopods to engulf small organisms and food particles by phagocytosis.

Cytoplasmic streaming: In plasmodium, the outer cytoplasmic region of a plasmodium becomes more fluid in places, and cytoplasm rushes into those areas, stretching the plasmodium. Microfilaments and a contractile protein called *myxomyosin* interact to produce the streaming movement. As it moves, the plasmodium engulfs food particles by endocytosis.

5. Many excavates, including the trypanosomes and the kinetoplastids, are parasitic and may be able to obtain energy without mitochondria. In addition, mitochondrial genes have been discovered in the nuclear genome, and so the lack of mitochondria may not reflect the lack of energy-generating reactions in the cell.
6. Ciliates in the genus *Paramecium* reproduce sexually in a process called conjugation, during which two single-celled organisms exchange genetic material. Although each *Paramecium* has both a macronucleus and many micronuclei, only the micronuclei are exchanged. The macronuclei then incorporate the new genetic material from the micronuclei. Genetic change is likely to occur during crossing over during meiotic divisions of a single micronucleus within each individual and also when the haploid micronuclei from each individual are exchanged and fuse into a novel diploid cell. The resulting daughter cells are not identical because their diploid micronuclei are the result of fusions of different haploid micronuclei.
7. Asexual reproduction results in clones of the original organism, and there is no genetic recombination or variation associated with the creation of offspring. Sexual reproduction allows for new genetic combinations. Protists undergo varied types of sexual reproduction, from fusing haploid myxamoebas in cellular slime molds, to alternation of generations in brown algae. Conjugation between paramecia is an example of sexual recombination without reproduction.
8. An organism that exhibits alternation of generations exists in a haploid gamete-producing form and a diploid spore-producing form. Prokaryotes have a single chromosome, so only in eukaryotes, which have multiple copies of chromosomes, is a diploid stage of a life cycle possible.
9. The protist that causes malaria is a member of the genus *Plasmodium*. *Anopheles* mosquitos are the vector for *Plasmodium* and transfer it to the human circulatory system through a bite. The parasites then move to the liver and lymph system and multiply. They then reenter the bloodstream and infect red blood cells, where they multiply again; the red blood cells burst, releasing more parasites. When another *Anopheles* mosquito bites the infected human, it takes in *Plasmodium* cells, which develop into gametes that produce a zygote. The zygotes move into the mosquitos salivary glands, ready to be passed to another human host.
10. The group apicomplexans (which is exclusively parasitic) is closely related to the dinoflagellates (which are photosynthetic), so it is not surprising that apicomplexans would have once possessed chloroplasts for photosynthesis. The question is why they would retain the chloroplast after adopting a parasitic lifestyle. One possibility is that the original photosynthetic endosymbiont provided other functions, such as synthesis of fatty acids, lipids, or other cellular molecules.

Test Yourself

1. **e.** Methods of reproduction are quite varied among the protists.
2. **c.** At the time the first eukaryotes evolved, the environment was becoming oxygen-rich. There was a switch from anaerobic to aerobic metabolism, not vice-versa.
3. **a.** Malaria is caused by *Plasmodium*, which is an apicomplexan. In fact, all apicomplexans are parasitic.
4. **d.** Trichocysts are defensive barbs ejected from ciliates when they are disturbed.
5. **d.** The vegetative (feeding) state of an acellular slime mold is called a plasmodium; it consists of multiple diploid nuclei enclosed in a single membrane. The vegetative state of a cellular slime mold is a myxamoeba with a single haploid nucleus. Although the myxamoebas of cellular slime molds do aggregate to form fruiting structures, individual myxamoebas never fuse into multinucleated structures.
6. **c.** Diatoms have a stiff cell wall and are composed of two halves that fit together like petri dishes. Neither half can grow, and the new cells must fit inside each parent cell half. Thus, the cells reduce in size each generation until sexual reproduction occurs, at which point the cell walls are shed during gamete production and the zygote is able to grow before a new cell wall is laid down.
7. **d.** Dinoflagellates are the cause of red tides.
8. **c.** The brown algae are multicellular protists notable for their organ and tissue differentiation and the presence of alginic acid in their cell walls.
9. **b.** Sleeping sickness, Chaga's disease, and Leishmaniasis are all caused by kinetoplastids, many of which are able to frequently change their cell surface recognition molecules.
10. **d.** Conjugation does result in genetic recombination but does not result in the production of clones or offspring.
11. **e.** Protists do not have a unifying characteristic other than being eukaryotic organisms that do not fit into the

kingdoms Plantae, Animalia, or Fungi. If an aquatic environment is considered to be a single droplet of water, however, protists would meet this criterion. For a single-celled organism, a droplet of water is a very large environment.

12. **a.** Alveolates are more closely related to the stramenopiles than to any other group.
13. **b.** Oomycetes, whose cell walls contain cellulose, are distantly related to fungi, whose cell walls contain chitin.
14. **c.** Protists are a very diverse group, but they do not release bases or neutralize environmental acids. Members of Archaea are likely to be found in an acidic environment.
15. **c.** The cytoskeleton, consisting of actin filaments, microtubules, and intermediate filaments, is a feature of eukaryotic cells. In protists, the cytoskeleton is important in anchoring organelles, movement of materials in the cell, movement of the cell in the environment, and controlling the shape of the cell.
16. **e.** Coral rely on the photosynthetic products of their dinoflagellate endosymbionts. Coral bleaching occurs when the dinoflagellates die or are expelled by the coral cells due to a change in conditions, such as warm temperatures or increased water turbidity.
17. **d.** These are the two characteristics that are common the amoebozoans, but are not characteristic of fungi.

The Evolution of Plants 21

The Big Picture

- Land plants are photosynthetic eukaryotes that utilize chlorophylls *a* and *b*, undergo alternation of generations, and develop from multicellular embryos that are protected by the parent plant. The ten extant (surviving) groups can be classified as nonvascular plants (those without highly developed vascular tissue) and vascular plants (those with highly developed vascular tissue). This chapter focuses on the nonvascular plants and the nonseed vascular plants.
- In order to colonize land, plants had to evolve strategies for coping with desiccation and gravity. This involved mechanisms for extracting water from soil, means of transporting water throughout the plant, methods of ensuring fertilization, and modes of protecting developing embryos.
- This chapter introduces and describes the liverworts, hornworts, and mosses, all of which are considered nonvascular plants, and the club mosses, horsetails, whisk ferns, and ferns, which are the nonseed vascular plants.

Study Strategies

- Rather than memorizing land plant types and characteristics, focus on the trends and relationships among them, particularly the evolutionary relationships.
- Plant terminology is probably less familiar than the vocabulary that refers to animals. Focus on the meaning of the word parts. For example, in the word "glaucophyte," the root of the word ("phyte") means plant and the prefix ("glauca") means bluish white or pale green, so unicellular glaucophytes are so named because of the distinguishing color of the peptidoglycan found between the inner and outer membranes. A gametophyte is the gamete-producing plant stage, and a sporophyte is the spore-producing plant stage. Understanding how the words are constructed will help you understand and retain their meanings.
- Focus on relationships among the land plants and how to differentiate one group from another. Many of the important distinctions between groups are related to adaptations for terrestrial life.
- When studying life cycles, be sure to note whether the sporophyte or the gametophyte is dominant and make note of the point at which mitosis and meiosis take place. Make diagrams of life cycles indicating haploid and diploid stages. Keep track of spore formation and the gametophyte in the seed plants.
- Go to yourBioPortal.com to review the following tutorials and activities:

  Animated Tutorial 21.1 Life Cycle of a Moss

  Animated Tutorial 21.2 Life Cycle of a Conifer

  Animated Tutorial 21.3 Life Cycle of an Angiosperm

  Web Activity 21.1 The Fern Life Cycle

  Web Activity 21.2 Homospory

  Web Activity 21.3 Heterospory

  Web Activity 21.4 Life Cycle of a Conifer

  Web Activity 21.5 Flower Morphology

  Working with Data 21.1 Genomes and Plant Phylogeny

Key Concept Review

21.1 Primary Endosymbiosis Produced the First Photosynthetic Eukaryotes

Several distinct clades of algae were among the first photosynthetic eukaryotes

There are ten major groups of land plants

The development of photosynthetic eukaryotes contributed greatly to the development of terrestrial life. Primary endosymbiosis is a shared derived trait of all members of the group Plantae, which includes many aquatic glaucophytes, red algae, and green algae in addition to the terrestrial organisms we commonly refer to as plants. The unicellular glaucophytes are distinguished by the small amount of peptidoglycan found between the inner and outer membranes of their chloroplasts—the same arrangement found in cyanobacteria. In contrast to the glaucophytes, red algae are mostly multicellular and are characterized by the photosynthetic accessory pigment phycoerythrin. Red algae also contain chlorophyll *a*, phycocyanin, and carotenoids. The ratio of phycoerythrin to chlorophyll *a* determines the color of red algae and depends on light exposure.

Green algae are the other algal members of Plantae. They are distinguished by the possession of chlorophylls *a* and *b* and their storage of photosynthetic products in their chloroplasts as starches. Three important clades of green algae include chlorophytes, coleochaetophytes, and charophytes. Together with land plants, the coleochaetophytes and charophytes form the "streptophytes."

All land plants develop from an embryo that is protected by tissues of the parent plant. This is a key shared trait, or synapomorphy, of land plants. For this reason some people refer to land plants as embryophytes. This book distinguishes green plants (including green algae) from land plants (embryophytes).

There are ten extant groups of land plants (see Table 21.1). Land plants as a whole are monophyletic, but the nonvascular plants as a group are paraphyletic.

Question 1. Chloroplast-containing organisms are often referred to as plants, and green plants, streptophytes, and land plants are sometimes referred to collectively as the "plant kingdom." What are the differences among them?
Textbook Reference: *21.1. Primary Endosymbiosis Produced the First Photosynthetic Eukaryotes, pp. 408–409*

Question 2. Differentiate between glaucophytes and red algae. In what ways are these two groups similar?
Textbook Reference: *21.1. Primary Endosymbiosis Produced the First Photosynthetic Eukaryotes, p. 409*

21.2. Key Adaptations Permitted Plants to Colonize Land

- Two groups of green algae share many features with land plants
- Adaptations to life on land distinguish land plants from green algae
- Life cycles of land plants feature alternation of generations
- Nonvascular land plants live where water is readily available
- The sporophytes of nonvascular land plants are dependent on the gametophytes

Molecular and fossil evidence indicates that land plants arose from a green algae ancestor. Both coleochaetophytes and charophytes retain the eggs within the parental organism. Land plants most likely arose from a group of green algae called the Charales (within the charophytes—see Figure 21.1) based on the following synapomorpies: (1) plasmodesmata that join the cytoplasm of adjacent cells, (2) branching and apical growth, and (3) similar peroxisome contents, chloroplast structure, and mechanics of cellular division.

To survive on land, aquatic organisms had to reduce their dependence on water and develop mechanisms to avoid lethal desiccation (drying). Terrestrial plants also needed support to resist gravity in order to grow upward, and they required some mechanism other than swimming gametes (through water) to move the gametes from one plant to another. The first land plants that met at least some of these challenges are the nonvascular land plants.

Many characteristics that distinguish land plants from green algae are adaptations to a terrestrial environment. The earliest land plants developed the following modifications for terrestrial life: a waxy cuticle, the presence of stomata, gametangia, embryos, protective pigments, thick spore walls, and mutualistic associations with fungi. Waxy cuticles (waxy lipid coatings of leaves and stems) prevent water loss from tissues exposed to dry air. The development of a cuticle was likely one of the earliest and most important adaptations to land.

Land plants undergo alternation of generations (see Figure 21.4). Haploid gametophytes, which grow from haploid spores, produce haploid gametes through mitosis; diploid sporophytes, which arise from the fusion of gametes, produce haploid spores through meiosis. The sporophyte and gametophyte generations differ genetically: the sporophyte has diploid cells while the gametophyte has haploid cells. In plant evolution, the trend is toward reduction of the gametophyte generation. The life cycle of nonvascular land plants is dominated by the gametophyte generation. The sporophyte is very tiny and is completely dependent on the gametophyte. Figure 21.6 illustrates the life cycle of a moss as an example of the nonvascular plant life cycle.

Land plants without vascular tissue depend on abundant water sources, thin tissues that can absorb water easily, mechanisms for capturing water vapor in the air, and capillary action to move water through the plant. They lack the support of lignin and hug the ground closely. The cuticle is either lacking or thin and ineffectual in reducing water loss. Mutualisms with fungi probably facilitate water and mineral absorption.

Liverworts (*Hepatophyta*) may be the oldest surviving land plant clade. The liverwort gametophyte can be either leafy or a flat plate of cells. Liverwort sporophytes remain attached to the larger gametophyte and rarely exceed a few millimeters in length. Liverworts can reproduce asexually as well as sexually.

Mosses (*Bryophyta*) are found in almost all types of terrestrial environments and are the most familiar and widespread nonvascular plants. The specialized cells called hydroids are functionally similar to the tracheids seen in vascular plants, but they lack lignin and the cell wall structure of tracheids. Like a tracheid, the hydroid cell dies and leaves a tiny channel that can transport water through the plant. Mosses (see Figure 21.6) are the sister lineage to vascular plants plus the hornworts and share with those groups the presence of stomata that liverworts lack.

Hornworts (*Anthocerophyta*) evolved simple stomata, which are pores that help with the exchange of gas and water vapor. Hornworts are distinguished from other nonvascular plants by two characteristics: the presence of a single platelike chloroplast in each cell, and sporophytes that are capable of indeterminate growth. The sporophyte produces new spore-bearing tissue from a basal region of cell division. In very moist environments, the sporophyte can be as tall as 20 centimeters; growth is limited only by the lack of a true vascular system. The hornwort sporophyte resembles a long, slender horn (see Figure 21.5C). Hornworts often have symbiotic relationships with cyanobacteria (contained in mucilage-filled internal cavities) that are able to fix atmospheric nitrogen and make it available to the hornwort.

In nonvascular land plants, the dominant photosynthetic and nutritionally independent generation is the haploid gametophyte. The diploid sporophyte may be photosynthetic, but it is always nutritionally dependent on the gametophyte and remains attached to it. The sporophyte produces unicellular haploid spores as products of meiotic division that occur within a sporangium. The germinating spore gives rise to a multicellular haploid gametophyte whose cells contain chloroplasts. The gametophyte forms specialized sex organs called gametangia, within which gametes form via mitotic cell division. The archegonium, a multicellular flask-shaped organ with a long neck and a swollen base, produces a single egg. The male sex organ, the antheridium, produces sperm in large numbers. Each sperm bears two flagella and swims to the archegonium and down the canal to fertilize the egg. The eggs release chemical attractants to guide them. Water is critical to all these events. Each gametophyte individual produces both archegonia and antheridia, but the density of gametophytes helps ensure some cross fertilization that maintains genetic variation. Once the sperm nucleus fuses with the egg nucleus, the diploid zygote produces a multicellular embryo that matures into a sporophyte that then repeats the cycle.

Question 3. Nonvascular plants faced many problems as they colonized land. Describe three problems and the mechanisms that land plants evolved to surmount those problems.
Textbook Reference: *21.2 Key Adaptations Permitted Plants to Colonize Land, p. 411*

Question 4. Explain why the largest mosses are less than a meter tall.
Textbook Reference: *21.2 Key Adaptations Permitted Plants to Colonize Land, pp. 412–413*

Question 5. Label the structures in the moss life cycle in the diagram below. Indicate the sporophyte generation, gametophyte generation, and diploid or haploid status. Also indicate, in the open boxes, if the process occurring is meiosis, mitosis, or fertilization.

Textbook Reference: *21.2 Key Adaptations Permitted Plants to Colonize Land, p. 414, Figure 21.6*

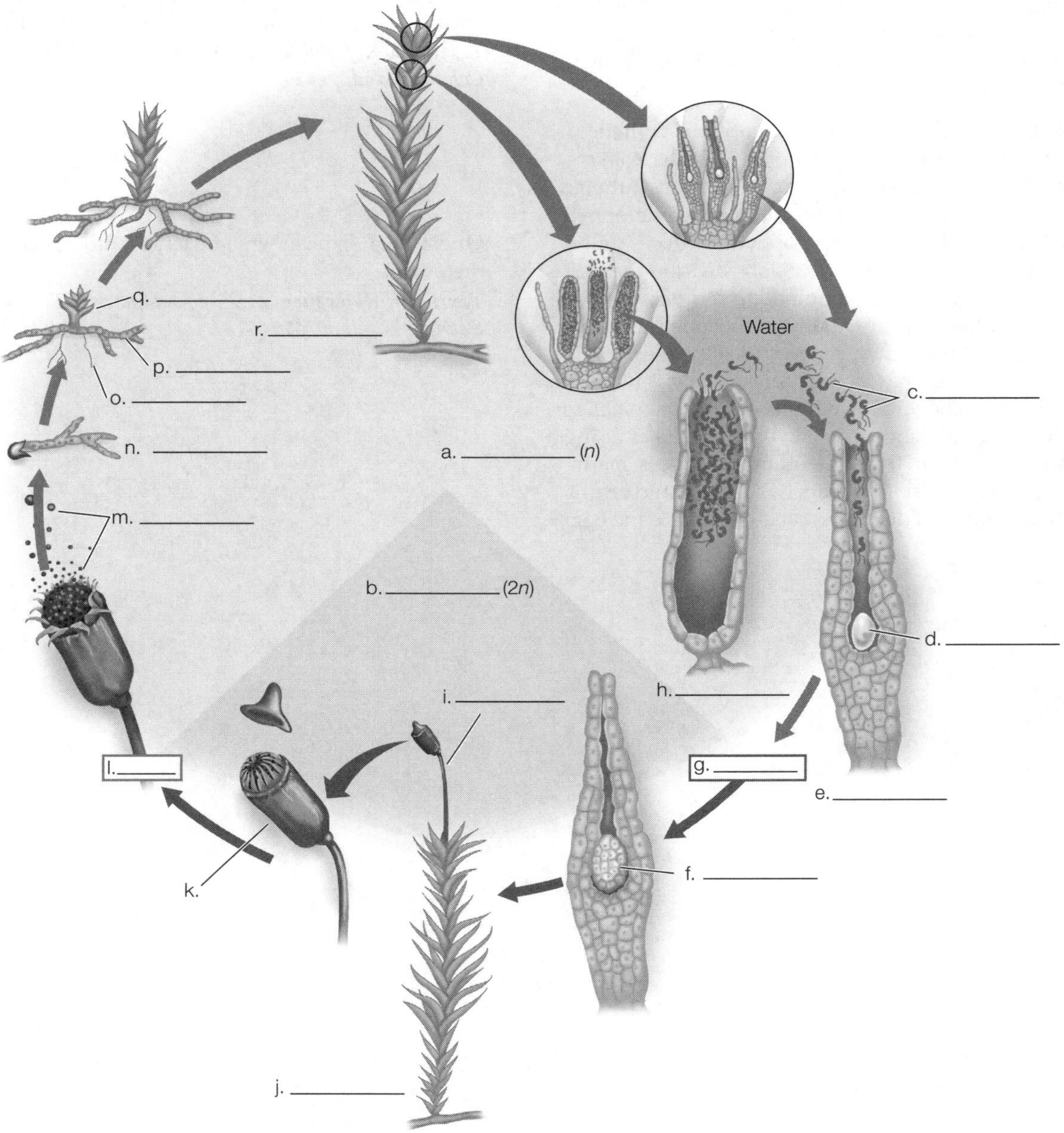

21.3 Vascular Tissues Led to Rapid Diversification of Land Plants

Vascular tissues transport water and dissolved materials

Vascular plants have been evolving for almost half a billion years

The earliest vascular plants lacked roots

The lycophytes are sister to the other vascular plants

Horsetails and ferns constitute a clade

The vascular plants branched out

Heterospory appeared among the vascular plants

The development of vascular tissue was key to the further adaptation to a terrestrial existence. Vascular tissue provides for the transport of water and food through a plant. Xylem is a vascular tissue responsible for the transport water and minerals from the soil to the aerial parts of the plant. Vascular plants also contain phloem, which conducts the products of photosynthesis from the location where they are produced to where they are stored.

Tracheids, which evolved about 430 mya, are strawlike cells that are the principal water-conducting elements of the xylem in all vascular plants except the angiosperms, although they do persist in angiosperms. The cell walls of tracheids provide a rigid structural support, which allows vascular plants to grow upward and compete for sunlight. Increased height also increases the distance over which spores can disperse.

Vascular plants are also characterized by branching, independent sporophytes that can produce more spores than an unbranched body and develop more complex architecture. The presence of terrestrial plants modified the environment to make it more hospitable to animals. By the Devonian period, trees appeared, including the giant lycophytes and horsetails that later dominated the landscape, including the swamps that would become the source of coal. About 200 mya, the climate dried and cooled and gymnosperms replaced the lycopophyte fern forests. About 65 mya, angiosperms became the dominant vegetation.

Significant new features that arose in early vascular plants included roots and true leaves. Early vascular plants, like rhyniophytes, did not have roots and were anchored by horizontal portions of stem called rhizomes. Their branching pattern was dichotomous. Lycophytes (club mosses and their relatives) have dichotomously branching true roots, simple vascular tissue, and simple leaflike structures called microphylls. Growth and branching occurs by apical cell division.

Heterospory appears to have evolved several times in the vascular plants (see Figure 21.11). Most early vascular plants exhibit homospory, in which spores (and the gametophytes that grow from them) are all of the same type; gametophytes produce both archegonia and antheridia. In heterospory, one spore type called a megaspore gives rise to the female, egg-producing megagametophyte; another spore type called a microspore develops into a male, sperm-producing microgametophyte. Land plants typically produce many more microspores than megaspores.

Extant nonseed vascular plants are varied and abundant. Club mosses (*Lycophyta*) diverged from the tracheid lineage relatively early and are the most "primitive" of the extant nonseed vascular plants. They have simple leaves arranged spirally on the stem. The sporangia of many club mosses are held in apical strobili, which are simple branching structures of fertile sporangia and sterile leaves (see Figure 21.8A). Other club mosses have sporangia called sporophylls on the upper surfaces of leaves. Club mosses may either be homosporous or heterosporous.

The horsetails, whisk ferns, and ferns were once thought to be only distantly related but are now in their own clade, the monilophytes. As with seed plants, there is differentiation between the main stem and side branches. Horsetails are represented by few extant species. They have true roots, their sporophytes are large and independent, and their gametophytes are highly reduced but also independent. The leaves of horsetails are simple, forming whorls around the stem (see Figure 21.8B). This group is one of the few plant groups that exhibits basal growth. Whisk ferns once were thought to be the "missing link" to *Rhynia,* but molecular evidence indicates that they evolved much later from more highly complex land plants. Their relatively simple body plan demonstrates that evolution does not necessarily move toward greater complexity. Wisk fern gametophytes live below the surface, lack chlorophyll, and depend on fungal partners for nutrition.

Ferns are the largest surviving group of nonseed vascular plants. They are characterized by relatively large, complex leaves with branching vascular strands. Like all other nonseed vascular plants and nonvascular plants, ferns continue to be dependent on water to carry motile sperm. They have advanced vascular structures and can reach great heights, but they do not produce true wood, and their root systems are poorly developed. The ferns consist of more than 12,000 species. The fern life cycle is dominated by the sporophyte, but the gametophyte is an independent photosynthetic structure (see Figure 21.9). In most species of ferns, the sporangia are found in clusters called sori. Most ferns are homosporous, but some fern groups are heterosporous. A few genera of ferns produce gametophytes that depend on a mutualistic fungus for nutrition.

True leaves are flattened photosynthetic structures with vascular tissue. Vascular plants have evolved two different kinds of leaves: microphylls and megaphylls (see Figure 21.10). Microphylls are present in the club mosses. They probably evolved from sterile sporangia. Most familiar leaves are megaphylls. These are thought to have evolved when photosynthetic tissue developed between the ends of small lateral branches.

Question 6. Describe the major consequences of the evolution of a branching sporophyte.
Textbook Reference: *21.3 Vascular Tissues Led to Rapid Diversification of Land Plants, pp. 417–418*

Question 7. Label the structures in the homosporous fern life cycle in the diagram below. Indicate the sporophyte generation, gametophyte generation, and diploid or haploid status. Also indicate, in the open boxes, if the process occurring is meiosis, mitosis, or fertilization.

Textbook Reference: *21.3 Vascular Tissues Led to Rapid Diversification of Land Plants, p. 418*

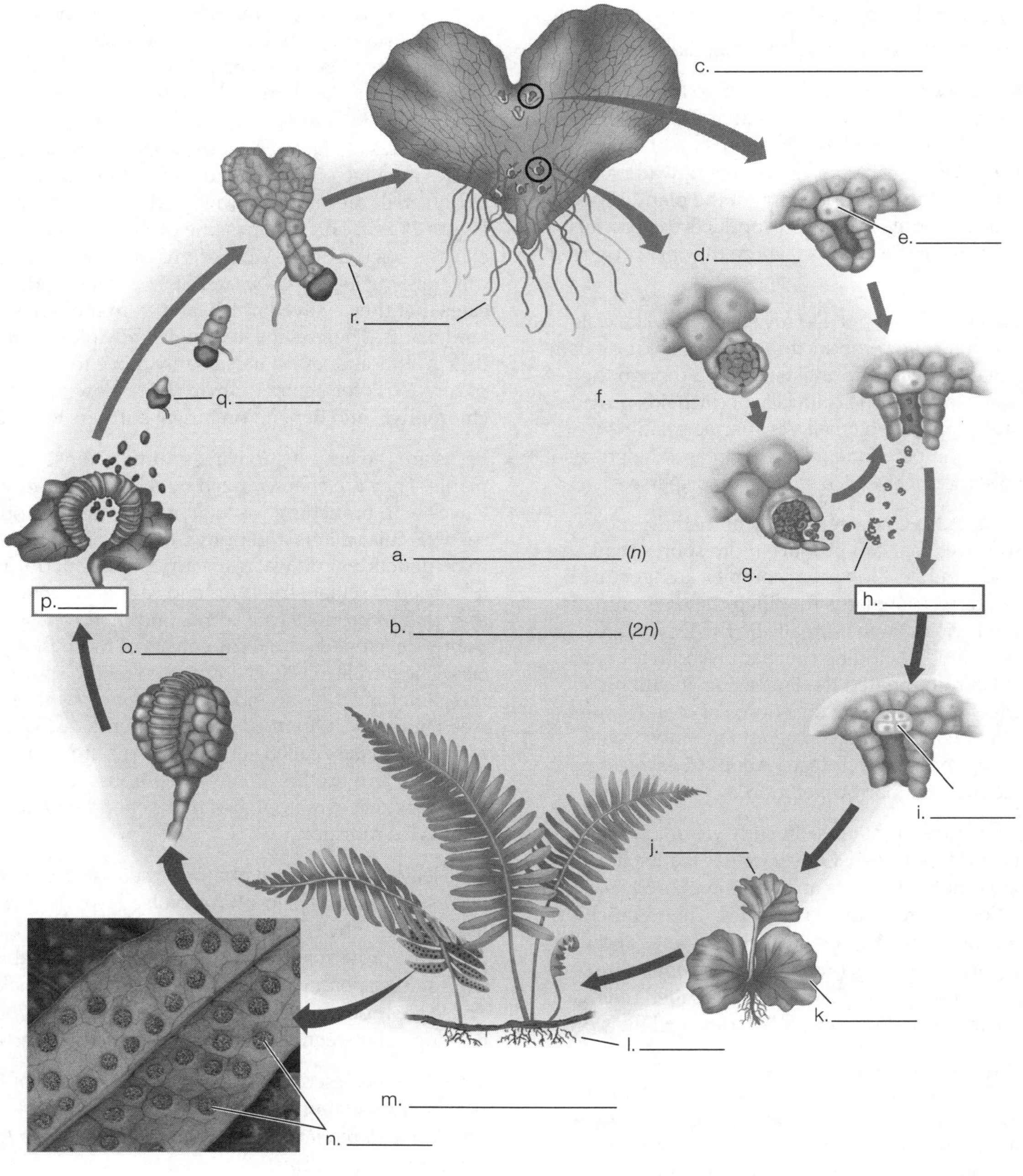

Question 8. Compare and contrast homospory and heterospory. Which reproductive structures result from meiosis in each type of life cycle, and which structures result from mitosis?

Textbook Reference: *21.3 Vascular Tissues Led to Rapid Diversification of Land Plants, pp. 418–420*

21.4 Seeds Protect Plant Embryos

Features of the seed plant life cycle protect gametes and embryos

The seed is a complex, well-protected package

A change in anatomy enabled seed plants to grow to great heights

Gymnosperms have naked seeds

Conifers have cones but no motile gametes

Seed plants have characteristics that set them apart from nonseed plants. They consist of two groups of vascular plants (tracheophytes): the gymnosperms and the angiosperms. Gymnosperms consist of four major phyla: Cycadophyta (cycads), Ginkgophyta (ginkgos), Coniferophyta (conifers), and Gnetophyta (gnetophytes). The angiosperms are the most diverse group and consist of a number of different phyla.

Seed plants have highly reduced gametophyte generations that are nutritionally dependent on the sporophyte for survival. This nutritional dependence is what sets the seed plants apart from the seedless vascular plants (see Figure 21.17). Very few seed plants (e.g., the cycads and ginkos) have retained swimming sperm; most have evolved other means for dispersing male gametes.

All seed plants are heterosporous in that they produce two types of spores. One becomes the female gametophyte and the other becomes the male gametophyte. Microspores and megaspores develop in specialized cones or in flowers. In the megasporangium, meiosis produces four megaspores, but in most seed plants only one of the four megaspores is retained. This megaspore divides by mitosis to form the multicellular (yet tiny) megagametophyte (female gametophyte), which produces the eggs. When the eggs are ready to be fertilized, they are surrounded by megagametophyte cells that are still within the megasporangium. The megasporangium is surrounded by sterile sporophyte tissues (the integument). Together, the megasporangium and the integument constitute the ovule.

Microspores develop into pollen grains, the male gametophytes. Pollen grains consist of sperm and supporting cells; they are dispersed by wind and animals. The wall of the pollen grain contains sporopollenin. Sporopollenin is one of the most chemically resistant biological compounds known, and represents a major advantage for land colonization by plants.

Because the female gametophyte is retained within sporophyte tissue, pollen grains do not have direct access to gametophytes and their eggs. The sporophyte housing a gametophyte creates tissue for receiving pollen grains; upon reaching this tissue, pollen produces pollen tubes that deliver the sperm through the sporophyte tissue to the eggs for fertilization. This process is known as pollination (see Figure 21.14). The embryo resulting from fertilization grows to a certain size and then becomes dormant within the surrounding tissues. This dormant protected embryo together with its surrounding tissues constitutes the seed.

Seeds are complex structures that may contain tissue from three plant generations. Recall that alternation of generations involves a multicellular sporophyte generation that alternates with a multicellular gametophyte generation. The embryo within a seed represents the beginning of a new sporophyte generation. It is still surrounded by the female gametophyte tissue, which will provide the embryo with nutrients to begin growth (particularly if it is a gymnosperm). The tough coat, or seed coat, that surrounds the seed consists of tissue provided by the embryo's sporophyte parent. This tissue is derived from the integument of the diploid parent. Seeds protect the embryo until conditions are right for germination. Seeds can remain viable or dormant for many years, waiting for favorable conditions to germinate. Many seeds have adaptations that allow dispersal by wind or another vector.

Wood provides support to many seed plants, allowing them to grow taller to capture more light for photosynthesis. The younger portions of wood consist of vascular tissue, allowing for water transport. As wood becomes older, it becomes clogged with resins or other materials and provides the plant with support. New layers of vascular tissue are produced underneath the wood. The existence of this secondary growth is one reason that seed plants became today's dominant vegetation.

Gymnosperms are seed plants that do not form flowers or true fruits. They are diverse and live in a variety of habitats, from sparse deserts to vast forests. There are four groups within the gymnosperms (see Figure 21.15), with the Coniferophyta group being the most abundant. Gymnosperms have significant vascular tissue and are capable of woody secondary growth. With the exception of the gnetophytes, their vascular tissue is less complex than that of the angiosperms. Tracheids are the only support and water-conducting cells within the xylem. The gymnosperm life cycle can be illustrated by that of the pine, a conifer (see Figure 21.17). Conifers differ from other gymnosperms in that they produce cones (megastrobili) and strobili (microstrobili), which are specialized structures for reproduction (see Figure 21.16). Cones house the megasporangia and produce megaspores, megagametophytes, and eggs; strobili house the microsporangia and produce microspores, microgametophytes (pollen), and sperm.

Pines do not have swimming sperm, and their pollen is modified for wind dispersal. Pollen lands on the female cone and lodges in the pollen chamber. The pollen tube grows through the maternal sporophyte tissue to the female gametophyte, where it releases two sperm. Only one of the two sperm will fertilize an egg. The resulting zygote develops into an embryo that remains encased in the tissues of the megasporangium and gametophyte. The seed is also protected by the scale of the cone (sporophyte tissue) until it is mature and ready for dispersal. Dispersal is aided by modifications of the seed coat.

Some conifer species have soft, fleshy modifications that surround the seed. Examples include the "berries" found on yew and juniper plants. These are not true fruits, however. The fruits of angiosperms are ripened ovaries, and gymnosperms do not have ovaries.

Question 9. Seed plants are heterosporous. Compare the protection provided to the microspores with the protection provided to the megaspores in seed plants.
Textbook Reference: *21.4 Seeds Protect Plant Embryos, p. 421*

Question 10. The sperm of seed plants is transported in the pollen grain. Describe in detail the two parts of the journey from the paternal parent to the egg in the maternal parent plant.
Textbook Reference: *21.4 Seeds Protect Plant Embryos, p. 421*

21.5 Flowers and Fruits Increase the Reproductive Success of Angiosperms

- Angiosperms have many shared derived traits
- The sexual structures of angiosperms are flowers
- Flower structure has evolved over time
- Angiosperms have coevolved with animals
- Fruits aid angiosperm seed dispersal
- The angiosperm life cycle produces diploid zygotes nourished by triploid endosperms
- Recent analyses have revealed the phylogenetic relationships of angiosperms

The production of flowers and fruits is the most obvious feature that distinguishes the angiosperms, which currently are the dominant plant form on Earth. Angiosperms have the most reduced gametophyte generation, with the female gametophyte consisting of just seven cells. Angiosperms differ from all other plants in that they have double fertilization, produce triploid endosperm (the nutritive tissue), have their ovules and seeds enclosed in a carpel, produce flowers and fruits, and have xylem and phloem with multiple modified cell types, including vessel elements, fibers, and companion phloem cells.

All flower parts are modified leaf structures (see Figure 21.12). Although flowers are very diverse, they all consist of the same small set of structures (see Figure 21.14B) that have evolved over time. The male structures in a flower are stamens. Each stamen consists of a filament and an anther; the anther contains microsporangia, which produce pollen. Flowers often have several stamens.

The female structure in a flower is the pistil, which consists of a stigma, a style, and an ovary. The stigma is modified to receive pollen. The style is a stalk that separates the stigma from the ovary. The ovary contains one or more ovules, each of which houses a megasporangium. Nonreproductive floral structures include petals, sepals, and the receptacle. Petals and sepals are often modified to attract animal pollinators. Sepals also serve to protect the developing flower bud. The receptacle is the attachment site for the sepals, petals, stamens, and carpels.

Flowers that have both megasporangia and microsporangia are called "perfect" or hermaphroditic. If either structure is missing, the flower is called "imperfect." Species that have both megasporangiate flowers and microsporangiate flowers on the same plant are called "monoecious." If they are on separate plants, they are called "dioecious." Flowers can be single or grouped together to form an inflorescence (see Figure 21.18).

Different phylogenetic approaches have led to different conclusions regarding angiosperm phylogeny. Investigators are still working on the question of how angiosperms first arose. Some angiosperms are neither monocots nor eudicots (see Figure 21.27). The overall relationship among the angiosperm clades is shown in Figure 21.26. Molecular and morphological evidence now points to *Amborella* as the living species most similar to the first angiosperms. *Amborella* has a variable number of carpels and stamens, and it lacks vessel elements.

Fruits develop from the ovary and its supporting tissues. Fruit types (see Figure 21.24) depend on the number of carpels associated with the fruit and the extent of support tissue incorporated into the fruit structure.

Many plants and animals have coevolved in such a way that the nutrition of the animal and pollination of the plant are interdependent. Some plants have evolved methods of limiting pollination to a single species of insect, but most can be pollinated by a range of species. Many flowers entice animals by providing food rewards such as nectar. Pollen grains themselves can be a food reward, and this pollen can be carried from one plant to another. Bee-pollinated flowers often have nectar guides that can be seen only in the ultraviolet spectrum visible to bees.

The angiosperm life cycle differs from that of all other plants in that double fertilization occurs (see Figure 21.25). In double fertilization, one sperm unites with the egg to produce a diploid zygote; the other sperm unites with two other haploid cells of the female gametophyte to produce a triploid cell. This triploid cell divides mitotically to create the (triploid) endosperm tissue. The endosperm provides nutrition for the developing embryo.

Angiosperm embryos have one or two seed leaves called cotyledons. The number of cotyledons distinguishes the two major clades of flowering plants. Monocots have one cotyledon; eudicots have two. A few other relatively small groups of angiosperms do not fit into these two lineages (see Figures 21.28 and 21.27).

Question 11. Describe the differences between the seeds of gymnosperms and angiosperms. For each difference, identify the advantage of the angiosperm characteristic that might contribute to the success of this group.
Textbook Reference: *21.5 Flowers and Fruits Increase the Reproductive Success of Angiosperms, pp. 426–430*

Question 12. In double fertilization, triploid endosperm results. Do the seed parent and the pollen parent contribute equally to the endosperm, and if not, how might this unequal contribution affect the offspring?
Textbook Reference: *21.5 Flowers and Fruits Increase the Reproductive Success of Angiosperms, pp. 431–432*

Test Yourself

1. A land plant may be reliably distinguished from green algae by which of the following characteristics?
   a. Chlorophyll type
   b. The presence of an embryo protected by parent tissue
   c. The presence of roots
   d. Swimming sperm
   e. All of the above
   Textbook Reference: *21.1 Primary Endosymbiosis Produced the First Photosynthetic Eukaryotes, pp. 409–410*
2. Which of the following characteristics was *not* necessary in order for plants to colonize land?
   a. Vascular tissue for moving water throughout the plant
   b. A waxy cuticle to reduce water loss
   c. The ability to screen ultraviolet radiation
   d. The development of thick spore walls to protect the spores from dehydration
   e. Development of embryos protected inside other tissues
   Textbook Reference: *21.2 Key Adaptations Permitted Plants to Colonize Land, p. 411*
3. The main difference between nonvascular plants and vascular plants is that vascular plants
   a. lack gametophytes.
   b. produce spores.
   c. have tracheids.
   d. reproduce sexually.
   e. All of the above
   Textbook Reference: *21.2 Key Adaptations Permitted Plants to Colonize Land, p. 412*
4. Land plants life cycles feature alternation of generations. Which of the following statements about alternation of generations is *false*?
   a. The life cycle includes both a diploid and haploid multicellular stage.
   b. Gametes are not produced by mitosis.
   c. Gametes fuse to form a zygote.
   d. The life cycle includes a multicellular spore producing generation.
   e. Sporangia undergo meiosis to produce haploid unicellular spores.
   Textbook Reference: *21.2 Key Adaptations Permitted Plants to Colonize Land, p. 411*
5. Two characteristics that help distinguish the mosses from the liverworts are _______ and _______.
   a. the presence of hydroids; gametophytic dominance
   b. gametophyte dominance; the presence of stomata
   c. sporophyte dominance; the presence of stomata
   d. the presence of stomata; hydroids
   e. sporophytic dominance; the presence of hydroids
   Textbook Reference: *21.2 Key Adaptations Permitted Plants to Colonize Land, p. 412*
6. During a plant's life cycle, meiosis takes place in the _______ and produces _______.
   a. gametophyte; haploid gametes
   b. sporophyte; haploid gametes
   c. sporophyte; haploid spores
   d. gametophyte; diploid spores
   e. gametophyte; haploid spores
   Textbook Reference: *21.2 Key Adaptations Permitted Plants to Colonize Land, p. 412, Figure 21.4*
7. Which of the following statements best represents the rationale for placing the charophytes as the sister group to the land plants?
   a. Retention of the egg within the parental organism and DNA evidence
   b. Apical branching growth, plasmodesmata, structural similarities, and DNA evidence
   c. Flattened growth form, plasmodesmata, and DNA evidence
   d. Presence of chlorophyll *b*, growth form, retention of the egg within the parental organism, and DNA evidence
   e. Structural similarities and DNA evidence
   Textbook Reference: *21.1 Primary Endosymbiosis Produced the First Photosynthetic Eukaryotes, pp. 408–411; 21.2 Key Adaptations Permitted Plants to Colonize Land, p. 411*
8. Which of the following enhances the ability of the hornworts to produce elongated sporophytes?
   a. A basal region where cells divide indefinitely
   b. An association with nitrogen-fixing bacteria

c. Developmental genes that increase growth of the sporophyte
d. Both a and b
e. Both a and c
Textbook Reference: *21.2 Key Adaptations Permitted Plants to Colonize Land, p. 413*

9. You are walking along a roadside and find a plant with the following characteristics: a very thin, waxy cuticle; stomata; simple leaves in whorls around a central stem; independent sporophytes and gametophytes; and sporangia in strobili. This plant is most likely a member of which of the following groups?
a. Bryophyta
b. Monilophyta
c. Anthocerophyta
d. Lycopodiophyta
e. Cycadophyta
Textbook Reference: *21.3 Vascular Tissues Led to Rapid Diversification of Land, p. 416*

10. Which of the following structures is common to all land plants *except* the seed plants?
a. Sporangium
b. Antheridium
c. Embryo
d. Sperm
e. Archegonium
Textbook Reference: *21.5 Flowers and Fruits Increase the Reproductive Success of Angiosperms, p. 431, Figure 21.25*

11. Which of the following changes contributed to the success of seed plants?
a. Development of secondary growth
b. Development of double fertilization
c. Development of the seed coat
d. Both a and b
e. Both a and c
Textbook Reference: *21.4 Seeds Protect Plant Embryos, p. 422*

12. The female cone in gymnosperms is known as the _______ and the male cone is known as the _______
a. megastrobilus; microstrobilus
b. microstrobilus; megastrobilus
c. megasporangium; microsporangium
d. modified leaves; modified branches
e. integument; micropyle
Textbook Reference: *21.4 Seeds Protect Plant Embryos, p. 424*

13. Species in which both megasporangiate and microsporangiate flowers occur on the same plant are called
a. perfect.
b. dioecious.
c. monoecious.
d. imperfect.
e. sterile.
Textbook Reference: *21.5 Flowers and Fruits Increase the Reproductive Success of Angiosperms, p. 427*

14. In which of the following groups are sporangia arranged in strobili?
a. Gingko
b. Whisk ferns
c. Hornworts
d. Club mosses
e. Ferns
Textbook Reference: *21.3 Vascular Tissues Led to Rapid Diversification of Land Plants, p. 416*

15. Which of the following groups has large leaves with branching vascular strands?
a. Horsetails
b. Whisk ferns
c. Hornworts
d. Club mosses
e. Ferns
Textbook Reference: *21.3 Vascular Tissues Led to Rapid Diversification of Land Plants, pp. 418–419, Figure 21.11*

16. Which pieces of evidence would be most helpful in determining the relationship between a newly discovered green aquatic organism and land plants?
a. Chlorophyll type
b. Presence or absence of chloroplasts
c. Whether or not they both exhibit alternation of generations
d. Retention of the egg in the parent organism
e. Both c and d
Textbook Reference: *21.1 Primary Endosymbiosis Produced the First Photosynthetic Eukaryotes, pp. 408–409, Figure 21.1A*

Answers

Key Concept Review

1. The presence of chloroplasts as a result of primary endosymbiosis is common to all photosynthetic eukaryotes, and the groups that evolved chlorophyll *b* with starch stored in chloroplasts are known as the green plants. Green plants that retain the egg within the parental organism are known as streptophytes. Streptophytes that exhibit a number of additional synapomorphies including a protected embryo, a cuticle, and a multicellular sporophyte are called land plants.

2. Glaucophytes are unicellular, microscopic, freshwater algae whose characteristic whitish-blue color is due to their retention of a small amount of peptidoglycan between the inner and outer membranes of their chloroplasts. Red algae are almost exclusively multicellular organisms that inhabit saltwater environments, lack peptidoglycan in their chloroplasts, and are usually red in color due to their characteristic pigment, phycoerythrin.

3. Land plants required protection against water loss, support against gravity, the ability to disperse in the absence of water, protection from ultraviolet radiation and dessication, and the ability to acquire water and

nutrients from a dry environment. Some of the characteristics that plants developed that address these problems include: the cuticle (a waxy coating) to protect against water loss; stomata (openings to regulate gas exchange and water loss); multicellular enclosures to protect gametes and embryos from drought; pigments for protection against ultraviolet radiation; thick-walled spores containing sporopollenin to protect spores against dessication and decay; and mutualisms with fungi to aid in the acquisition of resources.

4. Mosses have very rudimentary water transport cells called hydroids. Hydroids lack the waterproofing and support molecule lignin. Because of this, they can carry water only short distances and cannot support tall growth.

5.
   a. Haploid: Gametophyte generation
   b. Diploid; Sporophyte generation
   c. Sperm (*n*)
   d. Egg (*n*)
   e. Archegonium (*n*)
   f. Embryo (2*n*)
   g. Fertilization
   h. Antheridium (*n*)
   i. Sporophyte (2*n*)
   j. Gametophyte (*n*)
   k. Sporangium
   l. Meiosis
   m. Ungerminated spores
   n. Germinating spore
   o. Rhizoid
   p. Protonema
   q. Bud
   r. Gametophytes (*n*)

6. A branching sporophyte offers the opportunity for increasing structural complexity. Branching rhizomes or roots can provide increased anchoring for an independent sporophyte. A branching aerial portion of the sporophyte can support the production of an increased number of spores and the development of megaphylls, which can increase photosynthetic potential.

7.
   a. Haploid
   b. Diploid
   c. Mature gametophyte
   d. Archegonium
   e. Egg
   f. Antheridium
   g. Sperm
   h. Fertilization
   i. Embryo
   j. Sporophyte
   k. Gametophyte
   l. Roots
   m. Mature sporophyte
   n. Sori
   o. Sporangium
   p. Meiosis
   q. Germinating spore
   r. Rhizoids

8. In homospory, meiosis results in a single type of spore, but in heterospory, meiosis results in megaspores and microspores. Mitosis occurs throughout both types of life cycle as cells divide and plants grow, but the reproductive cells that result from mitosis in both homospory and heterospory are sperm and eggs. In homospory, the antheridium produces sperm by mitosis, and the archegonium produces eggs; in heterospory, the microgametophyte produces sperm and the megagametophyte produces eggs.

9. In seed plants, the microspores are protected by the spore wall which contains sporopollenin, the most chemically resistant biological compound known. The megaspore is enclosed in the megasporangium, which is enclosed by a layer of sporophytic tissue called the integument. Together, the megasporangium and the integument constitute the ovule, which after fertilization and maturation becomes a seed.

10. Most gymnosperms rely on wind or animals for transfer of the male gametes to the female parent plant. In the first part of the journey, the male gametes are transported from the microsporangia in the microstrobilus in the form of pollen grains. When the pollen grains reach the megastrobilus, they germinate, and the second part of the journey is accomplished via the growth of the pollen tube down through the maternal sporophytic tissue. When the pollen tube reaches the female gametophyte, it releases two sperm, one of which unites with the egg; the other sperm disintegrates.

11. Pollination in angiosperms involves double fertilization, which provides additional genetic variation in nutrition supplied to the offspring. Carpels in the angiosperm that enclose ovules and seeds provide protection as well as an opportunity for the maternal parent to control pollen access to the eggs, including selecting against self-pollination.

    In angiosperms, pollen germination on a stigma means that the male gametophyte journey is simplified. Pollen tube growth through the stigma provides an opportunity to screen out inappropriate matches (e.g., self-pollination or distantly related pollen). Pollen tube growth through the stigma is much faster than growth in gymnosperms, which requires traversing the megasporangium, the female gametophyte, and the archegonium before reaching the egg.

    Angiosperm flowers incorporate potential animal attractants such as color, shape, scent, and nutritional rewards that offer many options for animal-assisted pollination. The arrangement of flowers and their stamens and carpels offers opportunities to facilitate appropriate pollination.

    Fruits provide additional protection for the embryo and increase the likelihood of water, wind, or animal dispersal of seeds by providing morphological transport enhancement.

In angiosperms, the reduction of the microgametophyte to 2–3 cells and the megagametophyte to 7–8 cells that are completely dependent on the sporophyte means that the haploid stage of the angiosperm life history requires few resources and can be completed very quickly, reducing the generation time and decreasing the importance of environmental conditions for the gametophytes.

12. The seed parent contributes two copies of its n DNA, while the pollen parent contributes one copy of its haploid DNA. This means that the endosperm resembles the seed parent more closely than the pollen parent, so the seed parent characteristics may have more impact on the fitness of the offspring. The benefits could include (among others) longevity of the nutrients available to the embryo.

Test Yourself

1. **b.** All land plants produce embryos that are protected by tissue of the parent plant. Green algae and land plants make use of the same types of chlorophyll. Not all land plants have roots, so a plantlike organism lacking roots will not necessarily belong to the green algae.
2. **a.** Several successful groups of terrestrial plants lack vascular tissue.
3. **c.** Tracheids are found only in the vascular plants.
4. **b.** In alternation of generations gametes are produced by mitosis, not by meiosis.
5. **d.** Mosses possess hydroids and stomata, both of which are lacking in liverworts.
6. **c.** The outcome of meiosis is four cells, each of which has half the genetic material of the parent cell. A haploid cell already has only half the normal number of chromosomes of most eukaryotes, so a cell must be diploid (or have higher ploidy) to undergo meiosis. In all plant life cycles, the sporophyte is diploid and the gametophyte is haploid; therefore, only the sporophyte can undergo meiosis. Spores are the products of sporophyte meiosis.
7. **b.** Both coleochaetophytes and charophytes as well as land plants retain the egg on the parental plant; coleochaetophytes have flattened growth patterns similar to liverworts but not most land plants; and all green plants possess chlorophyll *b*. Apical, branching growth, plasmodesmata, and DNA evidence all support charophytes as the closest algal group to land plants.
8. **d.** Both the indefinitely dividing group of sporophyte-generating cells and the nitrogen-fixing cyanobacteria contained by the gametophyte contribute to the hornwort's ability to produce elongated sporophytes.
9. **b.** The plant is a horsetail, which is very common along roadsides in damp ditches, particularly in the Midwest.
10. **b.** In seed plants, the microgametophyte is reduced to 2–4 cells and no longer includes the sterile jacket of cells that protect the gametes in nonseed plants. The megagametophyte does produce an archegonium that protects the egg. In angiosperms, there is no longer an archegonium.
11. **e.** Development of secondary growth enabled seed plants to increase their diameter and reach greater heights, allowing them to become better competitors for light. The seed coat protects the embryo from dessication and herbivory, and also allows greater control over germination conditions. Double fertilization is very rare in gymnosperms.
12. **a.** The female cone is called the megastrobilus because it contains the megasporangia enclosed in the integument (together called the ovule). The smaller male cone is called the microstrobilus because it contains the microsporangia that bear pollen.
13. **c.** Monoecious means "one house" and refers to species that possess separate male and female flowers on the same plant.
14. **d.** The sporangia of many club mosses are aggregated in conelike structures called strobili.
15. **e.** Some ferns uncurl as their leaves grow.
16. **e.** Land plants exhibit alternation of generations and retain the egg in the parent organism, so these characteristics would provide insight as to whether the new plant belonged in that group. All photosynthetic eukaryotes have chloroplasts and the chlorophyll type may not be helpful.

The Evolution and Diversity of Fungi

22

The Big Picture

- Fungi are heterotrophic organisms that absorb nutrients from the environment. Fungi can be unicellular or multicellular. The mycelium is the body of a multicellular fungus and is composed of many tubular filaments known as hyphae. Many fungi form symbiotic relationships with photosynthetic organisms, creating mycorrhizae and lichens.
- Reproduction in the fungi occurs both sexually and asexually. Asexual reproduction involves the production of spores, breakage, fission, or budding. Sexual reproduction in multicellular fungi requires the union of hyphae with two different mating types.

Study Strategies

- You may be confused by the numerous ways that fungi reproduce. Create flowcharts to help you understand the ploidy level at each stage of the life cycle in the different phyla of fungi. Make sure to include both the sexual and asexual stages in your charts. Comparisons of your flowcharts for the different phyla will help you learn the differences among the groups.
- Organisms are placed in particular systematic groupings because they share unique characteristics. Look for patterns that distinguish the six major fungal groups from one another. Most fungi are grouped according to the reproductive structures that they produce.
- Go to yourBioPortal.com to review the following tutorial and activities:

  Animated Tutorial 22.1 Life Cycle of a Zygomycete

  Web Activity 22.1 Fungal Phylogeny

  Web Activity 22.2 Life Cycle of a Dikaryotic Fungus

Key Concept Review

22.1 Fungi Live by Absorptive Heterotrophy

Unicellular yeasts absorb nutrients directly

Multicellular fungi use hyphae to absorb nutrients

Fungi are in intimate contact with their environment

Fungi are grouped with animals and choanoflagellates in the opisthokonts and probably share a unicellular protist common ancestor with these two groups. The synapomorphies that unite fungi are absorptive heterotrophy and the presence of chitin in their cell walls.

Although most fungi are multicellular, unicellular fungi are found in almost all fungal groups and are commonly referred to as yeasts. Yeasts may reproduce by budding (see Figure 22.2), by fission, or sexually.

The body of a multicellular fungus is called a mycelium. A mycelium is made up of individual tubular filaments called hyphae. Hyphae grow rapidly into a substrate; the hyphae in a single mycelium may collectively grow as much as 1 kilometer per day. Hyphal cell walls are strengthened by the polysaccharide chitin. Hyphae may be septate (divided into compartments by chitinous walls) or coenocytic (continuous and multinucleate) (see Figure 22.3). Hyphae provide a large surface area-to-volume ratio, and thus a large surface area for absorption of nutrients from the substrate.

Most fungi are multicellular, but multicellularity in fungi is significantly different from the familiar multicellularity of plants and animals. In plants and animals, each cell is usually enclosed in its own membrane and has only one nucleus. In multicellular fungi, cells are separated by porous septa that do not completely block the movement of organelles (see Figure 22.3). Fungi are tolerant of hypertonic environments and temperature extremes.

Question 1. Early taxonomists considered fungi to be members of the plant kingdom. What evidence indicates that they are in fact more closely related to animals?
Textbook Reference: *22.1 Fungi Live by Absorptive Heterotrophy, p. 438*

Question 2. Label the diagram below using the following terms: hypha, nuclei, cell wall, septa.
Textbook Reference: *22.1 Fungi Live by Absorptive Heterotrophy, p. 439, Figure 22.3*

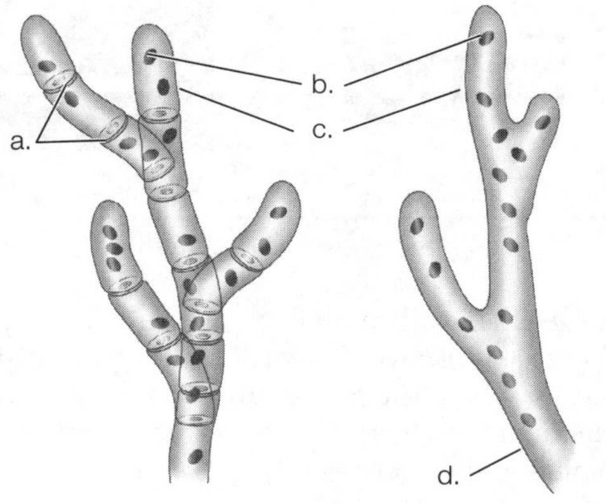

22.2 Fungi Can Be Saprobic, Parasitic, Predatory, or Mutualistic

Saprobic fungi are critical to the planetary carbon cycle

Some fungi engage in parasitic or predatory interactions

Mutualistic fungi engage in relationships beneficial to both partners

Endophytic fungi protect some plants from pathogens, herbivores, and stress

Fungi are important recyclers of elements in the ecosystem. They absorb the nutrition needed for their survival from dead matter (saprobes), from living hosts (parasites), and from mutually beneficial symbiotic relationships with other organisms (mutualists).

Adaptations of hyphae allow fungi to exploit different sources of nutrition. Some fungi that are plant parasites, for example, have hyphae that germinate from spores on the leaf surface and enter the leaf thorough stomata. The fungus forms a mycelium within the leaf (see Figure 22.5). Rhizoids are hyphae that anchor fungi to their substrate.

Parasitic fungi are either facultative (possessing the ability to grow independently) or obligate (dependent on their living host for growth). Some parasitic fungi, called pathogens, not only derive nutrition from their hosts, but also sicken or kill the host. Some fungi are active predators that trap microscopic protists or animals in a sticky substance; others form a constricting ring around their prey (see Figure 22.6).

Many fungi are beneficial to other organisms. Saprobic fungi secrete enzymes into the environment that help in the absorption of dead matter and the decomposition and recycling of elements—especially carbon—used by living organisms. As decomposers, saprobic fungi are essential to life on Earth.

Fungi also form two crucial types of symbiotic, mutualistic relationships: lichens and mycorrhizae. Lichens are associations of a fungus with a unicellular photosynthetic alga or cyanobacteria. The fungus provides minerals and water to its photosynthetic partner, and the photosynthesizing organism provides the fungus with organic compounds. Lichens are among Earth's hardiest organisms, and can thrive in barren and extreme environments such as Antarctica. Lichens are characterized by their appearance as either crustose (crusty), foliose (leafy), or fruticose (shrubby) (see Figure 22.7). Mycorrhizae are associations between fungi and the roots of plants in which a fungus obtains the products of photosynthesis from the plant and provides minerals and water to the plant. The mycorrhizal symbiosis is essential to the survival of most plants, and the evolution of this relationship may have been the most important step in allowing plants to colonize land. Ectomycorrhizal fungi wrap their hyphae around a plant root. The hyphae of arbuscular mycorrhizal fungi penetrate the root cell walls and form treelike (arbuscular) structures that provide the plant with nutrients (see Figure 22.9).

Endophytic fungi are symbionts living within the plant parts that are aboveground. They help certain plants (especially grasses) resist pathogens, herbivores, and stresses such as drought and salty soil; their role in other plants is not well understood.

Question 3. Explain how a haustoria, which does not enter the host's cells, obtains nutrients from the host.
Textbook Reference: *22.2 Fungi Can Be Saprobic, Parasitic, Predatory, or Mutualistic, p. 440*

Question 4. How can you distinguish between a mutualistic and pathogenic relationship between a fungus and its partner?
Textbook Reference: *22.2 Fungi Can Be Saprobic, Parasitic, Predatory, or Mutualistic, pp. 440–442*

22.3 Major Groups of Fungi Differ in Their Life Cycles

Fungi reproduce both sexually and asexually

Microsporidia are highly reduced, parasitic fungi

Most chytrids have an aquatic life cycle

Some fungal life cycles feature separate fusion of cytoplasms and nuclei

Arbuscular mycorrhizal fungi form symbioses with plants

The dikaryotic condition is a synapomorphy of sac fungi and club fungi

The sexual reproductive structure of sac fungi is the ascus

The sexual reproductive structure of club fungi is the basidium

Fungi are classified into six major groups on the basis of molecular evidence and type of life cycle. The groups are: microsporidia, chytrids, Zygomycota, Glomeromycota, Ascomycota, and Basidiomycota.

Fungi have many different life cycles and can reproduce both sexually and asexually. Fungi have several means of asexual reproduction: They can produce spores enclosed within sporangia or naked spores at the tips of hyphae known as conidia. Unicellular fungi can reproduce by fission or budding. Just about any part of a mycelium is capable of living independently of the rest; therefore, simply the division of a mycelium into two or more parts is a method of reproduction.

Members of the different fungal groups can be distinguished from one another by their mechanisms of sexual reproduction, which involves two or more mating types. Self-fertilization is prevented by the incapacity of individuals of the same mating type to mate with each other (see Figure 22.11). Sexual stages have not yet been identified in many fungi, so these species have been classified according to data from DNA sequence analysis.

Microsporidia are unicellular parasitic fungi and are among the smallest eukaryotes known. They lack mitochondria and are obligate intracellular parasites of animals. They use a polar tube projection from their cell to invade the host and convey the contents of the microsporidia into the host cell.

Chytrids are aquatic fungi, and are the only fungi with flagellated gametes. Chytrids likely resemble the common ancestor of all fungi more closely than any other fungal group. Chytrids have chitinous cell walls. Many chytrids display alternation of generations, which is not typical in other fungal groups. The genus *Allomyces* displays alternation of generations. A haploid zoospore becomes a small organism (the gametophyte) with both female and male gametangia that produce gametes (see Figure 22.13A). When the gametes fuse, they form a diploid zygote and a diploid organism (the sporophyte). The sporophyte produces sporangia that give rise to haploid zoospores via meiosis.

Terrestrial fungi (all except chytrids and microsporidia) do not have motile gametes; they reproduce sexually via fusion of the cytoplasms of opposite mating types (plasmogamy) followed by fusion of nuclei (karyogamy).

Zygomycota have coenocytic hyphae and (usually) no fleshy fruiting body. More than 1,000 species of Zygomycota have been described. The hyphae of a zygomycota spread randomly over a substrate, periodically producing stalked sporangiophores bearing sporangia (see Figure 22.14).

Sexual reproduction in Zygomycota occurs between adjoining individuals with different mating types (see Figure 22.13B). Branches from opposite mating types are attracted to each other by pheromones. The individuals grow toward each other until they produce gametangia. The gametangia fuse (plasmogamy), and a thick-walled zygosporangium containing many nuclei of opposite mating types forms at the fusion site. The opposite mating type nuclei fuse (karyogamy). A unique multinucleate zygospore forms within the zygosporangium. The zygospore nuclei undergo meiosis and a sporangium, containing haploid nuclei incorporated into new spores, sprouts from the zygospore. The spores disperse and germinate to form a new generation of haploid hyphae. The Zygomycota include black bread mold (*Rhizopus stolonifer*) and many other saprobic and parasitic species.

Glomeromycota form arbuscular mycorrhizae. Glomeromycota are entirely asexual and coenocytic. They are also entirely terrestrial; about half the fungi found in soils are Glomeromycota. The fewer than 200 species of Glomeromycota form arbuscular mycorrhizae with the roots of 80 to 90 percent of all plants (see Figure 22.9).

Ascomycota and Basidiomycota have a synapomorphy of a dikaryotic hyphal stage. This unique dikaryon condition is the first stage of sexual reproduction (see Figure 22.15). The cytoplasms of two individuals fuse, but the nuclei do not fuse immediately, and the resulting hyphae contain genetically different haploid nuclei. Two genetically distinct nuclei coexist and divide in each cell of the mycelium. This stage of the life cycle is called the dikaryon.. These nuclei eventually fuse in the "fruiting" structure of the fungus to form a zygote which undergoes meiosis.

The hyphae are neither haploid (n) nor diploid ($2n$), but dikaryotic ($n + n$). The dikaryon life history has no gamete cells, only gamete nuclei. The zygote is the only diploid cell. The dikaryotic mycelium can have characteristics not shared by the haploid or diploid stages.

Ascomycota (sac fungi) are found in marine, freshwater, and terrestrial habitats and include baker's yeast, truffles, and *Penicillium*. The Ascomycota are distinguished by their ascus, a sexual reproductive structure (see Figure 22.15A). Sexual reproduction of filamentous sac fungi involves two different mating types which move into a dikaryotic stage before forming a diploid cell population. A dikaryon is produced during sexual reproduction by the fusing of two mating structures (plasmogamy) (see Figure 22.15A). From the dikaryon hyphae ($n + n$), dikaryotic asci form where the nuclei fuse (karyogamy). The zygote ($2n$) undergoes meiosis and then produces an ascus with haploid ascospores (n). The ascospores germinate and grow into new haploid mycelia.

Asexual reproduction in Ascomycota involves the production of conidia at the tips of hyphae (see Figure 22.17). Ascomycota include molds and the cup fungi and have a number of uses for humans (see Figure 22.17). Mold from the genus *Penicillium* produces the antibiotic penicillin; other molds are important in making cheese. Brown molds are used in brewing sake (a Japanese alcoholic beverage) and soy sauce. The species of yeast used to make bread and alcoholic beverages, *Saccharomyces cerevisiae,* is a unicellular sac fungus. Reproduction in these yeasts is asexual and accomplished by budding.

The fruiting structures of Basidiomycota are familiar as mushrooms. Included in the 30,000 species of Basidiomycota fungi are puffballs and mushrooms, which produce the group's most spectacular fruiting structures. In Basidiomycota, the cap of a mushroom is the fruiting structure (see Figure 22.18).

Nuclear fusion (karyogamy) and then meiosis occur in the basidium on the underside of the mushroom cap along the gills (see Figure 22.15B). After meiosis, four basidiospores are formed on tiny stalks. The basidiospores are released into the environment and germinate to form haploid hyphae of opposite mating types. The hyphae grow and fuse (plasmogamy) with different mating types to form the dikaryotic mycelium. The fruiting structure, or basidiocarp, grows from the dikaryotic mycelium.

Question 5. Examine the four life histories shown in Figures 22.13 and 22.15 and describe one feature in each that is unique among these groups of fungi.
Textbook Reference: *22.3 Major Groups of Fungi Differ in Their Life Cycles, pp. 447, 449*

Question 6. Contrast the sporangiophore produced in the Zygomycota with the conidia produced by Ascomycota.
Textbook Reference: *22.3 Major Groups of Fungi Differ in Their Life Cycles, pp. 448, 451*

22.4 Fungi Can Be Sensitive Indicators of Environmental Change

Lichen diversity and abundance indicate air quality

Fungi contain historical records of pollutants

Reforestation may depend on mycorrhizal fungi

Although lichens and fungi can survive in extreme environments, their absorptive feeding leaves them vulnerable to environmental pollutants. Lichens are sensitive to environmental pollutants and can serve as air quality indicators. Collections of fungi in museums contain a historical record of environmental pollutants.

Forest restoration projects may depend on providing appropriate mycorrhizal symbionts for tree establishment. Fungi also continue to provide antibiotic and other disease-fighting weapons.

Question 7. What difficulties might you encounter when trying to reestablish trees and their mycorrhizal associate in a degraded environment?
Textbook Reference: *22.4 Fungi Can Be Sensitive Indicators of Environmental Change, pp. 452, 454*

Question 8. What characteristics might make fungi a good source of antibiotics?
Textbook Reference: *22.4 Fungi Can Be Sensitive Indicators of Environmental Change, p. 454*

Test Yourself

1. Many species of fungi can be placed into one of six main groups based on
   a. their methods of sexual reproduction.
   b. whether or not gametes have a flagella.
   c. the presence or absence of septa in the hyphae.
   d. DNA sequence analysis.
   e. All of the above
   Textbook Reference: *22.1 Fungi Live by Absorptive Heterotrophy, p. 438*
2. Which of the following would *not* be found in any of the typical fungal life cycles?
   a. Haploid nuclei
   b. Diploid nuclei
   c. Spores
   d. Chloroplasts
   e. A dikaryotic stage
   Textbook Reference: *22.3 Major Groups of Fungi Differ in Their Life Cycles, pp. 444–452*
3. Fungi are absorptive heterotrophs. Which of the following is an adaptation that greatly aids this mode of nutrient procurement?
   a. Dikaryosis
   b. A large surface area-to-volume ratio
   c. Conjugation
   d. A complex life cycle
   e. A small surface area-to-volume ratio
   Textbook Reference: *22.1 Fungi Live by Absorptive Heterotrophy, p. 439*
4. Assume that two normal hyphae of different fungal mating types meet. After a period of time, the cell walls between these hyphae will dissolve, producing a
   a. mycelium.
   b. fruiting body.
   c. zygote.
   d. spore.
   e. dikaryotic cell.
   Textbook Reference: *22.3 Major Groups of Fungi Differ in Their Life Cycles, p. 448*
5. Which of the following is the best way to represent the ploidy of dikaryotic hyphae?
   a. $1/2\ n$
   b. n/n
   c. n
   d. $n + n$
   e. $2n$

Textbook Reference: *22.3 Major Groups of Fungi Differ in Their Life Cycles, p. 448*

6. Which of the following statements about sexual reproduction in fungi is *false*?
   a. Motile gametes are present in all fungal species.
   b. An aquatic environment is not required for fertilization to occur in most fungi.
   c. There is no true diploid tissue in the life cycle of most sexually reproducing fungi.
   d. Sexual reproduction often begins with contact between hyphae of different mating types.
   e. Alternation of generations and sexual reproduction can occur in the same species.
   Textbook Reference: *22.3 Major Groups of Fungi Differ in Their Life Cycles, p. 446*

7. A mycorrhiza is
   a. a specialized type of lichen.
   b. the fruiting structure of a basidiomycota.
   c. a symbiotic association between a fungus and cyanobacterium or green algae.
   d. a reproductive stage of sac fungi.
   e. a symbiotic association between a fungus and a plant.
   Textbook Reference: *22.2 Fungi Can Be Saprobic, Parasitic, Predatory, or Mutualistic, p. 443*

8. Suppose a scientist investigating the classification of a fungus that has never been observed to reproduce sexually discovers that it has DNA sequences characteristic of basidiomycota. If this scientist could coax fungi of this species to reproduce sexually, which of the following would most likely be observed?
   a. Dikaryotic hyphae segmented by septa
   b. Dikaryotic hyphae without septa
   c. Asymmetrical cell division
   d. A dikaryotic ascus
   e. Flagellated gametes
   Textbook Reference: *22.3 Major Groups of Fungi Differ in Their Life Cycles, p. 452*

9. Which of the following fungi have coenocytic hyphae and stalked sporangiophores?
   a. Chytrids
   b. Zygomycota
   c. Ascomycetes
   d. Basidiomycota
   e. Glomeromycota
   Textbook Reference: *22.3 Major Groups of Fungi Differ in Their Life Cycles, p. 448*

10. Which of the following fungi display alternation of generations?
   a. Chytrids
   b. Glomeromycota
   c. Ascomycota
   d. Basidiomycota
   e. Zygomycota
   Textbook Reference: *22.3 Major Groups of Fungi Differ in Their Life Cycles, p. 446*

11. A saprobe is an organism that
   a. absorbs nutrients from the sap of a host plant.
   b. reproduces in the sap of a plant.
   c. undergoes asexual reproduction.
   d. is mutualistic.
   e. absorbs nutrients from dead organic matter.
   Textbook Reference: *22.2 Fungi Can Be Saprobic, Parasitic, Predatory, or Mutualistic, pp. 439–440*

12. Which of the following is *not* a form of asexual reproduction in fungi?
   a. Budding
   b. Formation of haploid spores in sporangia
   c. Formation of dikaryotic mycelia
   d. Fission
   e. Formation of haploid spores in conidia
   Textbook Reference: *22.3 Major Groups of Fungi Differ in Their Life Cycles, p. 449, Figure 22.15*

13. Which of the following statements about fungi is *false* (i.e., a common misconception)?
   a. Fungi grow only in warm, wet environments.
   b. Fungi lose water rapidly in a dry environment.
   c. Fungi can grow in environments too hypertonic to sustain bacteria.
   d Fungi are eukaryotes and have multiple mating types.
   e. Male and female fungi have no distinctive morphology.
   Textbook Reference: *22.1 Fungi Live by Absorptive Heterotrophy, p. 439*

14. Which of the following is the best explanation for why lichens can grow in extreme environments, but cannot tolerate poor air quality?
   a. Algae cannot get enough CO_2 when the air quality is poor.
   b. Lichens cannot excrete toxins.
   c. Fungal spores cannot germinate at low pH.
   d. The ascus of the fungal component is sensitive to pollution.
   e. Air pollutants are the most extreme toxins on Earth.
   Textbook Reference: *22.4 Fungi Can Be Sensitive Indicators of Environmental Change, p. 452*

15. Which of the following scientific groups is correctly matched with its correct common name?
   a. Chytrids – microspore fungi
   b. Zygomycota – sac fungi
   c. Glomeromycota – mycorrhizial fungi
   d Basidiomycota – sac fungi
   e. Ascomycota – club fungi
   Textbook Reference: *22.3 Major Groups of Fungi Differ in Their Life Cycles, p. 445, Table 22.1*

16. Suppose scientists are contending with a new fungal plant disease that has appeared on rhododendrons in Asia and in the United States. How would they best determine whether this was a recent spread of a disease or one that had not been previously noted?
   a. By attempting crosses among the various strains

b. By observing whether the various strains are resistant to a particular bacterium
c. By testing the fungus on European rhododendron to see if it can infect them
d. By sequencing the DNA of this and other strains and comparing them for similarities.
e. By testing the temperature tolerance of the various strains

Textbook Reference: *22.2 Fungi Can Be Saprobic, Parasitic, Predatory, or Mutualistic, p. 441*

17. Damp weather increases the incidence of mildew fungal disease in plants because
a. chitrids are flagellated and need water for motility.
b. fungal life histories require water for karyogamy to occur.
c. fungal sporangiophores require water for support.
d. basidiomycetes produce mushrooms in damp weather.
e. in damp conditions the stomata are open, giving the fungus easy access to plant leaf cells.

Textbook Reference: *22.2 Fungi Can Be Saprobic, Parasitic, Predatory, or Mutualistic, p. 441*

18. Which of the following fungi characteristics make it difficult to eradicate pathogenic fungi?
a. Lack of specificity to hosts
b. Similarity of pathogenic and beneficial fungal invasion processes
c. Copious spore production
d. None of the above
e. All of the above

Textbook Reference: *22.2 Fungi Can Be Saprobic, Parasitic, Predatory, or Mutualistic, p. 441*

Answers

Key Concept Review

1. Similarly to animals and choanoflagellates, fungi are believed to have had a unicellular flagellated protist ancestor with a posterior flagella. Additional evidence of the relatedness of these three groups are the similarities of many of their gene sequences.
2.
a. Septa
b. Nuclei
c. Cell wall
d. Hypha
3. Haustoria are branched hyphae that penetrate the target cell's wall, but not the cell's membrane. They invaginate into the membrane so that materials can transfer across the membranes of the cell and the hyphae.
4. Mutualistic fungi provide some benefit to the partner, while pathogenic ones lead to reduced growth and vigor of the host. In some cases this distinction is cloudy. For example, the photosynthetic partners in lichens sometimes grow faster on their own than they do when associated with the fungus in a lichen. However, the photosynthetic organism may not be able to survive a period of drought without the fungus. Endophytic fungi may or may not reduce growth in a plant while conferring resistance against herbivory and stress.
5. Chytrids have motile gametes and haploid zoospores. Zygospore fungi have a unique unicellular, multinucleate zygospore. Ascomycota have a unique microscopic spore bearing sac called an ascus. Basidiomycota bear their spores on unique clublike structures called basidia.
6. The sporangiophore is produced as part of sexual reproduction in the Zygomycota. It sprouts from the zygospore and contains numerous haploid nuclei that germinate to form haploid hyphae. The conidia produced by Ascomycota develop in chains at the tips of specialized hyphae. These contain spores that are produced through mitosis and are part of asexual reproduction.
7. It might be difficult to reestablish the mycorrhizae before there are trees, and vice-versa. Often bringing in some soil from a healthy forest can help. Until the mycorrhizae are established, addition of nutrients and water could provide trees without their ectomycorrhizal associates a temporary boost. While fewer than 200 species of arbuscular mycorrhizae have been identified, finding the appropriate associate might be difficult.
8. Because fungi have such a large surface-to-volume ratio and are in intimate contact with their environment, they are vulnerable to bacteria and other microbes. Their need for their own defenses means that they might be a source of substances that could fight diseases in plants and animals, including humans.

Test Yourself

1. **e.** Traditionally, a fungus's method of sexual reproduction was the primary criterion in its classification, but all of the traits listed are useful in classifying fungi.
2. **d.** Fungi have haploid and diploid nuclei at different stages of their life cycle, and many have a dikaryotic stage. They also produce spores. The chloroplasts are not found in the fungi, but in plants.
3. **b.** The large surface area-to-volume ratio of the hyphae increases the ability of a fungus to absorb nutrients.
4. **e.** When two hyphae of different mating types fuse, they form a dikaryotic hyphae.
5. **d.** Dikaryotic hyphae are neither truly diploid ($2n$) nor haploid (n). Because dikaryotic hyphae include genetic material from two haploid nuclei that remain separate, the best way to represent their ploidy is $n + n$.
6. **a.** Not all fungi have motile gametes; only the gametes of chytrids are motile.

7. **e.** Mycorrhizae are associations between fungi and the roots of plants. Lichens are symbiotic relationships between fungi and cyanobacteria or green algae.
8. **a.** If this fungus is indeed a basidiomycota, the fusing of hyphae of different mating types would most likely result in dikaryotic hyphae that are segmented by septa.
9. **b.** Coenocytic hyphae are characteristic of both the chytrids and the zygomycota, but only the zygomycota regularly produce sporangiophores.
10. **a.** Among the fungi, only the chytrids have a multicellular haploid stage and a multicellular true diploid stage.
11. **e.** Saprobes are organisms that absorb nutrients from dead matter. Some bacteria are saprobes.
12. **c.** The dikaryotic mycelium is a structure formed in the sexual reproduction life cycle.
13. **a.** Fungi are plentiful in warm, wet environments, but many also survive extreme temperatures and very dry conditions.
14. **b**. The fungal component of lichens surrounds the algal component. Fungi absorb material from their environment and are unable to excrete compounds that may be toxic.
15. **c.** The correct associations are as follows: Microsporidia/microsporidia; Chytrids/chytrids; Zygomycota/zygospore; Glomeromycota/mycorrhizae; Ascomycota/sac fungi; Basidiomycota/club fungi.
16. **d.** The more recently the fungi have spread, the more similar they are likely to be genetically.
17. **e.** Mildew enters the plant via stomates. In dry weather, plants close their stomates for at least part of the day. In damp weather, stomates may stay open longer, allowing greater access for the fungus.
18. **e.** Fungal spores are produced in large numbers and their intimate association with plants is similar whether they are pathogenic or beneficial, so interfering with pathogens can also disrupt relationships with, for example, mycorrhizae. There are some fungal pathogens that we might consider allies because they attack other disease organisms or weedy pests, but in some cases they may also attack beneficial organisms.

Animal Origins and Diversity 23

The Big Picture

- The animals are a monophyletic group, sharing a number of morphological and genetic traits. Animals are motile, multicellular organisms that must ingest nutrients. Animals are classified according to early embryonic patterns and body plan characteristics, which include symmetry, body cavity structure, segmentation, and type of appendages. The importance of acquiring nutrition (food) has led to the evolution of a variety of feeding strategies that maximize the available sources of nutrition. Life cycles also vary considerably among the animals, and involve a considerable number of trade-offs.
- The eumetazoans encompass all animals except the three groups of sponges. Ctenophores and cnidarians are diploblastic eumetazoans that are not bilaterally symmetrical. Bilaterians are divided into protostomes (in which the mouth develops first from the blastopore) and deuterostomes (in which the anus develops first from the blastopore).
- Protostomes include lophophores with wormlike body forms and echdysozoans with a rigid cuticle covering their bodies. The echdysozoan group of arthropods dominate life on Earth today, both in numbers of species and numbers of individuals.
- There are many fewer deuterostomes than protostomes, but many deuterostomes, including vertebrates, are large and ecologically important. Terrestrial vertebrate life is dominated by reptiles and mammals.

Study Strategies

- Many of the same evolutionary "themes" can be found in widely divergent species. Venomous or venom-producing structures, for example, are found in many different animal groups, from the cnidarians to spiders to snakes. As the different animal groups are described, create a table of the various characteristics or mechanisms that consistently appear. Use this table to help you organize your thinking about evolution and the creation of diversity.
- The phylogenetic trees in the textbook are good frameworks to which more information can be added. Doing so will provide a point of reference when studying the different groups. Remember that animals have a set of traits shared with and inherited from their ancestors, as well as distinctive derived traits.
- It is important to understand what features of an animal group may give it an advantage in its environment and the compromises (trade-offs) that are made in other areas. Don't forget that even though particular animals may be placed in a group or clade, not all species have all of the features that characterize the group. For example, think about the reasons that sponges are considered animals even though they lack many of the features of other animals.
- Sometimes learning the Latin root of a group can help you organize your thinking. Although the names are unfamiliar at first, they will make sense if their meanings are understood.
- Go to yourBioPortal.com to review the following tutorials and activities:

  Animated Tutorial 23.1 Life Cycle of a Cnidarian

  Animated Tutorial 23.2 Life Cycle of a Frog

  Interactive Tutorial 23.1 Reconstructing Animal Phylogeny

  Web Activity 23.1 Animal Body Cavities

  Web Activity 23.2 Sponge and Diploblast Classification

  Web Activity 23.3 Features of the Protostomes

  Web Activity 23.4 Protostome Classification

  Web Activity 23.5 Deuterostome Classification

  Web Activity 23.6 The Amniote Egg

Key Concept Review

23.1 Distinct Body Plans Evolved among the Animals

Animal monophyly is supported by gene sequences and cellular morphology

Basic developmental patterns and body plans differentiate major animal groups

Most animals are symmetrical

The structure of the body cavity influences movement

Segmentation improves control of movement

Appendages have many uses

Although there are exceptions, the following characteristics are commonly associated with animals: multicellurity, heterotrophy, internal digestion, and the capacity either to move to their food or to bring it to them. However, some animals have stages in their life cycle in which movement does not take place.

Phylogenetic analyses of animal gene sequences support the conclusion that animals are monophyletic (see Figure 23.1). Animals generally share the following morphological and genetic synapomorphies: tight junctions, desmosomes, and gap junctions between their cells; extracellular matrix molecules, including collagen and proteoglycans; and Hox genes that specify body pattern and axis formation. These traits were probably possessed by the common ancestor of all animals but have been lost in some groups.

The common ancestor of modern animals may have been a colonial flagellated protist such as a choanoflagellate (see Figure 23.2). Differences in patterns of embryonic development, including cleavage patterns, gastrulation patterns, and the number of cell layers present, show the evolutionary relationships among animals. On these bases, animals can be grouped as follows as diploblastic or triploblastic.

Diploblastic animals have two cell layers: the endoderm and the ectoderm. Triploblastic animals have three layers: the endoderm, mesoderm, and ectoderm. Triploblastic animals are divided further into protostomes ("mouth first"; the blastopore becomes the mouth and the anus forms later) and deuterostomes ("mouth second"; the blastopore becomes the anus and the mouth forms later). Sequencing data indicate that the protostomes and deuterostomes are two different animal clades. Together, they are known as the bilaterians (see Figure 23.1) and account for most of the animal species.

Basic developmental patterns and body plans differentiate major animal groups. The overall organization of an animal's body is known as its body plan. The features of an animal's body plan include symmetry or asymmetry, body cavity structure, support (skeletal) structure, segmentation, and the presence or absence of appendages.

Most animals are symmetrical. An animal that can be divided along at least one plane into similar halves is said to be symmetrical. Animals exhibiting bilateral symmetry can be divided into two mirror images by only one plane. Bilateral symmetry is common among animals that are able to move quickly. Bilaterally symmetrical animals often have sense organs and nervous tissue concentrated at the anterior end; this type of organization is known as cephalization (from the Greek word for "head").

Animals have one of three types of body cavities (see Figure 23.3). Acoelomates have no enclosed body cavity. Pseudocoelomates have a liquid-filled space known as the pseudocoel in which many of the internal organs are located. Coelomates have a true body cavity, a coelom, developing within the mesoderm. In coelomates, the internal organs are in pouches of the peritoneum. The structure of the body plan influences an animal's movement.

Fluid-filled body cavities act as hydrostatic skeletons for many animals. Other animals evolved rigid supportive skeletons that can be internal (bones or cartilage) or external (a shell or cuticle). Muscles attached to hard skeletons allow the animal to move.

Segmentation allows for specialization of the different body regions and can improve control of movement. In some animals, segments are not apparent (e.g., vertebrae column). In other animals, similar body segments are repeated many times, and in yet others the body segments differ (see Figure 23.4).

Appendages, especially jointed limbs, enhance locomotion. Jointed limbs in the arthropods and vertebrates are a major factor in their evolutionary success. Other appendages are specialized and can be used to sense the environment (e.g., antennae) or can be used to capture prey.

Question 1. Explain the fundamental difference between protostomes and deuterostomes. What evidence supports the division of organisms into these two clades?
Textbook Reference: *23.1 Distinct Body Plans Evolved among the Animals, pp. 457, 459*

Question 2. What are some of the limitations imposed by a hydrostatic skeleton? Are the limitations the same for terrestrial animals?
Textbook Reference: *23.1 Distinct Body Plans Evolved among the Animals, p. 460*

23.2 Some Animal Groups Fall Outside the Bilataria

Sponges and placozoans are weakly organized animals

Ctenophores are radially symmetrical and diploblastic

Cnidarians are specialized carnivores

The simplest animals are the sponges, which have no body symmetry and no distinct cell layers. Placozoans have only four cell types and weakly differentiated layers of tissue. All animals other than the sponges and placozoans make up the eumetazoans. Eumetazoans have body symmetry, a defined gut, a nervous system, and distinct organs.

The Bilateria are a monophyletic group that includes all of the eumetazoans except the ctenophores and the cnidarians. Bilaterian synapomorphies include bilateral symmetry, three cell layers, and the presence of at least seven Hox genes. The bilaterians have two major subgroups: the protostomes and the deuterostomes.

Sponges and placozoans are weakly organized animals. Sponges can be classified in three different groups based on molecular evidence and the morphology of their spicules. Sponges have differentiated cells but no true organs. Although the sponge body plan is relatively simple, the three sponge groups all have cells that are differentiated for specific functions. Most of the 8,000 species of sponges are marine filter feeders that remove small organisms and nutrient particles from seawater as it flows through pores in the walls of their inner cavity (see Figure 23.5). A few sponges are carnivores, and can trap prey on hook-shaped spicules that are on the outside of the body surface. Sponges have a supporting skeleton composed of branching spines (spicules). The spicules of glass sponges and demosponges (the largest sponge group) are made of silicon. The calcareous sponges take their name from their calcium carbonate spicules and are the sponge group most closely related to the eumetazoans. In addition to spicules, sponges also have an extracellular matrix composed of collagen, adhesive glycoproteins, and other molecules that hold the cells together.

The huge variety of sponge body sizes and shapes is a response to the different movement patterns of water, specifically tides and currents. Sponges that live in environments that have strong wave action tend to be firmly attached to substratum, while those that live in slowly moving water are generally flat and oriented at right angles to the current flow. Sponges reproduce sexually by producing both egg and sperm, and asexually by budding and fragmentation.

Placozoans are structurally very simple, with a diploblastic body plan. Placozoans do not have a mouth, gut, or true nervous system, and they consist of only four distinct cell types. They have upper and lower epithelial cell layers with contractile fiber cells between the layers. Based on phylogenetic analysis, their structural simplicity may have been secondarily derived (i.e., some of their common features were probably lost sometime during evolution from a common ancestor). The life cycle of placozoans remains relatively unknown because their transparency makes it difficult to observe them in nature (see Figure 23.6B). It is known that they have a pelagic (free-swimming) stage in the ocean and that they are capable of sexual and asexual reproduction.

Ctenophores are radially symmetrical and diploblastic. Ctenophores have two cell layers (ectoderm and endoderm) separated by mesoglea. The ctenophores (comb jellies) are marine diploblastic animals that are radially symmetrical (see Figure 23.7). There are 150 known species, and they are found primarily in the open ocean, where they feed on planktonic organisms that are filtered by sticky filaments on the tentacles. Although ctenophores are classified as eumetazoans, they lack most of the Hox genes. However, they have a complete gut (i.e., a gut with an entrance and exit, or mouth and anus). Ctenophores get their name from the eight rows of comblike plates of cilia, known as ctenes. The cilia are used to propel the animal through the water. Prey is caught on sticky filaments on the tentacles or body (see Figure 23.7). Ctenophores have a simple life cycle and reproduce sexually. In most species, the externally fertilized egg hatches into a miniature ctenophore (i.e., direct development).

The cnidarian life cycle has two stages: the polyp and the medusa. All cnidarians are diploblastic and have radial symmetry. They include sea anemones, corals, and jellyfish. The cnidarian gastrovascular cavity is a blind sac, so cnidarians do not have a complete gut. There is only one opening, which serves as both mouth and anus. The gastrovascular cavity functions in food digestion, respiratory gas exchange, and circulation. It also lends support as a hydrostatic skeleton. As in the ctenophores, a large amount of mesoglea is found between the two cell layers of cnidarians. Because of the inert nature of the mesoglea, cnidarians have low metabolic rates. Many species, however, are able to capture large prey; this combination of characteristics makes it possible for them to live in environments with little prey.

Cnidarian tentacles have specialized cells, called cnidocytes, which inject toxins into their prey with the help of stingers called nematocysts (see Figure 23.9).

The life cycle of most cnidarians includes a sessile polyp stage and a motile medusa stage (see Figure 23.8). The polyp stage usually reproduces asexually. The medusa stage reproduces sexually, with the fertilized egg becoming a planula larva that eventually develops into a polyp.

Cnidarians possess muscle fibers that enable them to move and simple nerve nets that integrate their activities. All but a few of the 11,000 or so species of cnidarians are marine. The smallest individuals are almost microscopic, while individual jellyfish can be quite large. Many cnidarians are colonial.

Anthozoans comprise about 6,000 species of sea anemones, sea pens, and corals.

Scyphozoans include the marine jellyfish. Hydrozoans include both freshwater and marine species.

Question 3. Discuss the two major strategies animals use to get food, and indicate which of these would most likely be used by sessile organisms.
Textbook Reference: *23.2 Some Animal Groups Fall Outside the Bilataria, pp. 461–465*

Question 4. How does the body structure of sponges reflect their mode of food acquisition?
Textbook Reference: *23.2 Some Animal Groups Fall Outside the Bilataria, p. 463*

23.3 There Are Two Major Groups of Protostomes

Cilia-bearing lophophores and trochophore larvae evolved among the lophotrochozoans

Ecdysozoans must shed their cuticles

Protostomes are highly diverse, but most are bilaterally symmetrical, possessing two major derived traits: an anterior brain surrounding the entrance to the digestive tract and a ventral nervous system consisting of paired or fused longitudinal nerve cords. Many species in both groups have a wormlike appearance. The blastopore of the embryo develops into the mouth in almost all protostomes.

Most protostomes belong to one of two major groups: the lophotrochozoans and the ecdysozoans. The arrow worms are not placed in either group; they may be sister to the protostomes as a whole, or they may be more closely related to the lophotrochozoans. The 100 or so species are small marine predators.

The common ancestor of the protostomes had a coelom (a fluid-filled cavity within the mesoderm). However, the protostomes include some groups that are coelomate and some that are pseudocoelomate. One important group, the flatworms, is acoelomate (lacks a coelom). Two prominent protostome groups, the arthropods and the mollusks, have had secondary evolutionary modifications of the coelom. In the arthropods, the coelom has become a hemocoel; the mollusks have returned secondarily to a virtually open circulatory system.

The lophotrochozoans get their name from two structures: the lophophore (a circular or U-shaped ring of ciliated, hollow tentacles used for feeding and gas-exchange found in a number of groups in this clade) (see Figure 23.10A) and the trochophore larvae (see In-Text Art, p. 466). A number of lophotrochozoan groups, including the annelids (segmented worms), and mollusks, undergo spiral cleavage during early development.

The 4,500 species of bryozoans are colonial; strands of tissue connect individuals in each colony, and in some species individuals are specialized for feeding, reproduction, defense, and support (see Figure 23.10B). Individuals have a great deal of control in manipulating their lophophores to increase contact with prey. Colonies grow via asexual reproduction of the founding members. Sexual reproduction also occurs; eggs are brooded internally and larvae emerge to seek suitable sites to form new colonies. Bryozoans can cover large areas of coastal rock and can even form small reefs in shallow seas.

Flatworms lack respiratory organs and have only simple cells for waste removal. The flat shape of the animal helps in oxygen transport and waste removal (see Figure 23.11A). The digestive tract consists of a mouth that opens into a blind sac, which has many branches that aid in nutrient absorption. Although there are some free-living flatworm species, most are parasites, including about 25,000 species of tapeworms or flukes. Parasitic flatworms feed on the nutrient-rich body tissues of their host animal and disperse their eggs in the host's feces.

Most of the 1,800 species of rotifers are very small (some smaller than single-celled protists), but they have specialized internal organs, including a complete gut and a pseudocoel that serves as a hydrostatic skeleton (see Figure 23.12). Cilia are used to propel the rotifers through the water. Most species live in freshwater habitats and feed using a ciliated organ called a corona. Some species have both males and females; some have only one sex and reproduce asexually. This is the only group of animals known to have existed for millions of years without the benefits of sexual reproduction.

Nemerteans, or ribbon worms, have a complete digestive tract with two openings. Small ribbon worms use their cilia for movement, and large ribbon worms move by means of muscle contractions. The feeding organ of the ribbon worms is a proboscis, which lies within a rhynchocoel, or fluid-filled cavity (see Figure 23.13). The proboscis has a sharp stylet and can be forcefully ejected from the body to catch prey. Most of the 1,000 or so species are marine, although a few species are found in fresh water or on land. Most are small, but some species can be up to 20 meters long.

The phoronids include 20 species of tiny sessile worms that live in chitinous tubes and extract food from the water with their lophophores (see Figure 23.14). While the lophophore is similar in function to that of the bryozoans, these groups are not closely related based on genetic analyses. This indicates that the lophophore structure has evolved more than once.

Brachiopods are solitary marine animals that live attached to the substratum. Their divided shell gives them a superficial resemblance to bivalve mollusks (e.g., clams), but the two halves are dorsal and ventral instead of lateral (see Figure 23.15). The lophophore is located in the shell, and cilia help draw water and food into the shell. More than 26,000 fossil brachiopod species have been described, but only about 335 species are known to exist today.

The annelids consist of approximately 16,500 species of segmented worms living in marine, freshwater, and moist terrestrial environments. Their thin body wall serves as a surface for gas exchange, and they are restricted to these environments because the thin covering causes them to lose moisture rapidly when exposed to air. A segmented body plan gives these worms extremely good control of their movement. A separate nerve center called a ganglion controls the movement of each segment, and in most cases each segment also contains an isolated coelom (see Figure 23.16). Many annelids have more than one pair of eyes and tentacles (see Figure 23.17A). Outgrowths called parapodia, used in gas exchange, extend from segments laterally over much of the body. Setae extending from the parapodia help attach the animal to the substrate and aid in movement.

Pogonophorans have secondarily lost their digestive tract and secrete tubes made of chitin and other substances that come from their surroundings (see Figure 23.17B). Pogonophorans are found in the deep ocean near hydrothermal vents. They harbor a number of endosymbiont bacteria in a specialized organ known as a trophosome, and the bacteria provide much of their nutrition.

The oligochaetes ("few hairs") include the most familiar annelids, called the earthworms. Oligochaetes live mainly in freshwater and terrestrial environments and are hermaphroditic, containing both male and female reproductive organs in the same individual. Sperm is exchanged between two individuals outside the worm's body in a cocoon secreted by the clitellum (see Figure 23.16).

Leeches are also hermaphroditic species that live either in fresh water or on land (see Figure 23.17C). The coelom of these parasitic annelids is not segmented but is composed of undifferentiated tissue. Clusters of segments at their anterior and posterior ends are modified into suckers, which the leech attaches to the substratum for movement, or to a host mammal from which it sucks blood (its nutritional source). To keep the blood from clotting, the leech secretes an anticoagulant.

About 100,000 species are included in the four main groups of mollusks: chitons, gastropods, bivalves, and cephalopods. Mollusks have evolved into this morphologically diverse group based on a distinctive three-part body plan including a foot, a visceral mass, and a mantle (see Figure 23.18A). All species have a large muscular foot. In some groups, such as the clams, this foot is a burrowing organ, while in squids and octopuses the foot has been modified and consists of arms and tentacles; the tentacles bear complex sensory organs. Organs such as the heart, the digestive tract, and the reproductive system are concentrated centrally in a visceral mass. A tissue fold, known as the mantle, covers a visceral mass of internal organs. In many species, the mantle secretes a hard, calcareous shell. In most species, the mantle is extended to create a mantle cavity holding the gills used in respiration. Mollusks have a secondarily reduced coelom (see Figure 23.18A). The blood vessels of the mollusks do not form a closed circulatory system. Instead, blood and other fluids empty into a hemocoel through which fluid moves around the animal to deliver oxygen to the internal organs. Mollusks have a heart that moves the blood back into the blood vessels.

Chitons are marine mollusks that feed on algae, bryozoans, and other organisms that they scrape off rocks using a razor-like body structure called the radula. A chiton's shell consists of eight overlapping plates that are surrounded by a girdle (see Figure 23.18B). They have simple internal organs, multiple gills, and bilateral symmetry.

Gastropods are the most species-rich and widely distributed of the mollusks, and are found in all environments. There are shelled and unshelled gastropod species, including the snails and slugs (the only terrestrial mollusks), as well as the marine nudibrachs (sea slugs), whelks, limpets, and abalones. Gastropods use their foot either to crawl or swim (see Figure 23.18C). Land snails and slugs are able to survive on land, as the mantle tissue is modified into a highly vascularized lung.

The 30,000 living species of bivalve mollusks include the familiar clams, oysters, scallops, and mussels. They are found in both salt and fresh water, but they are all aquatic. Bivalves use an opening called an incurrent siphon to bring water into their two-part hinged shells (see Figure 23.18D); they are filter feeders that extract foodstuffs from these water currents. Large gills inside the shell extract the food and also function as respiratory organs. Water and gametes exit from an excurrent siphon. Among the clams, the molluscan foot has been modified into a digging device that allows the animal to burrow into the mud or sand.

Cephalopods include the octopuses, squids, and nautiluses. The excurrent siphon is modified to give cephalopods the ability to control water movement into the mantle, allowing them to use ejected water as a jet for propulsion (see Figure 23.18E). They capture prey with their tentacles (see Figure 23.19B), and their greatly enhanced mobility makes them dominant ocean predators. They are also able to control gas movement in the mantle, which helps in buoyancy control. As is typical of active, rapidly moving predators, cephalopods have a head with complex sensory organs, most notably the eyes, which are in many ways comparable to those of vertebrates.

Nautiluses are the surviving cephalopods with external chambered shells. Shells have been lost by species in several molluscan groups, including gastropods (slugs and nudibranchs) and cephalopods (octopus)—shell protection in these groups is replaced by protective strategies including toxins and evasion tactics.

The ecdysozoans include more species than all other lineages combined. They are characterized by a rigid external covering (cuticle or exoskeleton), which must be molted periodically in order to allow the animal to grow (ecdysis is the Greek word for "shedding"). The new exoskeleton is temporarily soft after molting, leaving the animal vulnerable (see Figure 23.20B). Increasing molecular evidence, including a set of Hox genes shared by all ecdysozoans, supports the monophyly of these animals and suggests that cuticle molting is a lifestyle that may have evolved only once.

Wormlike ecdysozoans, which include the priapulids and the kinorhynchs, are unsegmented and have a thin cuticle that allows gas, mineral, and water exchange, but affords little protection and support and restricts the animals to moist environments.

Arthropods have a hard exoskeleton of chitin that provides protection, waterproofing, and muscle attachment sites. Their bodies are segmented, with individual muscles attached to the exoskeleton that operate each segment. The jointed appendages that give the group its name (arthros = joint, poda = limb) allow for a greater range of movement and the specialization of different appendages for different purposes. Chitin is waterproof and keeps the animal from dehydrating. They are Earth's dominant animals in both number of species and number of individuals.

Nematodes, or roundworms, range in size from microscopic to up to 9 meters in length. About 25,000 species of this diverse group have been described, and these may represent less than a quarter of extant nematode species. They live in soil, on the bottoms of lakes and streams, and in marine sediments. Nematodes use their gut for both gas exchange and nutrient uptake (see Figure 23.21). Many species prey upon protists and other microscopic organisms, but of most significance to humans is the large number of parasitic species; several of these, including Trichinella spiralis, are dangerous to humans. The free-living nematode Caenorhabditis elegans is a widely used model organism for geneticists and developmental biologists.

The 320 species of horsehair worms are long and very thin, as their name suggests. They live in fresh water or very damp soil near the edges of ponds and streams and feed in the larval stage as parasites of terrestrial and aquatic insects and crabs (see Figure 23.22). Adults have no mouth and have reduced guts. In some species the adults may not feed. The adults of other species continue to grow, so it is also possible that adult worms may absorb nutrients from the environment during the time between the molting of the old cuticle and the hardening of the new one.

Some marine ecdysozoans have a thin exoskeleton, called a cuticle, that is molted periodically. The cuticle allows for gas exchange with the environment, but it offers little protection. The priapulids, kinorhynchs, and loriciferans are groups of tiny, wormlike marine animals that live burrowed in ocean sediments (see Figure 23.23). Most priapulids have a larval stage; the kinorhynchs do not. The loriciferans were only recently discovered in 1983, and they are still being described.

Question 5. The diagram below shows a generalized mollusk body plan (A) and the body plans that are characteristic of gastropods (B) and cephalopods (C). Label each structure in the diagram.
Textbook Reference: *23.3 There Are Two Major Groups of Protostomes, p. 472, Figure 23.18*

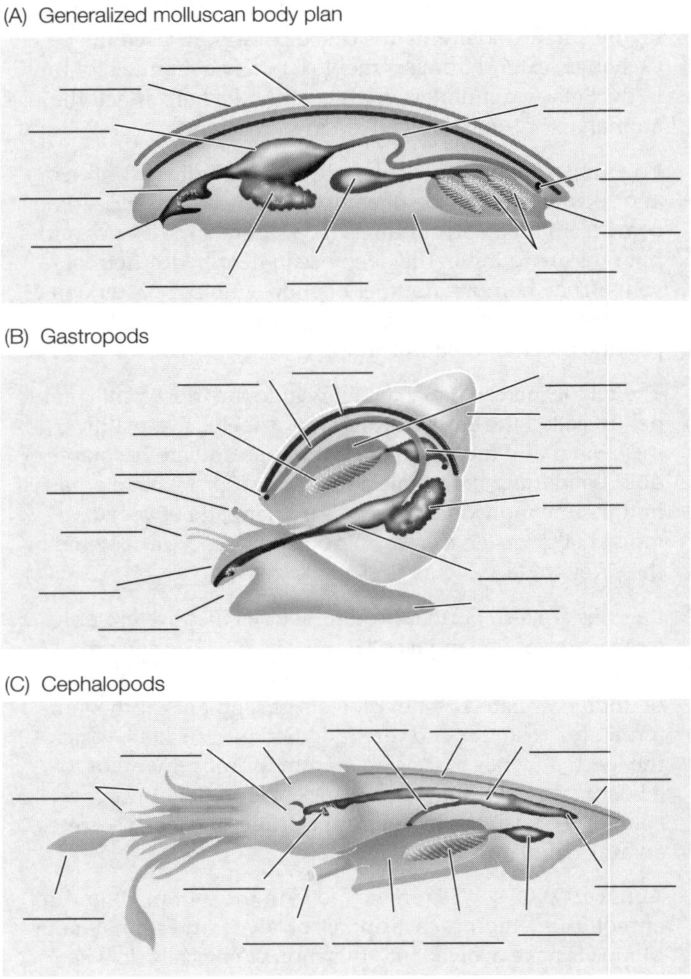

Question 6. What advantages did the incorporation of chitin into a cuticle have for ecdysozoans and what were some disadvantages?
Textbook Reference: *23.3 There Are Two Major Groups of Protostomes, pp. 473–474*

23.4 Arthropods Are Diverse and Abundant Animals

Arthropod relatives have fleshy, unjointed appendages

Chelicerates are characterized by pointed, nonchewing mouthparts

Mandibles and antennae characterize the remaining arthropod groups

Over half of all described species are insects

The arthropods and their relatives are ecdysozoans with paired appendages.

Today four arthropod groups—crustaceans, hexapods, myriapods, and chelicerates—are all numerous and are found in all environments. Several key features have contributed to their success. Their bodies are segmented, and their muscles are attached to the inside of their rigid exoskeletons. Each segment has muscles that operate that segment and the jointed appendages attached to it. Their rigid exoskeleton (composed of chitin) provides waterproofing and protection from predators. The relationships among the arthropod groups are currently being revised based on gene sequence data, and current data supports the idea that arthropods may be monophyletic.

Arthropod relatives have fleshy, unjointed appendages. The onychophorans (velvet worms) were once thought to be more closely related to the annelids, but recent genetic analysis links them to the arthropods. They have unjointed legs and thin cuticles composed of chitin (see Figure 23.24A). Onychophorans have probably changed relatively little from their common ancestor with the arthropods. The tardigrades have fleshy, unjointed legs and use their fluid-filled body cavities as hydrostatic skeletons (see Figure 23.24B). Tardigrades lack a circulatory system and do not have gas exchange organs. When their environment dries out, they shrink in size drastically and can survive in a dormant state for over a decade.

Four major arthropod groups survive: the myriapods (centipedes and millipedes), the chelicerates (including the arachnids—spiders, mites, ticks, etc.), the crustaceans (crabs, lobsters, scallops, barnacles, etc.), and the hexapods (insects and their relatives).

Chelicerates are characterized by pointed, nonchewing mouthparts. The 98,000 described species are placed in three major clades: pycnogonids, horseshoe crabs, and arachnids. All chelicerates have two body parts, and most have eight legs. The pycnogonids, or sea spiders, are a group of about 1,000 exclusively marine species. Most are very small and are rarely seen (see Figure 23.25A). There are only four living species of horseshoe crabs. These animals have changed so little in morphology over evolutionary time that they are often referred to as "living fossils" (see Figure 23.25B). The arachnids are the most prominent chelicerates and include the spiders, scorpions, mites, and ticks (see Figure 23.26). They have a simple life cycle, with young that resemble small adults. Spiders, the most familiar arachnids, build webs of protein threads that they use to capture prey. Spider webs are strikingly varied, often species-specific, and increase the predatory ability of spiders in many different environments. Spiders use hollow chelicerae to inject venom into prey. Mites and ticks are vectors for a variety of organisms that cause diseases in plants and animals.

Mandibles and antennae characterize the remaining arthropod groups. The myriapods include 3,000 described species of centipedes and 11,000 described species of millipedes. Members of both groups have similar body plans, consisting of a well-formed head and a long, segmented trunk. Centipedes have one pair of legs on each trunk segment (see Figure 23.27A); millipedes have two pairs on each segment (see Figure 23.27B). Crustaceans include many familiar animals (decapods, which include shrimp, lobsters, crabs, barnacles; see Figure 23.28A) as well as the less-familiar isopods (which include sowbugs; see Figure 23.28B), amphipods, ostracods, copepods (see Figure 23.28C), and branchiopods (see Figure 23.28D). About 50,000 species have been described so far. Barnacles are an unusual crustacean that is sessile as an adult (see Figure 23.28E).

According to recent gene sequencing, crusteans may be paraphyletic with respect to the hexapods. The crustacean body is divided into the head, thorax, and abdomen (see Figure 23.29A). In many species, a carapace extends dorsally from the head to protect and cover the body. The thorax and abdomen have one pair of appendages each; crustacean appendages are specialized for walking, swimming, feeding, sensation, and gas exchange. In some species, complex branched appendages have evolved. Fertilized eggs remain attached to the body of the female during the early stages of development. Upon hatching, the young are released as larvae in some species. With other species, the juveniles are similar in form to the adults. With some species, the fertilized eggs are released into the water or attached to an object.

Insects are six-legged terrestrial hexapods. More than half of all described species are insects. They are prominent occupants of terrestrial and freshwater environments. More than a million species of insects have been described so far, and many biologists believe this is only a small fraction of the species that actually exist. (See Table 23.2 and Apply the Concept, p. 480.)

Three groups of wingless hexapods—the springtails, two-pronged bristletails, and proturans—are related to the insects and probably resemble the insect ancestral form most closely. Members of these three groups differ from insects in having internal, rather than external, mouthparts. Springtails, which are probably the most abundant hexapod, have a very simple lifestyle in which the juveniles resemble the adults.

Like the crustaceans, the insect body has three parts: the head with a pair of antennae, the thorax with pairs of legs, and the abdomen. Unlike other arthropods, the abdominal segments do not bear appendages (see Figure 23.29B). Gas exchange occurs in a system composed of a series of air sacs and tubular channels, called tracheae, that extend from external openings called spiracles. Other distinguishing characteristics of insects include paired antennae with a sensory receptor known as Johnston's organs, three pairs of legs on the thorax, and external mouthparts.

There are two classes of insects: the wingless apterygotes and the winged pterygotes (some species of which have secondarily become wingless). The apterygotes, including the jumping bristletails and silverfish, have simple life cycles. Pterygotes typically have two pairs of wings attached to the thorax, but in some groups (e.g., parasitic lice and fleas, some beetles, and worker ants), one or both pairs of wings have been secondarily lost.

Hatching pterygotes do not look like the adults, and they undergo changes at each molt. Each stage between molts is called an instar. If the changes are gradual, the insect is said to undergo incomplete metamorphosis. If a drastic change occurs between some instars, then the insect is said to undergo complete metamorphosis. The most dramatic example is the change that occurs when a caterpillar transforms into a pupa in which the adult form develops. In insects that undergo complete metamorphosis, each stage is often specialized with regard to the environment and food source.

Pterygote insects were the first animals to evolve the ability to fly. Homologous genes control the development of insect wings and crustacean appendages and there is evidence that insect wings evolved from a dorsal branch of a crustacean appendage that functioned in gas exchange.

Most flying insects have two pairs of stiff, membranous wings attached to the thorax which can fold over their bodies, allowing them to fit into tight crevices. Those that can fold their wings over their bodies are collectively known as the neopterans. True flies have one pair of wings plus a pair of stabilizers called haltares. In beetles, one pair of wings forms a hardened cover. Insects that cannot fold their wings over their bodies include dragonflies and mayflies. The aquatic larvae can be predatory or herbivorous. This is an ancestral form for pterygotes. Dragonflies are active predators, while adult mayflies do not have a functional digestive tract.

Neopteran insects that undergo incomplete metamorphosis include grasshoppers (see Figure 23.30B), termites, stone flies, earwigs, thrips, true bugs (see Figure 23.30C), aphids, cicadas, and many others. In these groups, the hatchlings resemble small, usually wingless adults; as they molt from one stage to the next, they gradually acquire more adult characteristics.

More than 80 percent of neopterans undergo complete metamorphosis (see Figure 23.30D), including beetles (see Figure 23.30E), lacewings, caddisflies, butterflies, moths (see Figure 23.30F), sawflies and true flies (see Figure 23.30G), wasps, bees, and ants (see Figure 23.30H). In these groups, the larvae and adult forms are substantially different; the young pass through at least two stages (larva and pupa) before becoming adults. They form a subgroup of the neopterans called the holometabolous insects.

Question 7. What are some advantages to segmentation?
Textbook Reference: *23.4 Arthropods Are Diverse and Abundant Animals, p. 476*

Question 8. Distinguish the body plan of insects from that of crustaceans and other arthropods.
Textbook Reference: *23.4 Arthropods Are Diverse and Abundant Animals, pp. 479–480, Figure 23.29*

23.5 Deuterostomes Include Echinoderms, Hemichordates, and Chordates

- Echinoderms have unique structural features
- Hemichordates are wormlike marine deuterostomes
- Chordate characteristics are most evident in larvae
- Adults of most cephalochordates and urochordates are sessile
- A dorsal supporting structure replaces the notochord in vertebrates
- The vertebrate body plan can support large, active animals
- Fins and swim bladders improved stability and control over locomotion

The deuterostomes include the echinoderms, hemichordates, and chordates.

There are many fewer species of deuterostomes than there are of protostomes, but the deuterostomes are of special interest because mammals—a group that includes the largest living animals as well as the human lineage—are deuterostomes. Deuterostomes are united based on early developmental patterns.

Deuterostomes fall into three major clades: the echinoderms, the hemichordates, and the chordates. All are triploblastic, coelomate animals (see Figure 23.3C). Skeletal support features, when present, are internal. A few species have segmented bodies, but the segments are not visible. Echinoderms and hemichordates (collectively known as ambulacrarians) have bilaterally symmetrical, ciliated larvae (see Figure 23.31A). Adult hemichordates are bilaterally symmetrical, while echinoderm adults have unique pentaradial symmetry. Other deuterostome groups retain bilateral symmetry.

While 23 major groups consisting of about 13,000 species of echinoderms have been described from the fossil records, only six groups containing 7,000 species live today. All known species live in marine environments.

In addition to their pentaradial symmetry, echinoderms have two unique structural characters: A system of water-filled canals called a water vascular system that functions in feeding, gas exchange, and locomotion and an internal skeleton made up of calcified plates (see figure 23.31B).

Echinoderms are divided into two major clades. The crinoids includes about 80 species of stalked sessile animals called sea lilies and 600 species of feather stars, which have flexible appendages that allow for limited movement (see Figure 23.32A).

The remaining echinoderms are generally mobile and are divided into two main groups: the echinozoans (sea urchins and sea cucumbers; see Figures 23.32B and C) and the asterozoans (sea stars and brittle stars; see Figures 23.4D and E). Hemichordates are wormlike marine deuterostomes. The 100 species of hemichordates include the acorn worms and the pterobranchs. Both groups have a body plan composed of a trunk, a collar, and a sticky proboscis used for catching prey and digging (see Figure 23.33A). In the acorn worm, cilia move food from the proboscis to the mouth. Behind the mouth are pharynx and an intestine. The pharynx can open to the outside through pharyngeal slits that contain vascularized tissue for gas exchange. Acorn worms breath by pumping water through the mouth and out the pharyngeal slits. They burrow in sand or mud and extract food items from the substrate.

Pterobranchs are sedentary marine animals that are either solitary or colonial (see Figure 23.33B). Behind the proboscis are one to nine pairs of arms that are used to capture prey and also function in gas exchange. Chordate characteristics are most evident in larvae.

The features that reveal the evolutionary relationship between the echinoderms and chordates, as well among the chordates, are primarily observed in the larval stage or during the early states of development.

The three principle chordate clades are the cephalochordates, the urochordates, and the vertebrates. All three chordate clades share the following at some point during development (see Figure 23.34): (1) A hollow dorsal nerve cord; (2) a tail that extends beyond the anus; (3) a dorsal supporting rod called the notochord; and (4) pharyngeal slits.

The notochord is the main distinguishing characteristic of the chordates. In urochordates, the notochord is lost during the metamorphosis from larval to adult forms. In vertebrates, the notochord is replaced by skeletal structures called the vertebrae (spinal column).

Ancestral pharyngeal slits are generally lost in adults, but are retained in some urochordates and cephalochordates (see Figure 23.34). A pharynx develops around the slits and becomes large in some chordates, forming a pharyngeal basket.

Adults of most cephalochordates and urochordates are sessile. The 30 species of cephalochordates (lancelets) are small fishlike animals that retain the notochord for their entire life and use their pharyngeal baskets to catch prey (see Figure 23.34B). Fertilization of eggs takes place in the water.

The three major urochordate groups are the ascidians (sea squirts, also known as tunicates), thaliaceans, and larvaceans. All members of these three groups are marine, and more than 90 percent of urochordate species are ascidians. The body form of an adult ascidian is baglike in shape and surrounded by a tough tunic (hence the name "tunicate") (see Figure 23.35).

Ascidian larvae have pharyngeal slits, a nerve cord, and a notochord; it is the larva, not the adult form, that suggests their chordate ancestry (see Figure 23.34A). The adults lose the notochord and nerve cord but have an enlarged pharynx—the baglike pharyngeal basket—for catching food. Ascidians sometimes reproduce asexually by budding, thus forming colonies.

A dorsal supporting structure replaces the notochord in vertebrates. The vertebrates take their name from the unique jointed dorsal vertebral column that replaces the notochord during early development and provides support.

The vertebrates are divided into two groups: the jawless fishes and the gnathostomes. (See Figure 23.36 for the phylogenetic tree that shows the evolutionary relationships among the vertebrates.)

The elongate, eel-like hagfishes are the sister group of the remaining vertebrates (see Figure 23.37). These jawless fishes look superficially like the lampreys, which are also jawless, but hagfishes lack true vertebrae and have only a partial cranium (skull). Lampreys have true vertebrae (see Figure 23.37B). The hagfishes also have a weak circulatory system with three small accessory hearts; they lack a stomach, and their skeleton is composed of cartilage. Although they lack a jaw, they have a tonguelike structure with toothlike rasps that they use to capture prey. There is some debate as to the accuracy of placing the hagfishes with other vertebrates in the same phylogenetic tree. The analyses of gene sequences suggest that the hagfishes and lampreys are related, and it is possible that during evolution the ancestor of the hagfishes lost some of the structural characteristics that define vertebrates.

Lampreys resemble hagfishes, but they differ biologically. They have a complete braincase and distinct vertebrae made of cartilage. Unlike hagfishes, which undergo direct development, lampreys undergo complete metamorphosis. The larvae, called ammocoetes, are filter feeders, while the adult forms are often parasitic. The mouth is a rasping, sucking organ used by the lampreys to attach themselves to prey. The adults of a few lamprey species do not feed as adults, surviving only long enough to breed.

The four key features of the vertebrate body plan are: (1) an anterior head with a large brain and protective skull; (2) a rigid internal skeleton supported by a vertebral column; (3) a large coelom in which the internal organs are suspended; and (4) a closed circulatory system. This body plan is capable of supporting large-sized animals (see Figure 23.38).

After the Devonian period, jaws evolved from the skeletal arches that supported the gills. The jawed vertebrates fall into the category of gnathostomes, or "jaw mouths." The evolution of the jaw—and then of teeth—led to greatly enhanced feeding efficiency (see Figure 23.39), and jawed fishes were the major predators of the Devonian. Most aquatic gnathostomes have unjointed appendages, called fins, that help control movement through the water (see Figures 23.38 and 23.40).

The chondrichthyan fishes, including sharks, skates, rays, and chimaeras, have flexible and leathery skin and a skeleton composed of cartilage (see Figure 23.40A). Sharks move forward by laterally undulating their body and tail (caudal) fin; skates and rays move vertically by undulating their pectoral fins. Almost all chondrichthyans are marine, but a few live in estuarine waters or migrate to lakes and rivers.

Lunglike gas-filled sacs called swim bladders evolved in the ancestral fishes of the bony fishes. With the aid of the swim bladder and fins, many fish can control their position in the water column with minimal energy expenditure.

The ray-finned fishes and virtually all other vertebrates have a bony skeleton. The outer body is covered by thin, lightweight scales, which aid in movement and provide protection. A hard flap called the operculum, which covers the gills, allows for the movement of water over the gills and thus for gas exchange. The approximately 30,000 species of ray-fins encompass a remarkable variety of shapes, sizes, and lifestyles, and they exploit every kind of food source in the aquatic environment (see Figure 23.41).

Question 9. Ascidians, or sea squirts, are armless and legless organisms that spend their adult lives attached to a substrate under water. They feed by pulling water into one tube, filtering out planktonic organisms, and pushing the water out another tube. Grasshoppers have legs, move around freely on land, and feed by ingesting food through a mouth. What evidence has led biologists to believe that humans are more closely related to sea squirts than to grasshoppers?
Textbook Reference: *23.5 Deuterostomes Include Echinoderms, Hemichordates, and Chordates, p. 483*

Question 10. Suppose you discover the fossil of a new organism that has radial symmetry and was in a marine environment (as shown by examining the surrounding rock and adjacent fossils in the rock layer). What other features would you look for in order to determine if this fossil is an echinoderm?
Textbook Reference: *23.5 Deuterostomes Include Echinoderms, Hemichordates, and Chordates, p. 483*

Question 11. The evolution of a hinged jaw resulted in an extensive radiation of the fishes into the many modern jawed forms. What was the main advantage and significance of hinged jaws in this radiation?
Textbook Reference: *23.4 Arthropods Are Diverse and Abundant Animals, p. 488*

23.6 Life on Land Contributed to Vertebrate Diversification

Jointed fins enhanced support for fishes

Amphibians adapted to life on land

Amniotes colonized dry environments

Reptiles adapted to life in many habitats

Crocodilians and birds share their ancestry with the dinosaurs

The evolution of feathers allowed birds to fly

Mammals radiated as nonavian dinosaurs declined in diversity

Most mammals are viviparous

The evolution of lunglike swim bladders in some ray-finned fishes set the stage for the move to the land. Some fishes may have used these sacs to supplement their oxygen supply in low-oxygen freshwater environments, and the sacs may have allowed some fishes to survive out of water. The fins of these fishes were unjointed, so they could only flop around on land.

Two pairs of jointed fins evolved in the ancestors of the coelocanths and the lungfishes, which along with the tetrapods are known as sarcopterygians. The jointed fins of lungfishes are connected to the body by a single large bone, and they have lungs as well as gills. During dry periods when the water from ponds evaporates, they burrow deep in the mud and can survive in an inactive state for many months, breathing air.

Coelacanths were thought to have died out 65 million years ago, but two species were discovered in the twentieth century. One genus, *Latimeria*, is a predator and can weigh up to 80 kilograms (see Figure 23.42A). Unlike other fishes, *Latimeria* has a skeleton composed primarily of cartilage; this is thought to be a derived feature, since the ancestors to this group had bony skeletons. Only six species of lungfishes remain, all of them in the southern hemisphere (see Figure 23.42B).

The change in fin structure allowed these fishes to support themselves in shallow water and eventually to move onto land and exploit food sources there. These early land-dwelling fishes gave rise to the tetrapods—the four-legged vertebrates.

Amphibians have a wide variety of life histories, but most spend at least part of their life cycle in the water (see Figure 23.43). Many species return to water to lay their eggs, producing aquatic larvae. The skin surface is a common site of respiratory gas exchange in amphibians, but many also have lungs.

There are approximately 6,500 species of extant amphibians, falling into three groups. The caecilians are wormlike, legless, burrowing or aquatic amphibians found only in moist tropical regions (see Figure 23.44A). Anurans include the tail-less frogs and toads, and account for the vast majority of amphibian species (see Figure 23.44B). Anurans undergo metamorphosis from an aquatic to a terrestrial life form. The adults have a very short vertebral column and a pelvic region modified for hopping on the hind legs. Some species have adapted to life in very dry, even desert environments, while others have returned to an entirely aquatic lifestyle. Salamanders are tailed amphibians that exchange respiratory gases through both lungs and gills (see Figures 23.44C and D). One group relies on gas exchange through its skin and mouth as all amphibians do, but it does not have lungs. Paedomorphic evolution (retention of the juvenile form in the adult) has led to several entirely aquatic salamander species. Most species have internal fertilization. Sperm is transferred in a jellylike capsule called a spermatophore.

The social behaviors of many amphibians are quite complex; males make species-specific calls to females and can defend their own breeding territories. The amount of care given to the eggs varies. Some amphibians lay large numbers of eggs that are abandoned, while others guard a few fertilized eggs. A few species are viviparous, meaning that the adult gives birth to well-developed juveniles.

Amphibians are the subject of attention because globally over a third of all species has gone extinct or declined to endangered levels in the last few decades (see Figure 23.44B). Scientists are investigating several hypotheses about their decline including infection from a pathogenic chytrid fungus.

A key innovation that allowed colonization of dry environments was the amniote egg (see illustration, p. 493). The egg has a protective calcium-based shell that inhibits dehydration while allowing oxygen and carbon dioxide to pass through. Within the shell, extra-embryonic membranes further protect the embryo, and large quantities of yolk provide nutrition. The amniote animals—reptiles (including birds) and mammals—were thus freed from reliance on water to reproduce.

In several groups of amniotes the egg became modified to allow the embryo to grow inside the mother. In mammals, the egg has lost the shell and the embryonic membranes have been retained and modified.

Other evolutionary adaptations to life on land appeared among the amniotes, including a tough, impermeable skin covered with scales or modified scales (i.e., hair or feathers). The excretory systems of amniotes also evolved adaptations that allowed these animals to excrete nitrogenous wastes in the form of concentrated urine with a minimal loss of valuable water.

During the Carboniferous period, about 250 mya, the amniote animals diverged into two groups, the reptiles and the mammals. Birds, the only living members of the now-extinct dinosaurs, evolved from one of the reptilian groups (see Figure 23.45).

Reptiles adapted to life in many habitats. One relatively small reptilian group, the turtles and tortoises, has a unique body plan that has changed very little over the millennia (see Figure 23.46A). These animals are characterized by dorsal and ventral bony plates that evolved from the ribs to form a protective shell. The relationship of turtles and tortoises to other reptiles is not clear. For example, it is not known how, unlike in other vertebrates, the pectoral girdles evolved to be inside the ribs.

The lepidosaurs include the squamates (lizards, snakes, and amphisbaenians—another group of legless, wormlike burrowers), and the tuataras (see Figures 23.46B–23.46D). The body is covered in scales that greatly reduce the loss of water but make their skin surface unavailable for gas exchange. Gases are exchanged through the lung, which is larger in surface area compared to that of the amphibians. The lepidosaurs have a three-chambered heart that partially separates oxygenated from deoxygenated blood. The limbless condition of snakes is a product of secondary evolution. Most lizards and all snakes are carnivores, and adaptations to the jaws of snakes allow them to swallow prey much larger than themselves. Tuataras are represented by only two living species.

The archosaurs include the extant crocodilians (crocodiles and alligators), the extinct dinosaurs, and the "living dinosaurs" we know as birds (see Figure 23.47). The crocodilians—crocodiles, alligators, caimans, and ghalials—live only in tropical and warm temperate regions. They spend most of their lives in water but build nests on land. They are all carnivores, feeding on other vertebrates, including large mammals.

Dinosaurs arose among the reptiles around 215 mya and survived for about 150 million years. Both the fossil record and molecular evidence support the position of birds as a sister group of the saurischian dinosaurs.

Birds probably evolved from an ancestral theropod, a bipedal predatory dinosaur that appears to have had hollow bones, a furcula (wishbone), limbs with three digits, and a backward-thrusting pelvis.

Living birds diverged from the common flying ancestor and fall into one of two groups. The palaeognaths are flightless and include rheas, emu, kiwis, cassowaries, and the largest bird, the ostrich (see Figure 23.47B). The neognaths, most of which have retained the ability to fly, represent the remaining 9,600 species of birds.

Recent fossil discoveries have demonstrated that some dinosaurs had scales that were highly modified to form feathers. In one species, *Microraptor gui*, the feathers were structurally similar to those of modern birds (see Figure 23.48A).

Archaeopteryx lived 150 mya and represents the oldest known fossil of a bird. It was covered in feathers and had well-developed wings and a wishbone (see Figure 23.48B). It also had teeth, which were lost in later members of this lineage.

Feathers are complex and strong, but also lightweight. Their evolution was a major factor in diversification. The bones of dinosaurs and birds are hollow, and along with the development of feathers, they assisted in the evolution of flight.

Along with the ability to fly came a high metabolic rate needed to fuel flight. The metabolic rate of birds means that they generate a great amount of heat, and their feathers are adapted to allow for heat loss. The lungs of birds also function differently from those of other vertebrates, another adaptation to the needs of flight. Different bird groups feed on many different types of animal and plant material, including carrion. Birds that feed on fruits and seeds are major agents of plant dispersal.

Mammals radiated as nonavian dinosaurs declined in diversity. The earliest mammals lived side by side with reptiles from the first split of the two lineages early in the Mesozoic era. However, only after the large dinosaurs became extinct did mammals begin to flourish in size and number.

Four key features distinguish the mammals: (1) sweat glands in the skin that produce secretions that cool the body; (2) mammary glands that provide nutrient fluid for newborns; (3) external body hair that protects and insulates most mammals; and (4) a four-chambered heart that completely separates oxygenated from deoxygenated blood. The four-chambered heart also evolved convergently in the crocodilians and birds.

In most mammals, the amniote egg became modified for growth of the embryo within the mother's uterus. There are five species of egg-laying mammals. The prototherians exist today only in Australia and New Guinea (see Figure 23.50A). They supply milk to their young as do other mammals, but they do not have nipples on their mammary glands.

Most mammals are viviparous. The remaining mammals are classed as therians. Marsupial therians give birth to tiny underdeveloped young that they nurture externally, usually in a pouch on the mother's belly (see Figure 23.50B). Marsupials were once widespread, but today the approximately 330 marsupial species are largely found in Australia and South America, with minor representatives in North America.

The 4,500 species of eutherian mammals develop in the mother's uterus. This group is widely known by the name placental mammals, from the placenta that provides nourishment for the growing embryo. However, some marsupials also have placentas.

Grazing and browsing by many herbivorous eutherian groups have transformed the terrestrial environment. In plants, this lead to the evolution of spines and tough leaves. Despite such defenses, herbivores have developed adaptations to their teeth and digestive systems in order to consume these plants. This is an example of coevolution.

Eutherians exist in virtually all of Earth's environments and vary greatly in form (see Figure 23.50C–F). Several groups, most notably the cetaceans (whales and dolphins) returned to a marine lifestyle. Flight evolved in the bats (which represent the second largest number of eutherian species, after the rodents; see Table 23.3). The 235 species of primates are the best-studied eutherians because human beings belong to this group.

Question 12. Amphibians are dependant on water and moist environments for their life cycle. Which stages require an aquatic environment and why is such an environment necessary? Do any amphibians have strategies that allow them to avoid the aquatic stage? If so, describe them.
Textbook Reference: *23.6 Life on Land Contributed to Vertebrate Diversification, p. 491*

Question 13. How have mammals and reptiles adapted to some of the challenges of a terrestrial habitat?
Textbook Reference: *23.6 Life on Land Contributed to Vertebrate Diversification, pp. 492–497*

23.7 Humans Evolved among the Primates

Human ancestors evolved bipedal locomotion

Human brains became larger as jaws became smaller

The primate ancestor was a small, arboreal, insectivorous mammal. Primates are identifiable by the presence of opposable digits (i.e., thumbs). Early in their evolutionary history—about 65 mya—the primates split into two clades, the prosimians (lemurs, bush babies, and lorises) and the anthropoids (tarsiers, monkeys, apes, and humans; see Figure 23.51). Prosimian species were once found on all continents, but today they are restricted to Africa, especially the island of Madagascar.

Soon after the prosimian–anthropoid split, the anthropoids split further into the New World and Old World monkeys. The breakup of the African and South American continents meant that the two groups evolved in isolation. About 35 mya, an Old World lineage broke off that would lead to modern apes and humans. Another split occurred around 22 mya. The Asian apes (gibbons and orangutans) descended from two groups in this lineage (see Figure 23.51). A long prehensile tail, which is unique to the New World monkeys, allows them to grasp branches. All New World monkeys are arboreal.

About 6 million years ago, a split resulted in two primate lineages, one that led to chimpanzees and the other that led to the groups that gave rise to hominids.

Ardipithecines (the earliest protohominids) and their descendants, the australopithecines, were adapted for bipedalism, which freed up their hands for other tasks and elevated the eyes to enable seeing over tall vegetation. Bipedal movement also requires less energy. *Australopithecus afarensis* is currently regarded as the ancestor of the modern humans (genus *Homo*; see Figure 23.52). There is disagreement as to how many species are represented by australopithecine fossils. It is clear that these different groups of hominids coexisted in Africa several million years ago. The genus *Homo* arose from one of the smaller lineages of australopithecines. In parallel, the genus *Parathropus* arose from larger australopithecines and later went extinct.

The oldest known member of the genus *Homo, Homo habilis*, lived about 2 mya, and tools have been found with their bones. *Homo erectus* appeared about 1.6 mya. They were as large as modern humans, used stone tools, and cooked with fire. This is the first hominid species to leave Africa.

Larger brain size appears to have evolved concurrently with a smaller, less muscular jaw structure, suggesting that the two traits may be functionally correlated. A mutation in the myosin gene may have permitted changes in jaw size.

The rapid increase in brain size may have been favored by an increasingly complex social life that thrived on ever more sophisticated communication. Any trait that allowed more effective communication would have been favored in a society based on cooperative hunting and other complex social interactions.

A number of species of *Homo* existed simultaneously over evolutionary time between 1.5 mya and 250,000 years ago. One species, *Homo neanderthalensis,* appeared about 500,000 years ago and became widespread in Europe and Asia. Although they were short, they were powerfully built. Their skulls housed a brain that is somewhat larger than that of modern humans. They manufactured tools and were able to hunt large animals.

By 200,000 years ago, *H. sapiens* was predominant in Africa, and they began to extend their range around 60,000–70,000 years ago. Around 35,000 years ago they coexisted alongside *H. neanderthalensis.* It is likely that the two species interacted, but Neanderthals abruptly became extinct 28,000 years ago, possibly exterminated by *H. sapiens*. Genetic evidence suggests that there was some interbreeding between the two species.

By about 20,000 years ago, all other *Homo* species had been supplanted by *Homo sapiens* (modern humans), the only currently existing human species. It was about this time that *H. sapiens* reached North America.

Question 14. What major geologic change occurred during primate evolution and how did it alter the way primate evolution proceeded?
Textbook Reference: *23.7 Humans Evolved among the Primates, p. 499*

Question 15. What are three traits that evolved in primates and led to evolutionary success? Which ones seem to have had the greatest impact in hominid lines?
Textbook Reference: *23.7 Humans Evolved among the Primates, pp. 499–501*

Test Yourself

1. Which of the following statements about echinoderms is *false*?
   a. They have a water vascular system.
   b. They have an internal skeleton.
   c. They are protostomes.
   d. The larvae have bilateral symmetry.
   e. The adult form lacks a head.
   Textbook Reference: *23.5 Deuterostomes Include Echinoderms, Hemichordates, and Chordates, p. 483*
2. Which of the following is *not* a derived trait that is shared by all animals?
   a. Hox genes
   b. The extracellular matrix molecule collagen
   c. Tight junctions, desmosomes, and gap junctions
   d. Bilateral symmetry
   e. Absence of a cell wall
   Textbook Reference: *23.1 Distinct Body Plans Evolved among the Animals, p. 459*
3. Which of the following statements about a deuterostome is *false*?
   a. Three distinct layers of tissue were present during development.
   b. If a coelom is present, it formed within the embryonic mesoderm.
   c. Its early embryonic cleavage pattern was radial.
   d. It is diploblastic.
   e. Gastrulation occurred during development.
   Textbook Reference: *23.1 Distinct Body Plans Evolved among the Animals, p. 459*
4. Which of the following chordate groups evolved before the appearance of cartilaginous fishes?
   a. Ray-finned fishes and sea squirts
   b. Ascidians, lancelets, and hagfishes
   c. Ascidians, lancelets, and ray-finned fishes
   d. Lampreys and ray-finned fishes
   e. Coelacanths and ray-finned fishes
   Textbook Reference: *23.5 Deuterostomes Include Echinoderms, Hemichordates, and Chordates, pp. 487–488*
5. Which of the following statements about the body cavity of animals is true?
   a. The body cavity of coelomates develops from the embryonic ectoderm.
   b. The body cavity of acoelomates is filled with liquid.
   c. The pseudocoel of the pseudocoelomates has a peritoneum.
   d. The acoelomates do not have an enclosed body cavity.
   e. The coelomates have a body cavity surrounded by peritoneum.
   Textbook Reference: *23.1 Distinct Body Plans Evolved among the Animals, pp. 459–460*
6. Which of the following is *not* a feature of the vertebrate body plan?
   a. A ventral spinal cord
   b. An internal skeleton

c. A well-developed circulatory system
d. Organs suspended in the coelom
e. An anterior skull encasing a proportionally large brain.

Textbook Reference: *23.5 Deuterostomes Include Echinoderms, Hemichordates, and Chordates, p. 487*

7. The swim bladder of many fishes that evolved from a lunglike sac has the important function of
a. aiding in prey capture.
b. controlling swimming speed.
c. controlling buoyancy.
d. aiding in reproduction.
e. providing balance.

Textbook Reference: *23.5 Deuterostomes Include Echinoderms, Hemichordates, and Chordates, p. 489*

8. Which of the following traits is shared by the chondrichthyans and the ray-finned fishes?
a. Gills as the major site of gas exchange
b. A skeleton composed of cartilage
c. An outer surface covered with bony plates
d. A swim bladder
e. Propulsion by means of dorsal–ventral movements of their tails

Textbook Reference: *23.5 Deuterostomes Include Echinoderms, Hemichordates, and Chordates, pp. 488–490*

9. The transition from aquatic to terrestrial lifestyles required many adaptations in the vertebrate lineage. Which of the following is *not* one of those adaptations?
a. A shift from gills to air-breathing lungs
b. Improvements in the water resistance of skin
c. An alteration in the mode of locomotion
d. Development of feathers for insulation
e. Modifications to the nitrogen elimination system

Textbook Reference: *23.6 Life on Land Contributed to Vertebrate Diversification, pp. 490–495*

10. The amniotes evolved the ability to reproduce by laying eggs that have shells. The major advantage of shelled eggs is that
a. the embryo needs only a small amount of yolk for development.
b. they do not have to be laid in a moist environment.
c. the shells increase evaporation from the egg.
d. nitrogenous wastes can be excreted across the shell.
e. a shell contributes to the efficiency of gas exchange with the environment.

Textbook Reference: *23.6 Life on Land Contributed to Vertebrate Diversification, p. 493*

11. Which of the following is *not* a trait that identifies an animal as a mammal rather than an amphibian?
a. Mammary glands
b. Hair
c. Sweat glands
d. Kidneys
e. Four-chambered heart

Textbook Reference: *23.6 Life on Land Contributed to Vertebrate Diversification, p. 493*

12. A key difference between Old World and New World monkeys is that the latter
a. have a prehensile tail.
b. are arboreal.
c. have a placenta.
d. are less closely related to tarsiers.
e. All of the above

Textbook Reference: *23.7 Humans Evolved among the Primates, p. 499*

13. Which of the following is *not* part of the evidence that birds are closely related to dinosaurs?
a. The presence of feathers in representatives of both groups
b. The presence of hollow bones in representatives of both groups
c. DNA sequence data comparing birds to other living reptiles
d. A bipedal stance in fossilized therapods
e. The ability to fly

Textbook Reference: *23.6 Life on Land Contributed to Vertebrate Diversification, p. 495*

14. Which of the following are *not* deuterostomes?
a. Echinoderms
b. Hemichordates
c. Cephalochordates
d. Ecdysozoans
e. Chordates

Textbook Reference: *23.3 There Are Two Major Groups of Protostomes, p. 465*

15. Which of the following statements about sponge structure or function is *false*?
a. Choanocytes are flagellated cells that play a role in feeding.
b. Large species are found in areas of heavy wave action, where food is most abundant.
c. Individual sponges are both male and female.
d. Water enters a sponge through pores and exits via one or more oscula.
e. Sponges have an extensive extracellular matrix holding the cells together.

Textbook Reference: *23.2 Some Animal Groups Fall Outside the Bilataria, p. 463*

16. A restaurant appetizer of escargot (snails), clams on the half shell, and calamari (octopus) would contain which types of mollusks?
a. Chitons, bivalves, and gastropods
b. Bivalves, gastropods, and cephalopods
c. Chitons, gastropods, and cephalopods
d. Bivalves and gastropods
e. Chitons and cephalopods

Textbook Reference: *23.3 There Are Two Major Groups of Protostomes, p. 471*

17. Which of the following attributes is seen in the arthropod body plan?
a. Segmentation
b. Jointed appendages

c. Muscle attachments for the appendages inside the exoskeleton
d. A hard exoskeleton
e. All of the above

Textbook Reference: *23.4 Arthropods Are Diverse and Abundant Animals, p. 476*

Answers

Key Concept Review

1. Protostomes and deuterostomes differ in the structure that becomes the anus and the structure that forms the mouth. The sequence in which these structures develop is also different. In protostomes, the mouth develops from the blastopore; in the deuterostomes, the anus develops from the blastopore. In both, the opposite structure forms later. These groups were first described according to observations of their developmental patterns, but sequence data supports this classification.
2. Hydrostatic skeletons provide for controlled movement of body parts. In terrestrial organisms, the surrounding environment does not support the body, so most organisms with hydrostatic skeletons are very small and soft bodied. Larger bodies are possible with exoskeletons or hard interior skeletons that provide structural support and protection to soft tissues.
3. Animals can either move through the environment to where the food is located or they can move the environment and the food to themselves. Sessile organisms would most likely have adaptations to move the environment and food to themselves.
4. The sponge body plan consists of an aggregation of cells around a water canal system. Food particles, along with water, enter by way of small pores and are captured by choanocytes.
5.

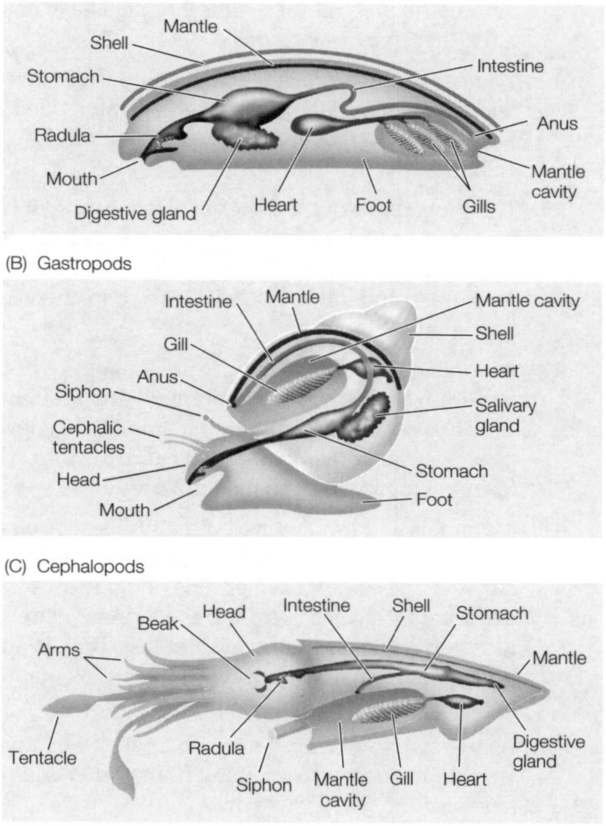

6. A cuticle reinforced with chitin provided support and protection from desiccation and predators. Its presence meant that gas exchange could not be accomplished through the body surface and that growth could not occur without molting of the rigid exoskeleton, leaving the temporarily soft-bodied animal vulnerable. Locomotion was also limited, since the animal could neither move in a wormlike fashion nor use cilia for propulsion.
7. Segmentation provides several advantages, including facilitating specialization of body regions and improved muscle coordination, as each segment can be manipulated independently.
8. Insects, like crustaceans, have three body parts: head, thorax, and abdomen. Unlike crustaceans, they have three pairs of legs attached to the thorax. They are distinguished from springtails and other hexapods by possessing external mouthparts, a pair of antennae on their heads with a motion receptor called a Johnston's organ, and tracheae.
9. Although adult ascidians and adult humans are very different animals, their embryonic stages share certain important characteristics, including the presence of a notochord (as in all deuterostomes) and a blastopore that develops into the anus. A look at embryonic grasshoppers, however, shows them to be fundamentally

different. The embryonic grasshopper's blastopore develops into the mouth, placing grasshoppers squarely among the protostomes. Grasshoppers have an external rather than an internal skeleton.

10. You would look for evidence of an internal skeleton and a vascular system composed of water-filled canals. These two traits are defining characteristics of echinoderms.
11. The hinged jaw of the fishes evolved during the Devonian period. The evolution of the jaw opened up a new food source for these animals. With a jaw, fish could now grasp and kill larger living prey and chew and tear body parts.
12. Amphibians generally require an aquatic environment for fertilization, larval development, and metamorphosis into a terrestrial adult. Some amphibians develop from eggs laid on land and move directly into adultlike forms, skipping the aquatic form entirely.
13. Mammals and reptiles are both amniotes. Their eggs minimize water loss while still permitting gas exchange. The skin of both groups of organisms is modified to reduce water loss, and the kidneys eliminate nitrogen while minimizing water loss. Both groups of organisms exchange gases with the atmosphere by means of lungs.
14. As the continental land masses separated 65 mya, New World monkeys were isolated from Old World monkeys and the two groups diverged.
15. Primates evolved opposable thumbs, bipedal upright locomotion, prehensile tails, and large cranium size. Opposable thumbs favored development of tool use. Bipedal locomotion increased the ability to see above vegetation and freed up hands to carry and manipulate objects. Prehensile tails increased the success of arboreal monkeys. Large cranium size permitted brain growth to accommodate language development and facilitated sophisticated social interactions.

Test Yourself

1. **c.** Species from the echinoderm clade are deuterostomes.
2. **d.** Not all animals have bilateral symmetry (e.g., sponges, cnidarians, ctenophores).
3. **d.** Deuterostome embryos have three layers (the ectoderm, the mesoderm, and the endoderm), making deuterostomes triploblastic, not diploblastic.
4. **b.** The ascidians, lancelets, and hagfishes all evolved before the cartilaginous fishes.
5. **d.** The body cavity of coelomates develops from the mesoderm and contains a peritoneum. The acoelomates lack a body cavity.
6. **a.** Along with an internal skeleton, well-developed circulatory system, and organs suspended in a coelom, the vertebrates have a dorsal spinal cord.
7. **c.** The swim bladder of modern-day fishes is involved in controlling buoyancy.
8. **a.** In both the ray-finned fishes and cartilaginous fishes, the major site of gas exchange is the gills. The ray-finned fishes have a skeleton of bone and a swim bladder. These traits are not shared with the cartilaginous fishes. Neither group has bony plates in its outer surface. They all move their tails laterally, in contrast to cetaceans, which move their tails vertically.
9. **d.** The move onto land did not require the development of feathers for insulation. Amphibians and reptiles do not have an insulation layer, and many mammals have hair for insulation.
10. **b.** The shelled egg of the birds and reptiles allowed them to occupy dry terrestrial habitats because the shell decreases water loss from the egg.
11. **d.** Kidneys are present in reptiles, birds, fishes, and mammals. The presence of kidneys is not a trait that identifies an animal as a mammal.
12. **a.** Only the New World monkeys have a prehensile tail.
13. **e.** Not all birds fly and not all dinosaurs were able to fly.
14. **d.** Ecdysozoans are a type of protostome (see Table 23.1).
15. **b.** Because they are not structurally robust, large, upright sponges would be destroyed by heavy wave action.
16. **d.** Snails are in the Gastropoda, clams are in the Bivalvia, and octopuses are in the Cephalopoda.
17. **e.** Arthropods have a hard exoskeleton, a segmented body, jointed appendages, and muscles for the appendages attached inside the exoskeleton.

The Plant Body

24

The Big Picture

- A plant can be thought of as having vegetative structures that carry out the major functions of day-to-day life and reproductive structures that are responsible for reproducing the plant. This chapter focuses on the vegetative plant body, consisting of the root system, which anchors the plant and absorbs water and nutrients, and the shoot system, which carries out photosynthesis and supports the plant against gravity. Modifications of the root and shoot systems lead to specialization of the plant.
- Plant cells are uniquely suited for support, transport, and the carrying out of cellular functions. Groups of cells form tissues that have specific roles within a plant. There are three basic tissue types in plants: dermal, ground, and vascular. Indeterminate growth and the modular organization of plants allow for regeneration of parts lost to damage and disease.
- Plant growth occurs from meristems. Apical meristems allow for elongation of the plant, whereas lateral meristems allow for secondary or woody growth. Not all plants exhibit secondary growth. All tissue types arise from the meristems and go through a process of elongation followed by differentiation.

Study Strategies

- The best study strategy for this material is to study the figures, pictures, and live material. Plant anatomy is the study of structure, and since structures are three-dimensional, they are best understood visually. Review stem, root, and leaf anatomy using the figures in the textbook.
- Basic plant anatomy is not difficult, but it is new to most students. The more time spent looking at diagrams and live specimens, the easier it will be to understand the anatomy. While studying, think about the function of each structure. You will find that "form follows function."
- This chapter exposes you to many new vocabulary words. A vocabulary list will help you organize your study, but do not try merely to memorize these terms. By looking at the words and understanding their roots, you will be better able to understand their meaning.
- Go to yourBioPortal.com to review the following tutorial and activities:

  Animated Tutorial 24.1 Secondary Growth: The Vascular Cambium

  Web Activity 24.1 Eudicot Root

  Web Activity 24.2 Monocot Root

  Web Activity 24.3 Eudicot Stem

  Web Activity 24.4 Monocot Stem

  Web Activity 24.5 Eudicot Leaf

Key Concept Review

24.1 The Plant Body Is Organized and Constructed in a Distinctive Way

Plants develop differently than animals

Apical–basal polarity and radial symmetry are characteristic of the plant body

The plant body is constructed from three tissue systems

Plants are composed of root systems and shoot systems (see Figure 24.1). Root systems are responsible for mineral and water uptake and support. Shoot systems consist of leaves (and leaf derivatives) involved in photosynthesis and supportive stems. Most plants are either broad-leaved eudicots or narrow-leaved monocots. The development of plants involves determination, differentiation, morphogenesis, and growth. The apical meristems allow for continual growth throughout a plant's life. Plant cells differ from other eukaryotic cells in that every plant cell is bounded by cellulose-containing cell walls. The plane of cell division controls the direction of plant growth. Plant cells are pluripotent or totipotent allowing for repair of damaged tissues.

The body plan of a plant is established in the embryo along a basal–apical axis and a radial axis. Meristems are undifferentiated cells found at the tips of the embryonic shoot and root that will become the organs of the plant as it grows. Two unequal daughter cells are produced as the zygote goes through a mitotic division. This results in the development of a thin suspensor and a globular embryo. The cotyledons (seed leaves) begin to form as the embryo enters the heart stage. As the cotyledons elongate, the embryo enters the torpedo stage and the internal tissues begin to differentiate. The root and shoot apical meristems develop between the cotyledons.

Tissues are composed of cells that function together. Different tissue types are grouped into the dermal, vascular, and ground tissue systems. The dermal tissue system makes up the outer covering of the plant and includes the epidermis and the layer of cuticle it secretes. Special epidermal cells include stomata, trichomes, and root hairs. The ground tissue system is found between the dermal and vascular tissue and is involved in storage, support, and photosynthesis. Thin-walled parenchyma cells have large central vacuoles and are frequently photosynthetic or used for storage. Collenchyma cells are support cells with special thickenings at the cell wall corners. Sclerenchyma cells have highly thickened cell walls and are either elongated fibers or variously shaped sclereids. Many undergo apoptosis and provide rigid support after they die. Fibers strengthen bark and woody stems. Densely packed sclereids are found in the shells of nuts and in some seed coats and produce the gritty texture of pears and other fruit.

The vascular tissue system is made of xylem and phloem, and is the conductive tissue of the plant. Xylem transports water and mineral ions from the roots to the rest of the plant. The tracheary elements involved in the water movement are functional after they die. Tracheids and vessel elements make up the cells involved in xylem transport. Living phloem moves carbohydrates and nutrients. Individual phloem cells are called sieve tube elements. Plasmodesmata enlarge where sieve tube elements join, making sieve plates.

Question 1. Draw a typical eudicot plant. Label the shoot system and root system. Indicate on your drawing where you would find an axillary bud and where you would find a terminal bud. Label the following structures: leaf blade, internode, petiole, taproot, and lateral roots.
Textbook Reference: *24.1 The Plant Body Is Organized and Constructed in a Distinctive Way, p. 507*

Question 2. The plant body is composed of three different tissue systems. While examining a plant you find a tissue within the plant composed of dead cells that have very thick cell walls and contain lignin. What types of cells are you likely seeing and why?
Textbook Reference: *24.1 The Plant Body Is Organized and Constructed in a Distinctive Way, pp. 509–510*

24.2 Meristems Build Roots, Stems, and Leaves

A hierarchy of meristems generates the plant body

The root apical meristem gives rise to the root cap and the root primary meristems

The products of the root's primary meristems become root tissues

The root system anchors the plant and takes up water and dissolved minerals

The products of the stem's primary meristems become stem tissues

The stem supports leaves and flowers

Leaves are the primary site of photosynthesis

Many eudicot stems and roots undergo secondary growth

Primary growth occurs as the shoots and roots lengthen and proliferate to produce the nonwoody primary plant body. Herbaceous plants go through primary growth only. Woody plants increase their girth during secondary growth and produce a secondary plant body composed of wood and bark.

Meristems are undifferentiated cells that are able to produce new cells indefinitely. These initial cells are similar to human stem cells. Meristems are classified as either apical or lateral, depending on whether they contribute to primary or secondary growth. Two types of apical meristems, shoot and root, give rise to the primary plant body and are located in buds and at the tips of stems and roots. They are responsible for primary growth—the elongation of the plant body. Apical meristems give rise to primary meristems that produce the primary plant body. These primary meristems are called the protoderm (dermal tissue system), the ground meristem (ground tissue system), and the procambium (vascular tissue system). All plant parts arise from division of the apical meristem cells.

Root apical meristems (protoderm, ground meristem, and procambium) produce root tissues. At the tip of a root, the root apical meristem forms a root cap and quiescent center. The zone of cell division includes the apical and primary meristems. The zone of elongation is found above this and is the site of new cell formation. The upper layer is the zone of maturation where the cells differentiate and take on special functions. The root has three primary tissue systems. The protoderm gives rise to the epidermis and root hairs, and both are involved in water and mineral ion uptake. The ground meristem gives rise to the cortex and endodermis. The endodermis is specialized with a waxy coating of suberin to assist with water and ion movement. The procambium gives rise to the stele, which houses three tissues: the pericycle, the xylem, and the phloem. In eudicots, the very center of the root is xylem, but in monocots the center is pith tissue (see Figures 24.8). The roots are the main site of water and nutrient entry into the plant. In eudicots, the taproot is the primary root with lateral roots radiating out. Monocot plants have a fibrous root system that is made up of more diffuse thin roots.

Shoot growth occurs as the plant adds repeating units of phytomers from the terminal and axillary buds. The shoot primary meristem also gives rise to three primary meristems that produce the three tissue systems. Shoot vascular tissues are arranged in vascular bundles containing both xylem and phloem. In eudicots, the vascular bundles are arranged in a cylinder, allowing for woody growth. In monocots the vascular bundles are scattered. Shoot apical meristems produce the leaves of the plant. Leaves arise from leaf primordia with bud primordia forming at each leaf base. Most eudicot leaves have two zones of photosynthetic cells called mesophyll. The upper level of cylindrical mesophyll is called palisade mesophyll, and the lower level is called spongy mesophyll. Air space around mesophyll cells is necessary for CO_2 to reach the photosynthesizing cells. Vascular tissue extends throughout leaves as a network of veins. Veins carry water to cells and transport carbohydrates to sink tissues. The entire leaf is covered by a protective epidermis and is waterproofed by a waxy cuticle. Gas exchange occurs through guarded stomata. Guard cells open and close stomata to control water loss and gas exchange.

Secondary growth in eudicots involves the laying down of wood and bark by the two lateral meristems, vascular cambium and cork cambium. Vascular cambium forms new secondary xylem (wood) and secondary phloem (bark). The cork cambium produces new dermal tissues to accommodate the increasing diameter and inhibits water loss with the production of periderm cells. The action of the vascular and cork cambiums is called secondary growth. Wood results from secondary xylem, and bark is made from the cork cambium, cork, phelloderm, and secondary phloem. Stretching, breaking, and flaking off of epidermis and cortex leaves the secondary phloem at risk. Cells at the surface of the phloem produce a protective layer of cork that is thickened and reinforced with waterproof suberin. New cork is produced as secondary growth proceeds. The annual rings seen in wood are a result of climate shifts in temperate zones, particularly water availability. Tropical trees do not undergo seasonal growth and do not lay down such visible rings.

Question 3: What are the differences between apical and lateral meristems? Do all plants have apical meristems? Do all plants have lateral meristems?
Textbook Reference*: 24.2 Meristems Build Roots, Stems, and Leaves, pp. 511–512*

Question 4: Draw a growing root. Label the primary meristems (protoderm, ground meristem, and procambium), root cap, cortex, stele, and epidermis. Discuss the function of each of these structures.
Textbook Reference: *24.2 Meristems Build Roots, Stems, and Leaves, pp. 513–514*

Question 5: Examine the leaf diagram in Figure 24.12 in the textbook. Which surface of that leaf faces the sun? How do you know? Why are the stomata opposite the palisade layer?
Textbook Reference: *24.2 Meristems Build Roots, Stems, and Leaves, p. 516*

24.3 Domestication Has Altered Plant Form

Humans have artificially selected plants to improve crop yield. This is possible because of the morphological variation found within wild plant species.

Question 6: How have humans domesticated plants? What are the benefits provided by domestication?
Textbook Reference: *24.3 Domestication Has Altered Plant Form, pp. 518–519*

Test Yourself

1. Which of the following is a *not* a component part of a phytomer?
   a. Leaf
   b. Root hair
   c. Axillary buds
   d. Internode
   e. All of the above are components of a phytomer.
   Textbook Reference: *24.1 The Plant Body Is Organized and Constructed in a Distinctive Way p. 507*
2. Suppose you are studying tropical plants in a Costa Rican cloud forest and find a tree that is a eudicot in the forest ecosystem. This plant most likely has a(n) _______ system.
   a. fibrous root
   b. taproot
   c. adventitious root
   d. rhizoid root
   e. terminal root
   Textbook Reference: *24.2 Meristems Build Roots, Stems, and Leaves, p. 514*
3. Some plants, such as sweet peas, will attach themselves to a fence by means of tendrils, which are modifications of
   a. stems.
   b. roots.
   c. branches.
   d. leaves.
   e. seeds.
   Textbook Reference: *24.2 Meristems Build Roots, Stems, and Leaves, p. 517*

4. Plant cells are easily distinguished from animal cells by their
   a. rigid cell walls.
   b. apical meristems.
   c. large vacuoles.
   d. chloroplasts.
   e. All of the above
   Textbook Reference: *24.1 The Plant Body Is Organized and Constructed in a Distinctive Way, p. 508*

5. Plant cells that are photosynthetically active are found in the _______ layer of the leaf and are called _______ cells.
   a. mesophyll; parenchyma
   b epidermis; parenchyma
   c. mesophyll; sclerenchyma
   d. epidermis; sclerenchyma
   e. xylem; mesophyll
   Textbook Reference: *24.2 Meristems Build Roots, Stems, and Leaves, p. 516*

6. Water is conducted in _______ tissue, and carbohydrates and nutrients are transported in _______ tissue.
   a. xylem; phloem
   b. phloem; xylem
   c. parenchyma; phloem
   d. parenchyma; xylem
   e. mesophyll; xylem
   Textbook Reference: *24.1 The Plant Body Is Organized and Constructed in a Distinctive Way, p. 510*

7. Plants are capable of indeterminate growth because of their
   a. regions of nondividing cells.
   b. meristem tissues.
   c. epidermis.
   d. xylem.
   e. All of the above
   Textbook Reference: *24.1 The Plant Body Is Organized and Constructed in a Distinctive Way, p. 507*

8. Which of the following best describes the developmental origin of wood?
   a. Xylem cells enlarge and deposit large amounts of lignin.
   b. Primary meristems increase the amount of xylem deposited.
   c. Lateral meristems contribute to continuous increases in vascular tissue.
   d. Spongy mesophyll cells become cork cambium.
   e. None of the above
   Textbook Reference: *24.2 Meristems Build Roots, Stems, and Leaves, p. 517*

9. Which of the following is *not* a ground tissue cell?
   a. Parenchyma
   b. Collenchyma
   c. Sclerenchyma
   d. Xylem
   e. All of the above are ground tissue cells.
   Textbook Reference: *24.1 The Plant Body Is Organized and Constructed in a Distinctive Way, p. 510*

10. Which of the following best describes the function of the cork cambium?
    a. It lays down a protective cork covering over exposed phloem tissue.
    b. It inhibits the sloughing off of epidermal tissue.
    c. It allows for diameter shrinking in stems and roots.
    d. It supplies the secondary xylem.
    e. All of the above
    Textbook Reference: *24.2 Meristems Build Roots, Stems, and Leaves pp. 517–518*

11. Sieve tube elements have sieve plates where they join other sieve tube elements. Which of the following statements about the sieve plates is true?
    a. Sieve plate pores are enlargements of meristems.
    b. They allow conduction between sieve tube cells through plasmodesmata.
    c. They allow for the joining of cytoplasm between adjacent stomata.
    d. They contain the organelles of the cell.
    e. None of the above
    Textbook Reference: *24.2 Meristems Build Roots, Stems, and Leaves, pp. 510–511*

12. Plants regulate gas exchange and water loss via
    a. the cuticle.
    b. xylem.
    c. coated pits.
    d. sieve plates.
    e. stomata.
    Textbook Reference: *24.2 Meristems Build Roots, Stems, and Leaves, p. 517*

13. The protoderm becomes the _______ tissue system.
    a. dermal
    b. ground
    c. vascular
    d. All of the above
    e. None of the above
    Textbook Reference: *24.1 The Plant Body Is Organized and Constructed in a Distinctive Way, p. 511*

14. Primary growth occurs at the
    a. lateral meristems.
    b. fruit.
    c. quiescent center.
    d. apical meristems.
    e. wood.
    Textbook Reference: *24.1 The Plant Body Is Organized and Constructed in a Distinctive Way, p. 511*

15. Vascular bundles are composed of _______ and _______.
    a. root hairs; xylem
    b. cork; phloem
    c. xylem; phloem
    d. wood; cork
    e. mesophyll; xylem
    Textbook Reference: *24.1 The Plant Body Is Organized and Constructed in a Distinctive Way, p. 515*

Answers

Key Concept Review

1.

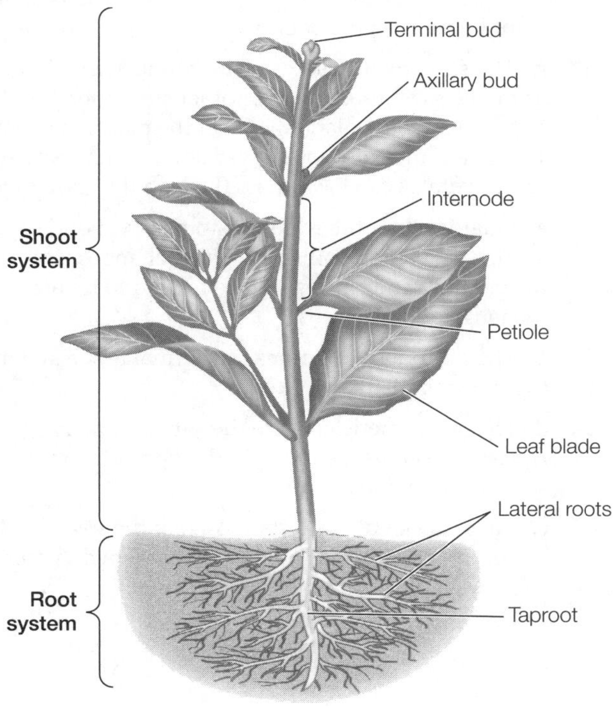

2. The three types of tissue are dermal, vascular, and ground. Dermal tissue acts as a covering for the plant and its cells are typically small. It provides protection and also functions in gas exchange and nutrient and water uptake. Vascular tissue, which is composed of cells that are either dead (xylem) or alive (phloem), is involved in transport of water and nutrients within the plant. Certain types of ground tissue are involved in photosynthesis, and others provide support for the plant body. The cells you found are probably scleren-chyma cells, which have thick cell walls that provide support.
3. Apical meristems are responsible for elongation of the plant body. Lateral meristems are responsible for an increase in girth. Lateral meristems are found only in woody eudicots and are responsible for creating wood. All plants have apical meristems.
4.

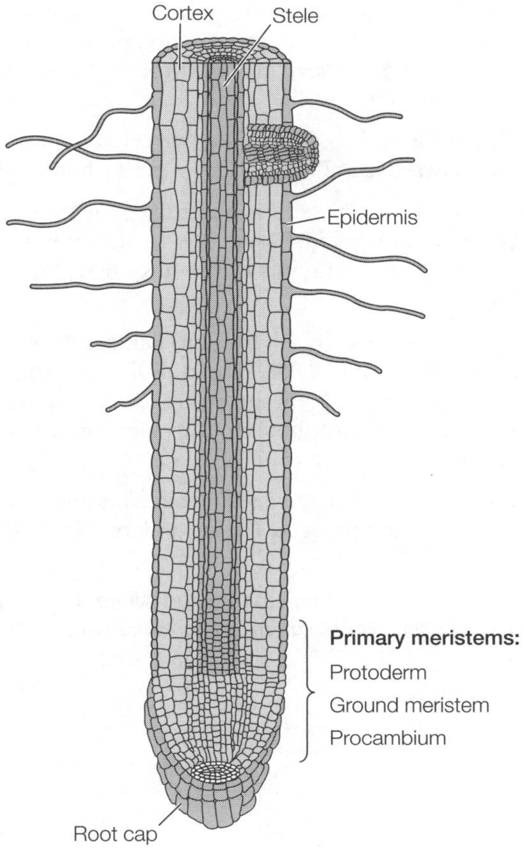

The protoderm gives rise to the epidermis for protection. The ground meristem gives rise to the cortex for storage. The procambium gives rise to the stele for transport. The root cap protects the meristem as it pushes through the soil.

5. The "top" of the diagram in the textbook would be the surface that faces the sun. This is the surface where photosynthetic cells, which need maximum sun exposure, are located. The stomata are on the opposite cooler side, which reduces water loss due to evaporation during photosynthesis.
6. Humans have taken advantage of the large variation in plant shape and size within a species. They have selectively used the seeds from plants with desirable characteristics, ensuring these characteristics remain in the crop. This has allowed humans to produce more productive crops.

Test Yourself

1. **b.** The phytomer is the repeating unit of a plant from node to node that includes the leaves, an internode, and one or more axillary buds.
2. **b.** A taproot would be necessary to anchor a plant of that size. Most eudicots have taproots.
3. **d.** Tendrils are modified leaves.

4. **e.** Plant cell shape is maintained by the cell walls. No animal cells have cell walls. Plants have vacuoles, chloroplasts, and apical meristems. Though some of these characteristics are seen in some protists, no animal cells have them.
5. **a.** Palisade and spongy mesophyll cells are photosynthetically active and derived from parenchyma cells.
6. **a.** Xylem tissue transports water from the roots throughout the plant. Phloem tissue transports carbohydrates and nutrients from source tissue to sink tissue.
7. **b.** The apical and (when present) lateral meristems are regions of continually dividing cells that can contribute to the growth of a plant throughout its life. A modular body plan also contributes to the ability of plants to grow continually.
8. **c.** Lateral meristems are responsible for the growth of new xylem and phloem. The secondary xylem gives rise to wood.
9. **d.** Parenchyma, collenchyma, and sclerenchyma are all ground tissue cells. Xylem is a vascular tissue cell.
10. **a.** The cork cambium is a meristematic region outside the secondary phloem. As girth increases and splits the epidermis, causing loss of those protective layers, the cork cambium produces new cells to cover the expanding vascular tissue. It can also produce cells inside the plant known as phelloderm.
11. **b.** The sieve plates allow for the conduction of sap from one sieve tube cell to another sieve tube cell. The pores are due to enlargements of the plasmodesmata. These cells may also lose nuclei and organelles so that carbohydrates can easily pass through the sieve tubes.
12. **e.** Guard cells at the edges of stomata respond to changes in osmotic pressure. These changes lead to the opening and closing of the stomata to regulate gas exchange and water loss.
13. **a.** The protoderm becomes the dermal tissue system of the growing plant.
14. **d.** The apical meristems are the site of primary growth. Apical meristems are found at the tips of the roots, in stems, and in buds.
15. **c.** Vascular bundles are composed of the vascular tissue of the plant, which includes the xylem and phloem.

Plant Nutrition and Transport

25

The Big Picture

- Plants require specific macro- and micronutrients. Deficiencies in any of these challenge the health of the plant. Essential nutrients must be available, and no substitutions will sustain the plant. Nutrients are procured from the soil solution that bathes the roots of a plant. The availability of nutrients depends on the quantity, solubility, and structure of soil. Many agricultural practices deplete soils of nutrients, and these must be replenished through fertilization.
- Some plants interact with fungi and bacteria that help them obtain needed nutrients. Some plants acquire the nitrogen essential to their growth by means of nitrogen-fixing bacteria in their root nodules, which use an enzyme called nitrogenase to convert nitrogen into ammonia. A small number of plants do not photosynthesize. These heterotrophic plants are often parasites of other plants and acquire nutrients solely through their hosts.
- In terrestrial plants, water must be acquired from the soil, transported through the plant, and used in the leaves for photosynthesis. At the same time, the nutritional products of photosynthesis must be transported throughout the plant to nonphotosynthetic tissues. This two-way transport is achieved through specialized cells that make up the vascular tissue of the plant. Water with dissolved mineral nutrients is absorbed through the roots and transported to cells in the xylem of the plant, where it is pulled up the stem to the leaves via the transpiration–cohesion–tension mechanism. Sugars and solutes are moved out of the leaves and to the rest of the plant through cells in the phloem.
- Water uptake is regulated by osmotic and water potentials in the root cells, and the rate of transport is controlled by the rate of evaporation at the leaf surface. Guard cell activity in the leaves regulates the opening and closing of stomata to match water availability, light, and drying conditions. Sucrose movement is regulated by active transport and facilitated diffusion in the phloem tissue. The rate of sucrose transport depends on the rates of loading and unloading at source and sink tissues.

Study Strategies

- The interactions between nitrogen-fixing bacteria and plants can be confusing. Remember that it is a mutualistic relationship and that the bacteria have the enzymes necessary to fix atmospheric nitrogen.
- It is easy to fail to look at the anatomy of the structures that are used in transport. Each structure is uniquely suited to its function. Think in terms of form coupled with function to understand this process.
- Water and sucrose transport are pathways that can be understood by visually tracing a molecule of water or sucrose through the plant. Use the figures in the textbook to follow these pathways.
- In order to understand water transport, it is important to understand osmosis and the properties of water molecules. Review these concepts from earlier chapters to make this easier.
- The basis of nutritional transport in plants is cell-to-cell transport. Review the sections in the textbook on diffusion, osmosis, and active transport.
- Be sure you understand the consequences of nutrient shortages in plants.
- Go to yourBioPortal.com to review the following tutorials and activity:

  Animated Tutorial 25.1 Nitrogen and Iron Deficiencies

  Animated Tutorial 25.2 Xylem Transport

  Animated Tutorial 25.3 The Pressure Flow Model

  Interactive Tutorial 25.1 Water Uptake in Plants

  Web Activity 25.1 Apoplast and Symplast of the Root

  Working with Data 25.1 Nickel Is an Essential Element for Plant Growth

Key Concept Review

25.1 Plants Acquire Mineral Nutrients from the Soil

Nutrients can be defined by their deficiency

Experiments using hydroponics have identified essential elements

Soil provides nutrients for plants

Ion exchange makes nutrients available to plants

Fertilizers can be used to add nutrients to soil

The basic nutrient requirements for all living things are carbon, hydrogen, oxygen, and nitrogen. These elements are the fundamental building blocks of all macromolecules. Plants are autotrophs and incorporate carbon through photosynthesis. Oxygen and hydrogen enter plants as water. Nitrogen's entry into plants is dependent on nitrogen-fixing bacteria in the soil. Mineral nutrients are essential to life. Sulfur, phosphorus, magnesium, and iron are all essential components of many macromolecules. Mineral nutrients enter biological organisms through soil solutions, which plants take up through their roots. Every plant requires specific essential nutrients for growth and development, and these nutrients cannot be replaced by another element and must be obtained directly for the day-to-day functioning of the plant. A deficiency in any essential nutrient leads to an unhealthy plant. Macronutrients are essential elements required at a rate of 1 g per 1 kg of dry plant matter, and micronutrients are essential elements required at a rate of 100 mg per 1 kg of dry plant matter. Though many nutrient deficiencies ultimately lead to plant death, specific symptoms of deficiency are evident in a plant before it dies. Deficiency can be corrected by the addition of fertilizers to supplement the soils. Scientists determine which elements are essential by growing plants hydroponically, which allows for manipulation of the nutrients that are available to the plant.

Soil provides an anchorage for plants, mineral nutrients, and water required for growth and survival, and oxygen for the roots. The living portion of soil contains roots, protists, bacteria, fungi, and many small animals. All soils have a soil profile consisting of two or more horizons (layers). Water-soluble nutrients are leached to deeper horizons through rainfall. Three major horizons (also called zones), termed A, B, and C, can be identified in soils.

The A horizon is topsoil that is organically rich and the most agriculturally important layer. A loam is a topsoil with an optimal mixture of sand, silt, and clay; it has high nutrient content, plentiful water, and adequate air spaces. Soils with too much sand typically do not hold nutrients or water well, and clays are too dense for the trapping of air. The B horizon is the subsoil that holds many leached nutrients. The C horizon is parent rock that gives rise to soil.

Soils form from mechanical and chemical weathering that breaks down rocks. Mechanical weathering is caused by freeze and thaw cycles, rain, and drying. Chemical weathering is caused by oxidation, water, and acids. Chemical weathering is very important in clay formation. Humus is formed from the breakdown of dead leaves and other organic matter.

Mineral nutrients are tied to clay particles in the soil. Because many nutrients are positively charged cations, clays with a negative charge can hold these nutrients and make them available to plants. Roots release protons into the soil that bind to clay particles, and the cation minerals are released. The CO_2 released from the roots can also form bicarbonate and free protons, which bind with clay. These processes are called ion exchange. Nutrients that are negatively charged and therefore do not participate in ion exchange are rapidly leached from soil. Thus, nitrate and sulfate are often not available to plants. In the past, nutrient content of soil was replenished by shifting agriculture and leaving the land unused for long time periods. Organic fertilizers such as compost or animal manure can be used to add nutrients. Inorganic fertilizers are used to quickly supply plants with nutrients

Agricultural fertilizers add nitrogen, phosphorus, and potassium to soils and are rated by their "N-P-K" percentages. A 10-10-10 fertilizer, for example, contains 10 percent nitrogen, 10 percent phosphate, and 10 percent potash (potassium source).

Question 1. Differentiate between micro- and macronutrients. Where are most of these nutrients acquired?
Textbook Reference: *25.1 Plants Acquire Mineral Nutrients from the Soil, p. 522*

Question 2. Explain how scientists determined which plant nutrients are the essential nutrients.
Textbook Reference: *25.1 Plants Acquire Mineral Nutrients from the Soil, pp. 522–523*

Question 3. Draw the different layers of the soil and describe the importance of each to a plants survival.
Textbook Reference: *25.1 Plants Acquire Mineral Nutrients from the Soil, pp. 523–524*

25.2 Soil Organisms Contribute to Plant Nutrition

Plants send signals for colonization

Mycorrhizae expand the root system

Rhizobia capture nitrogen from the air and make it available to plant cells

Some plants obtain nutrients directly from other organisms

Soils are full of bacteria and fungal hyphae, some of which interact with plants to help them obtain nutrients. Plants send out signals to attract these beneficial organisms. Fungi form associations with roots to form arbuscular mycorrhizae. Formation of arbuscular mycorrhizae is stimulated by the release of strigolactones by the root. A prepenetration apparatus helps guide the fungi as they grow into the root. Arbuscules in the cortical cells of the roots are the site of nutrient exchange. The mycorrhizae increase the surface area of the root, allowing for greater uptake of nutrients, primarily phosphorus. Bacteria known as rhizobia can form symbiotic relationships with legumes and live in root nodules. To establish the symbiosis, the plant releases flavonoids to attract the bacteria. In response to the flavonoids, the bacteria turn on the production of Nod factors, which cause formation of the nodule by the plant. Once housed in the nodule, the bacteria form swollen bacteroids capable of nitrogen fixation.

Nitrogen gas is readily available in the atmosphere, but plants are unable to break the triple bonds between the two nitrogens. A few bacteria species can fix nitrogen gas into biologically usable ammonia through nitrogen fixation using the enzyme nitrogenase to catalyze the reaction. Nitrogenase is inhibited by oxygen and therefore is active only under anaerobic conditions. The nitrogen-fixing bacteria are typically found in root nodules that maintain very low oxygen levels. The protein leghemoglobin binds with oxygen to keep oxygen levels low.

Some plants that live in nitrogen- or phosphorus-deficient soils are carnivorous. Carnivorous plants acquire nitrogen from the proteins of trapped decaying animals. Examples of carnivorous plants are Venus flytraps and pitcher plants. Some plants have lost the ability to photosynthesize and must acquire their nutrients from other sources. Some are parasitic and acquire some or all of their nutrients from a host plant at the host plant's expense. Hemiparasites obtain water and mineral nutrients from their host, but are able to photosynthesize. Holoparasites are unable to photosynthesize.

Question 4. Describe how plants and bacteria interact to form nitrogen-fixing root nodules. Why do many farmers plant crops such as alfalfa and soybeans without harvesting them?
Textbook Reference: *25.2 Soil Organisms Contribute to Plant Nutrition, pp. 526–527*

Question 5. Most carnivorous plants are found in boggy, wet, acidic environments. What is the effect of such an environment on nutrient availability?
Textbook Reference: *25.2 Soil Organisms Contribute to Plant Nutrition, p. 528*

25.3 Water and Solutes Are Transported in the Xylem by Transpiration–Cohesion–Tension

Differences in water potential govern the direction of water movement

Water and ions move across the root cell plasma membrane

Water and ions pass to the xylem by way of the apoplast and symplast

Water moves through the xylem by the transpiration–cohesion–tension mechanism

Stomata control water loss and gas exchange

The movement of water across a semipermeable membrane is a special type of diffusion known as osmosis (see Figure 25.8). The overall tendency of a solution to take up water across a membrane (called its water potential) is the sum of the negative solute potential and the positive pressure potential. Solute potential measures the influence of solutes on water movement, while pressure potential is the pressure inside the rigid cell (turgor pressure). All three parameters can be measured in megapascals (MPa). Water always moves to a region of more negative water potential. The structure of plants is maintained by osmotic phenomena. If a plant loses turgor pressure by a decrease in pressure potential, it wilts. Movement of water from cell to cell depends on the gradient of water potential. Specialized membrane channel proteins in plant cells called aquaporins can increase the rate of water movement by allowing water to cross the plasma membrane without interacting with the hydrophobic bilayer that slows water flow. Though aquaporins can increase the rate of osmosis, they cannot influence the direction of flow.

Mineral ion uptake from the soil solution requires active transport via proteins. When mineral concentrations are greater in the soil solution than in the plant, they are taken up by facilitated diffusion. If minerals are in smaller concentrations outside the plant than inside the plant, or if they must be moved against an electrochemical gradient, then the plant must rely on active transport. Plants rely on a proton pump for active transport of minerals into cells. Plants actively pump protons out of cells, causing the area outside the cell to become more positive. This assists facilitated diffusion of positive ions through protein channels. It also drives the movement of negatively charged ions into the cell by active transport (see Figure 25.10). The result of this pumping action is that the internal environment of the plant cell becomes highly negative compared to its environment. The proton gradient that develops across the membrane can also facilitate secondary active transport of ions such as Cl^-.

Minerals and water follow two paths for reaching the vascular tissue: the rapid apoplast or the slower symplast. The apoplast is formed by cell walls and intercellular spaces. Water and minerals may move unregulated through this space without ever having to cross a membrane. The symplast is the living portion of the plant and is enclosed in plasma membranes. Movement of water and minerals in the symplast is highly regulated. Water and minerals can travel through the apoplast as far as the endodermis. At the endodermis, water and minerals are stopped by the Casparian strips, which are waxy structures surrounding the endodermal cells. Because of this, water can reach the stele only via the symplast. The transport proteins in the endodermal cells determine which minerals enter the stele. Once past the endodermal barrier, water and minerals can again leave the symplast and move back to the apoplast with the aid of parenchyma cells. Ultimately, water and minerals from the soil solution end up in xylem cells and are referred to as xylem sap. Before the end of the nineteenth century, the movement of fluids in plants was thought to be controlled by upward pressure and capillary action. But a simple experiment in 1893, in which a sawed-off tree trunk was immersed in poison, showed that the poison continued to progress even as cells were killed along the way. Only when the poison reached the leaves did movement stop, showing that it was the leaves that were critical to transport.

The mechanism that pulls water up from the roots through the plant is known as the transpiration–cohesion–tension mechanism. This is a passive process requiring no energy input by the plant. Minerals are drawn passively along with the water column. In the process of transpiration, water evaporates from mesophyll cells of leaves, creating tension on the water associated with the mesophyll cell wall. Tension is a pulling force that drives water movement through the plant. Water molecules are cohesive, meaning that their hydrogen bonds stick together strongly enough to resist the tension, so they pull their neighboring water molecules toward the mesophyll cell walls in response to transpiration. Tension in the mesophyll in turn generates tension on the cohesive water molecules in the nearby xylem water column, which pulls water up from the roots and through the apoplast of the leaves. Transpiration also assists with temperature regulation through evaporative cooling of the leaves.

Leaf surfaces are covered with a waxy cuticle to prevent excessive water loss. However, the leaf must take up CO_2 for photosynthesis. Stomata with guard cells are pores that regulate gas exchange and water loss from a leaf. In response to osmotic differences, guard cells shrink and swell to open and close the stomata. Because closed stomata prevent water loss but also exclude CO_2, most plants open their stomata when light intensity is sufficient to maintain photosynthesis. Guard cells are also regulated by water potential. If the water potential in mesophyll cells is low, the cells release the hormone abscisic acid, which causes the guard cells to close. Specific wavelengths of light and low CO_2 levels stimulate a proton pump that helps regulate guard cell activity. Guard cells open when potassium ions diffuse into the cell as a result of the electrical gradient set up by the proton pump. High potassium levels cause water to move in by osmosis. Pressure potential builds in the guard cells, and they are pulled apart to reveal the stoma. For guard cells to shut, the proton pump ceases its activity, and potassium ions move back across the membrane. Water follows, and the cells go limp and seal off the stoma (see Figure 25.13).

Question 6. Create a flow chart of the path taken by a water molecule as it moves from the soil solution to the stele of a plant. Identify where the molecule is traveling through the apoplast and where it is traveling through the symplast.
Textbook Reference: *25.3 Water and Solutes Are Transported in the Xylem by Transpiration–Cohesion–Tension, p. 531*

Question 7. Under what conditions does transpiration occur most rapidly? What effect does increased transpiration have on water flow in a plant? What happens if adequate water for the plant is not available?
Textbook Reference: *25.3 Water and Solutes Are Transported in the Xylem by Transpiration–Cohesion–Tension, pp. 532–534*

Question 8. Explain how transpiration, cohesion, and tension work together to move water in a large plant.
Textbook Reference: *25.3 Water and Solutes Are Transported in the Xylem by Transpiration–Cohesion–Tension, pp. 532–534*

25.4 Solutes Are Transported in the Phloem by Pressure Flow

Sucrose and other solutes are carried in the phloem

The pressure flow model describes the movement of fluid in the phloem

Phloem moves materials from sources to sinks by translocation. Sources are organs that produce more sugars than are used by metabolism, storage, and growth. Sinks are organs that do not make enough sugar for their own growth or storage needs. Sucrose, amino acids, minerals, and other substances are translocated between sources and sinks in the phloem. Translocation proceeds in both directions along the stem. Translocation stops if phloem tissue is killed. Sieve tube elements are the cells of the phloem. They are connected end-to-end by sieve plates containing plasmodesmata, allowing for movement between cells. Because the sieve tube elements do not contain organelles, they rely on companion cells to provide them with all they need to survive.

The pressure flow model explains how materials move through the phloem. Once in the sieve tubes, sieve tube sap moves via bulk flow, which requires no energy input by the plant. Energy is required for the loading of the sieve tubes at the sources and the unloading of solutes when the sink is reached. Sucrose is actively transported into sieve tubes at the sources. This causes water to move into sieve tubes by osmosis, thereby increasing the pressure potential at the source end and pushing the contents toward the sink end.

Question 9. Differentiate between source and sink tissues. What happens relative to phloem in each?
Textbook Reference: *25.4 Solutes Are Transported in the Phloem by Pressure Flow, p. 536*

Question 10. Describe the pressure flow model of phloem transport. What would happen if the loading of sucrose were to be inhibited?

Textbook Reference: *25.4 Solutes Are Transported in the Phloem by Pressure Flow, p. 536*

Test Yourself

1. Which of the following nutrients is *not* considered essential for plant growth?
   a. Cadmium
   b. Nitrogen
   c. Manganese
   d. Potassium
   e. All of these nutrients are essential for plant growth.
   Textbook Reference: *25.1 Plants Acquire Mineral Nutrients from the Soil, p. 522*
2. Compared to micronutrients, macronutrients are
   a. larger.
   b. needed in greater quantities.
   c. more essential.
   d. of equal importance.
   e. less essential.
   Textbook Reference: *25.1 Plants Acquire Mineral Nutrients from the Soil, p. 522*
3. Nitrogen and potassium are acquired from
   a. the soil solution.
   b. heterotrophs.
   c. air.
   d. micronutrients.
   e. All of the above
   Textbook Reference: *25.1 Plants Acquire Mineral Nutrients from the Soil, p. 522*
4. Years of cotton farming in the South have stripped away much of the A horizon of the soils. The effects on agriculture have been _______ because _______.
   a. significant; the A horizon contains the most available nutrients
   b. negligible; the B horizon contains significantly more available nutrients
   c. negligible; the C horizon is most conducive to root growth
   d. significant; stripping of the A horizon leaches the C horizon of its nutrients
   e. All of the above
   Textbook Reference: *25.1 Plants Acquire Mineral Nutrients from the Soil, p. 523*
5. Clay particles in soils are important for
   a. holding the soil together.
   b. ion exchange.
   c. holding water.
   d. All of the above
   e. None of the above
   Textbook Reference: *25.1 Plants Acquire Mineral Nutrients from the Soil, p. 524*
6. Most clays form from the _______ of rock.
   a. mechanical weathering
   b. chemical weathering
   c. heaving
   d. grinding
   e. All of the above
   Textbook Reference: *25.1 Plants Acquire Mineral Nutrients from the Soil, p. 524*
7. The label "10-20-10" on a package of commercial fertilizer refers to the _______ the fertilizer.
   a. percentages of nitrogen, phosphate, and potassium in
   b. percentages of nitrogen, carbon, and oxygen in
   c. percentages of phosphate, iron, and potassium in
   d. rate at which nitrogen is released from
   e. ratio of organic to inorganic matter in
   Textbook Reference: *25.1 Plants Acquire Mineral Nutrients from the Soil, p. 525*
8. The relationship between rhizobium bacteria and the roots of legumes can best be described as
   a. parasitic.
   b. one-sided.
   c. mutualistic.
   d. carnivorous.
   e. detrimental.
   Textbook Reference: *25.2 Soil Organisms Contribute to Plant Nutrition, p. 527*
9. Nitrogen gas is reduced to ammonia by which of the following enzymes or processes?
   a. Rhizobium
   b. Nitrogenase
   c. Nitrification
   d. Denitrification
   e. Rhizobenase
   Textbook Reference: *25.2 Soil Organisms Contribute to Plant Nutrition, p. 527*
10. Plants are able to take up and use nitrogen in the form of
    a. ammonia.
    b. nitrogen gas.
    c. liquid nitrogen.
    d. nitrous oxide.
    e. All of the above
    Textbook Reference: *25.2 Soil Organisms Contribute to Plant Nutrition, p. 527*
11. Carnivorous plants are often found in acidic and nutrient-poor environments. The main selective pressure for carnivory is
    a. lack of nitrogen and phosphorus sources.
    b. lack of iron and calcium sources.
    c. incomplete ion exchange.
    d. lack of water sources.
    e. All of the above

Textbook Reference: *25.2 Soil Organisms Contribute to Plant Nutrition, p. 528*

12. Nitrate and sulfate tend to leach from the soil because
 a. they bind with ions such as K^+ and Mg^{2+}.
 b. the H^+ ions released by the roots push them out.
 c. they are unable to bind with the negatively charged clay particles.
 d. they bind with the positively charged clay particles.
 e. All of the above

 Textbook Reference: *25.1 Plants Acquire Mineral Nutrients from the Soil, p. 524*

13. A plant lowers the pH of soil by means of
 a. ion exchange.
 b. leaching.
 c. Na^+ pumping.
 d. proton pumping.
 e. All of the above

 Textbook Reference: *25.2 Soil Organisms Contribute to Plant Nutrition, p. 524*

14. The role of leghemoglobin is to maintain _______ levels in the root nodule.
 a. high O_2
 b. high CO_2
 c. low O_2
 d. low CO_2
 e. high N_2

 Textbook Reference: *25.2 Soil Organisms Contribute to Plant Nutrition, p. 528*

15. The function of the Casparian strips is to
 a. divert water and minerals through the membranes of endodermal cells.
 b. prevent water and minerals from entering the stele through the apoplast.
 c. provide regulation for water and mineral movement in the plant.
 d. All of the above
 e. None of the above

 Textbook Reference: *25.3 Water and Solutes Are Transported in the Xylem by Transpiration–Cohesion–Tension, pp. 531–532*

16. The primary difference between the apoplast and the symplast is that
 a. the apoplast consists of nonliving spaces and cell walls, whereas the symplast consists of living cells.
 b. apoplast movement is tightly regulated and symplast movement is not.
 c. the symplast consists of nonliving spaces and cell walls, whereas the apoplast consists of living cells.
 d. apoplast movement is slow and symplast movement is fast.
 e. the apoplast transports only ions and the symplast transports only water.

 Textbook Reference: *25.3 Water and Solutes Are Transported in the Xylem by Transpiration–Cohesion–Tension, p. 531*

17. Which of the following statements about water transport is true?
 a. Root pressure is sufficient to drive xylem sap movement.
 b. Bulk flow is not a mechanism by which water and minerals are transported.
 c. The cohesive nature of water is central to water movement in a plant.
 d. Water transport is an active process.
 e. None of the above

 Textbook Reference: *25.3 Water and Solutes Are Transported in the Xylem by Transpiration-Cohesion-Tension, p. 529*

18. Tension in the xylem is a result of
 a. transpiration at the leaf surface.
 b. the cohesive nature of water.
 c. the narrowness of the xylem tube.
 d. the surface area of the phloem.
 e. All of the above

 Textbook Reference: *25.3 Water and Solutes Are Transported in the Xylem by Transpiration–Cohesion–Tension, p. 532*

19. The fact that water transport continues as long as leaves are alive and active indicates that
 a. leaves pump water.
 b. leaves are necessary for transport of water.
 c. roots are active.
 d. water is not needed for leaves to remain alive.
 e. sieve tube elements are inactive.

 Textbook Reference: *25.3 Water and Solutes Are Transported in the Xylem by Transpiration–Cohesion–Tension, p. 532*

20. Which of the following statements regarding transport in phloem is true?
 a. It always moves in the direction of leaves to roots.
 b. It proceeds from source tissue to sink tissue.
 c. It requires no energy inputs from the plant.
 d. It is the same process as transport in xylem.
 e. None of the above

 Textbook Reference: *25.4 Solutes Are Transported in the Phloem by Pressure Flow, p. 536*

21. If the pressure potential of a plant's cells is 0.16 megapascals (MPa) and the solute potential is –0.24 MPa, then the water potential is _______ Mpa.
 a. 0.04
 b. 0.08
 c. –0.08
 d. –0.24
 e. –0.04

 Textbook Reference: *25.3 Water and Solutes Are Transported in the Xylem by Transpiration–Cohesion–Tension, p. 529*

22. Which of the following represents the correct ordering of the water potential of these root cells or regions, from least negative to most negative?

a. Xylem, cortex apoplast, stele apoplast, soil next to root
b. Soil next to root, xylem, stele apoplast, cortex apoplast
c. Xylem, stele apoplast, cortex apoplast, soil next to root
d. Stele apoplast, cortex apoplast, xylem, soil next to root
e. Soil next to root, cortex apoplast, stele apoplast, xylem

Textbook Reference: *25.3 Water and Solutes Are Transported in the Xylem by Transpiration–Cohesion–Tension, p. 531*

23. Which of the following statements about xylem transport and phloem transport is true?
a. Both are passive processes that do not require energy from the plant.
b. Both rely on living cells only.
c. Both rely on a water potential gradient.
d. The direction of flow can reverse in both.
e. The driving force for both is in the leaves.

Textbook Reference: *25.4 Solutes Are Transported in the Phloem by Pressure Flow, p. 536; 25.3 Water and Solutes Are Transported in the Xylem by Transpiration–Cohesion–Tension, p. 532*

24. Stomatal opening and closing are regulated by
a. oxygen.
b. sucrose.
c. carbon dioxide concentrations.
d. All of the above
e. None of the above

Textbook Reference: *25.3 Water and Solutes Are Transported in the Xylem by Transpiration–Cohesion–Tension, p. 534*

25. The opening and closing of the stomata are accomplished by the
a. sieve tube.
b. guard cells.
c. process of translocation.
d. aquaporins.
e. xylem.

Textbook Reference: *25.3 Water and Solutes Are Transported in the Xylem by Transpiration–Cohesion–Tension, p. 534*

26. Regulators of stomatal opening and closing work by activating the
a. proton pump in guard cells.
b. proton pump in stomata.
c. sodium–potassium pump in guard cells.
d. sodium–potassium pump in stomata.
e. All of the above

Textbook Reference: *25.3 Water and Solutes Are Transported in the Xylem by Transpiration–Cohesion–Tension, p. 534*

27. Mineral ions enter the cell due to the force of an electrochemical gradient set up by the pumping of _______ out of the cells.
a. K^+
b. Ca^{2+}
c. Na^+
d. H^+
e. Cl^-

Textbook Reference: *25.3 Water and Solutes Are Transported in the Xylem by Transpiration–Cohesion–Tension, p. 530*

Answers

Key Concept Review

1. Macronutrients are needed at a rate of 1 g/1 kg of dry plant tissue. Micronutrients are needed at a rate of 100 mg/1 kg of dry plant tissue. Most of these nutrients are in soil solution and are taken up as water is drawn into roots.

2. In a series of experiments, plants were grown in cultures that lacked specific nutrients. If a plant could not complete its life cycle, the missing nutrient was said to be essential. These experiments were well-controlled hydroponic studies because nutrient content can easily be manipulated in water.

3.

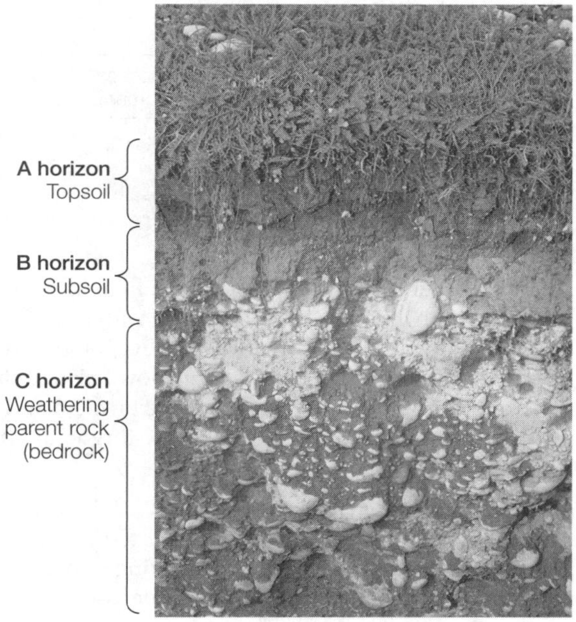

Soil is made up of three different layers: The top layer (topsoil), called the A horizon, supplies the plant's mineral nutrient needs and contains most of the soil's living and dead organic matter. The B horizon is the subsoil, which accumulates materials from the topsoil above and the parent rock below. The C horizon is the parent rock from which the soil arises.

4. See Figure 25.5. By rotating crops (and especially by rotating and plowing under legume crops) farmers can organically add nitrogen to depleted soils. This results in significantly larger yields of the harvested crops.
5. Acidic environments limit decomposition and thus nitrogen sources. They also limit ion exchange. Both conditions result in reduced nutrient availability that can be offset by carnivory.
6.

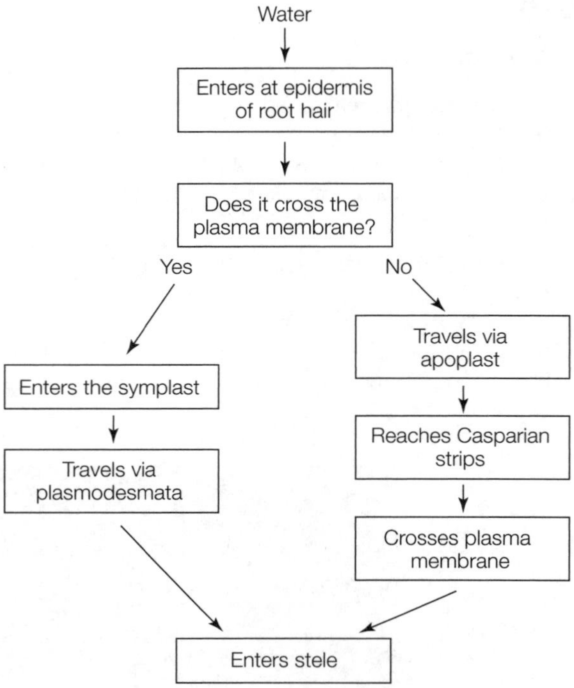

7. Transpiration occurs most rapidly in high light conditions when stomata are open, along with high wind conditions and low humidity when evaporation is greatest. This results in faster bulk flow through the xylem and increased water demands by the plant. If water is not available, plant cells lose turgor and the plant wilts.
8. The transpiration–cohesion–tension mechanism pulls water from the roots up through the plant. Water evaporates from mesophyll cells during transpiration. This puts tension on the film of water associated with the mesophyll cell wall. The tension at the mesophyll cell draws water from the xylem of the nearest vein. This creates tension in the entire xylem column, and the column is drawn upward from the roots.
9. Source tissues produce sugars in excess of what can be used and stored. Phloem loading occurs in source tissues and creates a pressure potential that results in bulk flow of sieve tube sap toward sink tissues. Sink tissues produce fewer sugars than can be stored or used and unload phloem through active transport.
10. The difference in solute concentration between sources and sinks creates a pressure potential along sieve tubes, resulting in bulk flow. For this to occur, sugars must be loaded at the source tissue and unloaded at the sink tissue through active transport, and the sieve plates must remain open and unclogged along the phloem column. If the loading of sugars is inhibited, then the plant will not be able to move fluid in the phloem and the plant will die.

Test Yourself

1. **a.** Cadmium is not one of the 14 essential micro- and macronutrients.
2. **b.** The main difference between micronutrients and macronutrients is in the quantity of each needed by a plant for survival.
3. **a.** Nitrogen and mineral nutrients are acquired in soil solution (dissolved in water).
4. **a.** Topsoil, or the A horizon, is most conducive to root growth. Ample nutrients are available in the topsoil, as are air spaces and water for ease of root growth.
5. **d.** Clay particles are critical for ion exchange. They are also important for retaining water and for the integrity of the soil.
6. **b.** Chemical weathering leads to clay formation.
7. **a.** The fertilizer is 10 percent nitrogen, 20 percent phosphorus, and 10 percent potash (a potassium source).
8. **c.** The relationship is mutualistic, in that both the plant and the bacteria benefit from the association.
9. **b.** Nitrogenase catalyzes the reduction of nitrogen gas to ammonia. This is an energy-expensive process.
10. **a.** Plants can take up and use nitrogen that is in the form of ammonia (or nitrate).
11. **a.** Carnivory supplements insufficient nitrogen and phosphorus availability.
12. **c.** Nitrate and sulfate are negatively charged anions. Without any positively charged molecules in the soil with which they can interact, they leach from the soil.
13. **d.** Plants decrease the pH in the surrounding soil by pumping protons out of the roots.
14. **c.** Leghemoglobin is a protein produced by the root nodule of plants to help maintain low levels of O_2 so that the nitrogen-fixing bacteria are in an anaerobic environment.
15. **d.** Not all minerals that enter the apoplast of a plant's root are beneficial to the plant. The Casparian strips prevent water and minerals from reaching the stele through the apoplast, diverting them instead through the plasma membranes of the endodermal cells. Channel proteins in these plasma membranes determine which minerals can enter the symplast, and from there, the stele. The Casparian strips thus contribute to the regulation of water and mineral movement in the plant.

16. **a.** The intercellular spaces and cell walls of the plant constitute the apoplast.
17. **c.** Water movement depends on the cohesive nature of water and its capacity to withstand the tension placed on the water column by transpiration.
18. **a.** Transpiration causes tension.
19. **b.** Leaves are necessary for transpiration to take place.
20. **b.** Transport in phloem does not always go from leaf to root, but it always proceeds from source tissue to sink tissue. The plant must contribute energy to create the water pressure gradient by pumping solutes into the phloem at the source and out of the phloem at the sink.
21. **c.** Water potential is equal to pressure potential plus solute potential.
22. **c.** The xylem would be the most negative, followed by the stele, then the cortex, then the area outside the root.
23. **c.** Both xylem transport and phloem transport depend on water potential.
24. **c.** Carbon dioxide levels are one of the regulators of stomatal opening and closing.
25. **b.** Guard cells are specialized epidermal cells that regulate the opening and closing of the stomata by covering the stomata opening.
26. **a.** Stomatal regulators work by activating and deactivating the proton pump in guard cells.
27. **d.** Cells pump H^+ ions out into the soil with the help of proton pumps.

Plant Growth and Development

26

The Big Picture

- Plant growth is controlled by interactions among a plant's environment, hormones, and genetic makeup. Changes in plant growth are dependent on the information transmitted by hormones to hormone receptors, the subsequent turning on of signal transduction pathways, and the eventual alteration of gene expression. Receptors are often highly specific and respond to select hormones. Hormones are produced in a specific region of the plant body and translocated throughout the plant; therefore, their effects are often concentration dependent. All growth in a plant is a result of changes in cell division, cell expansion, and cell differentiation.
- Seed germination, growth of the vegetative structures, reproduction, and senescence all proceed in defined patterns. Seed dormancy is broken and development begins when the seed coat is abraded, inhibitory chemicals are diluted, and the seed imbibes water. This begins a series of events that mobilize nutrients and induce growth. Once a seedling emerges from the soil, light begins to influence subsequent development under the control of hormones.
- Hormones exist in several classes, each with its own effects on growth and development. Some of these effects are antagonistic, and therefore control is maintained via relative concentration of several hormones. Hormones influence every step of development, from the breaking of seed dormancy to senescence.
- Light regulates plant processes through photoreceptors. Photoreceptors respond to very specific wavelengths and induce cascades that lead to changes in a plant.

Study Strategies

- Understanding how phytochrome shifts from the red to far-red forms is a difficult concept to understand—phytochromes have puzzled researchers for many years.
- There is tendency when studying hormones simply to memorize functions. You need to understand the effects in a plant of different ratios of hormones, not what one individual hormone does.
- The amount of information in this chapter may seem overwhelming. Take your time in learning how growth is regulated and the effects of the various hormones.
- Try to assimilate the big picture of how the environment, receptors, hormones, and genome interact before focusing on the details. Make sure the broad picture is clear before you memorize functions of hormones or specific sequences of development.
- Go to yourBioPortal.com to review the following tutorials and activities:

  Animated Tutorial 26.1 Tropisms

  Animated Tutorial 26.2 Auxin Affects Cell Walls

  Web Activity 26.1 Monocot Shoot Development

  Web Activity 26.2 Eudicot Shoot Development

  Web Activity 26.3 Events of Seed Germination

Key Concept Review

26.1 Plants Develop in Response to the Environment

The seed germinates and forms a growing seedling

Several hormones and photoreceptors help regulate plant growth

Genetic screens have increased our understanding of plant signal transduction

Plant development is influenced by multiple regulatory factors such as environmental cues, receptors, hormones, and its genome. Plant seeds are dormant and remain so until seed germination. Dormancy, which lasts for different periods of time depending on the plant, involves exclusion of water or oxygen from the embryo by the seed coat, mechanical restraint of the embryo by the seed coat, and chemical inhibition of the embryo. Dormancy ensures survival through adverse conditions. Germination is triggered by one or more mechanical or environmental cues. To germinate, a plant must imbibe water and draw polysaccharides, fats, and protein nutrients from the endosperm or cotyledons. The germination stage is complete once the embryonic root, known as the radicle, emerges from the seed. A plant is considered a seedling when its radicle breaks through the seed coat to end germination.

There are two types of plant growth regulators: hormones and photoreceptors. Hormones are regulatory compounds that are produced in one area of a plant and translocated throughout it. The effects of the hormones are determined by their relative concentrations, and they play multiple regulatory roles. Photoreceptors are pigment proteins that sense light and are altered by light quality to induce changes within a plant.

Our understanding of how regulatory signaling pathways function has been enhanced through genetic screening. Genetic screening involves mutating plants and comparing their phenotype with that of wild-type plants.

Question 1: Trace the basic steps that occur between the planting of a pea seed in a garden and the emergence of the pea plant.
Textbook Reference: *26.1 Plants Develop in Response to the Environment, p. 540*

26.2 Gibberellins and Auxin Have Diverse Effects but a Similar Mechanism of Action

Gibberellins have many effects on plant growth and development

The transport of auxin mediates some of its effects

Auxin plays many roles in plant growth and development

At the molecular level, auxin and gibberellins act similarly

Gibberellins and auxin are two important plant hormones. The first discovered gibberellin was isolated from a fungus that infects rice plants and causes them to grow tall and spindly. Gibberellins are produced in both plants and fungi. Charles Darwin and his son observed the function of auxin in the bending of canary grass seedlings toward the light. Asymmetrical movement of auxin in the canary grass simulates elongation of cells on the side opposite the light source. Mutant plants lacking either hormone result in short phenotypes that can be reversed by providing external sources of the hormones. Gibberellins are involved in stimulating elongation of the plant stem. Experiments with dwarf and normal plants showed that plants have innate gibberellins. Normal plants exposed to gibberellins showed no alteration in appearance, but dwarf plants exposed to gibberellins showed shoot elongation to near normal lengths. Gibberellins play a role in fruit development. Developing seeds produce gibberellins that enhance development of fruit tissue. It is common agricultural practice to spray seedless fruits (especially grapes) with gibberellins to enhance fruit growth. In the developing seed, gibberellins stimulate the aleurone layer, a tissue layer under the seed coat, to secrete enzymes that break down the seed coat.

Auxin movement in plant tissues is unidirectional and polar, from apex to base. Auxin enters the cell in its nonpolar acid form by passive diffusion, and proton pumps transport H^+ out of the cell, causing the nonpolar auxin to become an anion and producing an electrochemical gradient (see Figure 26.5). Auxin anion efflux carriers are carrier proteins at the basal end of the cell responsible for export of auxin anions from cells, and they contribute to the unidirectional movement of auxin. Lateral redistribution of auxin is responsible for phototropism and gravitropism. Under both conditions, higher auxin concentrations on one side of the plant cause increased rates of growth along that side and lead to bending. Auxin affects vegetative growth by initiating root growth in cuttings and promoting and maintaining the growth of a single main stem (apical dominance). Auxin inhibits abscisson (the dropping of leaves) and can stimulate unfertilized fruit to form (parthenocarpy).

The effects of auxin on growth are mediated by the cell walls, which determine the rate and direction of cell growth. Cells grow by taking up water. The amount of water that can be taken up is restricted by a rigid cell wall. Cell walls must loosen, stretch, and add polysaccharides and cellulose to maintain structure as the cell expands. According to the acid growth hypothesis of cell expansion, auxin stimulates the production and insertion of proton pumps into the plasma membrane (see Figure 26.7). The subsequent decrease in pH stimulates proteins called expansins to alter polysaccharide bonding so that they slide past one another during expansion.

Auxin and gibberellins work through similar transduction signaling pathways (see Figure 26.8). There is a receptor in the cell that binds with auxin or gibberellins. Simulation of the pathway by these hormones promotes gene expression of growth stimulating genes.

Question 2: Imagine that you are a plant physiologist interested in the hormones regulating development. Discuss an experimental approach you would use to try to determine the role of a hormone such as gibberellins in the regulation of development.
Textbook Reference: *26.2 Gibberellins and Auxin Have Diverse Effects but a Similar Mechanism of Action, p. 542*

Question 3: Explain how auxin distribution regulates phototropism and gravitropism.
Textbook Reference: *26.2 Gibberellins and Auxin Have Diverse Effects but a Similar Mechanism of Action, p. 545*

Question 4: Both gibberellin and auxin act in a similar fashion at the molecular level. Create a flow chart showing the

signal transduction pathways for gibberellin and auxin. Note their similarities and their differences.
Textbook Reference: *26.2 Gibberellins and Auxin Have Diverse Effects but a Similar Mechanism of Action, p. 548*

Question 5: Auxins and cytokinins appear to cancel out the effects of each other. Why is a hormone that affects bud growth produced in the roots, while the one that affects root growth is produced in the shoots? How does this relate to the polar distribution of hormones?
Textbook Reference: *26.3 Other Plant Hormones Have Diverse Effects on Plant Development, p. 549*

Question 6: Genetic engineering has produced fruits that are deficient in the ability to produce ethylene. How is this deficiency useful in the storage and marketing of fruits?
Textbook Reference: *26.3 Other Plant Hormones Have Diverse Effects on Plant Development, p. 549*

Question 7: Suppose that a scientist is experimenting in the laboratory with tissue culture methods. Pith tissue that has been isolated has grown an undifferentiated mass of cells (called a callus), and that tissue has been divided up and placed in the following culture media: (1) necessary nutrients plus auxin; (2) necessary nutrients plus zeatin (a cytokinin); (3) necessary nutrients plus equivalent concentrations of auxin and zeatin; (4) necessary nutrients plus an excess of zeatin and minimal auxin. Explain what happens to the mass of tissue under each condition.
Textbook Reference: *26.3 Other Plant Hormones Have Diverse Effects on Plant Development, p. 549*

Question 8: What hormonal influences affect a pea plant from the moment it is planted until its emergence?
Textbook Reference: *26.3 Other Plant Hormones Have Diverse Effects on Plant Development, pp. 549, 551; 26.2 Gibberellins and Auxin Have Diverse Effects but a Similar Mechanism of Action, p. 543.*

26.3 Other Plant Hormones Have Diverse Effects on Plant Development

- Ethylene is a gaseous hormone that promotes senescence
- Cytokinins are active from seed to senescence
- Brassinosteroids are plant steroid hormones
- Abscisic acid acts by inhibiting development

There are a number of additional hormones involved in the regulation of plant growth and development. Ethylene is a gaseous hormone that promotes leaf senescence and fruit ripening. In many instances it is given off by rotting fruit. The use of ethylene spray to promote fruit ripening is a common commercial practice. Ethylene scrubbers are used in fruit storage to prevent ethylene from accumulating and causing fruit to spoil. Other chemicals are used in the flower industry to inhibit ethylene's effects on flower senescence. Ethylene plays a role in maintaining the apical hook on emerging eudicots by inhibiting cells on the inner portion of the hook. Ethylene inhibits stem elongation, promotes lateral swelling of stems, and inhibits sensitivity to gravitropic stimulation—three responses that are known as the triple response.

Cytokinins are powerful stimulators of cell division and bud formation, aid in seed germination, inhibit stem elongation, and delay leaf senescence. Auxins and cytokinins regulate organ development based on the relative concentrations of the two. High auxin levels favor root formation, and high cytokinin levels favor bud formation. Cytokinins are synthesized primarily in the roots and are translocated throughout the plant. Cytokinins act through a two-component system similar to that found in bacteria. A receptor (AHK) phosphorylates proteins, and a target transcription factor (ARR) acts as an effector. An intermediate protein (AHP) transfers the phosphate from the receptor to the transcription factor (see Figure 26.9).

Brassinosteroids are involved in the response to light. They have been shown to stimulate cell elongation, pollen tube elongation, and vascular tissue differentiation and to inhibit root elongation, all of which are similar to the effects of auxin. The brassinosteriod receptor is found on the cell wall and initiates a transduction pathway that influences gene expression. Abscisic acid regulates the development of the seed and the production of proteins that will protect against desiccation; thus it is a stress hormone. Abscisic acid also inhibits germination of the seed on the plant. The condition of premature germination, in which a seed germinates while still on the plant, is called vivipary.

26.4 Photoreceptors Initiate Developmental Responses to Light

- Phototropin, cryptochromes, and zeaxanthin are blue-light receptors
- Phytochrome mediates the effects of red and far-red light
- Phytochrome stimulates gene transcription
- Circadian rhythms are entrained by photoreceptors

Light and photoreceptors interact to stimulate a variety of events in plants, known as photomorphogenesis. Photoreceptors in plants interpret intensity, duration, and wavelength of light. Light regulates a wide variety of plant processes, including germination, flower production, and shoot elongation. Phototropin, a blue-light receptor, is a protein kinase that stimulates cell elongation by auxin. Zeaxanthin and phototropin act together to regulate light-stimulated opening of the stomata. Cryptochromes absorb blue and ultraviolet light and influence seedling development and flowering. Red light stimulates developmental and physiological events that include germination, flowering, and production of chlorophyll in seedlings.

Exposure to far-red light reverses the effects of exposure to red light, and vice-versa. This "switching" occurs because phytochrome can be shifted from one isoform (red P_r) to the other isoform (far-red P_{fr}) upon absorption of light. When exposed to red light, P_r is converted to P_{fr}. When exposed to far-red light, P_{fr} is converted to P_r. Phytochromes are composed of a protein chain that interacts with transcription factors and the pigment chromophore. Red light changes the conformation of the protein from the P_r to P_{fr} form and exposes a nuclear localization sequence. The P_{fr} form then moves into the nucleus, where it stimulates gene expression through a transcription factor. It can also act as a kinase and phosphorylate other proteins (see Figure 26.12).

Biological organisms exhibit a daily cycle in their functioning known as a circadian rhythm. The phytochromes are probably involved in the circadian rhythms of plants.

Question 9: You have produced a plant with a mutation that has only one isoform for the photoreceptor phytochrome. This plant flowers continuously and develops more shoots than your unmutated control plants. What would happen if you were to expose this plant to either red light or far-red light? What isoform is this plant producing and how do you know this?
Textbook Reference: *26.4 Photoreceptors Initiate Developmental Responses to Light, p. 553*

Test Yourself

1. Plant growth is regulated by
   a. environmental cues.
   b. hormones.
   c. signal transduction pathways.
   d. the expression of the plant's genome.
   e. All of the above
   Textbook Reference: *26.1 Plants Develop in Response to the Environment, p. 540*
2. The mechanism by which gibberellins act involves
   a. the adding of proton pumps to the plasma membrane.
   b. the phosphorylation of proteins.
   c. binding with a transcription factor.
   d. the removal of a repressor from a transcription factor.
   e. None of the above
   Textbook Reference: *26.2 Gibberellins and Auxin Have Diverse Effects but a Similar Mechanism of Action, p. 547*
3. Which of the following triggers germination of a seed?
   a. The imbibing of water
   b. Its release from the fruit
   c. Chemical changes
   d. The exclusion of water
   e. All of the above
   Textbook Reference: *26.1 Plants Develop in Response to the Environment p. 540*
4. Which of the following may have the effect of breaking dormancy in seeds?
   a. Penetration of the seed coat
   b. Leaching of inhibitory compounds by water
   c. Exposure to fire
   d. Passing through an animal's digestive tract
   e. All of the above
   Textbook Reference: *26.1 Plants Develop in Response to the Environment p. 540*
5. Which of the following hormones is responsible for bud break in the spring in deciduous trees?
   a. Auxins
   b. Cytokinins
   c. Gibberellins
   d. Ethylene
   e. Brassinosteroids
   Textbook Reference: *26.2 Gibberellins and Auxin Have Diverse Effects but a Similar Mechanism of Action, p. 543*
6. Which of the following is *not* involved in the acid growth hypothesis for the regulation of cell expansion by auxin?
   a. The pumping of protons into the cell wall
   b. The pumping of protons into the cytosol
   c. Increased gene expression of the proton pump gene
   d. Increased insertion of proton pumps into the plasma membrane
   e. All of the above are involved.
   Textbook Reference: *26.2 Gibberellins and Auxin Have Diverse Effects but a Similar Mechanism of Action, p. 545*
7. Which of the following is *not* involved in the polar transport of auxin?
   a. Diffusion across a plasma membrane
   b. Membrane protein asymmetry of auxin transport carriers
   c. Proton pumping from the cytosol
   d. Ionization of auxin as a weak acid
   e. All of the above are involved.
   Textbook Reference: *26.2 Gibberellins and Auxin Have Diverse Effects but a Similar Mechanism of Action, p. 545*

8. A homeowner has installed an outdoor gas-burning grill on her back patio next to her favorite camellia bush. After the first few nights of using the grill, she notices that the camellia is beginning to lose its leaves. Which of the following is the best explanation for what is happening?
   a. The bush is getting too warm next to the grill.
   b. Ethylene is a by-product of the burning gas and is causing senescence in the plant.
   c. Abscisic acid is a by-product of the burning gas and is causing senescence in the plant.
   d. The plant is a biennial and is bolting.
   e. Auxin production is being inhibited.
   Textbook Reference: *26.3 Other Plant Hormones Have Diverse Effects on Plant Development, p. 549*
9. Cytokinins interact with which other hormone?
   a. Ethylene
   b. Abscisic acid
   c. Gibberellins
   d. Auxins
   e. Brassinosteroids
   Textbook Reference: *26.3 Other Plant Hormones Have Diverse Effects on Plant Development, p. 549*
10. Which of the following light receptors is responsible for absorbing blue and ultraviolet light?
   a. Phytochrome P_r
   b. Phytochrome P_{fr}
   c. Cryptochrome
   d. Phototropin
   e. Etiolatin
   Textbook Reference: *26.4 Photoreceptors Initiate Developmental Responses to Light, p. 522*
11. Etiolated seedlings are produced by germinating seeds that are kept in total darkness. Plants that are kept in the dark will begin to germinate after they are given a pulse of
   a. blue light.
   b. red light.
   c. red light followed by a pulse of far-red light.
   d. far-red light followed by a pulse of red light.
   e. ultraviolet light.
   Textbook Reference: *26.4 Photoreceptors Initiate Developmental Responses to Light, p. 553*
12. Ethylene is produced by what part of a plant?
   a. The seedling
   b. The leaves
   c. The fruit
   d. All of the above
   e. None of the above
   Textbook Reference: *26.3 Other Plant Hormones Have Diverse Effects on Plant Development, p. 549*
13. Auxin transport within a plant is said to be _______, and it is dependent on the action of _______ pumps.
   a. polar; proton
   b. nonpolar; potassium
   c. polar; potassium
   d. bidirectional; proton
   e. polar; sodium
   Textbook Reference: *26.2 Gibberellins and Auxin Have Diverse Effects but a Similar Mechanism of Action, pp. 544–545*
14. Which of the following is *not* initiated by auxin?
   a. Stimulation of root initiation
   b. Inhibition of leaf abscission
   c. Stimulation of leaf abscission
   d. Maintenance of apical dominance
   e. Auxin is involved in all of the above.
   Textbook Reference: *26.2 Gibberellins and Auxin Have Diverse Effects but a Similar Mechanism of Action, p. 546*
15. Red light activation of the phytochorome into the P_{fr} state leads to which of the following events?
   a. Inhibition of chlorophyll
   b. Leaf expansion
   c. Hook folding
   d. Hook unfolding
   e. Both b and d
   Textbook Reference: *26.4 Photoreceptors Initiate Developmental Responses to Light, p. 553*

Answers

Key Concept Review

1. The watering of the pea seed after planting promotes the leaching of germination inhibitors. At this point the seed also begins to imbibe water, resulting in metabolic changes. DNA synthesis is halted until the radicle emerges from the seed coat. Breakdown of starch and protein reserves in the cotyledons and the endosperm begins to provide nutrients for the developing embryo. The apical hook begins to push through the soil. Upon its emergence and exposure to light, chlorophyll synthesis begins.
2. To determine the role of a hormone on plant function, it helps to make a mutant plant that is unable to produce the hormone in question. In a controlled experiment, you would then grow the plant in the absence of external application of the hormone and in the presence of external application of the hormone. The differences in plant development and growth would be attributed to the action of the hormone.
3. Lateral distribution of auxin controls phototropism and gravitropism. Auxin accumulates in the shaded portions of a stem and stimulates cell growth. This uneven cell growth results in a bending of the stem toward the light. The same mechanism works in response to gravity. An accumulation of auxin occurs where the gravitational pull is the strongest. Cells grow in response to auxin, and stems bend upward, away from the gravitational force.

4.

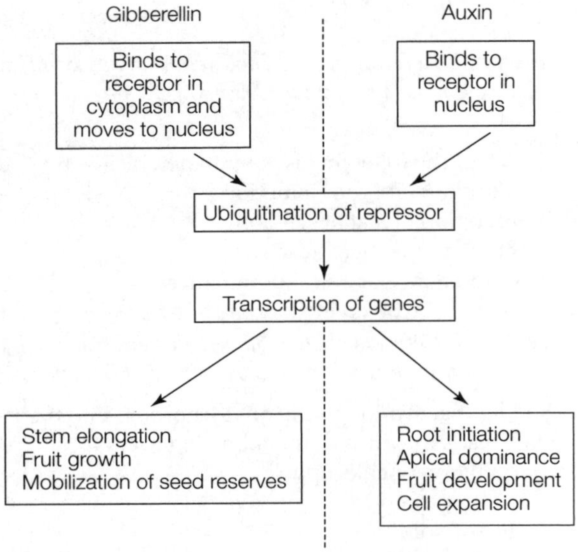

5. Roots need shoots, and vice-versa. Increases in roots require increases in shoots for photosynthesis. Regulation of the development of one by the other keeps growth in tandem. Because the distribution of the molecules is polar, a concentration gradient can be established. It is this relative gradient that controls development.
6. Fruits that are deficient in the ability to produce ethylene can be kept from ripening until they reach their destination. Once at market, they can be sprayed with ethylene to stimulate ripening. The result is fruit that can be shipped more easily and can still be ripe at any time in the market.
7. The results of the different treatments are the following: (1) roots will develop from the callus; (2) buds will develop from the callus; (3) both roots and buds will develop from the callus; (4) buds will develop from the callus.
8. Dormancy of the seed is maintained by abscisic acid until it is leached from the seed by water. Once water is imbibed, cytokinins begin to influence germination. Gibberellins assist with the mobilization of storage products to the growing embryo. Ratios of auxins and cytokinins balance root and bud formation. The apical hook is maintained by ethylene.
9. In a plant there are two isoforms of phytochrome that respond to red light and far-red light. Because your mutant plant only has one isoform for the phytochrome, exposure either to one type of light or the other will have no effect. The isoform that responds to far-red light is the active isoform that stimulates flowering and shoot development. Because your plant flowers continuously and has more shoots than the control, you can conclude that its phytochrome is the active isoform.

Test Yourself

1. **e.** The growth of a plant is regulated by all of these factors.
2. **d.** Gibberellin brings about a response in the plant cell by removing the repressor from a transcription factor, thus allowing transcription to occur.
3. **a.** The uptake of water by a seed begins the processes that lead to seed germination.
4. **e.** Mechanical abrasion, leaching of inhibitors by water, exposure to fire, and passing through a digestive tract may all trigger germination. Actual germination cannot begin until a seed imbibes water.
5. **c.** Gibberellins are responsible for bud break in deciduous trees.
6. **b.** The acid growth hypothesis involves an increase in the number of proton pumps in the plasma membrane and the pumping of protons into the cell wall.
7. **e.** All of the characteristics ensure that the transport of auxin is polar.
8. **b.** Ethylene gas promotes senescence and is one of the by-products of burning gas.
9. **d.** Cytokinins interact with auxins. The auxin-to-cytokinin ratio controls the bushiness of plants.
10. **c.** Cryptochromes respond to blue and ultraviolet light wavelengths.
11. **c.** The pulse of red light converts P_r to P_{fr}, which is the active form.
12. **d.** Ethylene can be produced by the fruit, seed, or leaves in most plants.
13. **a.** Auxin transport is a polar process, meaning that it moves in only one direction with the help of proton pumps and auxin anion efflux carriers.
14. **c.** Auxin inhibits leaf abscission rather than stimulating it. Auxin is involved in all of the other processes.
15. **e.** The P_{fr} phytochrome stimulates chlorophyll synthesis, hook unfolding, and leaf expansion.

Reproduction of Flowering Plants 27

The Big Picture

- Though plants can reproduce both asexually and sexually, maintenance of genetic variability depends on sexual reproduction. The flower is the basis of sexual reproduction in plants. The flower not only produces the necessary gametes, but it is also integrally involved in ensuring pollination. Eggs are produced through megasporogenesis, and pollen is produced through microsporogenesis. Angiosperms exhibit double fertilization, which results in a diploid embryo and a triploid endosperm.
- Hormones and signaling cascades are involved in a plant's transition from the vegetative state to the reproductive state. Meristem identity genes and floral organ identity genes code for genes necessary for flowering. The photoperiod sets the shoot apical meristem on the path to flowering. Some plants are short-day plants and flower when the day is shorter than a critical maximum while others are long-day plants and flower once the day length reaches a critical minimum. The key for the actual photoperiod is the length of the night (not day length). The gene FT codes for florigen and is involved in photoperiod signaling.
- Some angiosperms are also able to reproduce asexually. This occurs through vegetative reproduction, in which underground stems grow out from the plant. Apomixis is the asexual production of seeds without the necessity of fertilization.

Study Strategies

- Refer to the figures in the textbook to understand flower structure, megasporogenesis, microsporogenesis, and fertilization.
- Spend time thinking about the advantages and disadvantages of sexual and asexual reproduction, as related questions are often given in exams.
- Megasporogenesis is probably the most difficult concept in this chapter, but it is easier to understand when thinking about the source of each cell.
- Remember that flowering, like all other plant processes, is a result of environmental signals, receptors, enzyme cascades mediated by hormones, and alterations in gene expression.
- Be sure you understand the particular features of double fertilization.
- Go to yourBioPortal.com to review the following tutorials:

  Animated Tutorial 27.1 Double Fertilization

  Animated Tutorial 27.2 The Effect of Interrupted Days and Nights

Key Concept Review

27.1 Most Angiosperms Reproduce Sexually

The flower is the reproductive organ of angiosperms
Angiosperms have microscopic gametophytes
Angiosperms have mechanisms to prevent inbreeding
A pollen tube delivers sperm cells to the embryo sac
Angiosperms perform double fertilization
Embryos develop within seeds contained in fruits

Sexual reproduction is necessary for genetic recombination, and genetic variability provides plants with the ability to adapt to their environment. Plant reproduction involves alternation of diploid and haploid generations. The flower is the basis of sexual reproduction in angiosperms. The basic flower structures include carpels, stamens, petals, and sepals, all of which are modified leaves. The carpel and stamen are the male and female parts of the flower. Flowers with both a carpel and stamen are known as "perfect" and are found in monoecious species. Dioecious species have "imperfect" flowers that contain either the carpel or stamen, but not both. The flower produces haploid spores that develop into gametophytes. The female gametophytes, called embryo sacs, develop in the megasporangium. Male gametophytes, called pollen grains, develop in microsporangia.

Within the ovule, a megasporocyte produces four haploid megaspores through meiosis. Only one of these megaspores survives, and it divides mitotically to produce eight nuclei within a single large cell. The nuclei migrate to either end of the cell, but two remain in the middle. Cell walls form, isolating the three nuclei at either end into individual cells; the two nuclei in the middle remain together in one cell.

At one end, the three cells become two synergid cells and one egg cell; at the other are three antipodal cells, which degenerate. In the large central cell are the two polar nuclei. This seven-celled embryo sac is the megagametophyte (see Figure 27.2). Pollen grains develop when a microsporocyte undergoes meiosis. All products of meiosis are retained and undergo mitosis to form a two-celled pollen grain composed of the tube cell and the generative cell. Further development is halted until after the pollen is transferred from the anther to the stigma during pollination.

Pollen grains are carried to female flowers via wind, animals, or other vectors. Some plants are able to self-pollinate within the same plant or flower. However, most plants prevent self-fertilization by physical separation in either space or time of male and female gametophytes. Plants can also prevent self-fertilization by means of genetic self-incompatibility. A single gene, the *S* gene, regulates self-incompatibility and prevents self-fertilization in many plants (see Figure 27.3).

When a pollen grain germinates, a pollen tube grows down the style toward the embryo sac. Germination begins when the pollen begins to take up water from the stigma. Chemical signals, in the form of small proteins from synergids within the ovule, direct the growth of the pollen tube. The pollen grain consists of two cells at the time of pollination—the tube cell and the generative cell. The tube cell controls the growth of the pollen tube. As the tube is growing, the generative cell undergoes meiosis and produces two haploid sperm cells.

Once the pollen tube enters the embryo sac, the two sperm cells are released into a synergid that disintegrates and releases the sperm nuclei. One sperm cell fuses with the egg cell producing a diploid zygote, the other with the polar nuclei producing the triploid endosperm. All other cells disintegrate. This double fertilization resulting in a zygote and the nutritive endosperm is a characteristic feature of angiosperms (see Figure 27.4).

The embryos develop within seeds. The success of the embryo depends on its own development, the development of the endosperm, the integuments, and the carpel. Ultimately, a seed coat develops from the integuments and protects the dormant embryo. As the embryo and seed are developing, the ovary begins to form the fruit. Other parts of the flower and plant may be included in the fruit, but to be considered a fruit, only the ovary wall and the seed need be involved. The fruit disperses by various means, including by traveling on the coats of animals or by being consumed and later deposited. The fruit also protects the seed from animals and microbial diseases.

Question 1. In agricultural production, much effort is spent detasseling corn (removing male flowers). Why is it not possible to halt pollen production simply by spraying the corn plants with a meiosis inhibitor? (Hint: Male and female flowers occur on the same corn plant.)
Textbook Reference: *27.1 Most Angiosperms Reproduce Sexually, p. 559*

Question 2. Describe double fertilization. What is the ploidy level of the products of double fertilization?
Textbook Reference: *27.1 Most Angiosperms Reproduce Sexually, p. 560*

Question 3. Define a fruit. Which of the following are fruits: tomato, pear, potato, banana, cucumber, snow pea, peanut, sunflower seed?
Textbook Reference: *27.1 Most Angiosperms Reproduce Sexually, pp. 560–561*

27.2 Hormones and Signaling Determine the Transition from the Vegetative to the Reproductive State

Shoot apical meristems can become inflorescence meristems

A cascade of gene expression leads to flowering

Photoperiodic cues can initiate flowering

Plants vary in their responses to photoperiodic cues

Night length is the key photoperiodic cue that determines flowering

The flowering stimulus originates in the leaf

Florigen is a small protein

Flowering can be induced by temperature or gibberellins

Some plants do not require an environmental cue to flower

The beginning of flowering involves a reallocation of energy in the plant away from leaves and stem and toward flowers and gametes. Plants have one of three different life cycle patterns. Annual plants go from seed to seed set and die within one growing season. Biennial plants require one vegetative growing season before reproducing. Perennial plants flower repeatedly and live for many years.

The first transition from vegetative growth to floral production is the transition of the apical meristem into an inflorescence meristem. The inflorescence meristem can produce bracts and floral meristems. The floral meristem differs from the apical meristem in that growth is determined. The floral meristem is programmed to produce four consecutive whorls of flower organs. A gene cascade leads to flower formation, which begins with the activation of a set of meristem identity genes. Two genes important in switching from vegetative growth to reproductive growth are *LEAFY* and *APETALA1*. The expressions of floral organ identity genes specify successive whorls.

The gene cascade leading to flower development is controlled by environmental cues. Seasonal flowering is a result of changes in photoperiod, specifically the duration of continuous darkness. Experiments have shown that plants actually respond to the length of the night, rather than amount of daylight. In experiments in which daylight was interrupted, there was little effect on flowering, but in experiments in which dark was interrupted, there were significant effects on flowering. Each plant type has a critical day length corresponding to light availability that induces flowering. Short-day plants flower when the amount of light available is shorter than the critical maximum. Long-day plants flower only when the day is longer (more light available) than the critical minimum. Some plants require complex combinations of day lengths.

Plants, like all other organisms, have an internal mechanism for measuring the length of continuous dark periods. The duration of dark periods appears to be detected by special phytochromes that detect red light. The stimulus for flowering comes from the leaf of the plant. Florigen (FT) is thought to be the hormone responsible for flowering. The gene *FLOWERING LOCUS T* codes for the FT protein. High levels of FT induce the plant to flower. *CONSTANS* codes for a transcription factor (CO) that stimulates FT synthesis in the phloem cells. The *FLOWERING LOCUS D* gene codes for a transcription factor (FD) in the apical meristem that increases transcription of *APETALA1*.

Vernalization is the induction of flowering by low temperatures. A transcription factor (FLC) coded by the *FLOWERING LOCUS C* inhibits the FT pathway described above. Cold temperatures decrease the production of the FLC protein, leading to a functional FT pathway. Gibberellin can also stimulate a plant to flower. Those plants that do not require an environmental cue to flower rely on an "internal clock" to trigger flowering. This most likely works through changes in FLC concentrations in the plant, stimulating the FT–FD pathway described above.

Question 4. Draw a flow chart showing the mechanism of action of florigen, the plant hormone thought to be actively involved in the initiation of flower formation. Be sure to show the interaction of the three genes involved in initiating flower formation.
Textbook Reference: *27.2 Hormones and Signaling Determine the Transition from the Vegetative to the Reproductive State, p. 566*

Question 5. Flowering is stimulated when light sets off a gene cascade. Explain how this may be hormonally controlled.
Textbook Reference: *27.2 Hormones and Signaling Determine the Transition from the Vegetative to the Reproductive State, p. 566*

27.3 Angiosperms Can Reproduce Asexually

Angiosperms use many forms of asexual reproduction

Vegetative reproduction is important in agriculture

In certain conditions, asexual reproduction is advantageous to plants' success. Asexual reproduction occurs without genetic recombination. During the process of vegetative reproduction, asexual reproduction occurs through the modification of a vegetative organ. Stolons, tubers, rhizomes, and bulbs are all modifications of stems or roots that allow vegetative reproduction. Some plants, such as dandelions, reproduce asexually through seeds in a process called apomixis. Apomictic plants skip over meiosis and fertilization and produce diploid seeds with the identical genetic makeup of the maternal plant.

Asexual reproduction is used in agriculture. Crossing two inbred homozygous genetic strains can produce plants that are superior, resulting in hybrid vigor. Cuttings are commonly used in horticulture. Grafting is widely used in fruit crops in which the root-containing stock is grafted to a scion (shoot system). This allows for a hardy root stock to be combined with a good but often fragile fruit producer. Meristem culture (production of entire new plants from a small tissue sample) is leading to daily advances in agriculture.

Question 6. When is asexual reproduction beneficial to plants? Under what conditions is sexual reproduction beneficial?
Textbook Reference: *27.3 Angiosperms Can Reproduce Asexually, p. 568*

Question 7. Compare and contrast asexual and sexual reproduction in plants. In which category does self-fertilization belong?
Textbook Reference: *27.3 Angiosperms Can Reproduce Asexually, p. 568; 27.1 Most Angiosperms Reproduce Sexually, p. 559*

Test Yourself

1. Suppose that you manage a greenhouse that produces roses for Valentine's Day (in February). Roses normally bloom in June. Which of the following would be the best lighting schedule for inducing flowering in the plants?
   a. 16 hours of light, followed by 8 hours of interrupted dark
   b. 16 hours of light, followed by 8 hours of uninterrupted dark
   c. 10 hours of light, followed by 14 hours of uninterrupted dark
   d. 10 hours of light, followed by 14 hours of interrupted dark
   e. None of the above schedules would induce flowering.

   Textbook Reference: *27.2 Hormones and Signaling Determine the Transition from the Vegetative to the Reproductive State, p. 564*

2. After setting the correct photoperiod, the managers of a greenhouse still do not have blooming roses. Which of the following possibilities would most likely have contributed to the problem?
   a. The heating system allowed for fluctuations in temperature between 20°C and 25°C.
   b. The furnace mechanic accidentally turned off the lights for an hour two days in a row.
   c. The cleaning crew turned the lights on for an hour three nights in a row.
   d. All of the above
   e. None of the above

   Textbook Reference: *27.2 Hormones and Signaling Determine the Transition from the Vegetative to the Reproductive State, p. 564*

3. You have moved into a new house. During the first summer you notice many of the plants do not bloom. During the second summer your yard is a sea of blooms. It is now spring of the third year, and there are no plants. Which of the following is the best explanation for this observation?
   a. The plants are annuals.
   b. The plants are biennials.
   c. The plants are perennials.
   d. The plants are being affected by drought.
   e. The nights are too long for the plants to thrive.

   Textbook Reference: *27.2 Hormones and Signaling Determine the Transition from the Vegetative to the Reproductive State, p. 562*

4. You notice that a new houseplant sends out long stems with what look like "little plants" attached. You allow one of these to rest in a cup of water and note that roots form. What you are seeing is an example of
   a. asexual reproduction.
   b. apomixis.
   c. heterospory.
   d. parthenogenesis.
   e. vivipary.

   Textbook Reference: *27.3 Angiosperms Can Reproduce Asexually, p. 568*

5. Self-pollination in plants that produce both pollen and eggs is prevented by
   a. self-incompatibility genes.
   b. physical barriers.
   c. production of pollen and eggs at different times.
   d. All of the above
   e. None of the above

   Textbook Reference: *27.1 Most Angiosperms Reproduce Sexually, p. 559*

6. Most grapevines that produce wine grapes are grafted onto rootstock of another species. How does this practice increase grape yields?
   a. A hardy rootstock can replace a weak rootstock.
   b. A high-producing vine stock can replace a low-producing vine stock.
   c. It allows vintners to select for pest resistance without losing grape quality.
   d. All of the above
   e. None of the above

   Textbook Reference: *27.3 Angiosperms Can Reproduce Asexually, p. 570*

7. The induction of flowering by means of exposure to low temperature is called
   a. vernalization.
   b. frigidation.
   c. apomixis.
   d. viviparity.
   e. None of the above

   Textbook Reference: *27.2 Hormones and Signaling Determine the Transition from the Vegetative to the Reproductive State, p. 566*

8. The *LEAFY* and *APETALA1* genes are examples of _______ genes.
   a. floral organ identity
   b. meristem identity
   c. viviparity
   d. photoperiod
   e. inflorescence

   Textbook Reference: *27.2 Hormones and Signaling Determine the Transition from the Vegetative to the Reproductive State, p. 563*

9. The production of seeds without fertilization is called
   a. apomixis.
   b. parthenogenesis.
   c. conception.
   d. circadian rhythm.
   e. vernalization.

   Textbook Reference: *27.3 Angiosperms Can Reproduce Asexually, p. 569*

10. In the transition from vegetative growth to floral growth the _______ must be transformed into the _______. This involves a shift from _______ growth to _______ growth.

a. apical meristem; floral meristem; indeterminate; determinate
b. lateral meristem; floral meristem; indeterminate; determinate
c. apical meristem; floral meristem; determinate; indeterminate
d. apical cambium; floral cambium; determinate; indeterminate
e. floral meristem; apical cambium; determinate; indeterminate

Textbook Reference: *27.2 Hormones and Signaling Determine the Transition from the Vegetative to the Reproductive State, p. 562*

11. Which of the following best describes the fate of the generative cell of the pollen grain?
a. It coordinates growth of the pollen tube.
b. It divides by meiosis to produce two sperm nuclei.
c. It divides by mitosis to produce two sperm nuclei.
d. It forms the pollen tube.
e. None of the above

Textbook Reference: *27.1 Most Angiosperms Reproduce Sexually, p. 560*

12. Which of the following is *not* part of a megagametophyte?
a. Pollen grain
b. Synergids
c. Antipodal cells
d. Polar nuclei
e. None of the above is part of the megagametophyte.

Textbook Reference: *27.1 Most Angiosperms Reproduce Sexually, p. 558*

13. During double fertilization two sperm cells fuse (one each) with
a. an egg cell.
b. the two polar nuclei.
c. a pollen grain.
d. Both a and b
e. Both a and c

Textbook Reference: *27.1 Most Angiosperms Reproduce Sexually, p. 560*

14. Which of the following statements about plants and photoperiodic cues is true?
a. Short-day plants flower when the day is longer than a critical maximum.
b. Short-day plants only reproduce asexually.
c. Long-day plants flower when the dark night is longer than a critical maximum.
d. Long-day plants flower when the day is longer than a critical maximum.
e. Both a and c

Textbook Reference: *27.2 Hormones and Signaling Determine the Transition from the Vegetative to the Reproductive State, pp. 563–564*

15. Which of the following is *not* a part of the flower?
a. Carpels
b. Petals
c. Sieve Tube
d. Stamens
e. Sepals

Textbook Reference: *27.1 Most Angiosperms Reproduce Sexually, p. 557*

Answers

Key Concept Review

1. Meiosis occurs in the megasporocyte as well as the microsporocyte. Inhibition of pollen formation via a meiosis inhibitor will also inhibit egg formation.

2. Double fertilization occurs when the two sperm nuclei produced from the generative cell of the pollen grain unite with the egg nucleus and two polar nuclei, respectively. The resulting zygote is diploid and the endosperm is triploid.

3. A fruit is the seed and the ovary wall, and it may include other structures of a flowering plant. With the exception of the potato, all of the listed structures are fruits.

4.

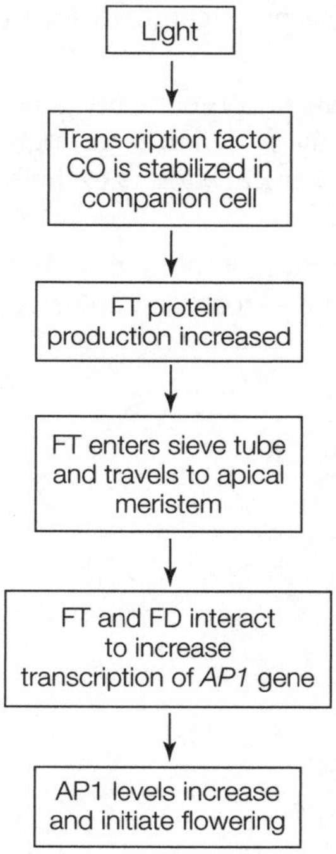

5. Light induction begins with the leaves. A single isolated leaf may stimulate floral production throughout the plant. In experiments in which leaves from one plant were grafted onto other plants, an "induced" leaf caused a plant to flower even though the plant had not been exposed to the required amount of darkness. Therefore, the leaf must send a signal (the protein florigen) that begins the gene cascade leading to flower formation.

6. Asexual reproduction is beneficial in stable environments in which many genetically identical plants are

sustainable. This process can help colonize a habitat or help a population of plants to spread. The disadvantage is lack of genetic diversity, which can be detrimental if the environment changes rapidly and adaptability of the population is required.

7. Asexual reproduction does not involve meiosis, fertilization, or genetic recombination. The offspring from asexual reproduction are genetically identical to the parent plant. Sexual reproduction requires a meiotic event and fertilization. Self-fertilization is sexual reproduction, even though genetic recombination is limited. This is because a meiotic event occurs, and fertilization is necessary for reproduction to take place.

Test Yourself

1. **b.** Flowering is regulated by darkness. Interruptions in darkness can prevent flowering.
2. **c.** The interruptions in the dark cycle could have affected the signals that tell the plant to begin flowering.
3. **b.** The plants are most likely biennials, which spend one season growing vegetatively and one season producing flowers before dying.
4. **a.** The runner (stolon) is propagating this plant asexually.
5. **d.** Some plants have self-incompatibility genes that prevent the growth of the pollen tube through the style. Others have mechanical barriers to their own pollen.
6. **d.** Grafting allows selection of hardy rootstock, produces well-producing vine stock, and contributes to disease resistance.
7. **a.** The induction of flowering by means of exposure to low temperature is called vernalization. This is the practice that induces flowering in winter wheat.
8. **b.** These two genes are meristem identity genes that code for proteins involved in the initiation of flower formation.
9. **a.** Dandelions and other plants produce seeds by apomixis, without meiosis and fertilization. These seeds are genetically identical to the parent plant.
10. **a.** Apical meristems and floral meristems differ in that growth from the apical meristem is indeterminate and growth from the floral meristem is determinate, leading to four whorls of floral structures.
11. **b.** The two sperm nuclei involved in double fertilization are derived from the generative cell of the pollen grain.
12. **a.** The megagametophyte is the female gametophyte. It is initially called the embryo sac and contains three antipodal cells, two synergid cells, and two polar nuclei at the seven-cell stage.
13. **d.** Double fertilization involves the fertilization of the egg cell and the two polar nuclei by two sperm cells.
14. **d.** The cue for flowering is the length of the night, so long-day plants flower when the day is longer than a critical maximum and when the night is shorter than a critical minimum.
15. **c.** The carpels, stamens, petals, and sepals are all part of the flower.

Plants in the Environment 28

The Big Picture

- Plants must respond to invasion by pathogens, physical damage by natural events and herbivory, variable water supplies, temperature fluctuations, and variable soil conditions. Natural selection has made plants uniquely suited to the areas they occupy. Some thrive in very harsh environments, but others cannot.
- Interactions between plants and pathogens stimulate a series of chemical changes that ward off further infection. Plant strategies to isolate pathogens include sealing plasmodesmata, releasing proteins that interact with pathogens, and signaling other parts of the plant. A plant produces chemicals to prevent herbivory and mounts defenses around the damaged tissues.
- Specific adaptations allow plants to withstand adverse conditions. Saline environments, water loss, and high temperatures can be survived if water uptake is maximized and water loss is minimized. Sunken stomata, reduction of surface area, and increased water potential of cells are adaptations to these conditions. Plants can survive extremes in temperature by means of heat shock proteins that stabilize metabolic proteins.

Study Strategies

- The best strategy is to study the various components of this chapter one at a time so that all of the defenses/modifications do not seem to run together. Focus first on pathogen interaction, then herbivory resistance, and so on.
- Think about how form follows function as you look at how plants are adapted to specific harsh environments.
- Gene-for-gene resistance can be difficult to understand, but remember that in any pathogen situation, the pathogen and the host are continually interacting; therefore, over time, interplay among the genomes has been selected for.
- There are many examples of the ways plants deal with pathogens and herbivory. Be sure to review them after reading the appropriate sections of the textbook.
- Go to yourBioPortal.com to review the following tutorial and activities:

  Animated Tutorial 28.1 Signaling between Plants and Pathogens

  Web Activity 28.1 Concept Matching

  Working with Data 28.1 Nicotine is a Defense against Herbivores

Key Concept Review

28.1 Plants Have Constitutive and Induced Responses to Pathogens

Physical barriers form constitutive defenses

Induced responses to pathogens may be genetically determined

The hypersensitive response fights pathogens at the site of infection

Plants can develop systemic resistance to pathogens

The action of a pathogen signals a plant to mount a defense. In turn, the plant's defense signals the pathogen. These signals go back and forth until one or the other "wins." The defense of a plant can either always be turned on (i.e., constitutive) or turned on in response to damage or stress (i.e., induced).

The first plant defense system is an attempt to prevent entry by pathogens. Cutin, suberin, and waxes cover the leaves and stems to exclude pathogens. Gene-for-gene resistance is a highly specific type of resistance and depends on a plant's having an allele for a gene that matches an allele in the pathogen. Dominant *R* genes code for resistance in plants and dominant *Avr* genes code for elicitors. If the receptor from a plant's *R* gene and the elicitor from a pathogen's *Avr* gene match, a defensive response is elicited in the plant (see Figure 28.2). Though this mechanism is not completely understood, it is thought to be mediated by nitric oxide and peroxide.

Invaded plants begin a three-pronged response to the pathogen called the hypersensitive response. Once invaded, plants produce chemical defenses against pathogens. Phytoalexins are nonspecific antifungal and antibacterial compounds produced by the plant. Pathogenesis-related proteins (PR proteins) are enzymes that function to break down the walls of pathogens or serve as signaling molecules to pathogen-free cells in the plant.

In the hypersensitive response, pathogen-containing tissue and surrounding tissues go through apoptosis upon infection and become necrotic lesions. This serves to contain and isolate the pathogen to the infected tissues. As these cells are dying, they release phytoalexins. Upon recognition of a pathogen, polysaccharides are deposited in the cell wall and seal off the plasmodesmata to form a barrier between cells.

Systemic acquired resistance protects the whole plant against further infection. Salicylic acid produced at the site of an infection stimulates PR protein production. Salicylic acid is also transported to other parts of the plant to signal the presence of an invader. Infected plants release methyl salicylate, which can become airborne and serve as a signal to other plant parts or to neighboring plants to mount a defense via PR proteins. Plants can use interference RNA (RNAi, also known as posttranscriptional gene silencing) to respond to attack by RNA viruses. A plant produces small pieces called small interfering RNA (siRNA) from interactions with the viral RNA and these siRNAs degrade the viral mRNA.

Question 1. It has long been a practice in rural areas to nail fencing to trees. Diagram how the act of nailing a fence to a tree trunk can introduce pathogens into the tree. Be sure to show what measures the tree takes to ward off disease.
Textbook Reference: *28.1 Plants Have Constitutive and Induced Responses to Pathogens, pp. 573–575*

28.2 Plants Have Mechanical and Chemical Defenses against Herbivores

- Constitutive defenses are physical and chemical
- Plants respond to herbivory with induced defenses
- Why don't plants poison themselves?
- Plants don't always win the arms race

Plants use a number of mechanical and chemical defenses against herbivores. In some plants, constitutive anatomical defenses provide protection, including morphological features such as thorns, spines, and hairs. Plants also produce secondary metabolites that act as a defenses against grazers. The effects of these chemicals are diverse, ranging from neurotoxins to hormone mimics (see Table 28.1). Canavanine is a defensive secondary metabolite that plays multiple roles in a plant, including assisting with nitrogen storage in the seed. Because it is similar to arginine, it is incorporated into the proteins made after consumption by an herbivore. Canavanine alters the tertiary structure of proteins in an herbivore and can be lethal.

Plants perceive the damage caused by herbivores through membrane signaling and chemical signaling. When an herbivore begins to eat the plant, changes in the plasma membrane electrical potential are passed along to every cell within the plant. Chemical signaling occurs when the herbivore's saliva combines with plant fatty acids to elicit a local and systemic response in the plant. The response of a plant to perceived damage stimulates a signal transduction pathway involving jasmonic acid (jasmonate). Wounding causes the release of an elicitor, which travels to other parts of the plant and induces membrane breakdown. A by-product of this breakdown is jasmonate, which enters the nucleus and activates production of a protease inhibitor that inhibits digestion in the predator (see Figure 28.6).

Plants use a variety of strategies to prevent harming themselves with their own defensive mechanisms. One way is by isolating toxic secondary metabolites into compartments. Water-soluble toxins are stored in vacuoles; hydrophobic poisons are stored in laticifers along with latex; other toxins are dissolved in the waxes covering the epidermis. Some toxic substances are produced only in already-damaged or dying tissue. The precursors of toxic material are stored separately from the enzymes that convert them into active poison, and these combine only once the cell is damaged. Insects sometimes find ways to get around the mechanisms used by plants to deter them. One example is a beetle that feeds on the leaves of milkweed.

Question 2. Describe some of the strategies by which plants can defend themselves against herbivory.
Textbook Reference: *28.2 Plants Have Mechanical and Chemical Defenses against Herbivores, pp. 576–578*

Question 3. Many plants produce toxins that damage eukaryotic predators. Considering that the cell structures of the predator and the plant are similar, why is the plant not affected by its own toxins?
Textbook Reference: *28.2 Plants Have Mechanical and Chemical Defenses against Herbivores, pp. 579–580*

28.3 Plants Adapt to Environmental Stresses

- Some plants have special adaptations to live in very dry conditions
- Some plants grow in saturated soils
- Plants can respond to drought stress
- Plants can cope with temperature extremes
- Some plants can tolerate soils with high salt concentrations
- Some plants can tolerate heavy metals

Plants have evolved the ability to live in harsh environments. In dry conditions such as deserts, some desert annuals are drought avoiders. They survive in these areas by completing their entire life cycle in the short period that water is available. Xerophytes are plants specifically adapted for year-round life in dry areas. They may have special modifications such as thickened cuticles, epidermal hairs, and stomatal crypts (see Figure 28.8) to prevent water loss. Trichomes help to decrease the intensity of light hitting the surface of the plant. Succulents have water-storing leaves or stems. Roots can also be adapted for drought conditions. Long taproots can reach water supplies far underground, and shallow fibrous root systems that grow only during rainy seasons are also adapted to desert growth. Xerophytic plants are able to extract more water when it is available because in response to the increased water the osmotic and water potential of their root cells changes. This is possible because proline is stored in their vacuoles.

Too much water can be as dangerous to plant survival as too little. Roots submerged in water do not get adequate oxygen to sustain respiration. Some plants that grow in standing water overcome this by producing pneumatophores, which are root extensions that grow above water and provide oxygen to the entire root system. Others have leaf modifications called aerenchyma for buoyancy and oxygen storage (see Figure 28.11B). Inadequate water supply results in changes in membrane integrity and in the three-dimensional structure of proteins.

Draught conditions stimulate the roots to release abscisic acid, which travels to the leaves. Abscisic acid closes the stomata and starts gene expression of late embryogenesis abundant (LEA) proteins. The LEA proteins stabilize membranes and proteins.

Temperature extremes pose yet another stress to plant survival. High temperatures denature proteins and destabilize membranes. Cold temperatures decrease membrane fluidity, and freezing causes rupturing of membranes if ice crystals form. Plants have evolved a number of adaptations to deal with heat, including hairs and spines to dissipate heat. Plants produce heat shock proteins that act as chaperonins and help stabilize protein structure against denaturation. Plants can adjust to cold through a process of cold-hardening, which involves repeated exposure to cool, nondamaging temperatures. This alters the saturated and unsaturated fatty acid composition of membranes, allowing them to remain fluid at cooler temperatures and making them more resistant to rupture. Heat shock proteins are also produced in response to cold temperatures. Some plants have antifreeze compounds that prevent ice crystal formation.

Some plants are adapted to survive in saline environments, however saline environments restrict the growth of angiosperms. Plants, such as halophytes (salt-loving plants), are adapted to living in saline environments and are exposed to an osmotic challenge. They are the only plants that accumulate sodium and chloride ions, which they store in leaf vacuoles. The increased salt concentration of the halophyte means it has a water potential that is more negative than that of the soil solution, allowing it to take up water from the saline environment. Some plants are able to excrete salt so that it does not reach toxic proportions. Salt glands move salt to the leaf surface, where it can be lost to wind or rain. Salt glands on the leaf assist with water procurement from the roots and with reduction of water loss to evaporation.

Heavy metals, such as chromium, mercury, lead, and cadmium in soils are poisonous to most plants. Some geographic areas are rich in these metals as a result of normal geological processes, and others have been contaminated by human activity. Plants living in these areas have adaptations that allow them to accumulate large quantities of heavy metals. These plants, known as hyperaccumulators, increase ion transport into the roots, increase translocation, accumulate ions in shoot vacuoles, and resist the toxin. These plants are important in bioremediation cleanup efforts.

Question 4. At a T-intersection in a small northern town, a local business plants several ornamental shrubs. During the winter, snowplows push snow from the intersection around the shrubs. In an attempt to save the shrubs, the business covers the shrubs with burlap. Will this save the shrubs? What conditions in the intersection are likely to cause the greatest damage to the shrubs?
Textbook Reference: *28.3 Plants Adapt to Environmental Stresses, p. 584*

Question 5. Pansies are planted in the winter in the southeastern United States. To ensure that the plants are ready to enter cool ground, nurseries cold-harden the plants for several days. Describe this process and explain why it is done.
Textbook Reference: *28.3 Plants Adapt to Environmental Stresses, p. 584*

Question 6. Imagine that you are put in charge of the bioremediation of a mine and ore refinery site in your city. How can plants help with the cleanup of sites contaminated with heavy metals?
Textbook Reference: *28.3 Plants Adapt to Environmental Stresses, p. 585*

Test Yourself

1. Defensive strategies in plants that are always turned on are called
   a. induced defenses.
   b. heat shock proteins.
   c. alternating defenses.
   d. constitutive defenses.
   e. All of the above
   Textbook Reference: *28.1 Plants Have Constitutive and Induced Responses to Pathogens, p. 573*
2. Plants acquire systemic resistance in much the same way that people acquire resistance to pathogens; however, the mechanism of systemic acquired resistance is quite different in plants. Which of the following does *not* have a role in acquired resistance in plants?
   a. Salicylic acid
   b. *R* genes
   c. Methyl salicylate
   d. PR proteins
   e. All of the above have a role in acquired resistance in plants.
   Textbook Reference: *28.1 Plants Have Constitutive and Induced Responses to Pathogens, p. 575*
3. Gene-for-gene resistances depend on which of the following?
   a. Compatible alleles in plant and pathogen
   b. Incompatible alleles in plant and pathogen
   c. Recessive *Avr* genes
   d. Recessive *R* genes
   e. PR proteins
   Textbook Reference: *28.1 Plants Have Constitutive and Induced Responses to Pathogens, p. 574*
4. Upon infection by a pathogen, plant cells increase synthesis of polysaccharides. The function of these polysaccharides is to
   a. destabilizes the cell walls.
   b. synthesize antibodies to the pathogen.
   c. isolate the pathogen in the invaded tissue.
   d. break down the wax barrier.
   e. communicate about the pathogen to the leaves.
   Textbook Reference: *28.1 Plants Have Constitutive and Induced Responses to Pathogens, p. 573*
5. Canavanine is toxic to many herbivores but not to plants. Which of the following statements regarding this differential toxicity is true?
   a. Canavanine is confused with arginine in plants but not in animals.
   b. Canavanine is incorporated into plant cells and causes them to fold properly.
   c. Plants are able to differentiate between arginine and canavanine, whereas animals cannot.
   d. Plants store canavanine in vacuoles, whereas animals metabolize it.
   e. None of the above
   Textbook Reference: *28.2 Plants Have Mechanical and Chemical Defenses against Herbivores, p. 576*
6. Which of the following best describes how plants produce their own insecticide?
   a. Wounded cells release an elicitor that directly induces synthesis of protease inhibitors, and protease inhibitors act as insecticides.
   b. Wounded cells release jasmonates, jasmonates stimulate elicitor synthesis, elicitor causes the production of protease inhibitors, and protease inhibitors act as insecticides.
   c. Wounded cells release an elicitor that causes membrane breakdown, membrane breakdown releases jasmonates, and jasmonates act as insecticides.
   d. Wounded cells release an elicitor that causes membrane breakdown, membrane breakdown releases jasmonates, jasmonates induce synthesis of protease inhibitors, and protease inhibitors act as insecticides.
   e. None of the above
   Textbook Reference: *28.2 Plants Have Mechanical and Chemical Defenses against Herbivores, p. 579*
7. Which of the following is *not* an adaptation to drought conditions?
   a. Water-storing leaves
   b. Leaf loss
   c. Sunken stomata
   d. Increased stomata number
   e. Water-storing stems
   Textbook Reference: *28.3 Plants Adapt to Environmental Stresses, pp. 580–583*
8. Halophytes are different from all other types of plants in that they
   a. can accumulate sodium and chloride ions.
   b. have a positive water potential.
   c. contain no sodium or chloride ions.
   d. contain no stomata.
   e. All of the above
   Textbook Reference: *28.3 Plants Adapt to Environmental Stresses, p. 584*
9. Which of the following conditions stimulates the production of heat shock proteins?
   a. Abnormally high temperatures only
   b. Abnormally low temperatures only
   c. Both abnormally high and abnormally low temperatures
   d. Heat shock proteins are continually available in a plant, no matter what the temperature.
   e. None of the above
   Textbook Reference: *28.3 Plants Adapt to Environmental Stresses, pp. 583–584*
10. The main function of heat shock proteins is to
    a. stabilize proteins necessary to a cell's survival.
    b. reinforce membranes that lose fluidity.
    c. cause the plant to enter dormancy.
    d. act as an antifreeze compound.
    e. stimulate bolting.
    Textbook Reference: *28.3 Plants Adapt to Environmental Stresses, p. 583*

11. Which of the following is *not* involved in the process of cold-hardening?
    a. Production of antifreeze proteins
    b. Increased production of saturated fatty acids
    c. Production of phytoalexins
    d. Production of heat shock proteins
    e. All of the above are involved in cold-hardening.
    Textbook Reference: *28.3 Plants Adapt to Environmental Stresses, p. 584*
12. Plants that produce toxins for defense do not also poison themselves because the toxic substances are stored in
    a. roots.
    b. laticifers.
    c. soil.
    d. the same place as enzymes that convert them to the active form.
    e. None of the above; toxic substances do slowly poison the plants.
    Textbook Reference: *28.2 Plants Have Mechanical and Chemical Defenses against Herbivores, p. 579*
13. Which of the following is *not* a secondary plant metabolite?
    a. Phenolics
    b. Alkaloids
    c. Terpenes
    d. Glucosinolates
    e. PR proteins
    Textbook Reference: *28.2 Plants Have Mechanical and Chemical Defenses against Herbivores, p. 577, Table 28.1*
14. The hypersensitive response involves the release of
    a. lignin.
    b. PR proteins.
    c. *R* genes.
    d. phytoalexins.
    e. secondary metabolites.
    Textbook Reference: *28.1 Plants Have Constitutive and Induced Responses to Pathogens, p. 574*
15. A plant's immune response to RNA viruses involves
    a. heat shock proteins.
    b. secondary metabolites.
    c. small interfering RNA.
    d. PR proteins.
    e. All of the above
    Textbook Reference: *28.1 Plants Have Constitutive and Induced Responses to Pathogens, p. 575*

Answers

Key Concept Review

1.

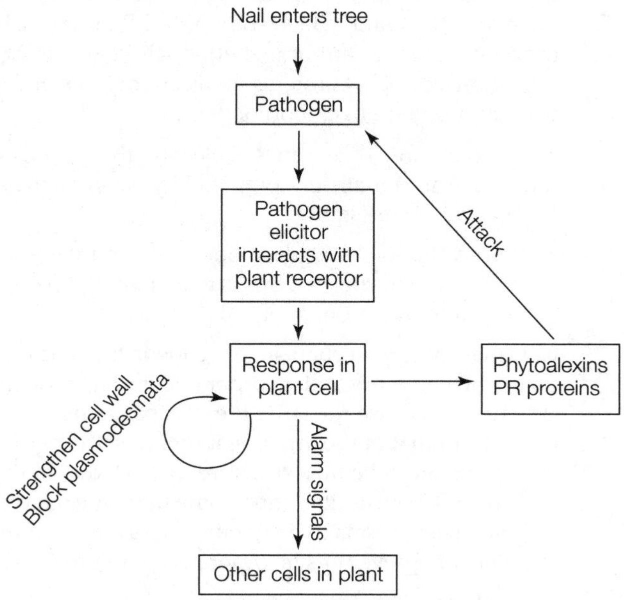

2. Plants prevent herbivory by producing secondary compounds that affect potential herbivores. These compounds may act as neurotoxins, inhibit digestion, or have a number of other consequences (see Table 28.1).
3. Plants may isolate the toxins in vacuoles within their cells so that the toxins do not come into contact with the active machinery of the cells. Plants may restrict toxin production to already-damaged tissue. Plants may also develop enzymes that differentiate toxins and render them harmless.
4. The burlap will most likely do little for the shrubs; the roots are in most danger. Because streets are heavily salted in the winter, the snow around them creates a saline environment. If the shrubs are not adapted to a saline environment, little will help them survive the salt accumulation around their roots.
5. Cold-hardening is gradual exposure to adverse temperatures. This allows for structural changes in cell membranes to help them cope with extreme temperatures. It also allows for the production of heat shock proteins and other compounds necessary for survival at temperature extremes.
6. Only plants that have evolved a strategy for taking up and coping with heavy metals can grow in these contaminated sites. Deliberately planting them at these sites initiates the slow process of removing the metals from the soil.

Test Yourself

1. **d.** If something is always turned on, it is said to be constitutive.
2. **b.** Plants that have been exposed to pathogens wall off invaded tissue. This tissue releases salicylic acid to the rest of the plant, which stimulates PR proteins. The tissue also releases airborne methyl salicylate, which causes release of PR proteins in distant regions of the plant and even in neighboring plants.
3. **a.** The plant must have an *R* allele, and the pathogen must have an *Avr* allele. Compatibility between these alleles leads to resistance.
4. **c.** Polysaccharides seal plasmodesmata and reinforce cell walls. The purpose of this is to contain the pathogen in a minimal number of cells.
5. **c.** Canavanine's mechanism of action is that it is taken up by the herbivore and incorporated into its protein structures. Because the structure of canavanine is different from that of the arginine it replaces, the herbivore's proteins take up a structure that is lethal to the herbivore. The plant itself and some herbivores have enzyme mechanisms to distinguish canavanine from arginine and prevent its incorporation into proteins.
6. **d.** See Figure 28.6 in the textbook.
7. **d.** Water storage tissues allow the plant to take advantage of any water that is available and hold on to it for lean times. Leaf loss, when photosynthesis cannot be sustained, prevents loss of resources. Sunken stomata prevent excessive water loss to evaporation.
8. **a.** They can accumulate sodium and chloride ions. This makes their water potential more negative, and they are able to extract more water from their surroundings than plants that do not store the ions.
9. **c.** Heat shock proteins are produced in response to extreme high and low temperatures.
10. **a.** Heat shock proteins are chaperonin proteins. They function to stabilize and prevent denaturation (unfolding) of essential proteins.
11. **c.** Cold-hardening involves production of antifreeze proteins, heat shock proteins, and an increase in saturated fatty acids in the membrane. The production of phytoalexins is involved in chemical defenses against pathogens.
12. **b.** Plants that produce toxins as a defense typically store them in special locations, such as laticifers, where they will not damage tissue.
13. **e.** Alkaloids, phenolics, terpenes, and glucosinolates are all examples of secondary plant metabolites. PR proteins are proteins that function in defense.
14. **d.** The hypersensitive response involves the release of antibiotic phytoalexins.
15. **c.** Plants that are invaded by RNA viruses use the RNA of the invading virus to interfere with and block viral replication.

Physiology, Homeostasis, and Temperature Regulation

29

The Big Picture

- One of the most important roles of tissues, organs, and organ systems is to help maintain homeostasis within the body's cells. Only by maintaining relatively constant intracellular conditions can basic metabolic reactions continue.
- Animals either generate their own body heat (endotherms) or rely on the environment to determine body temperature (ectotherms). Endotherms typically have more insulation and higher metabolic rates than ectotherms. The metabolic rates of endotherms and ectotherms respond differently to changes in environmental temperature. Endotherms increase metabolism as environmental temperature decreases; they also increase metabolic rate at high temperatures when energy is needed for sweating and panting. The metabolic rate of ectotherms is temperature-dependent and falls with decreases in environmental temperature.
- Control of body temperature can occur by behavioral and physiological mechanisms. In both endothermic and ectothermic vertebrates, the brain's hypothalamus is the thermostat for temperature regulation.

Study Strategies

- Remember that homeostasis does not necessarily mean holding every body variable absolutely constant, but rather making sure that body variables are held at the correct level for the circumstances. Even body temperature in humans, which we think of as being a constant 37°C, drops slightly in the early morning hours and rises considerably during extended vigorous exercise.
- Understanding the difference between ectotherms and endotherms can be difficult. These terms refer to the *source* of heat that determines an animal's temperature. Whereas body temperatures of ectotherms are determined primarily by external sources of heat, those of endotherms are determined by heat generated metabolically (i.e., within their bodies).
- Refer to Figure 29.3 to understand how homeostasis is maintained through regulatory systems.
- Review the organization of physiological systems, from cells to tissues to organs to organ systems. This will help you when you study each of the organ systems in later chapters.
- Go to yourBioPortal.com to review the following tutorial and activities:

  Animated Tutorial 29.1 The Hypothalamus

  Web Activity 29.1 Tissues and Cell Types

  Web Activity 29.2 Thermoregulation in an Endotherm

  Working with Data 29.1 A Hibernator's Thermostat

Key Concept Review

29.1 Multicellular Animals Require a Stable Internal Environment

Multicellular animals have an internal environment of extracellular fluid

Homeostasis is the process of maintaining stable conditions in the internal environment

Cells, tissues, and organs serve homeostatic needs

Most of the water within an animal is intracellular fluid (water within cells); the rest is extracellular fluid (see Figure 29.1). Some extracellular fluid is plasma (the liquid portion of blood) and the rest is interstitial fluid (fluid between cells).

Constancy of the internal environment is critical. Homeostasis is the state of having a constant, regulated internal environment. For example, temperature, pH, ion concentrations, and levels of glucose must be maintained within specific limits.

Question 1. Why is it important for an organism to maintain homeostasis?
Textbook Reference: *29.1 Multicellular Animals Require a Stable Internal Environment, p. 589*

A tissue consists of a group of similar cells with the same form and function. There are four types of tissue: epithelial, connective, muscle, and nervous. Epithelial tissues line the inner and outer body surfaces, such as the lining of the various organs and the skin. Connective tissue consists of an extracellular matrix (made of the proteins collagen and elastin) that holds together a dispersed group of cells. Cartilage, bone, adipose tissue, and blood are connective tissues. There are three types of muscle tissue: skeletal (attached to bone and responsible for body movement), smooth (generates force in blood vessels and internal organs, such as the stomach and bladder), and cardiac (generates the heartbeat and pumps blood). Nervous tissue consists of neurons, which generate and conduct electrochemical signals, and glial cells, which support and protect neurons.

Organs are groups of tissues that carry out a specific function in the body. Most organs are composed of all four of the major tissue types (see Figure 29.2). The integration of a number of organs to carry out a specific function is known as an organ system (e.g., the digestive system).

Question 2. Label the organs and four main tissue types in the diagram below.
Textbook Reference: *29.1 Multicellular Animals Require a Stable Internal Environment, p. 590*

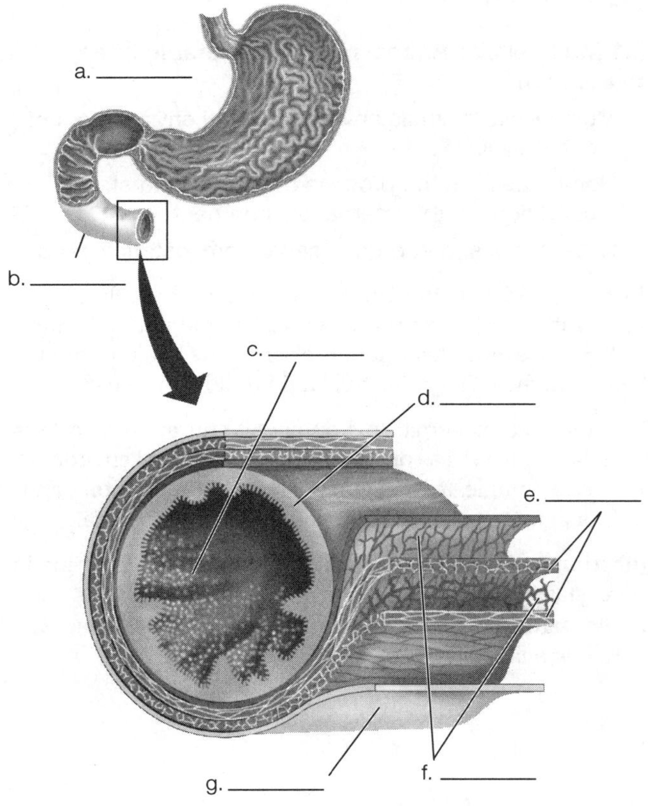

29.2 Physiological Regulation Achieves Homeostasis of the Internal Environment

Regulating physiological systems requires feedback information

Feedback information can be negative or positive

Physiological regulatory systems require both a set point at which the parameter is to be held, and feedback, which provides information about the state of the system (see Figure 29.3). Differences between the set point and the feedback information indicate to the regulatory mechanism which corrective measures are required to restore a parameter to its set point.

Physiological control systems must take information from the regulatory systems and effect changes. Negative feedback results in a slowdown or reversal of a process, returning the variable controlled by that process back toward its set point or regulated level. In contrast, positive feedback results in amplification of a response. Regulation by positive feedback is less common than regulation by negative feedback.

The set point maintained by an organism for a given physiological process can also be changed by feedforward information, which changes the set point.

Question 3. During childbirth, the pressure exerted on the mother's cervix by the emerging infant leads to increased contraction of the uterus. What type of feedback—negative or positive—is involved in this situation?
Textbook Reference: *29.2 Physiological Regulation Achieves Homeostasis of the Internal Environment, p. 592*

Question 4. In response to the smell of food, the stomach produces hydrochloric acid in anticipation of food. Is this an example of negative feedback, positive feedback, or feedforward information? Explain your answer.
Textbook Reference: *29.2 Physiological Regulation Achieves Homeostasis of the Internal Environment, p. 592*

29.3 Living Systems Are Temperature-Sensitive

Q_{10} is a measure of temperature sensitivity

Animals can acclimatize to seasonal temperature changes

Animals can regulate body temperature

Normal cellular functions are limited to a temperature range between 0°C and 40°C, with most organisms having much narrower limits in their range of survival temperatures. For example, the range for humans is between 35°C and 40°C.

All physiological processes in organisms are sensitive to temperature. Increases in temperature usually increase the rate of a given process. The Q_{10} is a measure of change in a physiological process as the temperature increases or decreases by 10°C (see Figure 29.4). Most biological Q_{10} values range from 2 to 3. A Q_{10} of 2 means that the reaction rate doubles when temperature increases by 10°C.

Acclimatization is the change in a physiological process that occurs over time in response to a change in the environment. For example, a fish experimentally exposed in summer to decreasing water temperatures will show slowed physiological functions. However, the physiological functions of the same fish in winter will not be as slow as they were when temperatures were experimentally reduced in summer because the physiology of the fish has acclimatized to seasonal changes in water temperature. Fish may catalyze reactions with one set of enzymes in the summer and a different set of enzymes in the winter.

The metabolic rate of an organism is the total energy used by the animal and is usually measured as oxygen consumption.

Animals can be classified as either endotherms ("heat from the inside"), which regulate their body temperature with internal heat production by means of a high metabolic rate and effective insulation, or ectotherms ("heat from the outside"), which rely on the environment to provide body heat.

Ectotherms have body temperatures that track environmental temperatures. In contrast, endotherms maintain a constant body temperature even as the environmental temperature changes (see Figure 29.5A). Also, in response to cold temperatures, metabolic heat production increases in endotherms, but decreases in ectotherms. The range of environmental temperatures at which the metabolic rate of endotherms does not change is known as the thermoneutral zone (see Figure 29.5B). Within the thermoneutral zone, metabolic rate is at the minimum level needed to maintain physiological functions essential to homeostasis in the resting animal; this level is called the basal metabolic rate.

Question 5. Having measured the metabolism of two animals of similar size at 15°C and 25°C, you find that the metabolic rate of Animal 1 is 115 ml of oxygen per hour at 15°C and 55 ml of oxygen per hour at 25°C. The metabolic rate of Animal 2 is 5.5 ml of oxygen per hour at 15°C and 11.5 ml of oxygen per hour at 25°C. Calculate the Q_{10} for the metabolic rate in both animals and determine if they are endotherms or ectotherms.
Textbook Reference: *29.3 Living Systems Are Temperature-Sensitive, pp. 592–593*

Question 6. Animals have the ability to undergo acclimatization. Describe this process in relation to temperature regulation.
Textbook Reference: *29.3 Living Systems Are Temperature-Sensitive, p. 593*

29.4 Animals Control Body Temperature by Altering Rates of Heat Gain and Loss

Mammals and birds have high rates of metabolic heat production

Basal metabolic rates are correlated with body size

Insulation is the major adaptation of endotherms to cold climates

Some fish can conserve metabolic heat

Evaporation is an effective but expensive avenue of heat loss

Some ectotherms elevate their metabolic heat production

Behavior is a common thermoregulatory adaptation in ectotherms and endotherms

Heat can be gained or lost to the environment through several routes (see Figure 29.6). Evaporation from the skin or respiratory tract cools the body. Solar radiation from the sun heats the body. Thermal radiation, convection, and conduction can either warm or cool the body, depending on the thermal environment.

The total balance of heat production and heat exchange for an animal is its energy budget. To maintain a constant body temperature, the heat entering an animal must equal the heat leaving. Generally, $heat_{in}$ is generated by metabolism (MR) and thermal radiation (R) and $heat_{out}$ occurs through convection, conduction, and evaporation.

Most of the energy expended by an endotherm goes into pumping ions across membranes to maintain proper concentrations inside and outside cells. Upper and lower critical temperatures are the environmental temperatures at which metabolic rate begins to increase. When the environmental temperature falls below the lower critical temperature, metabolic heat can be produced by shivering thermogenesis (skeletal muscles contract) and nonshivering thermogenesis (tissues other than skeletal muscle produce metabolic heat by uncoupling oxidative phosphorylation). Nonshivering thermogenesis involves a specialized adipose tissue, brown fat (see Figure 29.7).

For endotherms, the basal metabolic rate per gram of tissue decreases with increasing body mass (see Figure 29.8). The reason for this pattern is unclear.

Thermal conductance is a measure of how readily an animal loses heat. Insulation is a measure of how effectively an animal conserves heat. Fur and feathers provide insulation for mammals and birds in cold environments (see Figure 29.9). These structures retain heat produced in the body and help maintain a constant body temperature.

In most fish, heat picked up from metabolically active muscles is lost when blood is pumped from the heart to the gills, where it comes in contact with cold water (see Figure 29.10A). Some large fishes, such as bluefin tuna, have the ability to maintain a body temperature higher than the surrounding water. They possess special vascular countercurrent heat exchangers that retain metabolically produced heat in the muscle mass (see Figure 29.10B). In countercurrent heat exchangers, heat is exchanged between blood vessels carrying blood in opposite directions (see Figure 29.10C).

Evaporative cooling by sweating or panting provides an avenue for heat loss to the environment. However, evaporative cooling includes the costs of water loss and the expenditure of metabolic energy (sweating is an active process).

Some ectotherms can generate heat to raise body temperature. Some insects do so by contracting their flight muscles. Other insects, such as honeybees, engage in group thermoregulation whereby individuals form clusters in winter and regulate temperature by adjusting their own metabolic heat production and the density of the cluster (see Figure 29.11).

Both endotherms and ectotherms use behavioral mechanisms to regulate body temperature. For example, ectotherms shuttle between warm and cool environments during the day to maintain body temperature within a given range (see Figure 29.12).

Question 7. In the figure below, graph the lizard's predicted body temperature based on your knowledge of how ectotherms regulate their temperature with behavior.
Textbook Reference: *29.4 Animals Control Body Temperature by Altering Rates of Heat Gain and Loss, p. 599*

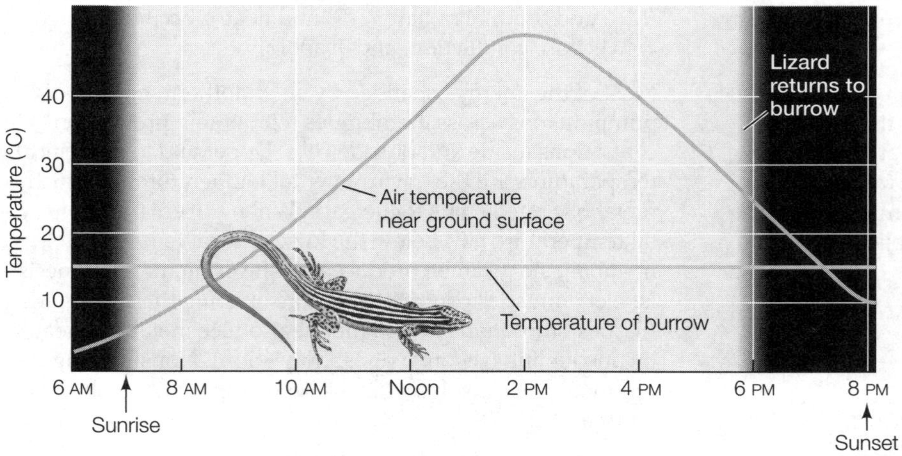

Question 8. Why would an endothermic animal's metabolic rate be higher at 40°C than at 30°C? What physiological phenomenon would explain why its metabolic rate rises when the environmental temperature is above the upper critical limit?
Textbook Reference: *29.4 Animals Control Body Temperature by Altering Rates of Heat Gain and Loss, p. 598*

Question 9. What types of insulation do mammals have, and what physiological controls exist to increase insulation?
Textbook Reference: *29.4 Animals Control Body Temperature by Altering Rates of Heat Gain and Loss, pp. 596–597*

29.5. A Thermostat in the Brain Regulates Mammalian Body Temperature

The mammalian thermostat uses feedback information

A thermostat can be adjusted up or down

The major integrative center for thermoregulation is the hypothalamus, a structure at the bottom of the brain. The hypothalamus has a set point temperature; changes in temperature of the hypothalamus result in thermoregulatory responses to offset any temperature change (see Figure 29.13). The hypothalamus also integrates thermal information received from the skin.

In mammals, hypothalamic set points are different at different environmental temperatures. Other factors, including the sleep–wake cycle and level of activity, also influence set points.

The thermostat can be adjusted up and down. The body has the ability to increase the set point temperature in response to infections, resulting in fever. A fever is a regulated rise in body temperature caused by a rise in the hypothalamic set point for the production of metabolic heat.

Hypothermia is a general state of below-normal body temperature. Some animals are able to decrease body temperature either on a daily scale (daily torpor) or over longer time periods (hibernation) (see Figure 29.14). Daily torpor and hibernation are examples of regulated hypothermia that allow animals to survive during periods of low temperature and scarce food.

Question 10. Complete the table below, which compares some general traits of endotherms and ectotherms.
Textbook Reference: *29.5 A Thermostat in the Brain Regulates Mammalian Body Temperature, p. 599*

	Endotherm	Ectotherm
Heat source		
Efficiency of energy usage		
Resting metabolic rate		
Temperature control center		
Insulation		

Question 11. Design a study to determine that the vertebrate thermostat is located in the hypothalamus.
Textbook Reference: *29.5 A Thermostat in the Brain Regulates Mammalian Body Temperature, pp. 599–600*

Test Yourself

1. Which of the following statements about interstitial fluid is true?
   a. It is found within cells.
   b. It represents most of the water in an animal's body.
   c. It makes up blood plasma.
   d. It does not exchange molecules with intracellular fluid.
   e. It is found between the cells of the body.
   Textbook Reference: *29.1 Multicellular Animals Require a Stable Internal Environment, p. 589*

2. In an environment with an ambient temperature lower than an animal's core body temperature, which of the following would be an *inappropriate* physiological or behavioral response if the animal needed to eliminate excess body heat?
   a. Wallowing in a pool of water
   b. Inhibiting blood flow to peripheral vessels
   c. Sweating
   d. Decreasing physical activity
   e. All of the above would be inappropriate responses.
   Textbook Reference: *29.4 Animals Control Body Temperature by Altering Rates of Heat Gain and Loss, pp. 594–598*

3. A lizard lives in a desert environment where the temperature is low at night and high during the day. What might it do to maintain the most stable body temperature?
   a. Stay in a burrow during the night and shuttle between the sun and shade on the surface during the day
   b. Stay on the surface during the night and move to a burrow during the day
   c. Increase metabolism and heat production during the night and decrease it during the day
   d. Decrease metabolism and heat production during the night and increase it during the day
   e. Move into a state of hypothermia during the night and a state of torpor during the day
   Textbook Reference: *29.4 Animals Control Body Temperature by Altering Rates of Heat Gain and Loss, pp. 598–599*

4. On a hot day, an active dog will pant heavily and visit his water bowl frequently. Which of the following statements about the dog's condition or behavior is *false*?
   a. He is lowering his core body temperature by means of evaporative cooling.
   b. He is operating at a basal metabolic rate.
   c. He is at the upper end of the thermoneutral zone.
   d. He is sending more blood to his skin.
   e. He is adjusting behavior in order to thermoregulate.
   Textbook Reference: *29.4 Animals Control Body Temperature by Altering Rates of Heat Gain and Loss, pp. 594–598*

5. Evaporative cooling is an effective way to increase heat loss, but it carries the physiological drawback of
   a. an increased use of ATP and substantial water loss.
   b. the lowering of the hypothalamic thermal set point.
   c. the decreased use of ATP and substantial water gain.
   d. the exhaustion of the supply of brown fat.
   e. a lowered basal metabolic rate.
   Textbook Reference: *29.4 Animals Control Body Temperature by Altering Rates of Heat Gain and Loss, p. 598*

6. Which of the following statements about fever is true?
   a. It is a higher-than-normal body temperature that is always dangerous.
   b. It decreases the metabolic rate of the body to conserve energy.
   c. It is a regulated resetting of the body's thermostat to a higher set point.
   d. It causes the liver to release large amounts of calcium, which seems to inhibit bacterial growth.
   e. It should always be controlled with fever-reducing drugs.
   Textbook Reference: *29.5 A Thermostat in the Brain Regulates Mammalian Body Temperature, p. 600*

7. Which of the following is a characteristic unique to connective tissue?
   a. Highly modified cells that show the special property of contractibility
   b. Diverse anatomy with distinct specializations for information transfer

c. Densely packed cells that cover inner and outer body surfaces
d. A loose array of cells embedded in an extracellular matrix
e. Possess cilia to move substances over surfaces.

Textbook Reference: *29.1 Multicellular Animals Require a Stable Internal Environment, p. 590*

8. Elephants use their ears to release heat to the environment. What mechanisms might they employ to increase heat loss from the ears?
a. Increased convection by means of ear flapping
b. Moving into the sun
c. Increased blood flow to the ears
d. Covering their ears with dust
e. Both a and c

Textbook Reference: *29.4 Animals Control Body Temperature by Altering Rates of Heat Gain and Loss, pp. 594–598*

9. Which of the following organ systems is not involved in the maintenance of homeostasis in humans?
a. Endocrine
b. Digestive
c. Muscle
d. Reproductive
e. All of the above organ systems are involved in maintaining homeostasis.

Textbook Reference: *29.1 Multicellular Animals Require a Stable Internal Environment, p. 591*

10. What is the difference between negative feedback mechanisms and positive feedback mechanisms?
a. Negative feedback mechanisms exist only in the circulatory system.
b. Negative feedback mechanisms return a system to a set point, whereas positive feedback mechanisms amplify a response.
c. Negative feedback mechanisms move a system away from a set point, whereas positive feedback mechanisms stabilize a system toward a set point.
d. Negative feedback mechanisms stabilize a system toward a set point, whereas positive feedback mechanisms reset the set point.
e. There is essentially no difference between the two feedback systems.

Textbook Reference: *29.2 Physiological Regulation Achieves Homeostasis of the Internal Environment, p. 592*

11. A number of physiological processes can undergo acclimatization. Which of the following statements about acclimatization is *false*?
a. It occurs in response to seasonal temperature changes.
b. Acclimatization of metabolic rate occurs because enzyme expression changes.
c. It involves the changing of a set point.
d. All multicellular animals are capable of acclimating to all the environmental changes they face.
e. Acclimatization is primarily seen in organisms with body temperatures tightly coupled to the environmental temperatures.

Textbook Reference: *29.3 Living Systems Are Temperature-Sensitive, p. 593*

12. In fast-swimming cool-water fishes such as sharks and tunas, which of the following contribute(s) to the generation and maintenance of core body temperatures in excess of ambient water temperatures?
a. Specializations in the size and arrangement of blood vessels
b. Low rates of activity in the swimming muscles
c. An ability to acclimatize rapidly to the surrounding water
d. Metabolic rates that are insensitive to temperature change
e. An insulating layer of fat located beneath the scaled skin.

Textbook Reference: *29.4 Animals Control Body Temperature by Altering Rates of Heat Gain and Loss, pp. 597–598*

13. Which of the following statements about hypothalamic function is *false*?
a. Circulating blood temperature is monitored by the hypothalamus.
b. Hypothalamic thermal set points never change.
c. The hypothalamus is a part of the central nervous system.
d. Different thermoregulatory responses have different hypothalamic set points.
e. The hypothalamus has an important function in fish, birds, amphibians, reptiles, and mammals.

Textbook Reference: *29.5 A Thermostat in the Brain Regulates Mammalian Body Temperature, p. 600*

14. Some animals use brown fat as a source of heat generation. Which of the following statements about brown fat is *false*?
a. Brown fat is involved in nonshivering thermogenesis.
b. Brown fat burns fuel without producing ATP.
c. Brown fat is found in some mammals.
d. Brown fat is highly vascularized.
e. Brown fat cells contain few mitochondria

Textbook Reference: *29.4 Animals Control Body Temperature by Altering Rates of Heat Gain and Loss, pp. 595–596*

15. Which of the following statements about nervous tissue is true?
a. It cushions internal organs and insulates the body.
b. Its glial cells fire nerve impulses.
c. It is contractile.
d. Its glial cells support and protect neurons.
e. Its glial cells are a type of connective tissue holding neurons in place.

Textbook Reference: *29.1 Multicellular Animals Require a Stable Internal Environment, p. 590*

Answers

Key Concept Review

1. Homeostasis is the maintenance of a constant internal environment.

   The body functions with the help of proteins and enzymes. Changes in, for example, pH, temperature, glucose level, and oxygen and carbon dioxide levels of the internal environment can affect cellular function. Loss of homeostasis can lead to improper function of proteins and cell membranes, and ultimately cell death.

2.
   a. Stomach
   b. Small intestine
   c. Epithelial cells
   d. Mucosa
   e. Smooth muscle
   f. Nervous tissue
   g. Epithelial cells and connective tissue

3. Increased uterine contraction in response to pressure on the cervix is an example of a positive feedback.

4. An example of feedforward information is when the smell of food prompts the stomach to produce hydrochloric acid in anticipation of food. Feedforward information is predictive of a change in the internal environment before that change occurs. Feedforward information functions to change the set point.

5. Metabolic rate can be calculated using the equation Q_{10} = (RT/RT – 10). The second animal shows a direct correlation between temperature and metabolism, and compared to the first animal it has a lower overall metabolic rate at both temperatures. This information can be organized into a table such as the following:

	Metabolic rate at 25°C	Metabolic rate at 15°C	Q_{10}	Response of metabolism to decreasing temperature	Type of animal
Animal 1	55 ml O_2/hour	115 ml O_2/hour	0.48 (=55/115)	Metabolism increases	Endotherm
Animal 2	11.5 ml O_2/hour	5.5 ml O_2/hour	2.1 (=11.5/5.5)	Metabolism decreases	Ectotherm

6. Animals undergo acclimatization in response to changes in seasonal conditions such as temperature. Often these changes are brought about by the production of enzymes that function better at the new temperature. Because of acclimatization, metabolic functions are less sensitive to long-term changes in temperature than to short-term changes.

7.

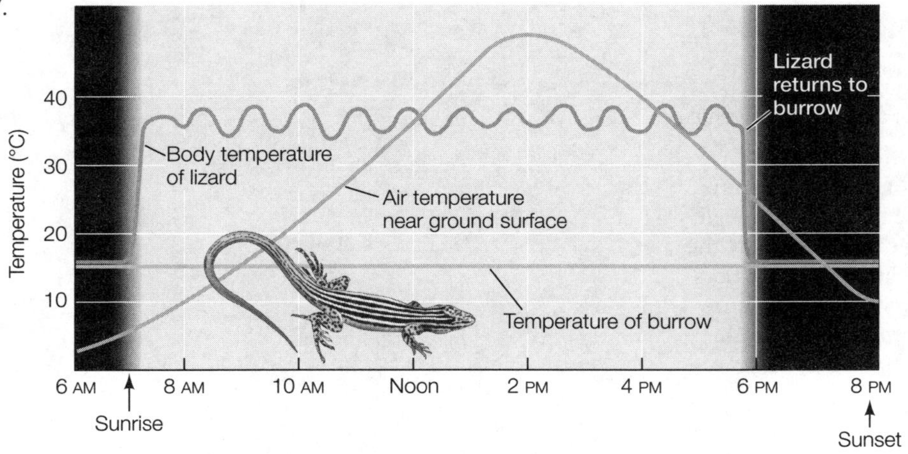

8. An endotherm in a hot environment will attempt to keep core body temperature from increasing. The heat-loss mechanisms available to an endotherm—sweating and panting—require the use of ATP and result in an increased metabolic rate. Endotherms will expend energy to protect body temperature and lose heat to the environment.

9. Fur and fat are effective insulators in mammals. Fur is an effective insulator because it traps still, warm air from the body; the longer the fur, the more effective it is as an insulator. Fat works in a similar manner because it has a low thermal conductance. Humans do not have sufficient hair on their bodies for it to provide adequate insulation; therefore, clothes are used as insulation. Physiologically, mammals can change their blood flow patterns to increase or decrease the amount of blood that is received by the skin. When they are active, they get rid of excess heat by transporting heat to hairless skin surfaces.

10.

	Endotherm	**Ectotherm**
Heat source	Behavioral and internal metabolism	Behavioral and environmental
Efficiency of energy usage	Leaky ion channels; inefficient	Non-leaky ion channels; more efficient
Resting metabolic rate	Higher	Lower
Temperature control center	Hypothalamus	Hypothalamus
Insulation	Fur, feathers, fat layer beneath skin	Little to none

11. Two possible approaches would be to conduct lesion experiments and thermal stimulation experiments. In lesion experiments, you would destroy the region of the hypothalamus thought to function as the thermostat and see what happens to the animal's ability to regulate body temperature. In thermal stimulation experiments, you could cool or warm the hypothalamus and monitor physiological and behavioral changes associated with thermoregulation (such as constriction or dilation of blood vessels in the skin, and sweating or panting).

Test Yourself

1. **e.** Interstitial fluid is found between the cells of the body.
2. **b.** If the environment is cooler than the animal, the response would be to lose heat to the environment across the skin. An increase in peripheral blood flow moves heat from the core of the animal to the surface, where it would be lost to the cooler environment.
3. **a.** For an ectothermic lizard in the desert to maintain the most stable body temperature, it should remain in a burrow during the night, where temperatures do not drop very much, and then shuttle between the sun and shade during the day.
4. **b.** After exercise, a dog will have a high metabolic rate and will pant in an attempt to cool down. Its body temperature will most likely be above the upper thermoneutral zone.
5. **a.** Evaporative cooling is a costly means to lower body temperature. ATP must be used either during sweating or panting. Water loss can also be high during evaporative cooling.
6. **c.** Fevers are brought about by a resetting of the thermoregulatory set point in the hypothalamus.
7. **d.** Connective tissues are blood, cartilage, and bone, all of which contain a loose population of cells embedded in an extracellular matrix.
8. **e.** Elephants use both physiological mechanisms (increased blood flow to the ears) and behavioral mechanisms (flapping of ears) to release heat into the environment.
9. **e.** All of these organ systems are involved in maintaining homeostasis. The endocrine system is a regulatory system, the digestive system acquires nutrients and raw materials for the entire body, the muscle system moves materials through the body and generates heat, and the reproductive system produces hormones that maintain secondary sex characteristics.
10. **b.** Negative feedback loops use information about the state of a system to bring the internal environment back toward the set point values. Positive feedback loops amplify the response of a system, moving the system away from the set point.
11. **d.** Physiological processes acclimate in response to a seasonal change in the environment. Acclimatization may involve changes in enzyme expression and a resetting of a physiological set point. Every organism has limits to its ability to acclimate.
12. **a.** Large fishes such as tuna are able to maintain core body temperature above ambient temperature by means of a number of mechanisms. They have a special countercurrent heat exchanger built into their blood vessels and high metabolic rates; thus they produce heat in their swimming muscles.
13. **b.** The hypothalamic thermal set points change daily and seasonally.
14. **e.** Brown fat, which is found in some mammals, burns fuel without producing ATP. This process of heat production is termed nonshivering thermogenesis. Brown fat is highly vascularized and rich in mitochondria.
15. **d.** Glial cells are a type of nervous tissue cell. They provide support and protection to neurons.

Animal Hormones 30

The Big Picture

- Hormones regulate functions ranging from growth to sexual maturity. In numerous ways, the endocrine system complements the nervous system, both of which are the major regulators of body function and homeostasis.
- Circulating hormones travel via the bloodstream to distant sites within the body to affect target tissues. Other endocrine signals, called paracrine signals, act locally. Autocrine signals act on the cells that secrete them. Cells that respond to particular endocrine signals are called target cells.
- Hormone action is controlled by several different mechanisms. In many instances there is a "hormone cascade," in which a hormone controls the release of another hormone, which controls the release of another hormone, and so on. In a situation of negative feedback, a released hormone inhibits (negatively feeds back on) the tissue that may have stimulated its release in the first place. Tropic hormones influence other endocrine glands.

Study Strategies

- The pituitary gland produces numerous regulatory hormones. The functions of the anterior and posterior pituitary are quite distinct, and keeping them separate is important for understanding the endocrine system.
- Remember that male sex steroids (androgens) and female sex steroids (estrogens and progesterone) are made and used by both sexes, but the relative levels vary in the two sexes.
- The many hormones may seem to constitute a long list of random compounds. Try to learn the function along with the name of each hormone, as this will help you remember the target tissues.
- Some hormones have extremely specific actions (e.g., follicle-stimulating hormone), whereas others have broad effects on many target tissues (e.g., epinephrine and thyroxine). Learning which hormones are "specialists" and which are "generalists" will help you better understand the endocrine system.
- Biologists studying the endocrine system frequently use abbreviations for the names of hormones. Learn these abbreviations along with the full names.
- Go to yourBioPortal.com to review the following tutorials and activities:

  Animated Tutorial 30.1 Complete Metamorphosis

  Animated Tutorial 30.2 The Hypothalamic–Pituitary–Endocrine Axis

  Animated Tutorial 30.3 Hormonal Regulation of Calcium

  Web Activity 30.1 The Human Endocrine Glands

  Web Activity 30.2 Concept Matching: Vertebrate Hormones

Key Concept Review

30.1 Hormones Are Chemical Messengers

Endocrine signals can act locally or at a distance

Hormones can be divided into three chemical groups

Hormonal communication has a long evolutionary history

Insect studies reveal hormonal control of development

Individual cells within a multicellular organism must be able to communicate with one another. Typically, one cell releases a chemical signal that travels to other cells with appropriate receptors to trigger a response.

The endocrine system consists of cells that make and release chemical signals called hormones. Endocrine cells may exist as single cells scattered within an organ or as large aggregates known as endocrine glands.

Hormones are released into the extracellular fluid, and then most diffuse into the blood. In contrast, secretions from exocrine glands are released into ducts that empty onto the surface of the skin or into a body cavity. Examples of exocrine glands are sweat glands and salivary glands.

Endocrine signals can be classified as hormones (which act on cells far from their release site), paracrine signals (which act on nearby cells), or autocrine signals (which act on the cells that secrete them). Cells that respond to endocrine signals are target cells. Target cells have receptors that recognize and bind to a signal, which eventually triggers a response by the cell.

Three classes of hormones include peptides or protein hormones, steroid hormones, and amine hormones (see Figure 30.1). Most hormones are peptides. These hormones are water-soluble and easily transported in the blood. Before release, peptide and protein hormones are packaged in vesicles and released by exocytosis. Because they do not pass easily through the cell membrane, their receptors are on the exterior of the target cell. Steroid hormones are derived from cholesterol and are lipid-soluble. Steroid hormones easily pass through membranes of the cells that make them, but they require carrier proteins to transport them in the blood. They also readily pass through the membranes of target cells, so receptors for steroid hormones are often inside target cells. Amine hormones are small molecules made from single amino acids. Some are water-soluble and some are lipid-soluble.

Chemical signaling between cells was critical for the evolution of multicellularity. Even sponges, the most primitive of multicellular animals, have forms of intercellular chemical communication. Hormone structure is highly conserved over evolutionary time, but hormone function may change. For example, prolactin stimulates milk production in female mammals, migration in salmon, and nest building in birds.

In insects, hormones control the processes of molting (shedding of the exoskeleton) and metamorphosis (transformation to the adult stage). Each growth stage between molts is called an instar.

In an elegant set of experiments, Sir Vincent Wigglesworth used the blood-sucking bug *Rhodnius* to examine the control of molting (see Figure 30.2). Decapitation of these bugs at various times after feeding showed that a substance secreted in the heads of the bugs stimulated molting. If the head was removed after a blood meal, molting did not take place. However, if the head was removed a week after a meal, molting did take place. Wigglesworth hypothesized that a substance produced in the insect's head shortly after feeding diffused slowly, and eventually triggered molting.

The brains of *Rhodnius* and other insects produce prothoracicotropic hormone (PTTH; previously called brain hormone), which is stored in the corpora cardiaca attached to the brain. After a meal, PTTH diffuses in extracellular fluid to the prothoracic gland. The prothoracic gland releases a steroid hormone called ecdysone, and ecdysone diffuses to the target tissues to stimulate molting. The control of insect molting by PTTH and ecdysone illustrates that the nervous and endocrine systems often act in concert to control growth and development; this link is also common in many vertebrates.

Juvenile hormone, which is secreted by the corpora allata, prevents premature maturation. In insects such as *Rhodnius* with incomplete metamorphosis, high levels of juvenile hormone result in the molting bug becoming a slightly larger juvenile. When levels of juvenile hormone drop, the bug becomes an adult. In insects such as butterflies or moths that undergo complete metamorphosis, the levels of juvenile hormone decrease as the larva molts a fixed number of times. When juvenile hormone falls below a certain level, the larva spins into a cocoon and molts into a pupa. No juvenile hormone is secreted during the pupal stage, so the pupa metamorphoses into the adult (see Figure 30.3).

Question 1. The hormone testosterone is produced and released mainly by the testes. Design a study to determine the effects of testosterone on the behavior of male rats.
Textbook Reference: *30.1 Hormones Are Chemical Messengers, pp. 605–606*

Question 2. Sir Vincent Wigglesworth conducted a number of studies on the insect *Rhodnius* to examine the control of molting. Describe what would happen to a fourth-instar *Rhodnius* if it were either partially or fully decapitated one week after a blood meal.
Textbook Reference: *30.1 Hormones Are Chemical Messengers, pp. 605–606*

30.2 Hormones Act by Binding to Receptors

Hormone receptors can be membrane-bound or intracellular

Hormone action depends on the nature of the target cell and its receptors

Hormone receptors are regulated

Water-soluble hormones (peptide and some amine hormones) cannot pass through the plasma membrane and thus act on receptors at the surface of target cells. These surface receptors are large transmembrane complexes with a binding domain that projects outside the plasma membrane, a transmembrane domain that anchors the receptor in the membrane, and a cytoplasmic domain that extends into the cytoplasm. Binding of a hormone changes the conformation of the receptor, and the cytoplasmic domain initiates the cell's response. This response usually involves activating protein kinases or phosphatases within the cell, which alter enzyme activity in the cytoplasm and lead to the cell's response (see Figure 5.17 for a review).

Lipid-soluble hormones (steroids and some amine hormones) are able to cross the plasma membrane and act on receptors inside the cell. Once a steroid hormone binds to a receptor, the hormone–receptor complex typically moves into the nucleus, where it ultimately alters gene expression.

A single hormone can act on many different types of cells to produce different effects; the response depends on the nature of the cell and its receptors. Epinephrine, secreted by the adrenal glands, activates the fight-or-flight response and causes changes in heart rate, blood flow, and the immune system. Epinephrine also stimulates the breakdown of glycogen in the liver and of fats in adipose tissue (see Figure 30.4).

Epinephrine and the related hormone norepinephrine bind to adrenergic receptors. This set of receptors contains at least five types with different affinities for epinephrine and norepinephrine (see Figure 30.5).

Downregulation is a decrease in the number of receptors of a target cell, often in response to continuous high levels of a hormone. Downregulation makes cells less sensitive to the hormone. Upregulation is an increase in the number of receptors of a target cell, often in response to continuous low levels of a hormone.

The human endocrine system consists of endocrine glands and organs with some endocrine tissue (see Figure 30.6).

Question 3. Describe the modes of action of lipid-soluble and water-soluble hormones. Which type of hormone would likely produce more rapid effects?
Textbook Reference: *30.2 Hormones Act by Binding to Receptors, p. 607*

Question 4. In Diagram A below, identify and match the gland with its location in the body. In the case of gonads, you may use the same location as long as you identify the sex of the individual. In Diagram B, label the structures indicated. In Diagram C, list the hormones secreted by each gland pictured.
Textbook Reference: *30.2 Hormones Act by Binding to Receptors, p. 610*

A.

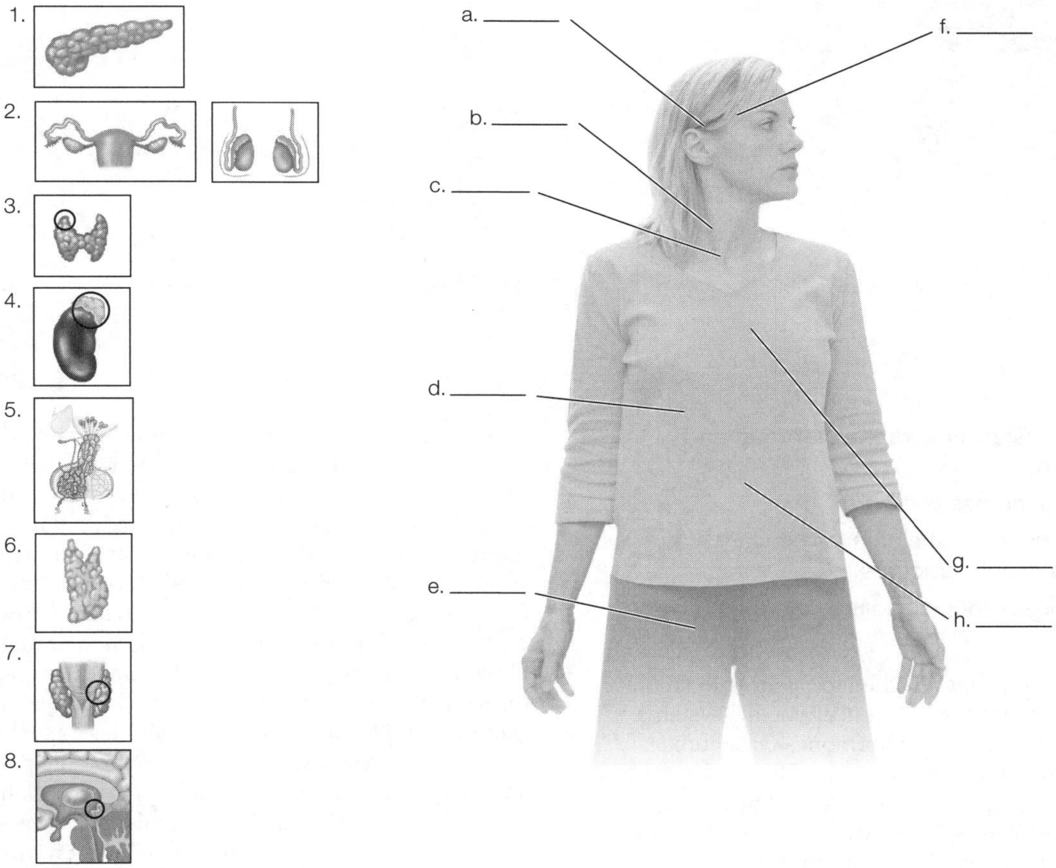

B.

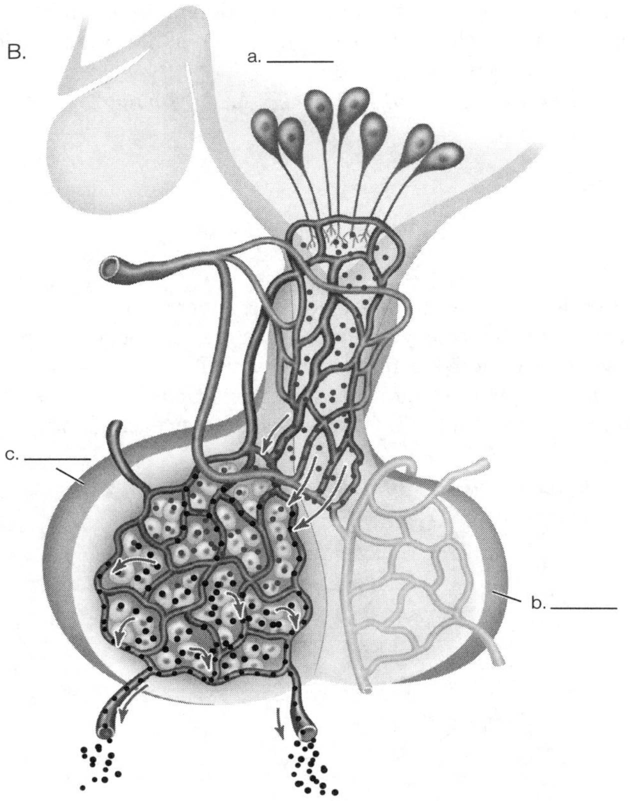

C.

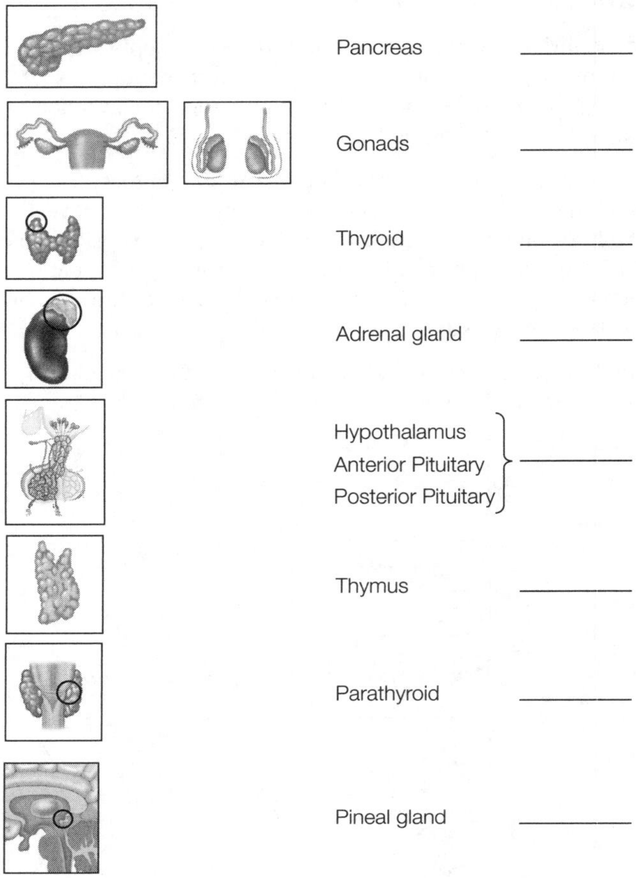

30.3 The Pituitary Gland Links the Nervous and Endocrine Systems

- The pituitary gland has two parts
- The neurohormones of the anterior pituitary are produced in minute amounts
- Negative feedback loops also regulate the pituitary hormones

The pituitary gland, which is attached to the hypothalamus, is a key meeting point for nervous and endocrine systems. The hypothalamus produces two hormones (antidiuretic hormone and oxytocin) that are stored and then secreted by the pituitary gland. Other hormones released by the hypothalamus influence release of pituitary hormones. Hormones released by neurons are called neurohormones.

The pituitary gland produces hormones that control many of the other endocrine glands in the body. The pituitary is made up of the anterior pituitary (derived from gut epithelium) and the posterior pituitary (derived from neural tissue). Each part uses different control mechanisms and releases different hormones.

Long axons from the hypothalamus extend into the posterior pituitary (see Figure 30.7). The posterior pituitary releases antidiuretic hormone and oxytocin, both of which are produced by neurons in the hypothalamus. Antidiuretic hormone (ADH, also called vasopressin) acts on the kidneys to stimulate reabsorption of water when blood pressure drops or salt increases in the blood. Oxytocin stimulates uterine contractions during childbirth and ejection of milk from mammary glands. It can be secreted simply in response to the sight and sound of a baby—a good example of how an external stimulus received by the nervous system controls a hormonal process. Oxytocin can also promote pair bonding and maternal bonding. The anterior pituitary gland releases four tropic hormones: thyroid-stimulating hormone (TSH), which controls thyroid function; luteinizing hormone (LH), which controls the function of the testes; follicle-stimulating hormone (FSH), which controls ovarian function; and adrenocorticotropic hormone (ACTH), which controls the adrenal cortex.

A different cell type produces each of the tropic hormones. The anterior pituitary also produces several other peptide hormones, including growth hormone and prolactin. Growth hormone (GH) stimulates the liver to release chemical signals called somatomedins, or insulin-like growth factors (IGFs), which stimulate bone and cartilage growth. GH also stimulates amino acid uptake by other cells. Over- or underproduction of GH can cause gigantism or pituitary dwarfism, respectively.

The anterior pituitary is composed of endocrine cells that also respond to neurohormones that are secreted by the hypothalamus and transported to the anterior pituitary by portal blood vessels (see Figure 30.8). One of these hypothalamic hormones is thyrotropin-releasing hormone (TRH), which causes the release of thyroid-stimulating hormone (TSH) by the anterior pituitary. TSH, in turn, stimulates the thyroid gland to release thyroxine. As with cortisol, these steps are controlled by a negative feedback loop.

Because portal blood vessels deliver molecules from the hypothalamus directly to the anterior pituitary, neurohormones from the hypothalamus can be produced in very small amounts. In addition, the anterior pituitary is under negative feedback control by hormones of the target glands it stimulates (see Figure 30.9). For example, cortisol is released by the adrenal glands in response to the production and secretion of adrenocorticotropic hormone (ACTH) from the anterior pituitary. As cortisol levels rise, release of ACTH is inhibited. Cortisol also inhibits release of corticotropin-releasing hormone (CRH) by the hypothalamus.

Question 5. Alcohol inhibits release of antidiuretic hormone (ADH). What effect would alcohol consumption have on urination?
Textbook Reference: *30.3 The Pituitary Gland Links the Nervous and Endocrine Systems, p. 611*

Question 6. Why does breast-feeding an infant soon after birth help a mother's uterus return to its prepregnancy size?
Textbook Reference: *30.3 The Pituitary Gland Links the Nervous and Endocrine Systems, p. 611*

30.4 Hormones Regulate Mammalian Physiological Systems

- Thyroxine helps regulate metabolic rate and body temperature
- Hormonal regulation of calcium concentration is vital
- The two segments of the adrenal gland coordinate the stress response
- Sex steroids from the gonads control reproduction

The thyroid gland wraps around the front of the trachea, expanding into a lobe on each side. It produces the hormones thyroxine and calcitonin. Thyroxine, also known as T_4 because it has four iodine atoms, is an amine hormone. The thyroid also produces the hormone T_3, which is nearly identical to thyroxine except that it has only three iodine atoms and is more active. Target cells have enzymes that can convert T_4 to T_3, thereby setting their own sensitivity to thyroid hormones.

Thyroxine raises metabolic rate. It is especially important during development and growth because it promotes uptake of amino acids and synthesis of proteins. Insufficient thyroxine during prenatal or early postnatal development can cause cretinism, a condition characterized by delayed physical and mental development.

Poor regulation of thyroxine production—either hyperthyroidism (thyroxine excess) or hypothyroidism (thyroxine deficiency)—can result in an enlarged thyroid gland, a condition known as goiter (see Figure 30.10). Hyperthyroidism is often the result of an autoimmune response that activates TSH receptors, causing uncontrolled production and release of thyroxine. In hypothyroidism there is insufficient circulating thyroxine to turn off TSH production. Insufficient iodine is the most common cause of this condition.

Careful regulation of calcium levels is essential because the flow of calcium ions across cell membranes is critical to muscle contraction and the release of neurotransmitters by neurons. Shifts below a narrow range can cause muscle spasms and seizures. Shifts above this narrow range can cause weak muscles and a depressed nervous system.

Most of the calcium in the body (nearly 99 percent) is in bone tissue. About 1 percent is found in cells, and only 0.1 percent is in extracellular fluid. The body has three sources for changing blood calcium levels: deposition or resorption of bone, excretion or retention by the kidneys, and absorption from the digestive system. These activities are controlled by calcitonin, calcitriol (a hormone synthesized from vitamin D), and parathyroid hormone (PTH).

Calcitonin, which is secreted by the thyroid gland, lowers blood calcium levels by stimulating osteoblasts (which deposit new bone) and inhibiting osteoclasts (which break down bone). In adults, bone turnover is not very high, so the other mechanisms of calcium regulation have greater importance.

Strictly speaking, vitamin D (also known as calciferol) is not a vitamin because the body can synthesize it. Cholesterol in skin cells is converted into vitamin D by ultraviolet wavelengths in sunlight. The liver and kidneys convert vitamin D into calcitriol, a hormone that stimulates cells of the digestive tract to absorb calcium from food.

The parathyroid glands consist of four small glands that are attached to the posterior surface of the thyroid gland (see Figure 30.11). They produce parathyroid hormone (PTH), which increases blood calcium by stimulating the bone-turnover activities of osteoclasts and osteoblasts, the reabsorption of calcium by the kidneys, and the absorption of calcium from food. Circulating calcium binds to a surface receptor on the parathyroid cells, which inhibits PTH synthesis and release.

The adrenal glands have two regions: the adrenal medulla makes up the core of the glands, and the adrenal cortex surrounds the medulla (see Figure 30.12). Epinephrine and norepinephrine are produced by the adrenal medulla and are involved in the fight-or-flight response. The release of these hormones by the medulla is controlled by the nervous system. The adrenal cortex produces mineralocorticoids, glucocorticoids, and relatively small amounts of sex steroids. The main mineralocorticoid is aldosterone, which stimulates the kidneys to reabsorb sodium and excrete potassium. The main glucocorticoid in humans is cortisol, which mediates our response to stress by stimulating cells that are not necessary for the fight-or-flight response to decrease their use of glucose and to use fats and proteins as energy sources instead.

In males, testes produce androgens (testosterone); in females, the ovaries produce estrogens and progesterone. Both sexes make and use androgens and estrogens, but the relative levels differ between the sexes.

Early in development, the sex organs of human embryos are similar. At about week 7, presence of the Y chromosome in male embryos causes the undifferentiated gonads to begin producing androgens. In response to androgens, the reproductive system develops into a male system. In the absence of androgens, the reproductive system develops into a female system (see Figure 30.13).

Sex steroids increase in concentration at the time of puberty. Both luteinizing hormone (LH) and follicle-stimulating hormone (FSH) from the anterior pituitary gland (collectively called gonadotropins) control the production of sex steroids. The production of these tropic hormones by the anterior pituitary is controlled by gonadotropin-releasing hormones (GnRH). At the onset of puberty, the GnRH-producing cells are no longer under negative feedback, so the level of GnRH increases, which stimulates the production of LH and FSH. Increased levels of LH and FSH in females and LH in males stimulate the gonads to produce more sex hormones. These sex steroids promote development of secondary sexual characteristics.

Question 7. Calcitonin is considered to be most important during childhood, and possibly important at certain times in adulthood, such as during the late stages of a pregnancy. Why might calcitonin be important at these particular times in life?
Textbook Reference: *30.4 Hormones Regulate Mammalian Physiological Systems, p. 615*

Question 8. Endocrine gland tumors often cause oversecretion of hormones. Why would weakened bones result from a tumor affecting one of the parathyroid glands?
Textbook Reference: *30.4 Hormones Regulate Mammalian Physiological Systems, pp. 615–616*

Question 9. Would the oversecretion or undersecretion of aldosterone produce high blood pressure? Explain.
Textbook Reference: *30.4 Hormones Regulate Mammalian Physiological Systems, p. 616*

Question 10. Would castration (removal of the testes) lead to an immediate and complete absence of testosterone in the person's bloodstream? Why or why not?
Textbook Reference: *30.4 Hormones Regulate Mammalian Physiological Systems, pp. 616–617*

Test Yourself

1. A paracrine signal
   a. circulates in the bloodstream and affects distant cells.
   b. always acts on a wide variety of target tissues.
   c. acts on nearby cells.
   d. acts on neuronal cells only.
   e. acts on glands only.
   Textbook Reference: *30.1 Hormones Are Chemical Messengers, p. 604*

2. A target cell's response to a hormone depends on
   a. the amount of hormone released.
   b. the number of receptors in or on that target cell.
   c. how well the hormone binds with the receptor.
   d. the presence of the proper receptor in or on the target cell.
   e. All of the above
   Textbook Reference: *30.2 Hormones Act by Binding to Receptors, pp. 607*

3. Which of the following is the definition of "upregulation"?
   a. An increase in hormone receptors in response to low hormone levels
   b. An increase in hormone receptors in response to high neurotransmitter levels

c. The increase in hormone levels produced by an increase in hormone receptors
d. The decrease in hormone levels produced by a decrease in hormone receptors
e. A change in hormone levels in response to a change in receptor levels

Textbook Reference: *30.2 Hormones Act by Binding to Receptors, p. 608*

4. Which of the following statements about hormonal functioning is *false*?
a. Hormones act in very low concentrations.
b. Many hormones act at sites distant from where they are produced.
c. Hormones are transported in the blood.
d. The same hormone can influence activities of several different target cells.
e. All of the above are true.

Textbook Reference: *30.1 Hormones Are Chemical Messengers, p. 604*

5. In insects, juvenile hormone
a. is produced by the corpora cardiaca.
b. is also known as brain hormone.
c. prevents maturation.
d. is produced only by those with complete metamorphosis.
e. All of the above

Textbook Reference: *30.1 Hormones Are Chemical Messengers, p. 606*

6. Which of the following statements about vitamin D is *false*?
a. It is a hormone secreted by skin cells.
b. It is activated as it passes through the liver and kidneys.
c. It decreases loss of calcium in the urine.
d. It can be acquired only from food.
e. All of the above are true.

Textbook Reference: *30.4 Hormones Regulate Mammalian Physiological Systems, p. 615*

7. Which of the following glands has one part derived from gut epithelium and a second part derived from neural tissue?
a. Adrenal gland
b. Thyroid gland
c. Hypothalamus
d. Pituitary gland
e. Parathyroid gland

Textbook Reference: *30.3 The Pituitary Gland Links the Nervous and Endocrine Systems, p. 610*

8. Which of the following glands regulates daily rhythms?
a. Pineal gland
b. Thyroid gland
c. Adrenal glands
d. Ovaries
e. Pituitary gland

Textbook Reference: *30.2 Hormones Act by Binding to Receptors, p. 610*

9. Which of the following represents the correct sequence of steps leading to the release of thyroxine by the thyroid gland?
a. Thyroid-stimulating hormone, thyrotropin-releasing hormone, thyroxine
b. Adrenocorticotropic hormone, thyrotropin-releasing hormone, thyroxine
c. Thyrotropin-releasing hormone, thyroid-stimulating hormone, thyroxine
d. Thyroid-stimulating hormone, aldosterone, thyroxine
e. Prolactin, thyroid-stimulating hormone, thyroxine

Textbook Reference: *30.3 The Pituitary Gland Links the Nervous and Endocrine Systems, p. 612*

10. The target tissues of hormones are those tissues that
a. can be penetrated by the particular hormones.
b. have specific enzymes with which the hormones interact directly.
c. have high concentrations of the "second messenger."
d. have receptors for the particular hormones.
e. have the particular genes that the hormones can express.

Textbook Reference: *30.2 Hormones Act by Binding to Receptors, p. 607*

11. Which of the following vertebrate hormones are produced in the anterior pituitary gland?
a. Somatostatin, antidiuretic hormone, and insulin
b. Prolactin, growth hormone, and enkephalins
c. Oxytocin, prolactin, and adrenocorticotropin
d. Estrogen, progesterone, and testosterone
e. Growth hormone, gonadotropin-releasing hormone, and thyroid-releasing hormone

Textbook Reference: *30.3 The Pituitary Gland Links the Nervous and Endocrine Systems, p. 611*

12. Hormones that are secreted by one endocrine gland and control the activities of another endocrine gland are called _______ hormones.
a. growth
b. obstructive
c. tropic
d. selective
e. paracrine

Textbook Reference: *30.3 The Pituitary Gland Links the Nervous and Endocrine Systems, p. 611*

13. Which of the following statements about parathyroid hormone (PTH) is true?
a. It increases calcium in the blood.
b. It decreases calcium in the blood.
c. It alters metabolic rate.
d. It increases blood sodium.
e. It initiates the fight-or-flight response.

Textbook Reference: *30.4 Hormones Regulate Mammalian Physiological Systems, p. 615*

14. Which of the following statements about sexual differentiation in human embryos is *false*?
    a. Estrogens are needed for development of female sex organs.
    b. Androgens promote the development of male sex organs.
    c. In the absence of androgens, female sex organs develop.
    d. Sexual differentiation begins at about the seventh week of development.
    e. Presence of a Y chromosome prompts production of testosterone by embryonic gonads.

    Textbook Reference: *30.4 Hormones Regulate Mammalian Physiological Systems, p. 617*

15. Which of the following statements about steroid hormones is true?
    a. They are water-soluble.
    b. They are produced by the thyroid gland.
    c. They are lipid-soluble.
    d. They are derived from the amino acid tyrosine.
    e. They are associated with second messenger cascades.

    Textbook Reference: *30.2 Hormones Act by Binding to Receptors, p. 607*

Answers

Key Concept Review

1. One method would be to remove the testes, the main glands that produce testosterone, and record the behavior of the castrated rats once they have recovered from surgery. Later, testosterone could be replaced through injections or implants and the behavior of the rats could be examined to see if it returned to presurgery levels. Another method would be to monitor levels of testosterone during one year, while simultaneously monitoring the behavior of the rats to see if there is a correlation between high levels of testosterone and certain behaviors, such as aggression. A third method would be to administer a drug that temporarily and reversibly blocks the effects of testosterone to see if it changed the behavior of the rats.

2. The insect *Rhodnius* can live for long periods of time after it has been decapitated. If the insect is partially decapitated, leaving the corpora allata, it will molt into a fifth-instar juvenile because the corpora allata produces juvenile hormone, which prevents maturation into an adult. An insect that is fully decapitated will molt into an adult rather than another juvenile instar. The fully decapitated animal has enough prothoracicotropic hormone (brain hormone) circulating one week after a meal to allow for molting; however, in the absence of juvenile hormone, it molts into an adult.

3. The plasma membrane of a cell is hydrophobic. Lipid-soluble hormones, which (as their name suggests) dissolve in lipids (fats), move through the plasma membrane and act on receptors inside the target cells. Water-soluble hormones cannot cross the lipid-based plasma membrane and act on receptors on the outside of target cells. Such binding by water-soluble hormones initiates changes in the cell by activating enzymes. Because lipid-soluble hormones cross the plasma membrane and enter the cell, they tend to take longer to act, but they also remain active for a longer time period than do water-soluble hormones.

4. A
    a. 8. Pineal gland
    b. 3. Thyroid gland
    c. 7. Parathyroid gland
    d. 4. Adrenal gland
    e. 2. Gonads (ovaries; testes)
    f. 5. Hypothalamus/Posterior pituitary/Anterior pituitary
    g. 6. Thymus
    h. 1. Pancreas

    B
    a. Hypothalamus
    b. Posterior pituitary
    c. Anterior pituitary

    C
    Pancreas: Insulin, glucagon, somatostatin
    Gonads:
        Ovaries: Estrogens, progesterone
        Testes: Testosterone
    Thyroid: Thyroxine (T_3 and T_4), calcitonin
    Adrenal gland:
        Cortex: cortisol, aldosterone, testosterone (in both sexes), estrogen (in both sexes)
        Medulla: epinephrine, norepinephrine
    Hypothalamus: antidiuretic hormone (ADH), oxytocin
    Anterior pituitary: Thyrotropin (TSH), follicle stimulating hormone (FSH; produced in both sexes), luteinizing hormone (LH, produced in both sexes), adrenocorticotropic hormone (ACTH), growth hormone (GH), prolactin, melanocyte-stimulating hormone (MSH), endorphins, enkephalins.
    Posterior pituitary: Releases (but does not make) ADH and oxytocin.
    Thymus: Thymosin
    Parathyroid: Parathyroid hormone (PTH)
    Pineal gland: Melatonin

5. Because alcohol temporarily inhibits secretion of antidiuretic hormone (ADH) by the posterior pituitary, it will increase the amount of urine produced. The main function of ADH is to conserve body water by decreasing urine output.

6. Suckling by a baby stimulates the release of oxytocin from the posterior pituitary, and this leads to milk ejection from the mammary glands. Oxytocin also causes uterine contractions, and after birth this helps the mother's uterus return toward its prepregnancy size.

7. When blood calcium is high, calcitonin stimulates the absorption of calcium by bone and inhibits the breakdown of bone. These activities would be especially important during periods of intense growth, such as childhood and during pregnancy when the fetus is growing rapidly.
8. A tumor on a parathyroid gland could cause oversecretion of parathyroid hormone (PTH). Because PTH pulls calcium from bone, weakened bones could result.
9. Aldosterone from the adrenal cortex stimulates the kidneys to conserve sodium and return it to the blood. Because water follows sodium, this will result in increased blood volume and blood pressure.
10. Castration would not lead to an immediate absence of testosterone in the bloodstream. Like many hormones, testosterone has a half-life of days to weeks. In addition, small amounts of the sex steroids are produced by the adrenal glands in both males and females.

Test Yourself

1. **c.** Paracrine signals act on nearby cells. Autocrine signals act on the same cells that secrete them. Circulating hormones enter the bloodstream and affect distant cells.
2. **e.** The response of target cells to a hormone depends on how much of the hormone is present and acting on the target, how many receptors are present on the surface of the cell that the hormone can act on, and how well the hormone binds with the receptor.
3. **a.** "Upregulation" of hormone receptors on a cell is the production of more receptors when a hormone is at low levels over time in the blood or in other fluids surrounding the cell.
4. **e.** All of the statements are correct. Hormones are found and act at low concentrations in the body. They are transported by the blood or diffuse through interstitial space to reach target cells that may be some distance from the secreting gland. The same hormone can influence several different target cells.
5. **c.** Juvenile hormone is secreted continuously by the corpora allata of insects and prevents maturation. Prothoracicotropic hormone (brain hormone) is secreted by the corpora cardiaca and is involved with ecdysone in molting.
6. **d.** Vitamin D is a hormone (not a true vitamin) secreted by skin cells, and it becomes more active once it passes through the liver and kidneys. Vitamin D raises levels of calcium in the blood by reducing loss of calcium in the urine.
7. **d.** The pituitary gland has an anterior lobe that develops from gut epithelium and a posterior lobe that develops from neural tissue.
8. **a.** The pineal gland secretes melatonin, a hormone that controls daily rhythms.
9. **c.** The hypothalamus releases thyrotropin-releasing hormone (TRH), which stimulates the anterior pituitary to release thyroid-stimulating hormone (TSH), which stimulates the thyroid gland to produce and release thyroxine.
10. **d.** For a hormone to act on a target cell, the target cell must have the receptors for the specific hormone to bind and trigger the hormonal action.
11. **b.** Prolactin, growth hormone, and enkephalins are produced in the anterior pituitary.
12. **c.** Hormones that control endocrine gland function are known as tropic hormones.
13. **a.** Parathyroid hormone increases the concentration of calcium in the blood.
14. **a.** Estrogens are not needed for the development of female sex organs. Female sex organs form in the absence of androgens. Male sex organs form when androgens are present.
15. **c.** Steroid hormones are lipid-soluble and are produced by the gonads and adrenal glands. They do not act through second messenger systems.

Immunology: Animal Defense Systems 31

The Big Picture

- Animals have two main ways to defend against pathogens: innate immune responses and adaptive immune responses. Innate immune responses are nonspecific and include barriers, phagocytic cells, specialized proteins, and inflammation. Adaptive immune responses are specific, slow to develop, and long-lasting. The two forms of adaptive responses are humoral (B cells and antibodies) and cellular (T cells).

Study Strategies

- The adaptive immune response may seem complex initially, but there are really only two major components: B cells, which secrete antibodies, and T cells, which regulate the activities of other white blood cells and thereby help direct the immune response.
- MHC proteins may also seem complicated at first, but they have only a few essential functions in the immune system. These proteins are specific for each individual and are essential for presenting antigens to the two different types of T lymphocytes: Class I MHC proteins (found on all nucleated cells in mammals) present antigens to T_C cells, while Class II MHC proteins (found on macrophages, B cells, and dendritic cells) present antigens to T_H cells.
- The specificity of the immune response is complex, but it can be understood by reviewing cellular communication. Antibodies of B cells bind with antigen on the surface of pathogens; T cell receptors bind antigen presented on MHC I and MHC II proteins on the surface of cells. Once these membrane proteins interact, cytokines are secreted, and cells begin to divide and proliferate. B cells secrete antibodies.
- The rearrangement of DNA in the nucleus of the B cell may seem foreign to students who think the integrity of the genetic information must be preserved. But this change in the genetic makeup of activated B and T cells is essential to providing the organism with a diverse and specific set of antibodies and cell receptors.
- List the innate (nonspecific) responses that a human body can make to an invading pathogen. Next, list cells that are part of the adaptive (specific) responses. Include proteins that are crucial to generating that specificity. Review how genetic rearrangements generate a diverse set of antibodies for B cells.
- Diagram an antibody and label the important components. Highlight those areas where mutations and alterations occur that change the amino acid sequence of the heavy and light chains.
- Review the expression and function of MHC I and MHC II proteins.
- Make a list of the functions of T_H cells and T_C cells.
- Draw a diagram of a chromosome and label it with 4 *V*s (*V*1–*V*4), 2 *D*s (*D*1 and *D*2), 3 *J*s (*J*1–*J*3), and 8 *C*s (*C*1–*C*8). Imagine that a B cell is maturing to produce antibody on its surface. Select one *V*, one *D*, one *J*, and one *C* segment to make an antibody. Repeat the process using different segments. Predict how many different antibodies can be generated from this set of gene segments.
- List some disorders of the immune system and their probable causes.
- Go to yourBioPortal.com to review the following tutorials and activities:

  Animated Tutorial 31.1 Cells of the Immune System

  Animated Tutorial 31.2 Humoral Immune Response

  Animated Tutorial 31.3 A B Cell Builds an Antibody

  Animated Tutorial 31.4 Cellular Immune Response

  Web Activity 31.1 Inflammation Response

  Web Activity 31.2 Immunoglobulin Structure

Key Concept Review

31.1 Animals Use Innate and Adaptive Mechanisms to Defend Themselves against Pathogens

White blood cells play many roles in immunity

Immune system proteins bind pathogens or signal other cells

Pathogens are harmful organisms and viruses that cause disease. Immunity is the ability to avoid disease once a pathogen has invaded.

Innate, nonspecific defenses are inherited mechanisms that act rapidly to protect the body from pathogens. Innate defenses include physical barriers and cellular and chemical defenses (see Table 31.1). Most animals have innate responses.

Adaptive, specific defenses are mechanisms aimed at particular targets. The specific defenses involve humoral (B cells and antibodies) and cellular (T cells) responses (see Table 31.1). These responses develop slowly, last a long time, and are found only in vertebrate animals.

There are two main types of white blood cells (leukocytes): phagocytes and lymphocytes (see Figure 31.1). Phagocytes, such as macrophages, engulf pathogens and are part of innate and adaptive responses. Lymphocytes, such as B cells and T cells, are involved in the adaptive response.

Cells and other components of the adaptive immune system interact with one another to fight off pathogens. There are four main players in the immune system response. Antibodies, which are produced by B cells, are proteins that bind to substances that are nonself. Binding can inactivate pathogens and toxins and mark them for attack by immune system cells. Major histocompatibility complex (MHC) proteins occur in two classes: MHC I and MHC II. MHC I proteins are found on the surface of most mammalian cells, and MHC II proteins are on the surface of immune system cells. These proteins are important self-identifying labels. T cell receptors are proteins on the surface of T cells that recognize and bind to nonself substances presented by MHC proteins on the surfaces of other cells. Cytokines are soluble signal proteins released by many cell types. They bind to cell surface receptors and can activate or inactivate B cells, macrophages, and T cells

Question 1. Deer ticks transmit the bacterium that causes Lyme disease. The test for Lyme disease involves drawing blood and looking for antibodies to the bacterium. Why would a doctor refuse to test for Lyme disease in a patient who found a deer tick attached to her body and asked to be tested immediately?
Textbook Reference: *31.1 Animals Use Innate and Adaptive Mechanisms to Defend Themselves against Pathogens, p. 621*

31.2 Innate Defenses Are Nonspecific

Barriers and local agents defend the body against invaders

Other innate defenses include specialized proteins and cellular processes

Inflammation is a coordinated response to infection or injury

Inflammation can cause medical problems

The innate defenses encountered first by a pathogen on the skin include the physical barrier of the skin, the saltiness of the skin, and the presence of normal flora (bacteria and fungi) with which the pathogen must compete.

Innate defenses encountered by pathogens that enter the nose or other internal organs include mucus (a secretion that traps microorganisms), lysozyme (an enzyme that causes bacteria to burst), harsh conditions (for example, acidic gastric juice in the stomach), and defensins (peptides that are toxic to many bacteria and enveloped viruses).

Phagocytes ingest cells and viruses, which are then killed by either hydrolysis or defensins inside the phagocyte. Natural killer cells are white blood cells that lyse virus-infected or cancerous cells. They also interact with the adaptive defense mechanisms by lysing antibody-labeled target cells.

About 20 complement proteins act as antimicrobial proteins in vertebrate blood. They function in the following characteristic sequence: one protein binds to an invading cell so that phagocytes can recognize and destroy it; another protein activates the inflammatory response; still other proteins lyse the invading cell. They are an important part of innate (nonspecific) and adaptive (specific) defense responses.

Interferons are small signaling proteins produced in response to viral infection. They bind to uninfected cells and stimulate signaling pathways that inhibit viral reproduction should the cells later become infected. They also stimulate cells to digest bacterial and viral proteins into smaller peptides; this is an important first step in adaptive immunity.

Inflammation defends the body against infectious agents through the release of tumor necrosis factor, prostaglandins, and histamine from mast cells (see Figure 31.3). Inflammation is an important defense response because it isolates the affected area to contain the damage, and it recruits cells such as mast cells to kill pathogens. The process of inflammation also promotes healing.

When the inflammation response is inappropriately strong it can result in allergic responses (histamine release and inflammation prompted by a nonself molecule that is typically harmless), autoimmune diseases (in which the immune system attacks tissues in the organism's own body), or sepsis (a severe bacterial infection that extends throughout the body). In the condition of sepsis, blood vessels throughout the body dilate, causing a severe drop in blood pressure.

Question 2. Pathogenic bacteria in the digestive tract often cause diarrhea. Why might diarrhea be considered a protective response by the body? Also, what other characteristic of the large intestines might help protect against invaders?
Textbook Reference: *31.2 Innate Defenses Are Nonspecific, p. 622*

Question 3. How might urine function as an innate defense?
Textbook Reference: *31.2 Innate Defenses Are Nonspecific, pp. 622–623*

31.3 The Adaptive Immune Response Is Specific

Adaptive immunity has four key features

Two types of adaptive immune responses interact

Adaptive immunity develops in response to a pathogen; after exposure, an individual develops antibodies against that pathogen. Passive immunity refers to the acquisition of immunity from antibodies received from another individual.

One feature of adaptive immunity is its specificity, which allows it to focus on pathogens that are present. T cell receptors and antibodies produced by B cells recognize specific antigens, which are nonself proteins or polysaccharides. More specifically, the immune cells recognize antigenic determinants, or epitopes (small portions of antigens), on invading pathogens. There can be many antigenic determinants on a pathogen surface.

A second feature of adaptive immunity is its diversity, meaning that it is able to respond to an enormous range of different pathogens. In order to do so, the body must generate diverse lymphocytes that are specific for particular antigens. The genetic changes resulting in diversity are produced by DNA changes, including chromosomal rearrangements and other mutations that occur as B and T cells are produced in the bone marrow. Each B cell can make only one type of antibody, and the T cells have a specific receptor to recognize an antigen. There are millions of different kinds of B cells and T cells with specific receptors. Essentially, the immune system has the machinery in place to recognize antigens before they are encountered. Also, when an antigen that fits the surface antibody binds to the B cell, that cell is activated and divides to produce clonal B cells, which secrete antibodies (see Figure 31.5). T cells are clonally selected in a similar manner (clonal selection).

A third feature of adaptive immunity is its capacity to distinguish self from nonself. Animals are able to tolerate their own antigens due to clonal deletion, a process that removes any B and T cells that show the potential to mount a strong response against self-antigens. Such cells undergo programmed cell death (apoptosis). Failure of clonal deletion leads to autoimmunity (an immune response within an individual to self-antigens, which may result in disease).

A fourth feature of adaptive immunity is its immunological memory (i.e., it "remembers" a pathogen), which allows it to respond even more effectively to subsequent encounters with the same pathogen. In the primary immune response, activated B and T cells produce effector cells and memory cells. Effector cells attack a pathogen by producing specific antibodies (in the case of B cells) or cytokines (in the case of T cells). Effector B cells are called plasma cells. This primary immune response occurs during a first encounter with an antigen. Effector cells generally live only a few days. Following the primary immune response, long-lived memory cells remain and divide at a slow rate. During subsequent encounters with the same pathogen, memory B and T cells divide rapidly to produce effector cells and more memory cells to serve as a more powerful immune response. This rapid response, the secondary immune response, provides a natural immunity to diseases caused by those pathogens. Vaccination triggers a primary response, so the body will be prepared to mount a stronger secondary response if it encounters the pathogen again.

There are three phases to the adaptive immune response: recognition (the organism detects a pathogen by distinguishing self and nonself); activation (cells and molecules mobilize to fight the invader); and effector (mobilized cells and molecules destroy the invader).

There are two interactive immune responses against invaders. The humoral immune response relies on B cells that make antibodies. The cellular immune response relies on cytotoxic T cells (T_C) that bind to and destroy self-cells that are mutated or infected by pathogens. These systems act in concert and share some of the same mechanisms (see Figure 31.6).

A key event in both the humoral and cellular responses is presentation of the antigen to the immune system. In the humoral immune response, presentation occurs when an antigen binds to a B cell that has an antibody on its surface that is specific for that antigen. In the cellular immune response, a T-helper cell (T_H) with a specific receptor binds to an antigen-presenting cell, which has a particular antigen inserted into its plasma membrane. In both responses, binding initiates the activation phase.

Question 4. Imagine that a particular bacterium has surface molecules that resemble membrane proteins found on heart valves. What would happen to an individual infected with this bacterium if the infection was not treated?
Textbook Reference: *31.3 The Adaptive Immune Response Is Specific, p. 627*

Question 5. In the diagram below, label each structure indicated by the letters "a" through "f." Also, at the bottom of the diagram (letters "g" and "h"), identify which represents the cellular immune response and which represents the humoral immune response.
Textbook Reference: *31.3 The Adaptive Immune Response Is Specific, p. 628*

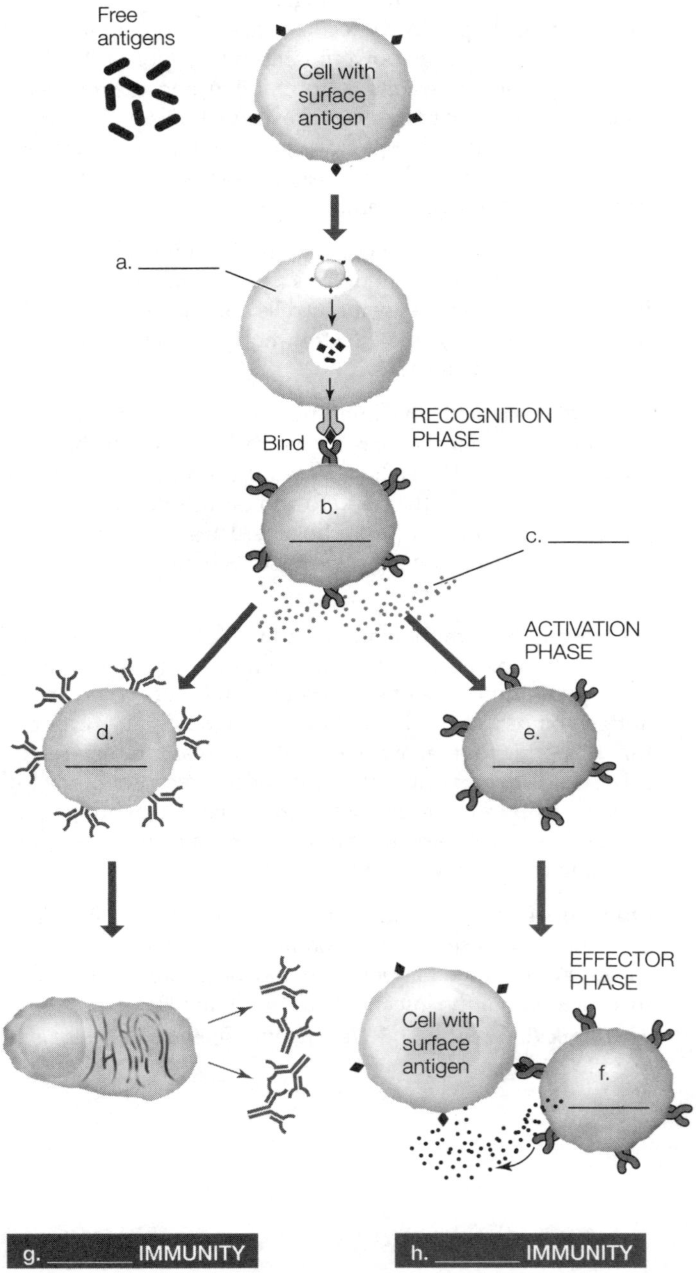

31.4 The Adaptive Humoral Immune Response Involves Specific Antibodies

Plasma cells produce antibodies

Antibodies share a common overall structure

Antibody diversity results from DNA rearrangements and other mutations

Antibodies bind to pathogens on cells or in the bloodstream

Activation of B cells results from the binding of an antibody to a particular antigenic determinant, and the arrival of a signal from a helper T cell that has bound the antigen on an antigen-presenting cell. Once this occurs, the plasma cell (effector B cell) is able to secrete antibodies. As the effector B cells proliferate, they produce antibodies that are specific for the antigen that bound to the parent B cell.

Antibodies (also called immunoglobulins) are proteins that can be grouped into five different classes, although they are all similar in structure. Antibody molecules consist of two identical heavy polypeptide chains and two identical light polypeptide chains held together by disulfide bonds. Each polypeptide consists of a constant region (which is similar in amino acid sequence from one immunoglobulin to another within a class) and a variable region that forms the antigen-binding site (see Figure 31.7). The variable region produces the specificity of the millions of antibodies.

The antigen binding sites on each immunoglobulin are identical, so antibodies are bivalent. Each antibody can recognize two antigen molecules, and because most antigens have multiple epitopes, antibodies can form large complexes with antigens. These complexes are easy targets for phagocytes.

The five classes of immunoglobulins are: IgG, IgM, IgD, IgA, and IgE. The different classes are based on differences in the constant region of the heavy chain. IgG is secreted by B cells and is the most abundant immunoglobulin.

The genome of the B cell undergoes genetic rearrangement during differentiation, and the combination of different alleles allows a B cell's genome to encode for a unique immunoglobulin. DNA fragments are rearranged and joined during B cell development to generate antibody supergenes (see Figures 31.8 and 31.9A). The variable region of the light chain originates from two families of genes (*V* and *J*), and the variable region of the heavy chain originates from three families of genes (*V*, *D*, and *J*). For the heavy chain in mice, there are multiple genes coding for each of the three parts of the variable region: 100 *V*, 30 *D*, and 6 *J* genes (see Figure 31.8). Each B cell randomly selects one gene from each of these multiple genes to make the final coding sequence, *VDJ*, for the heavy-chain variable region. Light chains are similarly constructed from DNA segments. Light and heavy chains are combined to create billions of possible antibodies. Additional diversity is achieved by imprecise recombination during DNA rearrangements, by the addition of extra nucleotides to cut DNA fragments as insertion mutations by the enzyme terminal transferase, and an increased spontaneous mutation rate in immunoglobulin genes.

Antibodies have two roles in B cells. First, because they are expressed on the cell surface, they can act as a receptor for an antigen in the recognition phase of the humoral response. Second, during the effector phase of the humoral response, antibodies are produced in large amounts by a clone of B cells. These antibodies enter the bloodstream, where they may bind to an antigen on the surface of a pathogen, stimulating macrophages or natural killer cells to dispense with the pathogen. Alternatively, if the antigen is free in the bloodstream, the antibodies may join to form large insoluble antibody–antigen complexes that are destroyed by phagocytic cells.

Question 6. What might happen to a person receiving a blood transfusion if the donor blood contained foreign antigens?
Textbook Reference: *31.4 The Adaptive Humoral Immune Response Involves Specific Antibodies, p. 633*

31.5 The Adaptive Cellular Immune Response Involves T Cells and Their Receptors

- T cell receptors specifically bind to antigens on cell surfaces
- MHC proteins present antigen to T cells and result in recognition
- Activation of the cellular response results in death of the targeted cell
- Regulatory T cells suppress the humoral and cellular immune responses
- AIDS is an immune deficiency disorder

T cells have specific glycoprotein surface receptors that are made up of two different polypeptide chains, both with a variable and a constant region (see Figure 31.10). The variable region provides the specificity of the T cell receptor. T cell receptors bind antigen fragments that are displayed on antigen-presenting cells in the presence of an MHC protein.

MHC proteins are plasma membrane glycoproteins that present antigens to the two different types of T lymphocytes (T_H and T_C). Class I MHC proteins are present on the surface of every nucleated cell in the mammalian body. They present antigens (fragments of virus proteins in virus-infected cells or abnormal proteins made by cancer cells) to T_C cells. Class II MHC proteins are found on the surfaces of B cells, macrophages, and dendritic cells. They present antigens to T_H cells (see Figure 31.11).

Activation of a T_H cell stimulates B cells to divide and produce antibodies against the antigen. Activation of a T_C cell results in the production of clones with the specific T cell receptor. These T_C cells bind to cells carrying the antigen and the Class I MHC protein. Once bound, the T_C cell kills the antigen-carrying cell; it accomplishes this either by producing perforins that lyse the cell or by stimulating the cell to undergo apoptosis.

Regulatory T cells (Tregs) are made in the thymus, express the T cell receptor, and are activated if they bind to antigen–MHC complexes. However, Tregs recognize self-antigens. When Tregs are activated, they secrete a cytokine called interleukin-10. This blocks the activation of T_C and T_H cells and causes them to undergo apoptosis (see Figure 31.12).

AIDS is caused by HIV, a virus that eventually destroys T_H cells. HIV is transmitted through body fluids, including blood and semen. HIV initially infects macrophages, T_H cells, and antigen-presenting dendritic cells. Initially there is an immune response to the viral infection and some T_H cells are activated. However T_H cells eventually decline: HIV itself kills them and T_C cells lyse infected T_H cells. Extensive production of HIV by infected cells activates the humoral response and antibodies bind HIV and the level of HIV in the blood decreases (see Figure 31.13). Nevertheless, the infection remains at a low level because of the depletion of T_H cells. A person may remain at this "set point" for 8 to 10 years. Eventually, the T_H cells are destroyed, and the patient becomes susceptible to opportunistic infections, including Kaposi's sarcoma and pneumonia caused by the fungus *Pneumocystis jirovecii*. Drug treatments generally focus on inhibiting processes associated with viral entry, assembly, and replication.

Question 7. If cytotoxic T cells were eliminated from a person's array of immune defenses, what kinds of disease would the person be susceptible to?
Textbook Reference: *31.5 The Adaptive Cellular Immune Response Involves T Cells and Their Receptors, p. 634*

Question 8. Why are organ transplants more successful when the donor is a relative of the recipient?
Textbook Reference: *31.5 The Adaptive Cellular Immune Response Involves T Cells and Their Receptors, pp. 634, 636*

Question 9. The polio vaccine comes in two forms: an injected dose of inactivated (dead) virus and an oral vaccine of attenuated virus. Which would be the safer vaccine for a child with an immunodeficiency disorder? Explain your reasoning.
Textbook Reference: *31.5 The Adaptive Cellular Immune Response Involves T Cells and Their Receptors, pp. 626–627*

Question 10. Draw a flowchart depicting the body's three lines of defense: first line of innate mechanisms; second line of innate mechanisms; and third line of adaptive mechanisms. Within each of the three lines of defense, include the component mechanisms.

Textbook Reference: *31.5 The Adaptive Cellular Immune Response Involves T Cells and Their Receptors pp. 633–635; 31.2 Innate Defenses Are Nonspecific, pp. 622–624; 31.4 The Adaptive Humoral Immune Response Involves Specific Antibodies, pp. 629–630, 633*

Test Yourself

1. Which of the following statements about phagocytes is true?
   a. They are derived from T and B cells.
   b. They present antigen on MHC I complexes.
   c. They digest nonself materials.
   d. They are a type of erythrocyte.
   e. They secrete antibodies.
   Textbook Reference: *31.1 Animals Use Innate and Adaptive Mechanisms to Defend Themselves against Pathogens, p. 621*
2. Which of the following statements about B cells is true?
   a. They secrete antibodies.
   b. They present antigen on MHC II.
   c. They ingest antigens.
   d. They are part of the humoral response.
   e. All of the above
   Textbook Reference: *31.3 The Adaptive Immune Response Is Specific, pp. 625, 628–629*
3. Which of the following is an innate, nonspecific defense of the immune system?
   a. Macrophages
   b. Natural killer cells
   c. Complement proteins
   d. Mucus
   e. All of the above
   Textbook Reference: *31.2 Innate Defenses Are Nonspecific, pp. 622–623; Table 31.1*
4. When the receptor of a T_H cell binds to a pathogen presented on a macrophage, it
   a. inactivates itself.
   b. secretes cytokines.
   c. inactivates B cells.
   d. inactivates the macrophage.
   e. becomes a T_C cell.
   Textbook Reference: *31.3 The Adaptive Immune Response Is Specific, pp. 628–629*
5. Part of the normal immune response includes
   a. the production of B memory cells.
   b. the production of memory macrophages.
   c. antibody secretion by eosinophils.
   d. the production of B cells that attack the individual's own cells.
   e. the production of complement proteins as a specified immune response.
   Textbook Reference: *31.3 The Adaptive Immune Response Is Specific, p. 627*
6. Antibody molecules are
   a. produced by B cells and have only a constant region.
   b. secreted by B cells once a signal (a cytokine) is received from a T cell.
   c. produced by T cells and have a variable and a constant region.
   d. proteins that induce B cells to form memory cells.
   e. proteins with antibody binding sites.
   Textbook Reference: *31.4 The Adaptive Humoral Immune Response Involves Specific Antibodies, pp. 629–630*
7. Which of the following statements about cytotoxic T cells is true?
   a. They release cytokines that activate B cells.
   b. They attack pathogens by binding to cell surface antigens on those pathogens.
   c. They destroy pathogens by engulfing them.
   d. They destroy host cells that are infected with virus.
   e. They stimulate the classical complement pathway.
   Textbook Reference: *31.5 The Adaptive Cellular Immune Response Involves T cells and Their Receptors, p. 634*
8. Which of the following statements about the humoral response is true?
   a. It involves the formation of antibodies.
   b. It occurs when T cells bind antigen-presenting cells.
   c. It is due to T cells' secretion of their receptors.
   d. It occurs when natural killer cells engulf cancer cells.
   e. It is initiated when macrophages engulf bacteria.
   Textbook Reference: *31.4 The Adaptive Humoral Immune Response Involves Specific Antibodies, p. 633*
9. An autoimmune disease is
   a. active when organ transplantation is successful.
   b. caused by viruses.
   c. a response in which the immune cells attack the body's own tissues.
   d. a result of the destruction of the immune system.
   e. inflammation that is always targeted to a specific location in the body.
   Textbook Reference: *31.2 Innate Defenses Are Nonspecific, p. 625*
10. Patients with HIV are susceptible to a variety of infections because
    a. the virus produces cell surface receptors that bind to pathogens, making it easier for those pathogens to be infective.
    b. the synthesis of a DNA copy of the viral genome makes a person susceptible to infection.
    c. HIV attacks and destroys the T_H cells (T-helper cells), which are central to mounting an effective immune response.

d. HIV destroys B cells, so antibodies cannot be made in response to invading pathogens.
e. HIV mutates the B cells, so they cannot make the huge array of antibodies needed for an effective immune response.

Textbook Reference: *31.5 The Adaptive Cellular Immune Response Involves T Cells and Their Receptors, pp. 635–636*

11. DNA rearrangements in the B cell
a. are responsible for generating single B cells that can express many different antibodies.
b. lead to mutations in T cells, resulting in the elimination of essential T cell genes.
c. occur only in B memory cells.
d. are responsible for generating many different antibodies, with each B cell expressing only one set of identical antibodies.
e. occur only in fully mature B cells.

Textbook Reference: *31.4 The Adaptive Humoral Immune Response Involves Specific Antibodies, pp. 630–631*

12. Major histocompatibility proteins function in the immune system by
a. engulfing pathogens.
b. presenting antigens to T_C and T_H cells.
c. generating antibodies to different pathogens.
d. presenting antigen fragments to B cells.
e. presenting macrophages to T_H cells.

Textbook Reference: *31.5 The Adaptive Cellular Immune Response Involves T Cells and Their Receptors, p. 634*

13. The process by which immature B or T cells with the potential to respond strongly to self-antigens undergo apoptosis is called
a. clonal selection.
b. immunological memory.
c. clonal deletion.
d. the primary immune response.
e. the secondary immune response.

Textbook Reference: *31.3 The Adaptive Immune Response Is Specific, p. 627*

14. Inflammation occurs when _______ release _______.
a. mast cells; histamine
b. neutrophils; toxins
c. B cells; histamine
d. mast cells; toxins
e. neutrophils; antihistamine

Textbook Reference: *31.2 Innate Defenses Are Nonspecific, pp. 623–624*

15. The process by which an antigen binds to a specific B cell and the B cell begins to divide is called
a. a nonspecific defense.
b. clonal selection.
c. normal flora.
d. meiosis.
e. a secondary immune response.

Textbook Reference: *31.3 The Adaptive Immune Response Is Specific, pp. 626–627*

Answers

Key Concept Review

1. Antibodies are part of the adaptive, specific immune response, and such responses typically take at least 96 hours to develop (see Table 31.1). Thus, it would not make sense to test immediately for antibodies to the bacterium that causes Lyme disease. Ideally, the test would be performed at least 96 hours after the deer tick was found.
2. Diarrhea helps move the pathogen out of the digestive tract quickly. The large intestine is home to many bacteria that keep invaders in check.
3. Urine flushes pathogens from the urinary tract. The chemical composition of urine (e.g., its acidity) makes the urinary tract inhospitable to some pathogens.
4. If the bacterial infection was not treated, the individual's immune response to this pathogen would include the secretion of antibodies. Such antibodies could cross-react with the individual's own membrane proteins on the cells of the heart valve, leading to an autoimmune response and possibly heart disease. This is an example of molecular mimicry inducing autoimmunity.
5.
a. Antigen-presenting cell
b. Helper T cell (T_H)
c. Cytokines
d. B cell
e. Cytotoxic T (T_C) cell
f. Cytotoxic T (T_C) cell
g. Humoral immunity
h. Cellular immunity
6. If a person receives a blood transfusion and the donor blood contains foreign antigens, then antibodies in the recipient's blood will cause the donor cells to clump or agglutinate, which can be damaging and possibly fatal. This is an example of antibodies forming large, insoluble antibody-antigen complexes when confronted with an antigen free in the bloodstream.
7. Cytotoxic T cells target virally infected cells and some cancer cells. If cytotoxic T cells were eliminated, the individual would be much more susceptible to viral infections and cancer.
8. Organ transplants are successful if the MHC proteins and other cell surface markers are similar in the donor and recipient, making the immune system of the recipient more tolerant of the foreign tissue. Since close relatives have more genes in common than unrelated people do, they are more likely to have surface anti-

gens on their cells that are similar. Organ transplants from unrelated individuals require the administration of immunosuppressive drugs to the recipient so that the patient's own immune system does not reject the foreign tissue. The drug cyclosporin, which inhibits T cell development, often is used.

9. The injected dose of inactivated poliovirus would be safer for a patient with a compromised immune system because an inactivated pathogen cannot cause disease. In contrast, the attenuated vaccine contains a mutant form of the virus that does reproduce itself in the patient, although at a very slow rate. However, it is possible for the virus to mutate back to a virulent form (which does happen occasionally) and cause disease.

10.

First line of defense (innate)
Physical barriers (e.g., skin, mucous membranes)
Chemical barriers (e.g., acid conditions, lysozyme)

↓

Second line of defense (innate)
Defensive cells (e.g., phagocytic cells, natural killer cells)
Defensive proteins (e.g., complement system, interferons)
Inflammation
Fever

↓

Third line of defense (adaptive)
Humoral immune response (B cells make antibodies)
Cellular immune response (cytotoxic T cells)

Test Yourself

1. **c.** Phagocytes are nonspecific cells (not B and T cells) that digest nonself materials and present protein fragments of those nonself materials on their surface via MHC class II complexes. They do not have antibodies.

2. **e.** B cells are antigen-presenting cells; they bind antigen, digest it, present fragments of that antigen on their MHC II proteins, and secrete antibodies as part of the humoral response.

3. **e.** Macrophages, natural killer cells, cilia on mucous membranes, and complement proteins are all part of the nonspecific response of the immune system. B and T cells are part of the specific response of the immune system.

4. **b.** When the T_H cell binds antigen being presented on a macrophage, it secretes cytokines, which stimulate other immune cells to divide.

5. **a.** In a normal immune response, B memory cells are produced, allowing the organism to mount a faster and more effective response to any subsequent encounter with the pathogen. Memory macrophages do not exist. Eosinophils do not secrete antibodies. In an abnormal immune response, B cells that attack the individual's own cells can be activated, as seen in autoimmune diseases. Complement proteins are part of the innate, nonspecific immune response.

6. **b.** T cells have cell surface receptors with a variable and a constant region and do not produce antibody molecules. Antibodies are not produced by macrophages. Only B cells produce antibodies in response to cytokines released from the T_H cell. Each antibody molecule has two identical heavy chains and two identical light chains, and each of these chains has a variable and a constant region. Antigens have "antibody binding sites" called epitopes.

7. **d.** Cytotoxic cells bind to virus antigen presented on MHC I protein and destroy those cells. T_H cells release cytokines to activate B cells. The antibodies of B cells bind cell surface antigens on pathogens. Cytotoxic T cells do not engulf pathogens. T cells have no involvement in the classical complement pathway.

8. **a.** The humoral response refers to that part of the specific response that releases antibodies (B cells) in the lymph and blood (the "humors" of the body). In the cellular response, T cell receptors bind antigen on antigen-presenting cells. T cells do not secrete their receptors. Macrophages are part of the nonspecific response.

9. **c.** Autoimmune diseases occur when the immune cells attack the body's own cells. Transplant rejection can occur if the transplanted tissue is recognized by the body as nonself, so this is not an autoimmune response. The immune system can be destroyed by an immune deficiency disorder.

10. **c.** An HIV-infected individual is more susceptible to a variety of infections because the virus destroys T_H cells, which are essential for mounting an effective immune response. HIV does not bind to pathogens and does not destroy B cells or cause mutations in their DNA that alter antibody production.

11. **d.** DNA rearrangements occur in B cell precursors and result in the expression of a unique kind of antibody by each mature B cell. DNA rearrangements also occur in T cells to generate T cell receptors. This rearrangement does not destroy essential T cell genes. Memory B cells are cells in which DNA rearrangement has already occurred.

12. **b.** MHC proteins present antigens to T_C and T_H cells. The cytokines of T cells (and not MHC proteins) activate B cells. MHC proteins do not generate antibodies. Macrophages are antigen presenting cells, but they are not themselves presented as antigens to T_H cells.

13. **c.** Clonal deletion is the process by which immature B or T cells that show the potential to mount a strong immune response to self-antigens undergo apoptosis.
14. **a.** Inflammation occurs in response to an infection or tissue injury. Once an infection or injury occurs, the local mast cells release histamine. The histamine in turn makes the capillaries leak, and fluid accumulating in the area produces the inflammation.
15. **b.** When an antigen binds to a B cell, the B cell begins to divide and make clone cells, which results in more B cells that recognize the antigen. This process is known as clonal selection. A secondary immune response occurs when memory cells encounter an antigen and begin to divide actively.

Animal Reproduction 32

The Big Picture

- Animals reproduce both asexually and sexually. Asexual reproduction includes budding, regeneration, and parthenogenesis. Sexual reproduction has an advantage over asexual reproduction in that it helps increase genetic diversity. Sexual reproduction has three main stages: gametogenesis, mating (or spawning), and fertilization.
- Gametogenesis is the process by which the sex cells are produced. In males, haploid sperm develop from diploid spermatogonia by the process of spermatogenesis. In females, the haploid egg develops from diploid oogonia by the process of oogenesis.
- The human female has two interrelated cycles, the ovarian and the uterine cycles, during which an egg is produced and released from the ovary and the uterus is prepared for implantation and pregnancy. These two cycles occur in parallel and are coordinated by changes in hormone levels. Luteinizing hormone, follicle-stimulating hormone, estrogen, and progesterone all play a role in these cycles.
- External fertilization is typical of many aquatic animals. Internal fertilization, often with the help of accessory sex organs to ensure fertilization, occurs in terrestrial animals. The eggs and sperm of a given species recognize each other.
- Several methods are available for preventing pregnancy and for overcoming infertility.

Study Strategies

- Understanding the ploidy level of the developing egg and sperm can be very confusing, but it is essential in determining how genetic inheritance works.
- Recognize that although sperm and egg formation are completely separate processes, each with its own characteristics, the same general events are happening in both. This will help you recognize and remember the stages of spermatogenesis and oogenesis. Be sure, though, to also understand the differences between spermatogenesis and oogenesis with respect to timing and number of gametes produced.
- Following the hormonal changes associated with human ovarian and menstrual cycles can be difficult. Create a chronological sequence of hormones and the events they trigger during both cycles.
- Go to yourBioPortal.com to review the following tutorials and activities:

  Animated Tutorial 32.1 Fertilization in a Sea Urchin Egg

  Animated Tutorial 32.2 The Ovarian and Uterine Cycles

  Web Activity 32.1 The Human Male Reproductive Tract

  Web Activity 32.2 Spermatogenesis

  Web Activity 32.3 The Human Female Reproductive Tract

Key Concept Review

32.1 Reproduction Can Be Sexual or Asexual

Budding and regeneration produce new individuals by mitosis

Parthenogenesis is the development of unfertilized eggs

Most animals reproduce sexually

A variety of animals, mostly invertebrates, employ asexual reproduction, which produces offspring that are genetically identical to one another and to the parent. Asexual reproduction has the following two advantages: (1) time and energy are not wasted on mating; and (2) every member of the population can produce offspring. The main disadvantage of asexual reproduction is that it does not generate genetic diversity, which can be essential if the environment changes.

New offspring can be produced either by budding outgrowths to produce a new individual or by regenerating a new individual from pieces of an animal (see Figure 32.1A and B). These methods rely on mitosis. Another way that animals produce new individuals asexually is through parthenogenesis—the development of unfertilized eggs into offspring. Sometimes sexual behavior is required for asexual reproduction; this is the case, for example, in a species of parthenogenetic whiptail lizard (see Figure 32.1C).

Most animals reproduce sexually, which involves the joining of two haploid gametes to form a diploid cell. Sexual reproduction has three stages: gametogenesis (producing gametes), mating or spawning (bringing gametes together), and fertilization (fusing gametes). Costs of sexual reproduction include searching for mates and engaging in complex and expensive courtship behavior. The main advantage to sexual reproduction is that it generates genetic diversity. Gametogenesis produces haploid gametes by meiotic cell division. During meiosis, crossing over between homologous chromosomes and independent assortment of chromosomes contribute to genetic diversity.

Question 1. Some animals can reproduce both sexually or asexually. When and why might these animals switch between sexual and asexual reproduction?
Textbook Reference: *32.1 Reproduction Can be Sexual or Asexual, pp. 639–640*

32.2 Gametogenesis Produces Haploid Gametes

Spermatogenesis produces four sperm from one parent cell

Oogenesis produces one large ovum from one parent cell

Hermaphrodites can produce both sperm and ova

Gametogenesis occurs in an animal's gonads. Male gametes—sperm—are produced in the testes, and female gametes—eggs or ova—are produced in the ovaries. Gametes are derived from germ cells, which migrate to the gonads of the embryo.

Spermatogenesis, the production of sperm, begins with a male germ cell ($2n$) that proliferates through mitosis to produce spermatogonia ($2n$; see Figure 32.2A). Spermatogonia mature into primary spermatocytes ($2n$) that enter meiosis. The first meiotic division produces secondary spermatocytes (n), and a second meiotic division produces four spermatids (n). The spermatids differentiate and mature into sperm cells, which are streamlined and have a flagellum.

Oogenesis, the production of ova, begins with a female germ cell ($2n$) that proliferates through mitosis to produce oogonia ($2n$; see Figure 32.2B). Oogonia mature into primary oocytes ($2n$). The primary oocytes enter prophase of the first meiotic division and in human females remain arrested in prophase for at least 10 years (until puberty is reached); some remain arrested for up to 50 years. No new primary oocytes will be produced during the lifetime of the female. Upon exiting prophase, the primary oocyte completes the first meiotic division to produce a secondary oocyte (n) and the first polar body. The second meiotic division produces an ootid (n) and the second polar body. This second meiotic division may not be completed until fertilization in some species. Polar bodies degenerate.

Two main differences between spermatogenesis and oogenesis are: (1) spermatogenesis continues to completion once the primary spermatocyte has differentiated, whereas oogenesis is characterized by a period of arrest during the first meiotic division; and (2) cytoplasm is apportioned equally to resulting cells in spermatogenesis and unequally to resulting cells in oogenesis (polar bodies receive much less cytoplasm than do secondary oocytes or ootids).

In most species, gametes are produced by individuals that are either male or female. In some species, however, a single individual may produce both sperm and eggs; such individuals are called hermaphrodites. Hermaphrodites can be either simultaneous (individuals are male and female at the same time) or sequential (individuals change sex at some point during their life span).

Question 2. How are spermatogenesis and oogenesis similar? How are they different?
Textbook Reference: *32.2 Gametogenesis Produces Haploid Gametes, pp. 640–641*

Question 3. To better understand the changes in ploidy that occur in gametes as they form, create a flowchart that shows how ploidy changes from germ cell to mature gamete in spermatogenesis and oogenesis. Be sure to label each stage of gamete formation (e.g., spermatogonium, primary spermatocyte, etc.) and include its ploidy.
Textbook Reference: *32.2 Gametogenesis Produces Haploid Gametes, pp. 640–641*

32.3 Fertilization Is the Union of Sperm and Ovum

Fertilization may be external or internal

Recognition molecules enable sperm to penetrate protective layers around the ovum

Only one sperm can fertilize an ovum

Fertilized ova may be released into the environment or retained in the mother's body

The major events of fertilization are: (1) sperm and ova recognize and bind to each other; (2) the sperm is activated, enabling it to gain access to the ovum; (3) once the plasma membrane of the ovum fuses with the plasma membrane of a single sperm, changes occur that block entry by additional sperm; (4) the ovum is stimulated to begin development; and (6) the nuclei of the ovum and sperm fuse to produce a diploid nucleus of the zygote.

Many aquatic animals have external fertilization of eggs. The eggs and sperm are released into the water (in a process called spawning), where fertilization may occur. Terrestrial animals employ internal fertilization, in which sperm and egg fuse within the female reproductive tract rather than in the external environment. Some aquatic animals have internal fertilization.

Organs other than gonads that are part of an animal's reproductive system are considered accessory sex organs. For example, in males of many species with internal fertilization, the penis is an accessory sex organ needed to deposit sperm in the female's reproductive tract. The vagina is an example of a female accessory sex organ. Copulation is the physical joining of male and female accessory sex organs.

For successful fertilization to occur, the sperm and egg must recognize each other. Specific recognition molecules on the gametes mediate recognition. These molecules prevent eggs from being fertilized by sperm from a different species. In sea urchins, the eggs release species-specific chemical attractants, and other recognition molecules are involved later in fertilization (see Figure 32.3B). In mammals, a glycoprotein in the zona pellucida (a layer surrounding the mammalian egg; see Figure 32.4) binds to recognition molecules on the head of the sperm and triggers the acrosomal reaction. During the acrosomal reaction, the plasma membrane covering the head of the sperm breaks down, as does the underlying acrosomal membrane, releasing enzymes that digest a path for the sperm through the protective layers surrounding the egg.

Fusion of the plasma membranes of the sperm and egg triggers blocks to polyspermy that prevent more than one sperm from entering the egg. Entry by more than one sperm usually means death of the embryo. Sea urchins have a fast block to polyspermy, which is based on a transient change in charge across the ovum's plasma membrane, and a slow block to polyspermy, during which the vitelline envelope becomes an impenetrable physical barrier. Mammals do not have a fast block to polyspermy, but they do have something similar to the slow block.

The two main modes of reproduction displayed by animals are oviparity (egg laying) and viviparity (live bearing). Oviparous animals deposit their eggs in the environment; the eggs of terrestrial animals have additional membranes and a protective shell to decrease water loss. Viviparous animals retain the developing embryo in the female reproductive tract. In the majority of mammals, the embryo develops in the female's uterus, and exchanges between mother and embryo occur via a placenta.

Question 4. Why is it problematic if more than one sperm enters an ovum at fertilization?
Textbook Reference: *32.3 Fertilization Is the Union of Sperm and Ovum, p. 644*

Question 5. Why might species-specific recognition molecules of sperm and ova be somewhat more important in species with external fertilization than in those with internal fertilization?
Textbook Reference: *32.3 Fertilization Is the Union of Sperm and Ovum, pp. 642–644*

32.4 Human Reproduction Is Hormonally Controlled

- Male sex organs produce and deliver semen
- Male sexual function is controlled by hormones
- Female sex organs produce ova, receive sperm, and nurture the embryo
- The female reproductive cycle is controlled by hormones
- In pregnancy, tissues derived from the embryo produce hormones
- Hormonal and mechanical stimuli trigger childbirth

Sperm are produced in the testes, which in most mammals are held in the scrotum (see Figure 32.5). This pouch of skin holds the testes outside the body cavity where temperatures are slightly lower than normal body temperature and optimal for spermatogenesis. Males produce semen, which contains sperm and fluids that support the sperm and facilitate fertilization. In the testis, sperm cells are produced by spermatogenesis in the seminiferous tubules (see Figure 32.6). Sertoli cells surround developing sperm cells and provide protection and nourishment. Leydig cells produce testosterone. Immature sperm move from the seminiferous tubules to the epididymis, where they mature, become motile, and are stored. The epididymis connects to the vas deferens, which connects to the urethra, which opens to the outside of the body at the tip of the penis.

In addition to sperm, semen contains fluids from accessory glands. The bulbourethral glands produce an alkaline and mucoid secretion that helps control pH in the urethra and lubricates the tip of the penis. The seminal vesicles produce seminal fluid that contains fructose (an energy source for sperm), mucus, and proteins. The prostate gland produces the milky prostate fluid and a clotting enzyme that causes the protein in seminal fluid to turn semen into a gelatinous mass, which aids its retention in the female reproductive tract. Another enzyme activated shortly after semen enters the female reproductive tract dissolves the clotted semen and liberates sperm. Prostate secretions are alkaline, so they also help neutralize the acidity of the male and female reproductive tracts.

During sexual arousal, the penis becomes engorged with blood, and contraction of smooth muscle in the ducts and at the base of the penis results in ejaculation of semen. The dilation of blood vessels in the penis during an erection is initiated by the neurotransmitter nitric oxide (NO) and its effects on the second messenger cGMP. After ejaculation, decreases in NO and cGMP lead to constriction of blood vessels, which results in loss of an erection. Erectile dysfunction (or impotence) is the inability to achieve or sustain an erection.

Testosterone, produced in the Leydig cells of the testes, is the male sex hormone that controls sexual function and sperm production (see Figure 32.7). Testosterone also prompts development and maintenance of secondary sexual characteristics, such as facial hair and a deep voice in males. The hypothalamus secretes GnRH, which prompts the anterior pituitary to release LH and FSH. LH then stimulates the Leydig cells to secrete testosterone, and FSH and testosterone stimulate the Sertoli cells to support spermatogenesis.

The ovaries release eggs into the body cavity. When an ovum matures, it enters one of the oviducts (also called the Fallopian tubes) and moves toward the uterus (see Figure 32.8A). At the bottom of the uterus is the cervix, which leads into the vagina. Sperm deposited in the vagina during copulation move through the cervix and uterus, and into the oviduct, where fertilization occurs. Two sets of skin folds, the labia majora and labia minora, surround the openings to the vagina and urethra. The clitoris, an erectile organ, is located at the anterior tip of the labia minora (see Figure 32.8B).

Mature eggs are produced during a 28-day ovarian cycle (see Figure 32.9). Initially, 6 to 12 primary oocytes begin to mature and are surrounded by follicle cells to produce individual follicles. After one week, one primary oocyte continues to grow while the others shrink. Prior to ovulation, the primary oocyte undergoes a meiotic division to become a secondary oocyte and is expelled from the ovary. The remaining follicle cells become the corpus luteum, a glandular structure that produces progesterone and estrogen.

The uterine cycle, in concert with the ovarian cycle, prepares the endometrium to receive the blastocyst. If a blastocyst does not implant by around day 14 of the uterine cycle, the endometrium is broken down and expelled in the process of menstruation. The uterine cycles of most mammals other than humans do not include menstruation. Other mammals resorb the endometrium; in these species, females display a period of sexual receptivity (termed estrus) around the time of ovulation.

Luteinizing hormone (LH) and follicle-stimulating hormone (FSH) control the timing of the ovarian and uterine cycles (see Figure 32.10). The first step of the uterine cycle is menstruation. Prior to day 10 of the uterine cycle, FSH and LH levels are kept low by negative feedback from estrogen. Around days 12 to 14, this changes, and estrogen exerts positive feedback, which causes a rise in FSH levels. Rising levels of FSH stimulate the follicles to mature. LH levels also increase, triggering ovulation, the release of the egg from the mature follicle. Following ovulation, the follicle cells form the corpus luteum. Estrogen and progesterone produced by the corpus luteum stimulate growth and maintenance of the endometrium. In the absence of fertilization, the corpus luteum degenerates, estrogen and progesterone levels decrease, and menstruation occurs to start the cycle again.

If an egg is fertilized, the initial divisions of the zygote produce a blastocyst, which moves from the oviduct into the uterus, where it burrows into the endometrium; this process is called implantation. In response to implantation, human chorionic gonadotropin (hCG) is produced, which keeps the corpus luteum functional. Continued production of estrogen and progesterone by the corpus luteum supports development and maintenance of the endometrium, which is needed for pregnancy. The blastocyst interacts with the endometrium to form the placenta, the organ that provides nutrients and removes wastes from the developing embryo. The placenta replaces the corpus luteum as the main source of estrogen and progesterone.

Late in pregnancy, estrogen increases contractility of uterine muscle. The uterine contractions of childbirth are triggered by pressure from the fetal head on the maternal cervix. This mechanical stimulus triggers release of oxytocin by the mother and fetus. Oxytocin stimulates increased strength and frequency of uterine contractions in a positive feedback cycle. After birth, oxytocin promotes bonding between mother and infant.

Question 6. What is human chorionic gonadotropin (hCG)? Why does its detection serve as the basis for pregnancy tests?
Textbook Reference: *32.4 Human Reproduction Is Hormonally Controlled, p. 651*

Question 7. Why might physicians recommend that men who want children but who have been diagnosed with low sperm counts avoid wearing tight-fitting shorts?
Textbook Reference: *32.4 Human Reproduction Is Hormonally Controlled, p. 646*

32.5 Humans Use a Variety of Methods to Control Fertility

Many contraceptive methods are available

Reproductive technologies help solve problems of infertility

The termination of pregnancy is known as abortion. A spontaneous abortion occurring early in pregnancy is commonly called a miscarriage. An abortion resulting from medical intervention may be performed for therapeutic purposes (e.g., to save the life of the mother) or fertility control.

Medical science has found ways to help couples overcome infertility, which is the persistent inability of a couple to conceive a child. Several problems may prevent couples from conceiving. A male may not produce adequate numbers of sperm or the sperm may lack motility. In the female, the environment of the uterus may be poor for hosting sperm or the fertilized egg, or the oviducts may be blocked, preventing passage of gametes. Artificial insemination places sperm in the female reproductive tract where fertilization can occur. Assisted reproductive technologies (ARTs) typically take eggs from a female, combine them with sperm outside the body and then place fertilized eggs or mixtures of sperm and eggs in the appropriate location in the female reproductive tract. In vitro fertilization (IVF) is an example of an ART.

Question 8. Given what you know about the hormones that regulate the ovarian and uterine cycles in human females, explain the mechanism by which birth control pills containing synthetic estrogen and progesterone prevent pregnancy.
Textbook Reference: *32.5 Humans Use a Variety of Methods to Control Fertility, pp. 652–653*

Question 9. Why are multiple births (for example, septuplets) associated with assisted reproductive technologies (ARTs)?
Textbook Reference: *32.5 Humans Use a Variety of Methods to Control Fertility, p. 653*

Question 10. Sperm sorting technologies are available that allow parents to choose the sex of their offspring. In what situations might parents want to select the sex of their offspring? What implications might gender selection have for society?
Textbook Reference: *32.5 Humans Use a Variety of Methods to Control Fertility, p. 653*

Test Yourself

1. Asexual reproduction is an effective strategy in stable environments because
   a. gametogenesis is most efficient under these conditions.
   b. the resulting offspring, genetically identical to their parents, are preadapted to their environment.
   c. asexual parthenogenesis produces a large amount of genetic diversity.
   d. animal cells tend to be more totipotent under stable conditions.
   e. sessile animals and sparse populations are more common under stable conditions.
   Textbook Reference*: 32.1 Reproduction Can Be Sexual or Asexual, p. 639*

2. An important difference between a sperm and an egg concerns
   a. their size.
   b. the amount of cytoplasm they contain.
   c. whether or not they are motile.
   d. the number produced over the course of a lifetime.
   e. All of the above
   Textbook Reference*: 32.2 Gametogenesis Produces Haploid Gametes, pp. 640–641*

3. External fertilization
   a. is typical of terrestrial animals.
   b. requires a penis.
   c. does not require species-specific recognition molecules of sperm and ova.
   d. occurs in some, but not all, aquatic animals.
   e. occurs before spawning.
   Textbook Reference*: 32.3 Fertilization Is the Union of Sperm and Ovum, p. 642*

4. Which of the following best represents the normal path of a sperm cell as it makes its way from the point of entry into a female's reproductive tract to the location where fertilization typically occurs?
   a. Cervix, vagina, ovary, oviduct
   b. Vagina, cervix, uterus, oviduct
   c. Uterus, cervix, vagina, oviduct
   d. Vagina, uterus, cervix, oviduct
   e. Cervix, vagina, oviduct, ovary
   Textbook Reference*: 32.4 Human Reproduction Is Hormonally Controlled, p. 648*

5. The fluid released by the prostate gland
   a. contains fructose, which serves as an energy source for sperm.
   b. neutralizes acidity of the male and female reproductive tracts.
   c. inhibits NO.
   d. lubricates the tip of the penis.
   e. facilitates spermatogenesis.
   Textbook Reference*: 32.4 Human Reproduction Is Hormonally Controlled, p. 647*

6. Which of the following is an example of positive feedback control in the reproductive cycle of males or females?
   a. The increased response of the hypothalamus and anterior pituitary gland in response to estrogen
   b. The decreased response of the hypothalamus and anterior pituitary gland in response to estrogen
   c. The inhibition of luteinizing hormone by high levels of testosterone
   d. The stimulation of luteinizing hormone by low levels of testosterone
   e. The inhibition of oxytocin release by increased uterine contractions during childbirth
   Textbook Reference*: 32.4 Human Reproduction Is Hormonally Controlled, pp. 648–650*

7. Which of the following statements about oogenesis is *false*?
   a. The polar bodies degenerate.
   b. The ovum produced is haploid.
   c. In many species, oocytes undergo developmental arrest in prophase I.
   d. The primary oocyte is haploid.
   e. The secondary oocyte is haploid.

   Textbook Reference: *32.2 Gametogenesis Produces Haploid Gametes, pp. 640–641*

8. Which of the following is the correct order of the layers of the mammalian ovum, from outermost to innermost?
   a. Plasma membrane, cumulus, zona pellucida
   b. Jelly coat, cumulus, plasma membrane
   c. Cumulus, zona pellucida, plasma membrane
   d. Zona pellucida, plasma membrane, cumulus
   e. Jelly coat, zona pellucida, plasma membrane

   Textbook Reference: *32.3 Fertilization Is the Union of Sperm and Ovum, pp. 642–644*

9. In the human female's menstrual cycle, when are progesterone concentrations high enough to maintain the uterus in a proper condition for pregnancy?
   a. For the entire duration of the cycle
   b. During no portion of the cycle
   c. During the first half of the cycle
   d. During the second half of the cycle
   e. During a 5-day window in the middle of the cycle

   Textbook Reference: *32.4 Human Reproduction Is Hormonally Controlled, p. 650*

10. Which of the following statements about the birth control pill is *false*?
    a. It works by preventing ovulation.
    b. It works by preventing implantation.
    c. It works by suspending the ovarian cycle.
    d. It contains low doses of estrogen and progesterone.
    e. It allows the uterine cycle to continue because the daily dose is suspended for one week every 21–28 days.

    Textbook Reference: *32.5 Humans Use a Variety of Methods to Control Fertility, p. 653*

11. Among the methods of contraception listed below, which has the lowest failure rate?
    a. Vaginal ring
    b. Rhythm method
    c. Diaphragm
    d. Coitus interruptus
    e. Condom

    Textbook Reference: *32.5 Humans Use a Variety of Methods to Control Fertility, p. 652*

12. If you compared the genetic makeup of a female animal produced by parthenogenesis with that of its mother, which of the following would you expect?
    a. About 100 percent genetic similarity
    b. About 50 percent genetic similarity
    c. No genetic similarity
    d. About 25 percent genetic similarity
    e. None of the above; parthenogenetic animals do not have mothers.

    Textbook Reference: *32.1 Reproduction Can Be Sexual or Asexual, p. 639*

13. Oviparous animals
    a. deposit fertilized eggs in the environment.
    b. retain the embryo in the mother's body.
    c. have a placenta.
    d. are described as live-bearing.
    e. are always aquatic.

    Textbook Reference: *32.3 Fertilization Is the Union of Sperm and Ovum, p. 645*

14. Spermatogenesis
    a. results in two sperm cells from each primary spermatocyte.
    b. involves a period of arrest during the first meiotic division.
    c. results in diploid gametes.
    d. apportions cytoplasm equally among resulting cells.
    e. apportions cytoplasm unequally among resulting cells.

    Textbook Reference: *32.2 Gametogenesis Produces Haploid Gametes, pp. 640–641*

15. Which of the following statements about fertilization is *false*?
    a. The egg permits several sperm to enter it.
    b. The plasma membranes of sperm and egg fuse.
    c. The egg is activated and stimulated to begin development.
    d. Species-specific recognition occurs between egg and sperm.
    e. During the acrosomal reaction, enzymes spill from a cap on the head of the sperm and digest a path through the protective layers surrounding the egg.

    Textbook Reference: *32.3 Fertilization Is the Union of Sperm and Ovum, p. 644*

16. Which of the following statements is *false*?
    a. The ovarian cycle produces mature ova and the uterine cycle prepares the uterus for the arrival of an embryo.
    b. If used correctly, condoms can prevent pregnancy and protect against sexually transmitted diseases (STDs).
    c. RU-486 blocks estrogen receptors.
    d. Men, but not women, have a refractory period following orgasm.
    e. Breastfeeding immediately after birth helps shrink the uterus.

    Textbook Reference: *32.4 Human Reproduction Is Hormonally Controlled, p. 648, 651; 32.5 Humans Use a Variety of Methods to Control Fertility, p. 652*

Answers

Key Concept Review

1. Asexual reproduction does not waste time and energy on mating, and every member of the population can produce offspring. However, asexual reproduction does not yield genetic diversity, which can be beneficial when environmental conditions change. Sexual reproduction generates genetic diversity. Thus, we would expect animals that can reproduce by either method to reproduce asexually when environmental conditions are stable and favorable, and to switch to sexual reproduction when environmental conditions are unstable and unfavorable.

2. Some similarities between spermatogenesis and oogenesis are that both processes occur in gonads and include the same general events. For example, once in the gonads, embryonic germ cells proliferate by mitosis to produce cells that eventually will enter meiosis, and the end result of both processes is the production of haploid cells. Two main differences between spermatogenesis and oogenesis are: (1) spermatogenesis continues to completion once the primary spermatocyte has differentiated, whereas oogenesis is characterized by a period of arrest during the first meiotic division; and (2) cytoplasm is apportioned equally to resulting cells in spermatogenesis and unequally to resulting cells in oogenesis (polar bodies receive much less cytoplasm than do secondary oocytes or ootids).

3. **Spermatogenesis:**

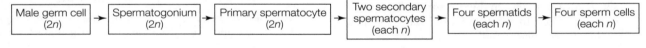

Male germ cell (2*n*) → Spermatogonium (2*n*) → Primary spermatocyte (2*n*) → Two secondary spermatocytes (each *n*) → Four spermatids (each *n*) → Four sperm cells (each *n*)

Oogenesis:

Female germ cell (2*n*) → Oogonium (2*n*) → Primary oocyte (2*n*) → Secondary oocyte, first polar body (each *n*) → Ootid, second polar body (each *n*) → Ovum (*n*)

4. More than one sperm entering an ovum would result in abnormal numbers of chromosomes in the zygote, which would make it incapable of normal development. One sperm entering the egg ensures equal genetic contributions by each parent.

5. Species with external fertilization release their gametes into the external environment where gametes from other species may be present, so recognition molecules are critical. In contrast, species with internal fertilization release gametes into the female reproductive tract where gametes of other species usually are not present because species-specific courtship and mating behaviors prevent this from occurring.

6. Human chorionic gonadotropin (hCG) is a hormone produced by the embryo once it implants in the endometrium of the uterus. Pregnancy tests make use of an antibody to detect its presence in urine.

7. The optimal temperature for spermatogenesis is slightly below normal body temperature, which explains why the testes are located outside the body cavity in the scrotum. Tight clothing pulls the testes close to the body, exposing them to higher than optimal temperatures for spermatogenesis.

8. During normal cycling, estrogen and progesterone serve as negative feedback signals (except at very high levels) to the hypothalamus and pituitary gland and

thus keep follicle-stimulating hormone (FSH) and luteinizing hormone (LH) at low levels. The low doses of estrogen and progesterone in birth control pills function in the same manner; these hormones prevent ovulation (and thus pregnancy) by keeping FSH and LH low.

9. Normally, a woman ovulates a single secondary oocyte each month; if this oocyte is fertilized, then a single birth may result. Many ARTs use hormones ("fertility drugs") to trigger superovulation—the release of several secondary oocytes from the ovary. If more than one of these oocytes is fertilized, then multiple births may result. Also, in some ARTs more than one embryo is introduced into the female reproductive tract in the hope that at least one will implant itself. If all become implanted, then multiple births can result.

10. Some couples want to select the sex of their child for therapeutic reasons (for example, to avoid certain X-linked genetic disorders). Others want to balance their families with respect to number of sons and daughters. Choosing the sex of offspring could result in more individuals of one sex present in populations. Such a gender imbalance means that the roles of males and females in society are likely to change. For example, a gender imbalance favoring males effectively makes females of reproductive age in short supply, potentially affecting population growth rates and gender roles in the work force.

Test Yourself

1. **b.** The parents that have survived to reproduce asexually are able to survive in the current stable environment. Therefore, the offspring should be preadapted for this stable environment.

2. **e.** There are many differences between eggs and sperm. Eggs are larger than sperm and contain more cytoplasm and organelles. Whereas eggs are immotile, sperm have a flagellum and are motile. Over the course of a lifetime, the number of sperm produced by a male far exceeds the number of eggs produced by a female.

3. **d.** External fertilization occurs in some, but not all, aquatic animals. It does not require a penis. It occurs after spawning and does involve species-specific recognition molecules.

4. **b.** A sperm is ejected by the male into the vagina. From the vagina the sperm move through the cervix into the uterus and finally into the oviduct, where fertilization occurs.

5. **b.** The fluid produced by the prostate gland neutralizes acidity in male and female reproductive tracts.

6. **a.** During days 12 through 14 of the ovarian cycle, estrogen exerts positive feedback control on the pituitary, prompting it to release LH and FSH.

7. **d.** During oogenesis, the primary oocyte is diploid; after the first meiotic division into the secondary oocyte the cell becomes haploid.

8. **c.** The correct order from outermost to innermost is cumulus, zona pellucida, and plasma membrane.

9. **d.** High levels of progesterone are needed to maintain the uterus in the proper condition for pregnancy. The levels of progesterone are high only during the second half of the uterine cycle.

10. **b.** The birth control pill interferes with the maturation of the follicles and the ova, inhibiting the release of an egg. Thus, it blocks ovulation, not implantation.

11. **a.** Among those methods listed, the vaginal ring has the lowest failure rate (<1%).

12. **a.** Parthenogenesis is a form of asexual reproduction in which offspring develop from unfertilized eggs; there is no sexual recombination of genes, so a female animal born parthenogenetically will be nearly identical to its mother genetically. In many hymenopterans (bees, wasps, and ants), only males are born from unfertilized eggs. Such males are haploid, so they are not genetically identical to their diploid mothers.

13. **a.** Oviparous animals deposit their fertilized eggs in the external environment, where the embryos develop.

14. **d.** During spermatogenesis, cytoplasm is apportioned equally among resulting cells. This differs from oogenesis, in which cytoplasm is apportioned unequally among daughter cells.

15. **a.** During fertilization, mechanisms in the egg that operate as blocks to polyspermy prevent more than one sperm from entering it.

16. **c.** RU-486 blocks progesterone (not estrogen) receptors and causes sloughing off of the endometrium and embryo.

Animal Development 33

The Big Picture

- Animal development involves several steps: fertilization, cleavage, gastrulation, and organogenesis (for example, neurulation).
- Fertilization is the joining of the egg and sperm to form a zygote. Following fertilization, the cells of the embryo begin to divide in a process known as cleavage. Eggs undergo either complete cleavage or incomplete cleavage, depending on the amount of yolk present. Those that undergo complete cleavage form a blastula, and those that undergo incomplete cleavage form a flat blastodisc.
- Cleavage in mammals is different from cleavage in other groups. It occurs more slowly and genes are transcribed earlier. Also, during the fourth division in mammals, cells separate into the inner cell mass (which will become the embryo proper and some extraembryonic membranes) and the trophoblast (which will prompt implantation and contribute to the chorion, another extraembryonic membrane).
- During gastrulation, the germ layers—endoderm, ectoderm, and mesoderm—form and move into specific positions. In the sea urchin, an archenteron that will become the gut forms, and specific cells form the germ layers. In frogs, a dorsal lip is formed, and the germ cells move into place. The dorsal lip is considered the primary embryonic organizer in amphibians. The blastodisc of reptiles, birds, and mammals goes through a very different pattern of gastrulation.
- Neurulation occurs early in organogenesis. The notochord induces overlying ectoderm to form a neural plate. The neural plate forms a neural tube that will become the brain and spinal cord. Somites produce the repeating segments of the vertebrate body plan. Several genes affect differentiation of tissues along the body axes. For example, Hox genes control differentiation along the anterior–posterior axis.
- Reptiles and birds have four major extraembryonic membranes: yolk sac, amnion, chorion, and allantois. Most mammals have a special placenta for the exchange of nutrients, gases, and wastes between the mother and the developing fetus.

Study Strategies

- All fertilized cells must divide to grow into adult animals. Sorting out the features of division that are common to all animals from those specific to certain taxonomic groups can be difficult and confusing until you begin to recognize the basic patterns.
- Gastrulation can be difficult to picture and understand, especially given the different ways in which gastrulation occurs in different organisms. Refer to the relevent figures in the textbook.
- The structure and role of the placenta is confusing. Is it embryonic? Is it maternal? And how do nutrients, gases, and wastes travel between the mother and embryo or fetus? Re-read appropriate sections of the textbook if needed.
- This chapter contains a great deal of terminology, yet many terms are common to many animals. Create a list of new terms from the chapter. Then determine which are the more general terms, applicable to numerous animals (e.g., "somites" and "gastrulation"), and learn these first. Then tackle those referring to development in specific groups of animals (e.g., "dorsal lip of the blastopore" and "primitive streak").
- A key point in understanding animal development is that all specialized tissues arise from the three germ layers: endoderm, ectoderm, and mesoderm. More important than just committing these three layers to memory is learning the specialized tissues that derive from them. This will enable you to make predictions about tissue functions and to compare body plans among groups of animals.
- Go to yourBioPortal.com to review the following tutorials and activity:

  Animated Tutorial 33.1 Gastrulation

  Animated Tutorial 33.2 Tissue Transplants Reveal the Process of Determination

  Web Activity 33.1 Extraembryonic Membranes

Key Concept Review

33.1 Fertilization Activates Development

The sperm and ovum make different contributions to the zygote

Rearrangements of egg cytoplasm set the stage for determination

Development begins when a haploid sperm joins with a haploid egg. However, fertilization does more than produce a diploid zygote; fertilization activates development. The egg contributes most of the cytoplasm and organelles (in addition to a haploid nucleus) to the zygote. The sperm contributes its haploid nucleus and a centriole that becomes the centrosome, which plays an essential role in microtubule organization.

In an unfertilized frog egg, nutrients are concentrated in the lower half of the egg, called the vegetal hemisphere, and the haploid nucleus is found in the upper half of the egg, known as the animal hemisphere. Upon fertilization, the cytoplasm of the frog egg undergoes rearrangement (see Figure 33.1). A sperm enters in the animal hemisphere at what will become the ventral side of the frog and causes rotation of the cortical cytoplasm to create the gray crescent, which will then become the dorsal region of the frog. This helps establish the left-to-right axis and anterior-to-posterior axis of the developing animal. During the movement of the cytoplasm, the transcription factor β-catenin is degraded in parts of the frog egg so that it becomes concentrated in the dorsal side of the embryo (see Figure 33.2). This happens because the protein GSK-3 moves to the ventral side of the zygote, where it targets β-catenin for degradation. On the dorsal side, proteins inhibit the action of GSK-3, and as a result, the concentration of β-catenin is higher there.

Question 1. What two organelles does the sperm contribute at fertilization and why is each organelle important?
Textbook Reference: *33.1 Fertilization Activates Development, p. 656*

33.2 Cleavage Repackages the Cytoplasm of the Zygote

Cleavage increases cell number without cell growth

Cleavage in mammals is unique

Specific blastomeres generate specific tissues and organs

In cleavage, the first cell divisions occur rapidly with little growth or differentiation. The embryo divides into smaller and smaller cells, producing first a solid ball of cells and then a blastula. The blastula has a central fluid-filled cavity called the blastocoel. Individual cells in the blastula are called blastomeres.

The amount of yolk in the egg determines the pattern of cleavage (see Figure 33.3). Cleavage furrows are impeded by yolk. Frog eggs have small amounts of yolk and complete cleavage in which early cleavage furrows completely divide the egg. The eggs of fishes, reptiles, and birds have large amounts of yolk and incomplete cleavage. In these organisms, the embryo develops from a disc of cells (known as the blastodisc) on top of the yolk mass. Some insects undergo superficial cleavage, a variation of incomplete cleavage in which a syncytium (single cell with many nuclei) is produced early in development when cycles of mitosis occur without cell division.

The early cell divisions of placental mammals occur more slowly than in other animals, and genes are transcribed earlier. Also unique to mammals is the separation of cells into the inner cell mass (which will become the embryo proper) and trophoblast (which adheres to the uterus during implantation). At this stage, the mass of cells is called a blastocyst (see Figure 33.4). Within the blastocyst is a blastocoel. The cells of the inner cell mass are pluripotent and called embryonic stem cells.

Cells of the blastula move and communicate with one another. Specific blastomeres will become specific tissues and organs, and this can be mapped out in the developing cells as a fate map of the blastula (see Figure 33.5). The blastomeres become determined (irreversibly committed to specific fates) at different times for different species. Animals with mosaic development have blastomeres that are set very early to contribute to specific parts of the embryo. In animals with regulative development, the blastomeres can be removed, and other cells will compensate for the loss.

Question 2. Explain how identical and non-identical twins form in humans. Are conjoined twins identical or non-identical twins?
Textbook Reference: *33.2 Cleavage Repackages the Cytoplasm of the Zygote, p. 659*

Question 3. Compare cleavage in birds and in mammals.
Textbook Reference: *33.2 Cleavage Repackages the Cytoplasm of the Zygote, pp. 657–658*

Question 4. Compare the trophoblast and inner cell mass of the human blastocyst with respect to function. From which of the two regions do embryonic stem cells come from?
Textbook Reference: *33.2 Cleavage Repackages the Cytoplasm of the Zygote, pp. 658–659*

33.3 Gastrulation Creates Three Tissue Layers

- Invagination of the vegetal pole characterizes gastrulation in the sea urchin
- Gastrulation in the frog begins at the gray crescent
- The dorsal lip of the blastopore organizes amphibian embryo formation
- Transcription factors underlie the organizer's actions
- Reptilian and avian gastrulation is an adaptation to yolky eggs
- Placental mammals retain the avian/reptilian gastrulation pattern but lack yolk

Placental mammals retain the avian/reptilian gastrulation pattern but lack yolk

During gastrulation, three germ layers form and position themselves in the embryo (see Figure 33.5). Gastrulation makes possible the inductive interactions between cells, which trigger differentiation and organ formation. The inner layer, or endoderm, will become the lining of the digestive tract, the lining of the respiratory tract, and some organs, such as the pancreas and liver. The outer layer, or ectoderm, will become the epidermis (outer layer of the skin), derivatives of the skin (hair, feathers, and teeth), and the nervous system. The middle layer, or mesoderm, will become the heart, blood vessels, muscle, and bone.

Gastrulation in sea urchins involves invagination of the vegetal pole (see Figure 33.6). The vegetal pole first flattens, and then moves inward to form the endoderm and primitive gut (archenteron). Some cells from the vegetal pole break free to become mesenchyme cells, which migrate between tissue layers. The opening of the archenteron or blastopore will become the anus, and the place where the tip of the archenteron meets the ectoderm will become the mouth.

The frog embryo has considerable yolk and differs from the sea urchin in being more than one cell layer thick. Initially, cells near the gray crescent begin to bulge into the blastocoel. These initial cells (bottle cells) move along the interior of the blastula and pull the outer surface of cells along with them, creating a dorsal lip; this process is called involution (see Figure 33.7). The first cells moving in are prospective endoderm, and they form the archenteron. As gastrulation continues, cells from the animal hemisphere move toward the site of involution in a process known as epiboly. As epiboly continues, the dorsal lip forms a complete circle around a plug of yolk-rich cells. At the end of gastrulation, the cells have become fate-determined and are layered as the ectoderm, endoderm, and mesoderm.

Hans Spemann examined the timing and fate of cells during gastrulation in salamanders. His research revealed that gastrulation and normal development depend on cytoplasmic factors in the gray crescent (see Figure 33.8). Spemann and Hilde Mangold, his student, found that when the dorsal lip was transplanted to another gastrula, the result was a second site of gastrulation and eventually two embryos attached at the belly. They concluded that the dorsal lip acts as the primary embryonic organizer (see Figure 33.9).

β-catenin, a transcription factor, and a complex series of interactions between growth factors and other transcription factors, create the organizer and lead to induction of the body plan. As the organizer migrates from the dorsal lip, it inhibits various growth factors along the way to achieve different patterns of differentiation along the anterior–posterior axis (see Figure 33.10).

In birds and reptiles, cleavage produces a blastodisc on top of the large amount of yolk in the egg. The blastula is a circular layer of cells composed of an outer epiblast layer and an inner hypoblast layer. The epiblast will form the embryo proper and the hypoblast will form extraembryonic membranes. There is a fluid-filled space between these two layers. The primitive streak is formed by the movement of cells in the epiblast toward the midline (see Figure 33.11). Along the primitive streak, a primitive groove forms, which functions as the blastopore. Cells move through this groove, into the blastocoel, becoming endoderm and mesoderm. In the chick embryo, there is no archenteron. Endoderm and mesoderm move forward and form gut structures. A group of cells at the anterior end of the primitive groove, known as Hensen's node, acts in the same manner as the dorsal lip of the frog blastopore. Cells that move over Hensen's node differentiate into the notochord and structures of the head.

In placental mammals, the inner cell mass splits into an epiblast and a hypoblast. The embryo and some extraembryonic membranes develop from the epiblast. An extraembryonic membrane also develops from the hypoblast and it interacts with cells of the trophoblast, eventually contributing to the placenta. Gastrulation in mammals occurs just as in birds, with the formation of a primitive groove through which epiblast cells migrate to form endoderm and mesoderm.

Question 5. In the diagram below of a sea urchin gastrula, label the ectoderm, endoderm, primary mesenchyme, secondary mesenchyme, blastopore, and archenteron.
Textbook Reference: *33.3 Gastrulation Creates Three Tissue Layers, p. 660*

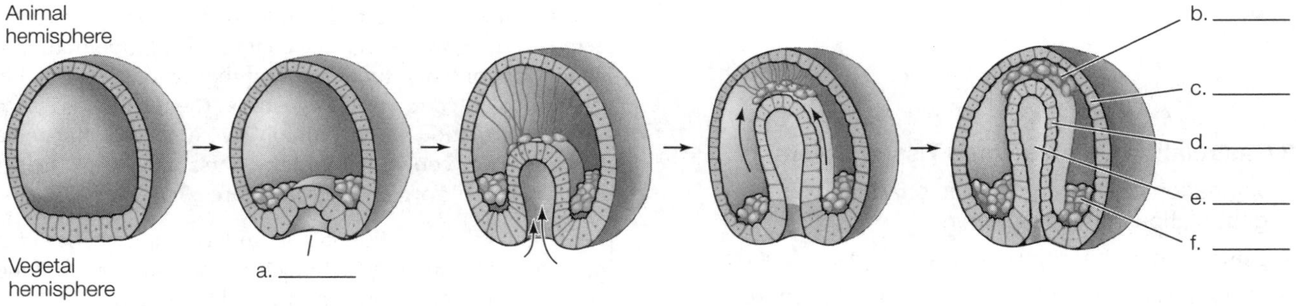

Question 6. Although details of gastrulation differ among sea urchins, frogs, and chickens, what is the common result in all of these animals?
Textbook Reference: *33.3 Gastrulation Creates Three Tissue Layers, p. 660*

33.4 Neurulation Creates the Nervous System

The notochord induces formation of the neural tube

The central nervous system develops from the embryonic neural tube

Body segmentation develops during neurulation

After gastrulation, the organs begin to develop through the process of organogenesis. Neurulation—the initial development of the nervous system—begins early in organogenesis. During neurulation, a rod of connective tissue, called the notochord, develops from chordamesoderm cells. The notochord provides structural support to the developing embryo and induces overlying cells to form the nervous system. A neural plate forms above the notochord from the ectoderm, which will begin to fold into a cylinder forming the neural tube (see Figure 33.12). The anterior end of the neural tube will become the brain, with the rest forming the spinal cord. Neural crest cells break off from the neural tube and migrate outward to prompt development of connections between the central nervous system and the rest of the body.

The hindbrain, midbrain, and forebrain form at the anterior end of the neural tube (see Figure 33.13). The hindbrain and midbrain give rise to structures collectively called the brain stem; these structures regulate critical physiological processes such as breathing. The hindbrain also gives rise to the cerebellum, which governs motor control. The cerebral hemispheres are major information processing centers and they develop from the forebrain. Structures such as the hypothalamus, thalamus, and pituitary also develop from the forebrain.

The repeating pattern of the vertebrate body plan forms from blocks of somite tissue derived from mesoderm and located on both sides of the notochord (see Figure 33.14). The somites will become ribs, vertebrae, and trunk muscles. Body segments differentiate as the embryo develops. Hox genes control this differentiation.

Question 7. In humans, spina bifida is a birth defect in which part of the spinal cord develops abnormally, as does the adjacent area of the spine. What process has been disturbed during prenatal development in individuals with spina bifida?
Textbook Reference: *33.4 Neurulation Creates the Nervous System, pp. 665–666*

Question 8. Design an experiment to determine which substance is responsible for the development of spinal cord circuits.
Textbook Reference: *33.4 Neurulation Creates the Nervous System, p. 666*

33.5 Extraembryonic Membranes Nourish the Growing Embryo

Extraembryonic membranes form with contributions from all germ layers

Extraembryonic membranes in mammals form the placenta

Extraembryonic membranes surround the embryos of reptiles, birds, and mammals. In birds, the yolk sac is derived from the hypoblast and nearby mesoderm, and it surrounds the entire yolk to help retrieve nutrients from the yolk and deliver them to the embryo via blood vessels (see Figure 33.15). Cells from the ectoderm and mesoderm form the amnion, which helps provide an aqueous environment, and the chorion, which develops just under the shell. The chorion regulates water, oxygen, and carbon dioxide exchanges across the shell. The allantois is derived from endoderm and adjacent mesoderm and is a membrane that forms a sac to store metabolic wastes.

In placental mammals, the trophoblast interacts with the endometrium to attach to the uterine wall and begin implantation of the blastocyst. The hypoblast cells form the yolk sac, but because there is no yolk in eggs of placental mammals, the yolk sac contributes to the mesodermal tissues, which interact with the trophoblast tissues to form the chorion. The chorion and tissues from the uterine wall form the placenta (see Figure 33.16). The amnion of the developing mammal surrounds the embryo to produce a closed, fluid-filled environment. The allantois of mammals has the function of removing nitrogenous wastes, but its function is relatively minor in many mammals, including humans. The tissues of the allantois help to form the umbilical cord, which carries major blood vessels that provide a route for exchanges of nutrients, wastes, carbon dioxide, and oxygen between the mother and fetus.

In humans, gestation can be divided into three trimesters, each about 12 weeks in length. During the first trimester of human development, cell division and tissue differentiation are rapid, and organ development begins. At this stage, the developing human is most sensitive to environmental disrupters. By the end of the first trimester, the embryo is considered a fetus. During the second trimester of development, organ systems continue to grow and mature. The third trimester is marked by extremely rapid growth of the fetus and further maturation of the internal organs. Development continues after birth.

Question 9. Why are the chorion, amnion, yolk sac, and allantois considered extraembryonic membranes?
Textbook Reference: *33.5 Extraembryonic Membranes Nourish the Growing Embryo, pp. 668–669*

Question 10. Chorionic villi sampling, one method of prenatal genetic testing, involves taking a small sample of chorionic villi from the placenta. Why would cells from the placenta reveal the genetic makeup of the fetus?
Textbook Reference: *33.5 Extraembryonic Membranes Nourish the Growing Embryo, p. 669*

Question 11. Why would exposure to a harmful chemical during the first trimester be more likely to produce a major birth defect than exposure to the same chemical during the last trimester?
Textbook Reference: *33.5 Extraembryonic Membranes Nourish the Growing Embryo, p. 669*

Question 12. Consider the following statement: "The placenta is the organ where fetal and maternal blood supplies mix." Is this statement correct or incorrect, and why?
Textbook Reference: *33.5 Extraembryonic Membranes Nourish the Growing Embryo, p. 669*

Question 13. In the diagram below of a 9-day chick embryo, label the yolk sac, embryo, allantoic membrane, allantois, amnion, and the chorion.
Textbook Reference: *33.5 Extraembryonic Membranes Nourish the Growing Embryo, p. 668*

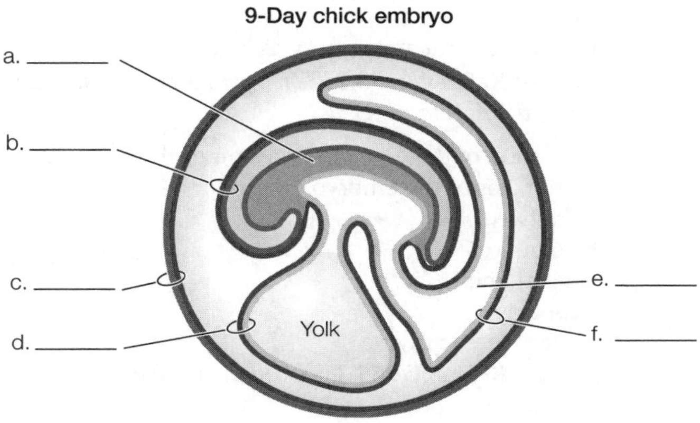

Test Yourself

1. Which of the following does *not* characterize the animal hemisphere of a frog egg?
   a. Contains dense yolk granules
   b. Contains site of sperm entry to which cortical cytoplasm rotates
   c. Heavily pigmented outer layer of cytoplasm
   d. Sperm-binding sites on the outside
   e. Contains the haploid nucleus
   Textbook Reference: *33.1 Fertilization Activates Development, p. 656*
2. In which of the following animals is the cleavage pattern superficial?
   a. Frog
   b. *Drosophila*
   c. Mammal
   d. Fish
   e. Bird
   Textbook Reference: *33.2 Cleavage Repackages the Cytoplasm of the Zygote, p. 658*
3. Which of the following statements about complete cleavage versus incomplete cleavage is true?
   a. Incomplete cleavage occurs in species with small volumes of cytoplasm.
   b. Complete cleavage is found in mammals and is the more evolved condition.

c. Incomplete cleavage occurs in species with large amounts of yolk.
d. Complete cleavage occurs only in eggs that have been fertilized by two sperm.
e. Incomplete cleavage occurs in species with small amounts of yolk.

Textbook Reference: *33.2 Cleavage Repackages the Cytoplasm of the Zygote, p. 657*

4. The tissues and organs that eventually will develop from specific germ layers can be depicted visually by means of a fate map. On the fate map, ectoderm will become
a. the lining of the gut.
b. the epidermal layer of the skin.
c. muscle.
d. the heart.
e. the lining of the respiratory tract.

Textbook Reference: *33.2 Cleavage Repackages the Cytoplasm of the Zygote, p. 659*

5. The location of the _______ determines the anterior–posterior axis of the embryo.
a. primitive streak
b. blastopore
c. vegetal hemisphere
d. hypoblast
e. tertiary mesenchyme

Textbook Reference: *33.3 Gastrulation Creates Three Tissue Layers, p. 661*

6. Which of the following represents the correct order of the germ layers, from the inside to the outside?
a. Mesoderm, ectoderm, endoderm
b. Endoderm, ectoderm, mesoderm
c. Ectoderm, mesoderm, endoderm
d. Endoderm, mesoderm, ectoderm
e. Mesoderm, endoderm, ectoderm

Textbook Reference: *33.3 Gastrulation Creates Three Tissue Layers, p. 660*

7. The dorsal lip of the blastopore organizes embryo formation in frogs. The equivalent structure in chickens is
a. the epiblast.
b. the hypoblast.
c. Hensen's node.
d. the bottle cell.
e. the gray crescent.

Textbook Reference: *33.3 Gastrulation Creates Three Tissue Layers, p. 664*

8. Birds develop extraembryonic membranes during development. Which of the following statements about avian extraembryonic membranes is *false*?
a. The yolk sac surrounds the yolk and provides nutrients.
b. The amnion and chorion are derived from ectoderm and mesoderm.
c. The allantois stores nutrients.
d. The chorion exchanges gases and water between the embryo and the environment.
e. The allantois is derived from endoderm and mesoderm.

Textbook Reference: *33.5 Extraembryonic Membranes Nourish the Growing Embryo, pp. 668–669*

9. Which of the following statements about the mammalian blastocyst is *false*?
a. The trophoblast gives rise to the embryo proper.
b. Genes are transcribed early during cleavage.
c. The blastocyst implants itself in the mother's uterus.
d. Early mammalian cleavage is relatively slow.
e. The trophoblast cells send out chorionic villi.

Textbook Reference: *33.2 Cleavage Repackages the Cytoplasm of the Zygote, pp. 658–659*

10. The primary embryonic organizer is most likely initiated by
a. the yolk.
b. Tcf-3 protein.
c. β-catenin.
d. cAMP.
e. None of the above

Textbook Reference: *33.3 Gastrulation Creates Three Tissue Layers, pp. 662–663*

11. The _______ eventually develop into vertebrae, ribs, and trunk muscles, and are found along the sides of the _______.
a. somites; neural tube
b. neural tube cells; notochord
c. blastopore cells; dorsal lip
d. neural crest cells; dorsal lip
e. neural plate cells; archenteron

Textbook Reference: *33.4 Neurulation Creates the Nervous System, pp. 666–667*

12. Hans Spemann called the dorsal lip of the blastopore the embryonic organizer because it
a. is the point where gastrulation begins.
b. becomes part of the nervous system.
c. becomes part of the notochord.
d. leads to the establishment of the embryonic axes.
e. is the location at which the sperm enters the egg.

Textbook Reference: *33.3 Gastrulation Creates Three Tissue Layers, p. 661*

13. During its development, the human embryo is contained within a fluid-filled chamber enclosed by the extraembryonic membrane called the
a. yolk sac.
b. amnion.
c. chorion.
d. allantois.
e. trophoblast.

Textbook Reference: *33.5 Extraembryonic Membranes Nourish the Growing Embryo, p. 669*

14. Impaired vision could be caused by a blockage of blood vessels in the
a. thalamus.
b. cerebellum.
c. pituitary.

d. hypothalamus.
e. spinal cord.

Textbook Reference: *33.4 Neurulation Creates the Nervous System, p. 666*

15. Which of the following statements about human development is *false*?
    a. At about 3 months gestation, the developing human is called a fetus.
    b. The allantois contributes to the umbilical cord.
    c. Substantial developmental changes occur in the brain after birth.
    d. Development ends at birth.
    e. The umbilical cord attaches the embryo and later the fetus to the placenta.

Textbook Reference: *33.5 Extraembryonic Membranes Nourish the Growing Embryo, p. 669*

Answers

Key Concept Review

1. The sperm contributes a haploid nucleus, which, when combined with the egg's haploid nucleus, produces a diploid zygote. The sperm also contributes a centriole. This is important because the centrosome of the egg deteriorates during oogenesis, so the sperm's centriole becomes the centrosome of the zygote. The centrosome functions as a microtubule-organizing center.
2. Identical (monozygotic) twins are formed during an early stage of cleavage when the mass of cells splits. Non-identical twins form when two secondary oocytes are released from the ovaries and are fertilized by two different sperm. Conjoined twins are identical twins formed when splitting of the mass of cells is incomplete.
3. Due to the large amount of yolk in the avian egg, cleavage in birds is incomplete, and the result is a blastodisc that sits on top of the yolk (see Figure 33.3B; although the egg pictured is from a zebrafish, bird eggs show a similar cleavage pattern). Mammalian embryos undergo complete cleavage, resulting in a structure called a blastocyst (see Figure 33.4A). The pace of cleavage is slower in mammals than in birds, and genes are transcribed earlier. Finally, mammalian cleavage is unique in that the cells separate into the inner cell mass and trophoblast. Mammalian embryos are similar to birds in that the blastula develops as a blastodisc.
4. The trophoblast secretes enzymes that allow the blastocyst to burrow into the uterine lining (endometrium) of the mother and to implant there. The trophoblast then forms chorionic villi, the embryo's major contribution to the placenta. In contrast, the inner cell mass forms the embryo proper and contributes to extraembryonic membranes (amnion, allantois, and yolk sac). Embryonic stem cells come from the inner cell mass; these cells are pluripotent.
5.
    a. Blastopore
    b. Secondary mesenchyme
    c. Ectoderm
    d. Endoderm
    e. Archenteron
    f. Primary mesenchyme
6. Although the details of gastrulation differ among sea urchins, frogs, and chickens, the common result is the formation of germ layers from which all tissues and organs will form.
7. Spina bifida is a neural tube defect characterized by failure of the neural tube to develop and close properly during prenatal development. It represents a disturbance in the process of neurulation.
8. The experiment should be directed at blocking Sonic hedgehog, the transcription factor released by the notochord. Sonic hedgehog diffuses into the ventral region of the neural tube, where it directs development of spinal cord circuits.
9. The chorion, amnion, yolk sac, and allantois are considered extraembryonic membranes because they lie outside the embryo. This name distinguishes them from membranes inside the embryo, such as mucous membranes that line the respiratory, digestive, urinary, and reproductive tracts.
10. The placenta is formed from the chorion of the embryo and the endometrium (uterine lining) of the mother. The chorion is formed when the yolk sac contributes mesodermal tissues that interact with the trophoblast. Thus, the chorionic villi are the embryo's contribution to the placenta and cells taken from them would have the same genetic makeup as cells of the fetus.
11. Exposure to a harmful chemical during the first trimester would be more likely to produce a major birth defect than exposure to the same chemical during the last trimester because tissues and organs are forming during the first trimester. The last trimester involves mostly rapid growth.
12. Under normal circumstances there is no direct mixing of fetal and maternal blood supplies at the placenta. Instead, all exchanges of materials (nutrients, gases, wastes) occur across fetal capillaries within chorionic villi.
13.
    a. Embryo
    b. Amnion
    c. Chorion
    d. Yolk sac
    e. Allantois
    f. Allantoic membrane

Test Yourself

1. **a.** Dense yolk granules are found in the vegetal hemisphere.

2. **b.** Cleavage in *Drosophila* is superficial.
3. **c.** Incomplete cleavage occurs because the cleavage furrows cannot completely penetrate the yolk. Incomplete cleavage occurs in animals, such as birds and reptiles, whose eggs have large amounts of yolk.
4. **b.** Ectoderm will form the epidermal layer of the skin.
5. **b.** The blastopore, which will eventually become the anus, marks the posterior of the embryo.
6. **d.** The germ layers that are formed during gastrulation are the inner layer of endoderm, the middle layer of mesoderm, and the outer layer of ectoderm.
7. **c.** In chickens, Hensen's node is the equivalent of the dorsal lip of the frog blastopore.
8. **c.** During development, an avian embryo produces wastes, but because birds develop in a shell, the wastes must be stored in the egg. The allantois forms a sac that is used for the storage of metabolic wastes.
9. **a.** In the mammalian blastocyst, the inner cell mass becomes the embryo proper.
10. **c.** β-catenin is thought to be the initiator of organizer activity during early development.
11. **a.** Somites are located along the neural tube in the developing vertebrate. These cells will develop into the vertebrae, ribs, and trunk muscles.
12. **d.** The dorsal lip leads to the establishment of the embryonic axes.
13. **b.** The amnion is the membrane that surrounds the developing mammalian fetus in its fluid-filled amniotic cavity.
14. **a.** Impaired vision could be caused by a blockage of blood vessels in the thalamus. The thalamus is a major relay station for sensory information, including vision.
15. **d.** Development does not end at birth. Growth continues until adult size is reached. Organs of the body continue to repair themselves.

Neurons and Nervous Systems

34

The Big Picture

- The neuron, with support from surrounding glia, is the functional unit of the nervous system. The neuron is composed of a cell body, dendrites, and an axon. Neurons generate membrane potential, which is the difference in charge across their membranes. Nerve impulses pass down the axon of a neuron as action potentials. An action potential is a temporary disruption of the "battery-like" state of the resting neuron membrane due to the opening and closing of sodium and potassium voltage-gated channels. The action potential is an all-or-none response that occurs when the depolarization of an axon reaches a threshold level.
- The axon divides into small nerve endings that each have an axon terminal. The axon terminals come very close to the membrane of the target cell to form a synapse, which is a tiny gap across which neurons communicate. There are many different neurotransmitters that transmit a nerve impulse from the presynaptic cell to the postsynaptic cell. The action of a specific neurotransmitter depends on the receptor to which it binds.
- The mammalian nervous system is divided both anatomically and functionally. The anatomical divisions are the central nervous system (CNS) and peripheral nervous system (PNS). The functional divisions are the sympathetic and parasympathetic nervous systems. Afferent nerves carry information from the PNS to the CNS, and efferent nerves carry information from the CNS to the PNS.
- The brain can be divided anatomically and functionally into many different regions, each responsible for its many vital actions. The "higher" brain centers of the cerebrum are responsible for conscious thought and deliberate (voluntary) movements. The "lower" brain centers such as the cerebellum, pons, and medulla regulate involuntary movements and are involved primarily in maintaining homeostasis throughout the body.

Study Strategies

- One of the most difficult challenges when learning about the nervous system is understanding how the resting membrane potential and action potential are produced. Remember that the neuron at rest is like a battery, and that the charge of the membrane is dependent on the ions that are on either side and that move across the membrane. Learn what each important anion and cation does at rest and during the action potential. Then, piece by piece, put together a sense of how the neuron membrane is charged at rest, and what happens when an action potential is generated.
- Make sure you understand the concept of "pre-" and "postsynaptic" neurons. Try to remember the sequence of events and relate them to the function of the neuron transmitting a stimulus from one cell to another.
- Examine the function of a synapse and how the action potential is transmitted across the synapse. Remember that in many cases the signal goes from an electrical signal, to a chemical signal, and back to an electrical signal.
- It is easy to confuse "afferent" and "efferent," and it is important in your understanding of the nervous system to differentiate them. Think of *a*fferent as *a*rriving, and *e*fferent as *e*xiting the reference point—usually the CNS.
- It is important to recognize that although the neurotransmitters acetylcholine and norepinephrine always have antagonistic effects on each other, there is no universal pattern as to which stimulates tissue and which inhibits tissue. For example, acetylcholine causes the smooth muscle in blood vessels in many regions of the body to relax, but it causes the smooth muscle in the stomach and intestines to contract.
- Sensory input to the brain can be a difficult notion. Specific regions of the body "map onto" specific regions of the cerebrum. Similarly, with regard to control of movements, specific regions of the brain "map onto" specific regions of the body. Moreover, the amount of brain matter devoted to a particular body region depends on the amount of muscle control and sensors contained in that body region. Thus, a relatively small area of the cerebral hemispheres is devoted to the upper leg (which has a relatively limited range of movement and sensation), whereas the tongue, with

its many sensory receptors and high degree of mobility, commands more of the tissue of the cerebrum.

- Several key terms in this chapter occur repeatedly: parasympathetic, sympathetic, afferent, efferent, preganglionic, and postganglionic. Until you master this vocabulary, it will be difficult to put together a comprehensive picture of the nervous system. Create your own list of the terms you see repeatedly, and make sure that you understand their meanings.
- Brain anatomy is complex. Start by dividing up the structures into forebrain, midbrain, and hindbrain regions. Then identify each of the subcomponents. Your understanding will be more complete if you learn the general functions of each brain section as you go along. That is, rather than learning the anatomy of the brain and then starting over to learn the functions, learn the two at the same time.
- Go to yourBioPortal.com to review the following tutorials and activities:

  Animated Tutorial 34.1 The Resting Membrane Potential

  Animated Tutorial 34.2 The Action Potential

  Animated Tutorial 34.3 Synaptic Transmission

  Animated Tutorial 34.4 Information Processing in the Spinal Cord

  Interactive Tutorial 34.1 Neurons: Electrical and Chemical Conduction

  Web Activity 34.1 The Nernst Equation

  Web Activity 34.2 The Human Cerebrum

  Web Activity 34.3 Structures of the Human Brain

Key Concept Review

34.1 Nervous Systems Consist of Neurons and Glia

Neurons transmit electrical and chemical signals

Glia support, nourish, and insulate neurons

Neurons are linked into information-processing networks

The nervous system is made up of two major types of cells: neurons and glia (also called glial cells). A neuron is composed of four main parts: the cell body, dendrites, axon, and axon terminals (see Figure 34.1). The cell body contains the nucleus and other organelles. Dendrites extend from the cell body and receive information from other neurons. The axon conducts action potentials away from the cell body. Axon terminals interact with the neuron's target cells to form a synapse, a tiny gap where information is passed from one neuron to another neuron. The first neuron is the presynaptic neuron, and the second is the postsynaptic neuron. Synapses can be either chemical or electrical.

Glia serve many functions in the nervous system, such as supplying nutrients to neurons, removing wastes, and helping neurons make the proper connections during development. Glia that insulate axons in the central nervous system are oligodendrocytes and those that insulate axons in the peripheral nervous system are Schwann cells (see Figure 34.2). The covering produced by Schwann cells and oligodendrocytes is called myelin; myelinated axons conduct action potentials more rapidly than do axons lacking myelin. Star-shaped glia, called astrocytes, contribute to the blood–brain barrier, which protects the brain from some toxins in the blood.

Neurons are organized into neural networks, with three functional categories of neurons. Afferent neurons carry sensory information into the nervous system (= input). Efferent neurons carry information from the nervous system to effectors, such as muscles or glands (= output). Interneurons facilitate communication between afferent and efferent neurons (= integration). Simple animals, such as the sea anemone, possess a nerve net (see Figure 34.3A). The nervous system of more complex animals, such as an earthworm, is made up of clusters of neurons (ganglia) distributed throughout the body (see Figure 34.3B). Animals with complex behavior have a central nervous system—the brain and spinal cord—and a peripheral nervous system—the neurons in the rest of the body (see Figure 34.3C).

Question 1. Why do certain substances, such as anesthetics and alcohol, have rapid effects on the brain, whereas others cannot reach the brain?
Textbook Reference: *34.1 Nervous Systems Consist of Neurons and Glia, p. 673*

Question 2. In Lou Gehrig's disease, motor neurons die and muscles no longer receive neural messages. Affected individuals gradually lose control over their limbs and body, and the eventual cause of death is respiratory failure due to deterioration of the muscles that control breathing. Sensory neurons and interneurons are not affected by the disease, and the individuals do not have loss of cognitive function. Explain in terms of the neurological processes why cognition is unimpaired.
Textbook Reference: *34.1 Nervous Systems Consist of Neurons and Glia, p. 674*

34.2 Neurons Generate and Transmit Electrical Signals

Simple electrical concepts underlie neural function

The sodium–potassium pump sets up concentration gradients of Na^+ and K^+

- The resting potential is mainly caused by K^+ leak channels
- The Nernst equation can predict a neuron's membrane potential
- Gated ion channels can alter membrane potential
- Graded changes in membrane potential spread to nearby parts of the neuron
- Sudden changes in Na^+ and K^+ channels generate action potentials
- Action potentials are conducted along axons without loss of signal
- Action potentials travel faster in large axons and in myelinated axons

Resting neurons have a negative charge inside and a positive charge outside, resulting in a difference in electrical charge across the membrane known as the membrane potential. In an unstimulated neuron, this voltage difference is called a resting potential. Electrodes can be used to measure resting potentials; such measurements show that the resting potential is typically between –60 and –70 mV (see Figure 34.4). The electrical charge across the membrane at rest is due to differences in concentrations of the charged ions sodium (Na^+), chloride (Cl^-), potassium (K^+), and calcium (Ca^{2+}). The lipid bilayer of the plasma membrane is impermeable to ions. Ions move across the plasma membrane through ion transporters and channels.

The sodium–potassium pump (also known as sodium–potassium ATPase) transports Na^+ out of the cell and K^+ into it, thereby maintaining higher concentrations of Na^+ ions outside the cell and higher concentrations of K^+ ions inside it (see Figure 34.5A). At rest, neurons have a specific charge due to K^+ movement to the outside of the cell, resulting in a resting potential. K^+ channels are the most common open (sometimes called leak) channels, allowing K^+ to diffuse out of the cell down the concentration gradient that has been set up by the Na^+–K^+ pump (see Figure 34.5B). K^+ leak channels are largely responsible for the membrane's resting potential.

Ion channels are selective pores in the plasma membrane that allow specific ions to diffuse across the membrane. Whereas some are always open (such as K^+ channels), others are gated (open under certain conditions and closed under other conditions). Ion channels can be voltage-gated (responding to changes in the voltage across the membrane), chemically gated (responding to the binding of a specific molecule), or mechanically gated (responding to mechanical force applied to the membrane).

Membranes can be depolarized or hyperpolarized (see Figure 34.6). Depolarization occurs when the inside of a neuron becomes less negative compared to the resting potential. Hyperpolarization occurs when the inside of a neuron becomes more negative compared to the resting potential.

Small, local changes in membrane potential are called graded membrane potentials. Action potentials are very short-lived, but large changes in membrane potential (see Figure 34.7). Action potentials are generated when the membrane reaches a threshold potential, Na^+ voltage-gated channels open, and Na^+ enters the cell to make the inside of the axon positive. Voltage-gated K^+ channels then open, allowing K^+ to leave the axons to help return the membrane potential back to the resting level. As the K^+ channels open, the Na^+ channels close and cannot be opened for a few milliseconds, which is called the refractory period. The Na^+–K^+ pump helps return the concentration of ions back to the resting levels.

Action potentials travel down an axon by a positive feedback mechanism that stimulates adjacent regions of an axon to generate the action potential. The Na^+ ions that enter during an action potential flow to adjoining regions of the axon, stimulating depolarization and the movement of the action potential along the axon. The refractory period, during which the Na^+ channels cannot act, can be explained by the presence of two gates in the channel, an activation gate and an inactivation gate. The refractory period keeps an action potential moving in one direction, away from the cell body.

An action potential is an all-or-none response; the depolarization must reach a threshold level for an action potential to occur. An action potential is also a self-regenerating response; once an action potential occurs at one location on an axon, it stimulates the adjacent area to generate an action potential.

In the nervous systems of invertebrates, the conduction velocity of axons increases with the increasing diameter of axons. In the nervous systems of vertebrates, conduction velocity of axons is increased by myelination, or the concentrated layers of myelin formed by glia that wrap themselves around the axons. The glia leave regularly spaced gaps called nodes of Ranvier, which are the sites where depolarization can occur. As action potentials jump from node to node, the speed of transmission increases in a process known as saltatory conduction (see Figure 34.8).

Question 3. In order to determine the role of the potassium channels in a neuron, a researcher has knocked out all the functional potassium channels and depolarized the membrane potential. What will happen to the membrane potential after depolarization?

Textbook Reference: *34.2 Neurons Generate and Transmit Electrical Signals, p. 676*

Question 4. Label the following structures on the myelinated motor neuron shown below: axon, axon hillock, axon terminals, cell body, and dendrites. What is the direction and manner in which an action potential is conducted along the neuron?

Textbook Reference: *34.2 Neurons Generate and Transmit Electrical Signals, p. 680 (see also p. 673)*

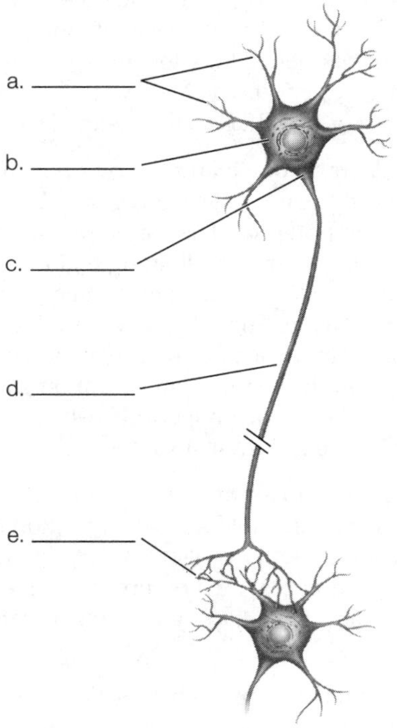

Question 5. Explain how an action potential travels more quickly down an axon wrapped in myelin than it does down an unmyelinated axon.

Textbook Reference: *34.2 Neurons Generate and Transmit Electrical Signals, p. 680*

34.3 Neurons Communicate with Other Cells at Synapses

- The neuromuscular junction is a model chemical synapse
- The postsynaptic cell sums excitatory and inhibitory input
- To turn off responses, synapses must be cleared of neurotransmitter
- There are many types of neurotransmitters
- Synapses can be fast or slow depending on the nature of receptors

Chemical synapses are more common than electrical synapses. At a chemical synapse, an action potential arriving at the axon terminals of a presynaptic neuron causes the release of neurotransmitter, which travels across the synaptic cleft to bind with receptors on the postsynaptic neuron. Electrical synapses are formed by direct contact between adjacent neurons; these synapses contain numerous gap junctions. Two neurons forming an electrical synapse are joined by connexons, which are tunnels (pores) between the two neurons that allow ions to pass between the two cells. Electrical synapses allow for rapid communication.

The neurotransmitter acetylcholine (ACh) is the chemical messenger carrying information between motor neurons and muscle cells at neuromuscular junctions (see Figure 34.9). Acetylcholine is enclosed in a vesicle at the presynaptic synapse, which fuses with the membrane to release the acetylcholine into the synaptic cleft. Ca^{2+} channels open when the action potential reaches the axon terminal, causing Ca^{2+} to rush in and regulate the fusing of the acetylcholine-containing vesicles to the presynaptic membrane. Acetylcholine released into the synaptic cleft binds with receptors in the postsynaptic membrane, the motor end plate, opening Na^{+} channels and resulting in depolarization of the motor end plate. The enzyme acetylcholinesterase breaks down acetylcholine in the synaptic cleft to halt the action of the released acetylcholine (see Figure 34.10).

Excitatory synapses between motor neurons and muscle cells in vertebrates depolarize the postsynaptic membrane, and inhibitory synapses between neurons hyperpolarize the postsynaptic membrane. Neurons may receive synaptic inputs from many neurons.

Excitatory and inhibitory postsynaptic potentials are summed by spatial summation (adding up of simultaneous potentials at different sites) or by temporal summation (adding up of the postsynaptic potentials generated at the same site in rapid sequence) (see Figure 34.11). The region of the cell body at the base of the axon, called the axon hillock, is the "decision-making" area of a neuron. If the axon hillock is depolarized, the axon will fire an action potential.

Neurotransmitters must be removed from the synapse in order for their action to be turned off. This can occur in several ways. First, enzymes may destroy the neurotransmitter. Second, the neurotransmitter may simply diffuse away from the synaptic cleft. Third, nearby cell membranes may take up the neurotransmitter by means of active transport. The drug Prozac, used to treat depression, acts by slowing the reuptake of the neurotransmitter serotonin, thus prolonging serotonin's action at the synapse.

There are more than 50 neurotransmitters, including amino acids, modified amino acid derivatives, and peptides. One neurotransmitter can act on several different receptors, and its particular action depends on the receptor to which it binds. Acetylcholine has nicotinic receptors, which tend to be excitatory, and muscarinic receptors, which tend to be inhibitory.

Two general categories of neurotransmitter receptors are ionotropic and metabotropic. Ionotropic receptors are ion channels on the postsynaptic membrane that are activated by binding of the neurotransmitter. They allow fast, short-lived responses. Metabotropic receptors are not ion channels. They act by initiating signaling cascades, which eventually cause changes in ion channels. When mediated by metabotropic receptors, postsynaptic cell responses are usually slower and longer-lived than those generated by ionotropic receptors.

Question 6. The active ingredients in many nerve gases belong to a class of chemicals called anticholinesterases (chemicals that block acetylcholinesterase). Explain a possible synaptic mechanism by which these chemicals can damage an animal's nervous system.
Textbook Reference: *34.3 Neurons Communicate with Other Cells at Synapses, p. 682*

Question 7. Clinical depression is thought to be due, in part, to insufficient levels of the neurotransmitter serotonin. Drugs known as selective serotonin reuptake inhibitors (SSRIs) can be used to treat depression. Taking the name of this class of drugs as a clue, propose a mechanism by which they may act.
Textbook Reference: *34.3 Neurons Communicate with Other Cells at Synapses, pp. 682–683*

34.4 The Vertebrate Nervous System Has Many Interacting Components

- The autonomic nervous system controls involuntary physiological functions
- The spinal cord transmits and processes information
- Interneurons coordinate polysynaptic reflexes
- The brainstem transfers information between the brain and spinal cord
- Deeper parts of the forebrain control physiological drives, instincts, and emotions
- Regions of the telencephalon interact to produce consciousness and control behavior
- Each cerebral cortex has four lobes

The brain and spinal cord together constitute the central nervous system (CNS). The part of the nervous system outside the brain and spinal cord is the peripheral nervous system (PNS). The PNS is composed of afferent nerves that carry information to the CNS and efferent nerves that carry information from the CNS to muscles and glands (see Figure 34.12). Efferent pathways can be classified as voluntary (executing our conscious movements) or involuntary (autonomic, controlling physiological functions). The CNS also receives chemical information from circulating hormones, and releases neurohormones.

The sympathetic and parasympathetic divisions of the autonomic nervous system have antagonistic effects on the organ systems they innervate and play a large role in the maintenance of homeostasis. The sympathetic division is involved in the fight-or-flight response of increased heart rate, blood pressure, and cardiac output. The parasympathetic division slows down heart rate and decreases blood pressure and cardiac output; however, it accelerates digestive activities. Every autonomic efferent pathway begins with a preganglionic neuron that has its cell body in the CNS. Axons of preganglionic neurons lead to a ganglion outside the CNS, where they synapse with postganglionic neurons. Norepinephrine is the neurotransmitter in postganglionic neurons of the sympathetic system, whereas acetylcholine is the neurotransmitter in postganglionic neurons of the parasympathetic system. Both acetylcholine and norepinephrine influence the pacemaker cells in the heart that control heart rate. The parasympathetic system has preganglionic neurons that come from the brain stem and the last segment of the spinal cord (the sacral region). In the sympathetic system, preganglionic neurons come from the middle (lumbar) and upper (thoracic) regions of the spinal cord (see Figure 34.13).

The spinal cord is a bidirectional neural pathway for information flow between the peripheral nervous system and the brain. Each spinal nerve has two roots: afferent (sensory) axons enter the spinal cord via the dorsal root, and efferent (motor) axons leave via the ventral root. Cell bodies are found in gray matter and axons are found in white matter of the nervous system. The white color is due to myelin.

The spinal cord converts some afferent information from the peripheral nervous system into efferent information sent back to the peripheral nervous system in a process known as a spinal reflex. A monosynaptic reflex, such as the knee-jerk reflex, involves only an afferent neuron, an efferent neuron, and one synapse (see Figure 34.14). Leg muscle stretch receptors trigger the sensory neuron to conduct action potentials to the spinal cord upon stretching. The action potential is passed through the synapse of the sensory and motor neuron located in the central gray matter of the spinal cord. More complicated reflexes involve interneurons and additional synapses. Flexor muscles, which flex limbs, and extensor muscles, which straighten limbs, control limb movement. The motor neurons that stimulate these muscles are antagonistic and polysynaptic, with each sensory neuron stimulating one motor neuron while inhibiting the other motor neuron at two distinct synapses.

All information traveling between the spinal cord and higher brain areas must pass through the brainstem, which includes the pons, medulla, and midbrain. Many of the connections involved in functional control of the body occur in the reticular system, along with the control of sleep and waking.

The limbic system is involved in instincts and emotions (see Figure 34.15). A portion of the limbic system, known as the hippocampus, helps transfer short-term memory to long-term memory. The amygdala, another part of the limbic system, is involved in fear reactions and fear memories.

The cerebral hemispheres are covered by a convoluted cerebral cortex, a sheet of gray matter that processes sensory information and higher-order information in the association areas. Each cerebral hemisphere consists of four main regions: the temporal lobe, the frontal lobe, the occipital lobe, and the parietal lobe (see Figure 34.16A). Each region has specific functions (see Figure 34.16B).

The temporal lobe processes auditory information, and the association areas are involved with recognition, identification, and the naming of objects. Damage to the temporal lobe results in conditions (agnosias) in which individuals are unable to identify a stimulus (e.g., a face), even though they can perceive it. Integration of spoken language can also be impaired by damage to the temporal lobe. The frontal lobe contains the primary motor cortex made up of axons that project to muscles in the body. The primary motor cortex can be mapped according to the locations that control movements of various body parts (see Figure 34.17A). The association areas of the frontal lobe are involved in planning and personality. The primary somatosensory cortex is located in the parietal lobe. As with the motor cortex, the location of the body that is sensed by the somatosensory cortex can be mapped on the brain (see Figure 34.17B). Visual information is processed by the occipital lobe of the cerebrum. The association areas for the occipital lobe translate visual stimuli into language.

Among vertebrates, humans (as well as porpoises) stand out as having larger brains than would be predicted by their body size. Degree of convolution of the cerebral cortex (a measure of the area of cortex) is greatest in humans, as is the percentage of cortex that is association cortex (which is devoted to the integration of information).

Question 8. Imagine that you have been eyeing candy in a dish and finally decide to unwrap a piece and eat it. As you begin to suck on the candy, your salivary glands begin to secrete saliva. What parts (divisions or branches) of the nervous system are involved in this sequence of events?
Textbook Reference: *34.4 The Vertebrate Nervous System Has Many Interacting Components, pp. 684–685*

Question 9. Label the following structures in the diagram of the knee-jerk reflex below: stretch receptors, sensory neuron, motor neuron(s), interneuron, dorsal root, ventral root, gray matter, and white matter. Indicate the direction of information flow along the neurons. Which pathway is polysynaptic?
Textbook Reference: *34.4 The Vertebrate Nervous System Has Many Interacting Components, pp. 685, 687*

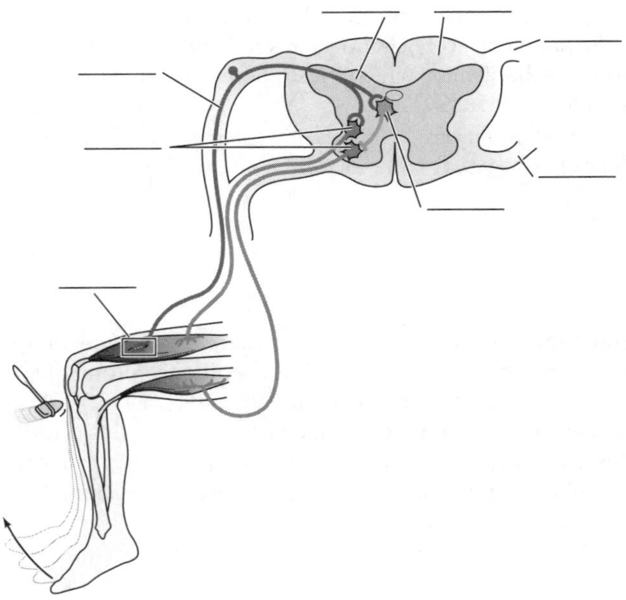

Question 10. During a boxing match, a sharp punch to the jaw of a boxer may cause loss of consciousness. Which area of the brain is likely to have been affected by such a knockout punch?
Textbook Reference: *34.4 The Vertebrate Nervous System Has Many Interacting Components, p. 687*

34.5 Specific Brain Areas Underlie the Complex Abilities of Humans

Language abilities are localized in the left cerebral hemisphere

Some learning and memory can be localized to specific brain areas

There are two different states of sleep

We still cannot answer the question "What is consciousness?"

In most people, the ability to produce and interpret language occurs in the left hemisphere of the cerebrum. Several areas have been located that are important for language (see Figure 34.16B). The frontal lobe contains Broca's area, which is involved in the production of language. Wernicke's area, which is located in the temporal lobe, is responsible for understanding language. Language ability involves the flow of information among these and several other areas of the left cerebral cortex, and damage to any area can result in aphasia, the condition of not being able to use or understand written or spoken words.

Learning occurs when behavior is modified as a result of experience. Long-lasting synaptic changes must occur for learning to take place. Long-term potentiation (LTP) occurs when high-frequency electrical stimulation makes certain circuits more sensitive to subsequent stimulation.

Associative learning in animals involves the linking of two unrelated stimuli; an example is the conditioned reflex of Pavlov's dogs, who were conditioned to associate eating with the ringing of a bell, so that even in the absence of food the mere ringing of a bell stimulated salivation. Observational learning is more complex than associative learning and has three elements: observation of another person's behavior; retention of a memory of what was observed; and the attempt to copy or use that information.

Memory is the phenomenon whereby the nervous system retains what has been learned or experienced. Declarative memory, which is memory of people, places, events, and things, lasts for varying amounts of time. Immediate memories are very short-term vivid memories of what has just occurred. Short-term memories contain less information than immediate memories but last between 10 and 15 minutes. Long-term memory can last for days, months, years, or even a lifetime. Repetition or reinforcement enhances the transfer of short-term memory to long-term memory. Procedural memory is memory of how to perform motor tasks, such as riding a bicycle.

Researchers who study sleep often use an electroencephalogram (EEG), which measures the electrical activity of entire brain regions, especially those in the cerebral cortex. There are two main states of sleep in humans: rapid-eye movement (REM) sleep and non-REM sleep (see Figure 34.19). During non-REM sleep, which always occurs first, the neurons in the thalamus and cerebral cortex become hyperpolarized. Eventually neurons fire action potentials in slow, synchronized bursts, which are recorded by an EEG as a large deflection with a slow-wave pattern. During REM sleep, dreams and nightmares occur along with near complete paralysis of skeletal muscles. Neurons that were hyperpolarized during non-REM sleep return to waking levels, allowing information to be processed. Afferent and efferent pathways are inhibited during REM sleep. During a night of sleep, the brain cycles between REM sleep and non-REM sleep, with 80 percent of sleep being non-REM sleep.

Consciousness requires a perception of self that can be integrated with information from the environment and past experience. The insular cortex (or insula) integrates information from all over the body and is greatly expanded with unique types of neurons in humans, great apes, dolphins, and elephants. The insula may be the area of the brain responsible for self-awareness and conscious experience.

Question 11. What are the differences between the brain functions involved in reading a written sentence out loud and those involved in repeating a sentence one has just heard?
Textbook Reference: *34.5 Specific Brain Areas Underlie the Complex Abilities of Humans, p. 690*

Question 12. Describe the neurological basis of the phenomenon of sleepwalking. A person who is sleepwalking is in which stage of sleep?
Textbook Reference: *34.5 Specific Brain Areas Underlie the Complex Abilities of Humans, pp. 690–691*

Question 13. Species that can recognize themselves in a mirror, including humans, great apes, dolphins, and elephants, have expanded insulas and are considered by some researchers to be self-aware. What do these species have in common, and how might the validity of the mirror test be challenged? What factors need to be considered when one is designing an experiment to test self-awareness in another species?
Textbook Reference: *34.5 Specific Brain Areas Underlie the Complex Abilities of Humans, p. 692*

Test Yourself

1. The extensions of postsynaptic neurons that provide the main receptive surface for presynaptic neurons are the
   a. nuclei.
   b. somas.
   c. axons.
   d. dendrites.
   e. glia.
   Textbook Reference: *34.1 Nervous Systems Consist of Neurons and Glia, p. 673*
2. The substance that wraps around the axon of many neurons and provides for increased conduction speed is
   a. dendrase.
   b. histamine.
   c. acetylcholine.
   d. myelin.
   e. microglia.
   Textbook Reference: *34.1 Nervous Systems Consist of Neurons and Glia, pp. 673–674*

3. The long extension from the cell body of a neuron that provides the pathway for action potentials to the synapse is the
   a. dendrite.
   b. Schwann cell.
   c. axon.
   d. presynaptic membrane.
   e. nerve net.
   Textbook Reference: *34.1 Nervous Systems Consist of Neurons and Glia, p. 673*

4. The threshold of a neuron is the
   a. amount of inhibitory neurotransmitter required to inhibit an action potential.
   b. membrane voltage at which an axon potential will be suppressed.
   c. amount of excitatory neurotransmitter required to elicit an action potential.
   d. membrane voltage at which the membrane potential develops into an action potential.
   e. closing of numerous sodium channels.
   Textbook Reference: *34.2 Neurons Generate and Transmit Electrical Signals, p. 678*

5. When a membrane is at the resting potential, the concentration of
   a. sodium and potassium ions is higher on the inside of its membrane than on the outside.
   b. sodium and potassium ions is higher on the outside of its membrane than on the inside.
   c. sodium ions is higher on the inside of its membrane and of potassium ions is higher on the outside.
   d. sodium ions is higher on the outside of its membrane and of potassium ions is higher on the inside.
   e. sodium equals the concentration of potassium inside the cell.
   Textbook Reference: *34.2 Neurons Generate and Transmit Electrical Signals, pp. 675–676*

6. Glia are specialized to do all of the following *except*
   a. generate neural impulses.
   b. insulate axons.
   c. supply neurons with nutrients.
   d. help maintain a proper ionic environment for the neuron.
   e. guide neurons to make proper contacts during development.
   Textbook Reference: *34.1 Nervous Systems Consist of Neurons and Glia, pp. 673–674*

7. The cells that create the blood–brain barrier, keeping some toxic substances from entering the brain, are _______ and belong to a type of neural tissue called _______.
   a. endothelial cells; Schwann cells
   b. astrocytes; glia
   c. glial fibers; axons
   d. dendrites; synapses
   e. oligodendrocytes; glia
   Textbook Reference: *34.1 Nervous Systems Consist of Neurons and Glia, p. 673*

8. A particular disease of the nervous system specifically involves the Ca^{2+} channels at the chemical synapses of motor neurons where neurotransmitter is stored and released. In other words, this disease affects the
   a. axon terminals of the presynaptic cell and the release of acetylcholine.
   b. axon terminals of the postsynaptic cell and the release of K^+.
   c. movement of Na^+ out of the postsynaptic cell.
   d. axon terminals of the presynaptic cell and the release of K^+.
   e. axon terminals of the postsynaptic cell and the release of Cl^-.
   Textbook Reference: *34.3 Neurons Communicate with Other Cells at Synapses, pp. 681–682*

9. Which of the following statements about the sympathetic division of the autonomic nervous system is *false*?
   a. It increases heart rate.
   b. It relaxes the urinary bladder.
   c. It stimulates digestion.
   d. It increases blood pressure.
   e. It relaxes airways.
   Textbook Reference: *34.4 The Vertebrate Nervous System Has Many Interacting Components, pp. 685–686*

10. Which of the following statements about neurotransmitter receptors is *false*?
   a. Ionotropic receptors are ion channels.
   b. The acetylcholine receptor of the motor end plate is a metabotropic receptor.
   c. Metabotropic receptors are not ion channels.
   d. Metabotropic receptors induce signaling cascades in the postsynaptic cell.
   e. Responses in the postsynaptic cell mediated by metabotropic receptors are usually slower than those mediated by ionotropic receptors.
   Textbook Reference: *34.3 Neurons Communicate with Other Cells at Synapses, p. 684*

11. The rapid depolarization of a neuron during the first half of an action potential is due to the
   a. exit of K^+ ions from the cell through gated potassium channels.
   b. rapid reversal of ion concentration caused by the action of the sodium–potassium pump.
   c. entry of Na^+ ions into the cell through gated sodium channels.
   d. movement of both Na^+ and K^+ ions through appropriate open channels.
   e. closing of sodium channels.
   Textbook Reference: *34.2 Neurons Generate and Transmit Electrical Signals, p. 678*

12. The refractory period of a neuron
   a. is the period when the sodium–potassium pump is nonfunctional.
   b. results from activation of voltage-gated chloride channels.

c. results from closing of inactivated voltage-gated sodium channels.
d. occurs when the action potential reaches the synapse.
e. lasts about a minute.

Textbook Reference: *34.2 Neurons Generate and Transmit Electrical Signals, p. 678*

13. Which of the following is *not* part of the central nervous system?
a. Brain stem
b. Spinal gray matter
c. Cerebellum
d. Neuronal cell body of a sensory afferent
e. Pons

Textbook Reference: *34.4 The Vertebrate Nervous System Has Many Interacting Components, pp. 684–687*

14. Which of the following statements about neurotransmitters is *false*?
a. Peptides can act as neurotransmitters.
b. Each neurotransmitter has a single type of receptor.
c. Amino acids and their derivatives function as neurotransmitters.
d. Neurotransmitters have different effects in different tissues.
e. Some neurotransmitters are cleared from synapses by enzymes that destroy them.

Textbook Reference: *34.3 Neurons Communicate with Other Cells at Synapses, pp. 682–684*

15. Which of the following statements about the knee-jerk reflex is *false*?
a. It is a monosynaptic reflex.
b. It causes the leg extensor muscle to contract.
c. Chemoreceptors sense a physician's hammer tap.
d. The afferent nerve travels from the receptor to the spinal cord.
e. The motor neuron leaves via a ventral root of the spinal cord.

Textbook Reference: *34.4 The Vertebrate Nervous System Has Many Interacting Components, pp. 685–687*

16. A man has damage to his brain that affects his ability to recognize the faces of people he knows. The damage must have occurred in the
a. hypothalamus.
b. temporal lobe.
c. parietal lobe.
d. frontal lobe.
e. occipital lobe.

Textbook Reference: *34.4 The Vertebrate Nervous System Has Many Interacting Components, p. 689*

17. The insular cortex is
a. located in the hindbrain.
b. greatly expanded in fishes.
c. most active during times of mild emotion in humans.
d. unrelated to perception of self.
e. greatly expanded in humans and great apes.

Textbook Reference: *34.5 Specific Brain Areas Underlie the Complex Abilities of Humans, p. 692*

18. When compared with brains of other vertebrates, the human brain
a. is larger than body size might lead one to predict.
b. has proportionately more of the cortex devoted to information integration.
c. has proportionately more cerebral cortex.
d. has more convolutions in the cortex.
e. All of the above

Textbook Reference: *34.4 The Vertebrate Nervous System Has Many Interacting Components, p. 688*

Answers

Key Concept Review

1. Astrocytes are glia that help form the blood–brain barrier by surrounding tiny, very permeable blood vessels in the brain. The barrier prevents most water-soluble substances and large molecules from reaching the brain. However, because the barrier is made of plasma membranes, fat-soluble substances such as anesthetics and alcohol can pass through it.

2. A person with Lou Gehrig's disease would have no cognitive deficits because only the motor neurons (concerned with output) are affected by the disease; sensory neurons (concerned with input) and interneurons (concerned with integration) are unaffected.

3. The potassium voltage-gated channels are responsible for setting up the resting potential of a membrane. Potassium ions have a tendency to diffuse out of the cell, leaving a negative charge inside. Knocking out the function of the potassium voltage-gated channels would result in the cell's being unable to maintain resting potential. If the cell was depolarized by the opening of sodium voltage-gated channels, then it might not repolarize because the potassium channels that help repolarize the membrane would not be functioning.

4.
a. Dendrites
b. Cell body
c. Axon hillock
d. Axon
e. Axon terminals

The action potential travels down the axon, away from the axon hillock.

5. The conduction of an action potential down a myelinated axon is called saltatory conduction. The myelin acts to insulate areas of the axon, preventing depolarization. The areas of the axon between the myelin sheaths are known as nodes of Ranvier. Depolarization can occur only at these nodes. As the action potential moves down a myelinated axon, the influx of sodium

ions at one node diffuses down the axon. This results in the depolarization of the next node of Ranvier. Depolarization can occur only in the downstream nodes because the upstream nodes are in a refractory period. As a result, the action potential moves quickly down the axon to the synapse.

6. Acetylcholine is the neurotransmitter used by all neuromuscular synapses in vertebrates. It transmits the action potential from a presynaptic cell to a postsynaptic cell. The enzyme acetylcholinesterase is found in the synaptic cleft, and it cleaves acetylcholine to help remove it from the synaptic cleft after an action potential. A nerve gas with components that block the action of acetylcholinesterase would cause acetylcholine to build up in the synaptic cleft. This buildup would mean that the receptors on the postsynaptic cell would remain bound with acetylcholine, resulting in prolonged muscle contraction.
7. Selective serotonin reuptake inhibitors, such as Prozac, increase the level of serotonin at the synapse by reducing its rate of removal.
8. The peripheral system contributed to both seeing the candy and the movements of the arms and legs that you used to pick it up. The parasympathetic branch of the autonomic nervous system stimulated salivation. The central nervous system was involved in the recognition and decision to unwrap and eat the candy.
9. The pathway that contains the neuron is polysynaptic.

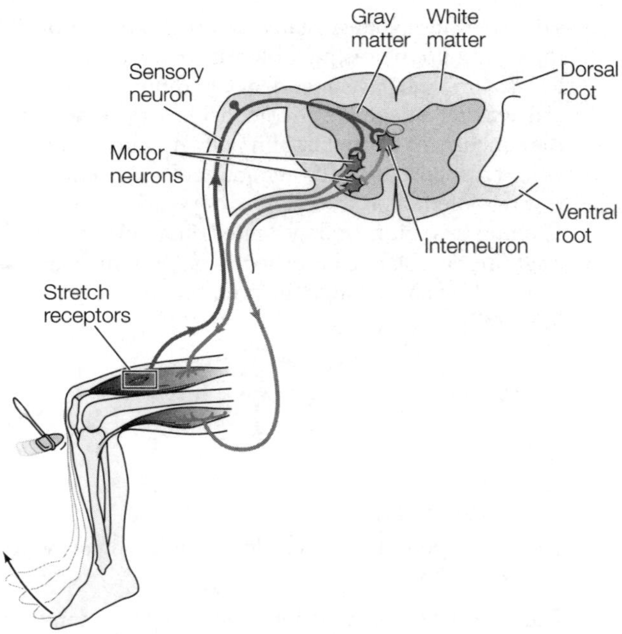

10. A sharp punch to the jaw will cause the head to turn sharply, and this is likely to twist the medulla and reticular activating system. The reticular system is a network of neurons in the brain stem that, unless inhibited by other regions of the brain, activates the cerebral cortex and causes consciousness. A sharp blow can affect the reticular system and cause temporary loss of consciousness.
11. Both speaking written language and repeating heard language involve similar pathways in the brain. The main difference has to do with the initial region of the brain perceiving the word. In reading a word, the area at the back of the cerebrum is used to visualize it. The spoken word activates an area of the cerebrum just behind the area used for speech. Once the word has been processed by the initial centers, the path used for speaking the word is the same. Wernicke's area is stimulated, followed by Broca's area, and then the motor area.
12. Sleepwalking takes place during non-REM sleep. During REM sleep the skeletal muscles of the body become paralyzed, and the sleeper is unable to move. A sleepwalker will not exhibit the eye movements typical of REM sleep.
13. Species such as humans, great apes, dolphins, and elephants are visually oriented animals. This raises the question of whether the mirror test is valid when applied to animals whose primary sensory modality is not vision, but perhaps olfaction or hearing. In designing an experiment to test self-awareness, a researcher should consider whether the mirror test is a valid test for the study species or whether the mirror test should be adapted to better match the sensory modality of the study species.

Test Yourself

1. **d.** The neuron is composed of a cell body, an axon, and dendrites. The dendrites form synapses with presynaptic cells to create the junction where information from one neuron is transferred to another neuron.
2. **d.** The glia that coat the axon of some neurons form myelin.
3. **c.** The neuron is composed of the cell body, the dendrite, and the axon. The axon carries action potentials away from the cell body to the synapses.
4. **d.** For an action potential to occur in an axon, the membrane must be depolarized above the level known as the threshold.
5. **d.** The resting potential of a neuron membrane occurs when the sodium ion concentration is higher on the outside and the potassium ion concentration is higher on the inside.
6. **a.** Glia perform many functions in the nervous system, but they do not generate neural impulses.
7. **b.** The blood–brain barrier is formed by astrocytes that wrap around the blood vessels traveling through the brain. Astrocytes are a special kind of glia.
8. **a.** If the disease acts on a chemical synapse where the neurotransmitter is stored and released, it is affecting

the axon terminals of the presynaptic cell. Ca^{2+} channels are involved in regulating the release of acetylcholine by allowing Ca^{2+} to enter the presynaptic cell and promoting the fusing of acetylcholine-containing vesicles to the membrane.

9. **c.** The sympathetic division of the autonomic nervous system inhibits digestion rather than stimulating it.
10. **b.** The acetylcholine receptor of the motor end plate is an ionotropic receptor.
11. **c.** The first step in an action potential is the influx of Na^+, leading to a depolarization of the axon membrane. Na^+ rushes into the cell due to the higher concentration outside of the cell and the negative membrane potential.
12. **c.** After the spike of the depolarization, the sodium voltage-gated channels close. One of the properties of these channels is that they will open again only after a short delay of a few milliseconds (known as the refractory period) when the sodium voltage-gated channels are inactive.
13. **d.** The cell bodies of the sensory neurons are located in the periphery and send their axons to the CNS.
14. **b.** Each neurotransmitter has multiple types of receptors.
15. **c.** The knee-jerk reflex is an example of a monosynaptic reflex. Stretch receptors (not chemoreceptors) sense the hammer tap on the tendon.
16. **b.** The temporal lobe is involved in recognition of people and objects. A person who has had damage to the temporal lobe will not be able to identify someone by face and must use other cues.
17. **e.** The insular cortex is greatly expanded in humans and great apes. This area of the forebrain may be involved with self-recognition and conscious experience.
18. **e.** Relative to the brains of other vertebrates, the human brain is larger than body size might lead one to predict, and it exhibits a greater degree of convolution of the cerebral cortex. The human brain also has proportionately more cerebral cortex and association areas within the cortex.

Sensors 35

The Big Picture

- Sensory structures work by converting some form of stimulus—mechanical, chemical, light—into action potentials in the nervous system, which then are interpreted by the central nervous system as a perceived sense.
- Receptors are named on the basis of their sensitivity. For example, chemoreceptors respond to chemical stimulation, mechanoreceptors respond to mechanical stimulation, and photoreceptors respond to light.
- Animals possess different types of senses, and differ in the acuity of their senses.

Study Strategies

- All of the senses have what appear to be very different mechanisms for the transmission of information to the brain. However, it is helpful to remember that there are only a few types of receptors that respond to stimuli and that they all generate action potentials. The steps in sensory transduction are also similar in all the different sensory systems.
- The neurons of the ear, eye, knee, stomach, or any other part of the body all fire action potentials. However, the action potentials are interpreted differently (e.g., action potentials coming from the eye are interpreted as light) because of the region of the brain that receives and analyzes them.
- Many of the receptors with complex structures have both neural and nonneural components. The nonneural components (e.g., the ear pinnae) help channel or otherwise alter or filter the stimulus that will arrive at the neural component of the receptor.
- The route by which sound travels in the ear can be very confusing. View the cochlea in the uncoiled form as in Figure 35.10. This will help you visualize how pressure waves of different wavelengths produce different sounds.
- Go to yourBioPortal.com to review the following tutorials and activities:

  Animated Tutorial 35.1 Sound Transduction in the Human Ear

  Animated Tutorial 35.2 Photosensitivity

  Interactive Tutorial 35.1: Sensory Receptors

  Interactive Tutorial 35.2: Visual Receptive Fields

  Web Activity 35.1 Structures of the Human Ear

  Web Activity 35.2 Structure of the Human Eye

  Web Activity 35.3 Structure of the Human Retina

  Working with Data 35.1 Rod Cell Response

Key Concept Review

35.1 Sensory Systems Convert Stimuli into Action Potentials

Sensory transduction involves changes in membrane potentials

Different sensory receptors detect different types of stimuli

Sensation depends on which neurons receive action potentials from sensory cells

Many receptors adapt to repeated stimulation

Sensory receptor cells (also called sensors or receptors) detect stimuli and respond with a change in membrane potential, which causes action potentials in the sensory receptor cells themselves or in neighboring cells. These action potentials transmit the information to the CNS.

Sensory transduction is the process by which a stimulus (e.g., mechanical, thermal, or chemical) received by a sensory cell is transformed ("transduced") into an action potential. Sensory transduction begins with a receptor protein that opens or closes ion channels in response to a particular stimulus. A receptor potential is a change in the resting membrane potential of a sensory receptor cell in response to a stimulus. A receptor potential is a graded membrane potential, meaning that it spreads over short distances only. Receptor potentials produce action potentials, either by causing the release of a neurotransmitter that induces an associated neuron to generate action potentials, or by generating action potentials within the sensory cell itself.

Even though all sensory systems process information as action potentials, we perceive different sensations (e.g., pain, light, sound) because messages from the different sensory systems go to different areas of the CNS. The frequency of action potentials encodes the intensity of sensation. Some sensory cells transmit information about internal body conditions, and we may not be consciously aware of such information.

Sensory organs are groups of sensory cells that, along with other cells, collect, filter, and amplify stimuli. Eyes, ears, and noses are examples of sensory organs. Sensory systems include the sensory cells, the associated structures, and the networks of neurons that process the information.

By means of a process known as adaptation, many sensory cells have diminished responses to a stimulus over time. Adaptation allows organisms to ignore background conditions and focus on new information.

Question 1. Although photoreceptors respond best to light, they can also respond to pressure. Explain why, when you press gently on your closed eyelids, you see spots of light.
Textbook Reference: *35.1 Sensory Systems Convert Stimuli into Action Potentials, pp. 696–697*

Question 2. The term "perception" describes the conscious awareness of sensations; it occurs when the cerebral cortex of the brain integrates sensory information. Are humans consciously aware of all information transmitted by sensory cells?
Textbook Reference: *35.1 Sensory Systems Convert Stimuli into Action Potentials, p. 697*

35.2 Chemoreceptors Detect Specific Molecules or Ions

- Olfaction is the sense of smell
- Some chemoreceptors detect pheromones
- Gustation is the sense of taste

Chemical stimuli in the external and internal environments stimulate chemoreceptors. Chemoreceptors are responsible for smell and taste, and for monitoring levels of particular chemicals (e.g., carbon dioxide) inside the body.

Olfaction is the sense of smell. The olfactory sensors of vertebrates are neurons with axons extending to the olfactory bulb of the brain; the dendrites of these neurons are exposed as hairs to the environment within the epithelium of the nasal cavity (see Figure 35.3). Olfactory receptor proteins are found on the hairs, and each receptor binds with specific odorants. Binding of the odorant generates action potentials, which are transmitted to glomeruli in the olfactory bulb. The ability to discriminate many different odors is due to the large number of specific receptors. Binding of an odorant results in depolarization of the cell through a G protein that activates a second messenger, which then opens sodium channels. The strength of a smell is related to the number of odorant molecules that bind to receptors.

Pheromones are chemicals involved in within-species communication. Among insects, pheromones attract mates by remotely stimulating their target's chemoreceptors (see Figure 35.4). The concentration of the pheromone released by a female creates a gradient that provides information about her specific location.

Some vertebrates have a vomeronasal organ, a paired structure located in the nasal epithelium.

Gustation, the sense of taste, relies on clusters of chemoreceptor cells called taste buds (see Figure 35.5). Binding of the stimulus to receptor proteins on the microvilli of sensory cells causes a change in membrane potential and the release of neurotransmitters that stimulate sensory neurons at the base of the taste bud. Humans can perceive five general tastes: sweet, sour, salty, bitter, and umami (savory).

Question 3. In snakes, the forked tongue presents odorant molecules from the environment to the chemoreceptors of the vomeronasal organ on the roof of the mouth. Is the tongue participating in smell or taste?
Textbook Reference: *35.2 Chemoreceptors Detect Specific Molecules or Ions, p. 699*

Question 4. One of the usual symptoms of the common cold is a diminished sense of smell. What is the cause of this loss of smell?
Textbook Reference: *35.2 Chemoreceptors Detect Specific Molecules or Ions, p. 698*

35.3 Mechanoreceptors Detect Physical Forces

- Many different cells respond to touch and pressure
- Mechanoreceptors are found in muscles, tendons, and ligaments
- Hair cells are mechanoreceptors of the auditory and vestibular systems
- Auditory systems use hair cells to sense sound waves
- Flexion of the basilar membrane is perceived as pitch

Various types of damage can result in hearing loss

The vestibular system uses hair cells to detect forces of gravity and momentum

Mechanical force causes distortion of the membranes of mechanoreceptors, which causes ion channels to open. Opening of ion channels creates a receptor potential that can lead either to the release of neurotransmitter or to the generation of an action potential.

The skin has several different types of mechanoreceptors (see Figure 35.6).

Meissner's corpuscles are very sensitive and rapidly adapting, and they provide information about changes in objects touching the skin. Merkel's discs adapt slowly and provide continuous information about objects touching the skin. Pacinian corpuscles and Ruffini endings are deeper in the skin; the former adapt rapidly and respond to high-frequency alternating stimuli, while the latter adapt slowly and respond to low-frequency vibrations.

Muscle spindles are mechanoreceptors (specifically, stretch receptors) in skeletal muscles that perceive muscle stretch. Golgi tendon organs are mechanoreceptors found in the tendons and ligaments that provide information about forces generated during muscle contraction. Collectively, these mechanoreceptors provide information on limb position, as well as stresses and strains on muscles and joints.

Hair cells are mechanoreceptors that have stereocilia projecting from their surface. Bending the stereocilia causes changes in the ion channels of the hair cell plasma membrane (see Figure 35.8). Bending in one direction opens the ion channels, causing depolarization and the release of neurotransmitters. Bending in the other direction closes ion channels. Hair cells are the mechanoreceptors for the vertebrate auditory and vestibular systems.

The auditory system takes in sound as pressure waves and transforms the waves into action potentials. The outer portion of the mammalian ear, called the pinna, collects sound waves and directs them into the auditory canal. At the end of the auditory canal is the tympanic membrane, which vibrates and transmits sound waves to tiny bones (ossicles) in the middle ear. The ossicles are the malleus (hammer), incus (anvil), and stapes (stirrup), and together they transmit and amplify vibrations to the membrane called the oval window.

Sound travels through the oval window into the fluid-filled cochlea (in the inner ear), where pressure waves are turned into action potentials. Movement of the oval window generates pressure waves in the cochlear fluid. The cochlea is a three-canal chamber with two membranes: Reissner's membrane and the basilar membrane. The pressure waves in the cochlear fluid cause the basilar membrane to vibrate. The organ of Corti is supported on the basilar membrane, and it contains hair cells with stereocilia that are in contact with the overhanging tectorial membrane. When the basilar membrane vibrates, the hair cell stereocilia of the organ of Corti are pushed against the tectorial membrane. Movements of the stereocilia are transduced into action potentials that are carried to the brain by the auditory nerve (see Figure 35.9). The round window functions to dissipate pressure created by movements of the oval window.

Different pitches of sound cause the basilar membrane to flex at different locations, stimulating different hair cells. The brain interprets input from hair cells in different areas as sounds of different pitch (see Figure 35.10).

The vestibular system of mammals consists of three semicircular canals and two chambers called the saccule and utricle; the entire system is filled with the fluid endolymph. Changes in position of the head cause shifts in the fluid within semicircular canals, which pushes on the gelatinous cupulae of hair cells, causing their stereocilia to bend (see Figure 35.11A). In the saccule and utricle, the stereocilia are bent by gravitational forces on otoliths (see Figure 35.11B). Otoliths are granules of calcium carbonate that sit on top of the gelatinous mass overlying the hair cells.

Question 5. An infection that causes vertigo (dizziness) would be located in which sensory organ and in which particular part of the organ?

Textbook Reference: *35.3 Mechanoreceptors Detect Physical Forces, pp. 704–705*

Question 6. In the diagram of the human ear below, label each of the following structures: tympanic membrane, malleus, incus, stapes, oval window, round window, cochlea, semicircular canal of the vestibular system, and auditory nerve.

Textbook Reference: *35.3 Mechanoreceptors Detect Physical Forces, p. 703*

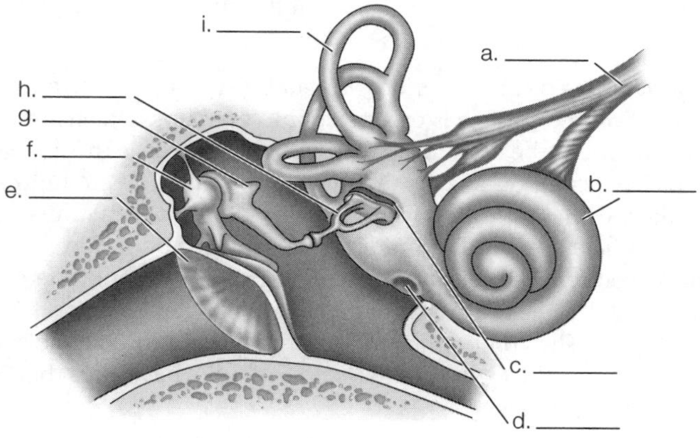

Question 7. Design a study to determine the lowest threshold of hearing for a mammal other than a human.

Textbook Reference: *35.3 Mechanoreceptors Detect Physical Forces, pp. 702–704*

Question 8. After a loud rock concert your hearing appears to be dampened. What portion of your ear might have been affected by the sounds of the concert? Will your hearing return to normal?
Textbook Reference: *35.3 Mechanoreceptors Detect Physical Forces, p. 704*

35.4 Photoreceptors Detect Light

Rhodopsins are responsible for photosensitivity
Rod cells respond to light
Animals have a variety of visual systems
Visual information is processed by the retina and the brain
Color vision is due to cone cells

Rhodopsins are a family of pigments responsible for light sensitivity. They are made up of the protein opsin and the light-absorbing nonprotein group 11-*cis*-retinal (see Figure 35.12). The 11-*cis*-retinal absorbs photons of light and changes conformation to all-*trans*-retinal, causing opsin to change conformation and become photoexcited rhodopsin. Photoexcited rhodopsin triggers a G protein cascade, which ultimately leads to changes in membrane potential and the photoreceptor's response to light.

In vertebrate eyes, rod cells are photoreceptor cells that contain an inner segment, a synaptic terminal, and an outer segment with many rhodopsin molecules (see Figure 35.13). Rods are found in the retina, along with a layer that transduces visual information into action potentials. Rod cells become hyperpolarized in response to light and respond by decreasing the levels of neurotransmitter released (see Figure 35.14).

Invertebrates display a variety of visual systems. Flatworms have photoreceptor cells organized into eye cups, which are used to orient the animal away from light sources. Arthropods have compound eyes that have from only a few to hundreds or even tens of thousands of ommatidia (optical units) per eye. The ommatidia contain light-sensitive photoreceptors called retinula cells (see Figure 35.15). The inner borders of the retinula cells are covered by microvilli that contain rhodopsin. The compound eye communicates a low-resolution image to the CNS. Cephalopod mollusks have eyes that form detailed images.

Cephalopod mollusks and vertebrates evolved image-forming eyes independently. The vertebrate eye is surrounded by the sclera (see Figure 35.16A). The cornea is the transparent sclera through which light passes. The iris controls the amount of light entering the eye through the pupil; it also gives the eye its color. The lens focuses images on the retina, the photosensitive layer at the back of the eye.

There are five layers of neurons in the retina, with the photoreceptive rods and cones in the last layer, farthest from the lens (see Figure 35.16B). The first layer of cells consists of ganglion cells (which create the action potential), the axons of which form the optic nerve. Bipolar cells are stimulated by neurotransmitters from the photoreceptors to transmit the signal from the photoreceptor to the ganglion cells by the release of a neurotransmitter. Thus, information flow in the retina is from photoreceptor cells at the back to bipolar cells to ganglion cells, which send the information to the brain. The two other layers in the retina are the horizontal cells (which connect adjoining groups of photoreceptors and bipolar cells) and amacrine cells (which connect adjoining groups of bipolar cells and ganglion cells). Horizontal and amacrine cells are interneurons responsible for lateral communication across the retina.

The human retina contains both rods (which are more light-sensitive) and cones (which absorb light of various wavelengths, allowing for color vision; see Figure 35.17A). Cones have different opsin molecules that absorb blue, green, yellow, or red (see Figure 35.17B). The center of the retina is the fovea, an area with the highest density of cone cells.

Question 9. In the diagram of the vertebrate eye below, label each of the following structures: sclera, cornea, iris, pupil, lens, retina, fovea, vitreous humor, and optic nerve.
Textbook Reference: *35.4 Photoreceptors Detect Light, p. 709*

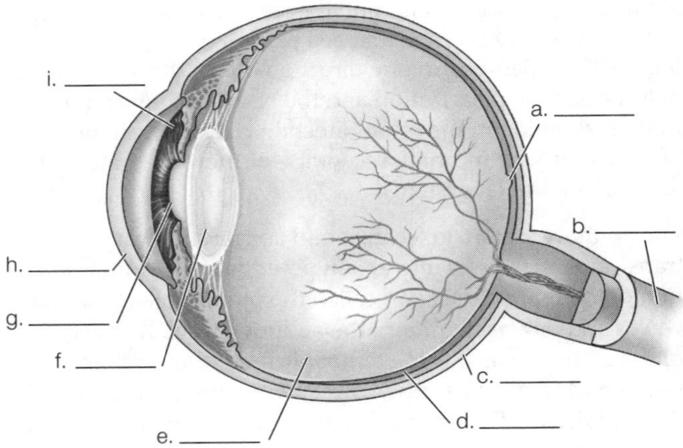

Question 10. You have just given a presentation in your biology class that had many elaborate red- and green-colored slides. Afterward, a male friend tells you that he could not see any of the differences you were reporting. Why could your friend not see the differences?
Textbook Reference: *35.4 Photoreceptors Detect Light, p. 710*

Test Yourself

1. An electrode is inserted into a chemosensory nerve leading away from a taste bud in the mouth of a dog. A mild acid solution is then flushed continuously over the sensors associated with this nerve. Initially, the nerve responds to this stimulation, but over time it ceases to carry action potentials. Which of the following processes would best explain this observation?
   a. Translocation
   b. Adaptation of the sensory cells
   c. Depletion of neurotransmitter in the sensory nerve
   d. Second messenger influences that increase cell membrane potentials
   e. Action potentials arriving at the wrong area in the CNS

   Textbook Reference: *35.1 Sensory Systems Convert Stimuli into Action Potentials, p. 697*

2. Which of the following statements about vertebrate auditory and vestibular systems is *false*?
   a. Otoliths are found in the saccule and utricle of the inner ear.
   b. Endolymph fills the canals and chambers of the vestibular system.
   c. The stapes transmits vibrations to the round window.
   d. Gravity and momentum are detected by the vestibular system.
   e. Different pitches of sound flex the basilar membrane at different locations.

   Textbook Reference: *35.3 Mechanoreceptors Detect Physical Forces, pp. 702–705*

3. Silkworm moths use chemosensory signals known as _______ for mate attraction.
   a. rhodopsin
   b. hormones
   c. pheromones
   d. G proteins
   e. locally acting chemical messengers

   Textbook Reference: *35.2 Chemoreceptors Detect Specific Molecules or Ions, pp. 698–699*

4. Stretch receptors in the aorta and carotid artery sense changes in arterial pressure. These receptors are therefore considered
   a. chemoreceptors.
   b. thermoreceptors.
   c. electroreceptors.
   d. mechanoreceptors.
   e. muscle spindles.

   Textbook Reference: *35.3 Mechanoreceptors Detect Physical Forces, p. 700*

5. Which of the following does *not* employ hair cells as its transducer?
   a. Meissner's corpuscle
   b. Lateral line
   c. Organ of Corti
   d. Semicircular canal
   e. Saccule

   Textbook Reference: *35.3 Mechanoreceptors Detect Physical Forces, pp. 700–705*

6. Which of the following statements about gustation is *false*?
   a. Some fish have taste buds on their skin.
   b. In humans, taste buds are confined to the oral cavity.
   c. Changes in the membrane potential of the taste bud sensory cells cause them to release neurotransmitter onto the dendrites of sensory neurons.
   d. Humans perceive only three categories of tastes: sweet, sour, and bitter.
   e. In humans, most taste buds are found on the papillae of the tongue.

   Textbook Reference: *35.2 Chemoreceptors Detect Specific Molecules or Ions, pp. 699–700*

7. Which of the following statements about the photosensitive molecule rhodopsin is *false*?
   a. Opsin is converted from the 11-*cis* to the all-*trans* form upon absorbing a photon of light.
   b. The retinal is the light-absorbing group.
   c. Photoexcited rhodopsin triggers a cascade of reactions that ultimately alters the membrane potential of a photoreceptor cell.
   d. Opsin is a protein; retinal is not a protein.
   e. 11-*cis*-retinal is covalently bonded to opsin.

   Textbook Reference: *35.4 Photoreceptors Detect Light, pp. 705–707*

8. In the human visual system, _______ send information directly to the brain.
   a. amacrine cells
   b. bipolar cells
   c. ganglion cells
   d. rods and cones
   e. horizontal cells

   Textbook Reference: *35.4 Photoreceptors Detect Light, pp. 708–709*

9. Through which of the following cell layers must a photon of light pass before striking a cone cell in the eye of a human?
   a. Amacrine
   b. Bipolar
   c. Ganglion
   d. Horizontal
   e. All of the above

   Textbook Reference: *35.4 Photoreceptors Detect Light, pp. 708–709*

10. Which of the following structures controls the amount of light entering the vertebrate eye?
    a. Cornea
    b. Retina
    c. Lens

d. Iris
e. Sclera
Textbook Reference: *35.4 Photoreceptors Detect Light, p. 708*

11. All sensory systems, no matter which type of stimulus they detect, convey information
a. from the CNS to the peripheral nervous system.
b. in the form of action potentials.
c. to the visual cortex.
d. to the CNS as receptor potentials.
e. to the CNS as graded membrane potentials.
Textbook Reference: *35.1 Sensory Systems Convert Stimuli into Action Potentials, p. 696*

12. Which of the following structures is *not* found in the inner ear?
a. Reissner's membrane
b. Tectorial membrane
c. Tympanic membrane
d. Basilar membrane
e. Semicircular canal
Textbook Reference: *35.3 Mechanoreceptors Detect Physical Forces, pp. 702–703*

13. Which of the following statements about receptor potentials is *false*?
a. They are changes in the resting membrane potential of a sensory cell in response to a stimulus.
b. They spread over short distances.
c. They must be converted into action potentials in order to travel over long distances.
d. They always prompt the release of a neurotransmitter that induces an associated neuron to generate an action potential.
e. They are graded membrane potentials.
Textbook Reference: *35.1 Sensory Systems Convert Stimuli into Action Potentials, p. 696*

14. Which of the following statements about the detection and integration of chemical stimuli is *false*?
a. In mammals, olfactory receptors communicate directly with the brain.
b. Many mammals have a vomeronasal organ to detect pheromones.
c. A greater frequency of action potentials is associated with perception of a more intense smell.
d. In mammals, sensors for the same odorant project to different areas of the olfactory bulb.
e. Chemoreceptors monitor aspects of the internal environment.
Textbook Reference: *35.2 Chemoreceptors Detect Specific Molecules or Ions, pp. 697–699*

15. Which of the following statements about visual systems is *false*?
a. Humans and a few other primate species have three kinds of cone cells, but most mammals have two kinds.
b. In cephalopod mollusks and vertebrates, image-forming eyes evolved independently.
c. Nocturnal animals have a high percentage of cones in their retinas, whereas diurnal animals have a high percentage of rods.
d. Rhodopsins are the basis for photosensitivity for all animals.
e. Compound eyes are found in arthropods.
Textbook Reference: *35.4 Photoreceptors Detect Light, pp. 705–710*

Answers

Key Concept Review

1. When you press gently on your closed eyelids, you see spots of light because the pressure stimulates photoreceptors, which send action potentials to the visual cortex. This is one illustration of sensation depending on the particular part of the brain that receives the nerve impulses.

2. Humans are not consciously aware of all information transmitted by sensory cells. Some sensory cells transmit information about internal conditions in the body (for example, about blood pressure and carbon dioxide concentration in the blood, or limb position) that is outside of conscious awareness.

3. In snakes, when the tongue presents odorant molecules to the vomeronasal organ, the tongue is being used in smell and not in taste.

4. Your sense of smell depends on olfactory cilia that line the surface of the nasal epithelium. The cilia's receptors bind with odorant molecules, triggering an action potential that is sent to the olfactory bulb of the brain. Usually this epithelium is covered with a thin layer of protective mucus. However, when you have a cold, the production of mucus increases and mucus covers the epithelium and the olfactory cilia, making it more difficult for odorant molecules to reach the cilia. Thus your sense of smell is decreased.

5. Vertigo is caused by infection in the ear, specifically the inner ear, which contains the organs of equilibrium (in addition to the cochlea).

6.
a. Auditory nerve
b. Cochlea
c. Oval window (under stapes)
d. Round window
e. Tympanic membrane
f. Malleus
g. Incus
h. Stapes
i. Semicircular canal of the vestibular system

7. The lowest threshold of human hearing can be determined by presenting auditory stimuli to subjects and asking them to respond with a yes or no answer as to whether a particular stimulus can be heard. In other mammals, the electrical activity of the auditory nerve and auditory regions of the brain can be monitored

directly. The lowest threshold of hearing can be determined by presenting sounds that stimulate the ear, auditory nerve, and parts of the brain involved with hearing and then recording, by means of electrodes placed strategically on particular locations of the head, the differences in electrical potentials elicited by the different sounds. Subjects are typically anesthetized during the procedure.

8. Sounds that are too loud can damage the hair cells of the organ of Corti, resulting in nerve deafness. At present, such damage to hair cells is considered irreversible. In other words, loud noise can cause permanent damage to hair cells, and once such damage occurs, your hearing will not return to normal.

9.
   a. Fovea
   b. Optic nerve
   c. Sclera
   d. Retina
   e. Vitreous humor
   f. Lens
   g. Pupil
   h. Cornea
   i. Iris

10. Your friend has red–green color blindness. The cones in our eyes allow us to see color. We have cones for red, green, and blue. A lack of one of type of cone, or a reduced number, can cause color blindness, as can problems in the functioning of one type of cone. Red–green color blindness is a sex-linked trait that is more common in men than in women.

Test Yourself

1. **b.** A sensor cell that is stimulated by an unchanging, steady-state stimulus will adapt to that stimulus. Adaptation allows the sensory system to ignore the unchanging stimulus while still being able to respond to new information.

2. **c.** The stapes transmits vibrations to the oval window (not the round window).

3. **c.** Pheromones are chemical signals used in communication within a species. The female silkworm moth releases a pheromone (bombykol) into the environment. The male uses chemoreceptors to follow the pheromone to the source.

4. **d.** The stretch receptors of the aorta, which detect changes in blood pressure, are examples of mechanoreceptors.

5. **a.** Meissner's corpuscle of the skin does not sense a stimulus by means of hair cells. The cell membranes of the Meissner's corpuscle deform in response to light touching of the skin.

6. **d.** Humans can perceive five tastes: sweet, salty, sour, bitter, and umami (a savory taste). The combination of taste and smell provides the complex subtle flavors of the food we eat.

7. **a.** Rhodopsin contains two groups: the protein opsin and the light-sensitive group retinal. Retinal, not opsin, is converted from the 11-*cis* to the all-*trans* form upon absorbing a photon of light. Opsin does change conformation in response to a change in the rhodopsin to signal the detection of light.

8. **c.** The ganglion cells transmit information from the bipolar cells to the brain. The axons of the ganglion cells connect with the optic nerve.

9. **e.** The photoreceptive cells are located at the back of the retina. Light must pass through a layer of ganglion cells, a layer of amacrine and bipolar cells, and a horizontal cell layer.

10. **d.** The iris controls the amount of light entering the vertebrate eye.

11. **b.** All sensory systems, no matter which type of stimulus they detect, convey information in the form of action potentials.

12. **c.** Although the tympanic membrane is found in the human ear, it is not found in the inner ear. The tympanic membrane is the membrane that transmits sounds from the auditory canal to the middle ear.

13. **d.** The receptor potential does not always prompt the release of a neurotransmitter to induce an associated neuron to generate an action potential. Sometimes the receptor potential generates action potentials within the sensory cell itself.

14. **d.** In mammals, sensors for the same odorant project to the same area of the olfactory bulb.

15. **c.** Nocturnal animals have a high percentage of rods in their retinas, and diurnal animals have a high percentage of cones.

Musculoskeletal Systems 36

The Big Picture

- Actin and myosin are the "universal" proteins for motion. Whether they are located in a unicellular animal or the leg muscle of a human, the molecular interactions of actin and myosin produce movement in the structures in which they reside.
- The sliding filament contractile mechanism of muscle contraction consists of actin and myosin filaments sliding past each other, resulting in the shortening (contraction) or lengthening (relaxation) of muscle cells. The whole process of actin and myosin interaction is tightly regulated by the movement of calcium ions into and out of the intracellular spaces of muscle cells, which in turn is activated by the arrival of action potentials in motor neurons. Muscle contraction requires energy in the form of ATP.
- In vertebrates, muscles act in concert with an internal skeleton made of bone. Bone is living tissue that is constantly remodeled. Bones are articulated, forming joints that provide for specialized directional movements of the tissues supported by the bones. Muscles controlling joint movement are often located in pairs acting antagonistically, with one set of muscles causing bending, or flexion, of the joint and the other causing straightening, or extension, of the joint. Some invertebrates have hydrostatic skeletons (a fluid-containing body cavity surrounded by muscles), while others have exoskeletons (rigid outer coverings).

Study Strategies

- The interactions of myosin, actin, troponin, tropomyosin, and calcium and the role of action potentials in stimulating muscle contraction constitute a complex, multistep process.
    - First, break the process down into its constituents and learn their locations and general structures.
    - Second, determine how actin and myosin move relative to each other through a series of power strokes.
    - Finally, understand how calcium ions released from the sarcomeres by action potentials initiate and maintain the whole process of muscle contraction.
- The sarcomere is the functional unit of the muscle cell, and until you understand its fine structure—Z lines, H lines, etc.—it will be difficult to appreciate how the sarcomere shortens through the actions of actin and myosin.
- Many people think of bone as tissue that is not living. In fact, bone is a living tissue that is constantly remodeled. Become familiar with the three types of living cells in bone: osteocytes, osteoblasts, and osteoclasts.
- The way that movements of multiple muscle groups cause both flexion and extension of a joint can be confusing. Thinking of joints in terms of levers and pulleys may help you understand their actions.
- Go to yourBioPortal.com to review the following tutorials and activities:

  Animated Tutorial 36.1 Molecular Mechanisms of Muscle Contraction

  Animated Tutorial 36.2 Smooth Muscle Action

  Web Activity 36.1 The Structure of a Sarcomere

  Web Activity 36.2 The Neuromuscular Junction

  Web Activity 36.3 Joints

Key Concept Review

36.1 Cycles of Protein–Protein Interactions Cause Muscles to Contract

Sliding filaments cause skeletal muscle to contract

Actin–myosin interactions cause filaments to slide

Actin–myosin interactions are controlled by calcium ions

Cardiac muscle is similar to and different from skeletal muscle

Smooth muscle causes slow contractions of many internal organs

Vertebrates have three types of muscle: skeletal, cardiac, and smooth. Skeletal muscle is responsible for voluntary movements and some unconscious movements such as breathing and maintaining posture. Cardiac muscle is responsible for the beating of the heart. Smooth muscle is controlled by the autonomic nervous system, and it is responsible for the contraction that occurs in many hollow organs, such as the bladder and the gut. In all three types of muscle tissue, contraction is due to the interaction between the contractile proteins actin and myosin.

Skeletal muscles are striated voluntary muscles made of large muscle fibers (see Figure 36.1). Muscle fibers have many nuclei and are composed of actin and myosin. Molecules of actin are organized into thin filaments, and those of myosin are organized into thick filaments. Bundles of actin and myosin filaments are arranged into myofibrils.

The contracting unit of myofibrils is the sarcomere, which contains actin and myosin filaments and has very distinct repeating patterns. In a sarcomere, the actin filaments are anchored by the Z lines, and myosin filaments are found at the center in the A band (see Figure 36.1). In relaxed muscle, the H zone and I band are the regions in which there is no overlap of actin and myosin. Within the H zone is the M band, which contains proteins that help hold myosin filaments in their regular arrangement. The protein titin runs from Z line to Z line and provides resistance to stretch in relaxed skeletal muscle. During muscle contraction, the Z lines move toward each other, and the H zone and I band shrink in size due to the sliding of actin filaments along the myosin filaments. This is the sliding filament contractile mechanism of muscle contraction (see Figure 36.2).

The proteins myosin and actin are the key to muscle contraction. Myosin molecules are made of two polypeptide chains wrapped around each other, each with a globular head at one end, much like two twisted golf clubs. A myosin filament is composed of many myosin molecules (see Figure 36.3). Actin filaments are composed of two monomer chains in a helical arrangement and look like two linear strings of pearls wrapped around each other. The proteins tropomyosin and troponin are associated with actin (see Figure 36.3).

Myosin heads change conformation when they bind to actin filaments at myosin binding sites, forming a cross-bridge connection. The conformational change in the myosin pulls the actin in toward the middle of the sarcomere. ATP then binds to an ATP binding site on myosin, resulting in the bound actin's release from the myosin and the myosin's return to the original conformation. Many myosin molecules cycle through binding with actin to shorten a sarcomere.

All the fibers activated by a single motor neuron constitute a motor unit. The strength of a muscle's contraction can be increased by an increase in the firing rate of an individual motor neuron or by the activation of more motor neurons. Action potentials spread deep into the sarcoplasm (cytoplasm) of the muscle through transverse tubules (T tubules) that are in contact with the sarcoplasmic reticulum (endoplasmic reticulum) throughout the sarcoplasm (see Figure 36.5). The sarcoplasmic reticulum takes up and releases Ca^{2+} ions into the sarcoplasm, thereby controlling relaxation and contraction of the myofibrils.

In relaxed muscle, tropomyosin and troponin cover the myosin binding sites on actin filaments, preventing muscle contraction. Calcium regulates contraction by binding with troponin, causing the tropomyosin to twist and expose the actin–myosin binding sites on the actin filaments (see Figure 36.6).

Cardiac muscle of the heart is composed of branched muscle cells that form a strong meshwork. Cardiac muscle cells are smaller than skeletal muscle cells and each cell has only a single nucleus. Intercalated discs add additional strength by holding the cells together. Gap junctions within the intercalated discs allow cardiac muscle cells to be electrically coupled.

Heartbeats originate at the pacemaker cardiac muscle cells and spread rapidly through gap junctions in the muscle. The heartbeat is described as myogenic because it is generated by the heart muscle itself; input from the nervous system is not necessary. The mechanism of excitation–contraction coupling in cardiac muscle cells is called Ca^{2+} -induced Ca^{2+} release.

Smooth muscles are involuntary muscles composed of long spindle-shaped cells, each with a single nucleus. Smooth muscles are controlled by acetylcholine and norepinephrine of the autonomic nervous system (see Figure 36.7). When smooth muscle is stretched, it contracts with strength that is proportional to the stretch of the muscle. Smooth muscle cells are arranged in sheets. Gap junctions allow electrical contact between the cells and promote coordinated contraction of cells in a sheet.

In smooth muscle, contraction is controlled by a calmodulin–Ca^{2+} complex. This complex activates myosin kinase, an enzyme that phosphorylates the myosin head to cause contraction. Myosin phosphatase works in the opposite direction by dephosphorylating myosin and stopping interactions between actin and myosin (see Figure 36.8).

Question 1. Label the following structures in the diagram below: muscle, tendon, single muscle fiber, single myofibril, sarcomere, actin filament, myosin filament, Z line, A band, H zone, I band, M band. Also label as many of the last seven terms as you can on the myofibril and on the enlargement of the sarcomere.

Textbook Reference: *36.1 Cycles of Protein–Protein Interactions Cause Muscles to Contract, p. 713*

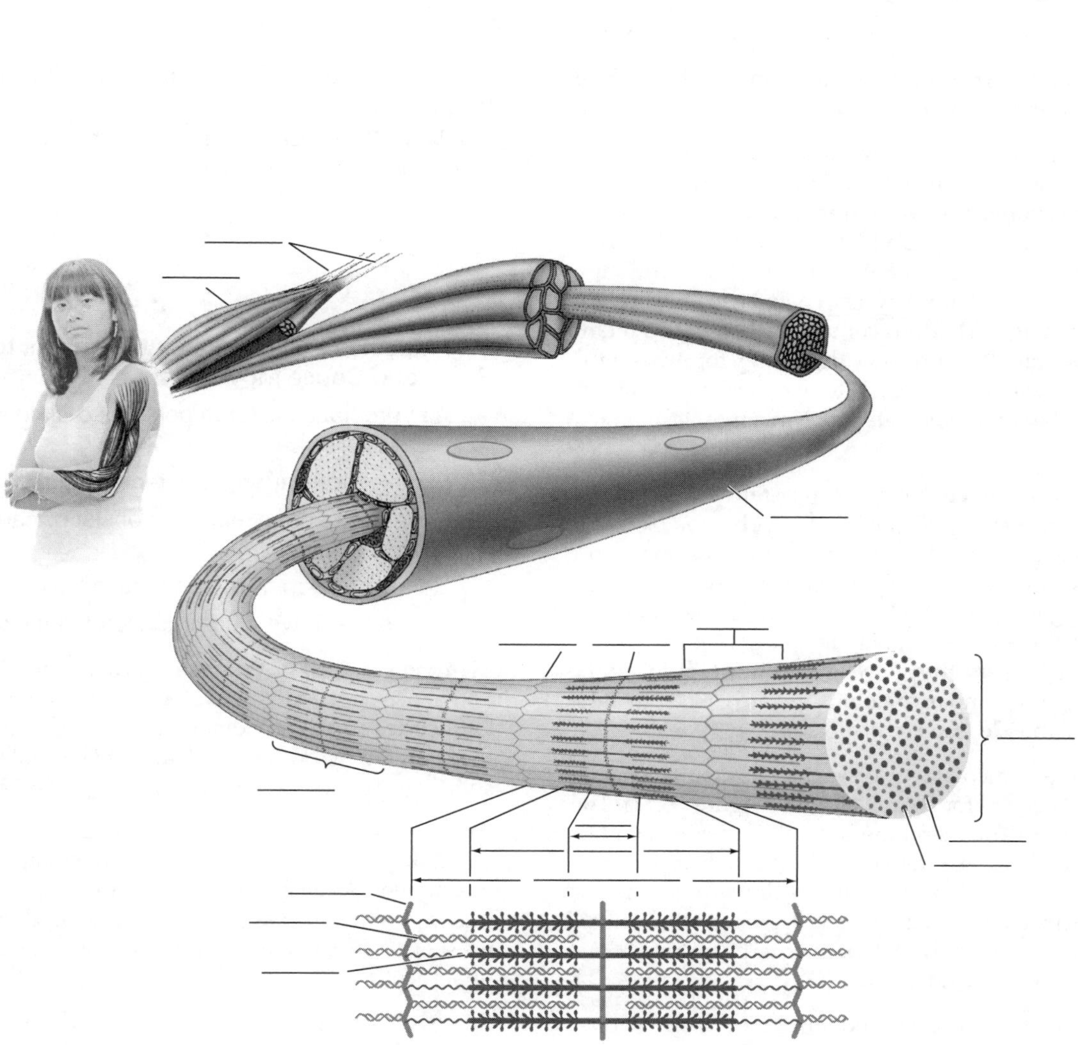

Question 2. Smooth muscle contracts involuntarily, whereas skeletal muscle contraction is under voluntary control. How do the mechanisms that control smooth muscle and skeletal muscle contraction differ?
Textbook Reference: *36.1 Cycles of Protein–Protein Interactions Cause Muscles to Contract, pp. 715–717, 720*

Question 3. What is the cause of rigor mortis, the stiffening of muscles after death?
Textbook Reference: *36.1 Cycles of Protein–Protein Interactions Cause Muscles to Contract, p. 715*

Question 4. Curare is a poison from South America that is applied to the tips of poison arrow darts. Mammals hit by the darts die by asphyxiation because their respiratory muscles cannot contract. Suggest a mechanism by which curare might work.
Textbook Reference: *36.1 Cycles of Protein–Protein Interactions Cause Muscles to Contract, pp. 715–716*

36.2 The Characteristics of Muscle Cells Determine Muscle Performance

- Single skeletal muscle fibers can generate graded contractions
- Muscle fiber types determine endurance and strength
- Muscle ATP supply limits performance

An action potential in a skeletal muscle fiber causes a twitch or contraction of the muscle. Twitches can occur as discrete contractions, or if they occur frequently enough, they can be summed together (see Figure 36.9A).

The level of tension generated by a muscle depends on the number of motor units activated and the frequency with which the motor units fire. Maximum muscle tension, or tetanus, occurs when there is a high rate of stimulation by action potentials (see Figure 36.9B). During tetanic contraction, actin and myosin bonds cycle to help keep a muscle fiber from stretching. ATP levels control the length of a tetanic contraction because ATP provides the energy for myosin to break the bond with actin. Muscle tone reflects the small but changing number of motor units active in a muscle at any given time.

Slow-twitch muscle fibers are highly resistant to fatigue because they are well supplied with myoglobin (an oxygen-binding protein similar to hemoglobin), mitochondria, and blood vessels. They are also called "red" or "oxidative" muscle (see Figure 36.10). Fast-twitch muscle fibers develop maximum tension rapidly but fatigue quickly. Compared with slow-twitch fibers, they have fewer mitochondria and blood vessels, little or no myoglobin, and are called "white" or "glycolytic" muscle (see Figure 36.10).

Muscles use the immediate, glycolytic, and oxidative systems to obtain ATP needed for contraction (see Figure 36.11). The immediate system utilizes preformed ATP and creatine phosphate. The glycolytic system follows the immediate system within seconds and metabolizes carbohydrates to lactate and pyruvate. The oxidative system is fully activated within about one minute and completely metabolizes carbohydrates or fats to water and carbon dioxide.

Question 5. White muscle and red muscle are found in different parts of the body and are used for different types of movement. What are the physiological and morphological characteristics that distinguish the two types of muscle?
Textbook Reference: *36.2 The Characteristics of Muscle Cells Determine Muscle Performance, pp. 721–722*

Question 6. In turkeys, the breast (flight) muscles are light in color while in mallard ducks they are dark. What does this tell you about the locomotor capabilities of these two birds?
Textbook Reference: *36.2 The Characteristics of Muscle Cells Determine Muscle Performance, pp. 721–722*

Question 7. Some athletes hoping to improve their performance take creatine as a dietary supplement. There are some reports that creatine supplements boost performance in activities that require short bursts of energy, such as sprinting, but not in those that require endurance. Why might this be the case?
Textbook Reference: *36.2 The Characteristics of Muscle Cells Determine Muscle Performance, p. 722*

36.3 Muscles Pull on Skeletal Elements to Generate Force and Cause Movement

- A hydrostatic skeleton consists of fluid in a muscular cavity
- Exoskeletons are rigid outer structures
- Vertebrate endoskeletons consist of cartilage and bone
- Bones develop from connective tissues
- Bones that have a common joint can work as a lever

Many soft-bodied invertebrates have a fluid-filled body cavity that acts as a hydrostatic skeleton. Earthworms have circular muscles and longitudinal muscles that oppose each other to act on the hydrostatic skeleton and control elongation and shortening of body segments (see Figure 36.12).

Arthropods have an exoskeleton, or cuticle, composed of chitin that offers protection and provides sites for muscle attachment. As an animal that has an exoskeleton grows, it undergoes the process of molting. With each molt the old exoskeleton is shed, revealing a new one that has developed underneath.

Endoskeletons are growing, living tissue that provide sites for muscle attachment and support for the body. The human skeleton is composed of a central axial skeleton (skull, vertebral column, sternum, and ribs) and an appendicular skeleton (pectoral girdle, pelvic girdle, arms, hands, legs, and feet) (see Figure 36.13).

The pliable portions of the endoskeleton, such as the framework of the nose and the surface of the joints of the endoskeleton, are composed of cartilage containing the protein collagen. The bone in the endoskeleton is strong and solid and is composed mainly of collagen and calcium phosphate. Bone is constantly being remodeled by two types of cells: osteoblasts, which lay down new bone, and osteoclasts, which break down bone (see Figure 36.14). When an osteoblast becomes enclosed by the matrix it is laying down, it stops forming matrix and exists within a lacuna; at this stage, the cell is called an osteocyte. Osteocytes communicate with one another and influence the activities of osteoblasts and osteoclasts.

Developing bone is created as membranous bone growing on a scaffolding of connective tissue or as cartilage bone hardening from an initial cartilage model. The long bones of the arms and legs are cartilage bones that ossify first at the center and then at each end (see Figure 36.15). Compact bone is solid, whereas cancellous bone is lightweight, with many cavities, but still rigid and strong.

There are six types of joints where bones meet: ball-and-socket, pivot, saddle, ellipsoid, hinge, and plane (see Figure 36.16). The muscles attached to bones at joints work antagonistically (see Figure 36.17). The flexor muscles bend the joints and the extensor muscles straighten them. Two types of connective tissues hold joints and bones together. Ligaments hold bone to bone, and tendons hold muscle to bone.

Question 8. Describe how the earthworm's hydrostatic skeleton is used to move the worm through the soil.
Textbook Reference: *36.3 Muscles Pull on Skeletal Elements to Generate Force and Cause Movement, p. 723*

Question 9. On the diagram below, label the four main components of the axial skeletal system.
Textbook Reference: *36.3 Muscles Pull on Skeletal Elements to Generate Force and Cause Movement, p. 724*

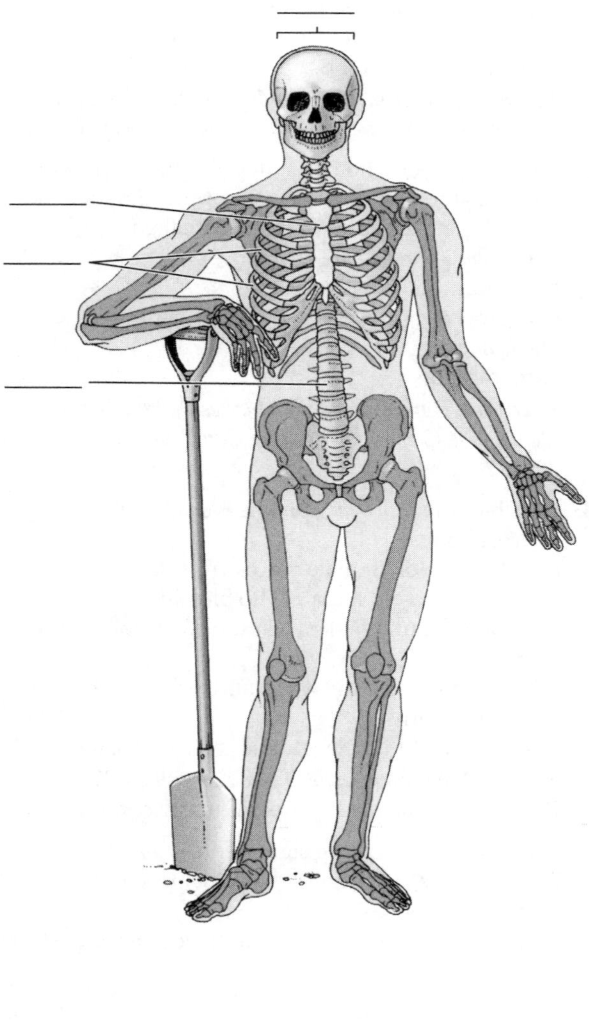

Question 10. Osteoporosis is a decrease in bone density that results when the destruction of bone by osteoclasts outpaces the formation of new bone by osteoblasts, leading to thin, brittle bones. Why does weight-bearing exercise help prevent osteoporosis?
Textbook Reference: *36.3 Muscles Pull on Skeletal Elements to Generate Force and Cause Movement, p. 724*

Question 11. Hydrostatic skeletons are found in annelids, such as earthworms, and cnidarians, such as hydras. Would a hydrostatic skeleton work for a terrestrial animal that moves by walking? Why or why not?
Textbook Reference: *36.3 Muscles Pull on Skeletal Elements to Generate Force and Cause Movement, p. 723*

Test Yourself

1. Which of the following would supply ATP during the last mile of a three-mile walk?
   a. ATP stored in muscle
   b. Glycolysis
   c. Creatine phosphate
   d. Oxidative metabolism
   e. Calcium ions
   Textbook Reference: *36.2 The Characteristics of Muscle Cells Determine Muscle Performance, p. 722*

2. A motor unit is best described as
   a. all the nerve fibers and muscle fibers in a single muscle bundle.
   b. one muscle fiber and its single nerve fiber.
   c. a single motor neuron and all the muscle fibers that it innervates.
   d. the neuron that provides the central nervous system with information about muscle contraction.
   e. a neuron that communicates information from sensory to motor neurons.
   Textbook Reference: *36.1 Cycles of Protein–Protein Interactions Cause Muscles to Contract, p. 715*

3. Which of the following statements is *false*?
   a. Cardiac muscle is striated.
   b. Actin is absent from smooth muscle.
   c. Skeletal muscle is considered voluntary.
   d. Smooth muscle is found in the digestive tract and the walls of the bladder.
   e. A single skeletal muscle cell has many nuclei.
   Textbook Reference: *36.1 Cycles of Protein–Protein Interactions Cause Muscles to Contract, pp. 713–714, 718*

4. The oxygen-binding molecule in skeletal muscle is
   a. myoglobin.
   b. hemoglobin.
   c. ATP.
   d. myokinase.
   e. creatine phosphate.
   Textbook Reference: *36.2 The Characteristics of Muscle Cells Determine Muscle Performance, p. 721*

5. The action potential that triggers a muscle contraction travels deep within the muscle cell by means of
   a. sarcoplasmic reticulum.
   b. transverse (or T) tubules.
   c. synapses.
   d. motor end plates.
   e. neuromuscular junctions.
   Textbook Reference: *36.1 Cycles of Protein–Protein Interactions Cause Muscles to Contract, pp. 716–717*

6. A sarcomere is best described as a
   a. moveable structural unit within a myofibril bounded by H zones.
   b. fixed structural unit within a myofibril bounded by Z lines.
   c. fixed structural unit within a myofibril bounded by A bands.
   d. moveable structural unit within a myofibril bounded by Z lines.
   e. collection of myofibrils.
   Textbook Reference: *36.1 Cycles of Protein–Protein Interactions Cause Muscles to Contract, pp. 713–714*

7. ATP provides the energy for muscle contraction by allowing for the
   a. formation of an action potential in the muscle cell.
   b. breaking of actin–myosin bonds.
   c. formation of actin–myosin bonds.
   d. release of calcium by the sarcoplasmic reticulum.
   e. formation of T tubules.
   Textbook Reference: *36.1 Cycles of Protein–Protein Interactions Cause Muscles to Contract, p. 715*

8. Ca^{2+} binds to _______ in skeletal muscle and leads to exposure of the binding site for _______ on the _______ filament.
   a. troponin; myosin; actin
   b. troponin; actin; myosin
   c. actin; myosin; troponin
   d. tropomyosin; myosin; actin
   e. myosin; actin; troponin
   Textbook Reference: *36.1 Cycles of Protein–Protein Interactions Cause Muscles to Contract, pp. 716–717*

9. Tropomyosin is moved by which of the following proteins?
   a. Calmodulin
   b. Acetylcholine
   c. Actin
   d. Troponin
   e. Titin
   Textbook Reference: *36.1 Cycles of Protein–Protein Interactions Cause Muscles to Contract, pp. 716–717*

10. Summation of frequent muscle twitches to give maximum contraction is called
    a. motor unit summation.
    b. twitch.
    c. facilitation.
    d. tetanus.
    e. muscle tone.
    Textbook Reference: *36.2 The Characteristics of Muscle Cells Determine Muscle Performance, p. 721*

11. _______ are responsible for the dynamic remodeling of bone that occurs continuously.
    a. Osteoblasts
    b. Osteoblasts and osteoclasts
    c. Osteoclasts and osteocytes
    d. Osteoblasts, osteoclasts, and osteocytes
    e. Myoblasts
    Textbook Reference: *36.3 Muscles Pull on Skeletal Elements to Generate Force and Cause Movement, p. 724*

12. Endoskeletons
    a. are characteristic of arthropods.
    b. are located on the inside of the body.
    c. lack joints.
    d. require molting as the animal grows.
    e. provide support to earthworms, along with their hydrostatic skeleton.
    Textbook Reference: *36.3 Muscles Pull on Skeletal Elements to Generate Force and Cause Movement, p. 723*

13. A soccer player who has suffered a knee injury that damages the tissue holding his upper and lower leg bones together has most likely damaged _______ tissue
    a. muscle
    b. tendon
    c. ligament
    d. cartilage
    e. membrane
    Textbook Reference: *36.3 Muscles Pull on Skeletal Elements to Generate Force and Cause Movement, pp. 725–726*

14. Which of the following people would likely have the densest bones?
    a. A 14-year-old female who swims daily
    b. An 20-year-old female who bicycles daily
    c. An astronaut who has just returned from 6 weeks in space
    d. A 30-year-old male who eats calcium-rich foods and jogs regularly
    e. A 60-year-old male who does sit-ups daily
    Textbook Reference: *36.3 Muscles Pull on Skeletal Elements to Generate Force and Cause Movement, p. 724*

15. Which of the following statements is *false*?
    a. In humans, cartilage is the principal component of the embryonic skeleton.
    b. Some vertebrates retain a cartilaginous endoskeleton into adulthood.

c. Cancellous bone has numerous cavities.
d. The outer bones of the skull are cartilage bone.
e. Calcitonin and parathyroid hormone regulate the deposition of calcium in bone.

Textbook Reference: *36.3 Muscles Pull on Skeletal Elements to Generate Force and Cause Movement, pp. 724–725*

Answers

Key Concept Review

1.

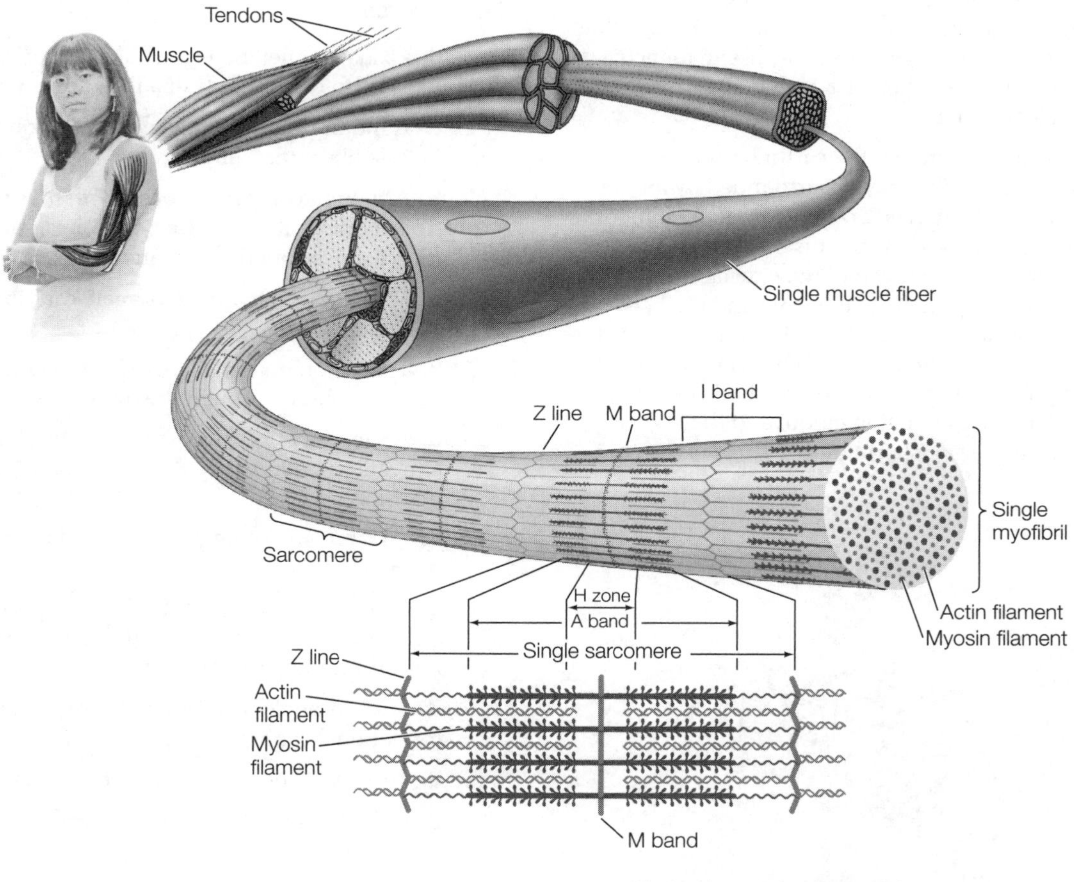

2. Smooth muscle contraction and skeletal muscle contraction are regulated by the presence of Ca^{2+}. In smooth muscle, Ca^{2+} joins with the protein calmodulin to activate myosin kinase in the sarcoplasm. Myosin kinase then phosphorylates the myosin head, allowing the myosin head to bind with actin. In skeletal muscle, Ca^{2+} binds with troponin on the actin filaments. This binding of Ca^{2+} causes a conformational change in tropomyosin and uncovers the myosin binding sites on the actin, allowing myosin to bind with actin.

3. ATP is needed to break the actin–myosin bonds. ATP production stops at death, so the actin–myosin bonds cannot be broken and muscles stiffen. Eventually, the proteins deteriorate and the muscles soften.

4. Skeletal muscles are used in breathing. Acetylcholine, released by motor neurons at the neuromuscular junction, initiates action potentials in skeletal muscle. Curare acts by preventing acetylcholine from binding to the postsynaptic membrane, where it would normally cause ion channels in the motor end plate to open.

5. Fast-twitch fibers are known as white muscle and have few mitochondria, small amounts of myoglobin, and few blood vessels. Muscles with many fast-twitch fibers are good for short-term work that requires maximum strength. Slow-twitch fibers are known as red muscle. Red muscle has many mitochondria, large amounts of myoglobin, and many blood vessels. Muscles with many slow-twitch fibers function well in endurance activities.

6. The breast muscles of turkeys are fast-twitch, or white, muscle (= "white meat"), and their leg muscles are slow-twitch, or red, muscle (= "dark meat"). This tells us that turkeys fly only in short bursts and typically walk or run. In contrast, the breast muscles of mallard ducks are slow-twitch, or red, muscle, indicating that they are capable of prolonged flight.
7. Taking creatine supplements may increase stores of creatine phosphate in muscle. Creatine phosphate stores energy in a phosphate bond, which it can transfer to ADP to form ATP. However, the energy is available immediately and the supply is quickly exhausted (creatine phosphate is part of the immediate system for supplying ATP to muscle). Thus, we would predict that creatine supplements would be most useful for activities such as sprinting when fast-twitch fibers quickly generate a lot of force.
8. The hydrostatic skeleton of the earthworm is an incompressible fluid-filled cavity surrounded by longitudinal and circular muscles. Contraction of the longitudinal muscles causes the segments to contract and the body to shorten. Contraction of the circular muscles causes the body segments to elongate and the body to lengthen. Alternating contractions between the longitudinal and the circular muscles move the animal in a push and pull manner. Bristles on the body help hold the animal in place after elongation, and the body is pulled forward during shortening.
9.

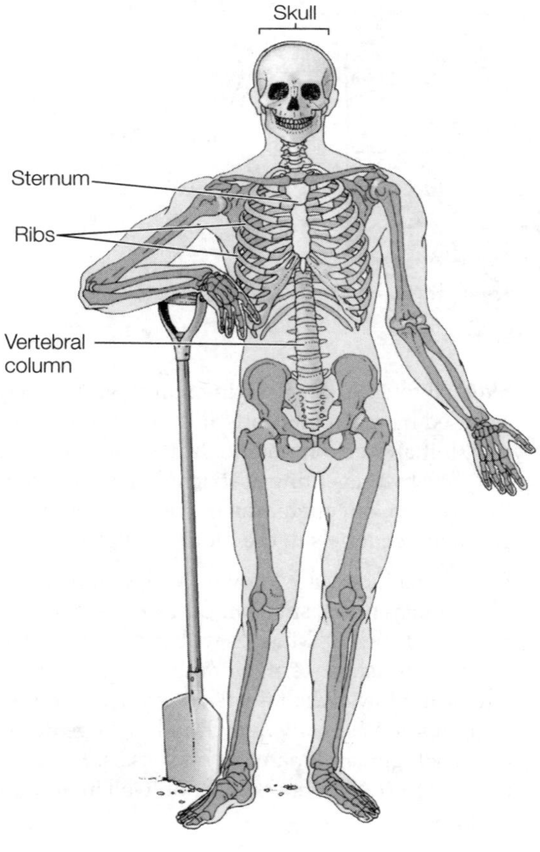

10. Weight-bearing exercises place stress on bone, ultimately altering the interplay of osteoblast and osteoclast activity to induce thickening of bone.
11. Hydrostatic skeletons work well for aquatic animals, such as hydras, and for terrestrial animals, such as earthworms that move by crawling through a substrate. However, a hydrostatic skeleton would not work for most terrestrial animals. Such skeletons provide little or no protection against drying out (dehydration is always a danger on land), and they do not provide sufficient support for a large animal that walks by holding its body off the ground.

Test Yourself

1. **d.** Oxidative metabolism would supply ATP needed by muscles in the last mile of a three-mile walk.
2. **c.** A motor unit is a single motor neuron and all the muscle fibers that it innervates.
3. **b.** Although contraction of smooth muscle is controlled differently from that of skeletal muscle, smooth muscle does contain actin and myosin.
4. **a.** Myoglobin is the main oxygen-carrying molecule in skeletal muscle.
5. **b.** The action potential arriving to the muscle travels into the muscle through the transverse (or T) tubules.
6. **d.** A sarcomere is a moveable structural unit within a myofibril bounded by Z lines; it contains actin and myosin.
7. **b.** ATP provides energy that is used to break actin–myosin bonds.
8. **a.** Calcium is released from the sarcoplasmic reticulum and binds with troponin, resulting in exposure of the myosin-binding site on actin.
9. **d.** The binding of calcium with troponin causes a conformational change in tropomyosin.
10. **d.** Tetanus is the maximum level of muscle contraction.
11. **d.** Osteoblasts, osteoclasts, and osteocytes are all involved in remodeling bone.
12. **b.** Endoskeletons (such as those of mammals) are found inside the body, and exoskeletons (such as those of insects) are found outside the body.
13. **c.** Ligaments hold bones together.
14. **d.** Placing stress on bones, which would occur through jogging, keeps them dense and healthy. Swimming, cycling, and a zero-gravity environment such as space, do not stress bones.
15. **d.** The outer bones of the skull are membranous bone, not cartilage bone.

Gas Exchange in Animals 37

The Big Picture

- Cellular metabolism requires O_2 and produces CO_2 as a waste product that must be eliminated. Animals have evolved diverse structures and mechanisms for exchanging these gases with the environment. Tracheal systems in insects, gills in fishes, and lungs in terrestrial vertebrates are three examples of gas-exchange organs. Very short diffusion distances and very large surface areas are adaptations designed to maximize gas exchange and characterize all gas-exchange organs.
- O_2 is transported by the protein hemoglobin in vertebrates. The P_{O_2} levels surrounding hemoglobin determine O_2 binding. If P_{O_2} is high, hemoglobin accepts O_2, but it readily gives up its O_2 when the P_{O_2} falls. These properties allow hemoglobin to bind O_2 in the gas-exchange organ, transport it to the metabolizing tissues, and then release O_2 for consumption in cellular metabolism.

Study Strategies

- Refer to Fick's law of diffusion as a guideline for understanding gas exchange. The various components of the law—distance for diffusion, partial pressure gradient, surface area over which gas exchange occurs—provide a good framework for understanding why gas-exchange organs have evolved with certain characteristics in common. For example, gas-exchange organs tend to have large surface areas and very short diffusion distances, and to experience large partial gradients for O_2 across their surfaces. All of this makes sense in the context of Fick's law of diffusion. Fick's law of diffusion also governs the movement of O_2 and CO_2 in the body, and it can help you remember where and why these gases are picked up and released in the body.
- Understanding how O_2 is picked up, transported by, and released from hemoglobin can be quite confusing. Think of the O_2 dissociation curve as a "tool." Think first of the situation at the top of the curve, where P_{O_2} is high (e.g., in the lungs or gills). If the P_{O_2} is high, hemoglobin will maximally bind O_2. Think, then, of transporting that blood to a region with a given P_{O_2}, and picture on the curve what must happen to O_2 saturation under the new P_{O_2} level. The O_2 from hemoglobin is now available for tissue respiration.
- A common error is the conception of O_2 and CO_2 exchange as a "two-way street," with O_2 taking the immediate place of CO_2 in the lungs and CO_2 then taking the place of O_2 in the tissues. This is incorrect. In fact, the mechanisms for exchange of these two gases are completely different.
- The structure of the avian and fish respiratory systems is complex and very different from the mammalian pattern. Study Figures 37.4 and 37.5. Follow the countercurrent exchange of gases in fish and be sure to remember that avian air sacs are not sites of gas exchange.
- Although one might expect that the O_2 content of the blood would regulate respiratory rate in mammals, including humans, it turns out that respiratory rate is primarily regulated by the CO_2 content of the blood.
- Go to yourBioPortal.com to review the following tutorials and activities:

  Animated Tutorial 37.1 Airflow in Birds

  Animated Tutorial 37.2 Airflow in Mammals

  Interactive Tutorial 37.1 Hemoglobin: Loading and Unloading

  Web Activity 37.1 The Human Respiratory System

  Web Activity 37.2 Oxygen-Binding Curves

  Web Activity 37.3 Concept Matching

  Working with Data 37.1 Describing Air Flow in Bird Lungs

  Working with Data 37.2 Calculating the Functional Residual Volume

Key Concept Review

37.1 Fick's Law of Diffusion Governs Respiratory Gas Exchange

Diffusion is driven by concentration differences

Fick's law applies to all systems of gas exchange

Air is a better respiratory medium than water

O_2 availability is limited in many environments

CO_2 is easily lost by diffusion

The respiratory gas oxygen (O_2) is required by cells to produce energy in the form of ATP. The respiratory gas carbon dioxide (CO_2) is one of the waste by-products of ATP production and must be eliminated from an animal's body. The transfer of these gases occurs by simple diffusion in the respiratory systems of animals.

Partial pressures are used to express the concentrations of gases in a mixture. The partial pressure gradient is the difference between the partial pressure of a gas at two locations. The partial pressure gradients of O_2 and CO_2 drive the movement of O_2 into the body and CO_2 out of the body.

The tendency for a gas to move by diffusion across gills, skin, or lungs depends on its partial pressure. The sum of all the gases' partial pressures in a gas mixture is the total partial pressure, which in the atmosphere equals the atmospheric pressure. O_2 makes up 20.9 percent of the atmospheric pressure. As elevation increases and atmospheric pressure decreases, the total amount of O_2 in air decreases.

Rate of diffusion, described by Fick's law of diffusion, depends in part on a diffusion coefficient that varies according to temperature, the medium, and the diffusing molecules. Rate of diffusion also depends on the cross-sectional area over which the gas is diffusing, and the partial pressure gradient.

There are major differences in the O_2 capacity of air and water. Water contains far less O_2 compared to air, and O_2 also diffuses far more slowly in water than in air. Temperature affects respiration of fishes because there is less O_2 in warm water than in cold, and a fish's need for O_2 increases with temperature. Animals that move water over their respiratory surfaces require more energy for gas exchange than those moving air, because water is more dense than air.

CO_2 diffuses across the respiratory organs in both air and water. In air, diffusion is rapid because of the large partial pressure gradient between the blood and the atmosphere. Getting rid of CO_2 typically is not a problem for water-breathing animals because CO_2 is much more soluble than O_2 in water. Lack of O_2 becomes a problem for water-breathing animals long before problems with CO_2 exchange occur.

Question 1. Why would a tropical fish in a fish tank face severe respiratory problems (and possibly death) if the heater in the tank malfunctioned and caused extremely high water temperatures?
Textbook Reference: *37.1 Fick's Law of Diffusion Governs Respiratory Gas Exchange, p. 731*

Question 2. Explain why getting rid of CO_2 is not typically a problem for aquatic animals.
Textbook Reference: *37.1 Fick's Law of Diffusion Governs Respiratory Gas Exchange, p. 732*

Question 3. How can an aquatic animal with no specialized respiratory surfaces get enough O_2 to all of its cells in order to survive?
Textbook Reference: *37.1 Fick's Law of Diffusion Governs Respiratory Gas Exchange, p. 731*

37.2 Respiratory Systems Have Evolved to Maximize Partial Pressure Gradients

Respiratory organs have large surface areas

Partial pressure gradients can be optimized in several ways

Insects have airways throughout their bodies

Fish gills use countercurrent flow to maximize gas exchange

Most terrestrial vertebrates use tidal ventilation

Birds have air sacs that supply a continuous unidirectional flow of fresh air

Animals maximize respiration by decreasing the distance that gases must diffuse between the blood and the external environment. They achieve this with very thin respiratory surface membranes. Animals also actively move the external medium over the respiratory surface (this is called ventilation) and perfuse the internal side of the respiratory organ with blood or another internal medium that carries the respiratory gases.

Some aquatic amphibians and insect larvae have external gills with large surface areas for exchange of respiratory gases with water (see Figure 37.1A). Internal gills, such as those in crayfish or fishes, have a large surface area and are protected from the environment by an animal's body cavity (see Figure 37.1B). They must be actively ventilated with water to achieve gas exchange. Other animals have lungs (see Figure 37.1C) or tracheae (see Figure 37.1D)

Insects have tracheae—air tubes that end at the tissue cells as air capillaries and open to the environment through spiracles. Gases diffuse through the tracheae into the air capillaries, but are also moved by movement of the animal's body parts (see Figure 37.2).

Fish have internal gills with a large surface area that they ventilate with unidirectional flowing water to maximize external P_{O_2} levels. Each gill has hundreds of gill filaments with folds, or lamellae, that act as the respiratory gas exchange surfaces. Countercurrent flow (blood in the lamellae flowing in the direction opposite to that of water flowing over the lamellae) maximizes the P_{O_2} gradient between the water and the blood (see Figures 37.3 and 37.4). As blood flows through the lamellae it is always in contact with water that has a higher O_2 level, resulting in a continuous O_2 gradient to maximize the uptake of O_2 into the blood.

Lungs evolved in fishes as outpocketings of the digestive tract. In mammals, lungs are dead-end sacs in which ventilation is tidal. In tidal ventilation, fresh air flows in and exhaled gases flow out by the same route. Tidal breathing limits the partial pressure gradient needed to drive the diffusion of O_2 from air into blood. Fresh air is not moving into the lungs during some of the breathing cycle, and when it does enter the lungs, it mixes with stale air.

Birds have a unidirectional flow of air through their lungs. Gas exchange occurs in the air capillaries that branch off the parabronchi of the lungs; gas exchange does not occur in the associated air sacs (see Figure 37.5). During inhalation, the posterior air sacs receive the incoming air, and the anterior air sacs receive the air that was in the lungs. During exhalation, the air in the posterior air sacs flows into the lungs, and the air in the anterior air sacs leaves the bird.

Question 4. The gills of fish employ a countercurrent exchange mechanism to ensure that O_2 is efficiently extracted from the water. Describe how the countercurrent exchange works.
Textbook Reference: *37.2 Respiratory Systems Have Evolved to Maximize Partial Pressure Gradients, pp. 733–734*

Question 5. Why is one more likely to find birds than mammals at high altitudes?
Textbook Reference: *37.2 Respiratory Systems Have Evolved to Maximize Partial Pressure Gradients, pp. 734–735*

Question 6. Why would the internal gills of fishes fail to work as respiratory organs on land?
Textbook Reference: *37.2 Respiratory Systems Have Evolved to Maximize Partial Pressure Gradients, pp. 733–734*

Question 7. On the diagram below of the avian respiratory system, label the following structures: anterior air sacs, posterior air sacs, trachea, lung, and bronchus.
Textbook Reference: *37.2 Respiratory Systems Have Evolved to Maximize Partial Pressure Gradients, p. 735*

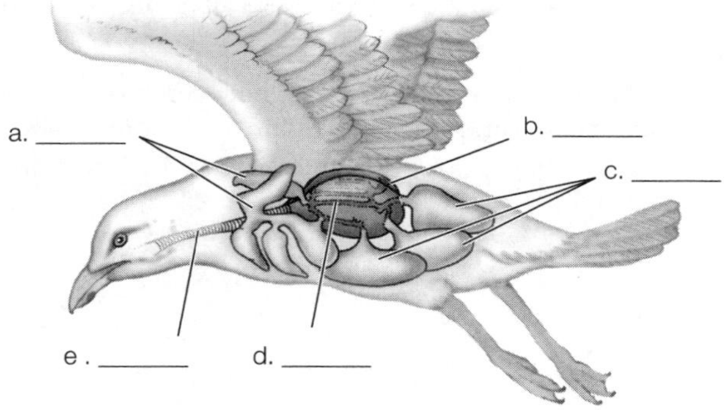

37.3 The Mammalian Lung Is Ventilated by Pressure Changes

At rest, only a small portion of the lung's volume is exchanged

Lungs are ventilated by pressure changes in the thoracic cavity

Mammalian lungs have a very large surface area and a very short gas diffusion pathway. Air enters at the oral cavity or nasal passage, travels down the pharynx through the larynx, and then through the trachea, which branches into two bronchi, one leading to each lung. In the lung, more branching occurs to produce bronchioles and finally the site of gas exchange—small air sacs, the alveoli (see Figure 37.6).

O_2 diffuses across the thin-walled alveoli (less than 2 µm) into many surrounding capillaries. The alveoli have surface tension due to an aqueous layer that makes inflation difficult. This is overcome by a surfactant produced by the cells forming the alveolar walls. The aqueous layer is needed to ensure the diffusion of gases across the membrane.

Mucus, produced along the airways, catches and removes inhaled dust and microorganisms. Cilia along the airways move the mucus up the airways and into the throat, where it can be swallowed.

The tidal breathing of mammals can be described in terms of a series of lung volumes measured with a spirometer (see Figure 37.7). The normal volume of air moved during one cycle is known as the tidal volume. The vital capacity consists of the resting tidal volume, plus the additional volume of air that can be taken in with a large breath (inspiratory reserve volume) and the additional volume of air that can be forced out with a large exhale (expiratory reserve volume). Not all of the air can be forced out of the lungs. Some air remains in the bronchi and trachea, making up the dead space; this air is called the residual volume. Increases in the residual volume are associated with certain respiratory diseases, such as emphysema.

Lungs are inflated by negative pressure created by contraction of the diaphragm muscle in the thoracic cavity. Closed pleural membranes line the thoracic cavity and allow for the development of negative pressure. During exhalation, the diaphragm relaxes and the thoracic cavity contracts (see Figure 37.8). Inhalation is an active process, requiring muscle contraction, and exhalation is normally a passive process, requiring no muscular activity. At times of strenuous exercise, however, intercostal muscles and abdominal muscles help change the volume of the thoracic cavity and exhalation becomes an active process.

Question 8. An asthma attack occurs when smooth muscles lining the bronchioles spasm, obstructing the flow of air and making breathing difficult. Why is this problem limited to the bronchioles?
Textbook Reference: *37.3 The Mammalian Lung Is Ventilated by Pressure Changes, p. 737*

Question 9. Are inhalation and exhalation active processes (requiring muscle contraction) or passive processes (not requiring muscle contraction)?
Textbook Reference: *37.3 The Mammalian Lung Is Ventilated by Pressure Changes, pp. 738–739*

37.4 Respiration Is under Negative Feedback Control by the Nervous System

Respiratory rate is primarily regulated by CO_2

CO_2 affects the medulla indirectly via pH changes

O_2 is also monitored

Breathing is controlled by the autonomic nervous system from neurons in the medulla. Brain areas above the medulla modify breathing in response to speaking, eating, and emotional states.

In humans and other mammals, chemoreceptors on the medulla are sensitive to CO_2 levels and the pH of cerebral spinal fluid. Breathing increases when CO_2 levels increase and pH decreases. Control of breathing is relatively insensitive to blood O_2 levels, but chemoreceptors in the arteries leaving the heart (carotid and aorta) can signal the medulla in response to low levels of O_2. In water-breathing animals, O_2 is the primary feedback stimulus for breathing.

Question 10. Explain why the immediate increase in respiratory rate at the start of aerobic exercise is an example of feedforward information.
Textbook Reference: *37.4 Respiration Is under Negative Feedback Control by the Nervous System, p. 739*

Question 11. During an episode of hyperventilation, a person breathes very deeply and rapidly, and large amounts of CO_2 are eliminated from the blood. What happens next and why?
Textbook Reference: *37.4 Respiration Is under Negative Feedback Control by the Nervous System, p. 739*

Question 12. Design a study to investigate whether brain areas above the medulla modify breathing patterns in rats. What three variables would you need to measure?
Textbook Reference: *37.4 Respiration Is under Negative Feedback Control by the Nervous System, pp. 739–741*

37.5 Respiratory Gases Are Transported in the Blood

Hemoglobin combines reversibly with O_2

Myoglobin holds an O_2 reserve

Various factors influence hemoglobin's affinity for O_2

CO_2 is transported primarily as bicarbonate ions in the blood

Gases diffuse between the respiratory organ and the circulatory system, where they are transported in blood. The liquid blood plasma transports only a small amount of dissolved O_2, with the vast majority of O_2 transported by the O_2-binding pigment hemoglobin found in the red blood cells of vertebrates.

Hemoglobin contains four polypeptide subunits, each with a heme group that reversibly binds O_2. The amount of O_2 bound to hemoglobin depends on the partial pressure of O_2. At the high blood P_{O_2} that is characteristic of blood leaving the lungs, hemoglobin is almost 100 percent saturated with O_2. At the tissues, the P_{O_2} is lower and hemoglobin releases approximately 25 percent of its bound O_2. The relationship between the P_{O_2} of the environment and the percentage of hemoglobin bound with O_2 is sigmoidal (S-shaped) (see Figure 37.11).

In muscle, myoglobin is the O_2-carrying molecule and functions primarily to store O_2. The O_2 affinity is higher in myoglobin than in hemoglobin, but myoglobin has only one unit to bind one O_2 molecule. Fetal hemoglobin has a different structure from the adult form, with a higher affinity for O_2 (see Figure 37.12). This allows the efficient movement of O_2 from maternal to fetal blood.

The Bohr effect is the shifting of the hemoglobin-binding curve to the right in response to a decrease in pH of the blood. This results in the release of more O_2 at a given environmental P_{O_2} at the lower pH. Hemoglobin's affinity for O_2 is lowered by 2,3-bisphosphoglyceric acid (BPG), a metabolite of glycolysis. BPG binds with hemoglobin to cause a conformational change in the hemoglobin, resulting in a right shift in the O_2-binding curve.

Tissues produce CO_2 as a by-product of metabolism. CO_2 must be moved from the tissues and excreted into the environment. Some CO_2 dissolves in the plasma and is carried in this form to the lungs; some is transported bound to hemoglobin. Most of the CO_2, however, is transported as bicarbonate ions. CO_2 reacts with water to make bicarbonate ions and a proton. Carbonic anhydrase speeds up the conversion of CO_2 to carbonic acid, which then dissociates into bicarbonate ions.

Question 13. What is the difference between the myoglobin O_2-binding curve and the hemoglobin O_2-binding curve?
Textbook Reference: *37.5 Respiratory Gases Are Transported in the Blood, p. 742*

Question 14. Why is carbon monoxide (CO) such a dangerous gas?
Textbook Reference: *37.5 Respiratory Gases Are Transported in the Blood, p. 742*

Test Yourself

1. Which of the following statements is *false*?
   a. Compared to a given volume of water, the same volume of air is moved more easily across a respiratory surface.
   b. Compared to water, air holds more O_2 per unit volume.
   c. Water breathers have a difficult time ridding themselves of CO_2 because CO_2 does not dissolve well in water.
   d. Temperature increases affect the O_2 content of water more than they do that of air.
   e. O_2 diffuses more rapidly in air than in water.
   Textbook Reference: *37.1 Fick's Law of Diffusion Governs Respiratory Gas Exchange, p. 732*
2. External gills, tracheae, and lungs share which of the following sets of characteristics?
   a. They are parts of gas-exchange systems, they exchange both CO_2 and O_2, and they increase surface area for diffusion.
   b. They are used by water breathers, their functioning is based on countercurrent exchange, and they make use of negative pressure for breathing.
   c. They exchange only O_2, they are associated with a circulatory system, and they are found in vertebrates.
   d. They are found in insects, they employ positive-pressure pumping, and their functioning is based on crosscurrent flow.
   e. They are parts of gas-exchange systems, they exchange both CO_2 and O_2, and they decrease surface area for diffusion.
   Textbook Reference: *37.2 Respiratory Systems Have Evolved to Maximize Partial Pressure Gradients, pp. 732–733*
3. Which of the following represents a larger volume of air than is normally found in the resting tidal volume of a human lung?
   a. Residual volume
   b. Inspiratory reserve volume
   c. Expiratory reserve volume
   d. Vital capacity
   e. All of the above
   Textbook Reference: *37.3 The Mammalian Lung Is Ventilated by Pressure Changes, pp. 737*
4. Both bird and mammal lungs
   a. have tidal ventilation.
   b. contain alveoli at the terminal ends.
   c. have an anatomical dead space.
   d. exchange O_2 and CO_2 with blood in capillaries.
   e. have a unidirectional flow of air moving through them.
   Textbook Reference: *37.2 Respiratory Systems Have Evolved to Maximize Partial Pressure Gradients, pp. 734–735*
5. According to Fick's law of diffusion, which of the following does *not* play a role in the diffusion of O_2 across a membrane?
   a. Surface area
   b. Volume
   c. Difference in concentration or partial pressure
   d. Diffusion distance
   e. Diffusion coefficient
   Textbook Reference: *37.1 Fick's Law of Diffusion Governs Respiratory Gas Exchange, p. 730*
6. Because of the relatively high altitude of Antonito, Colorado, the town has a normal barometric pressure of about 600 mm Hg rather than 760 mm Hg as at sea level. The partial pressure of O_2 in Antonito's air is approximately _______ mm Hg.
   a. 75
   b. 126
   c. 160
   d. 76
   e. 21
   Textbook Reference: *37.1 Fick's Law of Diffusion Governs Respiratory Gas Exchange, p. 730*
7. Air flows into the lungs of mammals during inhalation because the
   a. pressure in the lungs falls below atmospheric pressure.
   b. volume of the lungs decreases.
   c. pressure in the lungs rises above atmospheric pressure.
   d. diaphragm moves upward toward the lungs.
   e. internal intercostal muscles contract.
   Textbook Reference: *37.3 The Mammalian Lung Is Ventilated by Pressure Changes, pp. 738–739*
8. The movement of O_2 and CO_2 between the blood in the tissue capillaries and the cells in tissues depends most directly upon
   a. active transport of O_2 and CO_2.
   b. total atmospheric (barometric) pressure differences across the cell membranes.
   c. diffusion of O_2 and CO_2 down a concentration gradient.
   d. diffusion of O_2 and CO_2 down a partial pressure gradient.
   e. osmosis across cell membranes.
   Textbook Reference: *37.1 Fick's Law of Diffusion Governs Respiratory Gas Exchange, p. 731*
9. Although 20.9 percent of air is oxygen, the alveoli of the lungs do not contain air with this much oxygen because

a. we normally do not ventilate our lungs at a high enough rate.
b. the lungs have too many alveoli to ventilate them all at this level.
c. there is dead space in the trachea and bronchi.
d. the trachea and bronchi are too small to contain this much oxygen.
e. some O_2 has been exchanged before reaching the alveoli.

Textbook Reference: *37.3 The Mammalian Lung Is Ventilated by Pressure Changes, pp. 737–738*

10. Which of the following statements about hemoglobin is *false*?
a. Hemoglobin allows the blood to carry a large amount of O_2.
b. Hemoglobin contains a single polypeptide chain with very high affinity for O_2.
c. Hemoglobin is packaged inside red blood cells.
d. Fetal hemoglobin is structurally different from adult hemoglobin.
e. Compared to adult hemoglobin, fetal hemoglobin has a higher affinity for O_2.

Textbook Reference: *37.5 Respiratory Gases Are Transported in the Blood, pp. 741–743*

11. As blood becomes fully O_2 saturated, hemoglobin is combining with _______ molecule(s) of O_2.
a. 1
b. 2
c. 4
d. 8
e. 6

Textbook Reference: *37.5 Respiratory Gases Are Transported in the Blood, p. 741*

12. The Bohr effect describes the
a. outward movement of Cl^- from the blood cell in exchange for HCO_3^- moving into the cell.
b. leftward shift of the entire O_2 equilibrium curve when temperature rises.
c. rightward shift of the entire O_2 equilibrium curve when pH rises.
d. rightward shift of the entire O_2 equilibrium curve when pH falls.
e. leftward shift of the entire O_2 equilibrium curve when BPG increases.

Textbook Reference: *37.5 Respiratory Gases Are Transported in the Blood, p. 743*

13. The presence of CO_2 in blood will lower pH because CO_2 combines with _______, with the rate of reaction increased by _______.
a. H_2O to form H^+ and HCO_3^-; carbonic anhydrase
b. H_2O to form only HCO_3^-; carbonic anhydrase
c. H_2O to form only H^+; carbonic ions
d. H^+ to form HCO_3^-; oxyhemoglobin
e. H_2O to form H^+ and HCO_3^-; fetal hemoglobin

Textbook Reference: *37.5 Respiratory Gases Are Transported in the Blood, p. 743*

14. The largest proportion of CO_2 carried by the blood is in the form of
a. bicarbonate ions (HCO_3^-) carried in the plasma.
b. molecular CO_2 dissolved in the plasma.
c. bicarbonate ions (HCO_3^-) carried within the red blood cells.
d. molecular CO_2 chemically bound to hemoglobin.
e. molecular CO_2 chemically bound to myoglobin.

Textbook Reference: *37.5 Respiratory Gases Are Transported in the Blood, p. 743*

15. Which of the following is involved in the control of breathing?
a. Neurons in the medulla
b. Chemoreceptors on the surface of the medulla
c. Chemoreceptors in the aorta
d. Chemoreceptors in the carotid arteries
e. All of the above

Textbook Reference: *37.4 Respiration Is under Negative Feedback Control by the Nervous System, pp. 739–741*

16. Which of the following statements is *false*?
a. Mammals are remarkably insensitive to falling levels of O_2.
b. In water-breathing animals, O_2 is the primary feedback stimulus for breathing.
c. In humans, chemoreceptors for CO_2 are located in the medulla.
d. Areas of the brain above the medulla have no impact on breathing rate.
e. Breathing is an involuntary function of the central nervous system.

Textbook Reference: *37.4 Respiration Is under Negative Feedback Control by the Nervous System, pp. 739–741*

Answers

Key Concept Review

1. The tropical fish would be caught in the double bind of needing more O_2 as the water temperature increased (due to its increasing metabolic rate with increasing water temperature), and yet facing lower levels of O_2 in the warmer water (since warm water holds less dissolved O_2 than cold water does).
2. Compared to O_2, CO_2 is much more soluble in water, so getting rid of CO_2 is usually not a problem for aquatic animals. Such animals would first face problems concerning lack of O_2.
3. Since it has no respiratory surfaces, an aquatic animal must rely on simple diffusion from the environment into its cells. For simple diffusion to be an effective means of O_2 transport, the animal must be very small, with all of its cells only a few cell layers from the environment. The metabolic rate of this animal is most likely very low in order to accommodate the lack of any special respiratory surfaces.
4. In the countercurrent exchange in the gills of fish, water and blood move in opposite directions. This results in

blood with low O_2 content entering the gills and flowing past water with a higher O_2 content. O_2 diffuses down the partial pressure gradient into the blood. As blood moves along the lamella, it picks up O_2, but it always has less O_2 than the water moving in the opposite direction. In this way the blood is able to optimize O_2 uptake during the entire trip through the lamella.

5. Birds have an extremely efficient respiratory system characterized by the continuous unidirectional flow of air through the lungs. In contrast, mammals use tidal ventilation, a less-efficient method in which air flows in and exhaled gases flow out the same route. Tidal ventilation results in the mixing of fresh air with residual air in the lungs. O_2 availability decreases with altitude, and thus birds, rather than mammals, are likely to be found at very high elevations.
6. Like external gills, the internal gills of fishes consist of thin, delicate tissue. On land, such gills would collapse and clump together, significantly reducing the surface area for gas exchange.
7.
   a. Anterior air sacs
   b. Lung
   c. Posterior air sacs
   d. Bronchus
   e. Trachea
8. The problem occurs in the bronchioles because although the trachea and bronchi have cartilage supports to keep them open, the bronchioles do not.
9. Inhalation is always an active process because it involves contraction of the diaphragm. Exhalation is normally a passive process (the diaphragm relaxes), except during strenuous exercise when internal intercostal muscles and abdominal muscles contract.
10. Feedforward information is predictive of a change in the internal environment before that change occurs. Respiratory rate increases immediately at the start of aerobic exercise because information from receptors in the muscles and joints changes the respiratory center's sensitivity to CO_2 in anticipation of the upcoming rise in the partial pressure of CO_2.
11. Following hyperventilation, breathing temporarily stops because so much CO_2 has been removed from the blood that the respiratory control center temporarily stops sending signals to the diaphragm and intercostal muscles. Breathing begins again when the CO_2 builds to sufficient levels in the blood to prompt the respiratory center to send its signals.
12. Brain areas above the medulla can modify the basic breathing pattern to allow for activities such as eating and drinking (swallowing temporarily closes the opening to the respiratory system). This could be studied in rats by simultaneously monitoring the following: (1) the activity of neurons from higher brain centers that bring such information into the pons, which then communicates with the respiratory center in the medulla; (2) the behavior of the rats to keep track of when they are eating or drinking; and (3) the respiratory rates of the rats.
13. The O_2-binding curve of myoglobin is shifted to the left in relationship to the O_2-binding curve of hemoglobin. This allows myoglobin to pick up O_2 and hold it at lower partial pressures of O_2.
14. CO binds to hemoglobin much more readily than O_2 does, so CO prevents hemoglobin from binding and transporting oxygen. This can be fatal.

Test Yourself

1. **c.** Water breathing is more difficult than air breathing because the higher density of water makes it more expensive to move across the respiratory surfaces than air, it has less O_2 than air, and the O_2 content is very dependent on the temperature of the water. CO_2, however, dissolves easily in water and it is easily expelled, even in stagnant water.
2. **a.** The external gill, tracheae, and lungs are all examples of gas-exchange systems used by various animals to exchange both CO_2 and O_2. One of the special properties of all respiratory systems is that they increase surface area for diffusion.
3. **e.** All have a volume that is larger than the resting tidal volume.
4. **d.** The lungs of birds and mammals are both sites for the exchange of O_2 and CO_2 with blood in capillaries. The mammalian lung is the only one that ends in alveoli and has an anatomical dead space. Flow of air is unidirectional in birds, but tidal in mammals.
5. **b.** The volume of the gas is not part of Fick's law of diffusion. All the other parameters are important in determining the rate of diffusion of gases in the respiratory system of animals.
6. **b.** Air at sea level with an atmospheric pressure of 760 mm Hg has a partial O_2 pressure of 160 mm Hg (21% of 760). Therefore, the partial pressure of O_2 at Antonito, Colorado, is 126 mm Hg (21% of 600).
7. **a.** Inhalation in the mammalian lung occurs by means of negative pressure produced by contraction of the diaphragm. Thus, the pressure in the lungs falls below atmospheric pressure.
8. **d.** Movement of O_2 and CO_2 from the blood to the tissues always occurs by diffusion of O_2 and CO_2 down their partial pressure gradients.
9. **c.** The alveoli do not contain "air" with 20.9 percent O_2 because incoming air is mixed with air left in the dead space of the trachea and bronchi, which has had some of the O_2 removed by the lungs.
10. **b.** Hemoglobin is composed of four subunits. O_2-binding myoglobin found in muscle has only one unit.

11. **c.** Hemoglobin has four subunits, each of which binds to one molecule of O_2 for a total of four molecules of O_2 bound to one hemoglobin molecule.

12. **d.** The Bohr effect describes the action of pH on the O_2-binding curve. Decreases in pH result in a net rightward shift of the entire O_2 equilibrium curve.

13. **a.** CO_2 combines with H_2O in the plasma to form H^+ and HCO_3^-. The enzyme carbonic anhydrase catalyzes the reaction.

14. **a.** The majority of CO_2 is carried as bicarbonate ions (HCO_3^-) in the plasma.

15. **e.** All are involved in the control of breathing.

16. **d.** Areas of the brain above the medulla modify the breathing rate with respect to speech, eating, coughing, and emotional state.

Circulatory Systems 38

The Big Picture

- Cardiovascular systems are required when animals grow too big to acquire nutrients and eliminate waste by diffusion alone. Animals have evolved diverse cardiovascular systems, which can be classified as open or closed. An open system has relatively low pressure and large sinuses that bathe tissues and organs in circulating fluid. Closed systems have an interconnected series of vessels that keep blood separated from interstitial fluid but allow gases, nutrients, and wastes to diffuse through their walls.
- During the evolution of vertebrates, the heart has changed from the simple two-chambered heart of fishes to the complex four-chambered hearts of crocodilians, birds, and mammals. Turtles, lizards, snakes, and crocodilians often breathe intermittently; their hearts allow the shunting of blood away from the lungs and into the systemic circuit.
- The human heart is really two pumps in one. The left pump provides oxygenated blood under high pressure to the systemic tissues, and the right pump sends deoxygenated blood under low pressure to the lungs. A cardiac pacemaker sets the cardiac rhythm.
- The flow of blood through capillaries is highly regulated at the local level by changes in local surroundings; blood flow is regulated by the combined action of neural activity and circulating hormones.

Study Strategies

- It is frequently thought that the hearts of amphibians and certain reptiles such as snakes and lizards are primitive structures, awaiting "repair through evolution" to the more highly evolved bird and mammal heart. In fact, the hearts of amphibians and these reptiles are highly adapted to suit their particular lifestyles, specifically their intermittent breathing and relatively low metabolic rate. By being able to shunt blood from pulmonary to systemic circuits, these animals can save the energy that would otherwise be wasted in sending blood to the temporarily nonventilated lungs.
- People often fail to appreciate that the human heart is two distinct pumps—the right heart pumping blood to the lungs and the left heart pumping blood to the body. The two pumps just happen to be packaged in the same structure—the heart. In fact, cardiologists refer to the "left heart" and "right heart," emphasizing the separate nature of these two pumps. Understanding the flow of blood through the human heart becomes easier when it is considered from this perspective. Also, determining the pattern of blood flow through the human heart will be easier if you mentally "unfold" the circulation into a great circle, in which there are two pumps at opposite sides of the circle with a vascular bed intervening between each pump. Once you have mastered this concept, you will see that the only way to place the two pumps in proximity in the same structure (the heart) is to twist the circle into a folded figure eight—the typical diagram of the circulation in the textbook.
- Following from the preceding, learn the various chambers and valves in the context of a sequence of structures in each of the right and left pumps of the heart, rather than as a list of terms printed on a complex, anatomically accurate diagram of the heart.
- Go to yourBioPortal.com to review the following tutorials and activities:

  Animated Tutorial 38.1 The Cardiac Cycle

  Interactive Tutorial 38.1 Blood Pressure and Heart Rate Regulation

  Web Activity 38.1 Vertebrate Circulatory Systems

  Web Activity 38.2 The Human Heart

  Web Activity 38.3 Structure of a Blood Vessel

Key Concept Review

38.1 Circulatory Systems Can Be Open or Closed

Open circulatory systems move extracellular fluid

Closed circulatory systems circulate blood through a system of blood vessels

An animal's circulatory system transports nutrients and wastes to and from tissues and cells. In the smallest multicellular animals, every cell within the body is no more than one to two cells from the environment. Such small animals do not require a circulatory system because materials can move efficiently into and out of cells by diffusion alone. Some aquatic invertebrates have a gastrovascular system, which consists of a highly branched central cavity where exchange of gases and nutrients occurs. Large, active animals need a circulatory system to carry wastes, gases, and nutrients to and from all the cells in the body.

Circulatory systems may be classified as open or closed. In an open circulatory system, blood and other tissue fluids are not separated; the fluid is called hemolymph (see Figure 38.1A). A pump may be present to help move hemolymph through the various large compartments, assisted by movements of the animal. In a closed circulatory system, blood is pumped through a series of interconnected closed vessels and is kept separate from interstitial fluid (see Figure 38.1B). There are numerous advantages to closed systems compared to open ones: more rapid transport of fluid in the closed vessels, the ability to divert the flow of blood between different tissues, and the ability to direct hormone and nutrient transport to the tissues.

Question 1. What is the difference between interstitial fluid and extracellular fluid in a closed circulatory system?
Textbook Reference: *38.1 Circulatory Systems Can Be Open or Closed, p. 747*

Question 2. What are three advantages of a closed circulatory system over an open system?
Textbook Reference: *38.1 Circulatory Systems Can Be Open or Closed, p. 748*

38.2 Circulatory Systems May Have Separate Pulmonary and Systemic Circuits

- Most fishes have a two-chambered heart and a single circuit
- Lungfishes evolved a partially separate circuit that serves the lung
- Amphibians and most reptiles have partially separated circuits
- Crocodilians, birds, and mammals have fully separated pulmonary and systemic circuits

A fish pumps blood with a heart that has one atrium and one ventricle. Blood leaves the heart and enters the gills, where gas exchange occurs. Some arterial blood pressure is lost in passing through the gills. Blood leaving the gills collects in the dorsal aorta and travels to the systemic organs and tissues, eventually returning to the heart. African lungfish have a well-vascularized lung for gas exchange that is formed embryonically from an outpocketing of the gut. A heart with a partially divided atrium helps separate oxygenated blood returning from the lung and deoxygenated blood returning from the tissues as they enter the single ventricle. The lungfish heart and vascular system exhibit adaptations that partially separate systemic and pulmonary circuits.

Adult amphibians have two atria: one that receives deoxygenated blood from the tissues and one that receives oxygenated blood from the lungs. Anatomical features of the ventricle help direct deoxygenated blood preferentially to the lungs and oxygenated blood to the systemic body tissues. Even though the two circuits are not fully divided, blood tends to flow in parallel.

Turtles, snakes, and lizards have two atria and a ventricle that is internally divided into subchambers that allow for control of the distribution of blood to the lungs and body. During air breathing, blood from the right side of the ventricle is preferentially pumped to the lungs because there is a higher resistance in the systemic circuit than in the pulmonary circuit. When one of these reptiles stops breathing, vessel constriction in the lungs causes pulmonary circuit resistance to increase, causing blood from the right side to largely bypass the pulmonary circulation and flow to the systemic circuit.

Birds and mammals have a four-chambered heart that allows for complete separation and no mixing of systemic and pulmonary blood. This cardiovascular arrangement maximizes O_2 transport by eliminating mixing between pulmonary and systemic circulations. The pulmonary circulation operates at low pressure, protecting the delicate lung membranes, whereas the systemic circuit operates at higher pressure, allowing the perfusion of tissues and organs distant from the heart.

Alligators and crocodiles have a four-chambered heart, resembling that of a bird or mammal. However, alligators and crocodiles retain a left aorta that arises from the right side of the heart. Thus, crocodiles and alligators can operate their heart like that of a mammal, distributing blood equally to the pulmonary and systemic circulations, or they can partially bypass the lungs by directing blood from the left aorta into the right aorta and on to the systemic circulation.

Question 3. What are the important differences between the circulatory system of a fish and that of a turtle?
Textbook Reference: *38.2 Circulatory Systems May Have Separate Pulmonary and Systemic Circuits, pp. 748–749*

Question 4. In crocodilians, which circuit—pulmonary or systemic—would receive the most blood during a long dive?
Textbook Reference: *38.2 Circulatory Systems May Have Separate Pulmonary and Systemic Circuits, pp. 749–750*

38.3 A Beating Heart Propels the Blood

Blood flows from right heart to lungs to left heart to body

The heartbeat originates in the cardiac muscle

A conduction system coordinates the contraction of heart muscle

Electrical properties of ventricular muscles sustain heart contraction

The ECG records the electrical activity of the heart

The human heart has two atria and two ventricles (see Figure 38.2). Between each atrium and ventricle an atrioventricular valve stops backflow of blood from the ventricle into the atrium. The pulmonary valve resides between the right ventricle and the pulmonary artery and the aortic valve resides between the left ventricle and the aorta; these two valves help maintain one-way flow through the heart.

Deoxygenated blood flows from the body through the superior vena cava and inferior vena cava into the right atrium and the right ventricle by passive filling during heart relaxation. During heart contraction, blood flows from the right ventricle to the pulmonary artery and the lungs, where blood becomes oxygenated. Oxygenated blood from the lungs flows back to the heart via pulmonary veins into the left atrium, then into the left ventricle, and finally out through the aorta to the systemic tissues. Resistance in the blood vessels of the body's systemic tissues is higher than in the lungs, so the left ventricle must be larger and more muscular than the right ventricle to generate more force while pumping the same volume of blood.

Systole is the contraction of the ventricle, and diastole is the relaxation of the ventricle. At the very end of diastole, the atria contract. Together these two phases make up the cardiac cycle. During the cardiac cycle the pressure is highest during ventricle contraction (systole) and lowest during relaxation (diastole).

The human heartbeat is generated in the heart's pacemaker. The pacemaker, or sinoatrial node, is a small section of highly modified muscle tissue located at the junction of the superior vena cava and right atrium. The autonomic nervous system also plays a role in the control of the heartbeat. Sympathetic activity increases heart rate, and parasympathetic activity decreases heart rate (see Figure 38.4). The pacemaker rhythmically produces action potentials that pass on to the rest of the heart and stimulate a heartbeat (see Figure 38.5).

Gap junctions connect cardiac muscle cells and thus allow the action potentials to spread rapidly from cell to cell. The action potential moves from the atria to the ventricles through the atrioventricular node (AV node). The AV node slows down the transmission of the action potential from the atria to the ventricles. The action potential passes rapidly down the bundle of His and to the Purkinje fibers, fanning out in the ventricular tissue. From the Purkinje fibers the action potential spreads through both ventricles, causing contraction to occur slightly after the contraction of the atria.

The electrocardiogram (ECG) measures the electrical activity of the heart as a complex wave pattern (see Figure 38.6). The wave has five parts: P (depolarization and contraction of the atria), Q, R, and S (depolarization and contraction of the ventricles), and T (relaxation and repolarization of the ventricles).

Question 5. The human heart pumps blood to the pulmonary and the systemic circuits. Describe the path that blood takes as it moves through the circulatory system, starting at the right ventricle.
Textbook Reference: *38.3 A Beating Heart Propels the Blood, pp. 750–751*

Question 6. What happens if a problem arises with the heart's conduction system?
Textbook Reference: *38.3 A Beating Heart Propels the Blood, pp. 753–754*

Question 7. The diagrams below trace the course of the heartbeat. Label the following structures on the left diagram: sinoatrial node, atrioventricular node, bundle of His, Purkinje fibers, atria, and ventricles. On the middle and right diagrams, indicate with arrows the path of the action potentials at the particular stage indicated at the top of each diagram.
Textbook Reference: *38.3 A Beating Heart Propels the Blood, p. 754*

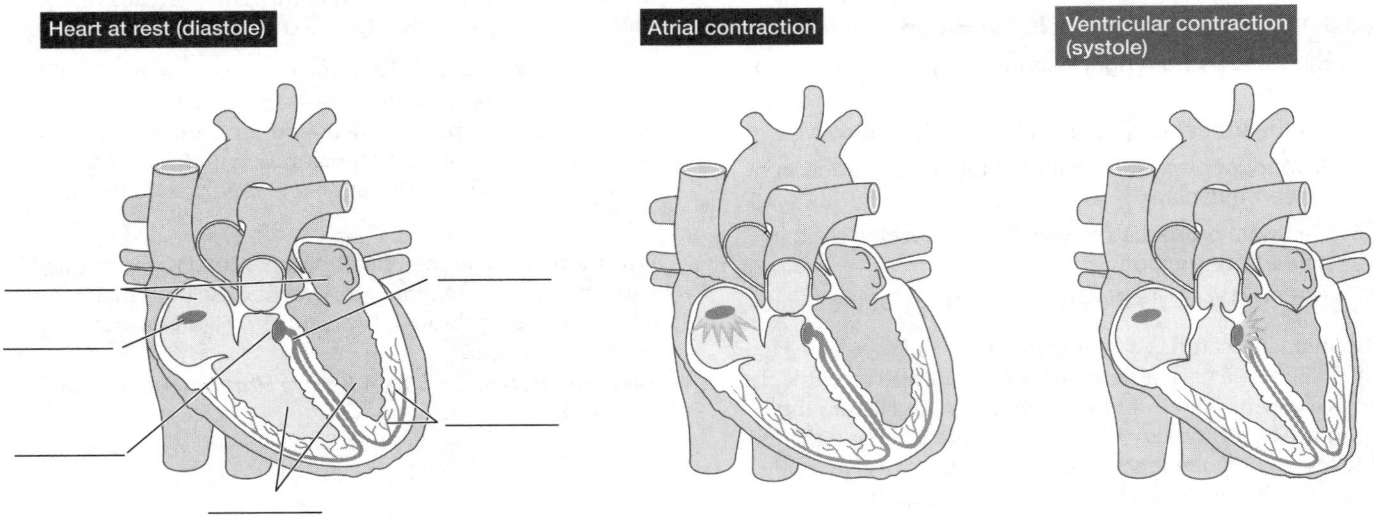

38.4 Blood Consists of Cells Suspended in Plasma

Red blood cells transport respiratory gases

Platelets are essential for blood clotting

The fluid matrix of blood is known as plasma and contains water, salts, and proteins. The hematocrit is the percentage of blood volume composed of cells. A normal hematocrit is about 42 percent for women and 46 percent for men. Plasma transports many gases, ions, nutrients, proteins, and other molecules. The plasma also contains the hormones that are circulated by the blood to target cells.

Several different cell types in blood carry out a wide range of functions (see Figure 38.7). The cell types include the erythrocytes, basophils, eosinophils, neutrophils, lymphocytes, and monocytes. Platelets are cell fragments found in the blood.

Erythrocytes, or red blood cells, are shaped like biconcave discs when mature. Their function is to transport respiratory gases. In mammals, but not other vertebrates, mature red blood cells lack a nucleus and other organelles. Erythrocytes are produced in stem cells located in the marrow of bone. Once released into the bloodstream, they survive about 120 days. The hormone erythropoietin is produced in the kidneys and regulates the production of red blood cells. The spleen serves as a major site of destruction of old red blood cells.

Platelets, which help to start blood clotting, are also found in blood and are produced by the breaking off of cell fragments from megakaryocytes in the bone marrow. Platelets bind to the edges of a break in a vessel wall, not only providing a mechanical plug but also starting a chemical chain reaction that leads to the conversion of the circulating protein prothrombin into thrombin. Thrombin stimulates the formation of sticky fibrin threads that form a meshwork and cover the hole in the vessel (see Figure 38.8).

Question 8. The rate of production of red blood cells in the body usually matches the rate of destruction, but under certain circumstances the body speeds up the rate of production. Under what circumstances would it do so, and how is this accomplished?
Textbook Reference: *38.4 Blood Consists of Cells Suspended in Plasma, pp. 755–756*

Question 9. A chronically fast heart rate is one of the symptoms that would lead a physician to test for anemia. Why?
Textbook Reference: *38.4 Blood Consists of Cells Suspended in Plasma, pp. 755–756*

Question 10. Some athletes engage in the practice of "blood doping," which is designed to increase the number of red blood cells and hence the blood's ability to carry oxygen. One approach is to remove blood and store it. The body responds to such removal by increasing its production of red blood cells, and the stored blood is then returned to the body. Design a test that would detect this practice.
Textbook Reference: *38.4 Blood Consists of Cells Suspended in Plasma, pp. 755–756*

38.5 Blood Circulates through Arteries, Capillaries, and Veins

Arteries have elastic, muscular walls that help propel and direct the blood

Capillaries have thin, permeable walls that facilitate exchange of materials with interstitial fluid

Blood flow through veins is assisted by skeletal muscles

Lymphatic vessels return interstitial fluid to the blood

Arteries carry blood away from the heart to the body under relatively high pressure. Veins carry blood toward the heart from the body under low pressure (see Figure 38.9). The very fine capillaries interconnect the arteries with the veins in the tissues, dissipate pressure, and are sites for all exchanges of gases, nutrients, and wastes.

Arteries and the smaller arterioles are elastic, allowing them to stretch during systole and rebound during diastole. In vertebrates, the walls of vessels contain smooth muscle that controls the vessel's resistance by causing changes in its diameter. Arteries and arterioles are known as resistance vessels. Blood pressure can be measured using a sphygmomanometer. A typical reading for a healthy young adult is 120 (systolic pressure) over 80 (diastolic pressure).

Capillaries connect the arterioles with the venules in the tissues. Capillaries have a large cross-sectional area compared with arteries and veins and extremely thin walls to allow the diffusion of gases, nutrients, and wastes (see Figure 38.9). Blood moves slowly as it enters the capillaries. In a process known as filtration, blood pressure from the arterial side forces fluids, ions, and small molecules out of the capillaries through small holes called fenestrations. Near the venule end of the capillaries, the difference in osmotic potential between the plasma inside the capillaries and the surrounding fluids creates an osmotic pressure resulting in fluid recovery back into the capillaries. Thus the net flow of water between the plasma and tissue fluid is determined by the difference in blood pressure and osmotic pressure (see Figure 38.11). The two opposing forces, blood pressure and osmotic pressure, are known as Starling's forces. Edema is tissue swelling that results from the accumulation of fluid in extracellular spaces. This condition is associated with protein starvation (which leads to low concentrations of proteins in the blood) and inflammation (which causes increased permeability of capillaries). Some capillaries are selective about which ions and molecules can pass through them, whereas others are less selective. The blood–brain barrier consists of highly selective brain capillaries.

Blood returns to the heart at low pressure in expandable veins. Because of the low pressure, veins that act against gravity have valves, and blood flowing toward the heart is helped by the movement of the surrounding skeletal muscles (see Figure 38.12). Gravity can cause pooling of blood in the veins if a person is inactive for a long period of time.

Once interstitial fluid enters the lymphatic system, it is called lymph. In humans, lymph is moved through the lymphatic vessels up the body by surrounding skeletal muscle contractions. Lymph is returned to the blood through the thoracic ducts that empty into large veins at the base of the neck. Lymph nodes exist in mammals and birds and contribute to the defenses of the body. Foreign materials and microorganisms are removed by the phagocytic action of white blood cells in the lymph nodes.

Question 11. Blood moves through the arteries and veins at a faster velocity than the blood moving through the capillaries. What is the cause and effect of blood slowing down as it moves through the capillaries?
Textbook Reference: *38.5 Blood Circulates through Arteries, Capillaries, and Veins, pp. 758–759*

Question 12. Cigarette smoke contains nicotine, which increases heart rate and causes blood vessels to constrict. Would these changes cause increased or decreased blood pressure?
Textbook Reference: *38.5 Blood Circulates through Arteries, Capillaries, and Veins, p. 757*

38.6 Circulation Is Regulated by Autoregulation, Nerves, and Hormones

Blood pressure is determined by heart rate, stroke volume, and peripheral resistance

Autoregulation matches local blood pressure and flow to local need

Sympathetic and parasympathetic nerves affect blood pressure

Many hormones affect blood pressure

Blood flow in the capillary beds of many tissues is controlled locally by autoregulatory mechanisms. Precapillary sphincters at the arterial end of the capillaries control the flow of blood, with contraction of these sphincters limiting or stopping blood flow through the capillary bed (see Figure 38.14). Low levels of O_2 and high levels of CO_2 can open the pre-capillary sphincters. The increased supply of blood brings in more O_2 and removes more CO_2.

Norepinephrine from sympathetic neurons and epinephrine released from the adrenal medulla cause arteries and arterioles to contract, reducing blood flow and increasing central blood pressure. Autonomic control of heart rate and constriction of blood vessels occurs from the medulla. Stretch receptors (known as baroreceptors) in the aorta and carotid arteries help regulate blood pressure. With rising blood pressure they inhibit sympathetic output and cause the heart to slow and arterioles to dilate. When pressure is low, sympathetic output is stimulated, increasing heart rate and constriction of arterioles. Chemoreceptors in the aorta and carotid arteries sense changes in blood composition.

In response to a fall in arterial pressure (signaled by decreased activity of baroreceptors), the posterior pituitary releases antidiuretic hormone (ADH, also known as vasopressin). ADH stimulates the kidneys to reabsorb more water, and this results in increased blood volume and pressure.

Question 13. You are about to take a final exam in your freshman biology course and are very nervous. What responses will your circulatory system have to your emotional state?
Textbook Reference: *38.6 Circulation Is Regulated by Autoregulation, Nerves, and Hormones, p. 762*

Question 14. How do precapillary sphincters and arterioles react to low oxygen, high carbon dioxide, and high lactate in tissue? How are total peripheral resistance and mean arterial pressure affected by the actions of the precapillary sphincters and arterioles?
Textbook Reference: *38.6 Circulation Is Regulated by Autoregulation, Nerves, and Hormones, pp. 761–762*

Test Yourself

1. Which of the following is *not* one of the reasons that closed circulatory systems are more efficient than open circulatory systems?
   a. Closed systems rely exclusively on simple diffusion for transport, whereas open systems rely on pumping mechanisms.
   b. Transport within closed systems is more rapid than in open systems.
   c. Blood can easily be directed to specific areas in closed systems, but not in open systems.
   d. Compared to open systems, closed systems operate better under higher pressure.
   e. In closed systems molecules and cells that transport hormones and nutrients can be kept in vessels until they unload their goods at specific tissues.
   Textbook Reference: *38.1 Circulatory Systems Can Be Open or Closed, p. 748*
2. When snakes, lizards, and crocodilians stop breathing,
   a. blood is shunted from the systemic circuit to the pulmonary circuit.
   b. blood is shunted specifically to the brain.
   c. blood is shunted from the pulmonary circuit to the systemic circuit.
   d. blood is shunted to the skin for gas exchange.
   e. the pattern does not change between breathing and nonbreathing periods.
   Textbook Reference: *38.2 Circulatory Systems May Have Separate Pulmonary and Systemic Circuits, p. 749*
3. In which of the following would the highest blood pressure be recorded?
   a. In the ventricle supplying blood to the gills of a fish
   b. In the anterior dorsal artery of an ant
   c. In the pulmonary vein of a frog
   d. In the ventricle supplying blood to the systemic circuit of a bird
   e. In the vessels leaving the gills of a mollusk
   Textbook Reference: *38.2 Circulatory Systems May Have Separate Pulmonary and Systemic Circuits, p. 749*
4. Blood consists of a fluid fraction of _______ and a solid fraction of _______.
   a. plasma; water, erythrocytes, and platelets
   b. erythrocytes; leukocytes, macrophages, and platelets
   c. plasma; erythrocytes, platelets, and leukocytes
   d. leukocytes; erythrocytes and platelets
   e. interstitial fluid; stem cells
   Textbook Reference: *38.4 Blood Consists of Cells Suspended in Plasma, p. 756*
5. Red blood cells are
   a. biconcave cells containing hemoglobin.
   b. spherical cells containing hemoglobin.
   c. spherical cells capable of amoeboid motion and containing hemoglobin.
   d. biconcave cells containing platelets.
   e. biconcave cells that function in the immune response.
   Textbook Reference: *38.4 Blood Consists of Cells Suspended in Plasma, p. 755*
6. The blood clotting cascade results in the conversion of _______ to _______.
   a. vitamin K; prothrombin
   b. fibrin; fibrinogen
   c. thrombin; prothrombin
   d. prothrombin; thrombin
   e. fibrinogen; thrombin
   Textbook Reference: *38.4 Blood Consists of Cells Suspended in Plasma, pp. 756–757*

7. Which of the following regions of the vascular bed is the actual site of gas exchange with surrounding tissue?
   a. Arteries
   b. Capillaries
   c. Lymphatic vessels
   d. Veins
   e. Venules

   Textbook Reference: *38.5 Blood Circulates through Arteries, Capillaries, and Veins, pp. 758–759*

8. Which of the following represents the correct sequence in which cardiac action potentials pass through the heart?
   a. Purkinje fibers, AV node, SA node, bundle of His, atrial fibers
   b. AV node, atrial fibers, SA node, bundle of His, Purkinje fibers
   c. SA node, bundle of His, atrial fibers, AV node, Purkinje fibers
   d. Purkinje fibers, bundle of His, AV node, atrial fibers, SA node
   e. SA node, atrial fibers, AV node, bundle of His, Purkinje fibers

   Textbook Reference: *38.3 A Beating Heart Propels the Blood, pp. 753–754*

9. The AV node _______, and the Purkinje fibers _______.
   a. creates simultaneous atrial and ventricular depolarization; speed up transmission of the cardiac impulse into the ventricle
   b. delays ventricular depolarization relative to atrial depolarization; insulate the cardiac impulse from the general ventricular fibers
   c. delays ventricular depolarization relative to atrial depolarization; ensure the rapid and even spread of the cardiac impulse throughout the ventricles
   d. delays atrial depolarization relative to ventricular depolarization; transmit the cardiac impulse to very small, localized groups of ventricular fibers
   e. initiates each heartbeat; transmit impulses to the atria

   Textbook Reference: *38.3 A Beating Heart Propels the Blood, pp. 753–754*

10. The atrial walls are _______ the ventricular walls, and pressure generated in the atrial chambers is _______ the pressure in the ventricles.
    a. thinner than; higher than
    b. thinner than; lower than
    c. thicker than; higher than
    d. thicker than; lower than
    e. the same thickness as; the same as

    Textbook Reference: *38.3 A Beating Heart Propels the Blood, pp. 750–751*

11. The left ventricle exceeds the right ventricle in the
    a. amount of blood that enters during heart contraction.
    b. volume expelled during contraction.
    c. pressure developed during contraction.
    d. speed with which it contracts.
    e. All of the above

    Textbook Reference: *38.3 A Beating Heart Propels the Blood, p. 750*

12. Which of the following structures of the human lymphatic system acts primarily as a filter for detecting and destroying microorganisms in lymph traveling through major lymph vessels?
    a. Lymph nodes
    b. Thymus
    c. Lymph capillaries
    d. Tonsils
    e. Thoracic ducts

    Textbook Reference: *38.5 Blood Circulates through Arteries, Capillaries, and Veins, p. 760*

13. The net loss of fluid from blood capillaries increases if
    a. plasma filtration decreases.
    b. the osmotic pressure of plasma increases.
    c. blood pressure increases in the capillaries.
    d. the osmotic pressure of interstitial fluid decreases.
    e. blood pressure drops below osmotic pressure.

    Textbook Reference: *38.5 Blood Circulates through Arteries, Capillaries, and Veins, p. 759*

14. Which of the following statements about the control of circulation in humans is *false*?
    a. Sympathetic nerve input to skeletal muscle causes the blood vessels in the muscle to dilate.
    b. Blood flow can be regulated by autonomic nerve signals emanating from cardiovascular control centers in the medulla of the brain.
    c. Carotid artery chemoreceptors detect low oxygen levels in the blood and promote increased blood pressure.
    d. Hormones such as angiotensin and vasopressin cause veins and venules to constrict.
    e. Autoregulatory mechanisms can cause constriction or dilation of arterioles.

    Textbook Reference: *38.6 Circulation Is Regulated by Autoregulation, Nerves, and Hormones, pp. 762–763*

15. Which of the following statements about vertebrate circulatory systems is *false*?
    a. Crocodilians have a four-chambered heart.
    b. In birds and mammals, pressures in the pulmonary circuit are lower than those in the systemic circuit.
    c. The ventricles of turtles, snakes, and lizards are partly divided by a septum.
    d. Amphibians have two atria and one ventricle.
    e. In fishes, blood passes through the gills, returns to the heart, and then is pumped to the body.

    Textbook Reference: *38.2 Circulatory Systems May Have Separate Pulmonary and Systemic Circuits, pp. 748–750*

Answers

Key Concept Review

1. Interstitial fluid is the fluid around the cells. Extracellular fluid includes the interstitial fluid plus the blood plasma.
2. Three advantages of closed circulatory systems over open systems are: (1) the circulatory fluid can flow more rapidly; (2) the diameters of vessels can be changed to control the flow of blood to particular tissues; and (3) specialized cells and molecules that transport oxygen, hormones, and nutrients are kept in the vessels and can be directed to particular locations.
3. Fish have a two-chambered heart (one atrium and one ventricle), with blood flowing from the heart of the fish to the gills and then directly to the body. Turtles have a more complex three-chambered heart (two atria and one ventricle). The ventricle, however, is partially divided, so outflow can be directed to the pulmonary circuit and to the systemic circuit. Thus, the circulatory system of a fish delivers blood to the respiratory organ and the body tissues in succession, whereas the circulatory system of a turtle delivers blood to the respiratory organ and the body tissues in parallel.
4. During a dive, crocodilians are not breathing, so they shunt blood away from their lungs (the pulmonary circuit) and toward the systemic circuit.
5. Blood from the right ventricle exits through the pulmonary valve into the pulmonary artery and then goes to the lungs. From the lungs, the blood returns to the left atrium via the pulmonary veins and passes through the atrioventricular valve into the left ventricle. Blood from the left ventricle is pumped to the systemic circuit through the aortic valve and into the aorta. From the aorta, blood moves to the tissues in arteries and arterioles, passes through tissues in the capillaries, and returns via the venules, veins, and ultimately the superior vena cava and inferior vena cava to the right atrium of the heart. From the right atrium, blood is pumped back where we started—into the right ventricle through the atrioventricular valve.
6. If the heart's conduction system is faulty, then cells can begin to contract independently. This cellular independence can cause rapid, irregular contractions of the ventricles, which in turn, could make the ventricles useless as pumps and circulation would stop.
7.

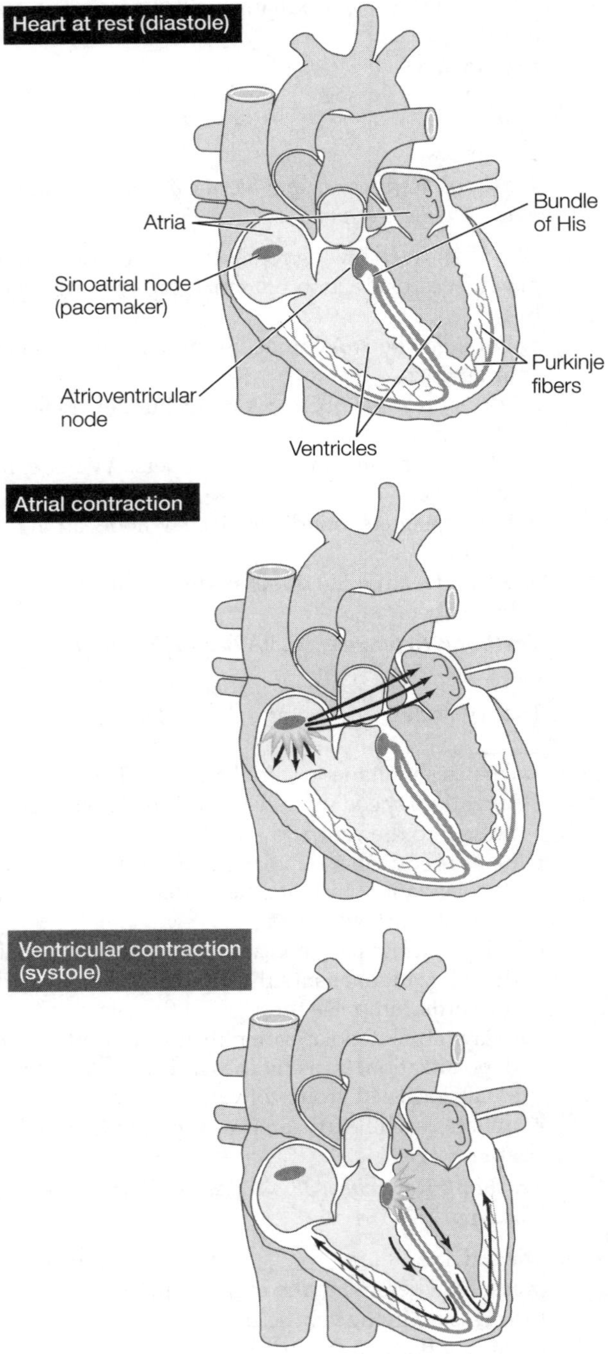

8. During times of blood loss, the body might speed its rate of red blood cell production. In response to a decreased supply of oxygen to the cells, the kidneys release erythropoietin, a hormone that extends the lives of mature red blood cells and stimulates production of new red blood cells in the bone marrow.
9. A person with anemia has an insufficient number of oxygen-transporting red blood cells. An anemic person's heart rate will therefore become faster to compensate for the blood's reduced ability to carry oxygen.
10. Removal of blood should prompt the kidneys to release erythropoietin, which stimulates production of red blood cells. Thus, one possible test for blood doping would be to test an athlete's urine for the presence of erythropoietin. Unusually high numbers of red blood cells would also suggest blood doping. This could be determined by measuring the hematocrit, the percentage of blood volume made up by cells (which are primarily red blood cells). The athlete's hematocrit could be compared to the normal hematocrit for women (42%) or for men (46%).
11. Blood is pumped from the heart into the arteries, where it has a relatively fast velocity. Upon entering the capillaries, much of the velocity is lost. The capillarie have a much higher total cross-sectional area than the arteries, and blood flow velocity is inversely proportional to the cross-sectional area; thus, the more area through which the blood flows, the more slowly it will flow. Just as water in a stream that widens to become a river flows more quickly in the stream than it does in the river, blood moves slowly through the capillaries and more quickly through the arteries and veins. The slow velocity of blood through the capillaries allows for greater exchange of materials with the tissues.
12. Increased heart rate and constriction of blood vessels would cause blood pressure to rise. Thus, cigarette smoking is associated with high blood pressure.
13. Stress and nervousness elicit the fight-or-flight response. Signals from the medullary cardiovascular control center stimulate sympathetic inputs from the autonomic nervous system. In response to sympathetic inputs, the adrenal gland releases epinephrine into the bloodstream. Epinephrine stimulates an increased heart rate and arterial pressure. The blood flow to the smooth muscles decreases due to constriction of blood vessels, and blood is diverted away from areas such as the digestive tract to the skeletal muscle needed for fight or flight.
14. In response to low oxygen, high carbon dioxide, and high lactate, precapillary sphincters would open, increasing blood flow to the area. Arterioles would dilate. These actions would decrease total peripheral resistance and mean arterial pressure.

Test Yourself

1. **a.** Both closed systems and open systems rely on pumping mechanisms (and not just simple diffusion) to distribute fluid throughout the body.
2. **c.** During nonbreathing periods, the blood is shunted from the pulmonary circuit to the systemic circuit. The resistance in the pulmonary circuit increases, resulting in the shunt.
3. **d.** Of the vertebrate groups, mammals and birds tend to have the highest blood pressure. Therefore, the ventricle supplying blood to the systemic circuit of a bird has the highest pressure.
4. **c.** The fluid portion of blood is the plasma; three components of the solid fraction are erythrocytes, platelets, and leukocytes.
5. **a.** Red blood cells are biconcave cells that contain the oxygen-binding hemoglobin.
6. **d.** Blood clotting involves the conversion of prothrombin to thrombin.
7. **b.** Gas exchange at the tissues occurs across the capillaries.
8. **e.** The cardiac action potential passes through the SA node, atrial fibers, AV node, bundle of His, and finally the Purkinje fibers.
9. **c.** The AV node delays the ventricular depolarization relative to atrial depolarization, so atrial contraction occurs before ventricular contraction. The Purkinje fibers ensure that the action potential spreads rapidly and evenly throughout the ventricles.
10. **b.** The atrium has thinner walls and generates lower pressure than the ventricles.
11. **c.** The pressure generated by the left ventricle in the blood flowing to the systemic circuit is greater than the pressure generated by the right ventricle in the blood flowing to the pulmonary circuit.
12. **a.** The lymph nodes filter and destroy microorganisms that are traveling through the lymphatic system.
13. **c.** Changes in blood pressure will change the amount of filtration that occurs at the capillaries. Thus, increases in blood pressure will result in greater rates of filtration.
14. **d.** Angiotensin and vasopressin control constriction of the arteries and arterioles, not the veins and venules.
15. **e.** In fishes, blood passes through the gills and directly out to the body; it does not return to the heart for additional pumping.

Nutrition, Digestion, and Absorption

39

The Big Picture

- Animals have evolved diverse mechanisms for acquiring needed energy and raw materials for metabolism. All animals are heterotrophs; they can achieve energy and nutrient input by eating plants, animals, or both.
- The digestive system is specialized for efficient digestion of the type of food it must break down. Carnivores have relatively short guts because meat is easier than plant material to digest. The longer guts of herbivores have numerous holding chambers for the laborious process of breaking down plant material. The teeth of carnivores and herbivores also reflect dietary differences.
- Food passing down the length of the gut is first broken down mechanically into small pieces and then broken down chemically into small molecules that can be absorbed across the gut. As nutrients and water are absorbed from ingested plant and animal material, the remaining material forms the feces, which are eventually eliminated from the body.
- The process of digestion is under both neural and hormonal control. The brain plays a role in regulating food intake, and the digestive tract has an intrinsic nervous system to regulate digestive functions. Numerous hormones control the peristaltic movements of the gut and the secretion of enzymes and other chemicals involved in digestion. Hormones also regulate food intake.

Study Strategies

- The study of digestive function may seem uninteresting or, worse, unpleasant. Yet, some of the best and clearest examples of negative feedback and homeostasis are to be found in the study of digestion.
- You may think that there is a single digestive mechanism for food and be surprised to learn of the complexity of the process. Be sure to recognize that different food components—proteins, carbohydrates, fats—are digested, absorbed, and transported by very different mechanisms.
- Although the guts of vertebrates are structurally complex, they can be divided functionally into a foregut, midgut, and hindgut. If you break down the structure of the gut into these sections and determine which physiological process occurs in each section, you will have divided the task into smaller pieces.
- Because different types of food are digested and absorbed by different mechanisms, learn how a particular type of food is digested and match the appropriate enzymes to that process. For example, it will be much easier to remember how proteins are digested, rather than to recount a long list of digestive enzymes and then try to pick the one that might break down protein. In short, learn the process, not the list!
- Go to yourBioPortal.com to review the following tutorials and activities:

  Animated Tutorial 39.1 The Digestion and Absorption of Fats

  Animated Tutorial 39.2 Insulin and Glucose Regulation

  Interactive Tutorial 39.1 Parabiotic Mice: Regulation of Food Intake

  Web Activity 39.1 Mineral Elements Required by Animals

  Web Activity 39.2 Vitamins in the Human Diet

  Web Activity 39.3 The Human Digestive System

Key Concept Review

39.1 Food Provides Energy and Nutrients

Food provides energy

Excess energy is stored as glycogen and fat

Food provides essential molecular building blocks

Food provides essential minerals

Food provides essential vitamins

Nutrient deficiencies result in diseases

Animals are heterotrophs, meaning that they must get their energy and nutrients from the fats, carbohydrates, and proteins of plants and other animals (see Figure 39.1). Energy is the ability to do work. The energy in food is in the form of chemical energy. The calorie and kilocalorie (Calorie, or Cal) are measures of heat energy.

Metabolic rate is a measure of the total energy used by an animal. Basal metabolic rate is the resting metabolic rate or the energy consumption needed for all essential physiological functions. Energy needed for metabolism can be obtained from food taken in or from stored food. Fats, carbohydrates, and proteins are the components of food that provide energy. Carbohydrates are stored as glycogen in the liver and muscles. Glycogen serves in short-term energy storage. Proteins can be broken down for energy as a last resort. Fat has more energy per gram than carbohydrates and can be stored more compactly, making it the most important form of stored energy in animals.

Essential molecules must be obtained from a food source. A nutrient that is required but that cannot be synthesized by the body is called an essential nutrient. The nutrients needed for survival are categorized as either macronutrients or micronutrients, depending on the quantity needed.

Amino acids are the building blocks of proteins. Some amino acids can be synthesized, but the essential amino acids can be obtained only from outside sources. In humans, the essential amino acids are isoleucine, leucine, lysine, methionine, phenylalanine, threonine, tryptophan, and valine. All eight essential amino acids can be found in meat, eggs, soybean products, and milk. Vegetarian diets should include complementary dietary mixtures, such as grains and legumes, to avoid protein malnutrition (see Figure 39.2). Human infants require four additional amino acids (histidine, tyrosine, cysteine, and arginine). Essential fatty acids, such as linoleic acid, must also be acquired in food and are necessary for producing membrane phospholipids.

An essential mineral is a chemical element required in the diet; examples include calcium, iron, and potassium (see Table 39.1). Calcium is a macronutrient needed for bones, teeth, and proper nerve function and muscle contraction. Insufficient calcium causes osteoporosis, a progressive thinning and weakening of bones. Iron is a micronutrient, an important component of hemoglobin, myoglobin, and certain enzymes in the respiratory chain. Insufficient iron in the diet can cause iron-deficiency anemia, a condition in which there are too few red blood cells.

Vitamins are carbon compounds that are required in small amounts and often function as coenzymes. Fat-soluble vitamins accumulate in the liver, whereas water-soluble vitamins are eliminated in urine. Humans require 13 vitamins (see Table 39.2). Vitamin D is needed to help in the uptake of calcium into the body. The body can synthesize vitamin D as long as the skin is exposed to sunlight; thus, this vitamin must be obtained in the diet only if sun exposure is inadequate. Shortages in vitamins can lead to disease. For example, a shortage of vitamin C can lead to scurvy, a progressive breakdown of connective tissue.

Malnutrition is caused by the lack of any essential nutrient in the diet. Chronic malnutrition leads to a particular deficiency disease. For example, lack of thiamin (vitamin B_1) causes beriberi, a nervous system disorder characterized by extreme fatigue.

Question 1. From a nutritional standpoint, what is the best way to cook a vegetable containing folic acid—boiling it in water or steaming it?
Textbook Reference: *39.1 Food Provides Energy and Nutrients, p. 769*

Question 2. What are some of the likely reasons that basal metabolic rate is higher in human males than in females?
Textbook Reference: *39.1 Food Provides Energy and Nutrients, p. 767*

Question 3. Why would taking mega doses of vitamin A be potentially more risky for your health than taking mega doses of niacin?
Textbook Reference: *39.1 Food Provides Energy and Nutrients, p. 769*

39.2 Digestive Systems Break Down Macromolecules

Simple digestive systems are cavities with one opening

Tubular guts have an opening at each end

Heterotrophs may specialize in different types of food

Hydrolytic enzymes break down proteins, carbohydrates, and fats into their component monomers. During hydrolysis, water is added to break a chemical bond. Digestive enzymes are classified according to the substances they hydrolyze: carbohydrases hydrolyze carbohydrates; proteases hydrolyze proteins; peptidases hydrolyze peptides; lipases hydrolyze fats; and nucleases hydrolyze nucleic acids.

In simple animals, a single opening, functioning as both a mouth and an anus, connects a gastrovascular cavity to the environment. More complex animals have tubular guts with a mouth that takes in food and an anus through which digestive wastes are eliminated (see Figure 39.3). The gut can be divided into three sections: foregut, midgut, and hindgut. In the foregut, food is physically broken up into smaller pieces in preparation for digestion by enzymes. Food enters the mouth, and in some animals it is broken up by grinding mechanisms that may include teeth or mandibles. After it is ground, food moves to a storage chamber, such as the stomach or crop, where digestion may or may not occur. It then enters the midgut or small intestine, which is the primary site of digestion and absorption. In many vertebrates, the parts of the midgut that absorb nutrients have increased surface area in the form of fingerlike projections known as villi, which in turn have microscopic projections called microvilli (see Figure 39.4). The hindgut absorbs water and ions and stores waste until it is expelled from the anus during defecation. The digestive tracts of most animals contain symbiotic bacteria that assist in the digestive process.

Two types of organisms feed on dead matter: saprobes (also called decomposers) absorb nutrients from decaying organic matter, whereas detritivores actively feed on dead matter. Predators feed on other organisms, and there are three main types: herbivores feed on plants, carnivores feed on animals, and omnivores feed on plants and animals. Filter feeders filter small organisms from water. Fluid feeders include mosquitoes (which feed on blood) and hummingbirds (which feed on nectar) as well as many others.

Mammals have four types of teeth: incisors, canines, premolars, and molars. The shapes and organization of mammalian teeth reflect different diets (see Figure 39.5). Because plant tissue is difficult to break down, herbivorous mammals have teeth that are good for clipping, crushing, and grinding. Because carnivorous mammals must capture, hold, and eat other animals, their teeth are adapted for stabbing, holding, ripping, and shredding their prey. Omnivorous mammals have teeth adapted for multiple purposes. Because plants are more difficult to digest than animals, the guts of herbivores tend to be relatively longer than the guts of carnivores and those of herbivores often contain cellulose-digesting bacteria. Characteristics of the human digestive system, including length of gut and types and appearance of teeth, indicate that we are omnivores.

Question 4. Some mammals have very high-crowned cheek teeth (premolars and molars). Would you expect these animals to be carnivores, herbivores, or omnivores?
Textbook Reference: *39.2 Digestive Systems Break Down Macromolecules, p. 772*

Question 5. Distinguish between the stomach and small intestine with respect to location in the gut and function in digestion.
Textbook Reference: *39.2 Digestive Systems Break Down Macromolecules, p. 771*

39.3 The Vertebrate Digestive System Is a Tubular Gut with Accessory Glands

- The vertebrate gut consists of concentric tissue layers
- Digestion begins in the mouth
- The stomach breaks up food and begins protein digestion
- Muscle contractions mix the stomach's contents and push them into the small intestine
- Ruminants have a specialized four-chamber stomach
- The small intestine continues digestion and does most absorption
- Absorbed nutrients go to the liver
- The large intestine absorbs water and ions

Four layers of different cell types form the wall of the gut (see Figure 39.7). The innermost layer is a layer of mucosa that surrounds the cavity of the gut, or lumen. Cells of the mucosal epithelium have secretory functions; some secrete mucus, while others secrete enzymes or hormones. The submucosa, lying under the mucosa, contains blood and lymph vessels that transport nutrients. This layer also contains a network of nerves that regulates gut activities. Smooth muscle surrounds the submucosa in both circular and longitudinal layers. These layers help move the contents of the digestive tract by peristalsis. The peritoneum is a membrane that lines the abdominal cavity and covers the gut and other abdominal organs.

In mammals, food enters the mouth where it is mixed with saliva and chewed; most other vertebrates swallow their food whole, without chewing it. The enzyme amylase, secreted by the salivary glands, begins the breakdown of starch in the mouth. Thus, in mammals, physical and chemical digestion of food begins in the mouth. Once food is swallowed, it passes through the pharynx into the esophagus and then into the stomach. Peristalsis moves food toward the stomach. The enteric nervous system coordinates peristalsis by creating an anticipatory wave of relaxation. As one section of the gut contracts, the smooth muscles of the area just beyond that section relax. The esophageal sphincter helps keep food from reentering the esophagus from the stomach.

The stomach holds food, physically breaks up food, and starts the digestion of proteins (see Figure 39.8). Gastric pits are infoldings of the stomach that contain three types of secretory cells. Chief cells secrete pepsinogen, which is the inactive form of pepsin, the enzyme that breaks down proteins into shorter peptides. Low pH activates the conversion of pepsinogen to pepsin through autocatalysis, a positive feedback process. The secretion of hydrochloric acid by the parietal cells of gastric glands maintains a low pH environment. A third type of cell secretes mucus, which protects the stomach from its other secretions. Ulcers of the stomach are locations where the mucosal lining has been damaged by the bacterium *Helicobacter pylori*. The acidic mixture of partly digested food and gastric juices is known as chyme, which slowly passes into the intestines through the pyloric sphincter.

Ruminants have a large four-chambered stomach (see Figure 39.9). The rumen and reticulum, the first two chambers, contain many microorganisms that break down cellulose. The contents of the rumen are periodically regurgitated into the mouth for rechewing; this process is often described as "chewing the cud." From the rumen, food passes into the omasum, where water absorption occurs, and then into the abomasum, or true stomach. In the abomasum, hydrochloric acid and proteases kill the microorganisms and digest them. Thus, ruminants acquire glucose from the breakdown of cellulose and protein from digesting the microorganisms.

In the small intestine, carbohydrate and protein digestion continue, the digestion of fats begins, and nutrients are absorbed. In humans, the small intestine can be more than six meters in length, with villi and microvilli contributing to a large surface area necessary for the absorption of nutrients. The small intestine is divided into the duodenum (where most digestion occurs), and the jejunum and ileum (where most absorption occurs).

Fat digestion begins in the small intestine with the help of secretions and enzymes from the liver and pancreas. The liver secretes bile through a side branch of the hepatic duct into the gallbladder, where the bile is stored. When lipids are present, bile is secreted by the gallbladder into the duodenum through the common bile duct (see Figure 39.10). Bile molecules have one lipophilic end and one hydrophobic end. The lipophilic ends attach to tiny fat droplets (micelles) leaving the hydrophobic ends sticking out. This increases the exposed surface area of the lipids for digestion by fat-digesting enzymes, the lipases (see Figure 39.11A).

The zymogen trypsinogen is secreted by the pancreas and is converted into active trypsin by enterokinase. The pancreas produces several zymogens that are released through the pancreatic duct, which joins the common bile duct and empties into the duodenum. The zymogens are converted to the active form by trypsin. The pancreas also helps maintain a slightly alkaline pH in the duodenum by secreting bicarbonate ions. This helps neutralize the acidic chyme entering from the stomach.

Final digestion of proteins and carbohydrates occurs among the microvilli of the small intestine. Proteins are broken down into amino acids, and disaccharides are broken down into monosaccharides. Absorption occurs via several methods, including diffusion, facilitated diffusion, active transport, osmosis, and secondary active transport. Fats move through the plasma membranes as fatty acids and monoglycerides (see Figure 39.11B). In the mucosal cells, they are converted into triglycerides and combine with cholesterol and phospholipids to form chylomicrons that pass into the lymphatic system. Bile is not absorbed across the membrane of the duodenum, but is actively reabsorbed in the ileum and returned to the liver.

The hepatic portal vein carries blood from the digestive organs to the liver, where nutrients are either stored or converted to needed molecules. The liver also breaks down toxins and drugs.

Material entering the large intestine has had most of the nutrients removed, but still contains important ions and water. The water and ions are absorbed across the large intestine, leaving behind semisolid feces. In the colon, absorption of too much water can lead to constipation, whereas absorption of too little water can produce diarrhea. The colon often houses symbiotic bacteria. Many vertebrates have a cecum, a dead end sac at the junction of the small and large intestines. In many vertebrates that eat plants, the cecum houses cellulose-digesting bacteria.

Question 6. Create a flow chart of the following organs, indicating the order in which they occur from the start to the end of the human gut: small intestine, anus, esophagus, large intestine, mouth, and stomach. Indicate in which of these organs chemical digestion occurs, and the particular nutrient (e.g., fat, protein) being broken down. Also indicate in which of the organs absorption occurs, and the particular nutrient or substance absorbed.
Textbook Reference: *39.3 The Vertebrate Digestive System Is a Tubular Gut with Accessory Glands, pp. 773–778*

Question 7. Suppose you have just eaten a small pizza with a lot of cheese. Because this food is high in lipids, your digestive system will have to perform specific tasks in order to digest and absorb the fats. What are those tasks, and in what order will they occur?
Textbook Reference: *39.3 The Vertebrate Digestive System Is a Tubular Gut with Accessory Glands, pp. 776, 778*

Question 8. In ruminants, such as cows and bison, microorganisms that break down cellulose are found in two chambers of the greatly enlarged stomach. In other herbivores, including rabbits, the microorganisms that break down cellulose are found in the cecum, a chamber off the large intestine. In which type of herbivore is the absorption of nutrients from cellulose more efficient?
Textbook Reference: *39.3 The Vertebrate Digestive System Is a Tubular Gut with Accessory Glands, pp. 775, 778*

39.4 Food Intake and Metabolism Are Regulated

- Neuronal reflexes control many digestive functions
- Hormones regulate many digestive functions
- Insulin and glucagon regulate blood glucose
- The liver directs the traffic of the molecules that fuel metabolism
- Many hormones affect food intake

The absorptive state is the period after a meal, when food is in the gut. The postabsorptive state is the period when the gut is empty and the body uses its energy reserves.

Neuronal reflexes regulate many aspects of digestion. Some reflexes, such as salivation, swallowing, and defecation, involve the central nervous system (CNS). The digestive tract also has an intrinsic nervous system, called the enteric nervous system, which coordinates movement of food through the gut. The enteric nervous system carries out its activities with limited involvement of the CNS.

Hormones control many digestive functions (see Figure 39.12). When food arrives in the stomach, cells in its wall secrete gastrin, which stimulates stomach secretions and movement. When the acidic chyme arrives in the small intestine from the stomach, the duodenum releases secretin, which stimulates the pancreas to release bicarbonate ions into the small intestine to control pH. Fats and proteins stimulate the mucosa of the small intestine to secrete cholecystokinin, which stimulates the release of bile and pancreatic digestive enzymes.

The pancreatic hormones insulin and glucagon control metabolic fuel use (see Figure 39.13). Glucose can be used as fuel by all body cells. When glucose is low, most cells switch to alternative fuels, such as fats. Cells of the nervous system, however, cannot switch to alternative fuels, so they always need an adequate supply of glucose to function properly. In response to high blood glucose levels, the pancreas releases insulin. Insulin facilitates the diffusion of glucose into cells where it can be used for energy and for the synthesis of fuel storage molecules—glycogen and fat. At postabsorptive blood glucose levels, insulin secretion is diminished, and glycogen and fat are broken down for fuel. At very low levels of blood glucose, the pancreas releases glucagon, which stimulates the liver to break down glycogen and produce glucose for release to the blood. Glucose is produced from lactate and certain amino acids in a process called gluconeogenesis. Diabetes mellitus results from a lack of insulin (type I diabetes) or an inadequate response to insulin (type II diabetes).

When fuel molecules are abundant in the blood (when an animal's body is in the absorptive state), the liver stores them as glycogen or fat. When fuel molecules are low in the blood (when the body is in the postabsorptive state), the liver delivers them back to the blood.

Lipoproteins are particles of fat and cholesterol covered with protein. The protein covering allows the particles to be suspended in water, which means they can be transported by the cardiovascular system. Chylomicrons are large lipoproteins produced by the mucosal cells of the intestine. The liver produces other lipoproteins. Very low-density lipoproteins (VLDLs) contain mostly triglyceride fats, which they transport to cells of adipose tissue. VLDLs are associated with excessive fat deposition and cardiovascular disease. Low-density lipoproteins (LDLs) carry cholesterol around the body for storage or use; they, too, are associated with cardiovascular disease. High-density lipoproteins (HDLs) remove cholesterol from tissues and carry it to the liver for bile synthesis; they are considered "good" lipoproteins.

Regulation of food intake involves the region of the brain known as the hypothalamus, which senses and responds to several digestive hormones. Leptin, a hormone produced by fat cells, provides information to the hypothalamus about the body's fat reserves. Ghrelin, a hormone produced by the stomach, acts on the hypothalamus to stimulate appetite. Peptide YY, produced by the gut, reduces appetite.

Question 9. When an animal's body is in the absorptive state, it digests and absorbs nutrients. When the body is in the postabsorptive state, it uses up stored fuel. What are the main differences in fuel use and control mechanisms during these two periods?
Textbook Reference: *39.4 Food Intake and Metabolism Are Regulated, pp. 778–780*

Question 10. A thin, diabetic child is brought to the emergency room suffering from weakness, anxiety, and disorientation. A blood test reveals extremely low blood glucose. What form of diabetes does this child likely have and how can the symptoms be explained?
Textbook Reference: *39.4 Food Intake and Metabolism Are Regulated, pp. 779–780*

Question 11. You have just discovered a new hormone that may control appetite. How would you experimentally determine its effects in rats?
Textbook Reference: *39.4 Food Intake and Metabolism Are Regulated, pp. 780–781*

Test Yourself

1. Certain amino acids are essential to the diet of animals because they
   a. prevent overnourishment.
   b. are cofactors and coenzymes that are required for normal physiological function.
   c. cannot be synthesized directly.
   d. are needed to make stored fats that are used during hibernation and migration.
   e. are the most important source of stored energy.
   Textbook Reference: *39.1 Food Provides Energy and Nutrients, p. 767*

2. Which of the following statements about sheep digestion is true?
   a. Sheep are saprobes; they engulf food and perform intracellular digestion.
   b. Sheep are autotrophs; they synthesize organic nutrients and perform extracellular digestion.
   c. Sheep are herbivores; they ingest food and perform extracellular digestion.
   d. Sheep are detritivores; they ingest food and perform intracellular digestion.
   e. Sheep are omnivores; they ingest plants and animals and exhibit intracellular digestion.

   Textbook Reference: *39.2 Digestive Systems Break Down Macromolecules, pp. 770, 772*

3. Which of the following protects the walls of the stomach against the action of its own digestive juices?
   a. An antienzyme chemical formed by the gastric glands
   b. The nervous reactions of the lining of the stomach
   c. Control by a center in the medulla of the brain
   d. Mucus covering the stomach's inner surface
   e. Alkaline secretions of the gastric glands that neutralize acidic secretions

   Textbook Reference: *39.3 The Vertebrate Digestive System Is a Tubular Gut with Accessory Glands, p. 774*

4. Chylomicrons are produced in the
   a. mouth.
   b. stomach.
   c. lumen of the small intestines.
   d. epithelial cells of the small intestines.
   e. liver.

   Textbook Reference: *39.3 The Vertebrate Digestive System Is a Tubular Gut with Accessory Glands, p. 778*

5. The gallbladder
   a. produces bile.
   b. is part of the liver.
   c. stores bile produced by the liver.
   d. produces cholecystokinin.
   e. is vestigial in humans.

   Textbook Reference: *39.3 The Vertebrate Digestive System Is a Tubular Gut with Accessory Glands, p. 776*

6. The pancreas
   a. is exclusively an endocrine gland responsible for the production and release of insulin and glucagon.
   b. is exclusively an endocrine gland that produces salivary amylase.
   c. contains villi to increase surface area.
   d. produces urobiligen (a bile pigment).
   e. produces exocrine products involved in chyme digestion.

   Textbook Reference: *39.3 The Vertebrate Digestive System Is a Tubular Gut with Accessory Glands, p. 777*

7. Hydrochloric acid
   a. is secreted by the gastric glands of the liver.
   b. is secreted by the gastric glands of the stomach.
   c. produces a low pH in the small intestine.
   d. promotes the growth of microorganisms in the stomach.
   e. is secreted by salivary glands.

   Textbook Reference: *39.3 The Vertebrate Digestive System Is a Tubular Gut with Accessory Glands, p. 774*

8. Bile produced in the liver is associated with which of the following?
   a. Emulsification of fats into tiny globules in the small intestine
   b. Digestive action of pancreatic amylase
   c. Emulsification of fats into tiny globules in the stomach
   d. Digestion of proteins into amino acids
   e. Emulsification of fats into tiny globules in the large intestine

   Textbook Reference: *39.3 The Vertebrate Digestive System Is a Tubular Gut with Accessory Glands, p. 776*

9. Most of the chemical digestion of food in humans is completed in the
   a. mouth.
   b. appendix.
   c. ascending colon.
   d. stomach.
   e. small intestine.

   Textbook Reference: *39.3 The Vertebrate Digestive System Is a Tubular Gut with Accessory Glands, pp. 774–777*

10. Which of the following does *not* contribute to the large surface area available for nutrient absorption in the small intestines?
    a. Villi
    b. Intestinal length
    c. Microvilli
    d. Bile duct
    e. Surface folds

    Textbook Reference: *39.2 Digestive Systems Break Down Macromolecules, p. 771*

11. Waves of muscle contractions that move the intestinal contents are
    a. caused by contraction of skeletal muscle.
    b. regulated by liver secretions.
    c. called peristalsis.
    d. voluntary.
    e. caused by the relaxation of smooth muscles of the gut in response to stretching.

    Textbook Reference: *39.3 The Vertebrate Digestive System Is a Tubular Gut with Accessory Glands, p. 773*

12. What is the function of enterokinase?
    a. It converts pepsinogen to pepsin.
    b. It converts trypsinogen to trypsin.
    c. It digests proteins.
    d. It activates HCl.
    e. It prompts opening of the lower esophageal sphincter.

    Textbook Reference: *39.3 The Vertebrate Digestive System Is a Tubular Gut with Accessory Glands, p. 777*

13. Cystic fibrosis causes the production of unusually thick mucus, which sometimes blocks the pancreatic duct. As a result, individuals with cystic fibrosis commonly
   a. experience heartburn caused by food backing up into the esophagus.
   b. suffer from disruption of the mechanical digestion of food in the stomach.
   c. are malnourished.
   d. are obese.
   e. have an abnormal secretion of insulin.

   Textbook Reference: *39.3 The Vertebrate Digestive System Is a Tubular Gut with Accessory Glands, p. 777*

14. Which of the following statements about the large intestine is true?
   a. It has almost no bacterial populations.
   b. It contains chyme.
   c. It absorbs much of the water remaining in waste materials.
   d. It is the site of most of the digestive processes.
   e. It receives blood from the digestive tract via the hepatic portal vein.

   Textbook Reference: *39.3 The Vertebrate Digestive System Is a Tubular Gut with Accessory Glands, p. 778*

15. Certain laxatives increase peristalsis in the large intestine, leaving other portions of the gut unaffected. Use of such laxatives
   a. results in loss of body fat.
   b. causes rapid passage of material through the colon.
   c. results in constipation.
   d. increases the absorption of water.
   e. increases the absorption of ions.

   Textbook Reference: *39.3 The Vertebrate Digestive System Is a Tubular Gut with Accessory Glands, p. 778*

16. Which of the following hormones stimulates gluconeogenesis?
   a. Glucagon
   b. Insulin
   c. Estrogen
   d. Secretin
   e. Gastrin

   Textbook Reference: *39.4 Food Intake and Metabolism Are Regulated, p. 780*

17. Which of the following statements about mammalian teeth and digestive tracts is true?
   a. All mammals have canines.
   b. Carnivores have longer guts than herbivores.
   c. The teeth of omnivores are more specialized than the teeth of carnivores and herbivores.
   d. Carnivores have large canines, whereas herbivores have large premolars and molars.
   e. The length of human teeth and the length of the human gut indicate that we are carnivores.

   Textbook Reference: *39.2 Digestive Systems Break Down Macromolecules, p. 772*

18. Which of the following statements about lipoproteins is *false*?
   a. Lipoproteins move fats from storage sites to sites where they are used.
   b. Lipoproteins have a hydrophobic core of cholesterol and fat.
   c. High-density lipoproteins are "good" lipoproteins.
   d. Low-density lipoproteins are "bad" lipoproteins.
   e. The liver is solely responsible for synthesis of lipoproteins.

   Textbook Reference: *39.4 Food Intake and Metabolism Are Regulated, p. 780*

Answers

Key Concept Review

1. Folic acid is a water-soluble vitamin found in vegetables. When vegetables are boiled, the folic acid is lost. Therefore, the best cooking method is steaming.

2. Human males have higher basal metabolic rates than females do because muscle uses more energy than fat does and a male's body typically has more muscle and less fat than a female's. The difference cannot be explained by any potential differences in activity, because the basal metabolic rate is a measure of the minimum energy required at rest.

3. Niacin is a water-soluble vitamin, so excesses are eliminated in the urine. Vitamin A, however, is fat-soluble and when taken in excess, it can accumulate in body fat and reach toxic levels in the liver.

4. Mammals with very high-crowned cheek teeth are likely herbivores, because such teeth are more adapted to the wear and tear of an abrasive diet. Carnivores and omnivores have low-crowned cheek teeth.

5. The stomach is in the foregut and functions primarily in the physical breakdown and storage of food. The small intestine is in the midgut and is the main site for enzymatic digestion of food and nutrient absorption.

6.

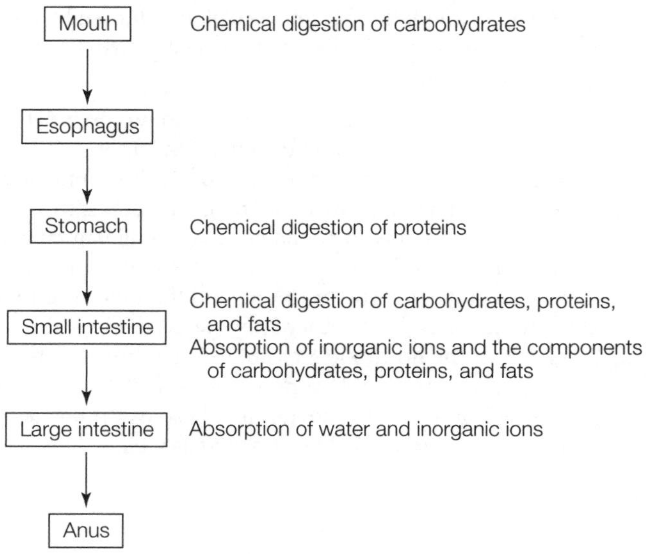

7. The digestion of fats begins in the small intestine. Lipases are water soluble, but lipids tend to aggregate into large droplets in an aqueous environment. Bile helps to increase the exposed surface area of the lipids by preventing the formation of large lipid droplets and promoting small micelles. This allows lipases to break down the fats into free fatty acids and monoglycerides. The fatty acids dissolve into the plasma membrane of the intestinal epithelial cells and are absorbed across this surface. In mucosal cells, the free fatty acids and monoglycerides are turned into triglycerides, which are incorporated into chylomicrons. The chylomicrons can then pass out of the mucosal cells and into the lymphatic system (see Figure 39.11).
8. In vertebrates, most nutrients are absorbed in the small intestine. Because ruminants break down cellulose in chambers of their stomach (which comes before the small intestine in position along the gut), the nutrients from cellulose are available for absorption in the small intestine, and absorption is relatively efficient. In nonruminant herbivores, however, cellulose is broken down in the cecum, after it has passed through the small intestine, so nutrient absorption is less efficient. Some nonruminant species produce two types of feces and eat the one containing cecal material to gain greater access to nutrients.
9. When an animal's body is in the absorptive state, it attempts to store nutrients and fuel. The preferred fuel source at this time is glucose, and insulin is released to help facilitate glucose uptake. When it is in the postabsorptive state, cells tend to metabolize fats, keeping the blood glucose reserves for the nervous system. Insulin levels are low during this period.
10. The child likely has type I diabetes, which is caused by a lack of insulin. Type II diabetes, a condition characterized by an inadequate response to insulin, typically develops later in life and patients tend to be inactive and overweight. The symptoms reflect the fact that the brain cannot switch to alternative fuel sources such as glycogen and fats. The very low blood glucose levels are affecting brain function and causing anxiety and disorientation.
11. There are several options for determining the function of the newly discovered hormone. One approach is to simultaneously monitor levels of the hormone and feeding behavior of the rats. The goal, here, is to look for changes in feeding behavior that parallel fluctuations in the level of the hormone. A second approach is to inject rats with the hormone and then monitor their feeding behavior. Another approach is to block the effects of the hormone, possibly with a drug that temporarily and reversibly suppresses its action, and then to monitor feeding behavior. In all three approaches, body mass would likely be tracked as well.

Test Yourself

1. **c.** Essential amino acids must be acquired through diet because an animal cannot directly synthesize all the amino acids needed for protein production.
2. **c.** Sheep are herbivores and ingest plant material for nutrition. Digestion occurs extracellularly in a sheep's digestive tract.
3. **d.** The stomach is protected from digestive enzymes and low pH by the mucus secreted over its inner surface.
4. **d.** The epithelial cells of the small intestine produce chylomicrons by combining triglycerides with cholesterol and phospholipids, allowing lipids to pass into the lymphatic system.
5. **c.** Bile is produced by the liver, stored in the gallbladder, and released into the small intestine to aid in lipid digestion.
6. **e.** The pancreas produces exocrine products such as lipases, nucleases, amylases, and trypsin, all of which are involved in chyme digestion.
7. **b.** Hydrochloric acid is a strong acid secreted by gastric glands in the lining of the stomach. It lowers the pH of the stomach fluid.
8. **a.** Bile aids in the digestion of lipids in the small intestine by changing fat droplets into small fat particles.
9. **e.** The small intestine is the main site for chemical digestion in humans.
10. **d.** The length of the intestines, surface folds, and the presence of microvilli and villi increase the surface area of the small intestine.
11. **c.** Food is moved through the digestive tract by wavelike contractions of smooth muscle called peristalsis.
12. **b.** Enterokinase converts the zymogen trypsinogen to trypsin.
13. **c.** People with cystic fibrosis often suffer from malnourishment because pancreatic enzymes cannot reach the small intestine to promote the breakdown of complex food molecules. As a result, nutrients are not absorbed in adequate amounts.
14. **c.** The large intestine is the site of water and ion absorption. The large populations of bacteria found in the large intestine contribute useful vitamins to their hosts.
15. **b.** Laxatives that enhance peristalsis in the large intestine cause rapid passage of material through the colon.
16. **a.** Glucagon increases the rate of gluconeogenesis (production of glucose from lactate and certain amino acids) and insulin reduces the rate of gluconeogenesis.
17. **d.** Carnivores have large canines for holding and killing prey animals. Herbivores have large premolars and molars for grinding plant material.
18. **e.** Chylomicrons are lipoproteins synthesized in the cells of the small intestine. Thus, the liver is not the sole producer of lipoproteins.

Salt and Water Balance and Nitrogen Excretion

40

The Big Picture

- Different animals experience different types of stresses related to water and salt balance. Animals living in marine environments have to overcome problems with water loss due to osmosis, whereas animals living in fresh water environments have to overcome problems with water gain due to osmosis. Terrestrial animals face scarcities of water and salts, so both must be conserved.
- All animals produce metabolic wastes (notably nitrogenous waste) as a by-product of metabolism. Invertebrates have evolved a variety of structures to eliminate wastes. Vertebrates have evolved variations on the kidney. The nephron is the "functional unit" of the vertebrate kidney. By eliminating wastes without losing valuable nutrients and salts, the kidneys cleanse the blood of metabolic waste products.
- Kidneys work by differentially processing nitrogenous wastes, ions, and water. In the kidney, large volumes of plasma are filtered from the blood into the interior of the nephron. As the fluid passes through the various specialized regions of the nephron, the desirable components to be retained (specific ions, nutrients, and water) are pulled back into the tissues, leaving behind the waste products in the more concentrated form of urine. As urine leaves the nephron, water content is adjusted in the collecting ducts through hormonal alteration of their water permeability. If the collecting ducts are highly permeable to water, it will be drawn by osmosis out of the collecting ducts into the region of high salt concentration created by the action of the loop of Henle.

Study Strategies

- Depending on whether they are living in fresh water or salt water, fishes may excrete copious amounts of dilute urine or small amounts of more concentrated urine. Because both environments create osmoregulatory problems, think about whether the tissues have higher or lower solute concentration relative to the surrounding water, and predict water and ion movements on that basis.
- Understanding how the nephron functions, and in particular how the loop of Henle works as a "countercurrent multiplier system," is one of the greater challenges in this chapter. This challenging topic can be understood more easily if you imagine a single Na^+ ion in the blood as it enters the nephron. Mentally follow its pathway—and its potential pathways—as it traverses the nephron, eventually ending up in the urine. Note especially that it may take the ion some time to pass through the loop of Henle as it leaves the ascending limb, only to recycle back into the descending limb—a process that can be repeated many times before it finally escapes from the countercurrent multiplier. Repeat this imaginary journey from the perspective of a water molecule.
- Go to yourBioPortal.com to review the following tutorials and activities:

  Animated Tutorial 40.1 The Mammalian Kidney

  Interactive Tutorial 40.1 Kidney Regulation

  Web Activity 40.1 Annelid Metanephridia

  Web Activity 40.2 The Vertebrate Nephron

  Web Activity 40.3 The Human Excretory System

  Web Activity 40.4 The Major Organ Systems

  Working with Data 40.1 What Kidney Characteristic Determines Urine Concentrating Ability?

Key Concept Review

40.1 Excretory Systems Maintain Homeostasis of the Extracellular Fluid

Excretory systems maintain osmotic equilibrium

Animals can be osmoconformers or osmoregulators

Animals can be ionic conformers or ionic regulators

Excretory systems perform the following four functions: (1) regulation of the volume of fluid in the body; (2) regulation of the solute concentration of the extracellular fluid; (3) maintenance of particular solutes at certain concentrations; and (4) elimination of nitrogenous wastes. The excreted water, solutes, and nitrogenous wastes form urine.

Osmolarity is the number of osmoles of solute particles per liter of solvent. Regulation of the osmolarity of extracellular fluid is critical. Differences in osmolarity between the extracellular fluid and the fluid inside cells can lead to changes in cell volume, ultimately causing cells to expand and burst or to shrink and die.

Osmoconformers do not actively regulate the osmolarity of their tissues; rather, they allow them to come to equilibration with their environment. Marine invertebrates tend to be osmoconformers unless the concentrations in the environment are extreme. Osmoregulators actively regulate the osmolarity of their tissues, even as environmental osmolarity changes. The brine shrimp is an osmoconformer over a wide range of salinities, but it becomes an osmoregulator in extremely low- or high-salt environments (see Figure 40.1).

Ionic conformers allow the ionic composition of their extracellular fluid to match that of the environment. Ionic regulators regulate the ionic composition of their extracellular fluid. Typically, hydrogen ion concentration is closely regulated because changes in pH can disrupt homeostasis. Buffers, such as bicarbonate ions found in mammalian blood, help to minimize changes in pH.

Question 1. Why are terrestrial animals always osmoregulators?
Textbook Reference: *40.1 Excretory Systems Maintain Homeostasis of the Extracellular Fluid, p. 785*

Question 2. Osmoconformers occur in marine environments, but not in freshwater environments. Why?
Textbook Reference: *40.1 Excretory Systems Maintain Homeostasis of the Extracellular Fluid, p. 785*

40.2 Excretory Systems Eliminate Nitrogenous Wastes

Animals excrete nitrogen in a number of forms

Most species produce more than one nitrogenous waste

The breakdown of carbohydrates and fats produces carbon dioxide and water; these two end products can be eliminated easily. Proteins and nucleic acids contain nitrogen, so their breakdown produces carbon dioxide, water, and nitrogenous wastes, which can be toxic.

Ammonia (NH_3) is a highly toxic nitrogenous waste that either must be excreted or converted to the less toxic urea and uric acid (see Figure 40.2). Ammonotelic animals excrete ammonia to the aquatic environment, usually across gill membranes. Examples include most bony fishes and aquatic invertebrates. Ureotelic animals excrete nitrogenous waste as urea. Examples include mammals, cartilaginous fishes, and most amphibians. Uricotelic animals, such as birds and other reptiles, excrete nitrogenous waste as uric acid. Insects also are uricotelic. Most species produce more than one nitrogenous waste.

Question 3. Discuss two treatment strategies that could be pursued by a pharmacology lab working on the development of a new medication for gout.
Textbook Reference: *40.2 Excretory Systems Eliminate Nitrogenous Wastes, p. 787*

Question 4. Why do water-breathing animals typically excrete ammonia as their primary nitrogenous waste?
Textbook Reference: *40.2 Excretory Systems Eliminate Nitrogenous Wastes, p. 787*

40.3 Excretory Systems Produce Urine by Filtration, Reabsorption, and Secretion

The metanephridia of annelids process coelomic fluid

The Malpighian tubules of insects depend on active transport

The vertebrate kidney is adapted for excretion of excess water

Mechanisms to conserve water have evolved in several groups of vertebrates.

Invertebrates have diverse excretory systems. Annelid worms, for example, filter blood across the capillaries into the coelom. Each segment of the worm contains a pair of metanephridia, and coelomic fluid enters the metanephridium through a funnel-like opening called a nephrostome. Tubules of the metanephridia actively secrete and absorb various ions and end in nephridiopores, which open to the environment to excrete dilute urine containing nitrogenous wastes and other solutes (see Figure 40.3). Malpighian tubules are the excretory organ of insects. Malpighian tubules actively transport uric acid and sodium and potassium ions from the tissue into the tubules, with water following passively (see Figure 40.4). The tubules lead into the hindgut, where uric acid precipitates and water is reabsorbed. Insects eliminate semisolid matter containing uric acid and other wastes.

Vertebrate excretory organs are kidneys, and nephrons are the functional units of kidneys. The nephron of the vertebrate kidneys is composed of the glomerulus, Bowman's capsule, renal tubule, peritubular capillaries, and other blood vessels (see Figure 40.5).

Three main processes lead to the formation of urine: filtration, tubular reabsorption, and tubular secretion. Filtration occurs at the glomerulus and tubular reabsorption and secretion occur along the renal tubule (see Figure 40.5). There are two capillary beds in a nephron. The glomerulus is the first capillary bed, with blood entering at the afferent arteriole and exiting at the efferent arteriole. From the efferent arteriole the peritubular capillaries of the second capillary bed emerge and surround the tubule component of the nephron. The renal tubule begins with Bowman's capsule, which encloses the glomerulus. Filtration occurs as blood pressure forces water and small solutes from the blood in the glomerulus to the space inside Bowman's capsule. Only ions, small molecules, and water can enter Bowman's capsule to form filtrate; large molecules and cells are prevented from entering. In order to reach the space inside Bowman's capsule, water and small solutes must pass sequentially through: (1) pores (fenestrations) in the endothelial walls of the capillaries in the glomerulus; (2) the basal lamina, a layer just outside the capillary walls; and (3) filtration slits formed by podocytes, which are specialized cells of Bowman's capsule that wrap around the capillaries of the glomerulus.

Arterial blood pressure is the driving force for filtration at the glomerulus. The porous capillary bed of the glomerulus also contributes to the high filtration rate. The filtrate entering the renal tubule is similar to blood plasma, but as the fluid moves through various sections of the tubule, ions and molecules are actively reabsorbed and secreted. This process controls the composition of the urine.

Vertebrates exhibit diverse excretory adaptations. In fresh water, water tends to move into the body by osmosis. To counter this influx, freshwater vertebrates produce large amounts of dilute urine. Marine bony fishes are osmoregulators, maintaining their extracellular fluids at one-third to one-half the osmolarity of seawater. They live in an environment with higher solute concentrations than their body fluids, so water tends to be drawn out of the body by osmosis. To cope with this environment, marine bony fishes produce small volumes of concentrated urine and actively secrete salts across the gills.

Cartilaginous fishes, almost all of which are marine, act as osmoconformers, allowing urea and trimethylamine oxide concentrations to increase, which increases the osmolarity of their tissues. Because the osmolarity of their body fluids is close to that of seawater, they do not lose body water to the environment. Cartilaginous fishes are not ionic conformers; they have a rectal gland to excrete salts.

Amphibians living in or near fresh water have a large water influx, and they produce large amounts of dilute urine in response. In contrast, amphibians living in dry environments have waxy skin that reduces water loss. Some also estivate (burrow underground and reduce metabolic rate) and use their bladders as canteens to store water for later use.

Reptiles, including birds, are amniotes; they lay shelled eggs, which allows them to reproduce in a fully terrestrial environment (i.e., they do not have to return to water to reproduce). They have scaly, dry skin to decrease evaporative water loss, and they excrete nitrogenous wastes as uric acid with little water. Mammals also are amniotes and they display several adaptations for water conservation, including their ability to produce urine that is more concentrated than their blood.

Question 5. The excretory systems of earthworms and humans share many characteristics. Describe some of the ways in which the excretory systems are similar in form and function.
Textbook Reference: *40.3 Excretory Systems Produce Urine by Filtration, Reabsorption, and Secretion, pp. 788–790*

Question 6. In the diagram below, label the following structures: afferent arteriole, renal tubule, Bowman's space, Bowman's capsule, collecting duct, efferent arteriole, peritubular capillaries, and glomerulus. Also label each of the three boxes to indicate where the following processes occur in the nephron: filtration; reabsorption and secretion; excretion.
Textbook Reference: *40.3 Excretory Systems Produce Urine by Filtration, Reabsorption, and Secretion, p. 789*

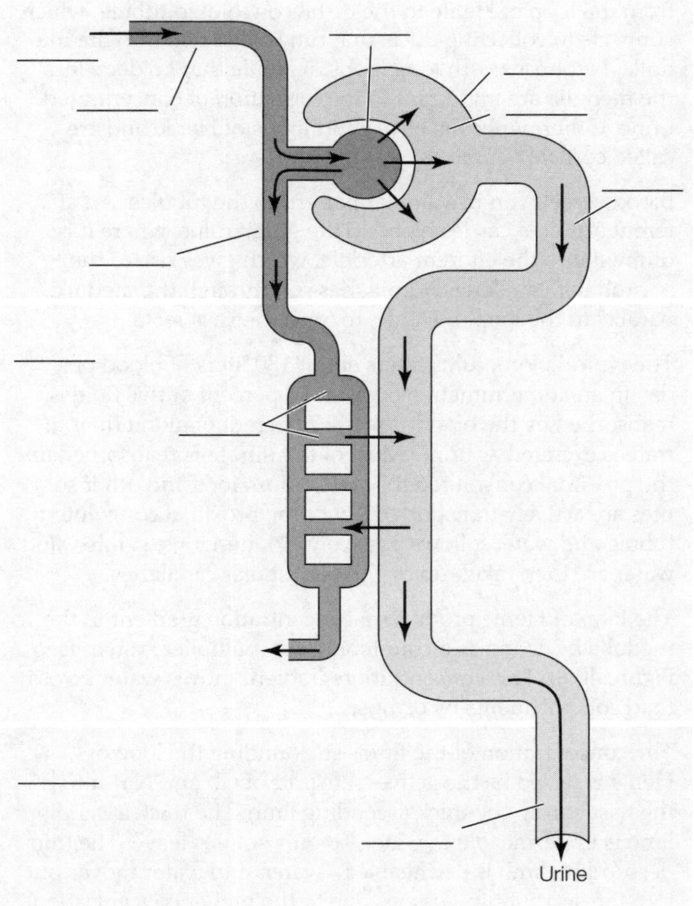

40.4 The Mammalian Kidney Produces Concentrated Urine

- A mammalian kidney has a cortex and a medulla
- Most of the glomerular filtrate is reabsorbed by the proximal convoluted tubule
- The loops of Henle create a concentration gradient in the renal medulla
- The distal convoluted tubule fine-tunes the composition of the urine
- Urine is concentrated in the collecting duct
- Kidney failure is treated with dialysis

Humans have two kidneys that filter blood and produce urine (see Figure 40.7). Urine exits the kidney through the ureter and travels to the bladder, where it is stored until released through the urethra.

The human kidney is shaped like a kidney bean, and has an internal structure consisting of a central medulla and a surrounding cortex. The glomerulus, Bowman's capsule, and adjoining proximal convoluted tubules are located in the cortex. The descending limb of the renal tubule runs down into the medulla, and the ascending limb returns to the cortex in what is known as the loop of Henle. Fluid moves from the loop of Henle to the distal convoluted tubule, which connects to collecting ducts that run back down into the medulla. Nephrons with long loops of Henle that go deep into the medulla are important in the formation of concentrated urine. Other nephrons have short loops of Henle and are called cortical nephrons.

Blood vessels run in a similar pattern to the tubules. An afferent arteriole carries blood to the glomerulus, where it is drained into the efferent arteriole, which gives rise to the peritubular capillaries. Capillaries run through the medulla, parallel to the loop of Henle, to form the vasa recta.

The typical glomerulus filters about 180 liters of blood per day in an adult human. However, 98 percent of this fluid is reabsorbed by the blood, so only 2 percent of glomerular filtrate is excreted as urine. Most of the filtrate is reabsorbed in the proximal convoluted tubule. Sodium ions and other solutes are actively transported out of the proximal convoluted tubule and water follows passively. Sodium ions, solutes, and water are then taken up by the peritubular capillaries.

The loop of Henle produces a concentration gradient in the medulla by acting as a countercurrent multiplier system (see Figure 40.8). This concentration gradient moves water across fluid compartments by osmosis.

The concentration of the fluids surrounding the loop of Henle is raised by the active transport of Cl^- and Na^+ into the tissue from the thick ascending limb. The thick ascending limb is impermeable to water, so only solutes leave. The thin descending limb is permeable to water, and water moves out into the surrounding tissues due to the higher concentration of Na^+ and Cl^- there. In the thin ascending limb, the filtrate is more concentrated than the fluid in the surrounding tissues, so Na^+ and Cl^- move passively out of the tubule. Water cannot move out because of the low permeability of the thin ascending limb.

The fluid reaching the distal convoluted tubule is less concentrated than blood plasma. The distal convoluted tubule fine-tunes the ionic composition of the urine. As the fluid leaves the distal convoluted tubule for the collecting duct, its major solute is urea.

The collecting duct passes through the medulla, which has the high solute concentration set up by the loop of Henle. Water moves out of the collecting duct and into the tissue, resulting in the production of concentrated urine. Some urea moves out of collecting ducts into the interstitial fluid of the medulla, adding to the concentration gradient there. This urea eventually diffuses into the loop of Henle and is returned to collecting ducts.

Kidney (renal) failure severely disrupts homeostasis, causing high blood pressure (due to the retention of salts and water), uremic poisoning (due to the retention of urea), and acidosis (due to decreasing pH). A dialysis machine can replace kidney function until a kidney transplant can be arranged.

Question 7. Based on your knowledge of how the nephron works, explain why mammals produce concentrated urine whereas nonavian reptiles do not.
Textbook Reference: *40.4 The Mammalian Kidney Produces Concentrated Urine, pp. 791–794*

Question 8. What would happen to the composition of excreted urine if all active transport processes in the nephron came to a halt?
Textbook Reference: *40.4 The Mammalian Kidney Produces Concentrated Urine, pp. 793–794*

40.5 The Kidney Is Regulated to Maintain Blood Pressure, Blood Volume, and Blood Composition

- The renin-angiotensin-aldosterone system raises blood pressure
- ADH decreases excretion of water
- The heart produces a hormone that helps lower blood pressure

Glomerular filtration rate (GFR) is a function of the blood flow and pressure to the kidneys. The kidneys have autoregulatory mechanisms that help maintain their high filtration rate. In response to a fall in blood pressure in the glomeruli, the afferent renal arterioles open, or dilate, to increase flow to the glomeruli. If this does not increase glomerular filtration rate sufficiently, the kidneys release the enzyme renin into the blood. Renin converts an angiotensin precursor into active angiotensin. Angiotensin constricts the efferent renal arterioles to increase pressure in the glomeruli. Angiotensin also constricts peripheral blood vessels, stimulates the release of aldosterone from the cortex of the adrenal gland, and stimulates thirst (see Figure 40.9). Aldosterone stimulates the reabsorption of sodium by the kidney, which increases water reabsorption because water follows sodium. These changes result in increased blood volume and pressure.

Antidiuretic hormone (ADH, also called vasopressin) controls the permeability of the collecting ducts to water by stimulating the insertion of aquaporins (water channels) into the plasma membranes of the cells in this region of the tubule. Osmoreceptors in the hypothalamus stimulate the release of ADH in response to increased serum osmolarity, resulting in increased water reabsorption and dilution of the blood, and the production of small amounts of concentrated urine.

Atrial natriuretic peptide (ANP) is released by the heart in response to high blood pressure. At the kidneys, this hormone decreases sodium reabsorption, resulting in decreased water reabsorption and a lowered blood volume and pressure. Under the influence of ANP, large amounts of dilute urine are produced.

Question 9. The drug urizadole inhibits antidiuretic hormone secretion. How does this inhibition affect glomerular filtration rate, urine volume, and urine concentration in patients taking this medication?
Textbook Reference: *40.5 The Kidney Is Regulated to Maintain Blood Pressure, Blood Volume, and Blood Composition, pp. 795–796*

Question 10. You have just eaten a large number of very salty potato chips, and the osmolarity of your blood has increased. What response will your body have in order to bring your blood osmolarity back to homeostasis?
Textbook Reference: *40.5 The Kidney Is Regulated to Maintain Blood Pressure, Blood Volume, and Blood Composition, pp. 795–796*

Test Yourself

1. In order to maintain homeostasis, a marine bony fish
   a. excretes only small amounts of water and pumps sodium out of its body at the gills.
   b. excretes large amounts of water and pumps sodium into its body.
   c. converts nitrogenous wastes to urea.
   d. excretes uric acid.
   e. None of the above
   Textbook Reference: *40.3 Excretory Systems Produce Urine by Filtration, Reabsorption, and Secretion, p. 790*
2. Which of the following statements about the excretory system of insects is *false*?
   a. Active transport moves materials from the coelomic fluid into the Malpighian tubules.
   b. The Malpighian tubules can produce a highly concentrated waste product, allowing insects to inhabit some of Earth's driest habitats.
   c. Reabsorption of salts takes place mostly in the gut.
   d. Water reabsorption takes place by osmotic movement only.
   e. Insects excrete urea.
   Textbook Reference: *40.3 Excretory Systems Produce Urine by Filtration, Reabsorption, and Secretion, pp. 788–789*
3. Which of the following represents the correct pathway of water and solutes traveling through a nephron?
   a. Glomerulus, Bowman's capsule, renal tubule, collecting ducts
   b. Bowman's capsule, glomerulus, renal tubule, collecting ducts
   c. Renal tubule, glomerulus, Bowman's capsule, collecting ducts
   d. Collecting ducts, glomerulus, Bowman's capsule, renal tubule
   e. Glomerulus, Bowman's capsule, collecting ducts, renal tubule
   Textbook Reference: *40.3 Excretory Systems Produce Urine by Filtration, Reabsorption, and Secretion, pp. 789–790*
4. Na^+ and Cl^- are actively transported out of the tubules to help set up the countercurrent multiplier. Sites of active Na^+ and Cl^- transport in the nephron are _______ and the _______.
   a. Bowman's capsule; thin descending limb of the loop of Henle
   b. the thin descending limb of the loop of Henle; thin ascending limb of the loop of Henle
   c. the thin ascending limb of the loop of Henle; proximal convoluted tubule
   d. the collecting duct; thin descending limb of the loop of Henle
   e. the proximal convoluted tubule; thick ascending limb of the loop of Henle
   Textbook Reference: *40.4 The Mammalian Kidney Produces Concentrated Urine, p. 793*

5.–9. Refer to the diagram below of the mammalian nephron to answer the questions that follow.

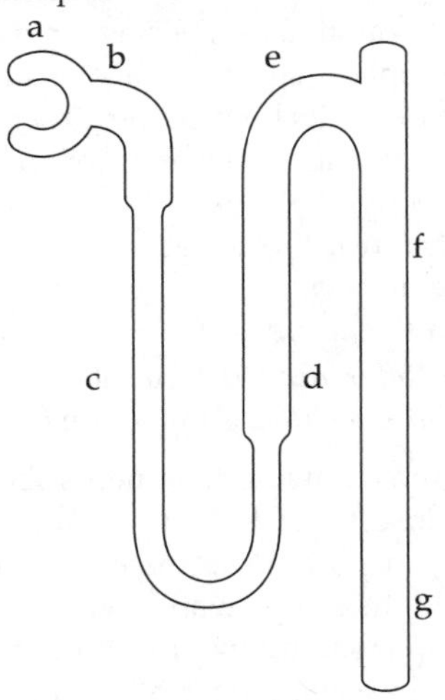

5. The composition of the filtrate would be most like plasma in the tubule next to letter _______.
6. The NaCl concentration in the extracellular fluid would be greatest in the area of letter _______.
7. The osmolarity of the filtrate next to letters _______ is similar to the osmolarity of blood plasma.
8. The urine would be most concentrated in the collecting duct next to letter _______.
9. Most of the glomerular filtrate is reabsorbed into the blood in peritubular capillaries next to letter _______.
   Textbook Reference: *40.4 The Mammalian Kidney Produces Concentrated Urine, pp. 793–794*

10. The sole mechanism for water reabsorption by the renal tubules is
    a. active transport.
    b. osmosis.
    c. cotransport with sodium ions.
    d. cotransport with bicarbonate ions.
    e. None of the above
    Textbook Reference: *40.4 The Mammalian Kidney Produces Concentrated Urine, p. 793*

11. Several hormones help regulate water and solute uptake and release in the nephron. Antidiuretic hormone (ADH) promotes _______ in response to _______.
    a. active transport of Cl^-; increased solute concentration
    b. active transport of Na^+; increased blood pressure
    c. increased permeability of the collecting duct to water; increased serum osmolality
    d. decreased permeability of the collecting duct to water; increased solute concentration
    e. decreased permeability of the collecting duct to water; decreased blood pressure
    Textbook Reference: *40.5 The Kidney Is Regulated to Maintain Blood Pressure, Blood Volume, and Blood Composition, pp. 795–796*

12. Which of the following is *not* a normal constituent of the glomerular filtrate?
    a. Red blood cells
    b. Urea
    c. Sodium ions
    d. Glucose
    e. Amino acids
    Textbook Reference: *40.3 Excretory Systems Produce Urine by Filtration, Reabsorption, and Secretion, p. 789*

13. If the afferent arteriole that supplies blood to the glomerulus becomes dilated, then
    a. the protein concentration of the filtrate decreases.
    b. hydrostatic pressure in the glomerulus decreases.
    c. the glomerular filtration rate increases.
    d. the glomerular filtration rate decreases.
    e. the water concentration of the filtrate decreases.
    Textbook Reference: *40.5 The Kidney Is Regulated to Maintain Blood Pressure, Blood Volume, and Blood Composition, pp. 795–796*

14. Which of the following is a hormone released by the heart that decreases sodium reabsorption by the kidneys?
    a. Aldosterone
    b. Antidiuretic hormone
    c. Renin
    d. Atrial natriuretic peptide
    e. Angiotensin
    Textbook Reference: *40.5 The Kidney Is Regulated to Maintain Blood Pressure, Blood Volume, and Blood Composition, p. 796*

15. As the glomerular filtrate passes through the renal tubule, about _______ percent is returned to the blood.
    a. 2
    b. 10
    c. 50
    d. 75
    e. 98
    Textbook Reference: *40.4 The Mammalian Kidney Produces Concentrated Urine, p. 793*

16. Which of the following statements about excretory systems is *false*?
    a. Osmoconformers are always ionic conformers.
    b. Osmoconformers are found only in marine environments.
    c. Terrestrial animals are always osmoregulators.
    d. Some marine vertebrates are osmoconformers.
    e. Most marine invertebrates are osmoconformers.
    Textbook Reference: *40.1 Excretory Systems Maintain Homeostasis of the Extracellular Fluid, pp. 785–786*

17. Which of the following end products of catabolism is most toxic?
    a. Water
    b. Ammonia
    c. Carbon dioxide

d. Urea
e. Uric acid
Textbook Reference: *40.2 Excretory Systems Eliminate Nitrogenous Wastes, p. 787*

Answers

Key Concept Review

1. On land, water and salts are usually in short supply, so terrestrial animals are osmoregulators, actively regulating the osmolarity of their extracellular fluid.
2. Osmoconformers allow their extracellular fluid to equilibrate with their surroundings. This can work in marine environments because there are many solutes in saltwater, some of which are necessary to support life. In addition, many marine osmoconformers have the ability to regulate the concentrations of certain ions in their extracellular fluids. In fresh water, however, there are few solutes, so the body fluids of an osmoconformer would be too dilute to support life.
3. The symptoms of gout are caused by the precipitation of uric acid crystals in joints due to high levels of uric acid in the extracellular fluid. Two potential approaches for new medications would involve blocking the production of uric acid by the body or improving the removal of uric acid from the body.
4. Water-breathing animals typically excrete ammonia as their primary nitrogenous waste because ammonia is highly soluble in water and diffuses rapidly. Ammonia excretion occurs continuously across the gills.
5. The metanephridia of the earthworm and the nephron of humans are both composed of tubules through which filtrate flows and in which certain molecules are actively transported out or secreted in. Both systems produce urine with an osmolarity different from that of the body fluid. The urine of the earthworm is hypotonic, whereas that of humans is usually hypertonic.
6.

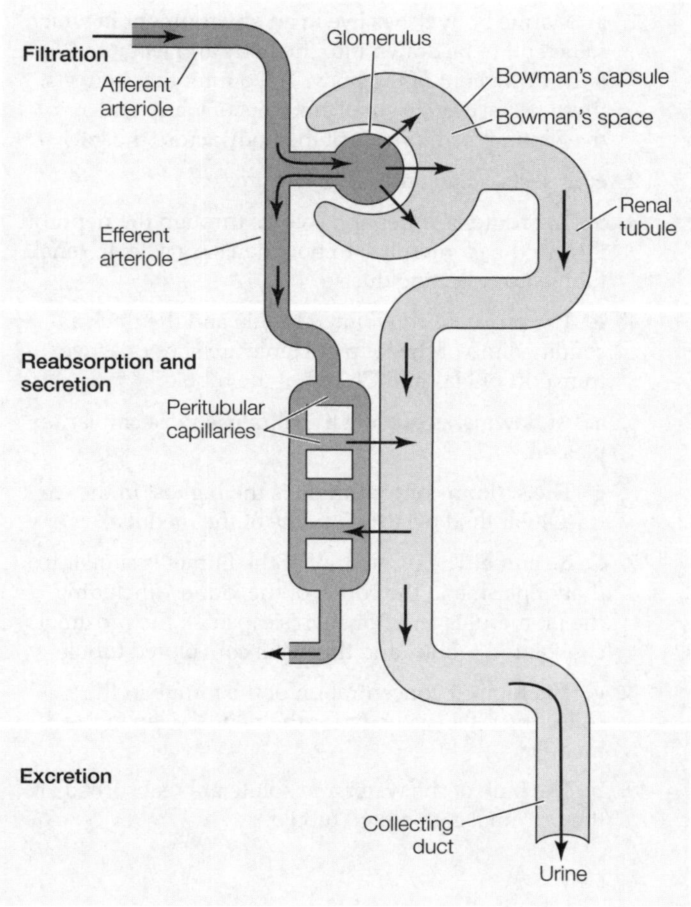

7. The loop of Henle acts as a countercurrent ion multiplier in the medulla of mammals. This allows mammalian kidneys to produce concentrated urine. Nonavian reptiles lack the countercurrent multiplier of the loop of Henle.
8. Active transport of Na^+ and Cl^- in the nephron provides the ions that set up the countercurrent multiplier, allowing for the production of concentrated urine in mammals. If active transport in the nephrons were to stop, the urine produced would eventually be isotonic with blood plasma because the countercurrent multiplier would disappear.
9. Inhibition of antidiuretic hormone by urizadole affects the permeability of the collecting ducts to water. This results in increased urine volume because water will not be reabsorbed across the collecting ducts. Blockage of ADH has no effect on the glomerular filtration rate.
10. In response to increased blood osmolarity, osmoreceptors in the hypothalamus stimulate the release of ADH. ADH acts to increase the permeability of the collecting ducts to water so that increased amounts of water can be reabsorbed to bring down the blood osmolarity. The osmoreceptors will also stimulate thirst, causing you to increase your water intake.

Test Yourself

1. **a.** Marine bony fishes live in an environment in which salts tend to be drawn into the body and water tends to be drawn out of the body. To counter these effects, these fishes excrete small amounts of water and actively pump sodium out of the body across the gills.
2. **e.** Insects excrete uric acid, not urea.
3. **a.** The route of water and solutes through the nephron is from the glomerulus, to Bowman's capsule, to renal tubule, to collecting ducts.
4. **e.** The proximal convoluted tubule and the thick ascending limb of the loop of Henle are sites of active transport of Na^+ and Cl^- out of the tubule.
5. **a.** At Bowman's capsule the filtrate is most similar to plasma.
6. **g.** The sodium concentration is the highest in the extracellular fluid near the middle of the medulla.
7. **a.**, **b.**, and **e.** The osmolarity of the filtrate is similar to that of plasma in the cortex of the kidney (including the glomerulus and Bowman's capsule), the proximal convoluted tubule, and the distal convoluted tubule.
8. **g.** The highest concentration of the filtrate in the collecting ducts will be near their ends, deep in the medulla.
9. **b.** The bulk of the water and solute are reabsorbed at the proximal convoluted tubule.
10. **b.** The sole mechanism for water reabsorption in the renal tubules is by osmosis.
11. **c.** Antidiuretic hormone acts on the collecting ducts by increasing permeability to water. Antidiuretic hormone secretion is stimulated by rises in serum osmolality.
12. **a.** Red blood cells are too large to be filtered out of the blood at the glomerulus and thus will not be found in the filtrate.
13. **c.** Changes in afferent arteriole pressure affect glomerular filtration rate. Dilation of the afferent arteriole increases pressure, which will increase filtration rate.
14. **d.** Atrial natriuretic peptide (ANP) is a hormone released by the heart in response to increased blood volume. ANP causes decreased reabsorption of sodium by the kidneys. This results in decreased water reabsorption and the production of large volumes of dilute urine.
15. **e.** About 98 percent of the filtrate is returned to the blood.
16. **a.** Osmoconformers can be either ionic regulators or ionic conformers. Many osmoconformers regulate the concentrations of certain ions in their extracellular fluids.
17. **b.** Ammonia is the most toxic end product of catabolism.

Animal Behavior 41

The Big Picture

- The distinction between proximate and ultimate explanations is of fundamental importance in the study of animal behavior and can be extended to many other fields of biology. Researchers concerned with proximate questions strive to unravel the underlying neural and hormonal mechanisms of behavior and to determine the relative roles of genes and experience in shaping behavior. Researchers asking ultimate questions attempt to understand how a particular behavior affects the animal's survival and reproductive success and how it might have evolved.
- Animals exhibit a wide range of species-specific behaviors that can be either genetically based or environmentally determined. Genetically based behaviors may require triggers or releasers to stimulate an animal. Nevertheless, patterns of behavior typically have genetic and environmental influences.
- Hormones are important in controlling the behavior of animals. Sex steroids are frequently involved in stimulating the different behaviors of males and females of a species. Sometimes the hormonal state during a brief period allows the acquisition of particular behaviors, such as imprinting. Biological rhythms coordinate behavior with environmental cycles, such as daily cycles of light and dark.
- Communication is an integral part of animal behavior. Communications occur in several modes, including chemical (involving pheromones), visual (involving displays of fins, feathers, fur, etc.), auditory (sounds created by either general or specialized structures and received by auditory sensors), tactile, and electric. Communication is an important component of territory-marking and reproductive behavior.
- As Chapter 15 (Mechanisms of Evolution) explained, the cost–benefit approach applied to the analysis of behavior can be extended to any kind of adaptation of organisms.
- Because standard Darwinian theory holds that only traits that contribute to individual fitness are favored by natural selection, explaining the evolution of altruistic behavior has presented a major challenge to evolutionary biologists. The concepts of inclusive fitness and kin selection have proved to be very fruitful in understanding altruism in animals.

Study Strategies

- It can sometimes be difficult to distinguish between behaviors founded in genetics (instinctive behaviors) and those that are learned during the course of an animal's lifetime. Further muddying the waters, the ability to learn new behavior is, of course, a heritable trait! As you learn about various patterns of behavior, take time to consider whether the behavior is primarily genetically determined or primarily environmentally determined.
- Hormones control many behaviors, and hormone production is often a function of the time of day or the time of year. Thus, a particular hormone that stimulates (or inhibits) behaviors under one condition at one point in time may not have the same effect at another time. Be sure to learn the conditions and caveats that accompany hormone actions.
- The categories of animal orientation—piloting, distance-and-direction navigation, and bicoordinate navigation—can be confusing. Refer to Figure 41.9 to consolidate your understanding of how the sun is used as a time-compensated compass.
- Learn to distinguish the various behavior cycles associated with internal controls via "biological clocks" as opposed to behaviors that are directly stimulated by an immediate event in an animal's surrounding environment.
- To study the properties of the various sensory modes of communication (visual, chemical, etc.), make a table that compares them with respect to characteristics such as cost of production, effective signaling distance (and

how signaling distance is affected by environmental conditions), durability, and information content.

- Go to yourBioPortal.com to review the following tutorials and activities:

  Animated Tutorial 41.1 Circadian Rhythms

  Animated Tutorial 41.2 Time-Compensated Solar Compass

  Animated Tutorial 41.3 The Costs of Defending a Territory

  Animated Tutorial 41.4 Foraging Behavior

  Interactive Tutorial 41.1 Time-Compensated Solar Compass

  Web Activity 41.1 Honey Bee Dance Communication

  Web Activity 41.2 Concept Matching

Key Concept Review

41.1 Behavior Has Proximate and Ultimate Causes

Biologists ask four questions about a behavior

Questions about proximate causes lead to mechanistic approaches

Questions about ultimate causes lead to ecological/ evolutionary approaches

Niko Tinbergen suggested that studies of animal behavior address the following four questions: (1) causation; (2) development; (3) function; and (4) evolution. Questions about causation and development address proximate causes of behavior while questions about function and evolution focus on ultimate causes.

The work of Ivan Pavlov and others inspired behaviorism, a school of animal behavior that focuses on proximate causes. Pavlov discovered the conditioned reflex through his experiments on the salivation response of dogs (see Figure 41.1). Before conditioning, food is an unconditioned stimulus (UCS) that elicits salivation, an unconditioned response (UCR). If the UCS is presented immediately after a neutral stimulus, such as a particular sound, the dog eventually salivates solely in response to the sound. The sound has become a conditioned stimulus (CS) and the salivation response to the sound is a conditioned response (CR). Salivation in response to the sound is thus a learned response.

Ethology, another school of animal behavior, arose around the same time as behaviorism. Ethologists are interested in fixed action patterns, which are patterns of behavior that are triggered by simple stimuli called releasers. For example, the red dot on the bill of an adult gull triggers pecking behavior by the chick (see Figure 41.2). Fixed action patterns are resistant to modification by learning.

Studies in behavioral ecology typically examine how certain patterns of behavior or choices affect an individual's survival and reproductive success.

Question 1. The phenomenon by which a young animal leaves its place of birth is called natal dispersal. Develop four research questions about natal dispersal in gray squirrels using Tinbergen's four questions.
Textbook Reference: *41.1 Behavior Has Proximate and Ultimate Causes, p. 800*

Question 2. Respond to the statement, "A particular behavior is either genetically determined or learned."
Textbook Reference: *41.1 Behavior Has Proximate and Ultimate Causes, pp. 800–801*

41.2 Behavior Can Have Genetic Determinants

Breeding experiments can reveal genetic determinants of a behavior

Studies of mutants can reveal the roles of specific genes

Gene knockouts can reveal the roles of specific genes

Behavioral geneticists conduct breeding experiments to analyze how particular behaviors (e.g., the hygienic behavior of honey bees) are inherited (see Figure 41.3). However, breeding experiments alone cannot identify the particular genes influencing a behavior or how they influence the behavior, so molecular genetic approaches are needed. Such approaches have been used to study the *Drosophila* mutant *fruitless*, in which mutant males are unable to tell the sexes apart, so they court both males and females. The *fruitless* gene has been cloned and sequenced, and experimentally altered. These studies have shown that the gene product is a transcription factor that controls the expression of many genes. This example illustrates that most behaviors are complex traits involving many genes that function in cascades and offer many points at which a change in a single gene can influence behavior.

Once a gene has been identified, a knockout experiment can be conducted in which biologists inactivate the particular gene and investigate the behavioral effects of its loss. A knockout mouse is one in which a particular gene is targeted and inactivated to eliminate the gene product, possibly a receptor. For example, inactivating particular receptors in the vomeronasal organ of male mice disrupts their ability to discriminate female from male conspecifics (see Figure 41.4).

Question 3. How can a change in a single gene have a major impact on behavior?
Textbook Reference: *41.2 Behavior Can Have Genetic Determinants, p. 802*

Question 4. The hormone progesterone is thought to mediate the aggression shown by male mice toward infant mice. Design a genetic knockout experiment to test this hypothesis.
Textbook Reference: *41.2 Behavior Can Have Genetic Determinants, pp. 802–803*

41.3 Developmental Processes Shape Behavior

Hormones can determine behavioral potential and timing

Some behaviors can be acquired only at certain times

Bird song learning involves genetics, imprinting, and hormonal timing

The timing and expression of bird song are under hormonal control

Hormones early in life can cause a behavioral potential to develop and then prompt expression of that behavior later in life. In rats, for example, exposure to testosterone early in development establishes the potential for a particular pattern of sexual behavior (mounting); testosterone secretion in adulthood prompts expression of that male sexual behavior (see Figure 41.5).

In the phenomenon of imprinting, an animal learns a set of stimuli during a limited sensitive (sometime called critical) period. Imprinting of offspring on parents or of parents on offspring is a learned response that helps in recognition. The critical period for imprinting is often determined by the developmental or hormonal state of the animal.

Birds use songs in territorial displays and in courtship. In some species, imprinting of the species-specific song is required in the nestling in order for it to sing the song as an adult, even though the juvenile bird never sings the song. Imprinting forms a memory of the song that is recalled as the bird approaches adulthood. The adults need both the initial imprinting of the song as a juvenile and the ability to match the song with auditory feedback in order to sing the correct song (see Figure 41.6).

The influence of hormones on behavior has also been investigated through experiments on song development in birds. In male birds, testosterone levels control singing by inducing certain regions of the brain to grow during the breeding season. During the nonbreeding season, the regions of the brain associated with singing are reduced in size. Female birds that are treated with testosterone during the spring will also develop the species-specific song. In these females, testosterone has stimulated growth in those areas of the brain associated with singing.

Question 5. Much like patterns of sexual behavior in rats (females display lordosis and males display mounting), urinary posture in dogs is sexually dimorphic (females typically squat and males lift their leg). In both sexual behavior in rats and urinary posture in dogs, testosterone early in development organizes the potential for particular behaviors in adulthood. Would testosterone have the same effect on the timing of expression of the behaviors in adulthood? (Hint: Testosterone is necessary for the expression of sexual behavior in adult male rats. Is testosterone necessary for the expression of the male urinary posture in adult male dogs?)
Textbook Reference: *41.3 Developmental Processes Shape Behavior, p. 804*

Question 6. Explain why deafening a young songbird affects song development differently than does deafening an adult songbird.
Textbook Reference: *41.3 Developmental Processes Shape Behavior, p. 805*

41.4 Physiological Mechanisms Underlie Behavior

Biological rhythms coordinate behavior with environmental cycles

Animals must find their way around their environment

Animals use multiple modalities to communicate

Control of behavior involves the nervous and endocrine systems. Execution of behavior typically involves the musculoskeletal system.

Responses to the environment must be timed appropriately. Circadian rhythms are daily cycles in activity, sleep, foraging, and other physiological processes and behaviors that are controlled by an endogenous clock (see Figure 41.7). The length of one cycle in a rhythm is defined as one period. Any point in the cycle is known as a phase. Two cycles can be in phase if the rhythms match or be phase-advanced or phase-delayed if they do not match.

Circadian rhythms can be reset by environmental cues, such as the light–dark cycle, during entrainment. In constant conditions an animal's circadian clock is said to be free-running and will have a natural period that is different from the 24-hour period of the day. In mammals, the master circadian "clock" is located in the suprachiasmatic nuclei (SCN) of the brain. The molecular mechanism of the circadian clock involves negative feedback control of certain clock genes that are expressed in SCN cells.

Animals find their way around their environment by several mechanisms, including piloting, distance-and-direction navigation, and bicoordinate navigation. Piloting is orientation using landmarks. It is the mechanism used by some species that migrate or that are capable of homing (the ability to return to a specific location). Homing and migrating species that are able to take direct routes to their destinations through environments they have never experienced must use mechanisms of navigation other than piloting. Examples of such species are homing pigeons and the many species of migrating birds that are able to fly great distances and return to the same breeding ground each season.

Many animals appear to have a compass sense, which allows them to use environmental cues to determine direction, and some appear to have a map sense, which allows them to determine their position. Distance-and-direction navigation requires knowledge of direction and distance to a destination. The position of the sun and stars can be a source of directional information. Pigeons, for example, have the ability to determine direction by means of a time-compensated solar compass (see Figure 41.9). The stars offer two sources of information about direction: moving constellations and a fixed point (the point directly over the axis on which Earth turns). Bicoordinate navigation (also known as true navigation) requires knowing the map coordinates of both the current position and the destination. The behavior of some species (such as albatrosses) suggests that they are capable of this type of navigation (see Figure 41.10).

Animals transmit information through communication. If the transmission of information benefits both the sender and the receiver, the behaviors of individuals may become elaborated through evolution into communication signals. Pheromones are chemical signals used to communicate among individuals of a species. Because of the diversity of their molecular structures, pheromones can communicate very specific, information-rich messages. Pheromones used in different types of communication vary in their volatility and diffusibility. Because pheromones remain in the environment for some time after their release, they are useful for such functions as marking territories, but they are unsuitable for the rapid exchange of information.

Visual signals provide rapid, directional communication over considerable distances. One drawback to visual signals is the need for light, except in the case of species that have evolved their own light-emitting mechanisms. Visual signals can also be intercepted by other species. Sound has advantages over sight in that it can travel in complex environments, such as a forest, and over long distances, such as the sound produced by whales in the ocean. Sound also communicates directional information. The characteristics of acoustic signals are typically adjusted to their function and the environment of the animal.

Animals in close contact with one another can use tactile communication. Honey bees dance to communicate the location of a food source in the environment (see Figure 41.11). The waggle dance is used to communicate distance and direction to food. Speed of the waggle indicates distance to the food source; direction of the straight run of the waggle dance indicates the direction of the food source relative to the sun.

The specificity of communication signals is enhanced by the use of multiple sensory modalities. Courtship behavior in fruit flies, for example, involves tactile, chemical, visual, and acoustic signals.

Question 7. A pigeon has been trained to feed at food bins at the eastern end of a circular cage from which it can see the sky. There are food bins at the N, NE, E, SE, S, SW, W, and NW ends of the cage. Based on this information, answer the following questions:

a. If the cage is covered and a fixed light source is presented at the east end of the cage at the time of sunrise, where will the pigeon search for food at noon?

b. If a mirror is used to shift the apparent position of the sun at noon from the south to the northwest, where will the pigeon search for food?

c. The bird has been placed in a light-controlled environment for three weeks and phase-delayed by six hours. If the pigeon is returned to the cage under natural lighting conditions at sunset, where will it search for food?

Textbook Reference: *41.4 Physiological Mechanisms Underlie Behavior, p. 809*

Question 8. What is the function of a honey bee's waggle dance, and what does it communicate to other bees in a hive?

Textbook Reference: *41.4 Physiological Mechanisms Underlie Behavior, p. 811*

Question 9. Draw five diagrams showing the orientation of the straight run of the waggle dance that would be performed on the vertical surface of a honeycomb by a foraging honey bee that has discovered a food source at the times and locations described below. Recall that in the northern hemisphere the direction (azimuth) of the sun is due south at noon; assume that the sun rises precisely in the east at 6:00 A.M. and sets precisely in the west at 6:00 P.M. As in Figure 41.11, let the top of the page indicate the "up" direction.

a. Food location: due south of the hive; time: noon

b. Food location: due north of the hive; time: 6:00 A.M.

c. Food location: due north of the hive; time: noon

d. Food location: due west of the hive; time: 9:00 A.M.

e. Food location: due east of the hive; time: 6:00 P.M.

Textbook Reference: *41.4 Physiological Mechanisms Underlie Behavior, p. 811*

41.5 Individual Behavior Is Shaped by Natural Selection

Animals must make choices

Behaviors have costs and benefits

The habitat of an animal is the environment in which it normally lives. In choosing their habitat, animals use cues that are good predictors of general conditions suitable for future survival and reproduction.

Natural selection molds behavior in accordance with costs and benefits. There are three aspects to the cost of behaving: (1) energetic cost (the amount of energy the animal expends during the behavior); (2) risk cost (the amount of risk the behavior entails for the animal); and (3) opportunity cost (the benefits the animal forfeits by not engaging in other behaviors instead).

The territory of an animal is an area from which other individuals of its own species (and sometimes individuals of other species) are excluded. By establishing a territory, an animal (usually a male) may improve its fitness by gaining exclusive use of the resources of part of its habitat. Cost–benefit analysis explains the diversity of territorial behaviors characteristic of different species (see Figure 41.12). Some animals defend all-purpose territories in which nesting, mating, and foraging take place. Other animals cannot establish feeding territories but defend nest sites or areas that provide access to females. Still other animals defend territories (leks) that are used only for mating.

Cost–benefit analysis can also be applied to foraging behavior (i.e., what food an animal selects and when and where it searches for it). More specifically, foraging behavior can be examined in terms of the energy and time expended (costs) as compared to the energy obtained (benefit). Because minerals and foods with medicinal value are important in the diets of many animals, they may sometimes forage in a way that deviates from energy-maximization.

Question 10. Why might animals preferentially settle in locations where conspecifics are present?
Textbook Reference: *41.5 Individual Behavior Is Shaped by Natural Selection, p. 812*

41.6 Social Behavior and Social Systems Are Shaped by Natural Selection

Mating systems maximize the fitness of both partners

Fitness can be enhanced through the reproductive success of related individuals

Eusociality is the extreme result of kin selection

Group living has benefits and costs

Social systems range from very simple (a single male and a single female) to very complex (such as honey bee colonies). Sociobiology proposes that the evolution of all variants of social behavior can be understood by asking how the behavior contributes to the fitness of the individuals involved.

Mating systems evolve to maximize the fitness of both partners. Because males produce vast numbers of sperm that contain almost no resources, and females produce relatively few eggs that are rich in resources, the energetic and opportunity costs for reproduction are greater for the female than for the male. This disparity in the investment in the young is particularly large in mammals because females bear the costs associated with gestation. For a female, the best way to maximize her fitness is to make sure her young are healthy and survive to pass on her genes.

Males have different options for maximizing their fitness, including monogamy (in which one male forms a pair bond with one female and both parents participate in rearing the young), polygamy (one male mates with many females), and polyandry (multiple males mate with one female). The mating strategy that maximizes his fitness is determined by environmental factors.

In species with polygamous mating systems, some males have high reproductive success while many males have none. As a rule, bigger, stronger males are the winners in the competition for females, and sexual dimorphism in body size evolves. Polyandry is a relatively rare mating systems that occurs in some birds and a few mammals in which paternal care for the young can have a large influence on fitness.

Natural selection sometimes favors altruistic acts—behaviors that reduce the reproductive chances of the individual performing the act but increase the fitness of the helped individual. Typically, such behavior is directed toward a relative of the altruist, with whom it shares alleles. By helping its relatives, an individual can increase the representation of some of its own alleles in the population. An altruistic behavior pattern can evolve if it increases the inclusive fitness of the altruist: the fitness derived from an individual's personal reproductive success (individual fitness) plus the fitness derived from the reproductive success of its relatives.

The maximization of inclusive fitness underlies kin selection, which is selection for behaviors that increase the reproductive success of relatives even when they have some cost to the performer. According to Hamilton's rule, for an apparent altruistic act to be adaptive, the fitness benefit of that act to the recipient times the degree of relatedness of the performer and the recipient has to be greater than the cost to the performer.

Eusocial species are those whose social groups include sterile individuals. In the Hymenoptera (ants, bees, and wasps), kin selection has probably facilitated the evolution of eusociality because of their sex determination system (haplodiploidy), in which males are haploid and females are diploid, with the result that sisters are genetically more similar to one another than to their own offspring. In eusocial species in which both sexes are diploid (naked mole-rats), the difficulty of establishing independent colonies has favored the evolution of eusociality.

Group living may benefit both predator and prey species. For predators, it may improve hunting efficiency or increase the size of the prey that can be captured. For prey, it may provide increased protection against predators. Living in a group imposes costs as well as benefits. Individuals in groups may compete for food, interfere with one another's foraging, injure one another's offspring, inhibit one another's reproduction, or transmit diseases to their associates.

Question 11. Why has the unusual sex determination system in Hymenoptera predisposed species in this group toward the evolution of eusociality?
Textbook Reference: *41.6 Social Behavior and Social Systems Are Shaped by Natural Selection, p. 816*

Question 12. For each of the two *y*-axes in the following graph, draw a labeled curve that correctly summarizes observations made on goshawks attacking wood pigeons, as described in the textbook.
Textbook Reference: *41.6 Social Behavior and Social Systems Are Shaped by Natural Selection, p. 817*

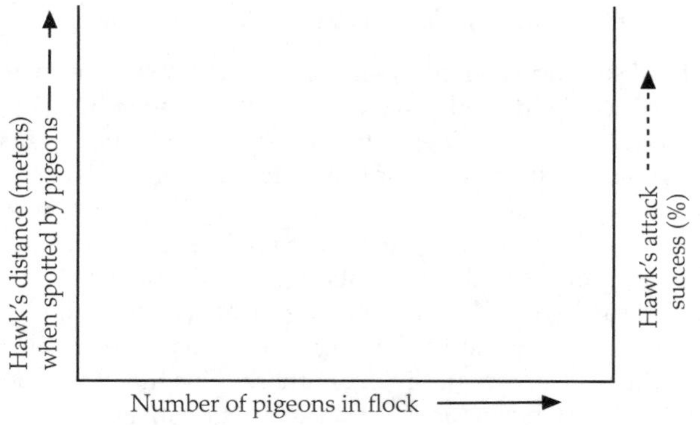

Test Yourself

1. On an April morning, you step outside your front door and notice the singing of a male robin. At that time you also observe that the female of the pair is building a nest in a nearby tree. A month later you observe the pair feeding their nestling offspring. Which of the following is a question about the ultimate cause of the behavior of these birds?
   a. Did the male bird begin to sing in April because the photoperiod had increased to a critical length?
   b. What were the relative roles of genes and experience in causing the female robin to build her nest out of particular building materials and to place it in a particular location?
   c. What combination of internal physiological factors and external cues stimulates the parents to feed their nestlings?
   d. Do robins that begin nesting in April raise more offspring than they would raise if they delayed nesting until June, and if so, why?
   e. None of the above
   Textbook Reference: *41.1 Behavior Has Proximate and Ultimate Causes, pp. 800–801*
2. The red dot on the bill of a gull parent that elicits pecking by a chick is considered a
   a. conditioned stimulus.
   b. fixed action pattern.
   c. unconditioned stimulus.
   d. conditioned response.
   e. releaser.
   Textbook Reference: *41.1 Behavior Has Proximate and Ultimate Causes, pp. 800–801*
3. Which of the following statements about the development of singing behavior in white-crowned sparrows is *false*?
   a. Social experience can profoundly affect song development.
   b. Female songbirds can be induced to sing by treatment with testosterone.
   c. Females learn their species song, but they do not normally express it under natural conditions.
   d. Testosterone is not necessary for singing in males once they have learned their song.
   e. To sing normally as an adult, a male must hear its species-specific song as a nestling.
   Textbook Reference: *41.3 Developmental Processes Shape Behavior, pp. 805–806*
4. Two species of mice live in the same geographical region, but Species 1 prefers open fields, whereas Species 2 lives in forests. When presented in an experiment with simulated "fields" and "forests," individuals of each species born in a laboratory prefer the environment in which they normally live. This experiment illustrates the concept of
   a. habitat selection.
   b. optimal foraging strategy.
   c. territoriality.
   d. imprinting.
   e. Both a and b
   Textbook Reference: *41.5 Individual Behavior Is Shaped by Natural Selection, p. 814*
5. The time that birds spend scanning the horizon for predators cannot be spent on foraging for food. This situation illustrates the phenomenon of
   a. cooperative hunting.
   b. energetic cost.
   c. risk cost.
   d. optimal foraging strategy.
   e. opportunity cost.

Textbook Reference: *41.5 Individual Behavior Is Shaped by Natural Selection, p. 812*

6. Which of the following is a cost of living in groups?
   a. Enhanced foraging efficiency
   b. Increased risk of disease transmission
   c. Reduced competition
   d. Predator deterrence
   e. None of the above

   Textbook Reference: *41.6 Social Behavior and Social Systems Are Shaped by Natural Selection, pp. 817–818*

7. Hygienic behavior in honey bees
   a. is adaptive because it helps prevent the spread of pathogens in the hive.
   b. has two components, each controlled by a separate gene.
   c. has been studied by hybridizing two closely related species of honey bees.
   d. has been studied using gene knockout experiments.
   e. Both a and b

   Textbook Reference: *41.2 Behavior Can Have Genetic Determinants, pp. 801–802*

8. Animals exhibit daily rhythms in their behavior and physiology. Which of the following statements about circadian rhythms is *false*?
   a. In mammals, the master circadian "clock" is located in the suprachiasmatic nuclei of the brain.
   b. A circadian clock can be entrained by environmental cues.
   c. Free-running circadian clocks are seldom exactly 24 hours long.
   d. Genetic mutations can cause changes in the length of the free-running circadian clock.
   e. Animals that are active at night do not display circadian rhythms.

   Textbook Reference: *41.4 Physiological Mechanisms Underlie Behavior, pp. 806–807*

9. If an animal with a compass sense, but not a map sense, is experimentally relocated, it will
   a. reach its goal.
   b. miss its goal by the extent of the displacement.
   c. head in a direction opposite to that of its goal.
   d. use bicoordinate navigation.
   e. use true navigation.

   Textbook Reference: *41.4 Physiological Mechanisms Underlie Behavior, p. 809*

10. Which of the following is an example of piloting?
    a. Silkworms following a trail of pheromones
    b. Marine migration over a featureless ocean
    c. Bees learning the direction of a food source from a dance
    d. Gray whale migration along the pacific coast of America
    e. None of the above

    Textbook Reference: *41.4 Physiological Mechanisms Underlie Behavior, p. 808*

11. There are several different navigational methods used by animals. Which of the following statements about navigational methods is true?
    a. Distance-and-direction navigation requires knowledge of latitude and longitude.
    b. Bicoordinate navigation requires knowledge of direction and of distance to a destination.
    c. The stars offer two sources of information about direction: a fixed point and moving constellations.
    d. Piloting involves the use of the sun as a compass.
    e. None of the above

    Textbook Reference: *41.4 Physiological Mechanisms Underlie Behavior, pp. 808–809*

12. Which of the following statements about pheromones is true?
    a. Pheromones are used only for communication between individuals of the same species.
    b. Mammalian pheromones can communicate information about the size, sex, and reproductive status of the signaler.
    c. Because of their durability in the environment, pheromones are unsuitable for rapid exchange of information.
    d. Pheromones differ in their volatility and diffusibility depending on their function.
    e. All of the above

    Textbook Reference: *41.4 Physiological Mechanisms Underlie Behavior, p. 810*

13. You observe an example of an apparently altruistic act by an animal that seems to reduce its short-term likelihood of reproductive success. What would be the *least* plausible explanation for this behavior?
    a. The act aids the reproductive success of individuals sharing a high proportion of genes with the altruistic individual.
    b. The act is only apparently altruistic; over the long term, the behavior actually contributes to individual fitness.
    c. The act increases the inclusive fitness of the animal performing it.
    d. The act is advantageous because it helps the species to survive and reproduce, even if the altruist itself does not.
    e. From the information provided, it is impossible to determine the least plausible explanation.

    Textbook Reference: *41.6 Social Behavior and Social Systems Are Shaped by Natural Selection, pp. 815–816*

14. Some birds give a species-specific vocalization called an "alarm call" when they see a predator, although this call may direct the predator toward them. Other members of their species respond to these calls by taking cover. This would be an example of altruistic behavior that is beneficial to the calling bird if
    a. the bird giving the vocalization survives the attack.
    b. the inclusive fitness of the bird giving the vocalization is increased.

c. all birds survive the attack.
d. the birds benefiting are offspring of the bird giving the vocalization.
e. All of the above

Textbook Reference: *41.6 Social Behavior and Social Systems Are Shaped by Natural Selection, pp. 815–816*

15. Which of the following characteristics is shared by all eusocial species?
    a. A sex determination system in which males are haploid and females are diploid
    b. Queens that mate with a single male
    c. The presence of sterile classes
    d. Both a and c
    e. All of the above

    Textbook Reference: *41.6 Social Behavior and Social Systems Are Shaped by Natural Selection, pp. 816–817*

16. Which of the following statements about territoriality is true?
    a. A male defending a territory in a lek is defending a space used only for mating.
    b. Territories do not always include foraging areas.
    c. Territorial defense is likely to impose energetic, risk, and opportunity costs.
    d. All-purpose territories provide food, shelter, and access to mates.
    e. All of the above

    Textbook Reference: *41.5 Individual Behavior Is Shaped by Natural Selection, pp. 812–813*

Answers

Key Concept Review

1. The question of causation would focus on what stimulates natal dispersal. For example, do peaks in certain hormones stimulate dispersal in young squirrels? The question about development would ask whether dispersal was prompted by changes in an individual's internal environment (e.g., does dispersal occur when a squirrel reaches a certain body mass?) or external environment (e.g., does dispersal occur due to increasing aggression from parents?). Questions about the function of dispersal would focus on how leaving home influences the survival and reproductive success of squirrels. Finally, to understand the evolution of natal dispersal, we might examine patterns of dispersal in other species within the squirrel family.

2. It is now recognized that most behaviors have genetic determinants and the potential for modification by experience. The balance between genetic determination and learning differs from one species to the next, and for different behaviors.

3. Because genes often function in gene cascades, a change in a single gene can cause major changes in behavior that impact fitness. For example, the product of a given gene may be a transcription factor that controls the expression of many other genes. Changes in any of these genes, or in their expression, can alter behavior. Thus, gene cascades create many opportunities for a change in one gene to cause a change in a behavior.

4. Male mice are typically aggressive toward infants. One way to test the hypothesis that progesterone mediates male aggression toward infants would be to develop progesterone receptor knockout mice. The gene for progesterone receptors would be targeted and inactivated. Thus, progesterone knockout mice would fail to respond to progesterone because they lack the appropriate receptors. If progesterone receptor knockout males were neutral or positive (i.e., not aggressive) to infants, we could tentatively conclude that progesterone mediates the aggression shown by typical male mice toward infants. We could increase our confidence in this conclusion by conducting additional studies.

5. Sexual behavior in male rats is organized by testosterone early in life and activated by testosterone in adulthood. Urinary posture in male dogs is organized by testosterone early in life, but testosterone is not necessary for the expression of urinary behavior in adulthood. This can be illustrated by the fact that neutered male dogs still lift a leg to urinate, even though their testes (the main source of testosterone) have been removed.

6. If a male songbird is deafened while young, before he has had time to match his song output to the song he memorized, then he will produce very abnormal song. If a male songbird is deafened in adulthood, the deafening will have little, if any, effect on song quality because his song has already crystallized.

7. a. The pigeon will eat from the northern bin at noon. Normally at noon the sun is in the south, and the bird would eat from the eastern bin that is 90° to the left of the sun. With the light in the east, the bird would eat in the north, which is 90° to the left of the light.

   b. The bird will go to the bin that is 90° from the sun at noon, and will feed from the southwestern food bin.

   c. At sunset the eastern bin with food would normally be 180° from the sun in the west. After the bird has been phase-delayed six hours, it will think that the setting sun is the noon sun. At noon the bird usually eats from the eastern bin, which is located 90° to the left of the sun. Therefore, the bird will eat from the southern bin.

8. Honey bees perform the waggle dance to inform other members of the hive about the location of a food source. The angle of the straight run indicates the direction of the food relative to the position of the sun projected down to the horizon. The duration of each waggle run communicates the distance of the food source from the hive, with a longer duration indicating greater distance.

9.

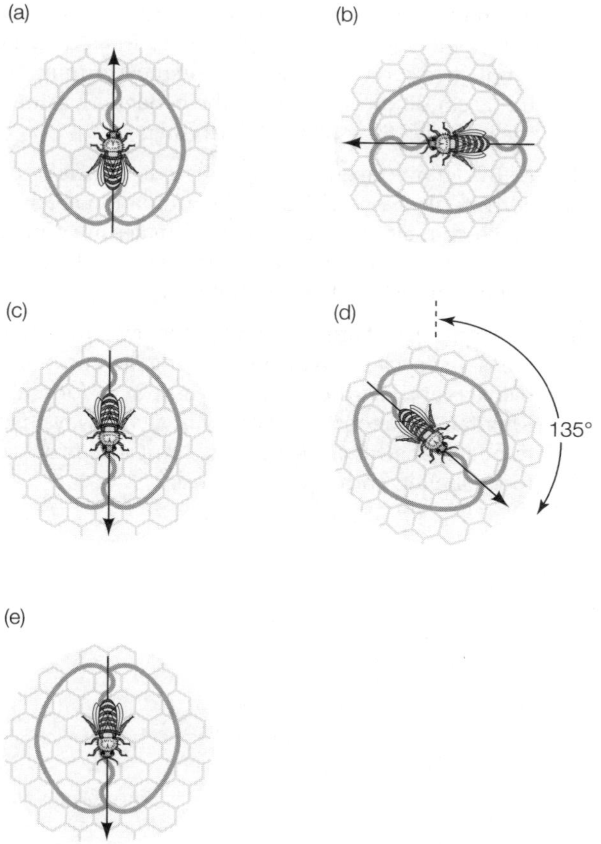

10. Animals looking for a place to settle might preferentially settle in areas where conspecifics are present because presence of conspecifics indicates that such locations are suitable places for individuals of this species to settle (i.e., potential mates may be present and the areas have the correct food, types of shelter, etc.).

11. All species of the Hymenoptera (ants, bees, and wasps) have a sex determination system in which males are haploid and females are diploid; thus females share 75 percent of their genes with their sisters but only 50 percent of their genes with any offspring they could produce. In species whose sexes are determined in this way, females may increase their inclusive fitness by foregoing reproduction and helping to raise sisters. This explanation does not apply to the evolution of eusociality in species without this mode of sex determination.

12. Your curves should show a positive relationship between the hawk's distance when spotted and pigeon flock size and a negative relationship between the hawk's attack success and pigeon flock size (see below).

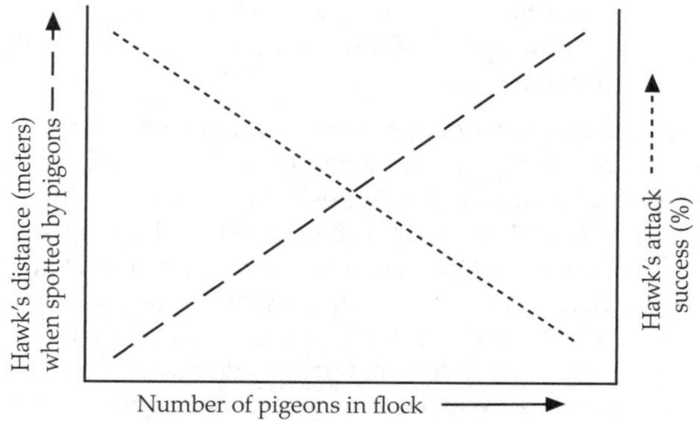

Test Yourself

1. **d.** The first three questions (answer choices a, b, and c) concern the proximate mechanisms that underlie a behavior. The fourth question is concerned with the ultimate cause of a behavior—the selection pressures that shaped its evolution.
2. **e.** The red dot on the bill of a gull parent that elicits pecking by the chick is a releaser.
3. **d.** In the adult, the presence of testosterone is needed to increase the size of regions in the brain associated with singing during the breeding season. Absence of testosterone will result in the inability to perform the correct song.
4. **a.** The environment in which a species normally lives is its habitat. During habitat selection, animals use cues that reliably predict good fitness outcomes.
5. **e.** The forfeited benefits of behaviors that could not be achieved as a result of performing a different behavior, like scanning, constitute the opportunity cost of the performed behavior.
6. **b.** Increased risk of disease transmission is a cost of group living.
7. **e.** Crosses of honey bees exhibiting hygienic behavior with other non-hygienic bees of the same species revealed that two genes control this behavior. The behavior decreases the probability that pathogens will spread to other larvae in the hive.
8. **e.** All animals exposed to a daily cycle display circadian rhythms, regardless of when they are active.
9. **b.** If an animal with a compass sense, but not a map sense, is experimentally relocated, it will miss its goal by the extent of its displacement.

10. **d.** Piloting is a means of navigation using landmarks. Following the coast during migration is considered piloting.

11. **c.** Many animals use the stars for navigation. The stars present two types of information: a fixed reference point (the point directly above Earth's axis of rotation) and moving constellations, which appear to revolve around the fixed point.

12. **e.** By definition, pheromones are chemicals used for communication among individuals of a single species. Mammalian pheromones can communicate information about several characteristics of the signaler, including all those mentioned in answer b. Though pheromones differ in their durability (as well as their volatility and diffusibility) in the environment depending on their function, they are all much more durable than most visual, auditory, or tactile signals and are therefore relatively poorly suited for rapid exchange of information.

13. **d.** For an altruistic behavior to evolve by natural selection, it must increase the inclusive fitness of the individual performing the behavior. Any explanation of an altruistic act based on its purported benefit to the species as a whole violates this principle.

14. **b.** Altruistic behavior is beneficial to the performer when the improvement in the reproductive success of kin (not including offspring) exceeds the reduced reproductive success of the individual performing the act. If this condition is met, then the behavior has improved the inclusive fitness of the performer.

15. **c.** By definition, eusocial species live in social groups with sterile classes. The sex determination mechanism in which males are haploid and females are diploid is found in the Hymenoptera (ants, bees, and wasps) but not in termites and naked mole-rats.

16. **e.** Territories of animals may not include foraging areas. In species in which males engage in communal displays at a lek, the territory is used only for mating. Maintaining a territory is costly in all the ways mentioned in answer c. All-purpose territories provide food, shelter, and access to mates.

Organisms in Their Environment

42

The Big Picture

- Much evidence supporting the theory of continental drift is based on the distributions of organisms on Earth. The acceptance by geologists and biologists of the reality of continental drift revolutionized the field of biogeography.
- The distribution of Earth's physical environments shapes the distribution of organisms. Biomes are distinct physical environments inhabited by ecologically similar organisms. Terrestrial biomes are distinguished largely by differences in vegetation, while aquatic biomes are distinguished by abiotic characteristics such as salinity, depth, and temperature.
- Human activities result in ecosystems that have fewer species and a more uniform physical structure than they would have in the absence of human activities.

Study Strategies

- To get an overview of terrestrial biomes, study Figure 42.10, which shows how terrestrial biomes are determined by average annual temperature and precipitation. These physical variables largely determine the type of vegetation in each biome.
- Also study Figure 42.11 to get a sense of the geographical location and extent of each biome. A useful exercise to help you learn the relative locations of biomes is to imagine a journey (e.g., from equatorial South America to Alaska, or from Maine to California), naming the biomes that you would pass through.
- Go to yourBioPortal.com to review the following tutorials and activity:

  Animated Tutorial 42.1 Rain Shadow

  Animated Tutorial 42.2 Biomes

  Web Activity 42.1 Major Biogeographic Regions

Key Concept Review

42.1 Ecological Systems Vary in Space and over Time

Ecological systems comprise organisms plus their external environment

Ecological systems can be small or large

Each ecological system at each time is potentially unique

Physical geography is the study of the distribution of Earth's climates and surface features. Biogeography is the study of the distributions of organisms. Ecology is the study of interactions among organisms and between organisms and their physical environment.

Ecological systems include organisms and the environment with which they interact. The environment has both abiotic factors (physical and chemical) and biotic factors (living organisms). Ecological systems can be studied from the following perspectives, listed in order of increasing spatial scale: individual organism; population (a group of organisms of the same species living in a particular area at the same time); community (an assemblage of interacting populations of different species within a particular geographic area); ecosystem (communities plus their physical environment); and biosphere (all the organisms and environments of Earth) (see Figure 42.1).

Although large ecosystems tend to be more complex than small ones, even small ecosystems can be very complex. For example, the microbial community of the human gut contains hundreds of species and is one of the most densely populated ecosystems on Earth. Further, this community varies from one person to the next and with diet.

Question 1. Distinguish between a community and an ecosystem.
Textbook Reference: *42.1 Ecological Systems Vary in Space and over Time, p. 824*

Question 2. Describe the human gut and its microbial inhabitants in terms of the following categories: organism, population, community, and ecosystem.
Textbook Reference: *42.1 Ecological Systems Vary in Space and over Time, pp. 824–825*

42.2 Climate and Topography Shape Earth's Physical Environments

Latitudinal gradients in solar energy input drive climate patterns

Solar energy drives global air circulation patterns

The spatial arrangement of continents and oceans influences climate

Walter climate diagrams summarize climate in an ecologically relevant way

Topography produces additional environmental heterogeneity

The climate of a region is the average of the atmospheric conditions found in that region over the long run. Weather is the short-term state of those conditions.

Solar energy varies with latitude. Regions near the poles receive less energy per unit of ground area than regions near the equator because of the lower angle of the sun at high latitudes (see Figure 42.3). The tilt of Earth's axis of rotation causes seasonality (see Figure 42.4).

Rising air expands and cools, releasing moisture, whereas descending air is compressed and warmed, taking up more moisture. Unequal heating of the atmosphere at low and high latitudes produces vertical and latitudinal movements of air masses. These movements lead to very moist climates at the equator and at 60°N and 60°S latitudes (where air rises) and to arid climates at about 30°N and 30°S latitudes and near the poles (where air descends) (see Figure 42.5).

The spinning of Earth on its axis causes air masses moving latitudinally to be deflected to the right in the Northern Hemisphere and to the left in the Southern Hemisphere. Thus, winds blowing toward the equator at low latitudes veer to become the northeast and southeast trade winds, whereas winds blowing away from the equator at mid-latitudes are deflected to become the prevailing Westerlies (see Figure 42.6).

Ocean currents are driven primarily by prevailing winds but are deflected by continents. The poleward movement of ocean water warmed in the tropics is a major mechanism of heat transfer to high latitudes. Winds can also cause upwellings, which are areas where colder water from deep below the water's surface rises to mix with and replace warmer surface water.

Heinrich Walter developed climate diagrams that plot average monthly temperature and precipitation throughout the year. These diagrams summarize climate in a particular location and illustrate when conditions allow terrestrial plant growth (see Figure 42.8).

Topography is variation in the elevation of Earth's surface; it influences physical conditions in terrestrial and aquatic environments. In terrestrial environments, for example, precipitation tends to be greater on the windward side of mountains (where rising air cools and releases moisture) than on the leeward side (where air descends, warms, and holds moisture) (see Figure 42.9).

Question 3. Evaluate the following remark: "Temperatures have been so cold over the last few days that global climate change—in particular global warming—cannot possibly be happening."
Textbook Reference: *42.2 Climate and Topography Shape Earth's Physical Environments, p. 825*

Question 4. If you were going to plant a crop that required a lot of water, would you be better off planting it on the windward or leeward side of a mountain? Explain your choice.
Textbook Reference: *42.2 Climate and Topography Shape Earth's Physical Environments, p. 830*

42.3 Physical Geography Provides the Template for Biogeography

Similarities in terrestrial vegetation led to the biome concept

The biome concept can be extended to aquatic environments

A biome is a distinct physical environment inhabited by ecologically similar organisms. In biomes that occur in several widely separated areas of the globe, species occurring in different locations are unlikely to be closely related phylogenetically, but they are likely to share many adaptations to their environment as a result of convergent evolution.

Differences in vegetation distinguish one terrestrial biome from another. These differences in vegetation reflect gradients in annual patterns of temperature and precipitation (see Figure 42.10). On a global scale, the distribution of terrestrial biomes reflects latitudinal and elevational gradients in temperature and precipitation. A particular biome can occur in widely separated regions (see Figure 42.11). Soil fertility also influences vegetation, and therefore biomes.

Biomes also occur in aquatic environments (see Table 42.1). However, unlike terrestrial biomes that are distinguished by their vegetation, aquatic biomes are not characterized by a structurally dominant group of organisms. Instead, aspects of the physical environment, such as the water depth, temperature, and salinity distinguish aquatic biomes.

Salinity is the prime characteristic used to distinguish aquatic environments because it is the most important determinant of the organisms that can live in a particular aquatic habitat. Freshwater biomes include streams, ponds, and lakes, and saltwater biomes include oceans and salt lakes. Estuarine biomes occur where freshwater and saltwater mix at river mouths.

Freshwater environments can be classified as running water environments (streams and rivers) and standing water environments (lakes and ponds). Bodies of standing fresh water and oceans can be divided into zones based on depth and light penetration (see Figure 42.13). The nearshore regions of lakes (littoral zone) and oceans (littoral or intertidal zone) are shallow and affected by wave action. The deeper, open-water areas of a lake are called the limnetic zone; the open-water areas of the ocean are the pelagic zone. The benthic zone is the bottom of a lake or ocean, and the abyssal zone is the deepest ocean environment. The photic zone (in both freshwater and saltwater environments) extends from the surface to the depth at which photosynthesis can no longer occur. Most aquatic life inhabits this zone. The aphotic zone occurs below the depth at which photosynthesis can occur.

Question 5. How does the basis for distinguishing terrestrial biomes differ from that used to distinguish aquatic biomes?
Textbook Reference: *42.3 Physical Geography Provides the Template for Biogeography, pp. 830–833*

Question 6. Construct a concept map whose theme is "Oceans." Include in your map the following terms: oceans, aphotic zone, currents, depth, direction of prevailing winds, distance from shore, intertidal zone, latitudinal differences in solar energy input, location of continents, pelagic zone, photic zone, photosynthetic organisms, rotation of Earth, and zones. Connect these concepts by verbs, phrases, or comparative terms to indicate the relationships among them.
Textbook Reference: *42.3 Physical Geography Provides the Template for Biogeography, pp. 833–834; 42.2 Climate and Topography Shape Earth's Physical Environments, pp. 828–829*

42.4 Geological History Has Shaped the Distributions of Organisms

- Barriers to dispersal affect the distributions of species
- The movement of continents accounts for biogeographic regions
- Phylogenetic methods contribute to our understanding of biogeography

Earth can be divided into several biogeographic regions, each containing characteristic assemblages of species. The biotas of the biogeographic regions differ because barriers such as oceans or mountains restrict the dispersal of organisms (see Figure 42.15).

Two scientific advances changed the field of biogeography: the acceptance of the theory of continental drift and the development of phylogenetic taxonomy. Continental drift has influenced the evolution and mixing of species throughout the history of life on Earth. It explains some discontinuous distributions that include several biogeographic regions (see Figure 42.16). In the process of biotic interchange, two different biota merge following the fusion of two formerly separated land masses. For example, following the formation of the Central American land bridge connecting North America (the Nearctic region) and South America (the Neotropical region), many species of mammals that had evolved on one continent colonized the other.

Biogeographers use phylogenetic information, together with the fossil record and geological history, to study modern distributions of organisms. For example, they can compare the sequence and timing of splits in a phylogenetic tree with the sequence and timing of movements and splits of geographic areas.

Question 7. South America is a center of diversification for many groups of freshwater fishes, including characins, the group that includes piranhas. Characins also occur in Africa. How would you explain the modern distribution of characins, given that these fishes cannot cross open salt water?
Textbook Reference: *42.4 Geological History Has Shaped the Distributions of Organisms, p. 836*

Question 8. As a biogeographer, what would you conclude if a phylogenetic split in your study organisms coincided in time with formation of a mountain range?
Textbook Reference: *42.4 Geological History Has Shaped the Distributions of Organisms, pp. 837–838*

42.5 Human Activities Affect Ecological Systems on a Global Scale

- Human-dominated ecosystems are more uniform than the natural ones they replace
- Human activities are simplifying remaining natural ecosystems
- Human-assisted dispersal of species blurs biogeographic boundaries

Humans have converted much of Earth's land into croplands, pasturelands, and urban areas. These human-dominated ecosystems are less complex than the natural ecosystems they have replaced. For example, when compared with natural ecosystems, agricultural areas have lower species diversity and are more spatially and physically uniform.

Human activities also influence surrounding natural ecosystems that remain. Examples of such activities include water control measures, the release of pollutants, and the introduction of new species. The deliberate or inadvertent introduction of nonnative species to areas is blurring biogeographic boundaries and homogenizing Earth's biota.

Question 9. At the end of their day of fishing at a local pond, some sportsmen have released a pail of leftover bait fish that are not native to the area. What potential repercussions could result from such an action?
Textbook Reference: *42.5 Human Activities Affect Ecological Systems on a Global Scale, p. 839*

42.6 Ecological Investigation Depends on Natural History Knowledge and Modeling

Models are often needed to deduce testable predictions with complex systems

Ecologists often use the tools of natural history and mathematical modeling. Natural history is the informal observation of nature. Such observations provide critical information at each step of a study, including hypothesis development, experimental design, and interpretation of results. Lack of natural history information limits our ability to answer questions. Models can be built on the basis of natural history information.

Question 10. In your ecology class, you have been given an assignment to conduct a field study concerning the effects on vegetation of excluding white-tailed deer. Although your instructor has suggested a period of informal observation of deer feeding in the study area, your lab partner suggests skipping that phase and going straight to developing an hypothesis and designing the study. Would you agree with your lab partner's plan? Why or why not?
Textbook Reference: *42.6 Ecological Investigation Depends on Natural History Knowledge and Modeling, pp. 839–840*

Test Yourself

1. Which of the following does *not* include multiple communities?
   a. Landscape
   b. Biosphere
   c. Ecosystem
   d. Population
   e. Both a and d
   Textbook Reference: *42.1 Ecological Systems Vary in Space and over Time, p. 824*
2. Which of the following is *not* a difference between the conditions on an acre of land in Colombia and those on acre of land in Michigan?
   a. The angle of the sun reaching the ground in the month of July
   b. The solar energy input in the month of July
   c. The annual solar energy input
   d. The total hours of daylight per year
   e. Both b and c
   Textbook Reference: *42.2 Climate and Topography Shape Earth's Physical Environments, p. 826*
3. If Earth did not spin on its axis, the northeast trade winds would blow from the
   a. northeast.
   b. south.
   c. north.
   d. east.
   e. southwest.
   Textbook Reference: *42.2 Climate and Topography Shape Earth's Physical Environments, pp. 827–828*
4. Which of the following statements concerning oceans is *false*?
   a. Prevailing winds drive ocean currents.
   b. Ocean currents are deflected by land masses.
   c. Oceans intensify Earth's terrestrial climates.
   d. Water circulation in oceans is three-dimensional.
   e. The Gulf Stream brings warm water from the tropical Atlantic Ocean and Gulf of Mexico northward.
   Textbook Reference: *42.2 Climate and Topography Shape Earth's Physical Environments, pp. 828–829*

5.–6. The diagram below shows a mountain with a sea breeze blowing in the direction indicated by the arrow.

5. Circle the letter for the area with air that would be *both* relatively warm and relatively dry.
   Textbook Reference: *42.2 Climate and Topography Shape Earth's Physical Environments, p. 830*

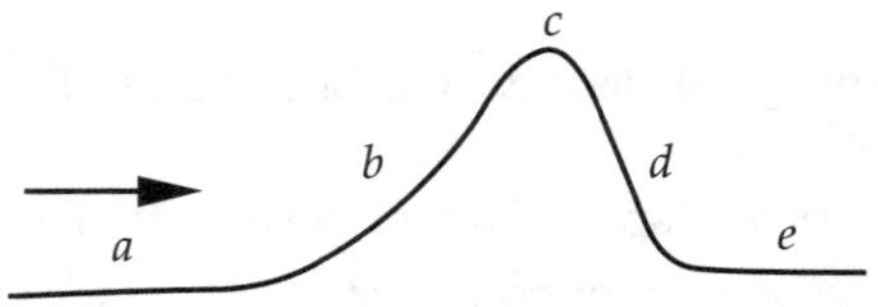

6. In what area of the diagram is a process occurring that is similar to the process that occurs in the region surrounding the equator?
   Textbook Reference: *42.2 Climate and Topography Shape Earth's Physical Environments, p. 827*

7. Which of the following terrestrial biomes is characterized by the highest average temperatures and the most precipitation?
   a. Boreal forest
   b. Tropical savanna
   c. Temperate seasonal forest
   d. Tundra
   e. Tropical rain forest
   Textbook Reference: *42.3 Physical Geography Provides the Template for Biogeography, p. 831*

8. Kangaroo rats of the deserts of the southwestern United States closely resemble jerboas, rodents that inhabit deserts in Asia and Africa. The two groups of rodents are not closely related, yet they both have small forelimbs, large hind limbs, and a long tail. These similarities are an example of
   a. competition.
   b. convergent evolution.
   c. dispersal.
   d. divergent evolution.
   e. None of the above
   Textbook Reference: *42.3 Physical Geography Provides the Template for Biogeography, pp. 830–831*

9. Which of the following statements about biogeography is *false*?
   a. North America and South America have always been connected by a land bridge.
   b. Gondwana was the southern supercontinent.
   c. The Nearctic and Palearctic are northern biogeographic regions.
   d. The Himalayas separate the Oriental and Palearctic regions.
   e. Laurasia was the northern supercontinent.
   Textbook Reference: *42.4 Geological History Has Shaped the Distributions of Organisms, pp. 835–837*

10. In comparison to natural ecosystems, human-dominated ecosystems
    a. have lower species diversity.
    b. have greater species diversity.
    c. are more spatially uniform.
    d. are structurally less diverse.
    e. a, c, and d
    Textbook Reference: *42.5 Human Activities Affect Ecological Systems on a Global Scale, p. 838*

11. Wallace's line separates the _______ and _______ biogeographical regions.
    a. Nearctic; Palearctic
    b. Oriental; Australasian
    c. Neotropical; Nearctic
    d. Oriental; Palearctic
    e. Ethiopian; Antarctic
    Textbook Reference: *42.4 Geological History Has Shaped the Distributions of Organisms, pp. 834–835*

12. Which of the following statements about natural history observations is true?
    a. They lead to new questions.
    b. They allow for the formulation of appropriate hypotheses.
    c. They help in the designing of appropriate experiments.
    d. They provide needed context for the interpretation of laboratory results.
    e. All of the above
    Textbook Reference: *42.6 Ecological Investigation Depends on Natural History Knowledge and Modeling, p. 839*

13. The region of the ocean that lies close enough to shore to be affected by wave action is the _______ zone.
    a. abyssal
    b. pelagic
    c. benthic
    d. aphotic
    e. intertidal
    Textbook Reference: *42.3 Physical Geography Provides the Template for Biogeography, pp. 833–834*

14. Which of the following statements about aquatic biomes is *false*?
    a. In freshwater biomes, slow-flowing water is associated with a soft bottom.
    b. Salinity is the primary characteristic that distinguishes aquatic biomes.
    c. Photosynthetic organisms are confined to the photic zone in both freshwater and saltwater biomes.
    d. In freshwater biomes, fast-flowing water is associated with a rocky bottom.
    e. The limnetic zone of lakes and the pelagic zone of oceans are near-shore environments.
    Textbook Reference: *42.3 Physical Geography Provides the Template for Biogeography, pp. 833–834*

15. An aquatic life zone in which fresh water and salt water mix is called a(n)
    a. estuary.
    b. intertidal zone.
    c. littoral zone.
    d. pelagic zone.
    e. photic zone.
    Textbook Reference: *42.3 Physical Geography Provides the Template for Biogeography, p. 833*

Answers

Key Concept Review

1. A community is an assemblage of interacting populations of different species in a particular geographic area. An ecosystem includes communities plus their physical environment.
2. A bacterium of the genus *Bifidobacterium* is one example of an individual organism inhabiting the human

gut. At a particular time, the gut of a person might have several hundred of these particular bacteria, which would constitute a population (i.e., individuals of one species living in the same area, at the same time, and interacting). There would also be many other species of microbes in this person's gut, and these interacting populations would constitute a community. This community plus the gut environment would make up the ecosystem.

3. The remark demonstrates confusion about what is meant by *weather* (conditions of the atmosphere over a short period of time such as days) and what is meant by *climate* (conditions of the atmosphere over relatively long periods of time).
4. It would be better to plant on the windward side of the mountain because rising air cools and releases moisture, which would be better for your crop. On the leeward side, air descends, warms, and little precipitation falls.
5. Terrestrial biomes are distinguished on the basis of vegetation. Aquatic biomes are not distinguished on the basis of a structurally dominant group of organisms, but instead on the basis of abiotic factors such as water salinity, movement, and depth.
6.

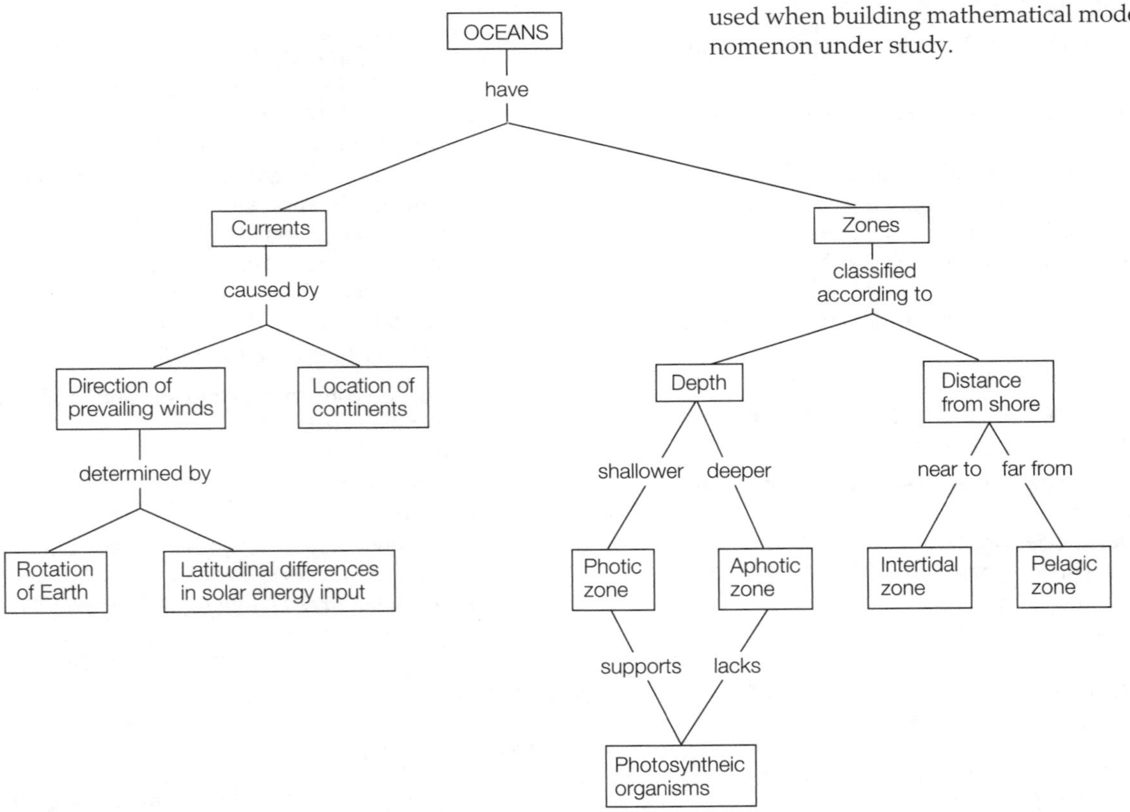

7. The characins existed before the continents split and drifted to their current locations.
8. If a phylogenetic split coincides in time with the formation of a barrier to dispersal (such as the uplift of a mountain range), it is reasonable to conclude that the barrier caused the phylogenetic split by subdividing the original range of the ancestral species.
9. Releasing a nonnative species of fish is a bad idea because the released individuals may survive and thrive in their new environment. If male and female individuals are released, then they may breed and produce offspring, thereby increasing the number of nonnative individuals in the pond. Individuals of a nonnative species may outcompete members of a similar native species or directly eliminate native species through predation. Sometimes nonnative species introduce diseases to which native species are susceptible. Introductions of nonnative species blur biogeographic boundaries and lessen spatial heterogeneity in Earth's species composition.
10. Informal natural history observations are critical to every stage of an ecological study. Such observations help in the generation of questions, development of hypotheses, design of experiments, and interpretation of results. Also, natural history information should be used when building mathematical models of any phenomenon under study.

Test Yourself

1. **d.** A population does not include multiple communities. A landscape, an ecosystem, and the biosphere all include multiple communities.
2. **d.** All areas on Earth receive equal hours of daylight per year. The seasonal distribution of those hours, the solar energy input, and the sun angle do vary latitudinally.
3. **c.** If Earth did not spin on its axis, air flowing south toward the equator would not be deflected to the right. Therefore, the northeast trade winds would blow directly from the North instead of from the Northeast.
4. **c.** Oceans do not intensify terrestrial climates; instead, oceans moderate them.
5. **e.** The air would have lost most of its moisture while rising on the windward side of the mountain, and in descending to area *e* it would have warmed again.
6. **b.** In regions surrounding the equator, warm air rises and loses much of its moisture. These same events occur when moist air rises over a mountain.
7. **e.** Tropical rain forests have the highest average temperatures and the most precipitation.
8. **b.** The resemblance between kangaroo rats and jerboas is an example of convergent evolution.
9. **a.** North and South America have not always been connected by a land bridge. The land bridge that formed about 6 million years ago resulted in the Great American Interchange.
10. **e.** When compared with natural ecosystems, human-dominated ecosystems have lower species diversity, are more spatially uniform, and are less structurally diverse.
11. **b.** Wallace's line separates the Oriental and Australasian regions.
12. **e.** Natural history observations are critical to all stages of an ecological study, including the generation of new questions, hypothesis development, experimental design, and the interpretation of results.
13. **e.** The intertidal zone is affected by wave action.
14. **e.** The limnetic zone of lakes and the pelagic zone of oceans are open-water, off-shore environments.
15. **a.** An estuary is the body of water found at the mouth of a river, where salt water mixes with fresh water.

Populations 43

The Big Picture

- Concepts of population ecology such as multiplicative growth and carrying capacity are fundamental to the study of human population growth. They also contribute to advances in the preservation of biodiversity and the control of undesirable species.
- Darwin's realization that all populations have the inherent capacity for multiplicative growth was crucial to the development of his theory of natural selection.

Study Strategies

- It can be difficult to interpret graphs depicting concepts of population ecology. Look carefully at how the axes of a graph are labeled. Ask yourself questions such as: Are the scales arithmetic or logarithmic? Does the graph trace the fate of a cohort of the population or the growth pattern of the entire population? Redrawing graphs from the textbook is a good way to reinforce your understanding of the concepts being presented.
- It is important to understand how an ever-faster per capita *growth* rate of the human population can be accompanied by a steadily *decreasing* doubling time. Review the definition of doubling time in the text (p. 851) and study Figure 43.9A.
- Take advantage of laboratory activities involving computer simulations of population growth or other aspects of population ecology.
- Go to yourBioPortal.com to review the following tutorials and activities:

  Animated Tutorial 43.1 Multiplicative Population Growth

  Animated Tutorial 43.2 Logistic Population Growth

  Animated Tutorial 43.3 Habitat Corridors

  Web Activity 43.1 Population Growth

  Working with Data 43.1 Habitat Corridors

Key Concept Review

43.1 Populations Are Patchy in Space and Dynamic over Time

Population density and population size are two measures of abundance

Abundance varies in space and over time

A population consists of the individuals of a species that interact with one another within a given area at a particular time. A species' role in a particular community is determined by characteristics of individuals as well by the relative abundance of the species, which is a population-level characteristic. Abundance can be measured as population density (the number of individuals per unit of area or volume) or as population size (the total number of individuals in the population). In particular areas, population densities change over time. The region in which a species is found is called its geographic range. Within the geographic range, a species occupies particular habitats. These may exist as habitat patches—areas of suitable habitat surrounded by unsuitable habitat.

Question 1. You have been live-trapping, marking, and releasing voles in a 1 hectare plot for several months. Based on your field work, you calculate the population density of the study plot to be 150 voles per hectare. The total area occupied by the population is 4 hectares. What is the total population size?
Textbook Reference: *43.1 Populations Are Patchy in Space and Dynamic over Time, p. 843*

Question 2. Timber rattlesnakes are found in the eastern half of the United States, north to southern Maine, south to northern Florida and west to central Texas. This species of snake prefers forested areas, particularly those containing wooded rocky ledges with southern exposures for basking. Areas with high rodent densities are ideal. Roads pose a great risk and agricultural fields are avoided. Explain the

following concepts and how they apply to this species: geographic range, habitat, and habitat patch.
Textbook Reference: *43.1 Populations Are Patchy in Space and Dynamic over Time, p. 843*

Question 4. Why is it unusual for population densities to remain unchanged over time?
Textbook Reference: 43.2 Births Increase and Deaths Decrease Population Size, p. 845

43.2 Births Increase and Deaths Decrease Population Size

Births add individuals to populations and deaths remove them. The "birth–death," or BD, model of population change states that the number of individuals in a population at some time in the future equals the number now, plus the number that are born, minus the number that die ($N_{t+1} = N_t + B - D$, where N is the population size, B is the number of births in the time interval from time t to time $t + 1$, and D is the number of deaths in that same time interval; equation 43.1). The growth rate of a population refers to the change in population size in a certain period of time ($\Delta N/\Delta T = B - D$, where the Greek symbol Δ means "change in"; equation 43.2).

It is often difficult to directly measure change in size of a total population of individuals, thus ecologists keep track of a sample of individuals, called a cohort. Using this sample, ecologists can calculate the per capita birth rate (b), which is the number of offspring produced by an average individual, and the per capita death rate (d), which is the average individual's chance of dying. The per capita birth rate minus the per capita death rate represents the average individual's contribution to the total population growth rate; this value is called the per capita growth rate and is symbolized by r. The following equation allows us to predict changes in population size: $\Delta N/\Delta T = rN$ (equation 43.4). If the per capita birth rate is greater than the per capita death rate, then r will be greater than zero and the populations will grow. If the per capita birth rate is less than the per capita death rate, then r will be less than zero and the population will shrink. If the per capita birth rate equals the per capita death rate, then r will equal zero and the size of the population will not change.

Question 3. An ecologist who has been monitoring a cohort of Couch's spadefoot toads in a locality in California describes the total population size as declining. What does this tell you about the per capita birth rates and death rates of the study cohort?
Textbook Reference: *43.2 Births Increase and Deaths Decrease Population Size, pp. 844–845*

43.3 Life Histories Determine Population Growth Rates

- Life histories are quantitative descriptions of life cycles
- Life histories are diverse
- Resources and physical conditions shape life histories
- Species' distributions reflect the effects of environment on per capita growth rates

The life history of a species includes information on the time course of an average individual's growth, development, reproduction, and death. This information can be summarized in a life table (see Table 43.1). Survivorship is the fraction of individuals that survive to different ages; it can also be expressed as mortality, (which equals 1 – survivorship). Fecundity is the average number of offspring each individual produces at each age. Survivorship and fecundity influence the per capita growth rate (r). Life histories vary considerably, both across species and within species.

Organisms need resources (materials and energy), and the time required to acquire them. They also need physical conditions they can tolerate. The distinction between these resources and conditions is that resources can be used up whereas conditions are experienced. The rate of resource acquisition increases with resource availability (see Figure 43.4). The principle of allocations states that a unit of an obtained resource can be used for only one function at a time. The resources obtained by organisms must be divided among competing functions, which include maintenance of homeostasis (usually the first priority), foraging, growth, reproduction, and defense (see Figure 43.5). Resource allocation to specific functions changes in relation to the abundance of resources and whether conditions are typical or stressful. Life-history tradeoffs are the negative relationships among growth, reproduction, and survival. For example, species that invest in reproduction have high fecundity but low survival. Species distributions can be predicted once we know how resource availability and physical conditions affect survivorship and fecundity (see Figures 43.6 and 43.7).

Question 5. Predict how higher survivorship and higher fecundity would influence per capita growth rate.
Textbook Reference: *43.3 Life Histories Determine Population Growth Rates, p. 846*

Question 6. Many small mammals produce large litters every few weeks during the breeding season. Given this aspect of their life history, what would you expect regarding their survivorship? This is an example of what kind of phenomenon?
Textbook Reference: *43.3 Life Histories Determine Population Growth Rates, p. 848*

43.4 Populations Grow Multiplicatively, but Not for Long

Multiplicative growth generates large numbers very quickly

Multiplicatively growing populations have a constant doubling time

Density dependence prevents populations from growing indefinitely

Variable environmental conditions cause the carrying capacity to change

Technology has increased Earth's carrying capacity for humans

For a period of time, populations can grow multiplicatively, which means that a constant multiple of the population size (*N*) is added during each time period. In contrast, when populations exhibit additive growth, a constant number of individuals is added during each time period. Multiplicative growth generates large numbers very quickly (the growth pattern is J-shaped) and has a constant doubling time, providing *r* does not change.

Populations cannot exhibit multiplicative growth indefinitely. Typically, such as during logistic growth, population growth slows and levels off at the carrying capacity (*K*), which is the number of individuals the environment can support indefinitely. Populations stop growing because *r* is density-dependent; *r* is highest when population densities are low, and *r* decreases as population densities increase (see Figure 43.8). At high population densities, each individual has access to fewer resources, and as a result, birth rates decrease and death rates increase, causing decreases in *r*.

The carrying capacity varies over space because resource availability varies over space, as do physical conditions that affect the costs of maintaining homeostasis. The carrying capacity also varies over time because resource availability and physical conditions vary over time, such as over years or across seasons.

Among populations of large animals, the human population stands out as one that has continued to grow at an ever-faster per capita rate, with its doubling time decreasing dramatically from the 1500s to the 1900s (see Figure 43.9A). The fast rate of growth of the human population has been made possible by technological advances that increase food production or improve health and therefore increase the carrying capacity (see Figure 43.9B). It is possible, however, that the human population has now exceeded its carrying capacity; the evidence for this includes the predicted decline in our finite fuel resources upon which the technological advances are based and environmental problems such as climate change and degradation of ecosystems. Change in size of the human population can occur through voluntary reductions in per capita birth rate or by the less appealing prospect of increasing mortality.

Question 7. Why is the per capita growth rate described as density dependent?
Textbook Reference: *43.4 Populations Grow Multiplicatively, but Not for Long, p. 852*

Question 8. Just as the per capita growth rate of a population can be described as density dependent, so too can other factors that influence population size. Of the following three factors, which would you describe as density dependent: (1) disease; (2) weather; and (3) starvation?
Textbook Reference: *43.4 Populations Grow Multiplicatively, but Not for Long, p. 852*

43.5 Extinction and Recolonization Affect Population Dynamics

Most populations are divided into geographically separated subpopulations that live in habitat patches—areas of suitable habitat that are separated from other patches by unsuitable environments. Individuals can disperse among the patches. The larger population to which the subpopulations belong is called the metapopulation.

Each subpopulation changes through births and deaths. Subpopulations also gain individuals through immigration and lose individuals through emigration. The BIDE model states that the number of individuals in a population at some time in the future equals the number now, plus the number that are born, plus the number that immigrate, minus the number that die, minus the number that emigrate.

Because individual subpopulations are much smaller than the metapopulation, they are prone to fluctuations leading to extinction. As long as subpopulations are connected by dispersal, however, they can be rescued from extinction by immigration.

Question 9. How does the division of some populations into discrete subpopulations relate to the possible "rescue" of a subpopulation from extinction? How would barriers to immigration influence the potential for rescue?
Textbook Reference: *43.5 Extinction and Recolonization Affect Population Dynamics, p. 854*

Question 10. Metapopulation A of crustaceans has 8 subpopulations, each with more than 50 individuals. Metapopulation B of crustaceans has 4 subpopulations, each with fewer than 30 individuals. In the face of environmental disturbance, which metapopulation would likely go extinct sooner and why?
Textbook Reference: *43.5 Extinction and Recolonization Affect Population Dynamics, p. 854*

43.6 Ecology Provides Tools for Managing Populations

Knowledge of life histories helps us to manage populations

Knowledge of metapopulation dynamics helps us conserve species

Life history information can identify particular life stages that are most important for reproduction, survival, and hence the population growth rate. This information, in turn, can be used to manage populations of species considered desirable or undesirable.

Conservation strategies also have been informed by knowledge of metapopulations. For example, once any remaining habitat has been identified and the risks to habitat patches have been evaluated, conservation efforts often focus on protecting as many habitat patches as possible, with priority given to the largest patches. Because opportunities for dispersal among patches are critical, corridors for dispersal may be protected or created.

Question 11. What are habitat corridors, and how have experiments demonstrated that they help species persist in patchy environments?
Textbook Reference: *43.6 Ecology Provides Tools for Managing Populations, pp. 856–857*

Question 12. Construct a concept map whose theme is "Metapopulation." Include in your map the following terms: metapopulation, subpopulations, births, deaths, immigrants, emigrants, recolonization, and dispersal corridor. Connect these concepts by verbs, phrases, or comparative terms to indicate the relationships among them.
Textbook Reference: *43.6 Ecology Provides Tools for Managing Populations, pp. 856–857*

Test Yourself

1. There are 150 golden shiners living in a pond. Which of the following would you need to know to calculate their population density?
   a. Their birth rate
   b. Their growth rate
   c. Their death rate
   d. The volume of water in which they live
   e. All of the above
   Textbook Reference: *43.1 Populations Are Patchy in Space and Dynamic over Time, p. 843*
2. *Per capita* means
   a. per population.
   b. per cohort.
   c. per individual.
   d. per family.
   e. per community.
   Textbook Reference: *43.2 Births Increase and Deaths Decrease Population Size, p. 845*
3. Which of the following statements about life histories is *false*?
   a. Life histories are essentially quantitative descriptions of life cycles.
   b. A species' life history reflects the timing of growth, development, reproduction, and death of an average individual.
   c. Life histories vary across species.
   d. Within species, life histories are invariant.
   e. A species' life history can be summarized in a life table.
   Textbook Reference: *43.3 Life Histories Determine Population Growth Rates, p. 846*
4. When resources such as food and roost sites are available at normal levels but environmental temperatures drop to stressful levels, a crow would likely
   a. allocate more resources to homeostasis.
   b. allocate more resources to reproduction.
   c. allocate more resources to growth.
   d. allocate more resources to defense.
   e. keep its resource allocation the same as when conditions were typical.

Textbook Reference: *43.3 Life Histories Determine Population Growth Rates, p. 848*

5. A positive per capita growth rate
   a. occurs when the per capita birth rate is greater than the per capita death rate.
   b. is expected to characterize populations that persist over time.
   c. occurs when the per capita birth rate is less than the per capita death rate.
   d. is expected to characterize populations that do not persist over time.
   e. Both a and b

   Textbook Reference: *43.2 Births Increase and Deaths Decrease Population Size, p. 845; 43.3 Life Histories Determine Population Growth Rates, p. 848*

6. Which of the following populations would grow most rapidly?
   a. A population exhibiting additive growth
   b. A population at its carrying capacity
   c. A population approaching its carrying capacity
   d. A population well below its carrying capacity and exhibiting multiplicative growth
   e. They would all grow at the same rate.

   Textbook Reference: *43.4 Populations Grow Multiplicatively, but Not for Long, pp. 850–852*

7.–8. Refer to the graph below.

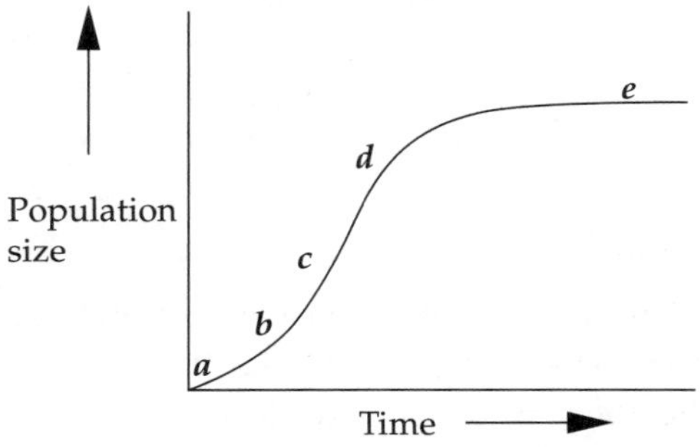

7. In the logistic population growth curve shown in the graph, the rate of growth is greatest at which point?

   Textbook Reference: *43.4 Populations Grow Multiplicatively, but Not for Long, pp. 851–852*

8. At which point in the graph would there be zero population growth ($r = 0$)?

   Textbook Reference: *43.4 Populations Grow Multiplicatively, but Not for Long, pp. 851–852*

9. A decreasing doubling time indicates
   a. an increasing per capita growth rate.
   b. a decreasing per capita growth rate.
   c. higher per capita death rates than per capita birth rates.
   d. that the population has reached equilibrium.
   e. logistic growth.

   Textbook Reference: *43.4 Populations Grow Multiplicatively, but Not for Long, pp. 850–853*

10. Which of the following statements about carrying capacity is *false*?
    a. Technology has increased Earth's carrying capacity for humans.
    b. At carrying capacity, $r = 0$ and populations stop growing.
    c. Carrying capacity is the number of individuals that the environment can support indefinitely.
    d. The carrying capacity of a particular environment can fluctuate over time.
    e. In a population exhibiting logistic growth, per capita growth rate is negative when the population density is below the carrying capacity.

    Textbook Reference: *43.4 Populations Grow Multiplicatively, but Not for Long, pp. 852–853*

11. The BIDE model of population growth differs from the BD model in that it
    a. includes immigration but not emigration.
    b. includes emigration but not immigration.
    c. includes both immigration and emigration.
    d. focuses only on births and deaths.
    e. does not account for the possibility of extinction.

    Textbook Reference: *43.5 Extinction and Recolonization Affect Population Dynamics, p. 854*

12. A metapopulation
    a. includes a cluster of distinct subpopulations linked by dispersal.
    b. has no carrying capacity.
    c. occurs when births equal deaths.
    d. has constant population density.
    e. None of the above

    Textbook Reference: *43.5 Extinction and Recolonization Affect Population Dynamics, p. 854*

13. Which of the following animals would likely need a continuous corridor of habitat for successful dispersal to a new patch of suitable habitat?
    a. Bat
    b. Bird
    c. Butterfly
    d. Salamander
    e. Dragonfly

    Textbook Reference: *43.6 Ecology Provides Tools for Managing Populations, pp. 856–857*

14. According to the BIDE model, the long-term fate of a population depends on
    a. immigration.
    b. emigration.
    c. births.
    d. deaths.
    e. All of the above

    Textbook Reference: *43.5 Extinction and Recolonization Affect Population Dynamics, p. 854*

15. Which of the following would aid in the persistence of subpopulations of a metapopulation?

a. Corridors for dispersal
b. Barriers to recolonization
c. Large size of subpopulations
d. Large number of subpopulations
e. a, c, and d

Textbook Reference: *43.6 Ecology Provides Tools for Managing Populations, pp. 856–857*

Answers

Key Concept Review

1. It is usually impossible to count all of the individuals in a population, so researchers typically measure population density and then multiply it by the area occupied by the population to calculate total population size. In this example, 150 voles per hectare × 4 hectares = 600 voles for the total population size.
2. A species' geographic range is the region where it is found. For timber rattlesnakes, it would be the eastern half of the United States, north to southern Maine, south to northern Florida, and west to central Texas. Within its geographic range, a species may be restricted to a particular type of environment, which is called its habitat. For timber rattlesnakes, the habitat would be forested areas with rocky ledges and high rodent densities. Sometimes suitable habitat occurs in patches surrounded by unsuitable habitat. Roads and agricultural areas may fragment the forested habitat of timber rattlesnakes into patches of suitable habitat.
3. If the total population size is described as decreasing then the per capita death rate must exceed the per capita birth rate of the cohort under study.
4. It is unusual for population densities to remain unchanged over time because this situation would only occur when the number of births exactly equals the number of deaths.
5. Higher survivorship and higher fecundity would typically result in a higher per capita growth rate.
6. Species such as small mammals that invest heavily in reproduction often do so at the expense of survivorship, so we would expect short life expectancies in these species. The negative relationship between reproduction and survival is an example of a life-history tradeoff.
7. As population density increases and resources become scarcer, birth rates decline and death rates increase. In other words, r decreases as the population becomes more crowded. Thus, the per capita growth rate is said to be density dependent.
8. Disease and starvation would be considered density-dependent factors because they have a greater impact as conditions become more crowded (i.e., as population density increases). Weather is not a density-dependent factor.
9. Rescue occurs if a subpopulation that has undergone a large decline is saved from extinction by the immigration of individuals from other subpopulations. As one would predict, barriers to immigration reduce the potential for rescue.
10. Metapopulation B will likely go extinct sooner because the time to extinction is shorter for the metapopulation with fewer and smaller subpopulations.
11. Habitat corridors are relatively thin strips of habitat of a particular type that connect larger patches of the same type of habitat. Their importance in permitting individuals to disperse from one patch to another was demonstrated in experiments described in Figure 43.11 of the textbook.
12.

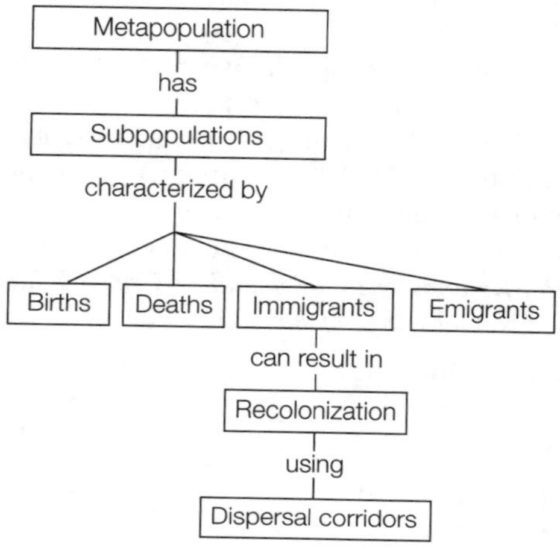

Test Yourself

1. **d.** Population density is the number of individuals per unit volume for organisms living in water.
2. **c.** Per capita means per individual.
3. **d.** Life histories vary across species and within species.
4. **a.** When resources are available at normal levels but environmental temperatures drop to stressful levels, a crow would likely allocate more resources to homeostasis.
5. **e.** A positive per capita growth rate occurs when the per capita birth rate is greater than the per capita death rate, and this would be expected to characterize populations that persist over time.
6. **d.** A population well below its carrying capacity exhibiting multiplicative growth would grow most rapidly.
7. **c.** The curve showing the relationship between population size and time is steepest at point c, so the growth rate would be greatest at that point.
8. **e.** The growth rate at e would be zero. This is the point called the environmental carrying capacity at which the birth and death rates are equal.

9. **a.** A decreasing doubling time indicates an increasing per capita growth rate. This is the case for the human population from the 1500s to the 1900s (see Figure 43.9A).

10. **e.** In a population exhibiting logistic growth, per capita growth rate is positive when the population density is below the carrying capacity.

11. **c.** The BIDE model of population growth differs from the BD model by including immigration and emigration. The BIDE model makes it possible for immigration to replace an extinct population.

12. **a.** A metapopulation has many subpopulations linked by dispersal.

13. **d.** A salamander is a terrestrial animal with limited dispersal abilities, so it would need a continuous corridor of habitat. Animals that fly can more readily reach patches separated by unsuitable habitat.

14. **e.** The BIDE model suggests that the long-term fate of a population depends on characteristics of the particular population, such as births, deaths, and emigration, and also on immigration from other populations.

15. **e.** Subpopulations would be more likely to persist if they have corridors for dispersal and many surrounding subpopulations of large size. Barriers to recolonization reduce the number of subpopulations that persist in a metapopulation.

Ecological and Evolutionary Consequences of Species Interactions

44

The Big Picture

- An individual's fitness is affected by both intraspecific interactions and interspecific interactions.
- Some mutualistic relationships have extraordinary evolutionary or ecological significance. Previous chapters have discussed the role of endosymbiosis in the evolution of eukaryotic cells and the association of nitrogen-fixing bacteria with their host plants. In this chapter we see that mutualistic relationships are diverse, often complex, and may involve cheating.
- The typically negative effect of invasive species on native species illustrates the complex web of interspecific interactions that characterizes ecological communities.

Study Strategies

- This chapter describes many different categories of interspecific interactions and introduces some terms relating to these categories that may be new to you. To learn the basic types of interaction, study the chart presented in Figure 44.1. Then try creating your own charts or concept maps to organize the material in more detail and to master the terminology. Focus also on how each type of interspecific interaction can modify per capita growth rate. Study the word equation concerning interspecific competition (p. 863) and develop similar word equations for the other interspecific interactions.
- The idea that intraspecific competition can increase carrying capacity through directional selection can be challenging. Review the example presented on pp. 865–866 of the textbook.
- Go to yourBioPortal.com to review the following tutorial and activity:

  Animated Tutorial 44.1 An Ant–Plant Mutualism

  Web Activity 44.1 Ecological Interactions

Key Concept Review

44.1 Interactions between Species May Be Positive, Negative, or Neutral

Interspecific interactions are classified by their effect on per capita growth rates

Many interactions have both positive and negative aspects

Interactions between individuals of different species are called interspecific interactions. Interspecific interactions influence population densities, species distributions, and lead to evolutionary change in one or both species.

Interactions between species in a community fall into five general categories that reflect whether the outcome of the interaction is positive (+), negative (–), or neutral (0) for each species' per capita growth rate (see Figure 44.1A). This system yields five broad categories of interspecific interactions: competition (–/–), consumer–resource (+/–), mutualism (+/+), commensalism (+/0), and amensalism (–/0).

Interspecific competition occurs when two or more species use the same resource. Limiting resources include food, shelter, water, space, and light (see Figure 44.1B).

In consumer–resource interactions, the consumer benefits and the consumed organism (the resource) loses (see Figure 44.1C). These interactions can be further classified as (1) predation, in which an individual of one species kills and consumes multiple individuals of another species; (2) herbivory, in which an animal consumes part or all of a plant; and (3) parasitism, in which an individual of one species consumes only certain tissues of its host species, typically without killing the host. Pathogens are parasites that cause symptoms of disease in their hosts.

Mutualism is a type of interaction between species that benefits both (see Figure 44.1D). Examples include leaf-cutter ants and their fungi, plants and the animals that pollinate them or disperse their seeds, and humans and their beneficial gut bacteria.

Commensalism is an interaction that benefits one participant while leaving the other unaffected. The interactions between cowbirds and ungulates are commensal because the cowbirds benefit from insects flushed from the grass as cows and bison walk, but the ungulates are unaffected by cowbird presence.

Amensalism is harmful to one participant while leaving the other unaffected. These interactions are often accidental, as when members of a herd of elephants crush insects and plants as they walk.

These five types of interaction are not always so easy to distinguish, and their outcomes depend on ecological circumstances (see Figure 44.2).

Question 1. Two of the main characters in the animated film *Finding Nemo* are anemonefish (also known as clown fish), which live inside of sea anemones and are unaffected by their stings. How is this relationship an example of the ways in which interactions between species are not always easily classifiable?
Textbook Reference: *44.1 Interactions between Species May Be Positive, Negative, or Neutral, pp. 861–862*

Question 2. You have discovered two individuals of different species living in close association. Design an experiment to determine whether the relationship is mutualistic.
Textbook Reference: *44.1 Interactions between Species May Be Positive, Negative, or Neutral, pp. 860–862*

44.2 Interspecific Interactions Affect Population Dynamics and Species Distributions

- Interspecific interactions can modify per capita growth rates
- Interspecific interactions can lead to extinction
- Interspecific interactions can affect the distributions of species
- Rarity advantage promotes species coexistence

Recall from Chapter 43 that per capita growth rates typically decrease with increasing population density; this reflects intraspecific interactions, and likely *intraspecific* competition for resources. The word equation on page 852 of the text can be extended to take into account *interspecific* competition: per capita growth rate (*r*) of species A = {maximum possible *r* for species A in uncrowded conditions – an amount that is a function of A's own population density} – {an amount that is a function of the population density of competing species B}. Interspecific interactions other than competition can be similarly described, with the effects of the other species either subtracted or added, depending on the nature of the interaction. For example, in a mutualistic interaction, the effect of each species would be added to the equation for the other because both species benefit.

Interspecific interactions have the following general consequences: (1) the per capita growth rate of each species is modified by the presence of the other species; (2) the average population densities of each species differ in the presence and absence of the other species; and (3) in certain interactions, such as consumer–resource interactions, extinction of one or both of the interacting species can occur. These consequences are illustrated in Figure 44.3 for competing species of *Paramecium*.

Interspecific interactions also influence the distributions of species. For example, competitive interactions can restrict distributions (see Figure 44.4).

Mathematical models of interspecific competition reveal that two competing species can coexist when intraspecific competition is stronger than interspecific competition. When species A is at low density and competing species B is at high density, species A gains an advantage, called the rarity advantage. This advantage prevents species A from decreasing to zero, and coexistence with species B results. Interspecific competition may be weaker than intraspecific competition due to resource partitioning, which occurs when competing species differ somewhat in their resource use (see Figure 44.5).

Question 3. How might the rarity advantage apply to a consumer–resource interaction, such as that between predators and their prey?
Textbook Reference: *44.2 Interspecific Interactions Affect Population Dynamics and Species Distributions, p. 865*

Question 4. Two species, A and B, have a commensal relationship in which species A benefits and species B is unaffected. Develop a word equation for species A that describes the effect of species B on its growth rate. As an example, see the word equation for interspecific competition on page 863 of the textbook. Do you need a reciprocal term in the word equation for species B in the commensal relationship?
Textbook Reference: *44.2 Interspecific Interactions Affect Population Dynamics and Species Distributions, p. 863*

44.3 Interactions Affect Individual Fitness and Can Result in Evolution

- Intraspecific competition can increase carrying capacity
- Interspecific competition can lead to resource partitioning and coexistence
- Consumer–resource interactions can lead to an evolutionary arms race
- Mutualisms can involve exploitation and cheating

Intraspecific competition can increase the carrying capacity of a population. This occurs when competition among individuals in the population results in directional selection for traits that allow individuals to use different resources or to be more efficient in their use of resources.

Interspecific competition can lead to resource partitioning, which in turn can allow competing species to coexist. Resource partitioning is thus an evolutionary response to interspecific competition (see Figures 44.6 and 44.7).

Consumer–resource interactions are those involving predators and their prey, parasites and their hosts, and herbivores and the plants on which they feed. In each of these three interactions, the species have opposing interests, and this can lead to an evolutionary arms race. For example, prey species continually evolve better defenses and predators continually evolve better ways to find, capture, and consume their prey. In an evolutionary arms race, neither prey nor predator has any lasting advantage.

Species do not display traits that have evolved solely to help another species. Instead, species A displays traits that have evolved because they benefited individuals of species A; these traits may happen to benefit individuals of species B as well. For example, pollinators visit flowers to get food and happen to pollinate the flowers when doing so.

Mutualisms involve the exchange of resources and services. The fitness effects of mutualistic relationships can vary with environmental conditions. For example, mycorrhizal fungi are beneficial to plants in nutrient-poor soils, but they are costly to plants in nutrient-rich soils. Cheating is characteristic of mutualisms.

Question 5. Evaluate the following statement: "Whereas plants sometimes benefit from herbivores, the prey of carnivores never benefit from being eaten." Explain why this statement is accurate or inaccurate.
Textbook Reference: *44.3 Interactions Affect Individual Fitness and Can Result in Evolution, pp. 866–869*

Question 6. Interspecific cleaning interactions occur in many natural systems. For example, some birds ride on ungulates and remove ectoparasites, benefitting both the birds and the ungulates. In other cases, animals come to a cleaning station for removal of ectoparasites. In a familiar example, large predatory fish come to cleaning stations because their gill cavities are infested with crustacean ectoparasites that interfere with respiration. The small cleaning fish swim into the mouths and gill cavities of the predators to eat the parasitic crustaceans. In this example, what is an actual or potential cost to each of the three interacting species (large predatory fish; small cleaner fish; parasite)? What is one benefit to the large predatory fish and small cleaner fish?
Textbook Reference: *44.3 Interactions Affect Individual Fitness and Can Result in Evolution, pp. 868–869*

44.4 Introduced Species Alter Interspecific Interactions

Introduced species can become invasive

Introduced species alter ecological relationships of native species

Introduced species that reproduce rapidly and spread widely are described as invasive. Invasive species usually affect native species negatively by outcompeting or eating them. Some introductions are accidental (for example, marine organisms are spread through the discharge of ballast water), while others are deliberate (for example, species are sometimes introduced to control other invasive species).

Invasive species alter interactions among native species. Introduced purple loosestrife has disrupted the relationship between a native plant and its pollinators (see Figure 44.10). Some invasive species alter ecological interactions by causing extinctions. Because species are parts of a web of interactions, impacts on one species can influence interspecific interactions throughout a community.

Question 7. After the European rabbit was introduced into Australia in 1859, it quickly became a devastating invasive species, causing the extinction or decline of many native animals and plants, major damage to crops, and soil erosion. To control the exponential growth of the rabbit population (which had grown to an estimated 600 million individuals from an original 24), biologists infected rabbits with the myxoma virus in 1950. It is estimated that in the first years after introduction of the virus, as many as 99 percent of infected rabbits died. Over a period of years, however, biologists observed an increase both in the average life span of infected rabbits and in the percentage of rabbits surviving infection. How would you explain these changes and what does this say about the effectiveness of this control measure?
Textbook Reference: *44.4 Introduced Species Alter Interspecific Interactions, pp. 869–870*

Question 8. The Great Lakes of North America have been subjected to many waves of introduced species as a result of canal construction in the eighteenth and nineteenth centuries. This lead to colonization by species from eastern North America, such as the sea lamprey. Subsequently, increases in international shipping traffic to the Great Lakes introduced species from other continents via discharge of ballast water, including the zebra mussel from Europe. Other species, such as coho salmon, a species native to the Pacific coast of North America, were deliberately introduced to enhance recreational fishing. Most recently, species of carp native to Asia have escaped from fish farms and now threaten to invade the Great Lakes via canals in the Chicago area that connect rivers of the Mississippi system to Lake Michigan. Draw a diagram indicating these four invasive pathways to the Great Lakes, specifying whether the species introductions were accidental

or intentional. Since there are different sources and pathways for these invasive species, and many invasive species now are well established in the Great Lakes, can you envision any way that the negative impacts of such species can be minimized without altering current Great Lakes ecosystems or the human infrastructure of canals, shipping pathways, or fish farms?
Textbook Reference: *44.4 Introduced Species Alter Interspecific Interactions, pp. 869–870*

Test Yourself

1. Which of the following does *not* affect the fitness of both participants?
   a. Sheep grazing on grass in a pasture
   b. A robin catching and eating a worm
   c. A flea biting a dog to obtain a blood meal
   d. Lions skirmishing with hyenas to prevent them from feeding from a zebra carcass
   e. All of the above affect the fitness of both participants.
   Textbook Reference: *44.1 Interactions between Species May Be Positive, Negative, or Neutral, p. 860*
2. Certain birds follow swarms of foraging army ants and prey upon the insects that the ants flush out. The relationship between these birds and the ants is an example of
   a. competition.
   b. commensalism.
   c. mutualism.
   d. amensalism.
   e. Both a and b
   Textbook Reference: *44.1 Interactions between Species May Be Positive, Negative, or Neutral, p. 860*
3. Parasitism
   a. is a form of competition.
   b. differs from predation in that parasites often do not kill their hosts, while predators do kill their prey.
   c. is a form of predation.
   d. does not affect the fitness of the host.
   e. does not affect the per capita growth rate of the host.
   Textbook Reference: *44.1 Interactions between Species May Be Positive, Negative, or Neutral, p. 861*
4. If a species of plant that is trampled by an animal eventually evolves sharp spines that prevent trampling, we can say that its association with the animal has changed from _______ to _______.
   a. amensalism; competition
   b. amensalism; commensalism
   c. commensalism; competition
   d. commensalism; mutualism
   e. commensalism; amensalism
   Textbook Reference: *44.1 Interactions between Species May Be Positive, Negative, or Neutral, pp. 860–861*
5. Models of interspecific competition indicate that
   a. presence of a competitor always reduces population growth rate.
   b. two species will coexist when intraspecific competition is stronger than interspecific competition.
   c. competition can cause one species to go extinct.
   d. coexisting species achieve equilibrium densities lower than either species would alone.
   e. All of the above
   Textbook Reference: *44.2 Interspecific Interactions Affect Population Dynamics and Species Distributions, pp. 863–865*
6. Species in a _______ interaction would increase the per capita growth rate of each other.
   a. commensal
   b. competitive
   c. mutualistic
   d. amensal
   e. consumer-resource
   Textbook Reference: *44.2 Interspecific Interactions Affect Population Dynamics and Species Distributions, pp. 863–864*
7. The phenomenon whereby two competing species differ somewhat in their use of resources is called
   a. resource partitioning.
   b. rarity advantage.
   c. extinction.
   d. a consumer–resource interaction.
   e. None of the above.
   Textbook Reference: *44.2 Interspecific Interactions Affect Population Dynamics and Species Distributions, p. 865*
8. Rarity advantage
   a. results in coexistence of competing species.
   b. protects prey from being driven to extinction by their predators.
   c. reflects the situation in which per capita growth rates of a species involved in an interspecific interaction may rebound at low density.
   d. may prevent the extinction of an interacting species.
   e. All of the above
   Textbook Reference: *44.2 Interspecific Interactions Affect Population Dynamics and Species Distributions, p. 865*
9. Evolutionary arms races are associated with
   a. predator–prey interactions.
   b. parasite–host interactions.
   c. herbivore–plant interactions.
   d. mutualisms.
   e. a, b, and c
   Textbook Reference: *44.3 Interactions Affect Individual Fitness and Can Result in Evolution, pp. 866–867*
10. Mutualism
    a. evolves so that one species can benefit another.
    b. evolves when both species benefit from the interaction.
    c. evolves when one species benefits at the expense of another species.

d. only occurs in plant–animal interactions.
e. None of the above

Textbook Reference: *44.3 Interactions Affect Individual Fitness and Can Result in Evolution, pp. 868–869*

11. Directional selection can result in changes in the average beak size of birds if
a. birds with small and large beaks feed equally efficiently on available seeds.
b. the size of available seeds is highly variable.
c. birds with small and large beaks differ in their efficiency for processing available seeds.
d. fitness is unaffected by beak size.
e. Both a and b

Textbook Reference: *44.3 Interactions Affect Individual Fitness and Can Result in Evolution, pp. 865–866*

12. A bird eats the fruit of a plant species. The seeds are not digested and germinate in the bird's excrement at some distance from the parent plant. This is an example of
a. predation.
b. competition.
c. commensalism.
d. mutualism.
e. amensalism.

Textbook Reference: *44.1 Interactions between Species May Be Positive, Negative, or Neutral, p. 860*

13. A nonnative, invasive wetland plant purple loosestrife (*Lythrum salicaria*) competes for pollinators with the native species of loosestrife, *Lythrum alatum*. Following the introduction of an insect that feeds specifically on the nonnative species, which of the following would you expect might happen?
a. The population of the purple loosestrife would decrease.
b. The population of the control agent (the insect) would first increase and then decrease as the purple loosestrife declined.
c. The native loosestrife species, *Lythrum alatum*, would receive more visits from pollinators.
d. The population of the plant might evolve defenses against the insect.
e. All of the above

Textbook Reference: *44.3 Interactions Affect Individual Fitness and Can Result in Evolution, pp. 866–867; 44.4 Introduced Species Alter Interspecific Interactions, pp. 869–870*

14. Which of the following is a possible outcome of competition involving two species?
a. Reduced growth and reproductive rates for some individuals of both species
b. Exclusion of one species from a habitat it would occupy in the absence of competition from the other species
c. Restriction of the geographic range of one species because of competition from the other species
d. Evolution of both species, allowing them to divide up the limiting resource and continue to coexist in the same locality
e. All of the above

Textbook Reference: *44.2 Interspecific Interactions Affect population Dynamics and Species Distributions, pp. 863–864*

15. In the nineteenth century, a few dozen European Starlings were deliberately introduced to New York City's Central Park in an effort to reconstruct familiar surroundings from Western Europe. The descendants of the initial introduction now number in the millions, and have spread widely across North America. The starlings found in North America today are thus _______ and the introduction was _______.
a. invasive; deliberate
b. native; deliberate
c. invasive; accidental
d. native; accidental
e. None of the above

Textbook Reference: *44.4 Introduced Species Alter Interspecific Interactions, pp. 869–870*

Answers

Key Concept Review

1. The anemonefish clearly benefit from the interaction with their host by gaining protection from their enemies. It is less clear that the sea anemones are unaffected (in which case the relationship would be a commensalism). The anemones may gain nutrients (especially nitrogen) from the feces of the fish, or the fish may occasionally steal prey from the anemones, or both. Thus the interaction could be mutualistic or competitive, or both simultaneously. The net effect of the interaction may depend on environmental conditions, such as the availability of nitrogen-rich food to anemones.
2. Both species benefit in a mutualistic relationship. One way to test for a mutualistic relationship is to establish two groups, one in which the two species are associated and another in which the species have been experimentally separated. Each species in the two groups should be monitored over time and the per capita growth rate determined. If both species display higher per capita growth rates when together than when apart, then a mutualistic relationship is suggested. Further experiments would likely be necessary to reveal the details of the relationship (i.e., what critical resource each species is obtaining from the interaction).
3. Once prey become rare, they may be harder to find because high-quality shelters are more available than when the prey population was at higher densities. Rare prey also may be in better physical shape (and therefore better able to escape or defend themselves) because their food resources are now shared among fewer conspecifics. Also, if finding rare prey becomes

too time-consuming and difficult, predators may switch to another prey species. For all of these reasons, the per capita growth rates of the prey may increase rather than decline to zero.

4. The following word equation describes the effect of species B on species A in this commensal relationship:

   per capita growth rate (r) of species A = {maximum possible r for species A in uncrowded conditions – an amount that is a function of A's own population density} + {an amount that is a function of the population density of species B}.

   By definition of a commensal interaction, species B is unaffected by the interaction, so the equation for species B does not have a reciprocal term (i.e., a term concerning the effect of species A on species B).

5. The prey of carnivores never benefit from being eaten. A plant may suffer a loss of fitness if an animal eats its roots, stems, or leaves, so these organs of the plant are often defended (for example, leaves may have spines and stems may have thorns). However, plants benefit when their pollen is transferred to another plant by a pollinator or when the seeds in their fruit are carried off to a new germination site. So, plants do sometimes benefit from the attentions of herbivores.

6. The large predatory fish benefits by having its ectoparasites removed. The cleaner fish benefits by getting a meal. A potential cost to the predatory fish is a cleaner that cheats and steals a bite, and a potential cost to the cleaner fish is being eaten by the fish it is cleaning. The parasite is consumed; there is no benefit.

7. The rabbits evolved resistance to the myxoma virus. Thus, the effectiveness of this control measure was limited over time. It did not solve the problems resulting from the introduction of rabbits, and it did not restore the affected communities to their original state.

8.

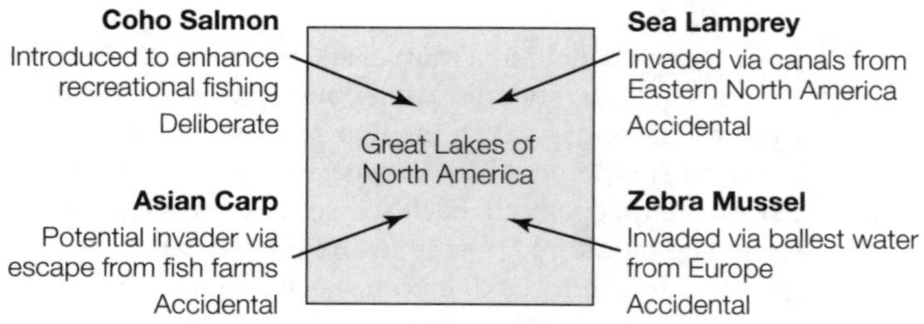

   There are no possible solutions that could return the Great Lakes to the previous uninvaded state, and it is likely that solutions for one invasive species would have impacts on many other species in the Great Lakes ecosystem.

Test Yourself

1. **e.** All of the interactions (competition and consumer-resource) affect the fitness of both participants.
2. **e.** Both ants and birds are competing for the same food supply; the birds also benefit from the ants' activity in flushing out the insects, but the ants might not be affected.
3. **b.** Parasitism is a consumer-resource interaction in which the parasite gains at the host's expense (both in terms of fitness and per capita growth rate). However, parasites often do not kill their hosts; predators do kill their prey.
4. **a.** Initially the plant is harmed and the animal is unaffected (amensalism). After the evolution of sharp spines, the two species are competitors for space.
5. **e.** Models of interspecific competition assume all of these factors.
6. **c.** Species in a mutualistic relationship will increase the per capita growth rates of each other.
7. **a.** The phenomenon whereby two competing species differ somewhat in their use of resources is called resource partitioning.
8. **e.** Rarity advantage describes the rebound of per capita growth rates at low density; this can result in the coexistence of competing species and the persistence of prey species. Thus, rarity advantage can prevent extinction of interacting species.
9. **e.** Evolutionary arms races are associated with consumer–resource interactions, which include predator–prey interactions, parasite–host interactions, and herbivore–plant interactions.
10. **b.** Mutualism evolves when both interacting species benefit.
11. **c.** Directional selection can result in changes in the average beak size if birds with small and large beaks differ in their efficiency for processing available seeds.
12. **d.** Both species have benefited; the bird dispersed and provided fertilizer for the plant's seed, and the plant provided food for the bird.
13. **e.** Following introduction of the insect, the population of *Lythrum salicaria* would likely decrease; the insect would initially increase and then decline as its food plant declined. Because plants often evolve defenses against herbivores, the purple loosestrife might evolve a defense, such as secondary compounds. The native plant, *Lythrum alatum*, would receive more pollination visits as the invasive plant declines.
14. **e.** It is generally true that competition, whether intraspecific or interspecific, causes reduced growth and reproductive rates of some individuals participating in the competitive interaction. Interspecific competition in particular may lead to exclusion from a habitat of one of the competing species, or to range restriction, or to resource partitioning.
15. **a.** The introduced European starlings are invasive because they have dramatically increased in numbers and spread widely. The introduction was deliberate.

Ecological Communities 45

The Big Picture

- In urging a cautious approach to human manipulation of the natural environment, ecologists like to say that "you can never do just one thing." In this chapter, the descriptions of human-induced trophic cascades and of other complex community interactions illustrate the meaning of this saying.
- The theory of island biogeography, which relates the area of an island to the equilibrium number of species inhabiting the island, has applications in conservation biology. For example, habitat destruction often leaves habitat islands that, if small, inevitably lose some of the species that existed in the previously more extensive habitat.
- Ecological communities provide goods and services to humans. The ability of a community to provide goods and services often depends on its diversity, which in turn is influenced by habitat complexity and rates of colonization and extinction.

Study Strategies

- The niche of a species is often confused with its habitat. The latter term refers to the preferred physical environment of a species (e.g., pond versus stream), whereas the former term refers to the entire set of physical and biological conditions a species needs in order to survive and reproduce. Thus the habitat of a species is just one of many attributes of its niche.
- The graphical representations of the concepts of island biogeography theory are sometimes hard to follow (see Figure 45.14). Try redrawing the graphs for yourself, paying particular attention to how the axes are labeled.
- This chapter contains descriptions of a number of studies of community structure, including several that relate species richness to other community characteristics. As you read about these studies, focus on the particular question that each study was designed to answer and how the results support the conclusions drawn from the study.
- Go to yourBioPortal.com to review the following tutorials and activities:

  Animated Tutorial 45.1 Succession on a Glacial Moraine

  Animated Tutorial 45.2 Biogeography Simulation

  Animated Tutorial 45.3 Edge Effects

  Web Activity 45.1 The Major Trophic Levels

  Web Activity 45.2 Energy Flow through an Ecological Community

  Working with Data 45.1 Habitat Fragmentation

Key Concept Review

45.1 Communities Contain Species That Colonize and Persist

A community consists of all the species that live and interact in a particular area. Thus, a community includes all the species that have colonized it minus those that have gone locally extinct, perhaps due to inappropriate environmental conditions, lack of a critical resource, or competition with resident species. Communities can be characterized by the particular collection of species present and the relative abundance of each species.

Question 1. A small lake contains vertebrates, such as snapping turtles, painted turtles, bluegill sunfish, and largemouth bass. What other information is needed to characterize the current species composition of this vertebrate community?
Textbook Reference: *45.1 Communities Contain Species That Colonize and Persist, p. 874*

Question 2. A landowner in Minnesota, hoping to develop a bass population for fishing, introduces six juvenile largemouth bass of unknown sex, each about 10 cm in length, into a small and shallow pond. Largemouth bass, a predatory species, do not reproduce until reaching about 25 cm

in length. The pond already contains herbivorous minnows and tadpoles and carnivorous salamanders. The population of largemouth bass fails to persist in the pond. What factors might have influenced its extinction?
Textbook Reference: *45.1 Communities Contain Species That Colonize and Persist, p. 874*

45.2 Communities Change over Space and Time

Species composition varies along environmental gradients

Several processes cause communities to change over time

Species composition changes along environmental gradients. When ecologists establish a transect (a straight line used when conducting surveys) along an environmental gradient and sample the species present, they find that certain species drop out and others appear. Thus, species turnover occurs when there is spatial variation in environmental conditions.

Extinction and colonization, disturbance, and climate change prompt changes in the species composition of communities through time. In a particular community, resident species may go extinct and new species may arrive. Sudden environmental change, whether caused by human activities or natural phenomena such volcanic eruptions or hurricanes, causes turnover of species. Following a disturbance, species often replace one another in a fairly predictable sequence called succession (see Figure 45.4). The precise sequence reflects species differences in colonizing ability and environmental tolerance. Early arriving species tend to have superior dispersal abilities and environmental conditions change over time as a result of physical processes and the activities of the early colonists. When a community is destroyed by disturbance, succession can lead to establishment of a community similar to the original one. Sometimes, however, an ecological transition to a different community occurs. Climate change causes temporal variation in the species composition of communities by influencing the geographic ranges of species (see Figure 45.5).

Question 3. White-tailed deer are often associated with wooded areas that contain a mix of trees, shrubs, forbs, grasses, and sedges. What might form the basis of this animal–vegetation association?
Textbook Reference: *45.2 Communities Change over Space and Time, p. 875*

Question 4. Forensic entomologists can estimate how many days a corpse has been dead by identifying the particular insects present on the body. This estimate is possible because different species of insects are found on corpses in a predictable sequence that reflects changes in physical conditions of the body over time (e.g., decreases in moisture). What ecological process forms the basis for the technique used by forensic entomologists to determine date of death?
Textbook Reference: *45.2 Communities Change over Space and Time, p. 876*

45.3 Trophic Interactions Determine How Energy and Materials Move through Communities

Consumer–resource interactions determine an important property of communities

Energy is lost as it moves through a food web

Trophic interactions can change the species composition of communities

The niche of a species is the set of physical and biological conditions necessary for its persistence. Within a community, each species has a unique niche.

The organisms in a community use diverse sources of energy, and organisms are grouped into trophic levels according to the number of steps through which energy passes to reach them. Energy passes, in sequence, from autotrophic primary producers (photosynthesizers) to heterotrophs, beginning with primary consumers (herbivores), followed by secondary consumers (which feed on herbivores), and so on. Detritivores (decomposers) feed on the dead bodies and waste products of other organisms. Omnivores obtain their food from more than one trophic level (see Table 45.1). A food web is a diagram depicting the linkages among interacting species, and organizing these linkages vertically by trophic level (see Figure 45.6).

Gross primary productivity (GPP) is the total amount of energy that primary producers capture and convert to chemical energy during a particular time interval. Net primary productivity (NPP) is the energy contained in the tissues that primary producers have made during that time interval; in other words, this is the portion of GPP that becomes available to consumers during the time period.

The ecological efficiency of energy transfer from one trophic level to the next is about 10 percent. Three factors account for such low ecological efficiency, and these are also the reasons that most communities have only three to five trophic levels. First, organisms use most of the energy they accumulate for respiration and other metabolic processes; this energy is ultimately dissipated as heat. Second, consumers do not ingest all of the biomass available to them (for example, herbivores may avoid plants with well-developed chemical defenses). Third, consumers cannot assimilate (digest and absorb) all of the biomass they ingest.

Recall that interspecific interactions influence per capita growth rates and population densities of the species involved. The elaborate web of interactions within a community means that change in one species—for example, the arrival of a new species or an increase or decrease in the population density of a resident species—can impact the entire community. In a trophic cascade, changes at one trophic level cause changes at other trophic levels (see Figure 45.8).

Question 5. Apply the information presented in the text on trophic levels to the ways in which humans eat. Do humans feed on one or more trophic levels? Give an example of what humans would eat at each trophic level you name.
Textbook Reference: *45.3 Trophic Interactions Determine How Energy and Material Move through Communities, pp. 878–879*

Question 6. Wolves were reintroduced to Yellowstone National Park in 1995. Describe the ecological effects within the park of this politically controversial decision.
Textbook Reference: *45.3 Trophic Interactions Determine How Energy and Material Move through Communities, p. 880*

45.4 Species Diversity Affects Community Function

The number of species and their relative abundances contribute to species diversity

Species diversity affects community processes and outputs

Species diversity has two main components, species richness and species evenness, both of which affect community function (see Figure 45.9). Species richness is the number of species in the community. Communities with many species are more diverse than those with fewer species, all else being equal. Species evenness considers the distribution of species' abundances within a community. The more even the distribution of species abundances, the more diverse the community. When quantifying diversity, ecologists use a mathematical index that incorporates both species richness and species evenness.

Community function can be measured by the outputs of species interactions. Such outputs increase with species diversity. For example, within a particular type of community, NPP is typically greater and more stable over time as species diversity increases. This association between species diversity and community function may reflect sampling (just by chance, communities with many species may be more likely to have some species with a strong influence on community output) or niche complementarity (communities with many species are more likely to have species with complementary niches, with the result that all available resources are used).

Question 7. Under what circumstances would a forest with many species have relatively low diversity?
Textbook Reference: *45.4 Species Diversity Affects Community Function, p. 881*

Question 8. Growing crops as monocultures is standard practice in modern agriculture. Why are monocultures unstable? What agricultural practices might result in more stable and more productive ecological communities?
Textbook Reference: *45.4 Species Diversity Affects Community Function, pp. 881–882*

45.5 Diversity Patterns Provide Clues to Determinants of Diversity

Species richness varies with latitude

Diversity represents a balance between colonization and extinction

For many taxa, species diversity is greatest in the tropics and decreases with increasing latitude (see Figure 45.11). Ecologists have suggested several explanations for latitudinal gradients in diversity. One possibility is that organisms in tropical regions have not been subjected to large-scale disturbance, such as the glacial cycles that affected temperate regions, and as a result the tropics have retained more species. Another possibility is that the greater productivity of the tropics (made possible by primary producers thriving in the warm, wet conditions) allows species with narrow, specialized niches to coexist. Greater niche specialization could interact with greater habitat complexity to amplify tropical diversity.

According to the theory of island biogeography, equilibrium species richness (the number of species living in an area) on islands is determined by the rate of arrival of new species (= rate of colonization) and the rate of extinction of species already present. The theory further predicts that the equilibrium number of species should increase with island size and decrease with distance from the mainland (see Figure 45.14). Scientists have confirmed the predictions of the theory both by observation and experiments (see Figure 45.15).

Question 9. How might the theory of island biogeography be applied to conservation efforts of mainland species?
Textbook Reference: *45.5 Diversity Patterns Provide Clues to Determinants of Diversity, pp. 883–884*

45.6 Community Ecology Suggests Strategies for Conserving Community Function

Ecological communities provide humans with goods and services

Ecosystem services have economic value

Island biogeography suggests strategies for maintaining community diversity

Trophic cascades suggest the importance of conserving critical components of food webs

The relationship of diversity to community function suggests strategies for restoring degraded habitats

Ecological communities provide humans with goods (for example, wood from trees and food from plants and animals) and services (for example, flood control, soil stabilization, and climate regulation) (see Table 45.2). These goods and services are called ecosystem services, since they often result from interactions between communities and the physical environment. The explicit acknowledgement that ecological systems have economic value is a relatively recent phenomenon.

Habitat fragmentation creates habitat islands—isolated patches of suitable habitat surrounded by areas of unsuitable habitat. Fragmentation is predicted to result in the loss of species, because as the total amount of habitat decreases, the average size of a habitat patch decreases, and patches become more isolated from one another. The theory of island biogeography can be applied to habitat patches. Steps that might enhance colonization and reduce extinction in fragmented habitat include keeping patches close to one another, providing dispersal corridors among patches, and retaining large patches of the original habitat.

Community function can be improved by targeting species with particularly important roles in their community and by restoring the original species diversity of communities.

Question 10. List five possible ecosystem goods or services.
Textbook Reference: *45.6 Community Ecology Suggests Strategies for Conserving Community Function, pp. 885–887*

Question 11. In the diagram below, the oval shape represents a large unbroken tract of forest. The four circles to the right represent habitat fragments: two large ones (signifying large area) and two smaller ones (signifying smaller area). The two filled in rectangles represent dispersal corridors. Select two of the four habitat fragments, along with the two dispersal corridors and the unbroken tract, and design an arrangement to achieve the goal of habitat fragments with the highest species richness.
Textbook Reference: *45.6 Community Ecology Suggest Strategies for Conserving Community Function, pp. 887–888*

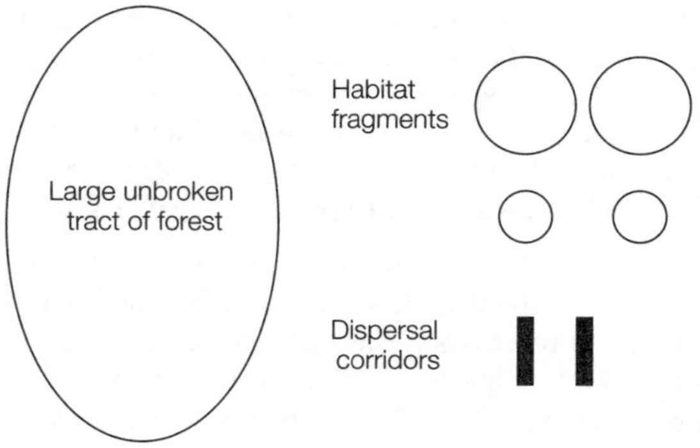

Test Yourself

1. The ecological community of a pond includes
   a. all the species living in the pond and the abiotic environment (water, sunlight, etc.) that supports them.
   b. all the species living in the pond.
   c. the invertebrates and vertebrate animals living in the pond.
   d. all of the producers and consumers living in the pond, but not the decomposers.
   e. None of the above
   Textbook Reference: *45.1 Communities Contain Species That Colonize and Persist, p. 874*

2.–3. Refer to the food web below to answer the questions that follow.

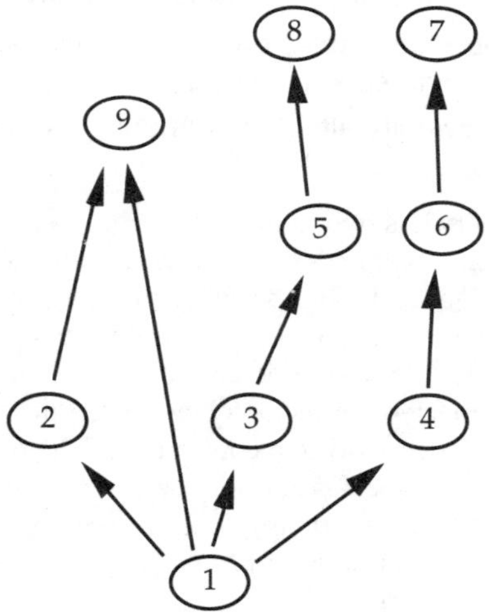

2. Organism 9 is a(n)
   a. herbivore.
   b. primary carnivore.
   c. secondary carnivore.
   d. primary producer.
   e. omnivore.
   Textbook Reference: *45.3 Trophic Interactions Determine How Energy and Material Move through Communities, pp. 878–879*

3. The food web has _______ trophic levels.
   a. two
   b. three
   c. four
   d. five
   e. six

   Textbook Reference: *45.3 Trophic Interactions Determine How Energy and Material Move through Communities, p. 878*

4. Which of the following statements about habitat fragmentation is *false*?
   a. Small isolated patches lose species more rapidly than larger isolated patches.
   b. Isolated patches lose species more rapidly than patches of similar size that are near other patches.
   c. Habitat fragmentation results in lower species richness in the fragments than in the original habitat.
   d. Human-dominated habitat surrounding patches increases the colonization rate of patches.
   e. Connecting fragments with dispersal corridors enhances colonization.

   Textbook Reference: *45.6 Community Ecology Suggests Strategies for Conserving Community Function, pp. 887–888*

5. Which of the following organisms is matched *incorrectly* with its trophic level?
   a. bison – primary consumer
   b. fungus – decomposer (detritivore)
   c. grass – primary producer
   d. parasite – secondary or tertiary consumer
   e. fungus – primary producer

   Textbook Reference: *45.3 Trophic Interactions Determine How Energy and Material Move through Communities, pp. 878–879*

6. If the net primary productivity of a community is 2,000 kcal/m^2/yr, what would be the best estimate of the productivity (in kcal/m^2/yr) of the secondary consumers in that community?
   a. 4,000
   b. 2,000
   c. 200
   d. 20
   e. 2

   Textbook Reference: *45.3 Trophic Interactions Determine How Energy and Material Move through Communities, pp. 878–879*

7. Experiments on wetland restoration have demonstrated that planting a rich mixture of species is associated with
   a. fast accumulation of nitrogen in roots.
   b. complex vegetation structure.
   c. rapid development of vegetation cover.
   d. rapid return to the community's original condition.
   e. All of the above

   Textbook Reference: *45.6 Community Ecology Suggests Strategies for Conserving Community Function, p. 889*

8. Which of the following communities of 24 individuals is most diverse?
   a. A community with 3 species, each of which has 8 individuals
   b. A community with 4 species, each of which has 6 individuals
   c. A community with 4 species; 1 species has 18 individuals and the remaining 3 species each have 2 individuals
   d. The communities in a, b, and c have equal diversity.
   e. It is impossible to determine from the information given.

   Textbook Reference: *45.4 Species Diversity Affects Community Function, p. 881*

9. Which of the following causes changes in species composition within communities over time?
   a. Climate change
   b. Local extinction of a species
   c. Disturbance
   d. Colonization
   e. All of the above

   Textbook Reference: *45.2 Communities Change over Space and Time, pp. 875–877*

10. The rate at which new species arrive on an island
    a. increases with closeness to the mainland.
    b. decreases with closeness to the mainland.
    c. increases as more species occupy the island.
    d. is lower than the extinction rate at the equilibrium species richness.
    e. None of the above

    Textbook Reference: *45.5 Diversity Patterns Provide Clues to Determinants of Diversity, pp. 883–885*

11. In the figure below, the species numbers of many islands of different sizes were plotted against their distance from the mainland, and data points for islands of similar size were connected to form the four curves.

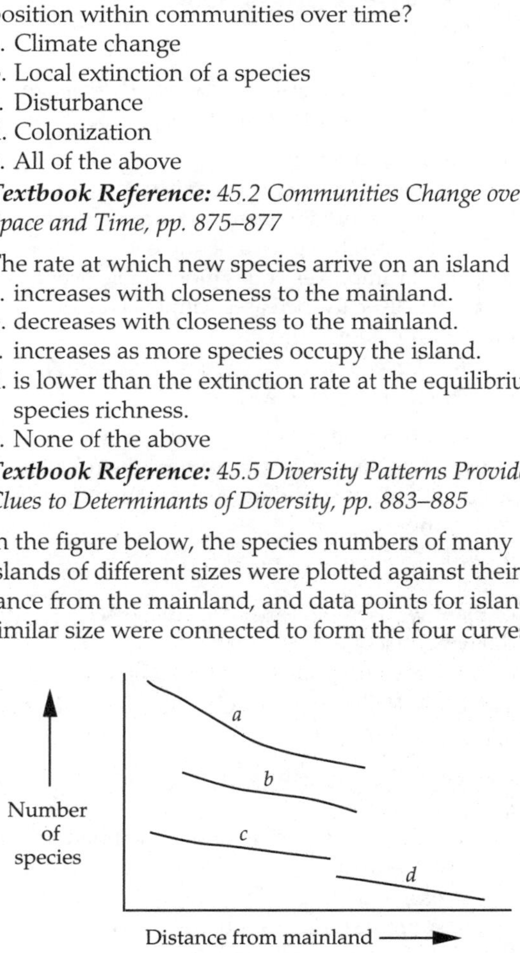

    Circle the letter of the curve corresponding to the group of islands with the smallest size.

    Textbook Reference: *45.5 Diversity Patterns Provide Clues to Determinants of Diversity, pp. 883–885*

12. The theory of island biogeography makes predictions about the effects of species number on the rate of extinction. Refer to the figure below, in which the solid curve shows this relationship for a large island. Which of the curves shows the expected relationship for a small island?

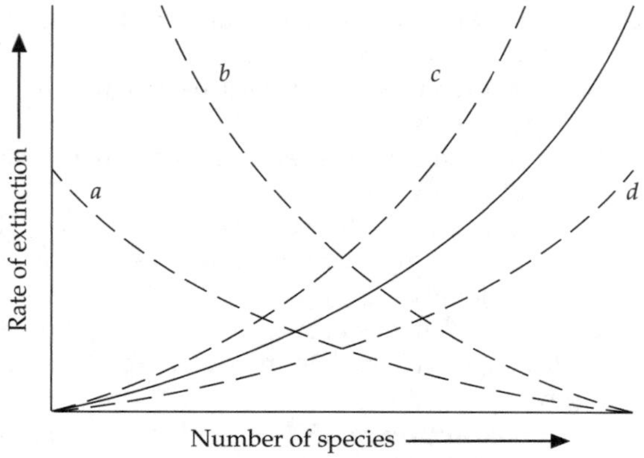

a. Curve a
b. Curve b
c. Curve c
d. Curve d

Textbook Reference: *45.5 Diversity Patterns Provide Clues to Determinants of Diversity, pp. 883–885*

13. Community function
 a. is influenced by species evenness.
 b. can be measured in terms of outputs, such as NPP.
 c. is influenced by species richness.
 d. generally improves and is more stable over time as species diversity increases within a type of community.
 e. All of the above

 Textbook Reference: *45.4 Species Diversity Affects Community Function p. 881*

14. Which of the following organisms would likely be one of the first colonists at a site where disturbance has wiped out all species that were members of the original community?
 a. Bird
 b. Bat
 c. Plant whose seeds are wind-dispersed
 d. Flying insect
 e. All of the above are good possibilities.

 Textbook Reference: *45.2 Communities Change over Space and Time, p. 876*

15. Which of the following statements about succession is *false*?
 a. Ecological transition describes the phenomenon whereby disturbance wipes out a community and succession leads to the eventual establishment of a community similar to the original one.
 b. Species that are early colonists may either facilitate or inhibit colonization by later-arriving species.
 c. Succession often leads to the establishment of a community that resembles the original one.
 d. Succession is the relatively predictable sequence of species replacements following a disturbance.
 e. Because environmental conditions at a site change over time, the successional sequence often reflects species differences in environmental tolerances.

 Textbook Reference: *45.2 Communities Change over Space and Time, p. 876*

16. Species richness
 a. is greatest in the tropics.
 b. decreases with increasing latitude.
 c. may increase when climates are more stable.
 d. typically increases with greater habitat complexity.
 e. All of the above

 Textbook Reference: *45.5 Diversity Patterns Provide Clues to Determinants of Diversity, pp. 882–883*

Answers

Key Concept Review

1. In order to characterize the current species composition of the vertebrate community of the lake, we would also need to know the relative abundance of each vertebrate species present.
2. Extinction of the introduced bass population can occur for many reasons. For example, the bass may not be able to tolerate the environmental conditions of the pond. More specifically, because the pond is small and shallow and located in Minnesota, it will be subjected to temperature extremes that the fish may be unable to tolerate. Related aspects of the physical environment, such as low oxygen levels in warm water during summer, could also prove problematic. Although food resources are initially present in the form of minnows, tadpoles, and salamanders, the prey supply may decrease to levels that cannot support a population of bass. Predatory birds, such as herons, may decrease the number of bass. Finally, the introduced population may simply be too small in size to successfully colonize the pond, and there is no guarantee that both males and females were part of the original introduction.
3. A certain species will be found in the plant communities that provide its preferred food. Animal–plant associations also may occur because plants modify physical conditions, such as temperature and humidity, and thereby make the environment more suitable for a particular animal. Plants also determine habitat structure, which influences an animal's ability to avoid detection by predators and to evade them once detected. In the case of white-tailed deer, their association with a habitat that contains this mix of vegetation centers on the species' need for year-round access to high quality food. Deer also use the vegetation for shelter and concealment.
4. The forensic entomologists are using the process of ecological succession to determine date of death. Ecological succession is the phenomenon whereby species often replace one another in a predictable sequence following a disturbance.
5. Humans feed on several trophic levels. Cows are primary consumers (herbivores), so we are secondary consumers when we eat beef. Tuna are secondary consumers (if they have eaten an herbivorous fish or crustacean) or tertiary consumers (if they have eaten another predatory fish), so we are tertiary or quater-

nary consumers, respectively, when we eat tuna. We also eat vegetables (primary producers), which makes us primary consumers.

6. In Wyoming's Yellowstone National Park, the reintroduction of wolves resulted in a reduction of the population of elk, the wolves' principle prey. In a trophic cascade, the decline in the number of elk led to increased tree reproduction of aspens and willows, both of which had been heavily browsed by elk. In turn, the increase in the population of streamside willows resulted in an increase in the number of beaver colonies in the valley. Thus, the reintroduction of wolves by humans caused changes in several trophic levels, illustrating the interconnectedness of species within communities.
7. A forest with many species could have relatively low diversity if it contained one or a few species that were especially abundant and the other species in the forest community had only a small number of individuals.
8. Community outputs are generally greater and more stable over time as species diversity increases. Monocultures are unstable because they are subject to attack by insect pests and pathogens that destroy or damage crops. More diverse systems in which two or more crops are grown on the same plot would be less subject to pest outbreaks, and NPP would be expected to be greater in these systems than in single crop systems.
9. Human activities have fragmented mainland areas, effectively establishing islands of suitable habitat surrounded by unsuitable habitat (for example, agricultural land). The theory of island biogeography predicts that we will lose species from habitat patches as the size of patches declines and as patches become more isolated from one another, making colonization more difficult.
10. Ecosystem goods and services may include: (1) food supplied by plants and animals; (2) wood from trees; (3) maintenance of fertile soil through decomposition; (4) waste treatment through decomposition; and (5) flood control. (See Table 45.2 for a more complete list.)
11. The general arrangement that would achieve the highest species richness in the habitat fragments is shown below.

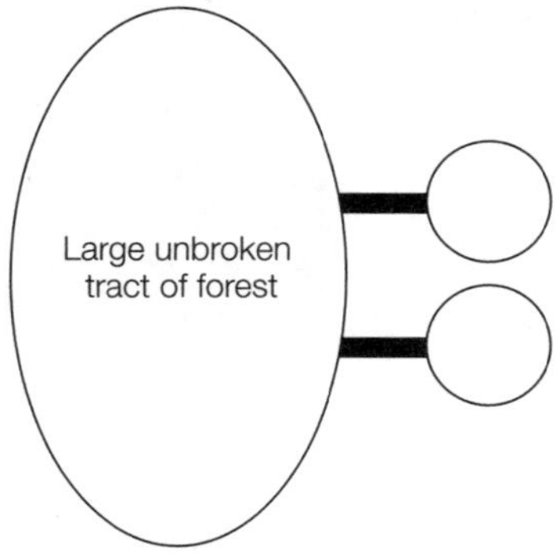

The essential elements are: (1) selecting the two largest habitat fragments from the four provided, because larger areas support larger populations, which are less prone to extinction; (2) arranging the two fragments and the large tract of unbroken forest as close together as possible to allow for the greatest dispersal among them, which increases colonization rates; and (3) placing the dispersal corridors between each fragment and the large unbroken tract of forest because this too would maximize colonization rates.

Test Yourself

1. **b.** An ecological community comprises all of the populations living in a defined area but excludes its nonliving components.
2. **e.** Because organism 9 eats from both the primary producer level (1) and the herbivore level (2), it is an omnivore.
3. **c.** Four trophic levels are depicted. The levels are primary producer (1), herbivore (also called primary consumer: 2, 3, 4), primary carnivore (also called secondary consumer: 5, 6), and secondary carnivore (also called tertiary consumer: 7, 8). The omnivore (9) occupies either the herbivore or primary carnivore level, depending on whether it is feeding on a primary producer or on an herbivore.
4. **d.** Human-dominated habitat surrounding patches acts as a barrier to dispersal, so it decreases the colonization rate of patches.
5. **e.** Fungi are decomposers (detritivores); they do not perform photosynthesis, so they cannot be primary producers.
6. **d.** Net primary productivity is the rate at which energy is incorporated into the bodies of primary producers through growth and reproduction. As a rule, only about 10 percent of the energy at one trophic level is transferred to the next. Because the secondary consumers are two trophic levels above the primary producers, their productivity would be expected to be only 1 percent of net primary productivity, which in the example equals 20.
7. **e.** Planting a mixture of species resulted in more rapid development of vegetative cover, more complex vegetative structure, greater accumulation of nitrogen in the roots of plants, and more rapid progress in restoring the wetland (see Figure 45.18).
8. **b.** The community with 4 species, each of which has 6 individuals, is the most diverse because it has the largest number of equally abundant species.
9. **e.** Temporal variation in species composition can result from climate change, disturbance, extinction, and colonization.
10. **a.** The rate at which new species arrive on an island increases with closeness to the mainland and declines as the island fills with species. The equilibrium species

richness occurs when the colonization rate equals the extinction rate.

11. **d.** Each curve corresponds to a group of similar-sized islands whose species number is plotted against their distance from the mainland. For each curve, species number decreases with distance from the mainland, and the curve that is lowest relative to the vertical axis would be the group of smallest islands.
12. **c.** With less space available, populations of the different species would be smaller. This would subject them to higher extinction rates than would be expected on larger islands.
13. **e.** Both species richness and species evenness influence community function, which can be measured in outputs such as NPP. Within a community type, community outputs (and hence function) are generally greater and more stable over time as species diversity increases.
14. **e.** Following environmental disturbance, the first species to arrive on a site are those with superior dispersal abilities. Animals that can fly and plants with wind-dispersed seeds are all good dispersers.
15. **a.** Ecological transition occurs when a distinctly different community emerges following disturbance, rather than a community similar to the original one. For example, following intensive grazing by cattle, certain grasslands in the United States/Mexico Borderlands changed to shrublands.
16. **e.** Species richness is greatest in the tropics and decreases with increasing latitude. Species richness may be enhanced by climatic stability and habitat complexity.

The Global Ecosystem 46

The Big Picture

- Net primary productivity (NPP) of terrestrial ecosystems varies with temperature and precipitation. NPP of aquatic systems varies with availability of light and nutrients. NPP is a measure of ecosystem function.
- Earth is an open system with respect to energy and a closed system with respect to matter. Photosynthetic organisms capture some of the energy provided by the sun. Energy flows from the primary producers to higher trophic levels and ultimately is dissipated as heat. In contrast, chemical elements, which are present in fixed amounts, come from within the system of Earth itself and continually cycle through ecosystems. Knowledge of how materials move through biogeochemical cycles is crucial for understanding and predicting the effects of human alterations of these cycles.
- Human activities have increased greenhouse gases in Earth's atmosphere and increases in global temperatures have followed. Climate change is altering the distributions, abundances, and interspecific interactions of species. International cooperation is needed to address the challenge of global climate change.

Study Strategies

- Eutrophication—the "enrichment" of a body of water with nutrients—often results in a "dead zone" because extra nutrients stimulate rapid growth of phytoplankton populations to a point at which consumers cannot eat them all. Respiration by the phytoplankton and by the decomposers that process their dead bodies depletes the water of oxygen, making it difficult for other aquatic organisms to survive.
- In studying the material on biogeochemical cycles, focus on the following basic questions: Where are the abiotic reserves of an element located? How does the element leave the reserve and enter living organisms? Why do living organisms need the element? How does the element return to its abiotic reserve?
- Go to yourBioPortal.com to review the following tutorials and activities:

  Animated Tutorial 46.1 The Global Water Cycle
  Animated Tutorial 46.2 The Global Nitrogen Cycle
  Animated Tutorial 46.3 The Global Carbon Cycle
  Animated Tutorial 46.4 Earth's Radiation Balance
  Web Activity 46.1 The Benefits of Cooperation
  Web Activity 46.2 Concept Matching

Key Concept Review

46.1 Climate and Nutrients Affect Ecosystem Function

NPP is a measure of ecosystem function

NPP varies predictably with climate and nutrients

An ecosystem is an ecological community plus the abiotic environment with which the organisms interact. Large-scale movements of organisms (e.g., migrations) and materials (e.g., by means of flowing water or circulating air) link ecosystems.

With very few exceptions, energy flow in ecosystems originates with photosynthesis. Net primary productivity (NPP) is the portion of assimilated energy left over after the energy used by primary producers for their own metabolism is subtracted, which takes the form of growth and reproduction of the producer organism. All of the other organisms in an ecosystem derive their energy directly or indirectly from net primary productivity. NPP is essentially a measure of the influx of energy and materials into a community, and can serve as a measure of ecosystem function.

NPP varies with type of ecosystem (see Figure 46.1). Within terrestrial ecosystems, tropical forests are the most productive area per unit of area, and tundra and deserts are the least productive. NPP varies with latitude, with productive areas in the tropics and less productive areas at high latitudes and in the dry regions around 30°N and S (see Figure 46.2). Terrestrial NPP generally increases with temperature, and increases to a certain level with precipitation and then declines (see Figure 46.3).

Within aquatic ecosystems, swamps, marshes, and coral reefs are the most productive and open ocean is the least productive. Nutrient availability is the main determinant of aquatic NPP. Nutrients are most abundant where land runoff brings nutrients into shallow coastal waters and where upwellings bring nutrients from the sediments.

Question 1. On the graph in part A, draw the general relationship between NPP and mean annual temperature for a terrestrial ecosystem. On the graph in part B, draw the general relationship between NPP and mean annual precipitation for a terrestrial ecosystem.
Textbook Reference: *46.1 Climate and Nutrients Affect Ecosystem Function, pp. 894–895*

(A) Temperature

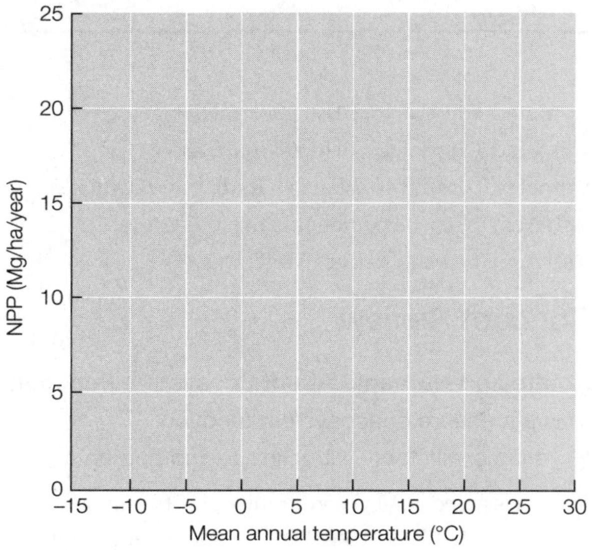

(B) Precipitation

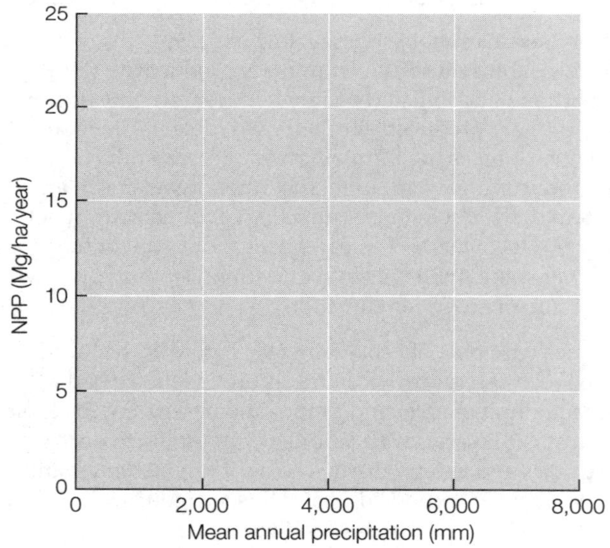

Question 2. Explain why marine NPP is highest in coastal areas.
Textbook Reference: *46.1 Climate and Nutrients Affect Ecosystem Function, p. 895*

46.2 Biological, Geological, and Chemical Processes Move Materials through Ecosystems

- The form and location of elements determine their accessibility to organisms
- Fluxes of matter are driven by biogeochemical processes

Earth is essentially an open system with respect to radiant energy from the sun, but it is a closed system with respect to matter. Nevertheless, the distribution of matter is not static because energy from the sun and heat from Earth's interior drive the processes that move materials around the planet.

Elements exist in different chemical forms and occur in different locations; some of these forms and locations are accessible to life while others are not (see Figure 46.5). These different forms and locations can be described as compartments, and biological, geological, and chemical processes cycle matter among compartments. The biosphere is the part of Earth where life exists; it is where atmosphere, land, and water come into contact. The pattern of movements of elements through organisms and compartments of the physical environment are called biogeochemical cycles. An accumulation of an element in a compartment is a pool; movement of elements into and out of a compartment is called flux.

Question 3. By what processes do inorganic compounds in rocks become accessible to organisms?
Textbook Reference: *46.2 Biological, Geological, and Chemical Processes Move Materials through Ecosystems, p. 896*

Question 4. Through what process do organic and inorganic compounds enter compartments where they are inaccessible to organisms?
Textbook Reference: *46.2 Biological, Geological, and Chemical Processes Move Materials through Ecosystems, p. 896*

46.3 Certain Biogeochemical Cycles Are Especially Critical for Ecosystems

- Water transports materials among compartments
- Nitrogen is often a limiting nutrient
- Carbon flux is linked to energy flow through ecosystems
- Biogeochemical cycles interact

Most of Earth's water (96.5%) is in the oceans, with the remainder divided among other compartments, such as ice and snow, groundwater, the atmosphere, and fresh water. Water moves between these compartments through gravity-driven flow when in its liquid state (e.g., when precipitation falls from the atmosphere to Earth's surface) or with changes in its physical state (e.g., when changing from liquid to gas during evaporation and transpiration). The sun powers the water cycle by causing evaporation, most of it from ocean surfaces (see Figure 46.6). Human activities that reduce vegetative cover disrupt the water cycle by increasing runoff and drying out local ecosystems.

Nitrogen is often in limited supply in biological communities. Although nitrogen gas (N_2) makes up 78 percent of Earth's atmosphere, most organisms cannot use this form of nitrogen. Some microorganisms, however, can reduce nitrogen gas to ammonium (NH_4^+) in a process called nitrogen fixation. Other microorganisms convert ammonium into nitrate (NO_3^-). Primary producers can take up ammonium and nitrate, thereby making nitrogen available to consumers. Microorganisms also perform denitrification, which returns nitrogen gas to the atmosphere (see Figure 46.7). Human activities have affected the nitrogen cycle. Industrial and agricultural nitrogen fixation now rival natural terrestrial nitrogen fixation. When nitrogen applied as fertilizer to cropland enters bodies of water, it can lead to eutrophication, a process in which high nutrient levels promote excessive algal growth, depleting oxygen in the water and creating a "dead zone" (see Figure 46.9). Human-caused perturbations of the nitrogen cycle have other adverse effects, including increased air pollution and acid rain.

The largest pools of carbon occur in fossil fuels and carbonate rocks; other pools include biomass and carbon dioxide (CO_2) and methane (CH_4) in the atmosphere (see Figure 46.10). Energy flow and the flux of carbon through biological communities are linked. Photosynthesis moves carbon from inorganic compartments in the atmosphere and water into the organic compartment, and respiration reverses this. When carbon moves from the atmosphere into surface waters, some of it is used by producers, some precipitates as carbonate (which forms rocks such as limestone), and some falls to the sea floor as detritus (which may eventually form fossil fuels). Burning fossil fuels increases the atmospheric pool of carbon dioxide. Livestock production increases the atmospheric pool of methane.

Biogeochemical cycles interact and perturbations in one cycle may affect others in ways that are difficult to predict.

Question 5. Analyze the changes to the water cycle in an ecosystem that has undergone deforestation.
Textbook Reference: *46.3 Certain Biogeochemical Cycles Are Especially Critical for Ecosystems, pp. 897–898*

Question 6. In what ways are farming methods that require the large-scale use of manufactured fertilizers contributing to eutrophication and other adverse environmental effects?
Textbook Reference: *46.3 Certain Biogeochemical Cycles Are Especially Critical for Ecosystems, pp. 899–900*

46.4 Biogeochemical Cycles Affect Global Climate

Earth's surface is warm because of the atmosphere

Recent increases in greenhouse gases are warming Earth's surface

Human activities are contributing to changes in Earth's radiation balance

About 49 percent of incoming solar energy is absorbed by Earth's surface. The rest is absorbed by the atmosphere or reflected back into space (see the left side of Figure 46.11). Earth's surface re-radiates some of the absorbed solar radiation in infrared wavelengths, and some of this is absorbed by molecules of gas in the atmosphere and radiated back to Earth's surface. The greenhouse effect describes the warming of Earth as a result of its atmosphere trapping heat and radiating it back to the surface. Greenhouse gases (such as carbon dioxide, water vapor, methane, and nitrous oxide) absorb strongly in the infrared wavelengths. Thus, these gases effectively trap outgoing infrared radiation emitted by Earth and radiate it back to the surface, raising surface temperatures (see the right side of Figure 46.11).

Greenhouse gases are increasing in the atmosphere (see Figure 46.12), and global temperatures also are increasing (see Figure 46.13). Higher global temperatures will affect climate, with wet regions expected to get wetter and dry regions drier. Storm intensity is predicted to increase. Human activities, including burning fossil fuels, deforestation, and livestock production, are adding greenhouse gases to the atmosphere.

Question 7. Why do humans deliberately set fire to natural vegetation? Analyze the ways in which human-set fires contribute to global warming.
Textbook Reference: *46.4 Biogeochemical Cycles Affect Global Climate, pp. 902–904*

Question 8. If you were asked to develop a computer model to predict climate change, what factors would you include in your model?
Textbook Reference: *46.4 Biogeochemical Cycles Affect Global Climate, pp. 902–904*

Question 9. Analyze how global warming could influence the spread of human disease.
Textbook Reference: *46.4 Biogeochemical Cycles Affect Global Climate, pp. 903–904*

46.5 Rapid Climate Change Affects Species and Communities

Rapid climate change can leave species behind

Changes in seasonal timing can disrupt interspecific interactions

Climate change can alter community composition by several mechanisms

Extreme climate events also have an impact

Changes in Earth's climate are occurring at a fast pace, which is challenging the ability of species to evolve in response to climate change. Climate change is altering the timing of some environmental cues but not others, so the temporal relationships among cues are shifting. Such mismatches in timing can disrupt interactions among community members who use different cues.

Climate change can alter the species composition of communities. For example, local populations unable to respond to changing environmental conditions may go extinct. Alternatively, changing conditions may favor one species over another and thereby alter the relative abundances of species. Climate change also can lead to shifts in the geographic ranges of species, which may lead to the formation of new communities. Finally, the geographic ranges of species and the composition of communities can be impacted by the increased frequency of extreme weather events predicted to accompany global warming.

Question 10. Many mammals exhibit seasonal reproductive rhythms that are largely determined by day length. Day length is an environmental cue that is independent of climate. Does this mean that mammals with seasonal reproduction will not be affected by climate change? Explain your answer.
Textbook Reference: *46.5 Rapid Climate Change Affects Species and Communities, p. 906*

Question 11. Speculate about the potential influence of climate change on bird migration and the impact this would have on communities.
Textbook Reference: *46.5 Rapid Climate Change Affects Species and Communities, pp. 906–907*

46.6 Ecological Challenges Can Be Addressed through Science and International Cooperation

Climate change has occurred in Earth's past. However, current climate change is unique because it has been precipitated by the diverse activities of a single species, *Homo sapiens*. Through science and cooperative interactions, humans are uniquely equipped to devise solutions and meet the challenges of developing sustainable economies and curbing the multiplicative growth of our population.

Question 12. Evaluate the following statement: "The current period of climate change is unprecedented."
Textbook Reference: *46.6 Ecological Challenges Can Be Addressed through Science and International Cooperation, pp. 907–908*

Test Yourself

1. Areas of upwelling in the ocean have high primary productivity because such areas
   a. bring nutrients from the seafloor up to the surface.
   b. are characterized by clear water that permits light to penetrate to an unusually great depth.
   c. bring warm water to the surface.
   d. trap nutrients washed into the ocean from nearby land masses.
   e. have water of lower salinity than other oceanic regions.
   Textbook Reference: *46.1 Climate and Nutrients Affect Ecosystem Function, p. 895*
2. Which of the following terrestrial ecosystems has moderate NPP?
   a. Tropical forest
   b. Cultivated land
   c. Desert
   d. Tundra
   e. None of the above
   Textbook Reference: *46.1 Climate and Nutrients Affect Ecosystem Function, p. 893*
3. The biosphere
   a. includes land and water, but not the atmosphere.
   b. is thick relative to Earth's radius.
   c. is where life exists on Earth.
   d. includes biotic but not abiotic components.
   e. None of the above
   Textbook Reference: *46.2 Biological, Geological, and Chemical Processes Move Materials through Ecosystems, p. 896*
4. Which of the following statements about water and its cycling is *false*?
   a. Most of Earth's water is in the oceans.
   b. Runoff is the gravity-driven flow of liquid water from land to oceans.
   c. Transpiration is the process by which water changes from gas to liquid.
   d. Water vapor is a greenhouse gas.
   e. Evaporation is the process by which water changes from liquid to gas.
   Textbook Reference: *46.3 Certain Biogeochemical Cycles Are Especially Critical for Ecosystems, pp. 897–898*

5. The largest pool of nitrogen is located in the
 a. benthic sediments.
 b. ocean waters.
 c. soil.
 d. vegetation.
 e. atmosphere.

 Textbook Reference: *46.3 Certain Biogeochemical Cycles Are Especially Critical for Ecosystems, p. 898*

6. The process by which nitrogen gas is reduced to ammonium by microorganisms is called
 a. denitrification.
 b. eutrophication.
 c. nitrification.
 d. nitrogen fixation.
 e. None of the above

 Textbook Reference: *46.3 Certain Biogeochemical Cycles Are Especially Critical for Ecosystems, pp. 898–899*

7. Terrestrial NPP
 a. generally increases with temperature.
 b. generally decreases with temperature.
 c. increases with latitude.
 d. does not vary with precipitation.
 e. None of the above

 Textbook Reference: *46.1 Climate and Nutrients Affect Ecosystem Function, pp. 894–895*

8. Which of the following statements about biogeochemical cycles is *false*?
 a. Burning fossil fuels increases the atmospheric pool of carbon dioxide, but it has no effect on the atmospheric pools of nitrogen oxides.
 b. Primary producers take up elements from inorganic pools and accumulate them as biomass.
 c. Some atoms in the human body may once have been part of a dinosaur.
 d. Biogeochemical cycles all include both organismal and nonliving components.
 e. Perturbations in one biogeochemical cycle can have major effects on other cycles.

 Textbook Reference: *46.3 Certain Biogeochemical Cycles Are Especially Critical for Ecosystems, pp. 898–901*

9. Which of the following processes releases carbon?
 a. Burning fossil fuels
 b. Erosion of sedimentary rock
 c. Decomposition of plants
 d. Decomposition of animals
 e. All of the above

 Textbook Reference: *46.3 Certain Biogeochemical Cycles Are Especially Critical for Ecosystems, pp. 900–901*

10. Which of the following statements about the greenhouse effect is *false*?
 a. Carbon dioxide and methane absorb strongly in infrared wavelengths.
 b. Perturbations to the nitrogen cycle have no effect on Earth's radiation balance.
 c. Earth's surface re-radiates some of the heat that is absorbed from the atmosphere.
 d. Livestock production increases methane in the atmosphere and therefore may increase warming of Earth's surface.
 e. Water vapor is a greenhouse gas.

 Textbook Reference: *46.4 Biogeochemical Cycles Affect Global Climate, pp. 902–903*

11. About _______ percent of incoming solar radiation is absorbed by Earth's surface.
 a. 49
 b. 31
 c. 20
 d. 80
 e. 90

 Textbook Reference: *46.4 Biogeochemical Cycles Affect Global Climate, p. 902*

12. Climate warming
 a. will likely affect different ecosystems in different ways.
 b. is occurring more rapidly in some areas than others.
 c. will likely be accompanied by more frequent extreme weather events.
 d. is currently occurring more rapidly than previous changes experienced by most species in their evolutionary histories.
 e. All of the above

 Textbook Reference: *46.4 Biogeochemical Cycles Affect Global Climate, pp. 903–904; 46.5 Rapid Climate Change Affects Species and Communities, pp. 904–907*

13. Which of the following statements about climate change and environmental cues is *false*?
 a. Temporal relationships among cues are shifting due to climate change.
 b. Climate influences day length.
 c. Mismatches in timing caused by climate change can disrupt species interactions within communities.
 d. Not all environmental cues respond in the same manner to climate change.
 e. None of the above

 Textbook Reference: *46.5 Rapid Climate Change Affects Species and Communities, p. 906*

14. Climate change can alter the species compositions of communities by
 a. causing local populations to go extinct.
 b. altering the relative abundances of species.
 c. causing shifts in geographic ranges.
 d. increasing the frequency and intensity of severe weather events.
 e. All of the above

 Textbook Reference: *46.5 Rapid Climate Change Affects Species and Communities, pp. 906–907*

15. Which of the following statements about the history of climate change on Earth is *false*?
 a. Asteroid impacts, continental drift, and volcanic activity have caused changes in temperature in the past.

b. Organisms have previously induced changes in the composition of the atmosphere.
c. Current climate change is unique because it has been precipitated by a single species, *Homo sapiens*.
d. Mass extinctions have never been associated with changes in temperature.
e. None of the above

Textbook Reference: *46.6 Ecological Challenges Can Be Addressed through Science and International Cooperation, p. 907*

Answers

Key Concept Review

1.

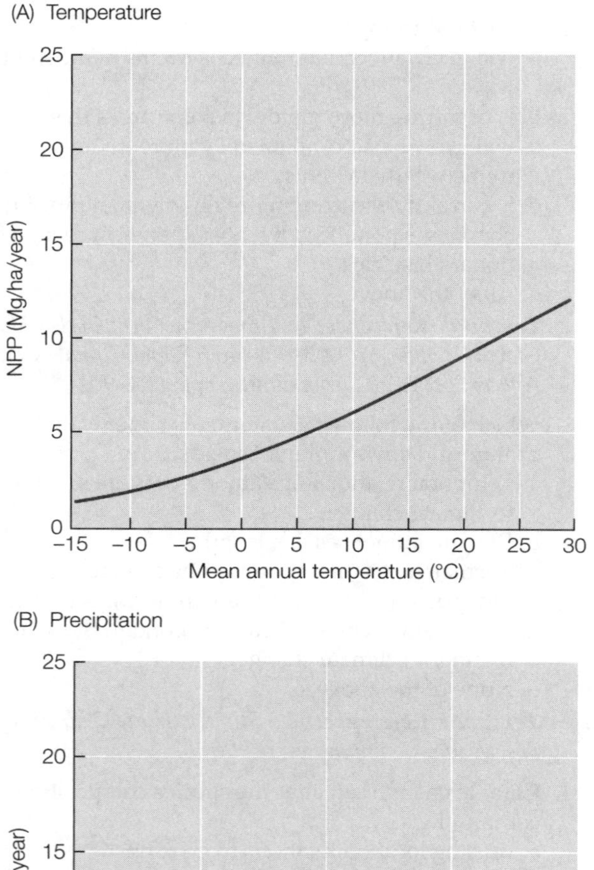

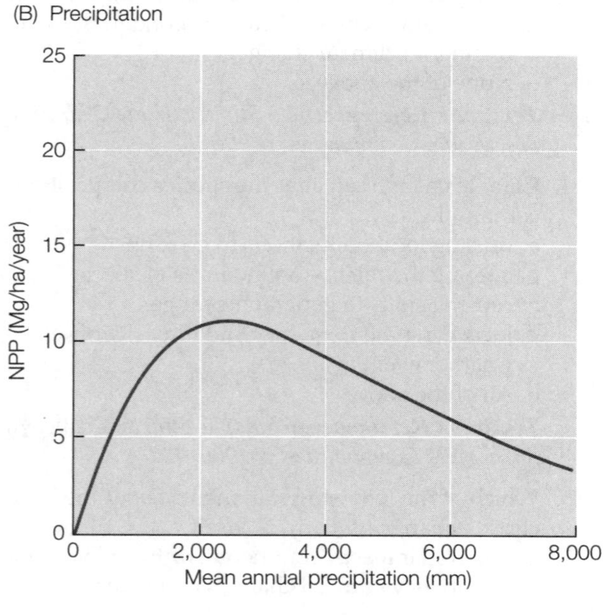

2. Marine NPP is highest in coastal areas because runoff from land brings nutrients into shallow waters. In these waters, photosynthesis occurs in the photic zone, where light can penetrate.

3. Inorganic compounds in rocks become accessible to living organisms when weathering and erosion release them to soil and water, where they can be taken up by autotrophs (see Figure 46.5).

4. Via sedimentation, organic and inorganic compounds enter compartments where they are inaccessible to organisms (see Figure 46.5).

5. Trees take up water through their roots and this water is released to the atmosphere via transpiration (the evaporation of water from plants). Transpiration also increases local humidity. Removal of trees disrupts the water cycle by decreasing transpiration, which can decrease precipitation and local humidity. Tree removal also lessens the ability of the land to retain water, so surface runoff increases. Overall, deforestation tends to dry out local ecosystems.

6. If farmers apply fertilizers to croplands in quantities that exceed the plants' abilities to absorb them, some of the excess nitrates leave the soil and enter bodies of water. This process contributes to eutrophication of lakes and the creation of "dead zones" in the sea, such as the one in the Gulf of Mexico around the mouth of the Mississippi River. In nutrient-poor terrestrial ecosystems, deposition of excess nitrogen can change the species composition of plant communities and convert species-rich communities to species-poor communities.

7. Most fires that occur in areas such as forests are deliberately set by humans to clear land for agriculture. These fires—together with other types of biomass burning, such as combustion of the wood—increase the influx of CO_2 into the atmosphere and contribute to the production of other greenhouse gases. Also, the removal of vegetation reduces the amount of CO_2 removed from the atmosphere for photosynthesis.

8. Factors that would be important to incorporate in a computer model of global climate change include incoming and outgoing radiation, patterns of circulation in the atmosphere and oceans, and interactions with Earth's surface (including those associated with human activities).

9. Global warming can affect human disease by increasing the population sizes and geographic ranges of vectors (such as insects and rodents) and pathogens (such as disease-causing viruses and bacteria) and the length of the transmission season. Cold temperatures in winter kill many pathogens. The milder winters associated with global warming will allow increasing numbers of pathogens to survive; hence the diseases they cause will become more common. Also, disease vectors such as mosquitoes, ticks, and rodents may thrive in a

warmer climate and possibly extend their geographic ranges; this could result in the spread of certain diseases to new areas.

10. Mammals whose reproduction is cued to day length can still be affected by climate change because the species with which they interact may be influenced by climate change. Different species make use of different cues, and so they respond in diverse ways to climate change. This makes possible the development of timing mismatches among species in a community, which could disrupt interactions between mammals and the organisms they feed on (e.g., trees may leaf out earlier or later, and prey species could go extinct locally). Timing mismatches could also disrupt interactions between mammals and their predators (e.g., a new species of predator could arrive through extension of its geographic range).
11. Climate change could affect bird migration in a number of ways. Species that typically migrate long distances could migrate shorter distances, leading to geographic redistribution and changes in the species composition of communities. Species that typically migrate short distances could become sedentary. Species could continue to travel their usual distance but arrive earlier on their breeding grounds. This could be problematic if the timing of their reproduction failed to coincide with an adequate supply of prey to feed their nestlings, leading to a decline in their population relative to resident bird species. Changes in their abundance would in turn affect other species in the community (e.g., predators, competitors, and parasites).
12. The current period of climate change is not unprecedented from the standpoints of changes in temperature and atmospheric composition. Over the course of Earth's history, temperatures have changed and extinctions have occurred as a result of asteroids, continental drift, and volcanic activity. Organisms have also changed the composition of Earth's atmosphere, such as when the first photosynthetic microbes increased atmospheric oxygen. What is unprecedented about current climate change is that it is being caused by the activities of a single species, *Homo sapiens*.

Test Yourself

1. **a.** Low concentrations of nutrients limit the primary productivity of most ocean waters. In zones of upwelling, nutrient-rich water is brought to the surface from the ocean bottom, thereby enhancing the growth of photosynthesizing plankton.
2. **b.** Cultivated land has moderate NPP. Terrestrial NPP is highest in tropical forests and lowest in deserts and tundra.
3. **c.** The biosphere is where life exists on Earth. It includes land, water, atmosphere, and organisms, and is thin relative to Earth's radius.
4. **c.** Transpiration is the process by which water evaporates (changes from a liquid to a gas) from plants. Plants move water from the ground to the atmosphere via transpiration.
5. **e.** The largest pool of nitrogen is nitrogen gas in the atmosphere.
6. **d.** Nitrogen fixation is the process by which nitrogen gas is reduced to ammonium by microorganisms. It is performed, for example, by certain bacteria in the soil and by symbiotic bacteria associated with plant roots.
7. **a.** Terrestrial NPP generally increases with temperature. It is highest in the tropics and declines at high latitudes. Terrestrial NPP increases with increasing precipitation to a point and then declines.
8. **a.** Burning fossil fuels increases the atmospheric pools of carbon dioxide and nitrogen oxides.
9. **e.** Burning of fossil fuels, decomposition of plants and animals, and erosion of limestone rock lifted to Earth's surface all release carbon.
10. **b.** Nitrous oxide is a greenhouse gas, so perturbations to the nitrogen cycle can affect Earth's radiation balance. Burning fossil fuels releases oxides of nitrogen.
11. **a.** About 49 percent of incoming solar radiation is absorbed by Earth's surface. Of the remaining amount, about 20 percent is absorbed by the atmosphere and 31 percent is reflected.
12. **e.** Climate warming will affect different ecosystems differently and is occurring more rapidly in some areas (e.g., the tropical Pacific and western Indian Oceans are warming more rapidly than other waters) than in others (e.g., some may become wetter, others drier). Extreme weather events are predicted to increase in frequency. Warming is occurring more rapidly than other changes most organisms have faced in their evolutionary histories.
13. **b.** Day length is an environmental cue that is independent of climate.
14. **e.** Climate change can directly alter species composition of communities by causing local extinctions, changes in geographic distributions, and changes in relative abundances. Climate change can also affect species composition through extreme weather events that cause sudden shifts in species distributions.
15. **d.** Temperature change has been linked to five mass extinctions in Earth's past.